Earth History

Providing a new approach to Earth history, this engaging undergraduate textbook highlights key episodes in the history of our planet and uses them to explain the most important concepts in geology. Rather than presenting exhaustive descriptions of each period of geological time, this conceptual approach shows how geologists use multiple strands of evidence to build up an understanding of the geological past, focusing on exciting events like the extinction of the dinosaurs and the formation of the Grand Canyon and the Himalaya. Beginning with an introduction to geology, tectonics, and the origin of the universe, subsequent chapters chronicle defining moments in Earth history in an accessible narrative style. Each chapter draws on a variety of sub-disciplines, including stratigraphy, paleontology, petrology, geochemistry, and geophysics, to provide students who have little or no previous knowledge of geology with a broad understanding of our planet and its fascinating history.

Peter Copeland is Professor of Earth and Atmospheric Sciences at the University of Houston, Texas. His expertise lies in thermochronology, geochemistry, and continental tectonics, with a particular emphasis on the evolution of the continental crust. In recent years, his research has focused on the formation of the Rocky Mountains and the Himalaya. From 2001 to 2004 he was co-editor of the *Geological Society of America Bulletin*.

Janok P. Bhattacharya is the Susan Cunningham Research Chair in Geology at McMaster University (Canada). His research interests are in sedimentary rocks of the western interior of North America. Prior to becoming a professor at the University of Texas at Dallas and subsequently at the University of Houston, Bhattacharya worked in the petroleum industry. He is an American Association of Petroleum Geologists (AAPG) Grover Murray Distinguished Educator (2007), AAPG Distinguished Lecturer (2005–2006), and the 2023 SEPM, Francis J. Pettijohn Medalist for excellence in sedimentology and stratigraphy.

"A novel, engaging approach to learning about Earth history. Instead of taking an exhaustive time-period approach that characterizes classic historical geology texts, key events are used to explore the planet's geologic evolution. Starting with general principles of geology, including planet formation, geologic time (fossil and radiometric), the unifying theory of plate tectonics, the origin of life and evolution, the text examines major events that shaped and changed our planet over its long history. Thematic events include Proterozoic Snowball Earth, the Cambrian Explosion of Life, the supercontinent Pangea, major extinction events (yes, dinosaurs too), the evolution of western North America and Southeast Asia, human evolution and modern Ice Ages, and more. The text closes with a brief exploration of today's environmental challenges, bringing ancient geologic history into today's conversation. The writing style is modern conversational, which will appeal to the college user, and the supporting illustrations are clear and informative. Instead of the past distinctions between physical geology and historical geology offerings, this text offers a fresh integration of disciplines, allowing instructors to amplify and cater to their and their students' interest. I heartily recommend this text."

Ben van der Pluijm, University of Michigan

"I am very pleased to recommend Dr. Copeland and Dr. Bhattacharya's textbook, *Earth History*, to the educational and academic community. Their book transcends the traditional, often dense, textbook format, presenting geologic science as a collection of exciting short stories in a narrative style that is both engaging and informative. Their writing style is playful yet analytical, data driven and entertaining. Using a coupled systematic and conceptual approach (with a very strong emphasis on conceptual), the authors make complex topics accessible and interesting. What sets this book apart is the fun! This book's unique blend of analytical rigor and rich storytelling is designed to capture student attention, which is no easy feat in the digital age. *Earth History* has significantly influenced the structure and tone of my own courses. I look forward to incorporating this book into my curriculum upon its release. I am confident that it will provide an enriching educational experience for my students."

Professor Jennifer Campo, Lone Star College, University Park

Earth History

Stories of Our Geological Past

Peter Copeland
University of Houston

Janok P. Bhattacharya
McMaster University, Ontario

CAMBRIDGE
UNIVERSITY PRESS

Shaftesbury Road, Cambridge CB2 8EA, United Kingdom

One Liberty Plaza, 20th Floor, New York, NY 10006, USA

477 Williamstown Road, Port Melbourne, VIC 3207, Australia

314–321, 3rd Floor, Plot 3, Splendor Forum, Jasola District Centre, New Delhi – 110025, India

103 Penang Road, #05–06/07, Visioncrest Commercial, Singapore 238467

Cambridge University Press is part of Cambridge University Press & Assessment,
a department of the University of Cambridge.

We share the University's mission to contribute to society through the pursuit of
education, learning and research at the highest international levels of excellence.

www.cambridge.org
Information on this title: www.cambridge.org/highereducation/isbn/9781108498524
DOI: 10.1017/9781108682800

© Peter Copeland and Janok P. Bhattacharya 2025

This publication is in copyright. Subject to statutory exception and to the provisions
of relevant collective licensing agreements, no reproduction of any part may take
place without the written permission of Cambridge University Press & Assessment.

When citing this work, please include a reference to the DOI 10.1017/9781108682800

First published 2025

Printed in the United Kingdom by CPI Group Ltd, Croydon CR0 4YY, 2025

Cover image: View of the Colorado River from Desert View Point, Grand Canyon National Park, Arizona,
USA. Alan Majchrowicz / Stone / Getty Images.

A catalogue record for this publication is available from the British Library

A Cataloging-in-Publication data record for this book is available from the Library of Congress

ISBN 978-1-108-49852-4 Hardback
ISBN 978-1-108-72415-9 Paperback

Additional resources for this publication at www.cambridge.org/earthhistory

Cambridge University Press & Assessment has no responsibility for the persistence
or accuracy of URLs for external or third-party internet websites referred to in this
publication and does not guarantee that any content on such websites is, or will
remain, accurate or appropriate.

CONTENTS

PREFACE

Geology is history, and history is a collection of stories. Rocks are books telling us about these geological stories, which are written in a variety of languages such as stratigraphy, petrology, paleontology, geochemistry, geophysics, and more. Naturally, the more languages you can speak, the more you can appreciate these stories, which is what makes the application of physics, chemistry, biology, and mathematics crucial to the understanding of the history, structure, and composition of Earth and other planets.

At the same time though, many geologists (including the authors of this book) have at least once met a curious non-geologist who has asked them questions such as: "Did the dinosaurs really die because of a meteor?," "Is California going to fall into the sea?," "Is it true that the rocks on the top of Mount Everest were formed in the ocean?," and "Are we going to run out of oil?"

The short answers to the above questions are: "It seems quite likely," "No," "Yes," and "Not really, but only because oil will get so expensive, we will have to find alternatives." In the response to such queries, some geologists make the mistake of going into way more detail than required about these events, perhaps describing the science behind them, and even resorting to the use of unnecessary jargon, therefore losing their interlocutor's interest.

This is why, in writing principally for undergraduate students taking courses on Historical Geology, Introductory Earth Science, or Earth through Time with no pre-requisites, we chose to simply tell stories. We introduce the details only in service of the story. In doing so, we hope to illustrate to the student the integrative nature of geoscience.

Our stories include highlights from all the major geologic eras from the Precambrian to the Holocene and include discussions of essential aspects of North America (the Appalachians and Rockies) and around the globe (Mediterranean salinity crisis and the Himalaya). Some stories are mostly tectonic whereas others emphasize the history of life, as recorded in the fossil record.

The stories are rarely simple, but they are usually fascinating. However, stories can change. Many of the ideas presented in this textbook as modern consensus were either derided by many experts or non-existent just a few decades ago. Yet, new data can come along (usually because of the invention of some new technology) that challenges old ideas. Therefore, we spend time in most of our stories to not just lay out the geologic evidence but describe the people involved with the big ideas in geoscience and how they struggled with the data and with each other. Such struggles are not over; very few places on Earth are left unexplored but new eyes bring new perspectives and we hope our discussions of how old ideas were replaced by new ones will inspire the next generation to look for yet newer and broader explanations of how our world came to be and where it might be going.

The telling of the history of Earth begins with these sorts of small-scale studies, often carried out by a small number of people over just a few years. This work is then often integrated into a larger picture to tell a broader story such as the evolution of a mountain belt over hundreds of millions of years, to the evolution of life across the planet over billions of years.

The stories we tell in the following chapters are of the broader sort and are therefore compilations of the work of many people (from dozens to thousands) over many years (decades to centuries). We hope to illustrate how broad insights into the history of our solar system, our planet, its diversity of life, and our own species, have come from being able to read many geologic languages.

Conceptual Approach

Rather than offering a linear overview of each time period in detail, we take key events in Earth history to demonstrate how geologists piece together different types of geological evidence to build up a clear picture of events in Earth history.

For example, in Chapter 13, when discussing the extinction of the dinosaurs, we begin with descriptions of some sedimentary rocks in Italy and the nuances of their chemical composition, but soon move on to geophysical discoveries in Mexico and giant tsunami deposits in Texas and North Dakota. We also mention that a large part of India is covered by kilometer-thick lava deposits that erupted right at the time the dinosaurs went extinct and may have played a role in their demise. Finally, we conclude with a consideration of the likelihood of another mass extinction caused by an asteroid impact. Or in Chapter 14, when addressing the question about the fate of California, we also discuss the details of Earth's interior and the processes that operate thousands of kilometers below the surface.

Every interesting thing you've ever heard about the history of our planet started with a geologist picking up a rock. Geologists are trained to notice different aspects of rocks that give clues to their genesis and subsequent history. Some of this information can be picked up in the field, whereas some can only be obtained by analyzing the rock in the lab. Regardless, the goal is to fully interpret this rock.

Understanding the formation of the rock (e.g., from a volcano, in a lake) is the starting point, but much of what we may want to know about a rock extends far beyond this initial step. If we are interested in finding oil, for example, we need to know how old the rock is, what kind of rocks were deposited above and below it, and how deeply it was buried by younger rocks. If, on the other hand, we want to uncover the processes associated with the formation of a particular mountain belt, we need to observe the faults and folds in rocks: when did they form, what is their orientation, and how much motion has occurred on these structures?

We might start a story by asking, "Did you know that some geologists think there was a time when the Earth was completely covered by ice?" From there, we can introduce aspects of glaciology, geochemistry, geophysics, and stratigraphy. This is the approach we've tried to take with chapters concerning the biologic and tectonic evolution of Earth.

Our goal, whether the topic is paleontologic or tectonic, is to try and present individual chapters as self-contained narratives. They really aren't totally self-contained (the order isn't random) but if a reader or instructor wanted to skip some chapters, we hope that most of what is needed to understand each chapter would be right there. Although we expect most people will encounter this book as part of a course often called Historical Geology, which usually assumes an initial course in Physical Geology, we have deliberately included enough foundational material in Chapter 1, so that a reader without any background can follow the stories and appreciate Earth history.

Key Features of this Book

- Accessible and concise, primarily aimed at non-majors, including broad introductory chapters for those with no prior knowledge of geology.
- Provides a far more conceptual approach to the topic, focusing on key events in Earth history, such as the extinction of dinosaurs and the formation of the Grand Canyon, to explain key geological concepts and how geologists use multiple strands of evidence to build up an understanding of the geological past.
- Uses an engaging, narrative style of writing to make the text more interesting to students.
- Limits the discussion of paleontology to the context of key events in Earth and life history, rather than as a detailed topic in its own right.
- Avoids large species lists, which can be overwhelming to students.

Pedagogical Features

Individuals learn in different ways. We therefore provide a variety of pedagogical aids to be used as the student chooses. Each chapter includes the following:

- Learning Objectives: to help students in clarifying, arranging, and prioritizing their learning.
- Introductions: to highlight the topic covered in each chapter and how it fits within the big picture, all whilst piquing the reader's interest.

- Color figures and photographs: to complement the text and provide a visual representation of the concepts discussed.
- KEY POINT boxes: to easily spot the main concepts covered in each section.
- Boxes: to highlight interesting and relevant side stories or information.
- Chapter Summary: to summarize the major points of the chapter and allow the reader to reflect on what they have learnt.
- Key Words: boldface within the text and listed at the end of each chapter for easy review. A full Glossary of key words appears at the end of the book.
- Further Reading and References: to help extend the coverage of topics that are not dealt with in so much detail within the chapter, or for students that need more information than can be presented in the chapter.
- Review Questions: to encourage students to think about what they have covered, and for guiding instructors' quiz or test questions.

Text Organization

The first six chapters provide much of the foundation for the remainder of the book, highlighting specific stories that we believe are both essentially interesting but also provide a systematic and sequential overview of most of the major events in Earth history calling on the variety of disciplines required to explain what we know about the subject. These chapters are organized in chronological order and cover most of the major tectonic and biologic events in Earth history that are traditionally covered in an Earth History course, but we emphasize the narrative versus a compendium of fact, events, and dates.

The remainder of the chapters discuss individual episodes in Earth history (restricted in both space and time). These in no way represent an exhaustive treatment but are chosen because of the way they illustrate the need to speak the several languages of geoscience in order to read the history of our planet.

Chapter 1 reviews the basic principles of geology and covers material usually taught in a Physical Geology course, followed by a review of the history of geology and the philosophies of geologic thought that have shaped geologic investigation over several centuries. Here, we take a fieldtrip (narrative-type) approach by using the Grand Canyon to illustrate many of the basic principles of physical geology. What can the reader learn about geology on a trip to the Grand Canyon, versus a systematic listing of the various types of rocks and minerals?

Chapter 2 covers the way that geological thinking developed and introduces the concepts of Neptunism, Plutonism, uniformitarianism, and actualism, using the Channeled Scablands as an example. The idea is for students to get a feel for how geologists think.

Chapter 3 tells the story of the origin of the universe, the formation of our solar system and the Earth–Moon system and Earth's early atmosphere and hydrosphere.

Chapter 4 uncovers how deep time was discovered and how we know that Earth is 4.56 billion years old.

Chapter 5 provides an overview of plate tectonics and how this theory was developed.

Chapter 6 concludes the first part of the book with a discussion of the process of natural selection that explains the observation that life has evolved through the immensity of deep time.

Chapter 7 discusses the biogeochemistry and stratigraphy that informs our understanding of the origin of life on Earth.

Chapter 8 reviews the paleomagnetism, geochemistry, glaciology, and stratigraphy that led to the snowball Earth hypothesis.

Chapter 9 describes the paleontological oddities of the Ediacaran fauna and the Cambrian explosion and what these fossils tell us about the mechanisms of evolution and the diversification of life.

Chapter 10 details the closing of Iapetus and the formation of Pangea from the perspective of the Appalachian Mountains.

Chapter 11 describes the changes in life on Earth during the Paleozoic and the paleontology, geochemistry, and volcanology that informs our understanding of the Great Dying at the end of the Permian.

Chapter 12 chronicles the rise of dinosaurs and their variety.

Chapter 13 details the geochemistry, geophysics, and stratigraphy that led to the giant impact hypothesis for the end-Cretaceous extinction.

Chapter 14 tells the tectonic history of western North America from Jurassic to present with reference to stratigraphy, magmatism, deformation, and geophysics.

Chapter 15 tells the story of the Himalaya using structure, geophysics, stratigraphy, and geochemistry.

Chapter 16 uses geochemistry, stratigraphy, and geophysics to tell the story of the drying of the Mediterranean during the Messinian salinity crisis.

Chapter 17 chronicles human evolution with paleontology and stratigraphy as well as a discussion of the cultural complications of these studies.

Chapter 18 describes the Pleistocene ice ages with glaciology, stratigraphy, and geochemistry.

Chapter 19 concludes the book by looking forward as well as to the past with a discussion of the effects of humans on the climate using stratigraphy, modeling, and geochemistry.

ACKNOWLEDGMENTS

We would first like to thank our editors at CUP, Ilaria Tassistro, Emma Kiddle, Matt Lloyd, and Rachel Norridge for their support of this book and all of their hard work in making it a reality. We thank reviewers Altaf Arain, Greg Benson, Ashok Bhattacharya, Steve Brusatte, Michael Happy, Rob Hatcher, Paul Higgs, Paul Hoffman, Mark Laflamme, Jonny Wu, and Jinny Sisson, as well as the anonymous reviewers solicited by CUP. David Hemsley at Aardwolf Books did a thorough job of copy editing and thanks too to our proofreader Julie Jackson. Discussions with Robert Stern helped with Chapter 5. Aidan Delorme helped compile research for Chapter 12. We appreciate the support in use of images and photos from collections at the Royal Ontario Museum, the Natural History Museum (London, UK), the Field Museum (Chicago), the Neanderthal Museum (Mettmann, Germany), the Houston Museum of Natural History, the Perot Museum (Dallas), Texas Christian University (Ft. Worth), and McMaster University. We also thank Laura Crossey, Bruce Damer, David Deamer, Karl Karlstrom, Bruce Watson, Alex Robinson, Michael Murphy, Steven Newton, and Chris Paola, all of whom provided images as credited throughout the book. We also thank Ron Blakey at Colorado Plateau Geosystems, Inc. for use of Deeptime Maps[TM], which we use throughout the text. Pat DeLuca, at McMaster, is thanked for helping find usable satellite images. We thank David Bonner, who compiled the Glossary. Finally, we thank our wives, Beth and Cyndy, for their patience as we worked on the book for the past eight years. Cyndy Bhattacharya joined a number of field excursions to the Grand Canyon, Rocky Mountains as well as driving to the Santerno Valley, Gubbio, and Sardinia, Italy to take photos featured in Chapters 1, 13, and 16 respectively.

Names and dates of units in the timescale figures for Chapters 7 to 19 are based on the Geologic Time Scale v. 6.0: Geological Society of America (Walker, J.D., and Geissman, J. W., compilers, 2022, Geologic Time Scale v. 6.0: Geological Society of America, https://doi.org/10.1130/2022.CTS006C), which is itself based on data from Cohen, K.M., Finney, S., and Gibbard, P.L., 2012, International Chronostratigraphic Chart: International Commission on Stratigraphy, https://stratigraphy.org/ICSchart/ChronostratChart2012.pdf (accessed Sept. 2022), Cohen, K.M., Finney, S.C., Gibbard, P.L., and Fan, J.-X., 2013 (updated), The ICS International Chronostratigraphic Chart: Episodes, v. 36, p. 199–204, http://www.stratigraphy.org/ICSchart/ChronostratChart2021-10.pdf (accessed Sept. 2022) and Gradstein, F.M, Ogg, J.G., Schmitz, M.D., et al., 2012, The Geologic Time Scale 2012: Boston, USA, Elsevier, https://doi.org/10.1016/B978-0-444-59425-9.00004-4. Colors follow the standard colors advised by the International Commission on Stratigraphy (administered by the International Geological Map of the World project in Paris - CCGM).

Aerial view of Colorado River, Grand Canyon, Arizona. Source: Barry Winiker / Getty Images.

Chapter 1

The Grand Canyon

Reading the Rocks

LEARNING OBJECTIVES

- Recall the main subdivisions (eons, eras, and periods) of the geological timescale.
- Define the main types of metamorphic rocks, the concept of a protolith, and explain how pressure and temperature control the formation of metamorphic rocks.
- Explain how igneous rocks form and how intrusive (plutonic) and extrusive (lava) igneous rocks can be distinguished.
- Discuss how light felsic magmas and associated rocks are produced during metamorphism by partial melting.
- Define Steno's laws of original horizontality, superposition, lateral continuity, and cross-cutting relationships.
- Define the main types of unconformities (non-conformities, angular unconformities, disconformities, and paraconformities).
- Define what a fossil is and explain how fossils are useful in environmental reconstructions and determining ages of rocks.
- Define a sedimentary basin and describe the kinds of evidence that can be used to interpret the environments of deposition of the sediments that fill basins, such as the difference between proximal and distal deposits.
- Explain how structural geology and the study of deformation can be used to interpret orogenic processes, such as compressional mountain building, and extensional breakup of continents.
- Explain the difference between a fold, fracture, and fault, and the difference between ductile and brittle deformation.

Introduction

Rocks are literally underfoot and can tell marvelous stories. The authors of this book have been spellbound by rocks for most of our lives. We have been astounded at the story of how the dinosaurs were killed by an asteroid crashing into Earth from outer space, and we smile at the idea that the rocks now at the top of Mount Everest were formed beneath the surface of an ocean. We have both spent time enjoying the splendor of Grand Canyon and pondering the incredible stories recorded in the rocks that are exposed in the cliffs and along the bottom (Figure 1.1).

The tiny details as well as the broad conclusions of these and thousands of other stories of our geologic past were put together when someone looked at a rock and began wondering about its history. To most folks, a rock may be just a rock, but to a geologist, every rock is like a page in a book. The pages are sometimes hard to read and incomplete, but by paying attention to various aspects of rocks, geologists can decipher the fascinating stories locked within. These stories are told in a book that is written in the "language" of geology. Sometimes, unwrapping the mystery requires **paleontology** – the study of fossil evidence of past

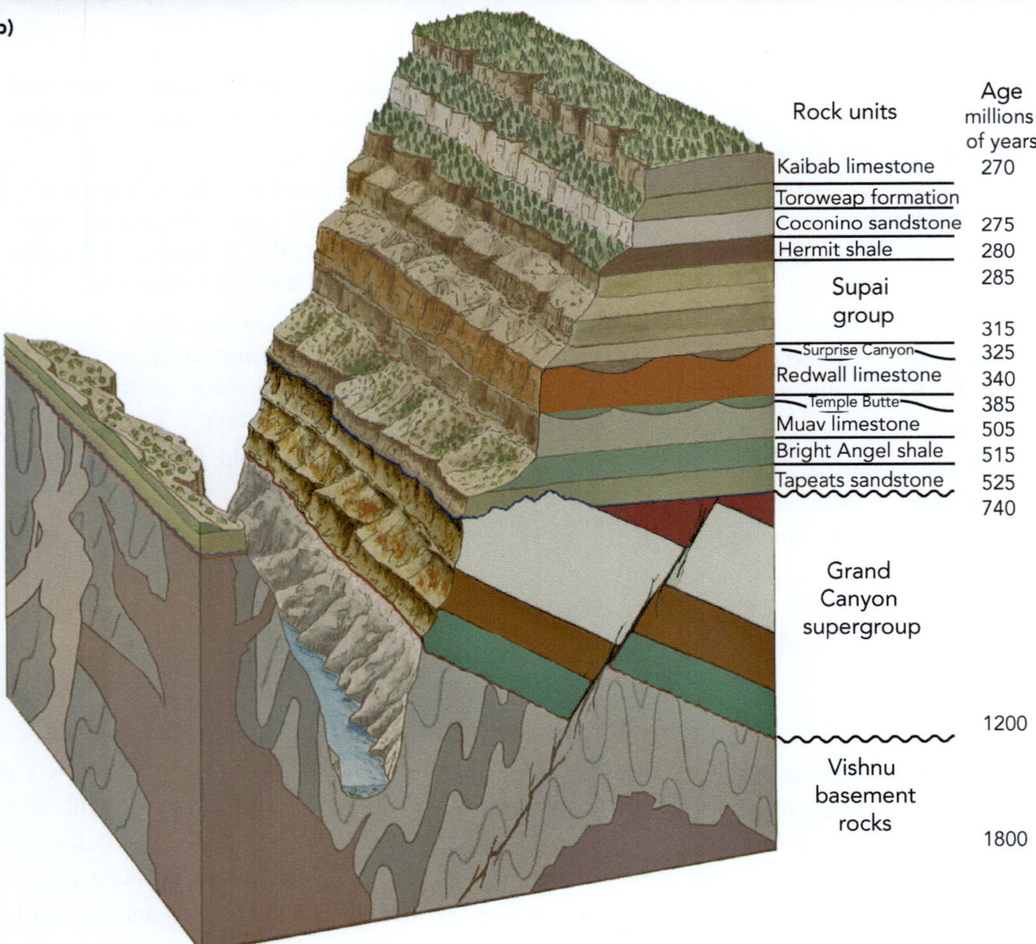

Figure 1.1 (a) Photo and (b) block diagram of the Grand Canyon. The older dipping layers of the Grand Canyon Group are separated from the younger horizontal Paleozoic layers by an unconformity surface, marked by the dashed line that separates them. The surface is actually continuous, but because it is exposed in different segments of the complex canyon, the line appears as different segments. The various formations are labeled: T – Tapeats Sandstone, BA – Bright Angel Shale, RW – Redwall Limestone. Source: (b) Karlstrom and Crossey (2019), used with permission.

life, or it could require **geochemistry** – the study of the chemical composition of rocks and minerals, or **structural geology** – the study of the deformation of rocks. In other cases, an understanding of **stratigraphy** – the study of rock layering – or Earth's geophysical properties, provide the key to unlock these stories. In this chapter we will use the example of the Grand Canyon to introduce some of fundamental concepts and terms that geologists use to read the stories that rocks tell us about how Earth and life on it have changed over its 4.5 billion years of history.

KEY POINT

Geological stories are told using the language of geology including paleontology, stratigraphy, geochemistry, and structural geology.

1.1 The Grand Canyon: Overview

The Grand Canyon (Figure 1.1), in the southwestern United States, is one of the seven natural wonders of the world. It reaches a depth of 1,857 meters, is 19 kilometers across at its widest, and stretches for about 446 kilometers. The canyon was carved over many millions of years due to a combination of the erosive power of the Colorado River, and regional uplift of the Colorado Plateau in response to **plate tectonic** processes that we will cover in more detail in Chapters 5 and 14.

The first humans to inhabit the Grand Canyon area were the native American ancestors of modern Puebloans about 3,000 years ago, and the first Europeans to see the Canyon were led by Spanish conquistador García López de Cárdenas y Figueroa, led by Hopi guides in 1540. In 1869 famed American geologist and explorer John Wesley Powell, who lost an arm in the American Civil War, conducted the first major geological expedition down the Colorado River. This expedition produced the first formal geologic reports of the Grand Canyon and surrounding Colorado Plateau, and the region has been the subject of geologic investigation ever since.

1.1.1 Age of the Grand Canyon

The details of the formation of the Grand Canyon are a subject of active investigation. The modern consensus is that the modern canyon system began forming around 70 million years ago, or 70 Ma. Throughout this book we will use the standard designations Ga for Giga-annum (billion years ago), Ma for Mega-annum (million years ago), and ka for kilo-annum (thousand years ago). The main phase of uplift and erosion occurred over the past six million years. One of the first concepts that can be deduced by considering the Grand Canyon is that slow erosion can result in the formation of vast geological features when operating over long times, such that the Colorado River was able to carve the deep canyon (see Chapter 2).

In addition to the idea that it has taken millions of years for the Colorado River to carve the canyon, the rocks that are exposed along the sides and at the bottom (Figure 1.1) record an even more impressive history that tells one of the most fundamental stories about Earth: that it is old – really old. Dating of the oldest rocks in the Grand Canyon show that they were formed about 1.8 billion years ago (1.8 Ga), but that is less than half of the age of Earth, which was formed around 4.56 Ga.

Geologists have developed the Geologic Time Scale (Box 1.1, Figures 1.2 and 1.3) to keep track of where rocks fit into Earth history. Much like a year is broken up into months, weeks, days, hours, and minutes, Earth time is broken up into **eons**, **eras**, **periods**, **epochs**, and **ages** (Figure 1.3). We will tell the story of how we came to know the age of the Earth in Chapter 4, but for now the key point is that over this immense stretch of time, many things have happened. We will start our story at the bottom of the Grand Canyon.

BOX 1.1 Core Knowledge: The Geologic Time Scale

Geologic time is divided into the longest eons, which are in turn divided into eras, periods, epochs, and ages (Figure 1.3). In the Phanerozoic Eon, the most recent eon, subdivisions are primarily based on variations in fossils. In older eons, which lack diagnostic fossils, other criteria may be used. Eons older than the Phanerozoic are collectively and informally referred to as Precambrian.

1.1.1.1 The Eons

Phanerozoic (542 to 0 Ma, i.e., to the modern day): The name means "evident life"; rocks of the Phanerozoic have abundant complex fossils.

Proterozoic (2,500 to 542 Ma): The name means "earlier life"; mostly simple fossils largely from single-celled life characterize Proterozoic rocks. The Proterozoic is divided into the *Paleoproterozoic* (ancient), *Mesoproterozoic* (middle), and *Neoproterozoic* eras (new), and these eras are in turn divided into a number of periods, discussed in Chapters 3, 6, and 8.

Archean (4,100 to 2,500 Ma): The name is from the Greek for "origin"; marked by the oldest rocks on Earth. It is divided into the *Eo-*, *Paleo-*, *Meso-*, and *Neoarchean* eras.

Hadean (4,567 to 4,100 Ma): The name comes from Hades, a reflection of the hellish conditions that existed at the time of formation of the planet. The Hadean is that eon of Earth history in which no rocks presently exist (in Chapter 3 we will see that they did exist once but have been destroyed by erosion, deformation, and metamorphism). By this definition, every time we find a new record-breaking oldest rock on Earth, the Hadean becomes a little shorter.

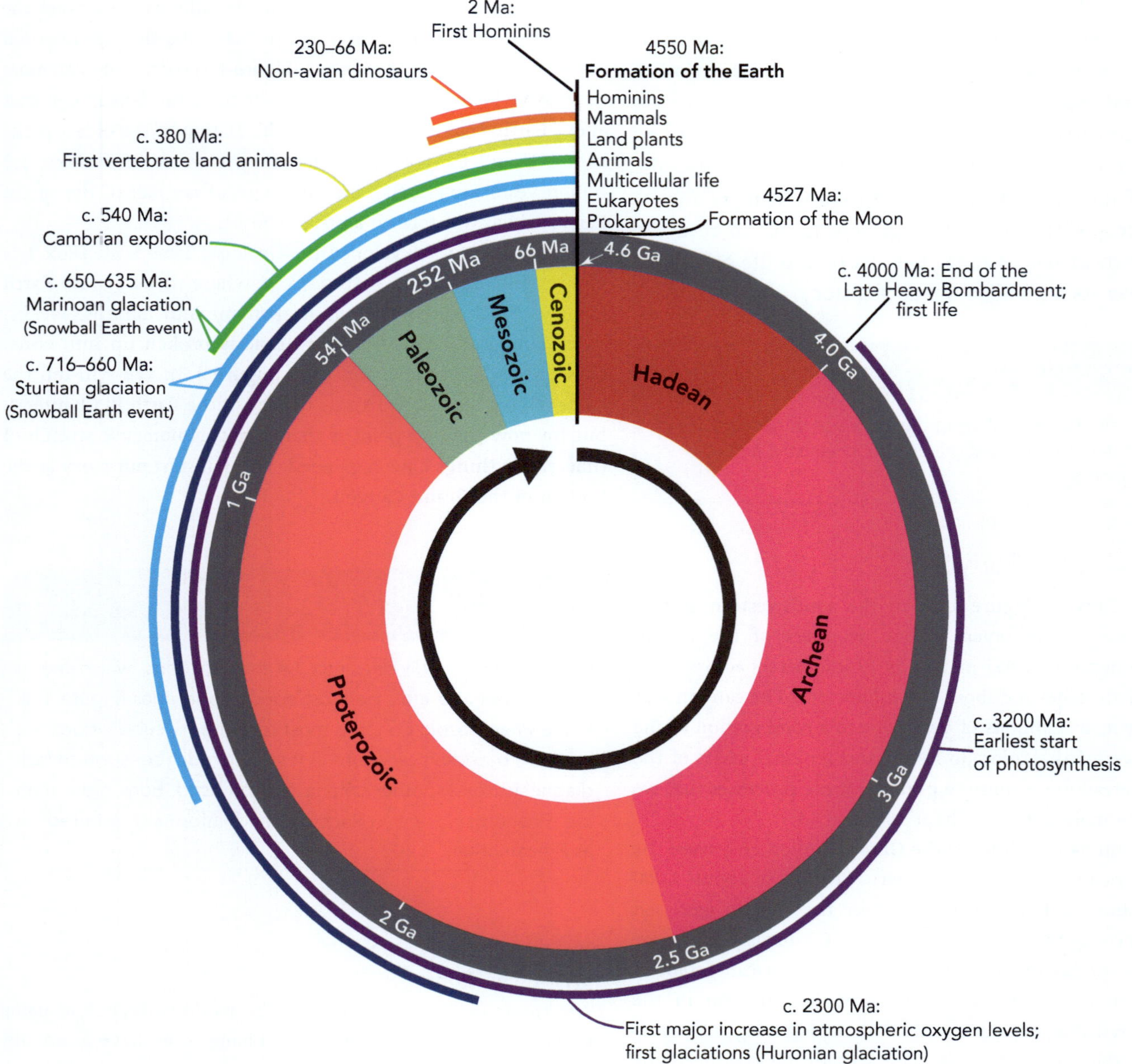

Figure 1.2 The Geologic Time Scale and key events in Earth history shown as a clock. Source: Adapted from WoudloperDerivative work: Hardwigg, Public domain, via Wikimedia Commons.

1.1.1.2 The Eras of the Phanerozoic Eon

Eras are the next level, or second-order division of geological time:

Cenozoic (65 to 0 Ma): Meaning "new life."
Mesozoic (251 to 65 Ma): Meaning "middle life."
Paleozoic (542 to 251 Ma): Meaning "old life."

1.1.1.3 The Periods of the Phanerozoic Eon

Eras are subdivided into periods, or third-order division of geological time:

Quaternary (2.6 to 0 Ma): This is the most recent period and extends to the present day. This name is a holdover from the four-fold division of rocks proposed by Alfred Werner in the late eighteenth century and discussed in Chapter 2.

Neogene (23 to 2.6 Ma): Neogene means "newborn." Fossilized organisms in this period are quite different to those from the preceding Paleogene period.

Paleogene (65 to 23 Ma): Paleogene means "ancient born" emphasizing the different nature of the fossils of this period compared to the succeeding Neogene period.

Cretaceous (145 to 65 Ma): One of the early sites of study of these rocks was southern England, where lots of the rocks are chalky limestones. The Latin word for chalk is *creta*, so these rocks were called Cretaceous.

Jurassic (199 to 145 Ma): These rocks are named from the Jura Mountains, in eastern France, where such rocks are found in abundance. This and the later Cretaceous period are famous for being the age of the dinosaurs, as discussed in Chapter 12.

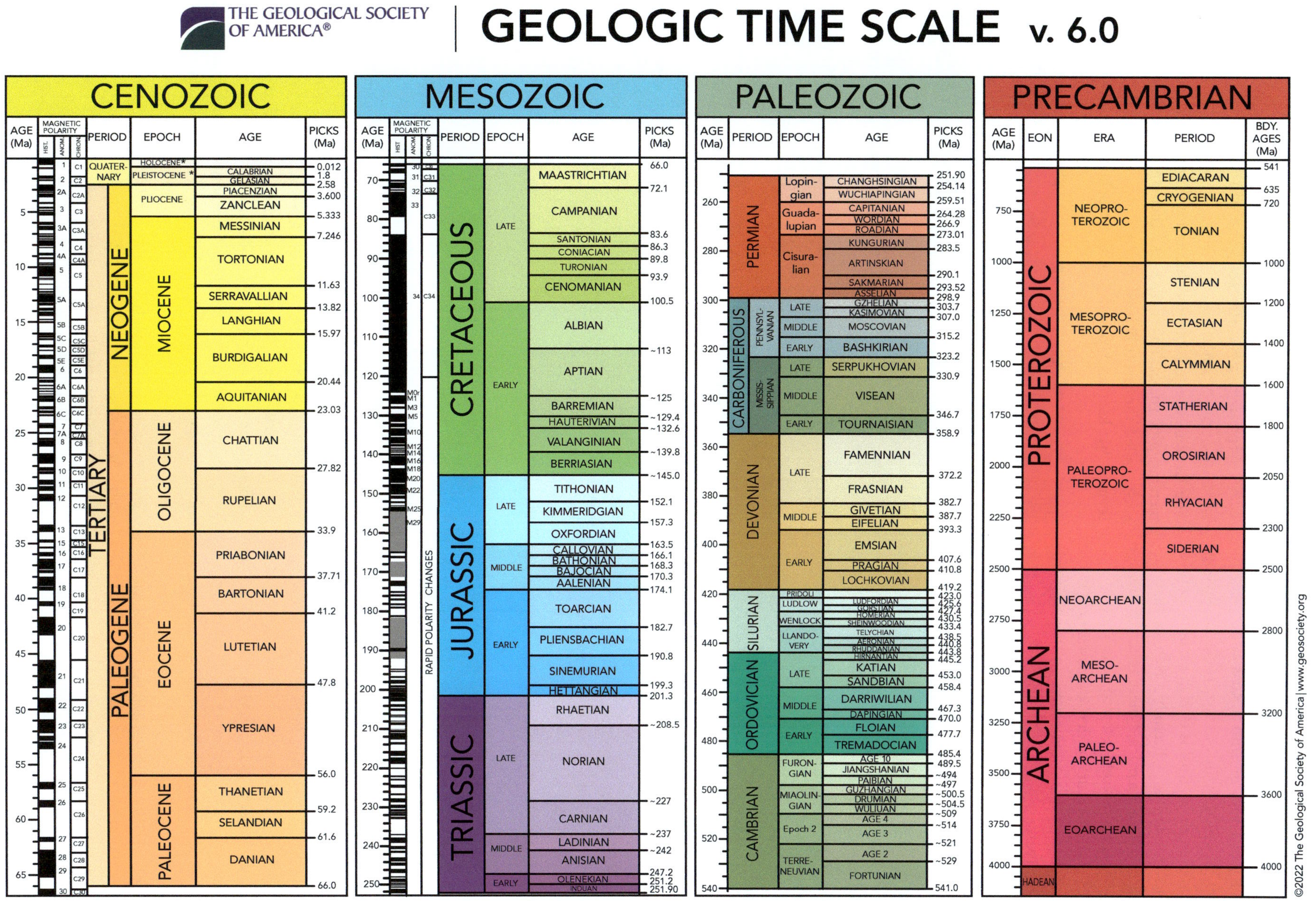

GEOLOGIC TIME SCALE v. 6.0

THE GEOLOGICAL SOCIETY OF AMERICA®

©2022 The Geological Society of America | www.geosociety.org

Figure 1.3 The Geologic Time Scale. Source: Used with permission of the Geological Society of America.

*The Pleistocene is divided into four ages, but only two are shown here. What is shown as Calabrian is actually three ages: Calabrian from 1.8 to 0.774 Ma, Chibanian from 0.774 to 0.129 Ma, and Late from 0.129 to 0.0117 Ma. The Holocene is divided into three ages: Greenlandian from 0.0117 to 0.0082 Ma, Northgrippian from 0.0082 to 0.0042 Ma, and Meghalayan from 0.0042 to present. The geologic community broadly recognizes the Anthropocene as a proposed new time interval of Earth history, partly coincident with the Holocene. Currently, the Anthropocene has an informal designation, with a proposed age span extending from the present to a beginning point between ca. 15,000 yr B.P. and as recent as 1960 CE.

The Cenozoic, Mesozoic, and Paleozoic are the Eras of the Phanerozoic Eon. Names of units and age boundaries usually follow the Gradstein et al. (2012), Cohen et al. (2012), and Cohen et al. (2013, updated) compilations. Numerical age estimates and picks of boundaries usually follow the Cohen et al. (2013, updated) compilation. The numbered epochs and ages of the Cambrian are provisional. A "~" before a numerical age estimate typically indicates an associated error of ± 0.4 to more than 1.6 Ma.

Walker, J.D., and Geissman, J.W., compilers, 2022, Geologic Time Scale v. 6.0: Geological Society of America, https://doi.org/10.1130/2022.CTS006C. (Walker—University of Kansas; Geissman—University of Texas-Dallas, University of New Mexico.)

REFERENCES CITED

Cohen, K.M., Finney, S., and Gibbard, P.L., 2012, International Chronostratigraphic Chart: International Commission on Stratigraphy, https://stratigraphy.org/ICSchart/ChronostratChart2012.pdf (accessed Sept. 2022).

Cohen, K.M., Finney, S.C., Gibbard, P.L., and Fan, J.-X., 2013 (updated), The ICS International Chronostratigraphic Chart: Episodes, v. 36, p. 199–204, http://www.stratigraphy.org/ICSchart/ChronostratChart2021-10.pdf (accessed Sept. 2022).

Gradstein, F.M., Ogg, J.G., Schmitz, M.D., et al., 2012, The Geologic Time Scale 2012: Boston, USA, Elsevier, https://doi.org/10.1016/B978-0-444-59425-9.00004-4.

Triassic (251 to 199 Ma): The rocks exposed in southern Germany have three distinctive layers, so these rocks were named Triassic – "tri" for three.

Permian (299 to 251 Ma): Rocks named for their prominent exposures near the city of Perm, in central Russia.

Carboniferous (359 to 299 Ma): In Europe, rocks of this age have many coal deposits in them. Coal is rich in carbon, so these rocks were named Carboniferous – literally "rich in carbon." This term is not used in North America, where rocks of this age are broken into two separate periods referred to as the:

Pennsylvanian (318 to 299 Ma) and older *Mississippian* (359 to 318 Ma), named for their prominent exposures in Pennsylvania and Mississippi respectively.

Devonian (416 to 359 Ma): Rocks named for their prominent exposures near the county of Devon, England, the largest deposits of which are the "Old Red Sandstones."

Silurian (444 to 416 Ma): These rocks were first studied in southern Wales and the time period is named after the Silures tribe, which occupied the region in Roman times.

Ordovician (488 to 444 Ma): These rocks were also first studied in Wales and named after the Ordovices tribe, which occupied the region in Roman times.

Cambrian (542 to 488 Ma): The name comes from Cambria, the Latinized version of Cymru, the name for Wales in the Welsh language.

1.2 The Basement: Metamorphic and Igneous Rocks

There are three main rock types, **igneous**, **sedimentary**, and **metamorphic**, and all of these are well represented in the Grand Canyon. The oldest rocks belong to the Vishnu basement rocks, so called because they lie at the base of the canyon and

include the Granite Gorge Metamorphic Suite and the Zoroaster Plutonic Complex. Box 1.2 summarizes the minerals that make up the majority of rocks on Earth.

1.2.1 Metamorphic Rocks (Granite Gorge Metamorphic Suite)

The Granite Gorge Metamorphic Suite (Figure 1.4) consists of metamorphic rocks. These are rocks that start off as one type of rock, called the **protolith** or original rock, and are later transformed by heat and pressure into a rock with new minerals

Figure 1.4 (a) The Vishnu Schist (dark) with pink granite intrusions is non-conformably overlain by the Tapeats Sandstone. Yellow arrows show onlap of Tapeats Sandstone against the Vishnu, recording deposition of shallow marine sands of the Tapeats during a time of rising sea level. The undulating top of the Vishnu would have created low and high areas that would have formed small islands as the area was flooded by the sea. (b) Banded dark schists and granite layers of the Vishnu schist.

BOX 1.2 Core Knowledge – Minerals: The Building Blocks of Rocks

Rocks mostly consist of minerals. A *mineral* is defined as a naturally occurring, inorganic solid with a regular internal structure and a narrow range of composition. Minerals are compounds made of elements. Although the periodic table includes more than 100 elements, most minerals can be described using only eight: oxygen (O), silicon (Si), aluminum (Al), calcium (Ca), sodium (Na), potassium (K), iron (Fe), and magnesium (Mg). Similarly, although more than 3,000 minerals have been identified, only about a dozen or so minerals are found in most rocks.

The crust of the Earth is about 47% oxygen and 28% silicon, so most common rock-forming minerals contain Si and O and are termed silicates. The simplest silicate is quartz, SiO_2; other common silicates include lighter-colored minerals, such as feldspars, which add potassium (K), calcium (Ca), and sodium (Na), and the so-called dark minerals, such as olivine, pyroxene, and

amphibole, which are rich in iron and magnesium. Of the common rock-forming minerals, only **calcite** is not a silicate, that is, it does not contain silicon. Calcite ($CaCO_3$, where C stands for carbon) is the most important constituent of the rock limestone.

The crystalline structure of minerals reflects the geometry of the molecules that they are constructed from (Figure 1.5). Although only a few elements make up the majority of most minerals, trace amounts of different elements can be found in many minerals, which leads to a wide range of different varieties. For example, emerald is the gem variety of beryl, which includes trace amounts of chromium (Cr), giving it its characteristic green color. As another example, zircon can include trace amounts of radioactive uranium (U), which decays over time to lead (Pb). As we will discuss in Chapter 4, the ratio of parent uranium versus daughter lead can be used to date zircons and the rocks that contain them.

(Figure 1.5) and new textures. The Vishnu Schist, found at the base of the Grand Canyon, started as sediments formed as muds and sands deposited in an ocean along the flanks of a chain of volcanic islands around 2.5 Ga. These muds and sands were buried to become **shales** (clay-rich sedimentary rock) and **sandstones** (sand-rich sedimentary rock) and later, around 1.7 Ga, they were buried further as a consequence of plate tectonics, which is the idea that the Earth's surface consists of a series of rigid plates that split, slide, and collide relative to each other to produce all the major and most of the minor features of the planet, such as mountains and oceans. The sedimentary rocks were squeezed and heated because of plate collision to form metamorphic rocks, including **schists** and **gneisses** (Figure 1.6). **Slates**, schists, and gneisses are common metamorphic rocks distinguished by the size and composition of crystals as well as the types of minerals they contain (Figure 1.6). Metamorphic rocks are reflective of the temperature and pressure of formation, or metamorphic grade. Low-grade metamorphosis of shales produces slate, at higher grades schists, and at even higher grades gneisses (Figures 1.6 and 1.8).

The protolith of a metamorphic rock can be deduced by the mineral composition. Quartz and mica-rich schists usually have

Figure 1.5 Examples of minerals. (a) White quartz and green emeralds. (b) Pyrite cube made of iron sulfide (FeS$_2$). (c) Inset shows that the molecular structure mimics the larger crystal shape. Source: (a) From Parent Géry, CC BY-SA 3.0 <https://creativecommons.org/licenses/by-sa/3.0>, via Wikimedia Commons.

Figure 1.6 Common metamorphic rocks: (a) slate roof shingles; (b) garnet schist, Franklin Mountains Texas; and (c) gneiss with folds, Ontario, Canada.

a sedimentary origin, and original sedimentary features may sometimes be observed (Figure 1.7). The forces required to heat and deform these basement rocks in the Grand Canyon are understood to have been created when several volcanic island chains called **island arcs**, such as seen in the modern-day Aleutian Islands between Alaska and Siberia, collided with the North American craton (Figure 1.9). The Brahma Schist consists of metavolcanics, the metamorphosed remnants of the island

Figure 1.7 Metasediments. (a) This schist shows alternating lighter and darker layers representing the original sand and mud interbedding. Undulating tops of red beds are interpreted as preserved ripple marks. (b) This metaconglomerate, technically a schist, shows well-rounded pebbles indicative of an origin as a gravel that are now aligned and flattened as a result of later metamorphism.

arcs around which the sediments of the Vishnu Schist were originally deposited.

Cratons are large areas of stable crystalline basement rocks that form the central parts of continents. In the North American continent, the craton is exposed in the **Canadian Shield**, an area covering most of Canada east of the Rockies (Figure 1.10). The North American craton extends farther south, but except for parts of Minnesota and Wyoming it is not well exposed in the United States because it is covered by younger sedimentary rocks.

As with most cratons, the North America craton is surrounded by mountain belts (Figure 1.10), the **Cordillera** to the west, which is active today and which we will discuss in Chapter 14, and the **Appalachians** on the eastern side, which were formed between 450 to around 300 million years ago, the story of which we will tell in Chapter 10. We commonly use the term **Laurentia** to refer to the older continent of North America, which formed long before the Appalachians and Rockies were attached to its margins. Like most continents, Laurentia was surrounded by oceans that opened and closed, because of plate

tectonic processes that we will introduce in Chapter 5. As the oceans surrounding Laurentia closed, volcanic island arcs (Figure 1.9) were accreted to its margins, increasing its size. Geologists like to give names to periods of mountain building, technically termed **orogeny**, and the one associated with the formation of the mountains associated with the formation of the Vishnu Schist, referred to as the Vishnu Mountains, which lie at the bottom of the Grand Canyon, is referred to as the **Yavapai Orogeny**.

Metamorphism (the transformation of one or a group of minerals into a new mineral assemblage) occurs entirely in the solid state, whereas igneous minerals that make up igneous rocks (described in a later section), form by crystallization from a liquid melt. Metamorphism happens because, given enough time, minerals subjected to temperatures or pressures different to those under which they formed change (often due to tectonic processes) because they are no longer stable under new combinations of temperature and pressure. Laboratory experiments can calibrate the pressures (P) and temperatures (T) at which diagnostic metamorphic minerals form, and observations of these minerals in metamorphic rocks can therefore be used to infer the conditions of formation (Figure 1.8). This can be used to build up a pressure–temperature (P–T) history, or "path" of a sequence of rocks, which can then be used to infer the sequence of burial, metamorphosis, and later uplift and erosion that they experienced, which in the case of the Vishnu basement rocks of the Grand Canyon record the formation and ultimate erosion of the ancient Vishnu mountain belt (Figure 1.9).

Analysis of key minerals show that the Granite Gorge Metamorphic Suite experienced maximum temperatures of 800 °C and pressures of 8 kilobars (8,000 times atmospheric pressure), indicating they formed at a depth of about 25 kilometers below the surface in the deep roots of the Vishnu Mountains (Figure 1.8). At these temperatures, in addition to metamorphism, some of the rocks also experienced partial melting, forming igneous rocks, discussed next.

KEY POINT

The oldest rocks in the Grand Canyon are metamorphic and igneous, formed 1.8 Ga during the Yavapai Orogeny at a depth of 25 km in the roots of the Vishnu Mountains.

1.2.2 Igneous Rocks: The Zoroaster Plutonic Complex

The compression and heating associated with the Yavapai Orogeny also caused some rocks to melt, which resulted in the formation of the rocks of the Zoroaster Plutonic Complex, although as we will explain below, the melting was not complete.

The term *igneous* literally means "fire-rock," from the root *ignis* from which we obtain the English word ignite. All igneous rocks start off as a liquid **magma**, which is called **lava** when it

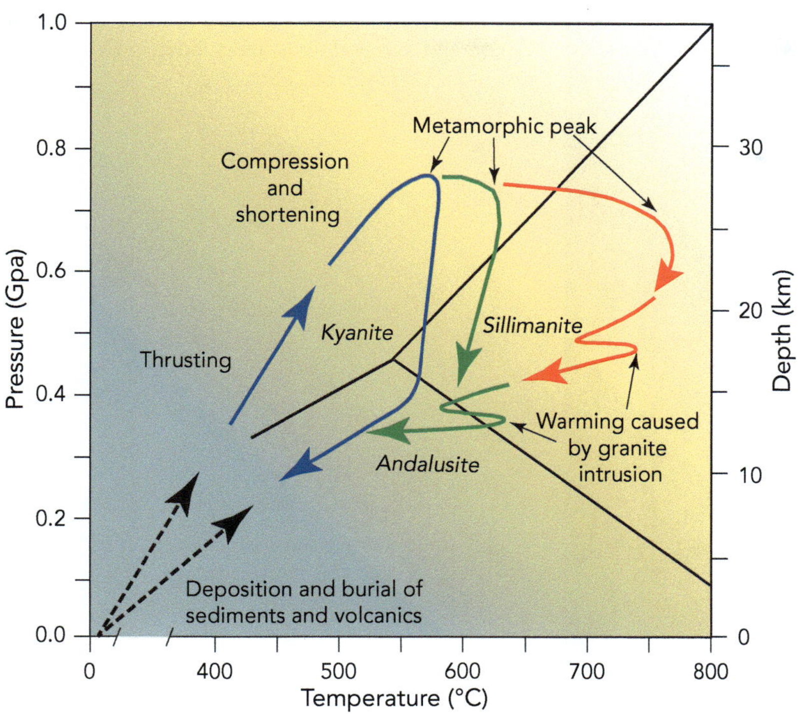

Figure 1.8 Pressure/temperature paths of different geological blocks, shown with different colored lines, of the Vishnu basement rocks, Grand Canyon. Different blocks experience different levels of metamorphism. Andalusite, Kyanite, and Sillimanite, as well as other mineral assemblages (not shown), are used to plot the PT pathways. Source: Modified from figure 10 in Karlstrom et al., in Timmons and Karlstrom (2012). Used with permission of the Geological Society of America.

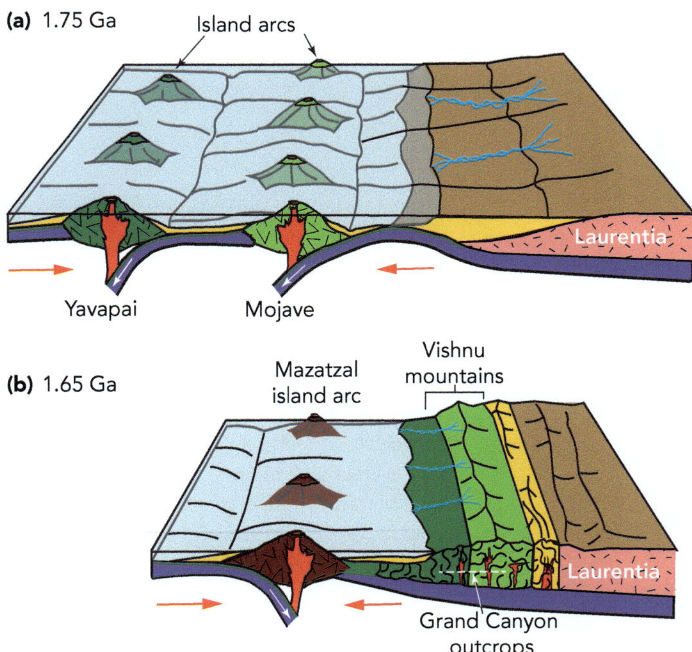

Figure 1.9 The Yavapai and Mojave volcanic island arc systems (a) eventually collided with Laurentia during the Yavapai Orogeny and built the Vishnu Mountains (b). The Vishnu basement rocks formed at a depth of about 25 kilometers below the surface in the mountain roots. Source: Modified from figure 6 in Karlstrom et al., in Timmons and Karlstrom (2012). Used with permission of the Geological Society of America.

flows at the surface. Igneous rocks that cool and solidify beneath the surface are called **plutonic** or **intrusive**, and when solidification happens on or above the surface, we call the resulting igneous rocks **volcanic** or **extrusive**.

Igneous rocks mostly contain minerals in the form of crystals (Figure 1.5). Crystal size primarily relates to how quickly a magma cools. The ability of magma or lava to flow means it behaves like a liquid. When lava is extruded at the surface, or underwater, it cools very quickly, and the liquid freezes without forming crystals, making volcanic glass (Figure 1.11a). Volcanic glass is therefore characteristic of extrusive rocks. Conversely, plutonic magmas that cool slowly below the surface allow more time for larger crystals to grow, eventually turning all that was once liquid into crystals (Figure 1.11c and d). Large crystals are therefore characteristic of plutonic rocks.

The Zoroaster Plutonic Complex includes a variety of igneous rock types, including **granites** (Figure 1.12), which have a distinctive pink color reflective of the dominance of the felsic minerals, potassium feldspar and quartz, and which show relatively large crystals, indicating that the magma cooled slowly, underground. It also includes darker-colored mafic **gabbros**, which have darker minerals, such as **pyroxene**, that are richer in magnesium (Mg) and iron (Fe). These different rock types record cooling of magmas that had different chemical compositions. Granites are formed from **felsic** magmas, rich in **feldspar** (fel) and silica (Si). Gabbros crystallize from **mafic** magmas, rich in magnesium (Mg) and iron (Fe).

But wait a minute, you might say, if plutonic igneous rocks form by cooling of magma, how do you get rocks with different composition? Isn't magma all one thing? The answer to this question is no. Magmas have variable compositions that relate to how magmas are produced in the first place. Most magma is produced by **partial melting** of older rocks. Compared to most people's everyday experience, the temperature needed to melt

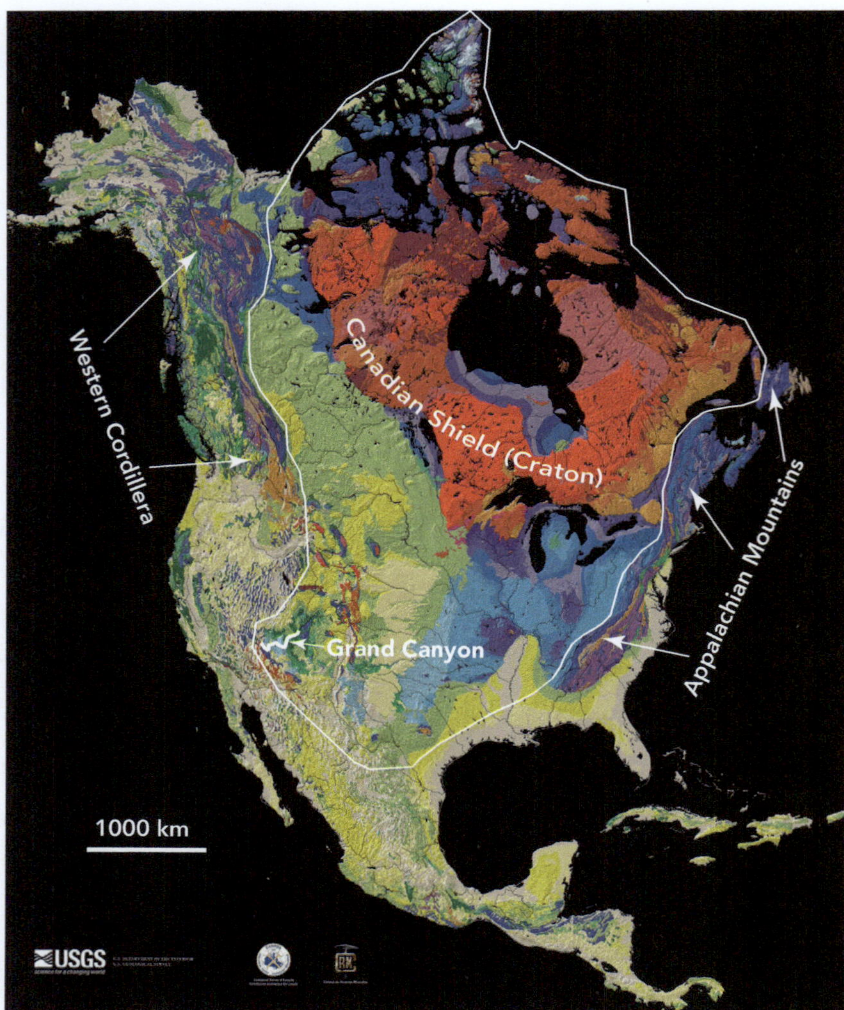

Figure 1.10 North American map showing Laurentian Craton and fold and thrust belts. Source: Original map from US Geological Society.

Figure 1.11 Examples of common igneous rocks: (a) obsidian (volcanic glass); (b) basalt with gas holes (vugs), some are filled with white calcite that formed after the rock cooled; (c) granite; and (d) pegmatite. In (c) and (d) the pink mineral is potassium feldspar and white crystals are quartz. Source: (a) Credit: V&G Studio via iStock / Getty Images Plus.

rocks is very hot, but rocks don't all melt at one temperature. The lowest temperature at which rocks begin to melt is about 650 °C (or about 1,200 °F), about three times hotter than a typical home oven. The important concept here is that when a rock begins to melt, minerals such as quartz and potassium-rich feldspar melt first, whereas mafic minerals like olivine and pyroxene, which contain iron and magnesium, melt at higher temperatures. Rocks that are richer in iron, such as volcanic **basalts** (Figure 1.11b) and plutonic gabbros, only begin to melt at about 1,300 °C (or about 2,300 °F). Therefore, if an area with variable rock types is being heated, due to metamorphism, and the temperature exceeds around 650 °C, the first magmas to be produced will have a felsic composition. If the temperature stays around there, no mafic magmas will be produced, but if the temperature increases to 1,300 °C, less-felsic magmas may be produced. Conversely, if new injections of magma from the

mantle occur, such as at mid-oceanic spreading centers (see Chapter 5), mafic igneous rocks are formed.

In the Grand Canyon, a lot of the darker-colored metamorphic rocks are cut by light-colored veins of granite, reflecting that while the darker rocks were experiencing a solid-state metamorphic change, partial melting produced a magma with a granitic (felsic) composition. Once the rock has partially melted, the remaining rock will have a different bulk (or overall) composition, and as temperature increases, the later produced magmas will have a more mafic composition. Because magma is usually less dense than the surrounding rock that produces it, the magma rises and may become separated from the parent rock. In this way, different types of rocks can be produced that vary from mafic to felsic in composition. In Chapter 3, we will show how partial melting produces mafic rock at oceanic spreading centers versus more-felsic rock typically forming continents, but for now the key point is that partial melting produces magmas of different composition, which in turn produces a variety of igneous rock types. Conversely, when minerals crystallize out of a magma, they do so in the exact opposite order to that in which they melt, in a process called **fractional crystallization** (Figure 1.13). As a magma cools, the first minerals to precipitate will be the darker mafic minerals, such as **olivine** or pyroxene.

This is somewhat like the formation of sea ice. When sea ice freezes, the salt in the water is unable to be incorporated in the ice crystals, so the ice is fresh, whereas the remaining seawater becomes saltier. Ice is unusual in being less dense than water so the freshwater icebergs float, but in magmas the opposite happens. The early formed crystals are denser than the surrounding magma, and these crystals will tend to sink leaving the remaining magma a bit more enriched in elements such as aluminum and potassium, which may later form potassium feldspars as cooling continues.

Figure 1.12 Granite overlain by the Tapeats Sandstone, Grand Canyon.

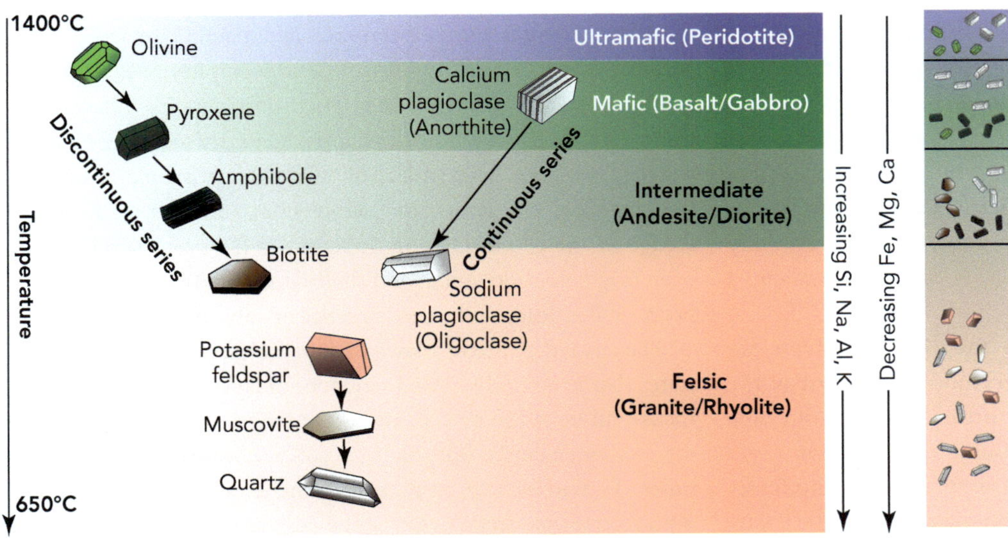

Figure 1.13 Fractional crystallization (Bowen's reaction series) comprises a discontinuous series, in which different minerals are precipitated as a magma cools, and the continuous series, in which individual plagioclase crystals become more sodium rich as the magma cools. The right column provides an idea of what the magma looks like as it cools and the composition changes. Late-stage residual melts produce felsic rocks. Alternatively, as you heat a rock and cause partial melting, the first magmas produced will be more felsic in composition.

Melting can also be enhanced by the addition of water. During metamorphosis, some minerals, such as **clays**, have water incorporated into their crystal structure, and as they are heated this water can be released. The released water can significantly lower the melting temperature of the surrounding rock. This process is somewhat like adding salt to ice, which in colder regions is used to melt ice on roads and sidewalks following winter storms. Fractional crystallization and partial melting are processes that cause differentiation of magmas that ultimately lead to the variable types of igneous rocks that we can see in the Grand Canyon, and play a major role in the evolution of oceans and continents that we will discuss in Chapter 3.

The other type of igneous rocks in the Grand Canyon are much younger basalts that lie at the surface. Basalts are much finer-grained igneous rocks formed from mafic lavas. There are numerous clusters of volcanoes in and around the Grand Canyon area, ranging in age from about six million years to less than a thousand years, which are the result of a geologic hot spot that upwells from deeper in the Earth.

KEY POINT

The different types of igneous rocks reflect whether they extruded at the surface as finely crystalline lavas or cooled within the Earth as coarser crystalline intrusive plutons. The compositional differences reflect either partial melting or fractional crystallization.

1.3 Sedimentary Rocks in the Grand Canyon and Principles of Stratigraphy

Most of the rocks exposed in the Grand Canyon are sedimentary (Figure 1.1). Sedimentary rocks form either as the accumulation and consolidation of pieces of older rocks, such as sediment accumulating on the seafloor or sediment and pebbles deposited by a river, or by the chemical precipitation of new minerals from seawater (often brought on via the metabolism of organisms), such as limestone or chalk. Sediments are mostly deposited in horizontal layers called **strata**. As sediment is buried it is cemented together by minerals, forming layers of sedimentary rock. Although the overwhelming proportion of rocks by volume on Earth are igneous or metamorphic, such as those in the basement rocks in the Grand Canyon, over three quarters of the Earth's continents are covered by a relatively thin layer of sediment and sedimentary rocks that lie above a crystalline basement.

The oldest sedimentary rocks in the Grand Canyon are referred to as the Grand Canyon Supergroup and consist of the Unkar and Chuar groups, which are further divided into nine formations ranging in age from 1.2 billion to 740 million years old (Figures 1.1 and 1.14). The Grand Canyon Supergroup forms a wedge of tilted sedimentary rocks, confined to the eastern part of the canyon, in contrast to the younger Paleozoic strata (541–250 Ma), which are mostly horizontal and extend across the entire Grand Canyon area.

1.3.1 Naming Layers

To keep track of sedimentary layers, such as seen in the walls of the Grand Canyon, geologists commonly assign specific names to specific layers, or strata. Stratigraphy is the branch of geology that is focused on the study and naming of regional-scale sedimentary layering. **Formations** are bodies of rock that have a consistent set of physical characteristics that can be used to distinguish them from adjacent rocks, and which occupy a particular position in a vertical sequence of layers. In the practice of formal naming of strata, the **formation** is the fundamental unit required to make geologic maps. Formations may be divided into smaller-scale units, called **members**, or lumped into broader units, called **groups** and supergroups, and the names are commonly based on the location where the units are best exposed or originally described. The place where a unit takes its name from and where it was first described in the geologic literature is called the *type section*. For example, the Coconino Sandstone (Figures 1.1 and 1.15), a Permian unit found near the top of the Grand Canyon, was named for outcrops where it is particularly well exposed in Coconino County, Arizona.

1.3.2 Steno's Laws

In addition to the absolute age of rocks, which we will discuss in Chapter 4, geologists can also determine the relative age of rocks using several principles that were introduced in 1669 by a Danish scientist, Nicolas Steno. Steno developed the basic principles for understanding the order of sedimentary layering, now known as **Steno's laws**. Steno realized that layered sedimentary rocks were originally deposited as loose sediment in a fluidized state that were laid down in successive layers and later **petrified** – literally "turned into rock" over time.

The first of Steno's stratigraphic laws is known as the **principle of original horizontality**. This simply acknowledges that, because sedimentary particles are deposited under the influence of gravity, sedimentary layers are laid parallel to the Earth's surface, which is broadly horizontal; this applies equally to volcanic rocks. The principle of original horizontality allows us to understand that something has happened to any sequence of layers that are now not horizontal. Following the principle of original horizontality, the now-tilted sedimentary layers that form the wedge of the Grand Canyon Supergroup (Figure 1.1) must have been originally horizontal, and their departure from their original horizontal attitude is the consequence of subsequent deformation.

The second of Steno's laws, the **principle of superposition**, states that in an undisturbed sequence of layered rocks, the oldest rocks will be at the bottom and the youngest rocks will be at the top. As an example, imagine that you are working at your computer. You are sitting on a chair, hunched over a desk or table, with a computer on top. Maybe the desk and chair are placed on a carpet, which is on the floor. Steno's law of superposition allows us

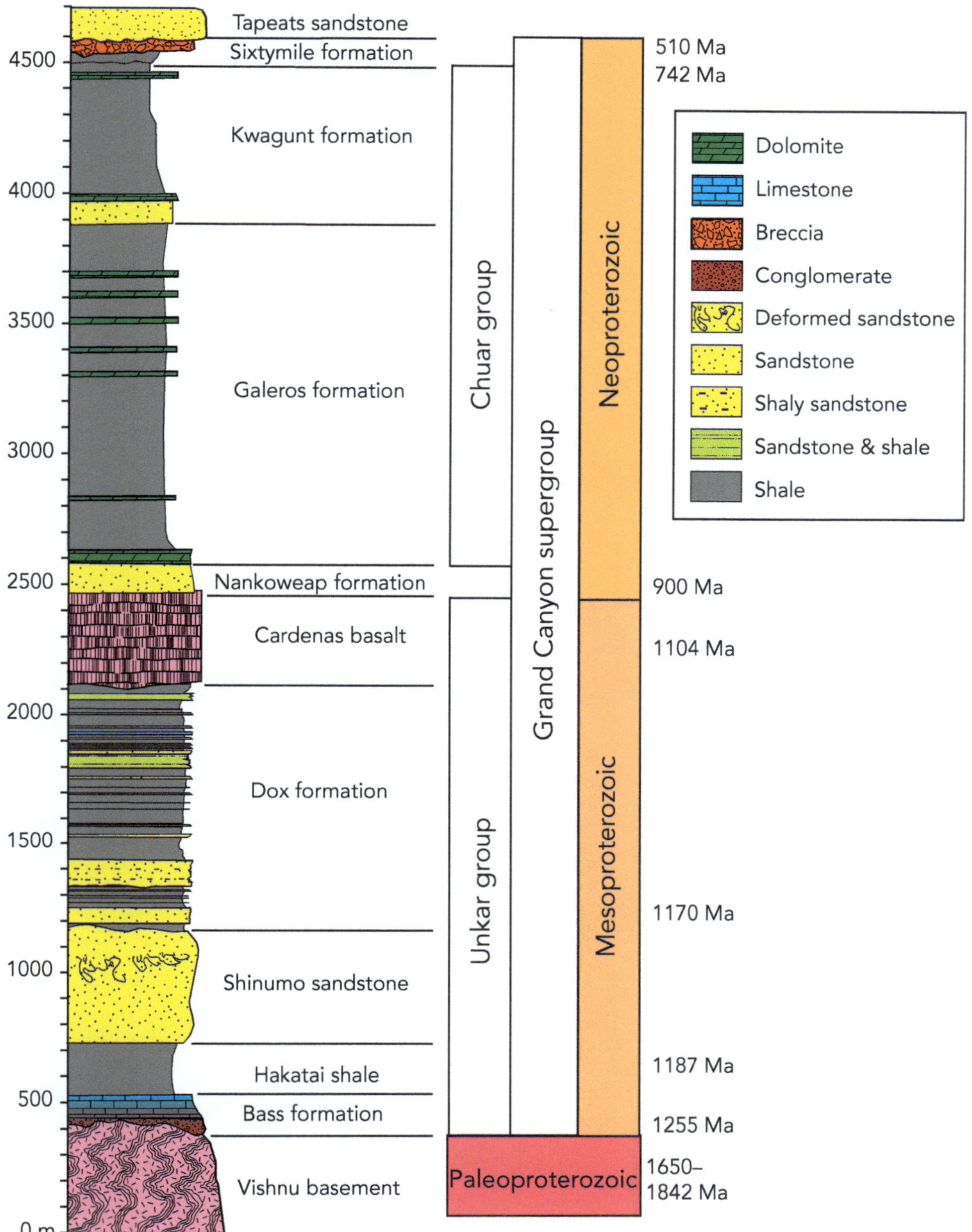

Figure 1.14 Stratigraphy of the Grand Canyon Supergroup. Source: Modified from Figure 3 in Dehler et al. (2012). Used with permission of the Geological Society of America.

to infer the order of things. The floor clearly must be there before the carpet. The carpet had to be laid down before the furniture was placed in the room. And obviously, you and your computer arrive later. The same reasoning can be used to understand the order of deposition of sedimentary rock layers in which the older layers are below and younger layers on top.

Applying the law of superposition to the Grand Canyon allows us to conclude that the rocks at the top of Grand Canyon are younger than those at the bottom (Figure 1.1). Within the Grand Canyon Supergroup, the oldest unit is the Bass Formation, which is overlain by the Hakatai Shale and progressively younger units upward (Figure 1.14). The youngest sedimentary formation exposed along the Grand Canyon rim is the Kaibab Limestone (Figure 1.15), about 270 million years old, deposited during the Permian Period. However, the law of superposition only holds for undisturbed sequences. Tectonic forces can deform, fold, and even overturn sequences of rocks. In such cases, independent methods of determining the order of deposition may be needed.

The third of Steno's laws is known as the **principle of lateral continuity** (Figures 1.16 and 1.17). To explain this, think of a cake, with alternating layers of cake, filling, and icing (Figure 1.16). As soon as a piece of cake is removed, there will be a gap in all the layers. The principle of lateral continuity allows us to infer that the icing and the layers of cake and filling were once continuous. If you look across the Grand Canyon, you will see similar looking rock layers on either side of the canyon walls (Figure 1.1). If you stand at the edge of the south rim of the

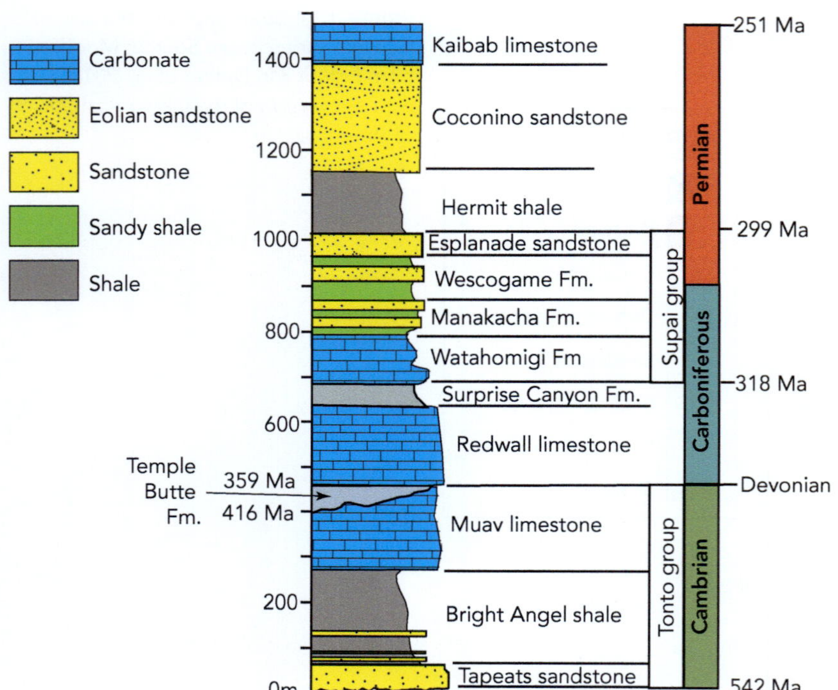

Figure 1.15 General Paleozoic stratigraphy of the Grand Canyon.

Figure 1.16 Photo of the Tapeats Sandstone, which originally covered the entire area but has later been eroded as the Colorado River incised the Grand Canyon, much like the layers in the cake. Source: Grand Canyon: Photo of cake from Scheinwerfermann, Public domain, via Wikimedia Commons.

Grand Canyon, you will be standing on the Kaibab Limestone. Looking to the North rim, you can see the same formation, but of course it has been eroded away in between as the Colorado River has incised downwards. The principle of lateral continuity generally states that once-laterally continuous layers, such as the Kaibab Limestone, are now discontinuous because of subsequent erosion, in this case by the Colorado River. This law also allows us to understand the order in which layers are deposited and later partly eroded. So long as the rock sequences are sufficiently similar, the principle of lateral continuity can be applied over distances much longer than across a modern river. Although we are focused on the Grand Canyon in this chapter, many of the formations identified in the canyon can be found in other places in North America and even globally. Correlation of strata around the world allows us to understand the history of how our planet has changed over geological time, and there are several chapters where we will discuss events for which the geological evidence can be found across the globe.

Steno also developed the idea of cross-cutting relationships, in the case where a body such as an igneous intrusion cuts

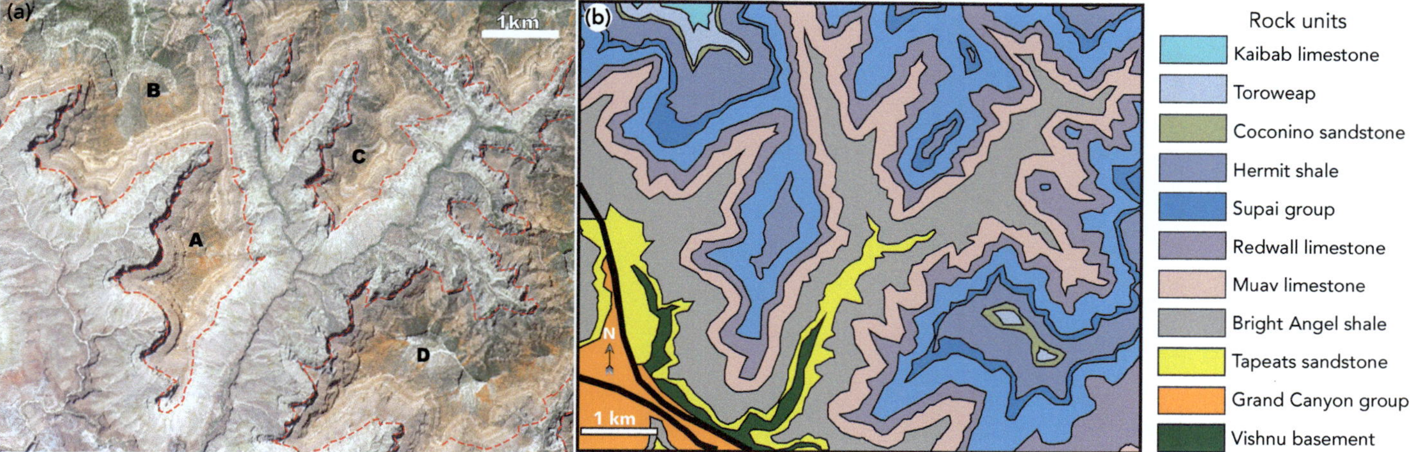

Figure 1.17 Aerial photo (a) and geological map (b) of part of the Grand Canyon. The dashed line marks the top of the Redwall Limestone. The areas labeled A to D are now physically separated as the Colorado River incised the canyon, but the layers were once contiguous across the entire area shown. The rock units are listed in the order they were formed with the Vishnu at the bottom and the youngest Kaibab limestone at the top. Source: (a) European Union; contains modified Copernicus Sentinel data 2023.

Figure 1.18 Older Hakatai Shale cut by a younger dike illustrates the principle of cross-cutting relationships.

across a layer, then the intrusion must have formed after the layer was deposited. If we go back to your desk example, if there are screws or nails in your desk that cut into the wood, Steno's law of cross-cutting relationships requires that the wood was there first, then the screws or nails were added later.

If we look lower in the Grand Canyon Supergroup, we can see younger **igneous dikes**, which formed as a crack in the rock filled with magma, cutting vertically into the Hakatai Shale (Figure 1.18). The law of cross-cutting relationships allows us to infer that the igneous dike is younger than the sedimentary rock that it cuts.

KEY POINT

Steno's laws of superposition, original continuity, and original horizontality, as well as cross-cutting relationships, are used to understand the order in which geologic features were formed and allow reconstruction of the geological history of an area.

1.3.3 Unconformities

Although rocks are essential in telling the story of our planet, they often don't record the full story and 'miss' significant periods of time. Stratigraphic successions, in which there is a relatively continuous record of deposition, are said to be **conformable**. In contrast, an **unconformity** is a surface between rock layers across which there is evidence of a significant break in the record of deposition, commonly marked by some erosion. This could be due to a period of time when little or no deposition took place, or more commonly, it could be due to subsequent erosion of the previous rocks.

In the Grand Canyon, the 540-million-year-old Tapeats Sandstone in some areas overlies the 1.8-billion-year-old igneous and metamorphic rocks of the Vishnu basement (Figure 1.4(a)) and elsewhere overlies various formations of the Grand Canyon Supergroup (Figure 1.1), forming an undulating unconformity. When sedimentary rocks overlie igneous or metamorphic rocks, this is termed a **non-conformity**. Where the Tapeats overlies the Vishnu the time gap at the contact is over 1.2 billion years or 25% of all Earth history (Figure 1.19). In other places, the Tapeats overlies tilted formations of the Grand Canyon Supergroup, forming an **angular unconformity** (Figure 1.1) characterized by the strata above the unconformity being tilted at a different angle to those below it. The time gap varies, as the Tapeats in some areas overlies the older Unkar Group and elsewhere the younger Chuar Group (Figure 1.19). In general, despite the outstanding exposures of rocks in the Grand Canyon, analysis of the time durations preserved by the rocks amount to about 400 million years, or about 10% of Earth's history. This exemplifies the incomplete nature of the geologic record, as well as the idea that unconformities commonly represent more "missing" time than that preserved by the rocks. Importantly, unconformities mark periods of erosion that may relate to orogenic events or times of major sea-level falls and can

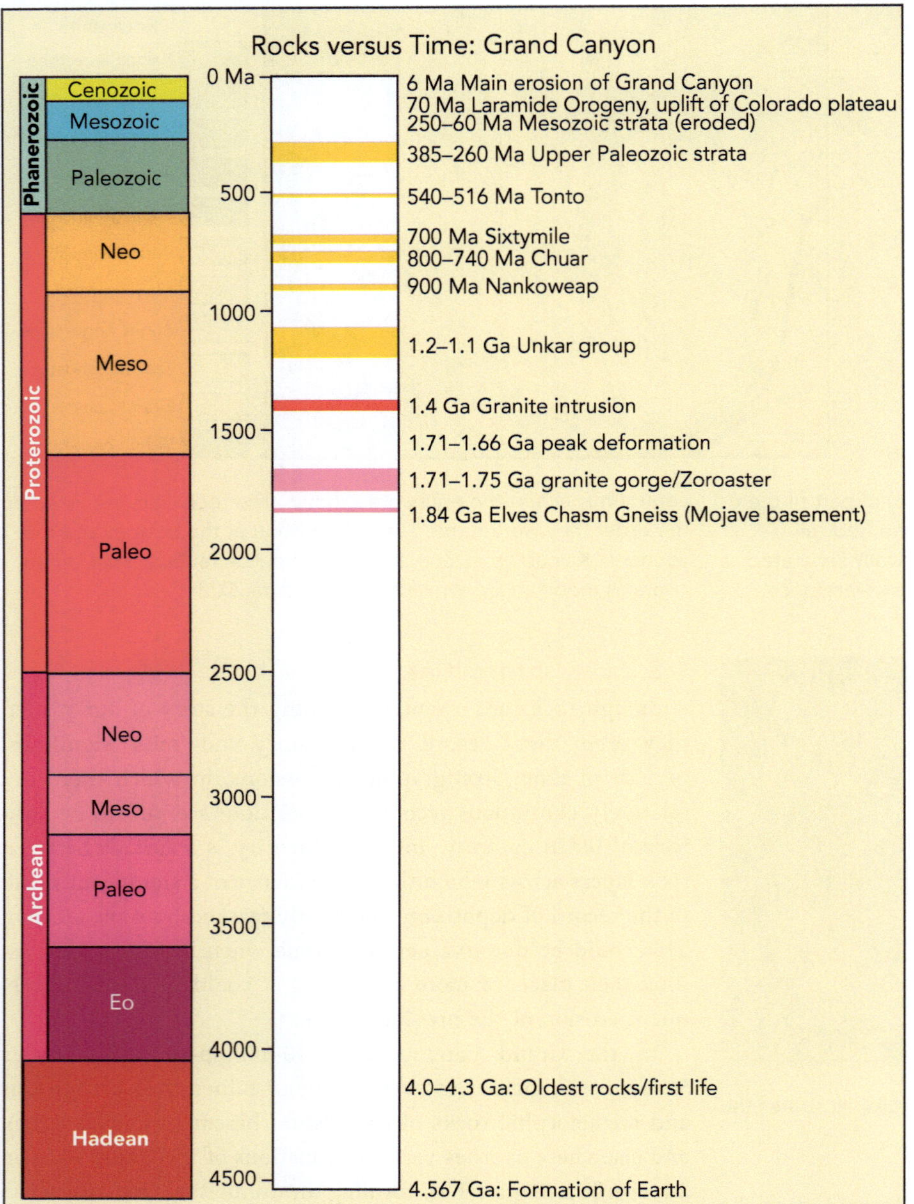

Figure 1.19 Time stratigraphy of the Grand Canyon.

be as important in deciphering Earth history as the rocks themselves.

The relatively flat-lying Tapeats Sandstone (Figures 1.1 and 1.4(a)) records the beginning of a long-lived rise of sea level or **transgression**. The contacts between the Tapeats, Bright Angel Shale, and Muav Limestone formations, which form the Tonto Group (Figure 1.15), are gradational and record an overall deepening of water in which the rocks formed, from sandstones, formed from sand deposited in shallow beach environments, into shales, formed in deeper oceans, and eventually the open marine limestones of the Muav. Tracking of the formations in the Tonto Group across the Grand Canyon (Figure 1.20) shows changes in the proportion and thickness of the limestones, sandstones, and shales that indicate progressive deepening

across the area as the old land surface was transgressed by an advancing sea. Tracking the occurrence of fossils like trilobites also shows that shales of the Bright Angel in the eastern sections are equivalent in age to the limestones of the Muav Formation to the west, and conversely, shales of the Bright Angel in the west are the same age as the Tapeats Sandstone to the east. The Tonto Group is one of several long-lived transgressive sequences capped by a major unconformity, representing a period of erosion that can be recognized across North America (Figure 1.21).

In some cases, major erosion results from a fall of sea level rather than tectonic uplift and tilting. In Chapter 16 we discuss the time, about 5.5 million years ago, when the Mediterranean Ocean completely dried out. The rivers that fed into the ocean

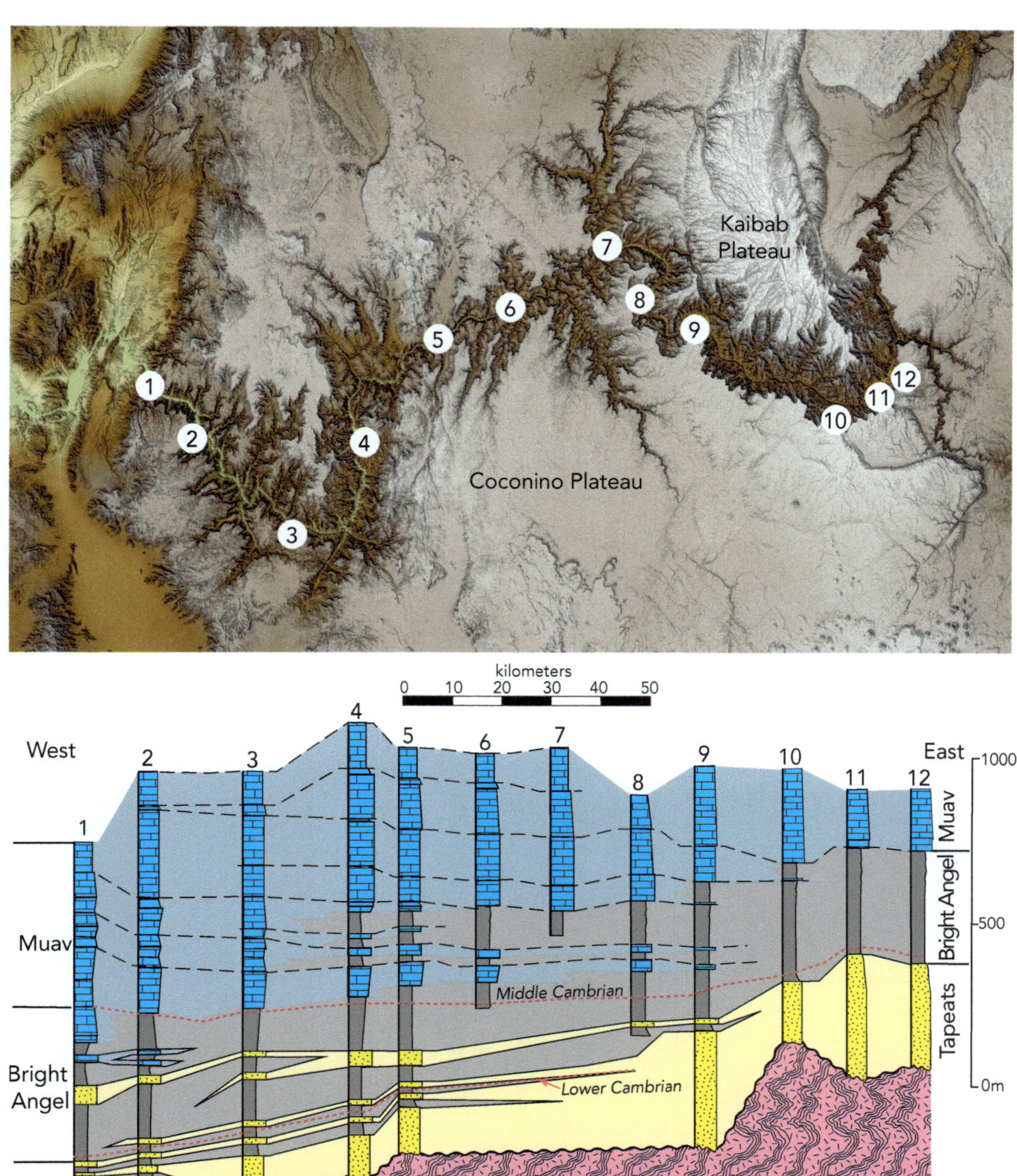

Figure 1.20 Stratigraphic cross section of the Cambrian Tonto Group with locations of measured sections along the Grand Canyon. The transition from sandstones through shales and into limestones indicates an overall rise of sea level during the Sauk transgression. The thicker sandstones to the east and thicker marine limestones to the west indicate that the transgression occurred from west to east and that an open ocean lay to the west. The red dashed lines mark fossil horizons of trilobites that are lower and middle Cambrian in age. For rock types, see key in Figures 1.14 and 1.15. Source: (top) FL-Maps / Shutterstock.

suddenly faced the sloping walls of kilometers-deep canyons and carved a series of side canyons, up to 1.5 kilometers deep, as they flowed down to the exposed ocean floor. These erosion surfaces lack the angular discordance that defines angular unconformities – strata above and below are both horizontal or tilted at the same angle, but they still represent missing time, and they are referred to as **disconformities**, which are more commonly diagnostic of sea-level fall than tectonic uplift.

The relatively continuous deposition of the Tapeats Sandstone, Bright Angel Shale, and Muav Limestone is punctuated by an undulating erosion surface that is filled with river valley deposits of the irregularly distributed Devonian Temple Butte Formation and overlain by the Mississippian Redwall Limestone (Figure 1.22). If we look at the stratigraphy of the Grand Canyon plotted in geological time (Figure 1.19), you can see that at this disconformity, the entire Ordovician, Silurian,

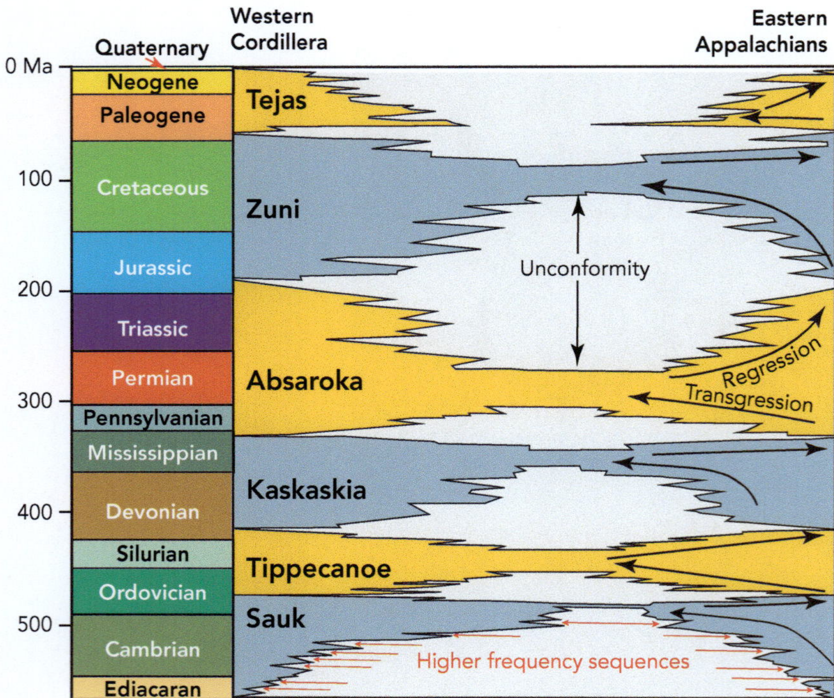

Figure 1.21 Sloss sequences. Regional compilations by Larry Sloss show that the stratigraphy of North America could be divided into six major sequences of rocks bounded by regional unconformities that show larger time gaps in the continental interior and more preserved rock records at the continental margins. Each sequence records a major transgressive-regressive couplet. The zig-zags in each sequence show higher-frequency sequences. Source: Adapted from Sloss (1963). © Geological Society of America. Used with permission.

Figure 1.22 Temple Butte disconformity. Source: Steve Newton, National Center for Science Education.

and much of the Devonian periods are missing, representing a gap of at least 66 million years. Although we appreciate the remarkable record of rocks in the Grand Canyon, in terms of geological time, the unconformities record more gaps than duration of time that the rocks record. Part of this reflects the fact that the Grand Canyon is found within the Laurentian continent, where even at the times of deepest sea level, the seas were never more than about 100 m deep. To find a more complete rock record, we must go elsewhere, such as the Appalachians (Figure 1.21), where a thicker section with fewer unconformities is preserved.

There is a fourth type of unconformity, called a **paraconformity**, marked where the layers above and below a contact appear to have no break, but in which examinations of fossils indicate a gap in time. Rather than erosion, these are surfaces across which there is a significant period of non-deposition, but they require careful analysis of the fossil content discussed in the next section.

KEY POINT

Unconformities are surfaces that mark periods of erosion, uplift, and non-deposition and can be as important in understanding the geological history of an area as the rocks themselves.

1.3.4 Fossils

A fossil is evidence of past life and can be either the physical remains of once living organisms, or impressions or traces made by them in sediments (now preserved in rocks). In some cases, hard parts such as shells or skeletons are preserved as constituents of sedimentary rocks (Figure 1.23), and in other cases, parts of the organism (originally soft or hard) are "mineralized," replaced by different minerals, such as is observed in petrified wood where the woody structure of trees is commonly replaced by microscopic quartz. Sometimes, the impression, or cast, of the hard part is all that remains, but in rarer cases, impressions of soft parts can be found, such as tracks and trails. Many sedimentary rocks also preserve **trace fossils** (Figure 1.24), disruption of the sediment by organisms that live on or in the sediment, recording animal behavior, such as burrowing, even though no parts of the organism that did the burrowing remain.

The oldest fossils in the Grand Canyon are **stromatolites** (Figure 1.25). These are found in the limestones of the Bass Formation of the Unkar Group, and consist of thinly laminated, bulbous mounds, interpreted to be formed by colonies of sticky bacterial mats that covered the ancient seafloor. We will discuss the origin and history of early life in detail in Chapter 7, but for now we will make the point that for most of Earth history, the dominant life forms were single celled. **Metazoans** – multi-celled animals – appeared around 570 Ma, 170 million years after deposition of the youngest formations in the Grand Canyon Supergroup. The younger Paleozoic layers contain a much more diverse range of fossils.

Figure 1.24 Short and longer sand-filled tubes made by burrowing animals (probably worms), Tapeats Sandstone.

Figure 1.25 Stromatolites, Bass Formation, Grand Canyon. Field of view is about 1 meter.

Figure 1.23 Examples of Cambrian trilobites: (a) *Olenellus sp.*, Lower Cambrian; (b) *Olenoides superbus*, Middle Cambrian; and (c) *Kochina sp.*, Middle Cambrian. Source: Photos by JPB. With permission from Houston Museum of Natural Science (HMNS).

Some fossils, such as those in carbonate reefs, are found in the same place in which the organisms lived, and thus provide direct information about the local environment. Fossils may be used to determine whether an environment was on land (terrestrial) or in the ocean (marine). Fossilized plants, for example, may indicate that the environment was terrestrial, whereas fossil coral would indicate a marine environment. Fossils are often essential in the assessment of the environment in which the sediment formed because many organisms that are preserved as fossils (or that produce traces that become preserved), only like to live in certain environments – fossil coral, for example, is indicative of a shallow tropical sea.

However, fossils, like any other particle, may be transported. Rivers may wash plant material out to the sea, such that the observation of broken bits of fossilized plant material, commonly observed in marine clays and silts, does not require that environment to be terrestrial. However, terrestrial plants are a relatively recent addition to our planet and are not found in rocks older than the Paleozoic (see Chapter 11). It can be harder to interpret the environment of pre-Paleozoic rocks (often referred to as *Precambrian*), such as the Grand Canyon Supergroup, because they generally have fewer diagnostic fossils.

Fossils also provide critical information regarding the age of deposition of sedimentary rocks. In Chapter 6 we will explain evolution of life on Earth, but for now the critical point is that over geologic time, new species appear, and older species become extinct. Extinction is a one-way process. Once a species goes extinct it never reappears. Therefore, specific fossils are diagnostic of the particular geological time in which the organisms lived. Indeed, the very names we use to designate the great geologic eras emphasize the life and fossils found within. The term **Paleozoic** means ancient life, **Mesozoic** means middle life, and **Cenozoic** means recent life (see Box 1.1) and the name of the eon that that encompasses all these eras, *Phanerozoic*, means "evident life." For example, trilobites (Figure 1.23) are only found in Paleozoic rocks as they went extinct at the end of this era, so if you find a trilobite in a rock it must be Paleozoic in age. Conversely dinosaurs are only found in Mesozoic rocks (the exception being birds, which we will discuss in Chapters 12 and 13). **Biozones** are periods of time characterized by age-diagnostic fossil species or fossil assemblages (i.e., several species). Where undisturbed and apparently conformable surfaces show missing biozones, paraconformities can be identified.

> **KEY POINT**
>
> In addition to providing the record of the history of life on Earth, fossils provide critical information on the age of rocks as well as environment of deposition.

1.3.5 Environments and Sea Level

In reading sedimentary rocks, geologists are interested in determining the **depositional environment** (Figure 1.26). This includes deciphering the physical, biological, or geochemical *processes* by which sediment was transported and deposited as well as the *place* of deposition. In general, sedimentation occurs

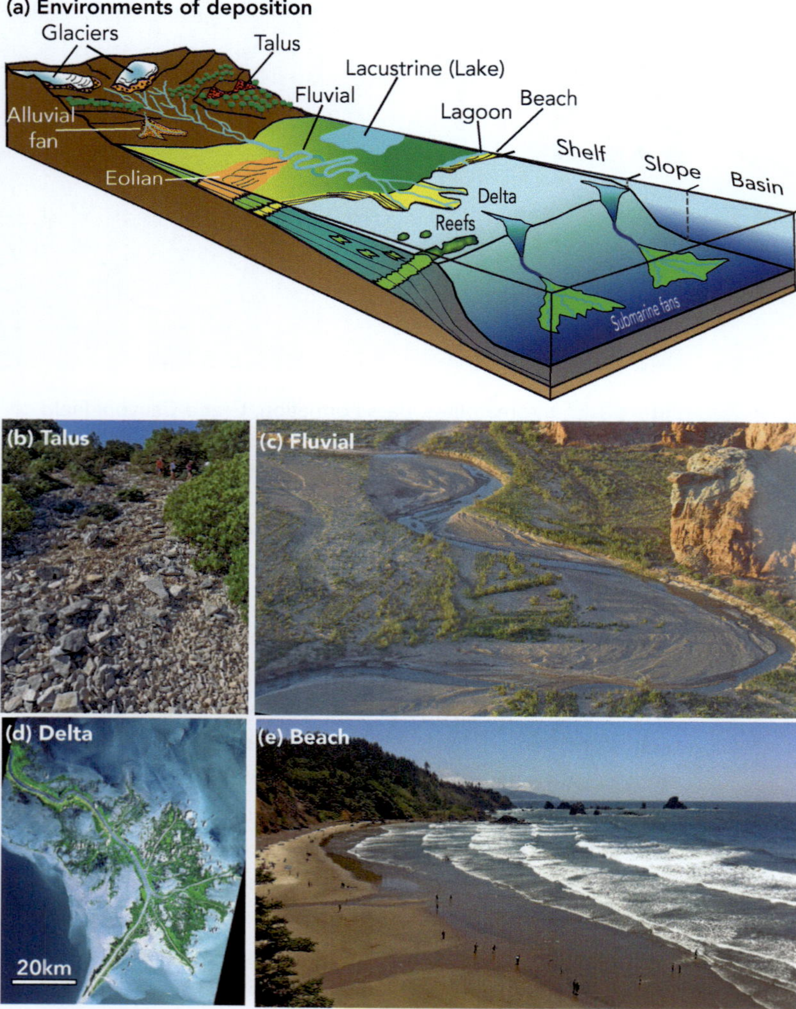

Figure 1.26 (a) General environments of deposition of sediments. (b) Rock fall forming a talus cone at the base of a scarp, Sardinia, Italy. (c) Gravel and sand bed, Fremont River, Utah. (d) Satellite image of the Mississippi Delta, USA. (e) Sand beach along the Oregon Coast fed by small rivers. Source: (d) Satellite image of the Mississippi Delta, USA (PD NASA).

in **sedimentary basins**, low areas on the surface of the Earth, which can include oceans, lakes, or any areas where **subsidence**, or sinking of the land surface, occurs. The Grand Canyon Supergroup is thought to have been deposited in a variety of lowland and marine areas, even though it overlies rocks that originally lay 25 km below the surface of a now-vanished mountain chain. In addition to formation of low areas by subsidence, such as can happen during rifting, when tectonic plates split apart (more on this in Chapter 5), deposition in sedimentary basins is also controlled by sea level, which rises and falls over time, and which we will examine in greater detail in Chapter 18. Of course, tectonic movements can also cause uplift, tilting, folding, and erosion, resulting in the various unconformities discussed above.

Figure 1.27 Unkar conglomerates. Source: NPS / Alamy Stock Photo.

The oldest sedimentary rocks in the Grand Canyon are conglomerates of the lower Bass Formation of the Unkar Group (Figure 1.27). Conglomerates originate as gravels, and the **sorting** (defined as the range of different particle sizes), size, and angularity versus roundness of particles provide clues as to how far sediment has traveled from the source area (Figure 1.28). Imagine rolling a car over a cliff. As it rolls, bits break off and the car becomes smashed and bent. Boulders may be stiffer than cars, but as boulders roll down a hill, edges are broken off, and if they are transported far enough, they may become smooth and rounded, and smaller in size as a result of bits breaking off. The coarse (large) nature of the Bass conglomerates may indicate deposition closer to, or **proximal** to, the sediment source area, in contrast to the overlying Hakatai Shale, which consists of mud and fine (small) sand, indicating deposition farther, or more **distal**, from the source area (Figure 1.28). However, the roundness of the **clasts** in the Bass indicate that the clasts tumbled and rolled, forming well-rounded boulders and cobbles, likely deposited by flowing water. The composition of clasts in the Bass Formation include rocks similar to the metamorphic and igneous Vishnu basement rocks, which are therefore likely to be locally derived (i.e., proximal), but they also contain quartzite clasts that are not local, indicating that some of the sediment was transported from source areas much farther away.

The Bass conglomerates are gradually overlain by limestones rich in stromatolites (Figure 1.25), which represent deposition in a shallow sea and indicate an overall increase in relative sea level through time. We say it is relative, because in this case we cannot be sure if global sea level rose, or the basin experienced a local increase in subsidence, or a bit of both. Regardless of

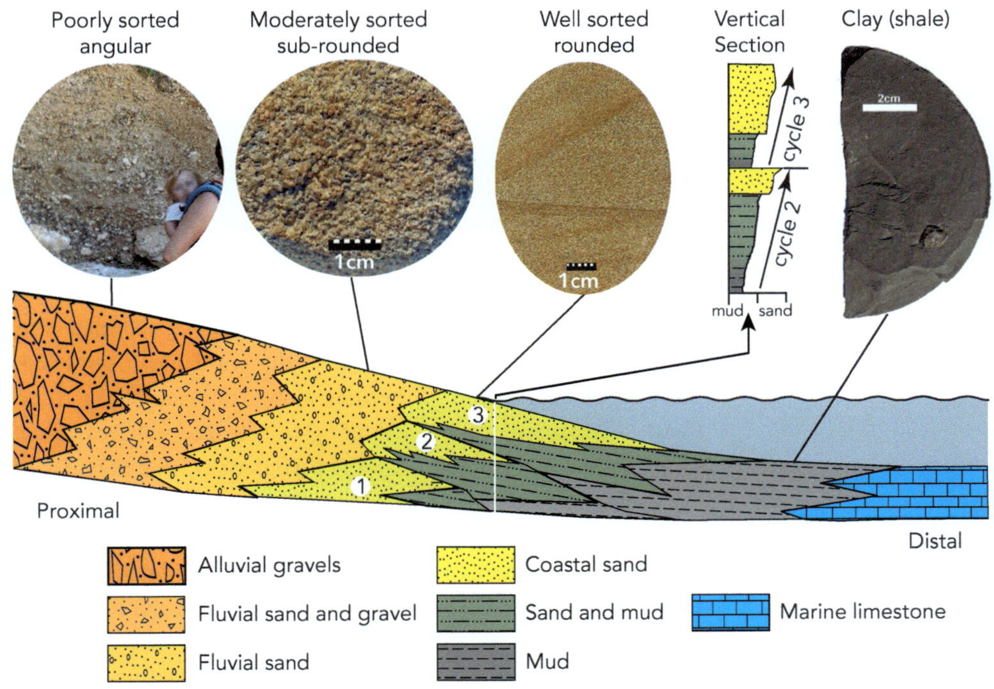

Figure 1.28 Sorting and rounding increase the farther sediments are transported from the proximal source area. In contrast to Figure 1.20, this shows an overall regression where more proximal sediments build (prograde) seaward and overlie more distal sediments as the shoreline retreats seaward. Three cycles of sediment progradation can be seen, but only cycles 2 and 3 are found in the measured section. The cycles of progradation show up as coarsening-upward sequences, where shallow marine coastal sands, such as the beach and delta deposits shown in Figure 1.26 overlie muds deposited in deeper environments. These upward-coarsening cycles are called **cyclothems** and are commonly the result of short-term climate cycles.

which process was more important the net effect is the same: the sediments record an overall change from shallower more energetic proximal environments, associated with deposition of the conglomerates, to a more distal environment characterized by limestones. The overlying Hakatai Shale then shows evidence that it was deposited in a somewhat shallower water environment. This includes wave ripples (Figure 1.29(a)), indicating that it was deposited in a standing body of shallow water where wind blew across the sea and the surface waves molded the loose sediment on the seafloor into the ripple forms, and mudcracks (Figure 1.29(b)), which indicate that the area occasionally dried out. Mudcracks are one of these diagnostic features that provide direct evidence that a mud has been exposed to the air and therefore the environment of deposition could not have been deep. However, the lack of conglomerates in the Hakatai Formation indicates that it was still a more distal environment than the Bass conglomerates.

In interpreting sedimentary rocks, geologists use observations of sedimentary structures, such as the ripple marks mentioned above, to indicate the presence of standing water, such as an ocean or lake. Much higher in the Grand Canyon in the Coconino Sandstone, we see large-scale cross beds (Figure 1.30 (a)). They are called cross beds because the sandy layers (i.e., the

Figure 1.30 (a) Large-scale cross beds indicate eolian dunes in an arid desert, in the Permian Coconino sandstone, Grand Canyon. (b) Trackways (*Chelichnus gigas*) of five-toed reptiles (highlighted with dashed white lines) can be seen covering the surfaces of beds.

Figure 1.29 (a) Wave-rippled and (b) mud-cracked sandstones in the Hakatai Shale. Source: (a) and (b) NPS photo by Erin Whittaker from the Grand Canyon Trail of Time, an interpretive walking trail that focuses on Grand Canyon's vistas and rocks.

beds) dip at a steep angle compared to the top and base of the formation, which are relatively flat. The cross beds are interpreted to have been deposited in an **eolian** or wind-blown desert environment, in which the cross beds were formed into dunes by the wind (Figure 1.31). As the wind blew, sand was moved in the direction of the flow along the back of the dune (stoss side). Larger grains roll along the surface, whereas finer sand and silt are suspended by the turbulence of the flow. Once the grains reach the top part of the dune (termed the crest), the larger grains flow down the lee side, whereas the finer sand and silt continue to be carried by the flow and eventually settle further out as the speed of the flow is slower on the lee versus the stoss side. In this way, dipping layers called cross beds are formed. The observation of cross beds can be used to infer the wind direction at the time of deposition (Figure 1.31).

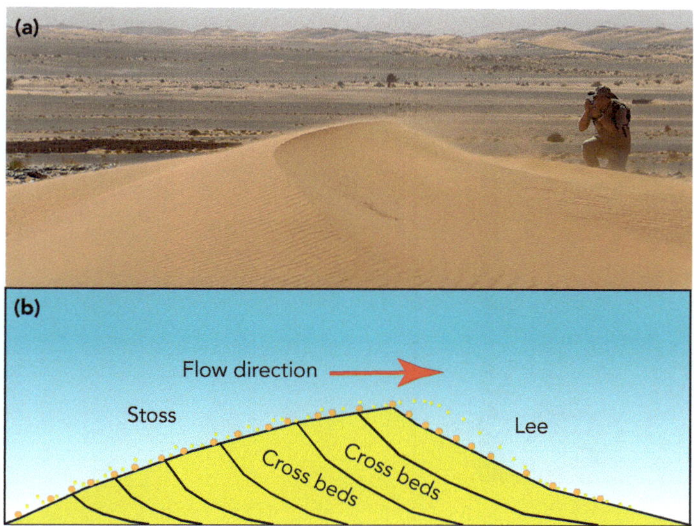

Figure 1.31 (a) Larger-scale dune in Morocco is covered by smaller-scale ripples, both formed by wind. You can see some sand flying off the crest of the dune. (b) Schematic cross-section of a dune, which can form in wind or water when coarser sand grains (orange dots) are dragged along the surface, whereas the finer silt grains (yellow dots) fly off in suspension but eventually fall back to the surface. The difference in the movement of different size grains sorts the grains and makes the cross beds visible.

Very similar dune features form underwater in rivers and can be distinguished from their eolian counterparts by typically having a coarser grain size and being smaller in scale. Also, in the case of the Coconino, tracks of animals, probably a type of reptile, are seen on the exposed surfaces of these ancient cross beds (Figure 1.30(b)). The combination of sedimentary structures and fossils allow a robust interpretation of the Coconino as being deposited in a relatively dry desert.

Combining evidence based on sedimentological observations, such as diagnostic rock types and sedimentary structures, as well as paleontological observations, such as body and trace fossils, allow details of the environments of deposition and how they change to be deciphered.

In interpreting ancient sedimentary environments, geologists commonly use observations of present-day environments of deposition to infer past conditions (Figure 1.26). For example, most limestones today, and particularly coral reefs, grow in shallow, warm, tropical to subtropical environments, away from places where rivers feed mud and sand into the coast. These are the kinds of tropical places like the Bahamas that many people like to visit during the cold winter. The Bass Limestone, in the Unkar Group, is inferred to have been deposited close to the equator in a subtropical environment. We can test this by looking at the remnant magnetic field of associated volcanic rocks, in which magnetic minerals, such as magnetite and hematite, can lock in the ancient **paleomagnetic field** of Earth as they cool and crystallize. As we will discuss in later chapters, Earth's magnetic field has switched from North to South as well as wandered around over time, and **paleomagnetism**, the study of ancient magnetic fields locked into

rocks, can be used to decipher how the Earth's continents have moved over time and the paleo-latitude at which sedimentary rocks were deposited. Of course, in Unkar time there were no land plants, much lower atmospheric oxygen, and nothing living more complicated than simple bacterial mats, so it would be a very different experience sun-tanning on an Unkar Beach a billion years ago, versus a modern-day tropical paradise with coral reefs, fish, and lush palm trees. Clearly, Earth has undergone many remarkable changes over time.

As we continue up section in the Unkar Group (Figure 1.14), we observe alternations of shales and sandstones of the Shinumo Quartzite and Dox Formation, mostly deposited in shallow marine deltaic, intertidal, and **fluvial** (river) environments capped by the 1.1-billion-year-old Cardenas basalt. The sandstones can be correlated farther to the east to Texas, where they correlate with terrestrial (alluvial) conglomerates and cross-bedded eolian sandstones of the Hazel Formation. This would have looked more like the deserts of the Sahara or Death Valley, in California, but again without any plants or animals. Several stories can be told here. The eolian sandstones in Texas indicate a relatively arid climate. But where does all the sand come from?

Another tool that can be used to determine where sediment was originally sourced from are sand grains comprised of the mineral **zircon**. As we will discuss in more detail in Chapter 4, zircons contain small amounts of radioactive uranium that decays to lead at a predictable rate, and it is possible to determine the age of zircon grains by measuring the contained uranium/lead ratio. Doing so reveals a high proportion of zircon grains that are around 1,100–1,170 Ma. The broader geology of North America shows that there was a major mountain chain building event along the eastern margin of Laurentia at this time, referred to as the **Grenville Orogeny**, which reflected the collision, and assembly, of a super-continent named **Rodinia** (Figure 1.32). These uplifted mountains shed their debris westward, and this sediment came to rest in Texas and Arizona as the Hazel and Dox formations. Throughout this book we will see examples of sedimentary wedges deposited adjacent to mountain chains, even though the mountains may be long gone, and pieces of the puzzle and pages of the book removed.

The 1,600 m-thick Chuar Group lies unconformably over the Unkar Group. The Chuar contains meters-thick alternating layers of shale, dolomite (a magnesium- and calcium-rich carbonate that represents a lightly altered limestone), and sandstone (Figure 1.33). The shales were deposited in water a few tens of meters deep and the dolomite and sand units show evidence such as wave ripples indicating deposition in shallower water, as well as evidence of exposure to the air, such as mudcracks. These define shallowing-upward cycles of deposition, commonly referred to as *cyclothems*. Over 300 of these cyclothems have been identified in the Chuar Group. The Chuar Group is thought to represent a duration of a few tens of millions of years and the number of cyclothems indicates that they have a frequency of several tens of thousands to a hundred thousand years. These rapid alternations of sea level are likely

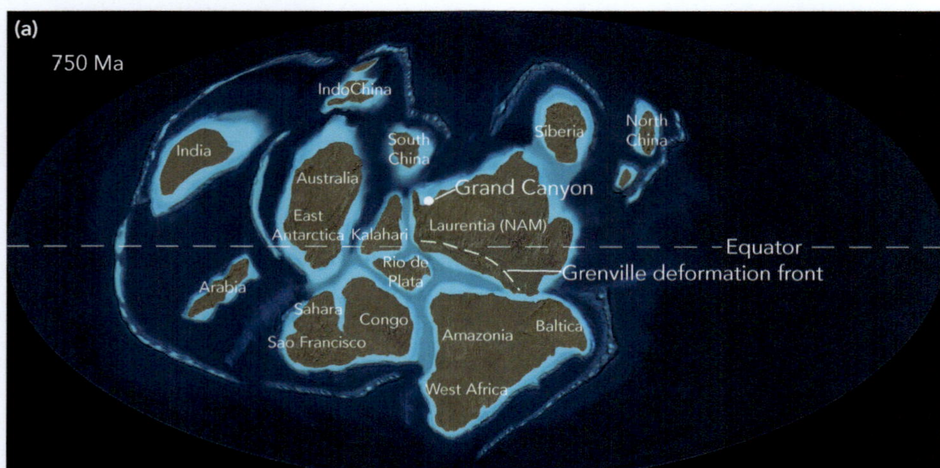

Figure 1.32 (a) Paleogeographic map of Earth at 750 Ma showing supercontinent Rodinia. Note that Laurentia is rotated 90° compared to today and lies on the equator. (b) Close up of Laurentia showing the Grenville mountain belt and the distance to the Grand Canyon. Source: © DeepTimeMaps.

Figure 1.33 The arrows mark meters-thick alternations of shale and dolomite which represent shallowing-upward cyclothems that record rapid (10–100 thousand year) climate-driven cycles of sea-level rise and falls in the Chuar Group.

the result of glaciations caused by rapid climate changes that we will discuss in more detail in Chapters 8 and 18.

The oldest Paleozoic layer in the Grand Canyon is the Cambrian Tapeats Sandstone, a cross-bedded sandstone containing marine trace fossils that we attribute to the overall sea-level rise associated with the Sauk transgression. The Tapeats is overlain by the Bright Angel Shale, which also contains cyclothems of interbedded shales and sandstones that coarsen upwards. The shale indicates continued deepening of the water, but the coarsening-upward cyclothems represent shorter-term regressions, times when shorelines were retreating as sea level fell. The Bright Angel in turn grades into the Muav Limestone, indicating the peak of a major transgression (Figures 1.15 and 1.20). The Muav Limestone is about 505 Ma old and is unconformably overlain by Mississippian rocks of the 385 Ma old Temple Butte and 340 Ma Redwall limestones. This unconformity records a time gap of over 100 million years and represents a period of major exposure and erosion, followed by resumption of marine conditions in the Mississippian Period. As discussed above, the Tonto Group records an overall deepening referred as the Sauk transgression (Figures 1.20 and 1.21). We also see something quite remarkable in the Tonto Group: the first clear evidence of complex animals. Although plants had yet to appear, meaning that terrestrial environments were very unlike today, the Earth's seas were now teaming with a diverse biota, including the trilobites that the Cambrian period is best known for. Although you still would not have found a palm tree to relax under, a time traveler or visitor from outer space could go fishing in the Cambrian seas. In Chapter 9 we will tell the remarkable story of the explosion of multi-celled life, known as the Cambrian explosion.

> ## 🔑 KEY POINT
>
> Sediment composition, textures, such as grain size, sorting, and roundness, as well as key physical and biogenic structures, can all be integrated to interpret ancient depositional environments. Examination of changes in successive layers can be used to infer transgressions and regressions of the shoreline related to sea-level changes. Sedimentary rocks can also provide evidence of tectonic processes and the presence of ancient mountain belts.

1.4 Deformation

In reading the pages of geologic history it's not always possible to tell the whole story just by identifying the type of rock and how it formed. In many cases, we also need to examine the structural geology by studying the deformation the rock has experienced. Deformation of rocks can include folds (Figure 1.34) and faults (Figure 1.35). In Figure 1.34 we show a simple example of folding a flat piece of paper by pushing on one side.

Deformation of rocks may occur at a very wide variety of scales, from slippage along individual flakes of clay in a shale that might be so small it can only be clearly seen in a microscopic **thin section** (Figure 1.36), to regional folds and faults that can form mountain chains. Structural geologists routinely integrate observations at a wide variety of scales to understand

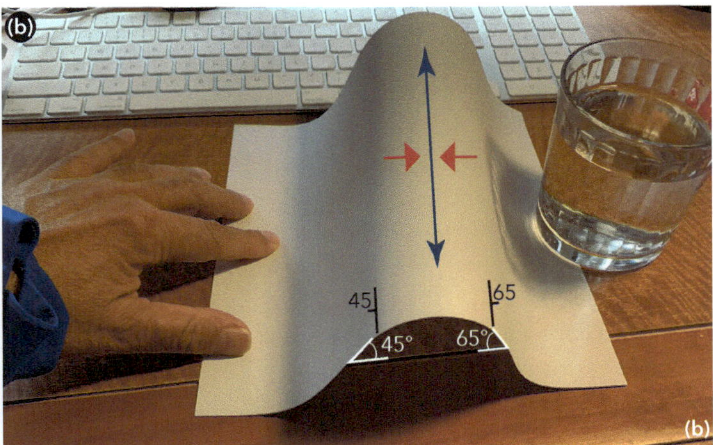

Figure 1.34 Formation of a fold by compression. The orientation of the dipping sides of the folded paper can be measured by the orientation of the plane (called strike marked by the black line with arrows, and the angle of inclination (45°) called the dip that is measured perpendicular to the strike.

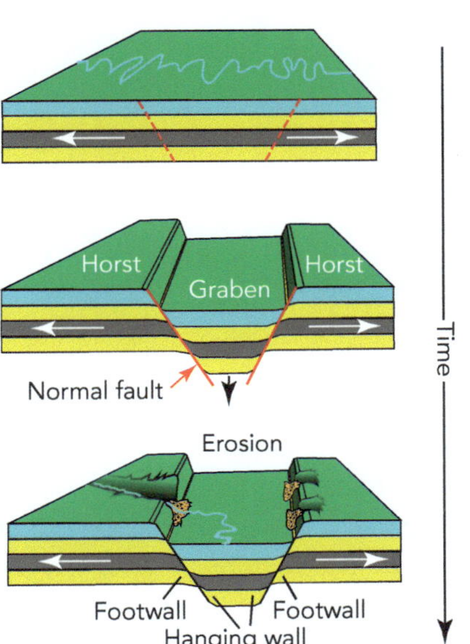

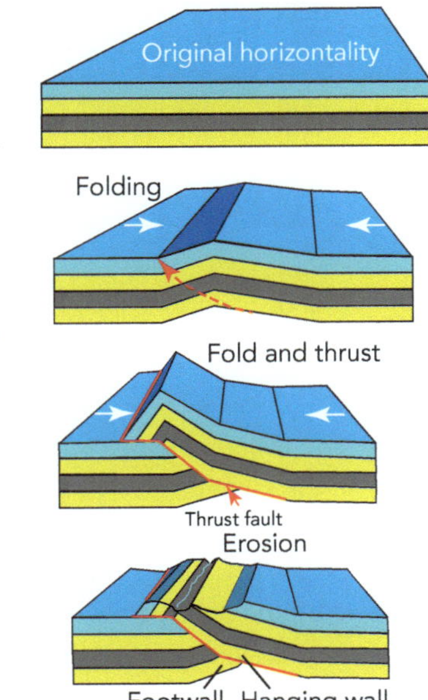

Figure 1.35 Development of folding and faulting in (a) extensional and (b) compressional regimes through time. Note bending of layers along the normal faults in (a). The thrust fault, labelled in (b), is a type of reverse fault and is also associated with folding of the rocks below and above the fault.

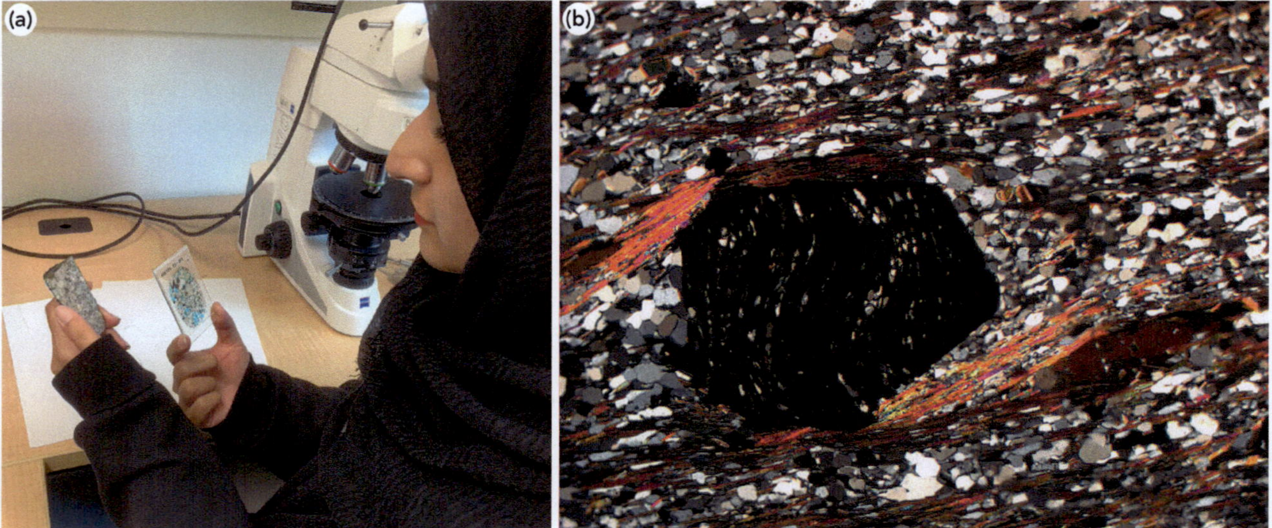

Figure 1.36 (a) Student holding a thin section and hand sample of conglomerate. (b) Thin section of a mica-garnet schist. The garnet is the black crystal in the middle flanked by brightly colored elongate micas and white to gray quartz crystals. The garnet rotated clockwise as it grew, marked by the s-shaped inclusions of quartz and mica. Source: (b) From Jackdann88, CC BY 3.0 <https://creativecommons.org/licenses/by/3.0>, via Wikimedia Commons.

the deformational history of an area. Studies at the smaller scale commonly provide insight into the larger scale and vice versa. Physical deformation may also be accompanied by metamorphism, thus allowing both pressure and temperature regimes to be estimated.

In the Grand Canyon, we can see intense deformation of the Vishnu basement rocks, shown by the vertical layering (Figure 1.4). We see significant tilting of the Grand Canyon Supergroup, forming the angular unconformity with the overlying Tapeats Sandstone (Figure 1.1) and, with more regional observations, subtle deformation of the younger Paleozoic rocks. The type of deformation a rock experiences is a function of the temperature and pressure, and the rate at which it is being strained. When rocks are hot (more than about 250 °C or at depths greater than at least 7 km), they are more likely to bend without breaking, forming folds. This kind of deformation reflects **ductile** behavior, in which the rocks can essentially flow rather than break or crack while still in the solid state (albeit quite slowly). It is similar to the kind of deformation that happens when you play with modeling clay, or for those of you who like to bake, dough. The deformation seen in the Vishnu basement rocks occurred at temperatures quite a bit higher than 250 °C and indicates intense ductile deformation with associated metamorphism. In contrast, when deformation occurs at lower temperatures, rocks will fracture in a brittle manner, rather than bend. More like what happens when you drop a glass or ceramic plate or break a piece of wood at a campfire.

If movement occurs along a fracture, such that the rock on either side is displaced, the break is referred to as a **fault** (Figures 1.35 and 1.37). Where faults are shallow, the rocks lock up as the stress builds up on either side of the fault. At a critical threshold, the rocks fail (or break) and move, so that the rocks

Figure 1.37 Examples of (a) normal faults and (b) reverse fault. Source: (b) Photo from Mike Murphy.

on either side of the fault can become displaced, often in just a few seconds, producing an earthquake.

Tectonic forces can pull rocks apart (extension or rifting, Figure 1.35(a)), push them together (compression, Figure 1.35 (b)), or cause them to slide past each other. The type of faulting produced by these different motions can be diagnostic of different tectonic settings, as is discussed in more detail in Chapter 5.

For example, if layers are offset downwards with respect to each other, these are called **normal faults** (Figures 1.35(a) and 1.37(a)) and typically indicate extension. When layers are moved upwards with respect to each other, these form **reverse** or **thrust** faults (Figures 1.35(b) and 1.37(b)).

The different types of stresses can be recorded by the type of folds and faults observed in rocks. Rocks can be folded and deformed during both compression and extension, but the geometry will be distinctive, and geologists are usually able to decipher the mechanism and timing of deformation by analysis of the types of folds and faults. The orientation of folds and faults, and associated metamorphic fabrics, or textures, such as **foliation**, the alignment of planar minerals such as the micas that are found in slates and schists predominantly in one plane, and **lineation**, a preferred orientation of elongate minerals, provide information about the orientation of the deforming forces, which can in turn be used to interpret larger-scale plate-tectonic forces. Foliation can also be readily seen in thin sections (Figure 1.36).

Analysis of the pressure and temperature paths of the Vishnu basement rocks indicate that they formed at a depth of about 25 km but were brought to the surface as the overlying rocks were eroded away before the Unkar Group rocks were deposited. A major mountain belt, the Vishnu Mountains, had gone and only the once deeply buried roots remain.

The Unkar Group is broken by NW-directed reverse faults and folding, indicating compression likely associated with the Grenville Orogeny formed during assembly of the Rodinian supercontinent. The Chuar Group is clearly associated with an extensional rift, based on the composition of the sandstones and the presence of some large normal faults, which appear to have formed at the same time as deposition of the sediment. This records extension due to the breakup of Rodinia. The subtle deformation of the younger rocks in the Grand Canyon is mostly related to the **Laramide Orogeny**, which started about 90 Ma and is discussed in more detail in Chapter 14.

KEY POINT

Structural geology is the analysis of the deformation of rocks, recorded by folds and faults as well as the orientation of metamorphic fabrics, such as foliation due to the alignment of micas. This can help reconstruct the forces that caused the deformation and the tectonic processes that caused them.

1.5 Historical Summary of the Grand Canyon

Our discussion of the various rock types and how rocks might be deformed is just a brief introduction to how geologists study rocks to tell the story of Earth and other planets. Because of the excellent exposures, the Grand Canyon offers an ideal place to illustrate how many lines of geologic evidence, including the simple concepts found in Steno's laws, a basic understanding of the importance of unconformities, and the observations of cross-cutting relationships, have been combined to tell the geologic story of a region. Applying the principle of superposition, we know that the oldest rocks in the Grand Canyon region are the igneous and metamorphic rocks of the Vishnu basement Complex. The type of rocks in this group and their deformational and metamorphic history tells us they were initially formed in island arcs formed in a now-vanished closing ocean (Figure 1.38(a)) that collided with the ancient Laurentian craton during the Yavapai Orogeny to form the Vishnu Mountains (Figure 1.38(b)). Metamorphism and partial melting produced metamorphic rocks intruded by igneous plutons. Subsequent erosion removed the mountains and exposed these basement rocks that were previously buried up to 25 km (Figure 1.38(c)). The Vishnu Complex was non-conformably overlain by a sequence of sedimentary rocks of the Unkar Group, deposited as far-traveled sediment eroded from mountains formed during the Grenville Orogeny farther southwest, linked to the collision of plates and assembly of the supercontinent Rodinia around 1 Ga. Later, starting around 750 Ma, Rodinia began to break up, creating fault-bounded rifts into which the Chuar Group sediments were deposited (Figure 1.38(d) and (e)). The region was then again subjected to widespread erosion (Figure 1.38(f)), producing an undulating surface onto which the Cambrian Tapeats Sandstone was eventually deposited as the area once again subsided and as sea levels rose (Figure 1.38(g)). The Paleozoic records a history of cyclic changes in environment, from marine to non-marine and back again, punctuated by several regional disconformities.

Regional considerations show that the deposition of sediments did not end in the Paleozoic and the region was later covered by several kilometers of younger Mesozoic sedimentary rocks (Figure 1.38(h)). These are no longer present in northern Arizona but are found in southern Utah. In the approximately 66 million years since the deposition of these Mesozoic rocks, tectonic processes have uplifted the Colorado Plateau, causing the Colorado River to erode down to form the modern-day canyon (Figure 1.38(i) and (j)).

In recounting this brief history of the Grand Canyon, we have relied on information gleaned from the geological languages of stratigraphy, sedimentology, paleontology, geochemistry, and structural geology. The story of the Grand Canyon is well known because of interest generated from the undeniable grandeur of the setting and the excellent exposure of the rocks. The approaches that geologists have used to understand the natural history of the Grand Canyon have been applied to many other areas of our planet, not all of which contain such spectacular exposures. In all cases, the telling of these ancient yet fascinating stories begins when a geologist picks up a rock and ponders its origin.

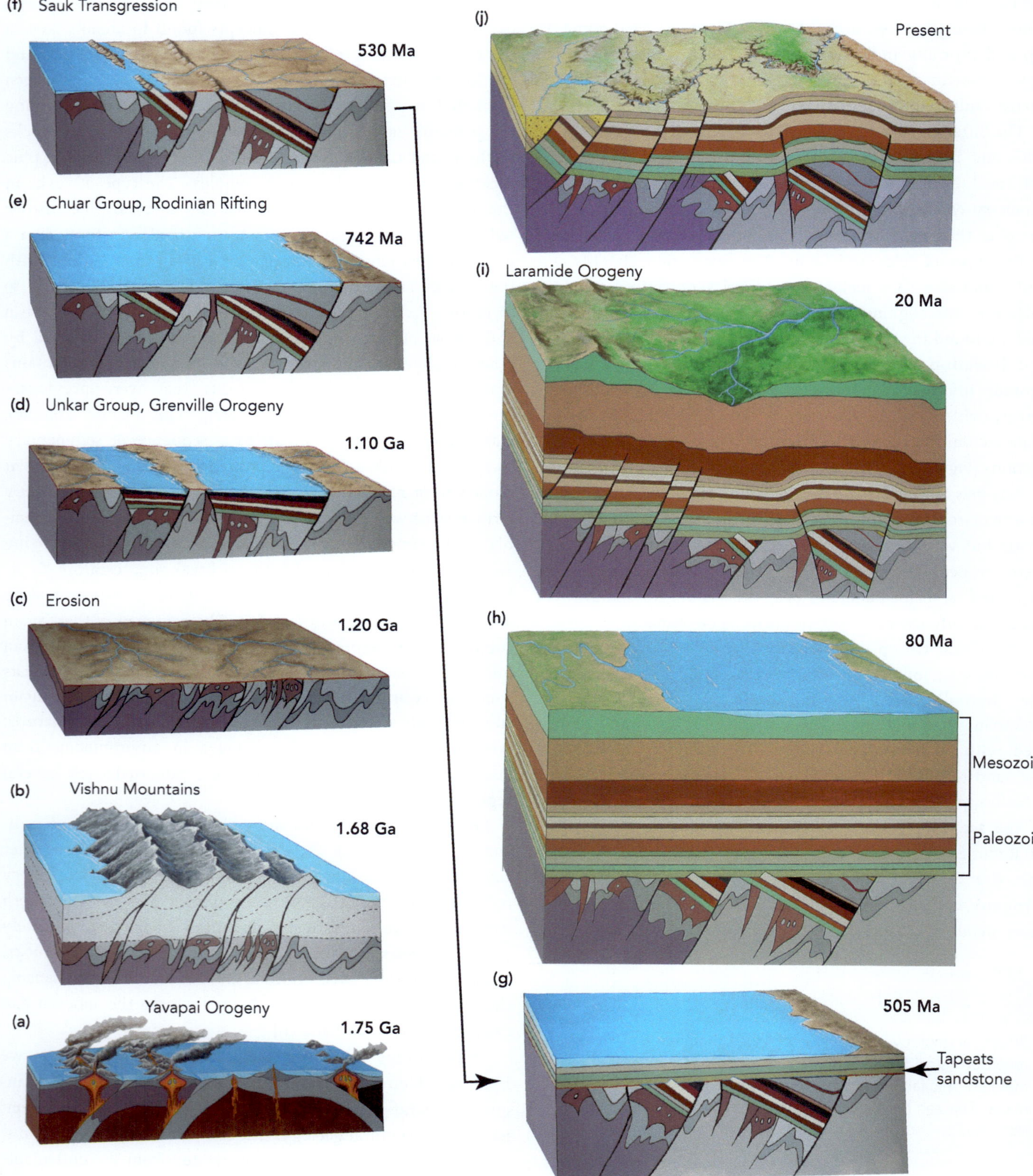

Figure 1.38 Grand Canyon history. Details are discussed in the text. Source: Courtesy of Karlstrom and Crossey (2019).

Key Words

- paleontology
- geochemistry
- structural geology
- stratigraphy
- plate tectonics
- eons
- eras
- periods
- epochs
- ages
- igneous
- sedimentary
- metamorphic
- shales
- sandstones
- schists
- gneisses
- slates
- calcite
- island arcs
- cratons
- Canadian Shield
- Cordillera
- Appalachians
- Laurentia
- orogeny
- Yavapai Orogeny
- magma
- lava
- plutonic
- intrusive
- volcanic
- extrusive
- granites
- gabbros
- pyroxene
- felsic
- feldspar
- mafic
- partial melting
- basalts
- fractional crystallization
- olivine
- clays
- strata
- calcite
- formations
- members
- groups
- Steno's laws
- petrified
- principle of original horizontality
- principle of superposition
- principle of lateral continuity
- igneous dikes
- conformable
- unconformity
- non-conformity
- angular unconformity
- transgression
- disconformities
- paraconformity
- trace fossils
- stromatolites
- Metazoans
- trilobites
- Paleozoic
- Mesozoic
- Cenozoic
- biozones
- depositional environment
- sedimentary basins
- subsidence
- sorting
- proximal
- distal
- clasts
- cyclothems
- eolian
- paleomagnetic field
- paleomagnetism
- fluvial
- Grenville Orogeny
- Rodinia
- thin section
- ductile
- brittle
- fault
- normal faults
- reverse faults
- thrust faults
- foliation
- lineation
- Laramide Orogeny

Further Reading and References

Dehler, C. M., Porter, S. M., Timmons, J. M., and Karlstrom, K. E., 2012, The Neoproterozoic Earth system revealed from the Chuar Group of Grand Canyon, in: Timmons and Karlstrom (eds.), *Grand Canyon Geology: Two Billion Years of Earth's History,* Geological Society of America Special Paper 489, pp. 49–72.

Karlstrom, K., and Crossey, L., 2012, *The Grand Canyon, Trail of Time Companion: Geology Essentials for your Canyon Adventure,* US National Park Service.

Karlstrom, K., and Crossey, L., 2019, *The Grand Canyon, Trail of Time Companion: Geology Essentials for your Canyon Adventure,* US National Park Service.

Karlstrom, K.E., Ilg, B.R., Williams, M.L., Dumond, G., Mahan, K., and Bowring, S., 2012, Vishnu basement rocks of the Upper Granite Gorge: Continent formation 1.84 to 1.66 billion years ago, in: Timmons and Karlstrom (eds.), *Grand Canyon Geology: Two Billion Years of Earth's History*, The Geological Society of America Special Paper 489, pp.7–24.

Sloss, L. L., 1963, Sequences in the Cratonic Interior of North America, *GSA Bulletin*, 74(2), 93–114, https://doi.org/10.1130/0016-7606(1963)74[93:SITCIO]2.0.CO;2.

Timmons, M., and Karlstrom, K., 2012, *Grand Canyon Geology, Two Billion Years of Earth's History*, Geological Society of America, Special Paper 489.

Review Questions

1. What are the main orders of subdivisions of the Geologic Time Scale? Can you name some of the specific units?
2. What are the three types of rocks?
3. What is a protolith?
4. What is a metamorphic rock and how do pressure and temperature control metamorphism?
5. What is the difference between plutonic and volcanic igneous rocks?
6. Explain the concepts of partial melting and fractional crystallization and how they yield magmas and igneous rocks of different compositions.
7. What are the main types of sediment and sedimentary rocks?
8. What are the main principles of stratigraphy (especially Steno's laws) that help understand the order in which layers are deposited?
9. Explain cross-cutting relationships and how they allow us to interpret the order of geological events.
10. What are the different types of unconformities and how do they form?
11. What are fossils and how are they used to interpret environments of deposition and the age of a given rock?
12. How do geologists determine the processes and environments in which sediments and sedimentary rocks form?
13. How do rocks deform and what are the main types of structures?
14. How can you tell if a rock was deformed in a ductile or brittle fashion?
15. Give a brief history of the Grand Canyon.

Palouse Falls at sunset, Palouse Falls State Park, Washington. Source: Mark Lee / 500px / Getty Images.

Chapter 2
The Philosophies of Geology

Assumptions Steer Interpretations

LEARNING OBJECTIVES

- Discuss the development of geological thinking, starting with the recognition of fossils as the ancient remains of once living organisms, versus Renaissance ideas that the fossils were inorganic counterparts of the living world produced by the latent plastic virtue.

- Understand Steno's reasoning that fossils must have originated when the sedimentary rocks that contain them were liquid and record a complex history.

- Recall the observations of William "Strata" Smith that led to the principle of faunal succession and its importance in understanding modern stratigraphy.

- Discuss how observations of apparently sudden extinctions led to the idea of *catastrophism*.

- Define *Neptunism* and how it was used to explain the origin of the so-called primary, secondary, and tertiary rocks.

- Define *Plutonism*, and how this concept was used to explain the origin of igneous rocks and deformed rocks.

- Define *uniformitarianism* and discuss how it relates to the concept of gradualism.

- Describe some of the observations that led J Harlan Bretz to propose a catastrophic origin for the formation of the Channeled Scablands and how this led to the modern concept of *actualism*, a modification of uniformitarianism.

- Understand how geology works as a modern science and the challenge of geology as a historical science.

Introduction

It was during the time of the Greek philosopher Aristotle (384–322 BC) that someone may have first argued that rocks can be viewed as recorders of ancient environments, although just how ancient was still not clearly understood in his time. By walking along the seashore and observing the types of sediments present and the various organisms living there, philosophers as early as Aristotle were able to deduce that the rocks which made up the landscape shared many characteristics of the modern ocean. From this, Aristotle would argue that "where there is sea, there is at another time land." This was perhaps the first application of the now-foundational dictum of geology: "the present is the key to the past" – the idea that we can apply our understanding of the natural world today to gain an understanding of the past. This chapter reviews some of the historical milestones and key people in the development of geological reasoning and concludes with a discussion of how modern geological science works as an academic and scientific discipline.

2.1 The Meaning of Fossils

For thousands of years, it had been noted that some rocks contained objects that had an obvious resemblance to modern plants and animals, an example of which is shown in Figure 2.1.

Figure 2.1 Fossil crinoids, an ancient, stalked echinoderm, related to the starfishes and sea urchins. In medieval times we might have asked whether this is evidence of past life now encased in stone. Was it a plant or an animal, or was it the product of spontaneous formation after deposition by some modifying force? Source: Photo by JPB. With permission from Houston Museum of Natural Science (HMNS).

Before the seventeenth century, observations of rocks that contained bits that had some resemblance to organisms, or their constituent parts, were called "figured stones." But the significance of what we now call fossils was long debated. These "figured stones" had the physical characteristics of the rocks in which they were found, so some observers had trouble suggesting they had an organic origin. Furthermore, some of the forms in these rocks were broadly similar to modern organisms, but others had no recognizable counterpart in the modern world. Even when the forms were similar, some observers were vexed by the fact that many rocks that had forms like modern marine organisms were found far away from the sea and sometimes at great elevation.

To deal with these apparent paradoxes, many argued that the forms that resembled living things were not related to ancient life but were rather the product of a force called the **latent plastic virtue**; a more contemporary English translation is perhaps "inherent formative power." This power was thought to be an intrinsic property of Earth that could cause the transformation of material in the rocks into the forms, such as the fossil in Figure 2.1, that we now see. Part of the enthusiasm for this idea came from the similar notion of the spontaneous generation of life, in which Greek philosophers attempted to give natural explanations for phenomena that had previously been said to be the work of the gods. This archaic idea was that the mineral domain must have symmetry with the animal domain and in this view "figured stones" were thought to represent the inorganic counterparts of living things created by "modifying forces" within the rocks themselves. In fact, the entire planet was considered to be analogous with the living human body, in which the water and magma in the interior moved and flowed like the pumping of blood and other bodily fluids.

Danish clergyman, scientist, and world-class anatomist Nicolas Steno, who we first met in Chapter 1, was one of the first Renaissance thinkers to realize that fossils were the remains of past life and to understand their relationship to the rocks in which they were found. Originally known as Niels Steensen, he was born in Copenhagen in 1683 and later moved to Italy, where he took on the Italian version of his name.

The idea that fossils are the remains of the hard parts of dead animals presented a conundrum. How could hard fossils be encased in hard, solid rocks? Steno's novel theory was that the rocks must at one time have been in a fluid state and later solidified. He reasoned that the layers of sedimentary rock that contain fossils must have been deposited as fluidized layers on top of which shells or skeletal remains of dead animals or plants were deposited, and which later were solidified to form rocks. He also reasoned that where sedimentary rock layers stacked on top of each other, the lower layers were deposited first, trapping fossils, and later new layers were deposited above, forming the stratigraphic record that we see today. Steno's ideas eventually became the foundational laws of stratigraphy, which we reviewed in Chapter 1.

Steno was also interested in how crystals formed in igneous rocks, and he further hypothesized that igneous rocks must also have been originally liquid. We now know that solid crystals are precipitated out of liquid magmas and as they cool the magmas experience a phase in which solid crystals "float" in the remaining liquid before the entire magma becomes solid (see also Chapter 1), and this concept can also be traced back to Steno.

At the end of the eighteenth century, about a hundred years after Steno, William "Strata" Smith played a major role in deciphering the relationships of fossils and their associated sedimentary layers. Smith was a geologist, hired to survey the geology of central England to help the engineers of his day to build canals that facilitated the shipping of coal across the country during the industrial revolution. In his compilation of the data, he noted the "wonderful order and regularity with which Nature has disposed of these singular productions [fossils] and assigned to each its class and peculiar stratum." In other words, he was struck by the observation that different sedimentary rock layers contained specific fossil types that were different from those found in layers above and below. The fossils seemed diagnostic of the age of the layer. At the same time, work on fossils was also being done by French scientists Georges Cuvier and Alexandre Brongniart, both at the vanguard of the Enlightenment movement that ushered in the era of modern science. They showed that it was possible to not only understand that fossils are evidence of past life but that they also provide information about the environment and time of deposition.

The observation that fossils succeed each other vertically in a specific and reliable order, and can furthermore be identified over wide horizontal distances, was identified by Smith as the **principle of faunal succession**. This concept also allowed geological units to be correlated and mapped across large areas. To that end, in the process of surveying, Smith is also credited with making the first published geological map (Figure 2.2). Unfortunately, Smith had financial troubles, spending some time in a debtor's prison, but he was eventually awarded the Wollaston Medal by the Geological Society of London and finally recognized as the father of English Geology.

KEY POINT

Fossils allow geologists to understand the relative age of sedimentary rocks, the environment in which they formed, and the relationship of far-distant outcrops.

Smith's understanding allowed outcrops of rocks to be understood not in isolation but in a regional and ultimately global context by the process of correlation of strata from one location to another using fossils as a guide (Figure 2.2). If, through analysis of fossils, rocks can be understood to have been deposited at the same time, paleogeographic maps can then be constructed, showing the distribution of various environments (e.g., beach, deep sea, lakes, see Figure 1.26) present at a particular time. This then allowed the understanding that these environments are not static, but rather can change through time. And finally, if rocks that were originally formed in the ocean (based on the fossils we now see in them) are today found at the top of mountains, great forces must have been at play that probably required vast amounts of time (we will return to the concept of geologic time in Chapter 4).

But even if a consensus could be found on the meaning of fossils and their utility in interpreting layered rocks, a variety of philosophical and methodological approaches needed to be reconciled before modern geology could come into being.

2.2 Catastrophism

The observation that some fossils are unlike any modern forms led to considerable debate during the seventeenth century.

Through careful examination of the fossil record, Cuvier showed that there were indeed many fossil forms that had no modern counterpart, suggesting that they had gone extinct, never to return. This caused many people of the time some theological discomfort: if God had created Nature according to some divine plan, why did not all creatures survive? Three arguments were used to confront the concept of extinction. First, it was argued that the differences between modern and fossil forms could be ascribed to the normal variation in any group. Second, some thought the modern knowledge of all life on Earth was insufficient at the time to rule out the possibility that fossil forms were the ancestors of a modern form that was yet undiscovered. The third criticism was that perhaps the fossil and modern forms were representatives of the same group (i.e., no extinction) but had undergone some transformation over time. Eventually, through the work of Cuvier, Brongniart, and Smith, the concept of extinction became widely accepted.

This, of course, leads to the question, "Why did they go extinct?" Cuvier surmised that Earth was vastly old and that, for most of that time, conditions were like those we live in today. He reasoned that it would take violent geologic events to cause the removal of a species from the face of the Earth. Cuvier noted that these extinction events commonly occurred at marked points in sedimentary successions that we now know are unconformities, introduced in Chapter 1 (Figure 2.3), and referred to these events as "revolutions." These were eventually called "catastrophes" (Figure 2.3). As an Enlightenment scientist, Cuvier disliked this term because it implied a supernatural cause. Many others, however, were quite comfortable with this nomenclature, precisely because of the way it appeared to reinforce the biblical story of Noah's flood. This was particularly in line with the views of early English naturalists, who assumed the history of Earth was short and marked by biblical catastrophes rather than the product of gradual changes over long timespans (we shall see in Chapter 4 how the Church of England promoted the idea that the Earth was but 6,000 years old, a conclusion Cuvier was not in favor of). Eventually, the term **catastrophism** evolved to refer to the idea that major, short-lived events are responsible for many of the major features visible on Earth.

The concept of extinction is clearly one that scientists are very comfortable with today, but the idea of catastrophism is much more problematic. In the catastrophist view, whenever there is a break in the fossil record (or even a significant change in a stratigraphic succession of rocks without fossils), the change can be explained as being due to a period during which the normal scientific rules that govern nature (e.g., the laws of physics) were not in effect. In this view, if extinctions are caused by supernatural interventions, there is no need to understand scientific principles or apply them to adjacent sequences of rocks.

Cuvier's view was that no supernatural causes were necessary to produce his "revolutions." As we shall see in later

© The Trustees of the Natural History Museum, London

Figure 2.2 Geologic map of William Smith (1815). This is sometimes considered the first geologic map, which Smith produced by comparing and correlating rock types and fossil assemblages in England and Wales. Comparison to work by Cuvier and Brongniart in France showed that this approach could be used over long distances. Source: © The Trustees of the Natural History Museum, London.

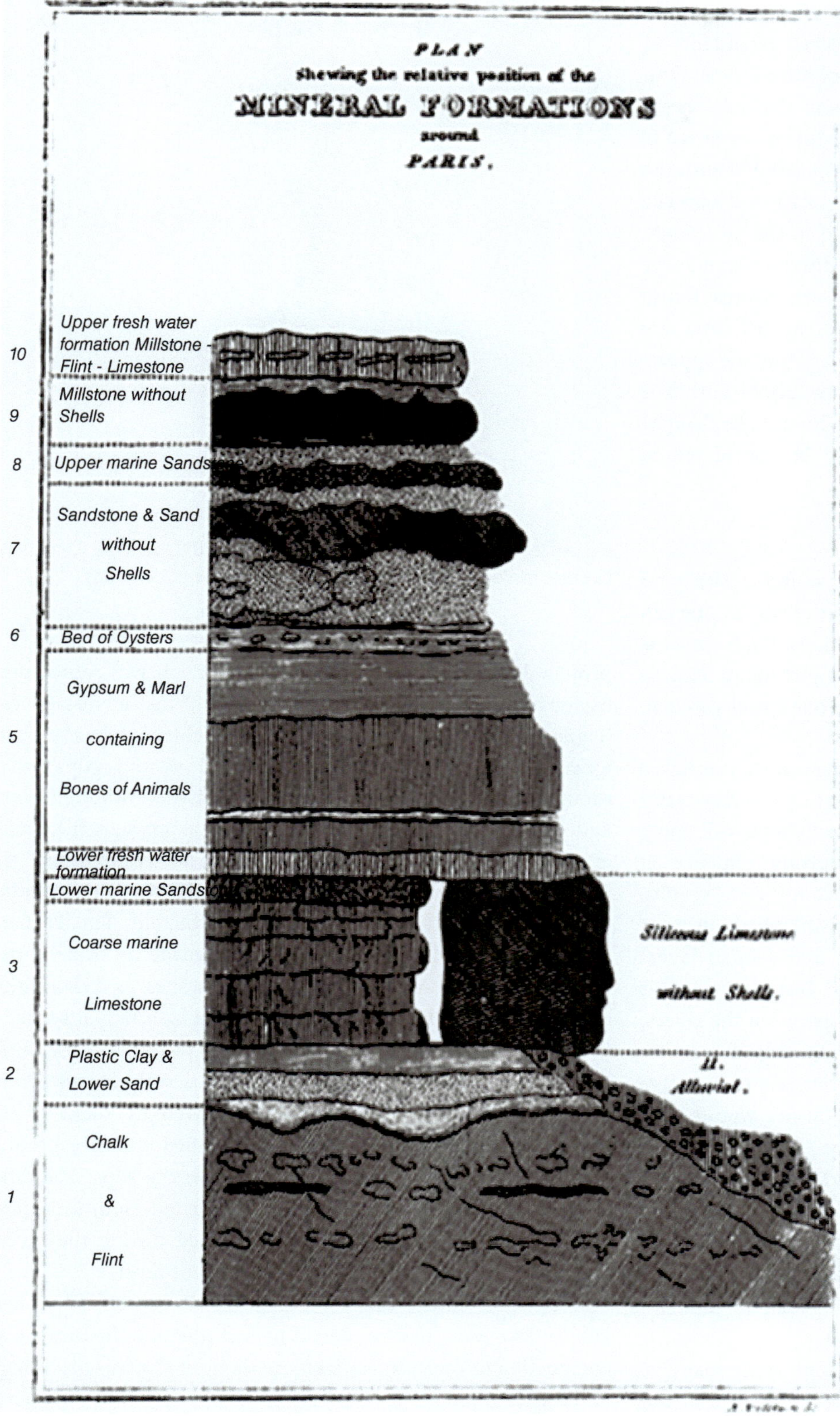

Figure 2.3 Stratigraphic column of the rocks in the Paris Basin by Cuvier and Brongniart. He suggested major breaks in rock type were the consequence of "revolutions," during which the normal rules of science might not have applied. Source: From Cuvier (2010). With permission of Cambridge University Press.

chapters, the greatest extinction events in Earth history are now understood to be the result of unusual conditions. These extinctions do not, however, mark times in which the normal scientific rules of nature, as we understand them today, did not apply. The importance of this view for the succession of rocks carries over to the interpretation of the succession of fossils. If nothing out of the realm of science was happening at the time when there are drastic changes in the rocks (unusual perhaps, but not supernatural), then the evolution of fossils, which can appear gradual or abrupt, must also have occurred without the intervention of supernatural forces. In this view, later organisms (i.e., fossils found in layers above) were descendants of those that came before. Thus, the evolution of life on Earth is an observation, not a theory (see Chapter 6), and the only methodological underpinning necessary to repeat this observation is Steno's principle of superposition. Anyone comfortable with the notion that younger rocks lie above older rocks will observe that life has evolved over time by reference to the systematic changes in the different fossils contained in different layers. For example, rock layers containing trilobites are always found below rocks containing dinosaurs, because the trilobites are not only considerably older, but went extinct before the appearance of dinosaurs. Two essential questions associated with this pattern of fossil distribution in the rocks – Why did life change? and How much time was involved? – will be considered in subsequent chapters.

2.3 Neptunism

In today's world, the barrier to a geologist seeing a variety of rocks is rather low. If one can afford the cost of travel, one can visit essentially any rock outcrop in a few days. Such travel is considered essential for the professional geologist today because geologic *provinces* are so large that staying in one's backyard can limit one's point of view.

During the Renaissance (fourteenth–seventeenth centuries) and ensuing Enlightenment (eighteenth century), when many ideas regarding the fundamental nature of rocks were still being hashed out, where one lived could have a strong influence on one's interpretation. For example, Cuvier's concept of "revolutions" was highly influenced by the Paris Basin, where he spent most of his time. The Paris Basin contains five distinct layers that can be observed throughout the area. Distinctive fossils occur in each of the layers and the contacts between the layers, which we now know are unconformities, mark the "revolutions" between the layers. When viewed from the perspective of the Paris Basin alone, one can see why Cuvier would have been drawn to the concept of revolutions. However, when extended beyond northern France, the same stratigraphic breaks and layers may not be present, and the revolution concept may not be as clear.

Next, we will consider how location influenced two important eighteenth century contributors to geologic thought: Abraham Gottlob Werner and James Hutton.

Abraham Gottlob Werner (Figure 2.4) lived in Germany in the last half of the eighteenth century where he was a professor at the prestigious Freiberg Mining Academy, and by 1800 he was one of the world's most famous geologists. Werner advanced a theory of Earth history in which almost all the rocks and minerals of the crust were precipitated from a universal ocean. This theory became known as **Neptunism** (named after Neptune, the Roman god of the sea).

Based on observations in and near the Alps of southern Germany, where Werner lived and worked, he and his students

Figure 2.4 Abraham Gottlob Werner, 1749–1817. Source: Christian Leberecht Vogel, Public domain, via Wikimedia Commons.

proposed that the geology of the entire Earth was represented by four formations related to the receding of the universal ocean (Figure 2.5). The oldest *Primary Series*, which are made up of igneous and metamorphic rocks of the central Alps, were thought to have been the first to precipitate. Although Steno had established that igneous rocks form as crystals (i.e., minerals) precipitated from a liquid, it was not realized that the liquid was hot magma, so Werner thought that precipitation from colder ocean water was quite reasonable. The Primary Series are overlain by a *Transition Series* made up of deformed non-fossiliferous limestones, muddy sandstones, as well as igneous sills. Werner thought this marked the beginning of sea-level fall that exposed and weathered the rocks of the Primary Series forming the first land, and that storms caused these rocks to be deformed. The Transition Series is followed by rocks of the *Secondary Series*. These are clearly stratified and fossiliferous sedimentary rocks, found in the foothills of the Alps. The third *Tertiary Series* is made of loosely consolidated mud, sand, and gravel, which he interpreted to be formed during the recent past as the universal ocean fell to its present level.

Werner also noted that the Tertiary Series were locally overlain by low-lying basalts, which he considered to be formed by combustion of coals. In contrast, he thought that basalts at high elevation, found in his Primary Series, were the precipitates from the universal ocean. Both ideas were controversial in their time and were attacked by those geologists who thought that the formation of volcanic rocks must be related to heat in the Earth's interior, leading to the idea of Plutonism that we will consider in the next section. Werner's views were likely a consequence of his provincialism: there are many coal beds in Germany but no active volcanoes. Had Werner had the opportunity to travel to sites of active volcanism, such as in southern

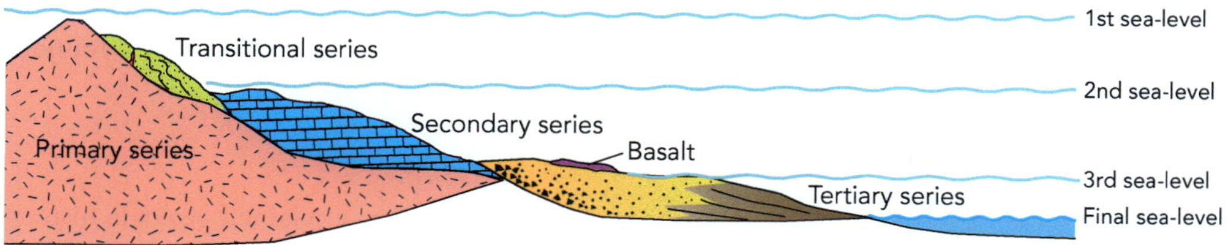

Figure 2.5 Werner's Neptunian classification of rocks.

Italy, he may have realized that coal has nothing to do with the formation of volcanoes.

One of the other difficulties with Neptunism was accounting for the fate of the water that originally covered the entire globe. Werner needed some explanation as to where all this water could have gone as the universal ocean receded. Werner rejected the suggestion that the water was somehow drained into the Earth's interior as too conjectural, and preferred the idea that a passing celestial body somehow drew away the water. We now know that any passing planet big enough to Earth to draw away most of the ocean's water would have had significant effects on the solid Earth as well, for which there is no evidence.

Although from our perspective, some 250 years later, Neptunism is clearly nonsensical, Werner's attempt to view stratigraphy in a global perspective was important because it led to a way of thinking in which local observations are used to invoke global causes. Despite abandoning Neptunism, Werner's legacy lives on. Until the mid 1990s, Werner's Tertiary Period was still formally recognized in the Geologic Time Scale. Also, in 1829, 12 years after Werner died, French geologist Jules Desnoyers named the *Quaternary Period*, building on Werner's scheme, and that name persists today even though we long ago abandoned the idea of a Primary, Secondary, and Tertiary Series formed by withdrawal of a universal ocean.

2.4 Plutonism

One of the key thinkers offering an alternative to Neptunism in the late eighteenth century was the Scottish naturalist James Hutton (Figure 2.6). As Werner was influenced by the geology of Germany, Hutton took many of his ideas from observations of rocks in Scotland and England, which are quite different from those in Germany. In many outcrops, Hutton and his colleagues observed veins of igneous granite intruding into metamorphic rocks on a variety of scales. He concluded that the only way for this to come about would be for the granite to have once been in a liquid state, again following Steno, but Hutton further surmised that the only agent that could turn rocks into liquid was great heat in the interior of the Earth. Although there are no active volcanoes in Scotland, Hutton was influenced by reports of recent volcanism in France and Italy. Hutton also thought that the consolidation of loose sediment into sedimentary rock was primarily due to internal heat, although we now

Figure 2.6 James Hutton, 1726–1797, geologist, painted in 1776 by Henry Raeburn, Scottish National Gallery. Source: Henry Raeburn, Public domain, via Wikimedia Commons.

understand that cementation of sediment to sedimentary rock occurs at relatively low temperatures (typically $<100\,°C$) compared to the temperatures required for metamorphism or melting, and that pressure and the presence of water are also important considerations.

Hutton and his followers further ascribed that the deformation of rocks (something easily observed in Scotland) and the uplift required to produce mountains was driven by subterranean heat. Calling on the effects of internal heat to explain this wide range of geologic activity became known as **Plutonism**, after Pluto, the Roman god of the underworld. Plutonism is sometimes also called Vulcanism (for the Roman god of fire, Vulcan).

Many essential extensions of Plutonism are in direct opposition to the ideas of Neptunism. In Plutonism, granites are often younger than the rocks they are in contact with, whereas Neptunism places granites in the category of the oldest Primary rocks. Neptunism invokes processes that only occurred once, as a single great ocean retreated, whereas Plutonism is an

Figure 2.7 An example of three generations of dikes (white rocks) cross cutting older (gray) rocks. By examining the geometry of the younger dikes, one can discern the order of their emplacement (1, then 2, then 3). Examples such as this led Hutton and colleagues such as James Hall to reject the tenet of Neptunism that all rocks are formed by precipitation from a global ocean.

explanation that allows multiple recurring events. Hutton saw the deformed rocks of Scotland to be the product of deposition, burial, heating, deformation, and uplift, which we now know as the rock cycle. He saw that rocks at the surface were being eroded and that the fragments from those rocks would eventually be buried as sediment and the cycle would begin again.

KEY POINT

Neptunism and Plutonism were both important in the history of thinking about the origin of rocks because they both suggested global causes for local observations.

Thus, in 1800, two starkly different views of how Earth history operated were competing in the marketplace of ideas. Hutton died in 1797 so it was his colleagues and protégés who carried on as champions of Plutonism. One of the most notable of these was James Hall. In 1790, Hall noted that,

> wherever the junction of the granite with the schists was visible, veins of the former, from fifty yards to the tenth of an inch in width, were to be seen running into the latter, and pervading in all directions, so as to put it beyond all doubt, that the granite in these veins … must have flowed in a soft or liquid state into its present position.

The features Hall saw are similar to the examples shown in Figure 2.7. Hall referred to experiments in which mixtures of quartz and feldspar were melted in coal-fired furnaces to produce a flowing glass. Thus, through a combination of experimental and field observations, the merits of Plutonism were gradually appreciated and the concepts of Neptunism faded in popularity during the first half of the nineteenth century.

Many of the nuances – and even some not-so-fine details – of how we understand Earth history were unknown or misunderstood by Hutton and his followers, but the major tenet of Plutonism, that heat from the Earth's interior is an important driving force of geologic change, continues today as a foundation for plate tectonics (see Chapter 5).

2.5 Uniformitarianism

Although Neptunism and Plutonism explained geologic history in fundamentally different ways, they shared a similar point of view regarding the pace of geologic processes. Werner said, "Our Earth is a child of time and has been built up gradually." Plutonists disputed the idea that rocks were formed by past catastrophic processes that no longer operate today and in general favored the idea that most rocks form gradually, whether it be the grain-by-grain deposition that builds sedimentary layers, or the slow cooling of magmas that yields igneous rocks. Of course, even Hutton and his followers realized that sometimes rocks can be formed quickly, but by fast natural processes, such as volcanic eruptions, as opposed to supernatural catastrophes delivered by the hand of an angry god.

Fundamental to Hutton's analysis is the gradual change of Earth over long periods of time. Hutton is perhaps most famous for writing, "the present is the key to the past," and his observations of the present indicated that most changes occur very slowly. Hutton realized that, given enough time, the slow and gradual work of wind and water will ultimately wear down entire mountains, but the effects of such erosion may be imperceptible over the span of a human lifetime.

Hutton's appreciation of the vastness of geologic time was captured in his observations of one of the most famous angular unconformities on the southeast coast of Scotland at a place called Siccar Point (Figure 2.8 and see other examples in Figure 1.1). Hutton observed tilted sedimentary rocks overlain by flatter lying sedimentary rocks. Hutton realized that a great deal of time was required to deposit the first set of layers as flat-lying sediment (recall Steno's law of original horizontality). They were then buried and lithified, and later uplifted, tilted, and eroded. Then another lowering of the land surface was required for the next set of sedimentary strata to be deposited over the unconformity.

We now know that the oldest rocks at Siccar Point are Silurian sandstones. These are overlain by Devonian sandstones and conglomerates, which are about 65 million years younger. Thus, the Silurian sediment experienced burial, transformation into sedimentary rock, and later deformation and uplift. As these rocks were exposed at the surface, they were buried by horizontal layers of younger Devonian strata, forming an angular unconformity. These sediments were again buried and lithified into sandstone. Finally, we can see that the Devonian sandstones are no longer horizontal, indicating that they too have experienced a period of deformation, which we now know is associated with the tectonic assembly of the Pangean supercontinent in the Paleozoic Era (see Chapter 10).

In Hutton's time, one of the fundamental questions of Natural Science being debated was the age of the Earth: is the Earth relatively young or do the rocks and structures seen on its

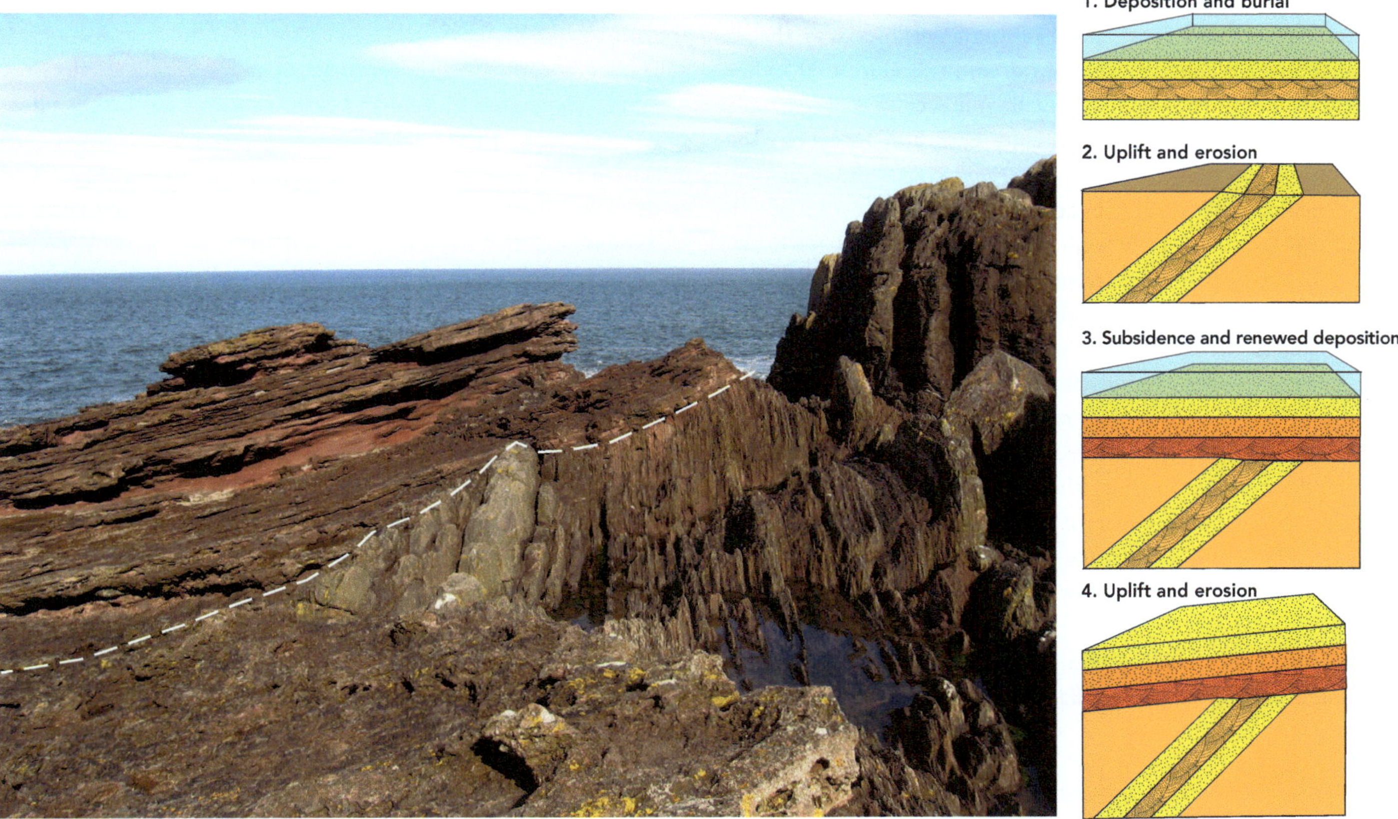

Figure 2.8 The angular unconformity exposed at Siccar Point, Scotland. Source: Dave Souza, BY CC 4.0, via Wikipedia Commons.

surface indicate great antiquity? Hutton advanced the idea that we could best understand the past by better understanding geologic processes that can be observed today. So, with his understanding of modern erosion and sedimentation, when he came upon the angular unconformity at Siccar Point, Hutton understood that a very great deal of time was involved. He couldn't say exactly how long this took, but this location played an important role in his understanding of the great antiquity of Earth. Emphasizing the importance of this field relationship, one of Hutton's companions, John Playfair, later recorded of their visit, "The mind seemed to grow giddy by looking so far into the abyss of time." Indeed, based on the discovery of the unconformity at Siccar Point, Hutton made his oft-quoted comment on the immensity of geologic time by concluding that for the history of Earth he found "no vestige of a beginning, no prospect of an end."

Without the present being a guide to our interpretations of the past, the time associated with geologic observations — even those as dramatic as can be seen at Siccar Point (Figure 2.8) and the Grand Canyon (Figure 1.1) — cannot be deciphered. The idea that understanding modern processes is critical to understanding Earth's past came to be known as **uniformitarianism**, and one of its foremost advocates was English geologist Charles Lyell, who was born in 1797, the year Hutton died. In 1830, Lyell published what was to become the most influential textbook in geology during the nineteenth century, *Principles of Geology*. The subtitle of the book — *Being an attempt to explain the former changes of the Earth's surface by reference to causes now in operation* — clearly sets out the philosophy of his book. One of the sources from which uniformitarianism can be traced was John Playfair's *Illustrations of the Huttonian Theory of the Earth* (1802), in which he stated,

> amid the revolutions of the globe the economy of Nature has been uniform, and her laws are the only things that have resisted the general movement. The rivers and the rock, the seas and the continents have been discharged in all their parts; but the laws which direct those changes, and the rules to which they are subject have remained invariably the same.

Compared to his predecessors, Lyell was relatively well-traveled, having conducted fieldwork in Scotland, England, Wales, France, and Italy. Illustrations from his travels filled his book. With a relatively widespread and interprovincial set of observations as his foundation, Lyell argued that one should assume that processes that we can observe today, such as erosion, are of the same kind and magnitude as those processes that acted in the past. Thus, the products (i.e., rocks) reflect formation by similar processes that have occurred over geological time. Uniformity had to be assumed, he argued, for the study of the Earth to be a scientific enterprise.

Lyell's approach influenced many natural scientists of the nineteenth century, perhaps most notably Charles Darwin.

Darwin who, in an 1844 letter to fellow geologist Leonard Horner, wrote: "the great merit of the *Principles* was that it altered the whole tone of one's mind, and therefore that, when seeing a thing never seen by Lyell, one yet saw it partially through his eyes." Darwin took a copy with him on his circumnavigation of the globe on the HMS *Beagle* (1831–1836) and this view of geology was critical in enabling Darwin to formulate his theory of natural selection, which we discuss in Chapter 6.

Uniformitarianism was embraced because at its core it invokes simplicity. No extra – that is, supernatural, or divine – causes need to be called upon to explain observations if presently known causes are sufficient. This, of course, stands in contrast to the approach of catastrophism, in which unusual (even supernatural) causes were invoked to explain major, or even modest, transitions in the geologic record. Explaining difficult-to-understand observations with a supernatural cause might be considered intellectually lazy. However, insisting that what we observe today is the only tool available to understand the billions of years of Earth history that has preceded us is nevertheless a shortcoming of uniformitarianism. Lyell was – at least sometimes – guilty of conflating uniformity of process with uniformity of rates.

KEY POINT

"The present is the key to the past" is a key tenet of geology but strict uniformitarianism can take this too far if events in "the present" are not representative of processes that have acted in the geologic past.

2.6 Uniformitarianism vs. Actualism: Channeled Scablands

The problem with strict uniformitarianism is that "today" cannot fully be the key to the past because today is not eventful enough. Even if we expand our definition of "today" to the expanse of human civilizations, a few thousand years still isn't enough to experience the complete variety of geologic phenomena that have operated throughout time. If one concentrates on "today," however broadly we define "today," one is likely to lean towards gradualism to explain the past because much of what we can see around us today is happening gradually.

At the beginning of the twentieth century, the champions of uniformitarianism had fought so hard to place this idea at the pinnacle of geologic philosophy that they allowed the simplicity of the catch phrase, "The present is the key to the past" to cloud their judgment. Anything that strayed from a strictly gradualist dogma was considered not just wrong but unscientific. In philosophy, claims about the natural world that can be disproved by observation or experiment, for example, "the Sun revolves around the Earth," are said to be substantive. Uniformitarianism is now recognized as a methodological assumption rather than a substantive claim about the world around us.

During the nineteenth century, so much intellectual effort had been expended in refuting catastrophism (and its Biblical overtones), that at the beginning of the twentieth century any deviation from gradualism was likely to be met with skepticism or even hostility, even if the motivation was entirely secular. As we will see in Chapter 6, this also affected views of evolutionary rates and led to debates about whether evolution is primarily gradual or episodic.

Into this environment appeared J Harlan Bretz. In the 1920s, Bretz was a young geology professor from the University of Chicago. Along with his students, Bretz spent several field seasons studying the unusual landscapes of eastern Washington – an area known as the **Channeled Scablands** (Figure 2.9). The Scablands are a vast area with unusual surface features including large, braided channels that cut through basalt bedrock (Figure 2.10(a)), gravel deposits as much as 15 m high, giant ripples 100 m in wavelength (Figure 2.10(b)), cliffs over which waterfalls plunge tens of meters, and large bowl-like depressions formed in the basalt.

Bretz was sure that all these geomorphic features were related to the glaciers that extended from Canada southward into northern Washington as recently as 18,000 years ago. However, the lack of *glacial till* (unsorted and unstratified material deposited,

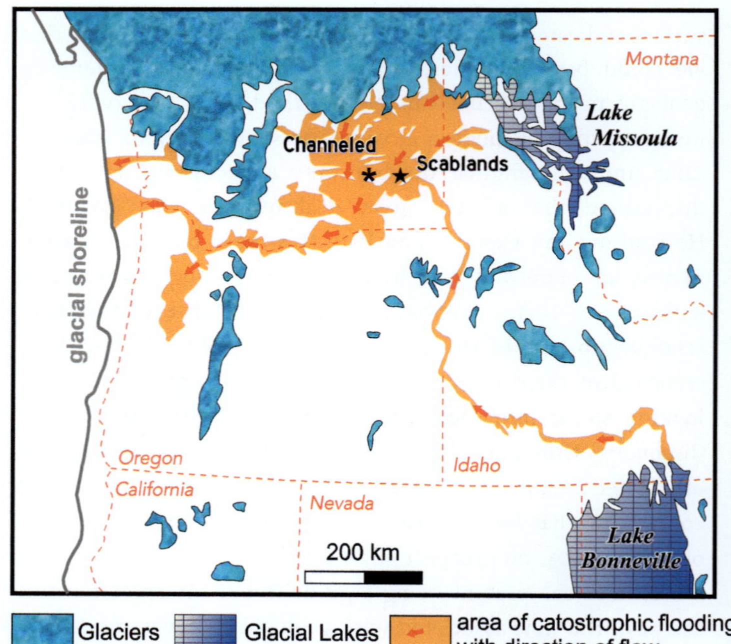

Figure 2.9 A schematic map of the northwestern USA at the end of the Pleistocene ice age showing the ice sheet, and Glacial Lake Missoula, and Glacial Lake Bonneville (later to become Great Salt Lake). When the ice dam broke on Glacial Lake Missoula, the lake floodwaters flowed eastward at up to 100 km/hr and carved the Channeled Scablands, creating gigantic sedimentary structures, showing the direction of flow. Star and asterisks show locations of Figure 2.9(a) and (b), respectively.

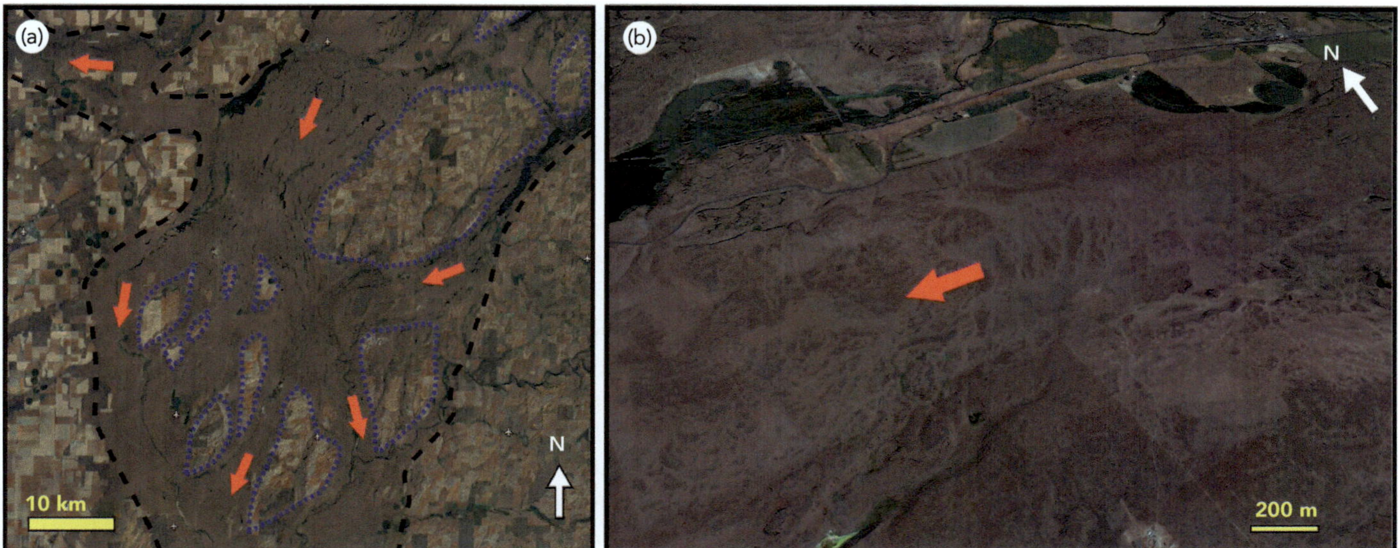

Figure 2.10 (a) Satellite image of a region approximately 75 km SE of Spokane, Washington (location of star on Figure 2.9) showing geomorphic features associated with mega-floods from Glacial Lake Missoula. Black lines show approximate edges of channels of floodwater (up to 35 km across) and red arrows show direction of flow. Blue lines show outlines of islands in the braided stream. (b) Satellite image of region near Marlin, Washington (location of star in Figure 2.9) showing giant ripples associated with floodwaters draining Glacial Lake Missoula. Red arrow shows direction of flow. The prominent current ripples are defined by a series of regularly spaced linear ridges seen to the right of the arrow. These ripples are up to 400 m long, 60 m wide, and 3 m high, but have been documented as much as 20 meters in height. Geometries for such ripples suggest a depth of water of tens to hundreds of meters and a flow velocity of up to 120 km/hour.

for example, in front of a moving glacier as if by a bulldozer, see Chapters 8 and 18) led him to realize that these features could not have been produced by simple glacial erosion. So, glaciers could not have produced these landscapes, but the scale of the features were inconsistent with normal erosion by rivers.

Bretz concluded that the Channeled Scablands must have been formed by flowing water but at a scale not known in human history; he called for flooding of water up to 200 m deep moving at speeds approaching 100 km/hour and peak discharge of at least 1 million cubic meters per second. Remarkable claims such as this will always have an uphill climb towards acceptance, but Bretz faced particular trouble, given that this hypothesis amounted to a slap in the face of gradualism. It also didn't help that when he began his study of the Scablands, Bretz was a relatively young man telling his seniors that they were looking at the world incorrectly.

In 1927, Bretz gave a presentation at the annual meeting of the Geological Society of America (GSA) in which he outlined his ideas. He argued that the features of the Channeled Scablands could only be formed in a great flood. This was met with a very hostile reception for two reasons. First, Bretz's interpretation flew in the face of contemporary geologic thinking: "great flood" sounded too catastrophic, even Biblical, and the concept of uniformitarianism had achieved such a strong purchase on the philosophy of early twentieth-century geologists that catastrophic events were considered unsuitable for scientific study. Second, at the beginning of his study, Bretz did not have a good explanation for the source of the more than 2 billion cubic meters of water that would be necessary to produce what he had observed (greater than the amount of water in today's Lake Ontario). But, during Bretz's 1927 presentation, a geologist from the United States Geological Survey, J. T. Pardee, is supposed to have leaned over to a colleague and said, "I know where Bretz's water came from." This was, at the time, a conjecture, so he did not come to Bretz's defense publicly because he didn't want to become part of the controversy without firm data.

After this now-famous GSA meeting, Bretz continued to insist, for several decades (almost alone), that the cause of the landscapes of eastern Washington was a massive flood. Research during this time, much of it by Pardee working independently from Bretz, showed that at the end of the last ice age, around 18,000 years ago, there existed a massive lake near the site of modern Missoula, Montana. **Glacial Lake Missoula** (Figure 2.9) is recognized by shoreline deposits now high in the valleys of western Montana, up to 400 km to the east of the Channeled Scablands. Today water flows from the mountains of western Montana, but Pardee and others argued that the southern edge of the continental glacier would have acted as a dam on the northern and western edges of Glacial Lake Missoula. The modern Flathead Lake in western Montana (approximately 450 km^2 in area with depths up to 100 m) is the largest remnant of Glacial Lake Missoula, which was approximately 7,700 km^2 in area (Figure 2.9), sequestering the volume of water proposed by Bretz.

A dam made of ice is inherently fragile and would be even more so in a time of global warming, as was the case around 15,000 years ago. The last ice age was ending, and continental glaciers were receding, rapidly (see also Chapter 18). In such circumstances, the dam holding back the water in Glacial Lake Missoula could have failed catastrophically. The pressure of the deep water is also thought to have caused some of the ice at the bottom of the dam to transform into liquid water, further reducing the strength of the dam. This event would have sent tens of millions of cubic meters of water rushing westwards towards eastern Washington. This was consistent with Bretz's idea that the landscape of eastern Washington was formed not by gradualist processes, but by a significant – or even a catastrophic – event. Based on studying modern ice dams in Iceland and hydrodynamic modeling in the last half of the twentieth century, it became clear that the giant sand bars and wide, braided canyons were formed on a timescale measured not in thousands of years or even years but rather hours. Later fieldwork in Washington and Montana at the end of the twentieth century suggests that there may have been several floods, not just one, as Glacial Lake Missoula filled and then breached the ice dam dozens of times, each time resulting in a massive flood.

In 1979, 52 years after he received a hostile reception at the GSA meeting in Washington, the Geological Society of America gave Bretz the society's highest award, the Penrose Medal (the medal is given annually to a single individual in "recognition of eminent research in pure geology, for outstanding original contributions or achievements that mark a major advance in the science of geology"), completely vindicating Bretz's efforts and ideas. It was a timely award, as the rules of the society state that the Penrose Medal may not be given posthumously and Bretz died only two years later at the age of 98. After he received the award, he told his son, "All my enemies are dead, so I have no one to gloat over."

The example of the Channeled Scablands shows that strict uniformitarianism (especially gradualism) cannot account for the full range of geologic processes that have operated in the past. The modern philosophy of geoscience that evolved out of uniformitarianism is now called **actualism**. This is a point of view that advances the idea that we need to understand what happened without the blinders of strict gradualism constraining us.

Actualism can seem like a blending of uniformitarianism and catastrophism but only in the sense that it rejects strict gradualism. The day the dam broke on Glacial Lake Missoula, the day a 10 km wide asteroid slammed into the Yucatan Peninsula, causing the extinction of the dinosaurs at the end of the Cretaceous (see Chapter 13), and the day in 1812, when movement on the New Madrid fault in southern Missouri caused the Mississippi river to flow backwards for several hours were all very unusual, but none were the consequence of special laws of Nature unrelated to everyday ordinary geologic processes, nor were they acts of a vengeful God.

2.7 The Scientific Method and Geology's Place in It

The modern practice of geology is a science, just like physics, chemistry, and biology. Nevertheless, geologists often face challenges many physicists, chemists, and biologists rarely worry about. Geologists commonly must deal with an incomplete record of events long past, and sometimes these events may be rare, catastrophic, and outside the experience of human civilizations, but nevertheless part of the natural world. These events may involve long-extinct organisms, about which we have no modern counterparts. Many geologists must reconstruct a world that looked very different from what we see today. Geologists also routinely have to work simultaneously at a wide range of scales that can include remote geophysical sensing of large swaths of the Earth's surface to microscopic examination of rocks, minerals, and fossils, to the smallest scale lab analysis of the very atoms themselves that can reveal much about the origin of the rocks that contain them. Of course, we can now also sample the geology of moons and planets, such as was done during the Apollo Missions and the more recent NASA missions to Mars.

2.7.1 Observation, Hypothesis, Test

A particular way of dealing with observing and explaining natural phenomena is called the scientific method or hypothesis testing. This is a way of making observations, proposing ways to explain these observations, and testing the explanations. The basic workflow of this approach is outlined in Figure 2.11.

In any study of the natural world interesting observations are bound to be made but for any given observation to lead to anything truly valuable, an explanation, or hypothesis needs to be proposed. To be useful, these explanations need to make predictions about other observations, either from the field or the lab, and these predictions should guide further work. That is, new experiments or fieldwork can be specifically designed to test the predictions and therefore the hypothesis.

Our flow chart in Figure 2.11 shows that in such instances, where new data are not consistent with the predictions of existing explanations, the hypothesis must be rejected, and we need to come up with another one. But what if the new data are in line with the prediction? Does that mean we are done? Have we proven the hypothesis to be correct? The answer is always, "NO." Scientists are in the testing business, not the proving business. As shown in Figure 2.11, if the new data are consistent with the predictions of the hypothesis, we return to make more observations. We are never done, even if we are very confident in our explanation; there is no end to the workflow in Figure 2.11. One classic example of this will be discussed in detail in Chapter 5: until after World War II many people thought the main reason for high mountains and deep oceans was predominantly vertical motions of the crust. After new data was obtained, primarily from geophysics, it is now understood from the theory of plate tectonics that mountains and oceans primarily form due to horizontal motion of Earth's outermost shell. The strength of an idea is not measured by the confidence of the researcher proposing it but rather by its ability to withstand scrutiny in light of new data.

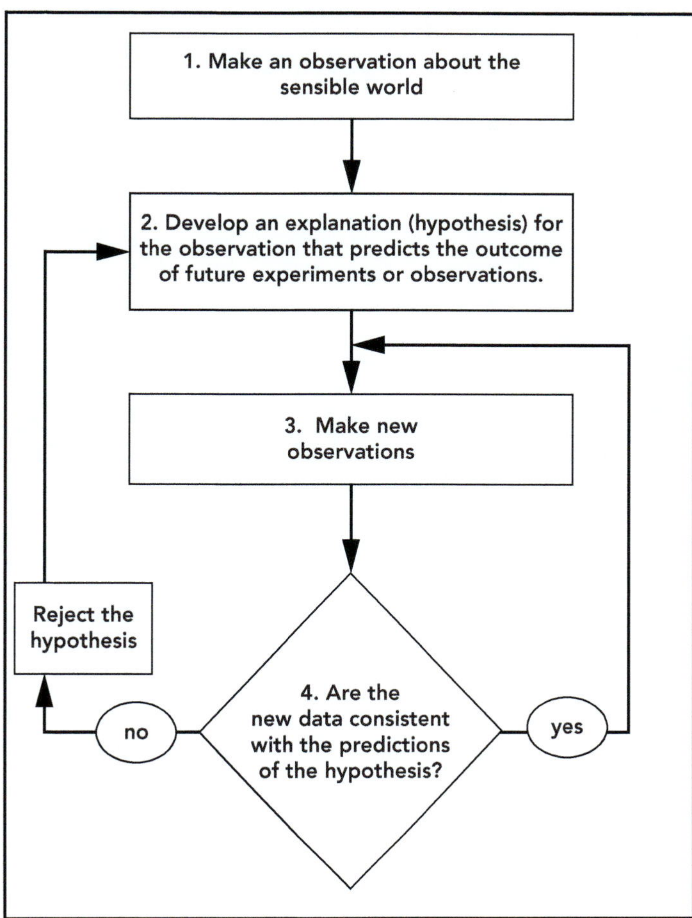

Figure 2.11 Schematic flow chart of the scientific method. Note that there is not a point on this diagram labeled "finished" or "proven." All ideas that today seem good are still regarded as provisional and need to be evaluated when new data are obtained.

2.7.2 The Problem of Geologic Experiments

Like all scientists, geologists gather data to test their ideas about the natural world. But unlike many other scientists, geologists face problems in conducting experiments because the systems of interest are usually very big and operate over very long timescales.

We may be interested in the interplay between climate, tectonics, and erosion in the Himalaya. For example, does a change in the sedimentary deposits weathered from the mountain range reflect a change that occurred in the annual rainfall, the frequency of earthquakes, both, neither, or some other factor? A truly experimental approach would be to construct several 2,000 km-long mountain ranges and vary the rainfall in different ways over a time span of 20 million years and then see how the different climates affected the sedimentation patterns. After that we would probably run the experiment again, holding rainfall constant and varying the timing and magnitude of earthquakes, and compare the sedimentation patterns with those in the previous experiments to determine whether climate or tectonics makes a bigger impact to erosion and sedimentation.

Well, obviously, we can't do that.

But if we can't build mountain ranges and subject them to known rates of faulting and rainfall, can geologists really participate in the scientific method? Yes, they can, but only within the constraints that time and space impose. Geologists cannot do experiments as described above but they can look at Earth as it is and deduce which processes were operating in the past to produce the features we see today. From an experimental perspective, geologists look at rocks and try to decipher which processes produced them. Conversely, we propose geologic experiments, predict their outcome, and go into the field (and sometimes later the lab) and see if the rocks are as predicted. For example, let's consider some of the stories we will discuss in detail in forthcoming chapters. In Chapter 8 we will see that some geologists have suggested there was a time (around 650 Ma) when almost the entire surface of the planet was covered by ice. Well, if this did happen, rocks of that age should show evidence of worldwide glaciation.

In Chapter 13 we will discuss the hypothesis that the dinosaurs (and many other life forms) went extinct at the end of the Cretaceous period because of an impact with a comet or meteor. Again, if this happened, the products of this impact should be observed regionally or even globally in rocks deposited across the boundary, and there could be evidence of the actual impact crater. In the case of the Cretaceous event, an impact was hypothesized, and the actual impact crater discovered later. This led to a further hypothesis that maybe other mass extinction events were also triggered by impacts, but the search for other major impacts associated with these other extinction events has not been very fruitful and that hypothesis has been largely abandoned. In many ways the geologist is like the police detective who arrives at the scene of the crime and needs to decipher what happened.

For these reasons, geology is sometimes referred to as a historical science, but geologists are not completely removed from experimental work. Much of what we know about the melting and crystallization of igneous rocks and the pressure–temperature significance of metamorphic mineral assemblages (introduced in Chapter 1) comes from heating up rocks in the lab or measuring chemical concentrations or isotopic ratios (Figure 2.12). Sedimentologists use flume labs to control the rate of water flow, the size and composition of sediment, the slope of transport and other factors to produce sedimentary deposits in the lab (Figure 2.13). They then can examine the types and sizes of sedimentary features produced under known conditions. Work in these labs provides insight that could not be gained by fieldwork alone.

Sometimes compromises are made, such that laboratory conditions may be hotter than expected in nature (because chemical reactions generally go faster at higher temperatures); or in flumes, flows are much shallower than in natural systems. Despite the limitations of modern experiments, we are commonly able to extrapolate the data obtained at high temperatures or shallow flows done on smaller-scale experiments to the larger and longer timescale over which many Earth processes operate.

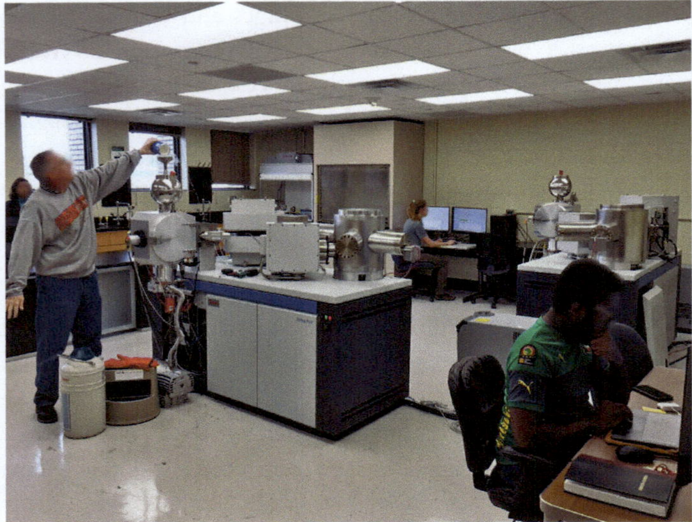

Figure 2.12 A modern geochemistry lab. The machine in the center of the photo is a mass spectrometer used to isolate individual elements that make up the minerals in rocks. Source: Courtesy of Alan Brandon, University of Houston.

Figure 2.13 Experimental flume in which water and sediments are delivered into a standing body of water creating channels and delta sands. (Note ladders in background for scale.) Source: Image courtesy of St Anthony Falls Laboratory, University of Minnesota, Minneapolis.

Of course, all experiments must be conducted under reasonable space and time constraints. Although the time limit on such experiments might go as high as 20 or 30 years in some instances, because most research gets done at universities by graduate students, the true time limit on most experiments is commonly the time it takes to get a MS or PhD degree, usually two to five years. Experiments that take longer than that are not likely to be started.

In some cases, experimentation is simply not possible and some type of numerical modeling is employed, typically using computers. Models are numerical approximations of Earth.

Modelers take observed conditions as the input to their data and then evolve their system computationally. Models can either look backwards, to describe the previous geologic events that occurred to bring us to a current state, or look forwards to predict how things will look in the future, such as the climate models that we will discuss in Chapter 19. These models take advantage of the extreme processing power of modern computers that can simulate, in a few hours or days, the long-term behavior of any given Earth process under investigation that may take millions of years.

2.8 Summary

- Initially, fossils were thought to represent inorganic counterparts of life on Earth. Renaissance scientists realized that fossils were the remains of once living organisms and theorized that they recorded a complex history of the formation of the rock from sediment that was once fluidized.
- *Neptunism* is the idea that Earth history largely recorded the evolution and disappearance of ancient oceans, with precipitation of a Primary, Secondary and Tertiary series, whereas Plutonism is the hypothesis that internal heat drives Earth processes.
- *Catastrophism* is the idea that Earth history was driven by sudden supernatural events. This was replaced in the Enlightenment by the idea of *uniformitarianism*, which is the theory that processes acting today operated over vast geological time and account for most of the features observed in the geological record.
- Uniformitarianism was also linked to the idea of *Gradualism*, which states that these processes operate very slowly.
- J Harlan Bretz interpreted the Channeled Scablands as a result of catastrophic flooding, which was later recognized as due to massive release of waters stored in a glacial lake, and led to the concept of *actualism*, which states that some events in Earth history may be so rare that they have not been observed in the present era and not all geological features are the result of gradual processes.
- Geology is a science and is governed by the same scientific principles as physics, chemistry, and biology, but one challenge is testing processes and events that occur at temporal and spatial scales far greater than those easily replicated in a laboratory.

Key Words

- latent plastic virtue
- principle of faunal succession
- catastrophism
- Neptunism
- Plutonism
- uniformitarianism
- Channeled Scablands
- Glacial Lake Missoula
- actualism

Further Reading and References

Cutler, A., 2003, *The Seashell and the Mountaintop*, Dutton.

Cuvier, G., 2010, *Essay on the Theory of the Earth*, 2nd ed., Cambridge University Press.

Repchek, J., 2003, *The Man Who Found Time*, Perseus Publishing.

Winchester, S., 2001, *The Map that Changed the World: William Smith and the Birth of Modern Geology*, Harper Collins.

Review Questions

1. What was "latent plastic virtue" and how did Greek philosophers use this concept to explain what we now call fossils?
2. How did Steno reinterpret the origin of fossils, now contained in rocks, as actual remains of once living organisms?
3. How did William Smith, Georges Cuvier, and Alexandre Brongniart advance the understanding of regional geology?
4. What is the principle of faunal succession and how is it used to understand the relationships and correlations of strata in different regions?
5. How does Catastrophism contrast with the concept of uniformitarianism? Who were the principal champions of these ideas?
6. Explain Hutton's statement "The present is the key to the past" and how it relates to the concept of uniformitarianism.
7. What is Neptunism and what are its chief flaws?
8. How does Neptunism contrast with the concept of Plutonism? Who were the principal champions of these ideas?
9. How did the debate regarding the genesis of the landforms in the Channeled Scablands in eastern Washington play into the consideration of the appropriate place of uniformitarianism, gradualism, and actualism in the views of geologic thought in the twentieth century?
10. What are some of the challenges in interpreting geology, as a historical science, particularly the fact that the record is incomplete and fossils may have no modern counterparts?

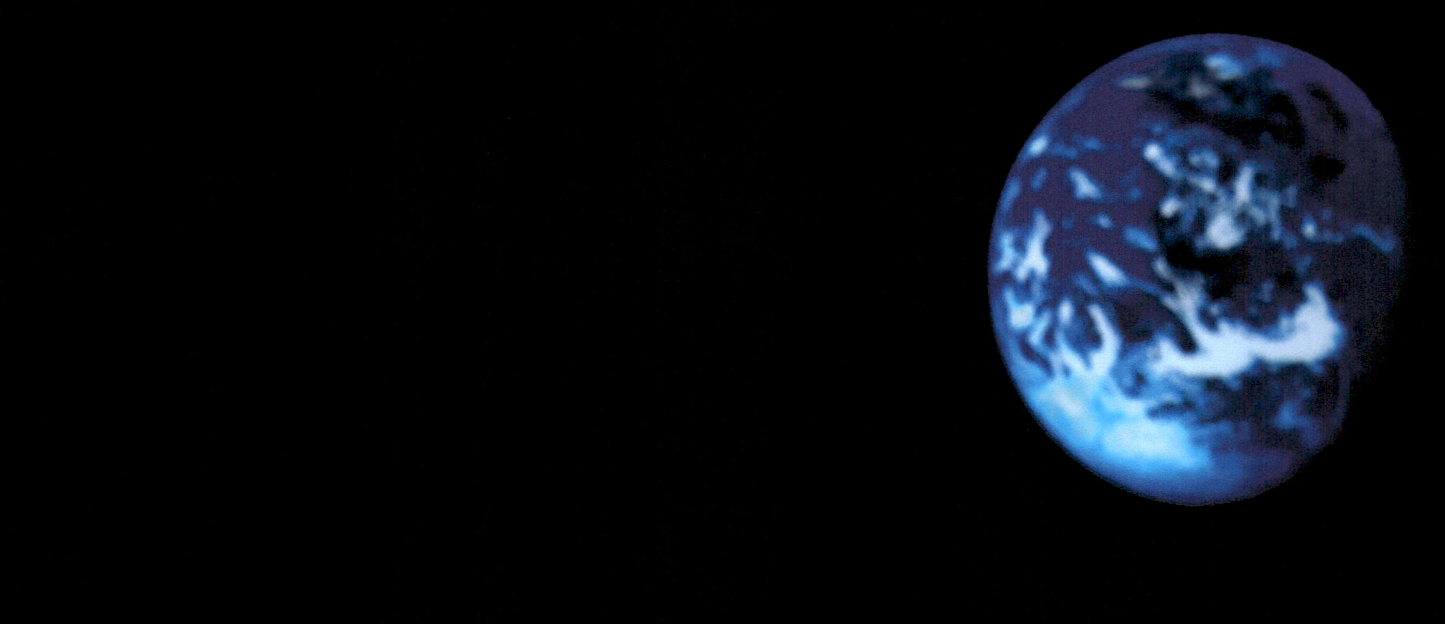

Photo of Earth taken from the Moon by Apollo astronaut. Source: Caspar Benson / Getty Images.

Chapter 3

The Origin of Earth

From the Beginning of the Universe to the Early Earth

LEARNING OBJECTIVES

- Describe how we know that the universe began with the Big Bang, which unevenly dispersed the first-formed matter and energy.
- Illustrate the origin of early formed hydrogen and helium to heavier elements formed by fusion in stars and how elements heavier than iron were formed through the collapse of supergiant stars and subsequently dispersed by supernovae.
- Define the nebular hypothesis, which explains how our own solar system formed and why the outer solar system consists of gas giants and the inner solar system the rocky planets.
- Discuss the impact hypothesis, which explains how the collision with the planet Theia early in its history resulted in the Earth–Moon system and the off-axis rotation of Earth.
- Explain how Earth became a differentiated planet, with the formation of a core, mantle, and crust.
- Examine the concept of isostasy and how it explains the elevation of continents.
- Describe the origin of the atmosphere and why planetary atmospheres such as Venus and Mars are so different from Earth.

Introduction

We are literally stardust. Any understanding of the history of our planet must begin with a discussion of the origin of the elements that make up our solar system. This chapter reviews the origin of the universe and subsequent atomic elements within stars and how the growth and death of stars ultimately allowed solar systems and planets to form. We then focus on the formation of Earth and its component rocks and minerals, previously introduced in Chapter 1, as well as its hydrosphere and atmosphere, which define the Earth systems that initiate the rock cycle.

3.1 The Big Bang

Following the discoveries of relativity by Albert Einstein in the 1920s and improved telescopes in the twentieth century, astronomers realized that the universe is expanding (see Box 3.1). If the universe is expanding, it is possible to extrapolate backwards to determine when this expansion began. Current data suggest that the universe formed some 13.77 billion years ago as a consequence of what is termed the **Big Bang** from an initial exceedingly hot and dense state, referred to as a *singularity*: defined as the point where the universe can be shown to have entered into a state where the laws of physics as we know them can be applied (Figure 3.1).

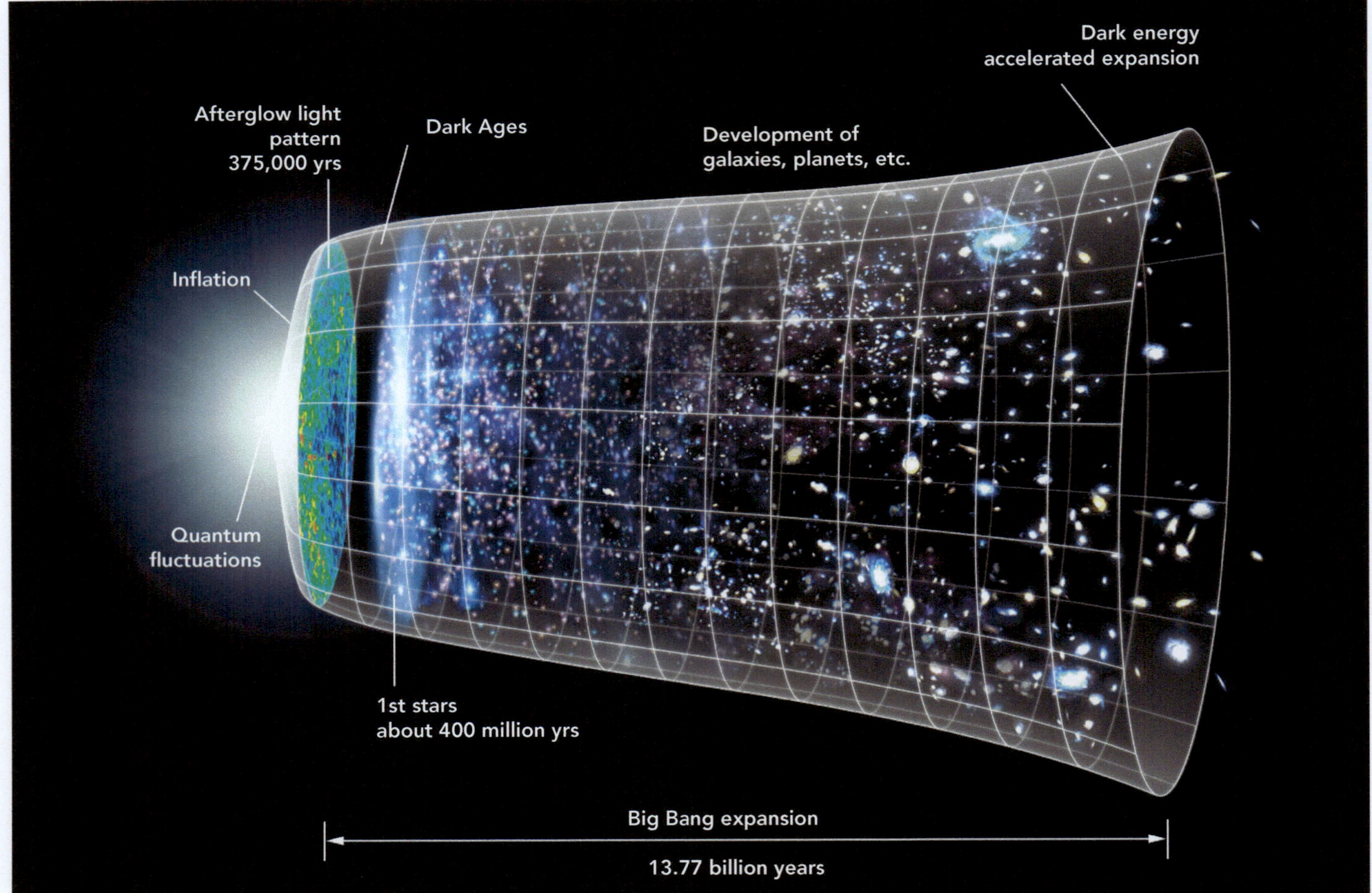

Figure 3.1 Evolution of the universe from the Big Bang to today. Source: NASA/WMAP Science Team – original version: NASA; modified by Cherkash.

BOX 3.1 How Do We Know the Universe is Expanding?

One of the key observations noted by astronomers is the stretching of the wavelengths of electromagnetic waves, including light rays, X-rays, and gamma rays, emitted by objects such as stars. If a star is moving away from you or if stars are moving away from each other (i.e., expanding) then the electromagnetic radiation shifts to longer red wavelengths, causing a *redshift*, as red is at the longer-wavelength end of the light spectrum. This is an example of a common phenomenon referred to as the Doppler effect. You may be familiar with this when you hear a siren go past, such as a firetruck, ambulance, or police vehicle. As the vehicle moves towards you the sound is higher, reflecting a shortening of sound waves as the velocity of the sound is added to the speed of the vehicle. This results in a shorter wavelength. If the waves are electromagnetic, rather than sound, we call this a *blueshift*, as blue is at the short-wavelength end of the light spectrum. Once the vehicle passes, the sound becomes lower as the sound waves are stretched. The redshift that is routinely observed in stars and between galaxies indicates that the universe is expanding. A collapsing universe should exhibit a *blueshift* or shortening of wavelengths.

Events happened incredibly fast in the initial expansion. For example, during what is known as the **inflationary epoch**, which lasted from 10^{-36} to 10^{-32} seconds after the Big Bang, the universe (i.e., everything, everywhere) expanded in scale by a factor of 10^{78} (Figure 3.1). To give a sense of the fantastic magnitude of such inflation, consider the size of a proton, which has a diameter of 8×10^{-16} m. The diameter of our **Milky Way** galaxy is about 10^{18} km, or 10^{33} times bigger than a proton, so something that is 10^{78} times bigger than a proton is a million, billion, billion, billion, billion, billion times bigger than the Milky Way. During the first 20 minutes after the Big Bang, fusion of protons, neutrons, and other subatomic particles produced a universe composed of about 75% hydrogen (H), 25% helium (He), and trace amounts of a few other light elements.

The unfathomable energy of the Big Bang sent this matter out in all directions but not equally spaced. Initially all this material would have been moving in straight lines, outward, from the singularity, but gravitational attraction gradually forced this material to move in a spiral motion. Spiral galaxies (Figure 3.2), which include our own Milky Way, formed as a consequence of the gravitational attraction and turbulence driven by uneven mixing of matter and energy in the early universe. Galaxies can be thus thought of as large-scale turbulent eddies.

Similar spiral patterns, such as hurricanes, are turbulent eddies in our atmosphere, but these only last a few weeks or days as they make their way across the world's oceans. The great red spot of Jupiter (Figure 3.3), which is larger in diameter than Earth, is an atmospheric storm on Jupiter's equator that has lasted for as long as we have been looking at it through telescopes, which is several hundred years, but it is a dynamic feature that clearly moves around. In contrast, spiral galaxies (Figure 3.2) may persist for many billions of years and of course this much longer timescale has allowed entire solar systems and planets to be formed, and on at least one planet in our own Milky Way, for life to evolve.

Figure 3.2 The M101 Pinwheel Galaxy. Source: Hubble image: NASA, ESA, K. Kuntz (JHU), F. Bresolin (University of Hawaii), J. Trauger (Jet Propulsion Lab), J. Mould (NOAO), Y.-H. Chu (University of Illinois, Urbana), and STScI; CFHT image: Canada-France-Hawaii Telescope/ J.-C. Cuillandre/Coelum; NOAO Image: G. Jacoby, B. Bohannan, M. Hanna/NOAO/AURA/NSF.

> ### 🔑 KEY POINT
>
> The universe started with the Big Bang, 13.77 billion years ago. Turbulence resulted in an uneven distribution of matter and energy that was concentrated by gravity into galaxies.

3.2 Stars and the Formation of the Elements

Small differences in the initial distribution of mass resulted in volumes of space that were slightly denser. All matter experiences gravitational attraction and eventually this gravitational attraction causes matter initially dispersed in the Big Bang to slowly coalesce to form gas clouds, larger nebular galaxies in

Figure 3.3 Jupiter, Earth, and Mars. The Great Red Spot of Jupiter is an equatorial cyclonic storm that has persisted for about 1,000 years. Earth also shows cyclonic storms, but these storms typically only persist for several days to a week. Source: (Earth photo) NASA/GSFC/ Reto Stöckli, Nazmi El Saleous, and Marit Jentoft-Nilsen, Public domain, via Wikimedia Commons; (Jupiter photo) NASA/JPL-Caltech/ SwRI/MSSS/Kevin M. Gill, Public domain, via Wikimedia Commons; (Mars photo) ESA / Michael Benson via The Image Bank Unreleased / Getty Images

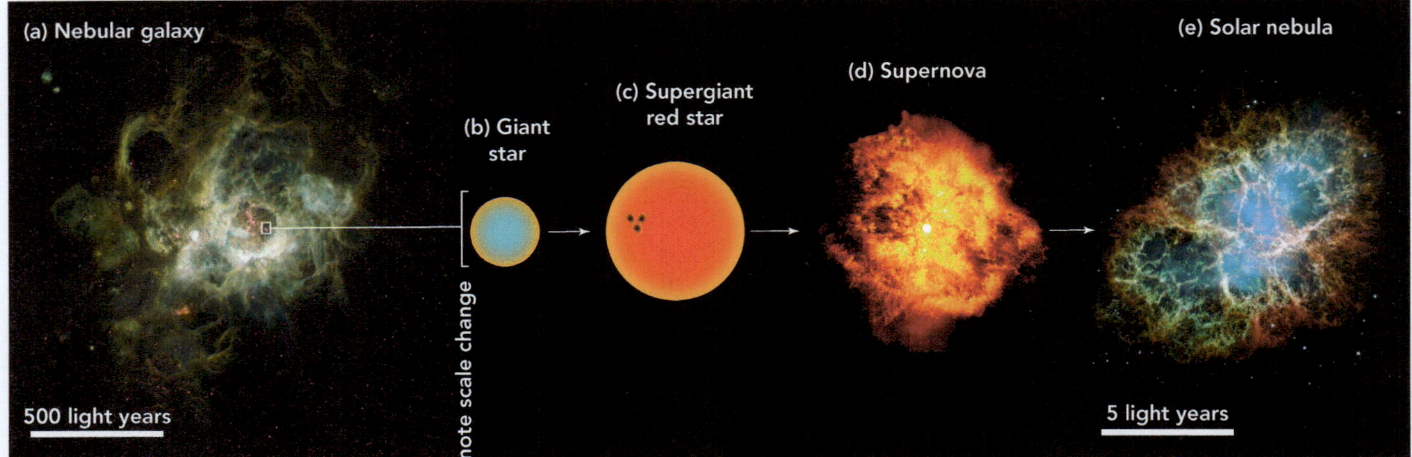

Figure 3.4 Evolution of giant stars. (a) Stars form within a gigantic nebular galaxy. This example (NGS 604) is 500 light years across. (b) A giant blue star is an example of one of the larger stars formed in the nebular galaxy. (c) The giant star expands into a supergiant red star and eventually collapses and (d) subsequently explodes, forming a supernova. (e) The resulting solar nebula (e.g., Crab Nebula) is much smaller than the nebular galaxy and may eventually coalesce to form a solar system (see Figure 3.6). Source: (a) Hui Yang (University of Illinois) and NASA/ESA, Public domain, via Wikimedia Commons; (d) Source: Yves Grosdidier (University of Montreal and Observatoire de Strasbourg), Anthony Moffat (Universitie de Montreal), Gilles Joncas (Universite Laval), Agnes Acker (Observatoire de Strasbourg), and NASA, Public domain, via Wikimedia Commons; (e) Source: NASA, ESA, J. Hester and A. Loll (Arizona State University).

which stars are formed (Figures 3.1 and 3.4). Sometimes these clouds became dense enough that the material at the center began to heat up and conditions became so dense and hot that hydrogen atoms were fused together to make helium (Figure 3.5). This *nuclear fusion* generates even more heat and eventually a star is born (Figure 3.4). The first stars likely formed about 400 million years after the Big Bang and these in turn coalesced to form the oldest observed galaxies around 350 million years later (Figure. 3.1).

As stars increase in heat and density, fusion of the earliest-formed helium and hydrogen (Figure 3.5) produces heavier elements, such as lithium (Li) and beryllium (Be). These further combine, again by fusion, to form carbon (C), oxygen (O), and so on. Each of these fusion reactions release energy, allowing yet more fusion. This fusion process continues to produce heavier and heavier atoms, with iron (Fe) being the heaviest, with 26 protons. Elements heavier than iron cannot be made by this process because fusion of Fe and heavier elements consumes more energy than it releases. Once a star starts making Fe, it causes its own destruction in this way, and eventually dies out. But, if this is the case, how are elements heavier than Fe produced?

Stars experience stages of evolution that largely depend on their size (Figure 3.4). Our own Sun, for example, is relatively small on a galactic scale, and relatively long-lived, with a lifespan of about 10 billion years (of which we are about halfway through). Larger stars burn out much faster than smaller stars, and the biggest stars (i.e., those 10 times larger than our Sun) only last for a few tens of millions of years. Our Sun is presently in the *main sequence* of its life and is primarily fusing hydrogen; however, when it finally exhausts its supply of hydrogen, in about another five billion years, it will expand into a *red giant*.

In this phase of growth, the Sun will be so large that it will engulf Mercury, Mars, Venus, and possibly Earth. Following the red giant phase, the Sun will eventually shrink and end its life in a white dwarf stage. Stars also experience changes in luminosity and their ability to radiate heat as they change from fusing hydrogen to helium and so on. In the early history of our Sun, it produced only about 70% of its current radiative heat, but 3.5 billion years later, in the Neoproterozoic Era, it was at about 94% (see Chapter 8). This is termed the *faint Sun hypothesis*, which has led to questions about how Earth was able to hold liquid water in its early stages, and we will discuss this further in Section 3.5.7.

Larger stars (between 5 and 250 times the mass of the Sun), in contrast end their existence in astronomical explosions known as supernovae that restart the stellar cycle (Figure 3.4(d)). The mass of large stars produces a gravitational attraction that draws all material to the center of the star but during the phase where nuclear fusion occurs (i.e., its main sequence), the fusion balances the gravitational attraction, maintaining the star. However, once fusion can no longer proceed (i.e., the fuel has been used up), gravitational forces dominate. In very massive stars, the sudden collapse at this terminal stage brings all the mass towards the center in less than one minute. This extremely energetic and rapid collapse causes the formation of elements heavier than Fe, by fusion of atoms made in the star up to that moment, and is followed by a **supernova explosion** (Figure 3.4 (d)). The debris of these supernova explosions eventually are scattered within a solar nebula (Figure 3.4(e)), resulting in dispersion of elements, including the ones heavier than iron made during the supernova. The solar nebula is orders of magnitude smaller than a nebular galaxy (compare Figure 3.4(a) and (e)) and can subsequently collapse to form a new solar system (Figure 3.6). Repetition of these stellar cycles over billions of years has increased the proportion of heavy elements, and eventually these heavy elements have coalesced to be incorporated, along with lighter elements, into solar systems with planets including our own solar system (see Table 3.1).

KEY POINT

All elements are formed by nuclear fusion in stars. Elements heavier than iron form when giant stars collapse and are dispersed in supernovae.

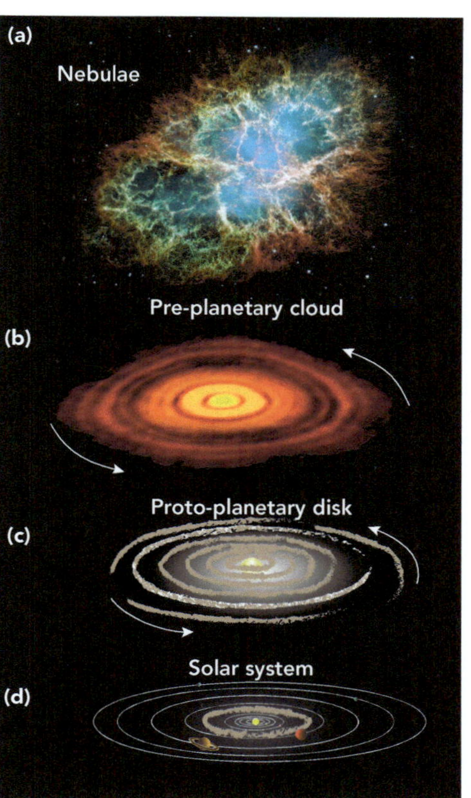

Figure 3.6 An illustration of the nebular hypothesis for the formation of our solar system. (a) Solar nebulae (as formed following a supernova of a giant star, as shown in Figure 3.4(e)); (b) spinning and gravity of the solar nebula forms a pre-planetary cloud with initial fusion at the center and formation of a protosun (HL Tauri photo taken by ALMA); (c) proto-planets begin to form around the newly formed sun; (d) planets form a solar system. Source: (a) NASA, ESA, J. Hester and A. Loll (Arizona State University); (b) ALMA, CC BY 4.0 <https://creativecommons.org/licenses/by/4.0>, via Wikimedia Commons.

3.3 Forming Our Solar System

Our solar system (Figure 3.7) formed from the coalescence of an interstellar cloud of hydrogen and helium, with lesser amounts of carbon, silicon, and iron, and even smaller amounts of heavier elements from previous supernovae (Figure 3.6). As the cloud, known as a nebula, collapsed, the center attracted the main mass, forming a star, surrounded by a proto-planetary disk of material that would eventually form the planets (Figure 3.6). The widely accepted **nebular hypothesis** suggests that planets grow around eddies in a flat proto-planetary disk that rotates around the evolving star at the center (Figure 3.6). Small bodies, a few meters across, began colliding and coalescing with each other to make larger bodies several kilometers across called **planetesimals** (literally "tiny planets"). After many collisions and widespread gravitational attraction these grew into **proto-planets** (primitive or original planets).

These protoplanets were all large balls that consisted of mostly gases, including hydrogen (H), helium (He), water (H_2O), carbon dioxide (CO_2), methane (CH_4), and ammonia

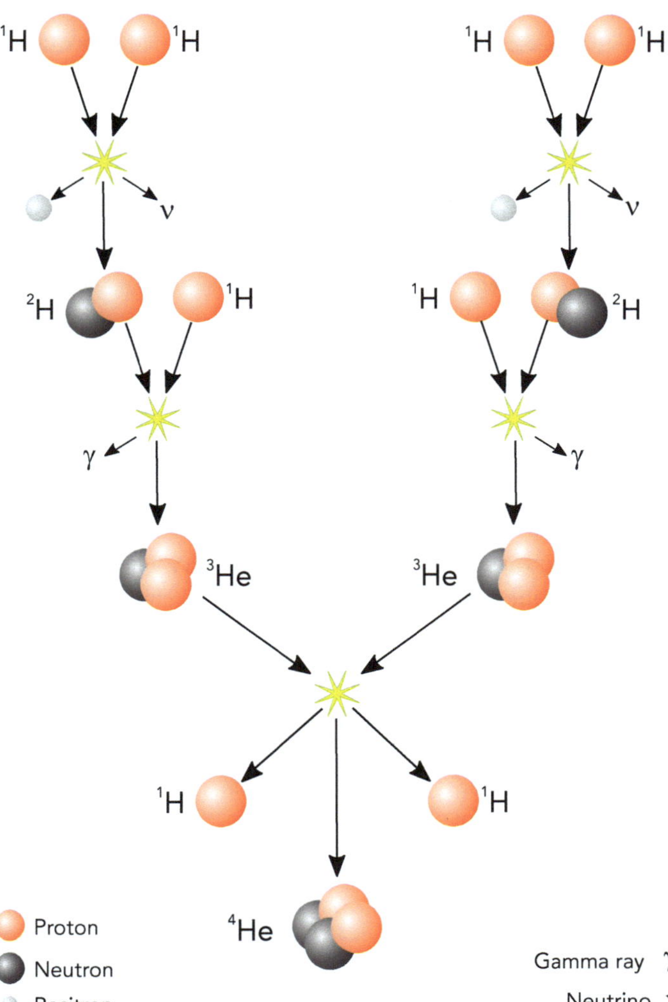

Figure 3.5 Fusion of two protons (^{1}H) forms hydrogen atoms (^{2}H), which include a neutron and one of the original protons. The process of fusion releases a positron and a neutrino. The hydrogen atom is then able to fuse with another proton to form a helium atom and releases gamma radiation in the process. Fusing of ^{3}He releases two protons and results in formation of ^{4}He. As this process evolves, heavier elements can fuse with other atoms or with protons to form heavier elements. Source: Sarang, Public domain, via Wikimedia Commons.

Table 3.1 Weight percent of elements that form the bulk Earth, core, and continental crust (not including atmosphere and hydrosphere)

Element	% In bulk Earth	% In core	% In continental crust
Si	17.22	7.4	27.72
O	32.46	4.1	46.60
Fe	28.18	79.4	5.00
Mg	15.87		2.09
Ca	1.60		3.63
Na	0.25		2.83
K	0.01		2.59
Al	1.51		8.13
S	0.70		0.05
Ni	1.61	4.8	<0.01

Source: Allegre et al. (1995), Haynes (2016), and Kring (1995).

(NH_4), which formed a volatile envelope surrounding a smaller rocky core. Much like a centrifuge, the heaver rocky materials settled to the interior during the formation of these protoplanets. At this early stage, all planets were much like the modern outer planets (Jupiter, Saturn, Neptune, and Uranus). But when the temperature and density of the Sun allowed fusion of H into He (Figure 3.5), the Sun began to generate large amounts of heat that radiated outwards. The inner planets, Mercury, Venus, Earth, and Mars, lost much of their volatile envelope due to heating from the Sun and its solar wind. The outer planets are too far from the Sun for this solar wind to be sufficient to remove the thick volatile envelope, so these planets still look much as they did at the dawn of the solar system (Figure 3.7).

Whether or not a planet retained its initial atmosphere depends on the size of the planet and the distance to the Sun, which controls the thermal energy at the planet's surface. The velocity needed to overcome the gravitational attraction of a planetary body is called the **escape velocity**, and for atmospheric gases this is controlled by the gravitational force. Therefore, smaller bodies have lower escape velocities than larger planets and thus are more likely to lose their atmosphere. This is illustrated well in a comparison of Earth and Mars (Figure 3.3). Mars is 1.5 times farther from the Sun than Earth so one might expect that this extra distance would make the planet colder, allowing it to retain a thicker atmosphere than that of Earth. However, Mars has only about 38% of the mass of Earth so its gravitational pull on gases in the atmosphere is significantly less and it is easier for the gases to escape from the influence of Mars' gravity. Therefore, although Mars is much colder than Earth, it has a much thinner atmosphere because it is much smaller. In this case, gravity is more important than temperature.

The differences in the sizes of planets are the consequence of small random variations of density and velocity in the disk that formed around the Sun, all of which came from the initial nebula. The size of planets varies from the gas giant Jupiter (Figures 3.3 and 3.7), which has a volume more than 1,340 times that of Earth, to Mercury, which is only slightly larger than Earth's moon. However, despite its greater volume, the mass of Jupiter is only about 300 times greater than Earth because most of Jupiter consists of low-density gases; here temperature is more important than gravity and Jupiter retains its gases because it is so far away from the heat of the Sun.

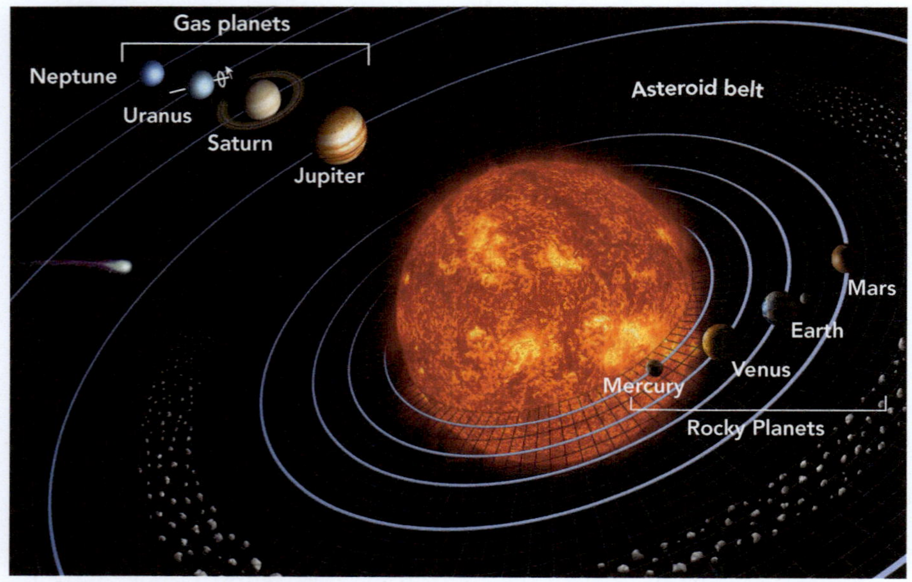

Figure 3.7 The solar system (not to scale). Note that Uranus rotates at a high angle to the solar plane. Source: Harman Smith and Laura Generosa (nee Berwin), graphic artists and contractors to NASA's Jet Propulsion Laboratory, with Pluto removed by User: Frokor, Public domain, via Wikimedia Commons.

The gravity from Jupiter, the largest planet, made it difficult for any additional planets to form between Mars and Jupiter, and today that space is occupied by the **asteroid belt**, a collection of millions of rocky fragments that are the remains of the material that formed the proto-planetary disk (Figure 3.7). Because these fragments never formed into a large planet, their composition can provide information on what the early solar system looked like prior to the formation of the planets. Material from the asteroid belt occasionally falls to Earth as meteorites and the study of meteorites is thus key to deciphering the composition and origin of the solar system.

The eight planets revolve about the Sun in the same direction but rotate about an axis to the plane of revolution that varies from essentially perpendicular (Mercury, Venus, and Jupiter) to 23–29° (Earth, Mars, Saturn, and Neptune), to 98° (Uranus). If all the planets were formed from the same nebula, they should have all started with an axis of rotation perpendicular to the plane of revolution about the Sun. The variation from this presumed initial condition for five of the eight planets is due to the effect of the pull of gravity from the other planets plus significant collisions. The favored hypothesis is that the tilt of Earth and Uranus were caused by major impacts that knocked these planets on their sides. These impacts tilted the planets as well as caused the formation of their moons. Uranus has rings as well as 27 moons, which range in size from 1,579 km in diameter to only 18 km in diameter, whereas Earth has a single moon. Venus is odd in that it is the only planet to rotate clockwise (as seen from above). It could be that Venus experienced an even more energetic impact than experienced by Earth and Uranus, knocking Venus upside down or perhaps the orbit of Venus slowed down and reversed from tidal influences from the gravity of the Sun.

3.4 Formation of the Earth–Moon System

Before the **impact hypothesis**, a range of explanations were proposed for the origin of our Earth–Moon system. Important observations any model must explain include the 23° tilt of Earth relative to the plane of orbit around the Sun, as well as the size, relative position, bulk composition, and the age of the two bodies.

One idea is that Earth and Moon evolved together from the early nebula. This model is consistent with the Earth and Moon being very close in bulk composition and could explain their relative size and position, but this model predicts that both Earth and Moon should have axes of rotation perpendicular to the plane of revolution about the Sun rather than being tilted at 23°. It also requires that the Moon should be the same age as the oldest meteorite (see more about this in Chapter 4), but the oldest rocks sampled from the Moon by the Apollo missions are no more than 4.53 Ga and the oldest meteorites are 30 million years older at 4.56 Ga. Of course, we have only visited a small part of the Moon so there may be older, yet undiscovered areas. However, some of the Apollo mission landing sites were chosen specifically to try and find the oldest rocks, and 4.53 Ga is the oldest moon rock recovered to date.

A second model is the capture hypothesis. In this model the Moon was formed in another part of the solar system and was later captured by Earth's gravity as the Moon passed by. Recall in Chapter 2 that Abraham Gottlob hypothesized that a passing celestial body removed excess water from his proposed global ocean. One problem with this idea is that Earth and the Moon are so similar in bulk composition that it would be unlikely for them to have formed in different parts of the solar system and this hypothesis was also successfully tested by the Apollo samples. However, a much bigger problem is that Earth's gravity would not be of sufficient magnitude to capture a body the size of the Moon unless the velocity of the Moon relative to the Earth was quite slow, and this is considered unlikely. Astronomical calculations show that a Moon-sized planet would be very likely to collide with Earth rather than move into an orbit around it and any path that did not include a collision would have a Moon-sized object just pass by and not take up an orbit around Earth. This model also predicts an older age for the Moon than has been observed.

The model favored today is the collision hypothesis (Figure 3.8). In this model, early in the formation of the planets from the nebular disk (Figure 3.6), Earth collided with another protoplanet approximately the size of Mars. This body has been given the name *Theia*; in Greek mythology, Theia was the mother of Selene, the goddess of the Moon. The Theia hypothesis proposes a giant impact that would have reformed both planets with mixing of the two bodies into a new Earth and a disk of material that eventually coalesced to become the Moon. The impact would have literally melted the surface of the Earth, creating a magma ocean (Figure 3.8(b)). This model is favored because it explains the close but not complete match between the composition of the Earth and Moon. It also explains how the rotational axis of Earth was tilted because of it being knocked on its side from the glancing blow.

One of the main differences in composition between Earth and the Moon is that the Moon has lower density. The giant impact hypothesis predicts that most of the Fe on both the original Earth and Theia would have been drawn into the center of Earth, with much less being thrown out into the surrounding cloud. Melting due to the impact would have aided the migration of the heavier iron and nickel toward the center of the rotating early Earth. This explains that, although the Moon has an Fe core, it has a radius about 25% of the radius of the entire Moon, in contrast to 50%, which is typical of the other inner planets in the solar system.

Another difference in composition between Earth and the Moon is that the Moon is dry, not only on its surface but in its mineralogy. On Earth some minerals, such as micas, have water bound in their mineral structure. This is not "water" exactly but hydrogen in the form of the hydroxyl ion (OH^-) bound in the mineral lattice. This bound water is extremely rare

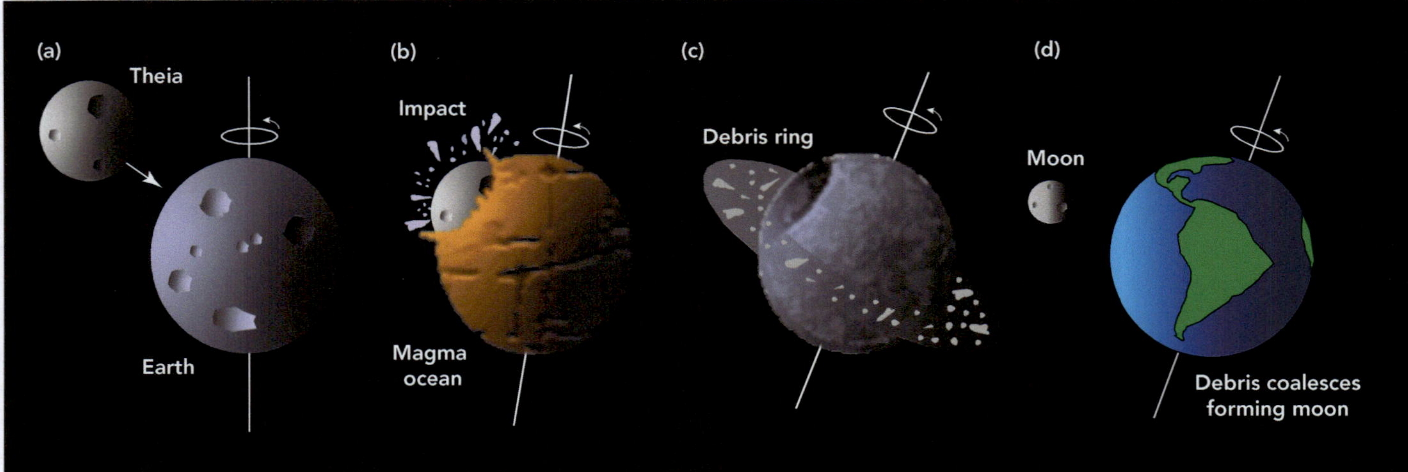

Figure 3.8 A model for the formation of Earth's moon.

in the rocks returned by the Apollo missions. This is consistent with volatile components, such as water and elemental hydrogen, being lost into space in the extremely hot conditions following the impact. There is some evidence for ice at the surface of the Moon in the deepest parts of craters near the poles that never receive sunlight, but this ice is thought to represent water added to the surface possibly by comets long after the formation of the Moon.

Finally, Earth spins quite fast compared to some other planets. This higher rate of spin is also consistent with momentum being added to the Earth–Moon system during the impact with Theia and is inconsistent with the Earth and Moon forming together out of the nebula or by capture.

Rocks returned by the Apollo astronauts indicate the Moon was volcanically active in its early history (Figure 3.9), as indicated by igneous rocks yielding crystallization ages from 4.5 to 3.2 Ga. It appears that this igneous activity, and any related tectonic activity, stopped by about 3.1 Ga. The reason that vigorous tectonics continued on Earth while the Moon has been tectonically inactive for the past three billion years is that the Moon has about one-sixth the mass of Earth and it cooled much faster. After the interior of a planet cools below a critical temperature, motion of material in the interior will stop. If the Moon ever had an atmosphere, it was lost due to insufficient gravitational force to hold on to whatever gases were once present because the escape velocity of the Moon is low. Thus, because of its small size, the Moon quickly lost the aspects that make Earth a dynamic planet, particularly plate tectonics and weather. This difference in geological activity between the active Earth and moribund Moon during the previous several billion years makes the Moon an ideal place to study the early solar system. Rocks on the Moon have remained essentially unmodified since their formation, whereas rocks of the early

Figure 3.9 (a) Apollo 17 astronaut Harrison Schmidt, trained as a geologist, examines a boulder; (b) basalt sample collected from the Moon. Source: (a) Avalon / Contributor via Hulton Archive / Getty Images; (b) Jstuby at English Wikipedia, CC0, via Wikimedia Commons.

Earth are very hard to find (see Chapter 4). The Moon, and other planets that have no plate tectonics, and are pockmarked by craters (Figure 3.10). Dating of moon rocks showed a clustering from 4.1 to 3.8 Ga and led to the *Late Heavy Bombardment*

(LHB). This may reflect an episode of more-frequent collisions or the end of prolonged period. In either case, the apparent slowing of the frequency of large collisions has important implications for the habitability of Earth with the likely extinction of life after many of the largest impacts (see also Chapter 7).

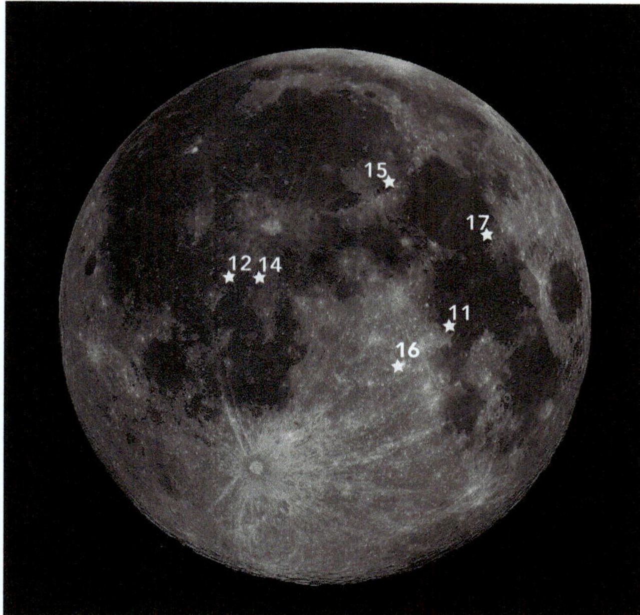

Figure 3.10 Photo of the Moon and Apollo landing sites. The circular features are craters caused by impacts. The dark areas mark ancient basaltic lava flows that filled some of the lager craters. Source: Christophe Lehenaff via Moment / Getty Images.

Much of what we know about the Moon is the result of the six Apollo missions that returned samples to Earth (Figures 3.9 and 3.10). Between July 1969 and December 1972, 12 moonwalkers spent about 80 hours on the surface of the Moon. Despite the brief fieldwork on the Moon, each of these missions was carefully targeted to investigate parts of the Moon that are most likely to reveal its history, including searching for the oldest rocks. Few rocks have been as carefully studied as these Apollo samples, and lunar fieldwork is another example of how much one can learn from the simple act of picking up a rock (Figure 3.9).

3.5 Early Earth Evolution

We next will consider the evolution of our planet during approximately the first two billion years. This takes us through the time periods known as the Hadean and Archean eons (Chapter 4). Earth began as a chemically undifferentiated planet, but over time it developed a **core**, **mantle**, and **crust** (Figure 3.11) through the process of physical and chemical differentiation described below.

3.5.1 Thermal Evolution

To understand our planet's journey from initial formation to the present day, it is helpful to understand the thermal state of the planet. On Earth, temperature increases with depth below the surface (Figure 3.11). It increases rapidly at first then more slowly with increasing depth. In the upper ~40 km, the temperature increases about 25–30 °C for every additional kilometer in depth, so the *geothermal gradient* in the crust is about 25 °C/km.

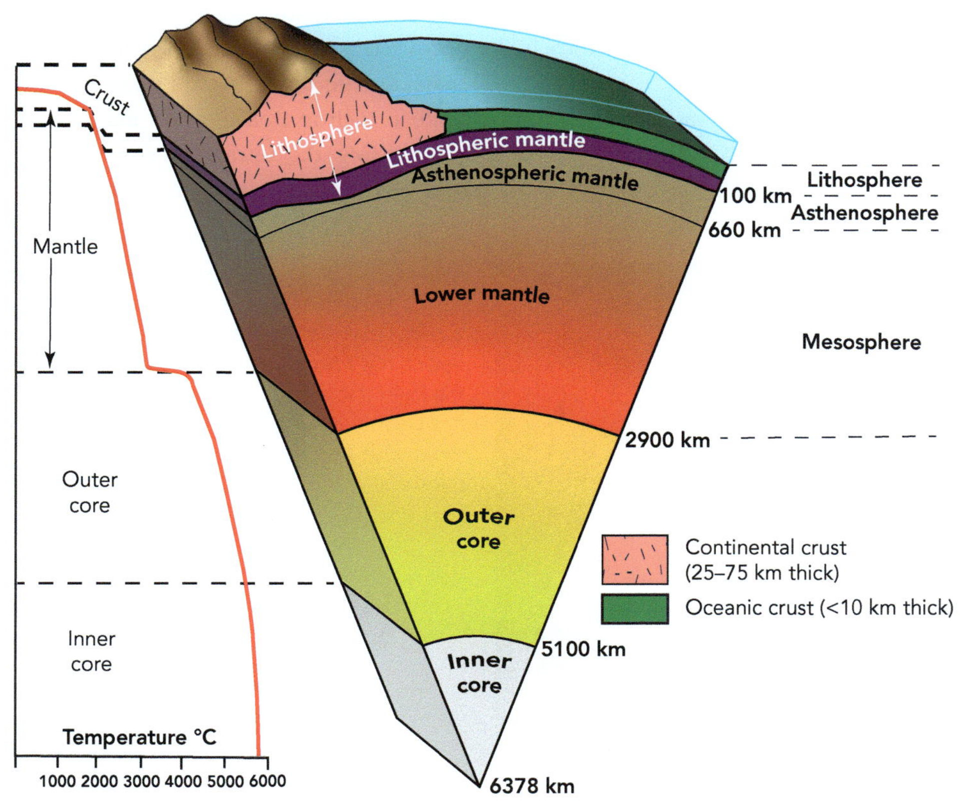

Figure 3.11 Earth layers and temperature gradients. The figure is not to scale as the thickness of lithospheric layers is exaggerated. The correct scales are shown in Figure 3.12.

At a depth of about 660 km, the temperature is about 1600 °C, indicating a gradient of about 2.5 °C/km, and in the inner core, temperatures are estimated to be about 5,700 °C (Figure 3.11), which is as hot as the surface of the Sun. The radius of Earth is 6,378 km (Figure 3.11), so the overall gradient is about 1 °C/km. The higher gradient of ~25 °C/km near the surface reflects the fact that these rocks are not subject to the insulating effect of overlying rocks, which are poor conductors of thermal energy. We will return to an in-depth discussion about the age of Earth in Chapter 4, but a fundamental pair of questions is: (1) Where does the thermal energy inside Earth come from? and (2) If the planet is billions of years old then why hasn't it cooled off more, much like a pie cools off after it is removed from the oven?

As we discussed in Section 3.3, the collisions that built planetesimals and protoplanets generated heat that raised the temperature of these bodies. Gravitational pull also caused compression within the developing protoplanets, further raising the temperature. Thus, Earth formed with a high degree of *primordial* (i.e., initial) heat. The analogy to a pie cooling on the countertop is not actually very good because rocks are poor conductors of thermal energy, which means that they can retain heat for much longer than even a planet-sized pie. Unlike our pie, in addition to the thermal energy present at the beginning, all rocky planets in our solar system have an additional heat source. This comes from the process of radioactive decay, which we will discuss in more detail in Chapter 4. Some atoms are unstable and from time to time they will decay into other atoms, releasing energy in the form of heat. This internal source of radioactive heat occurs in all the rocky planets, including the Earth and Moon, and explains why Earth has not cooled down like our pie, even given the enormous span of time involved (see Chapter 4).

3.5.2 Earth Structure

At the beginning of the transformation of the nebula into the Sun and discrete planets (Figure 3.6), the distribution of chemical elements in the planets was fairly homogeneous, but heat generated by collisions and radioactive decay allowed protoplanets to become molten, initiating the first igneous processes. These igneous processes allowed chemical and physical differentiation of Earth, forming a layered planet (Figure 3.11).

When rocks in planetesimals grew to larger protoplanets and were heated to the point of partial melting, the melts were higher in silica (SiO_2) and lower in iron and magnesium oxide (FeO and MgO) than the solid material remaining, resulting in chemical differentiation. The liquids are also less dense than the rocks they were derived from and therefore this liquid rose, moving towards the surface (Figure 3.12). Over time, these lighter, more silica-rich melts cooled to create Earth's first crust. At the same time, recall that the planets were spinning, and as they spun, heavier liquids, and particularly iron and nickel, sank to the center. These processes ultimately formed planets with a metallic iron and nickel *core* surrounded by a rocky *mantle* capped by a thin solid *crust* (Figures 3.11 and 3.12). It is likely that the collision of Earth and Theia at 4.5 Ga (Figure 3.8) imparted enough energy to melt the early Earth lithosphere, resulting in a *magma ocean*. This impact homogenized the combined mass but eventually two distinct bodies coalesced out of the debris. After this, Earth heated up due to internal radioactivity and a few hundred million years later, parts of the interior were hot enough to produce molten iron and nickel. Because of the enhanced mobility and high density of the metallic liquid this material migrated to the center of the planet to produce Earth's core (Figure 3.12).

The mantle is about 2,900 km thick (Figure 3.11), comprising about 84% of Earth's volume. The mantle is mostly made up of *peridotite*, an **ultramafic** rock that has a low silica and high magnesium and iron content. Peridotite refers to the mineral *peridot*, which is the gem-variety of the common mineral olivine, named after its characteristic olive color. Peridotites are mostly olivine and pyroxene, rich in iron (Fe) and magnesium (Mg) and low in SiO_2, reflecting the ultramafic composition. Partial melting of ultramafic peridotites releases mafic melts that

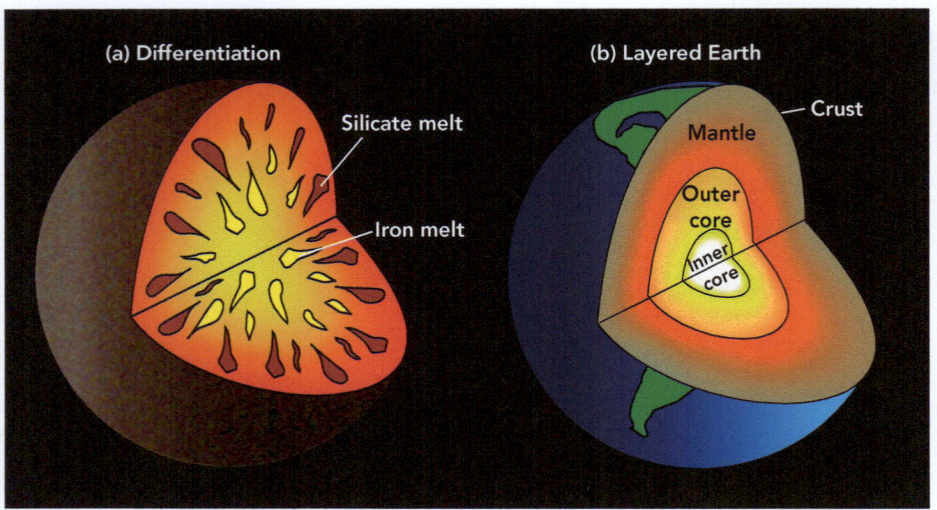

Figure 3.12 (a) The young hot Earth was largely liquid and heavy, where iron-rich liquids separate and sink while lighter silicate liquids rise forming (b) a layered planet. In (b) the layer thicknesses are shown to scale, and it is clear that the crust is much thinner than the mantle and core. We exaggerate the thickness of the crust and lithosphere in Figure 3.11.

cool to form basalts, which form the crust, and which have a higher ratio of silica to iron and magnesium.

We know about the chemical composition of the crust and mantle by direct observation (see Table 3.1). The crust is, of course, at the surface, making for easy sampling. The mantle is not usually present at the surface but sometimes volcanoes bring up pieces of the mantle during eruptions, and plate tectonic processes sometimes uplift and expose mantle rocks (see Chapter 10 for examples). We also can determine the thicknesses and composition of the core, mantle, and crust from seismic studies (see Box 3.2). We don't have any samples of the core but, based on the size, we know the mass and density necessary in the core to explain the rate of revolution around the Sun and an Fe-rich core provides the necessary mass. Moreover, samples of meteorites, which can be rich in iron, give us an idea of what a planet would have looked like before any geochemical differentiation, and an Fe-rich core matches the composition of the most primitive meteorites.

BOX 3.2 Understanding the Earth's Interior

Much of what we understand of the details of Earth's interior comes from seismology. The geometry of this process is llustrated in Figure 3.13. When an earthquake occurs several different kinds of energy radiate out in waves from the source. For understanding Earth structure, the two most important types of waves are called primary (P) waves and secondary (S) waves. They are given these names because of the relative velocity of the waves and differences in the way the waves affect rocks as they pass through them. P waves are faster and vibrate in the direction of travel, causing compression of the rocks. S waves are slower and vibrate perpendicular to the direction of travel, causing rocks to shear. Compressional waves can travel through solids and liquids, but shear waves can only pass through solids; when an S wave encounters a liquid zone, such as a magma or water, it dies.

The velocity of these waves is dependent on the material through which they are traveling and when the material changes the speed and direction will change. Layers in Earth can change their properties either gradually, because of continuous pressure changes, or abruptly, such as where two different rock types are juxtaposed.

Seismometers arrayed all around the world record the arrival times of P waves and S waves. From the various arrival times at the many locations, seismologists have been able to construct models for the composition and thickness of the layers of Earth's interior.

Because of increasing pressure with depth, the ray paths of both P and S waves curve as they leave the source of an earthquake, but S waves from any given earthquake are only received on about two-thirds of Earth's surface. This occurs because any wave that encounters the surface of a region at a depth of about 2,885 km stops. The observation that S waves are unable to penetrate the outer core indicates that it is mostly or entirely liquid. P waves pass through the outer core but do so in a way that requires another major change at a depth of about 5,155 km. This tells us that the inner core is made up of the same material as the outer core, mostly iron with a small concentration of nickel, but the inner core is solid rather than liquid.

P waves also abruptly change their direction of travel as they pass from the mantle and into the outer and inner core. These changes in direction are caused by refraction of the waves and indicate that the core is made of much denser material than the mantle, which helps to determine its composition.

Also, during the generation of P and S waves by a given earthquake, there are areas on the other side of the globe that will not receive any signal of the earthquake. The size and position of these shadow zones are a consequence of the thickness and composition of Earth's layers and the behavior of waves as they interact with these layers.

By observing the location of earthquakes, the timing of arrival of the various seismic waves around the globe, as well as the various ways that P and S waves behave, including the shadow zones, seismologists have been able to refine our knowledge of the Earth's interior and allow us to understand the composition and thickness of these layers.

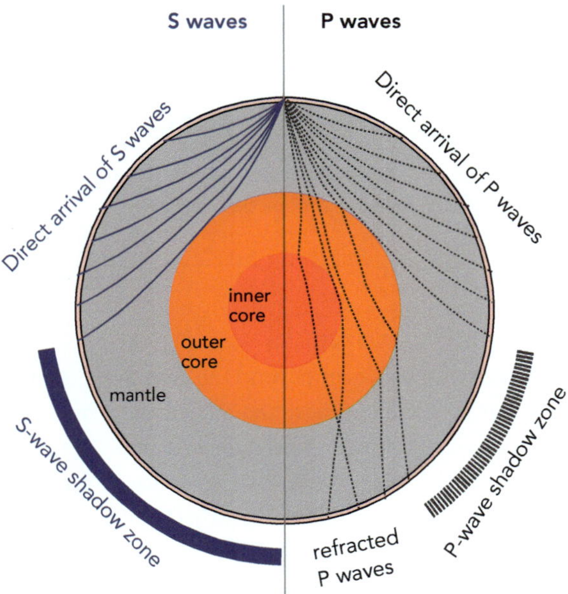

Figure 3.13 An illustration of the passage of seismic waves from a large earthquake with the focus a few kilometers beneath the North Pole; S waves are illustrated on the left and P waves are illustrated on the right, but both types of waves move out in all directions.

3.5.3 Core Formation

All the rocky planets (Figure 3.7) are thought to have iron-nickel cores, and Mercury, the innermost planet, has the largest. Iron and nickel cores are important in explaining why rocky planets have a magnetic field, as observed on Earth. In addition, because of the excessive temperature and pressures deep within planets, some planetary layers are liquid and others solid. Liquids can move about quickly but even solid rock (if it is hot enough) can flow and move, albeit very slowly on timescales of tens to hundreds of millions of years, and this is a key in allowing plate tectonics, as we will discuss further in Chapter 5.

The metallic core of Earth is divided into a solid inner core (about the size of the moon at 2,440 km in diameter) and a 2,400 km-thick liquid outer core (Figures 3.11 and 3.12). The temperature of the inner core is estimated to be about 5,700 °C, and the pressure is 330 to 360 gigapascals (3,300,000 to 3,600,000 atm). Despite the heat, the extremely high pressure allows the iron and nickel to exist in a solid state. The outer core is around 2,300–5,700 °C but experiences a lower pressure than the inner core, and at these conditions the iron and nickel maintain a liquid state and can flow in a slowly turbulent fashion. These slow turbulent eddies generate our magnetic field, and as they shift around so the magnetic field enveloping the Earth has changed over time.

The interior of the Earth has been cooling at a rate of about 0.1 °C per million years, and because of this cooling the inner core has grown by about 1 mm per year. At the time of formation of the Earth, the core is thought to have been entirely liquid and the inner core is thought to have started growing about 1 to 1.5 billion years ago.

KEY POINT

The layers of the Earth were formed by chemical and physical differentiation, such that metallic fluids moved to form the iron- and nickel-rich core and the silica-rich lighter fluids moved upward to form the mantle and crust. The liquid outer core causes Earth's magnetic field.

3.5.4 Crustal Growth

Early in Earth history the crust would have been entirely mafic in origin (i.e., rich in iron and magnesium with a lower proportion of SiO$_2$), but it is very likely that convection in the mantle would have produce upwelling, marked by volcanoes, similar to those that we see on Mars. As introduced in Chapter 1, Earth's outermost crust can be subdivided into felsic continental and mafic oceanic crust (Figure 3.11), but the earliest Earth's crust was almost certainly mafic in composition, inviting a key question of how and when the first continental felsic crust was formed. Our present configuration of felsic continental and mafic oceanic crust is the consequence of the ratio of formation of crust versus recycling back into the mantle by the process of plate tectonics (see Chapter 5). As we will discuss in greater detail in Chapter 5,

oceanic crust is relatively easy to make and recycle (also see Chapter 10). Mafic oceanic crust is generated by partial melting of the ultramafic mantle, but continental crust usually requires further differentiation by partial melting of previously formed mafic oceanic crust to generate a more felsic melt. It likely took several cycles of partial melting before a truly granitic type of continental crust could be produced on Earth. Once continental crust is made, it is much harder to recycle than oceanic crust because it is quite buoyant, so the question remains, when did this first occur? There is significant debate as to when plate tectonics began and whether plate tectonics is required to produce felsic melts, and we will discuss this further in Chapter 5.

Compilations of the ages of continental rocks show a peak around 2.7 Ga, suggesting that this represents the first major peak of formation of continental crust. However, the oldest continental rocks in the Acasta Gneiss in northern Canada are about 4.1 Ga, indicating that here was continental crust farther back in time, and this is discussed in more detail in Chapter 4. However, geochemical analysis of even older rocks suggest they could not have formed by primary partial melting of the mantle, which would yield a basalt, but were more likely formed by reworking of older continental material. Some of the oldest rocks on Earth also contain a small proportion of individual grains of zircon, dated at 4.4 Ga. These zircons are particularly important because zircons are found almost exclusively in granites, which are almost exclusively found in continents. There are thus several lines of evidence from geochemistry and observations of zircons that indicate some granitic crust had begun to form within the first few million years of Earth history, although the volume of continental crust didn't grow to something close to modern levels for billions of years.

KEY POINT

Some continental crust was formed early in Earth history, as indicated by 4.4-billion-year-old zircons that require a continental origin.

3.5.5 Isostasy and Topography

The significance of continents goes beyond the large-scale chemical stratification of Earth. This is because the difference in density in these chemical layers resulted in physical differentiation. Being made of mostly quartz and feldspar, continental crust has a density of about 2.8 g/cm^3 (water has a density of 1 g/cm^3). Oceanic crust has a higher density of about 2.9–3.0 g/cm^3 and the upper mantle has a density of around 3.3 g/cm^3. The mantle is split into two layers (Figure 3.11). Above about 100 km depth the upper mantle is rigid but below this depth the mantle has a density of about 3.4 g/cm^3 or more but is not as strong because it is hotter. This transition, at about 100 km, marks the boundary between the rigid *lithosphere* above and the weaker asthenosphere below (Figure 3.11). As discussed above, although the asthenosphere is technically a solid, it can flow very slowly by a process termed convection and as we will see in Chapter 5 is a key component in driving plate tectonics.

The differences in density produce topographic variation, a feature on Earth fundamental to a variety of processes. The density differences between the lithosphere versus the underlying asthenosphere affects topography. Lower-density lithosphere, such as found in continents, effectively "floats" as it is buoyed up on the higher-density asthenosphere. The density difference and thickness of the floating mass determines the height that the lithosphere rises as it rides above the asthenosphere (Figure 3.14), a property referred to as **isostasy**.

A good illustration of the concept of isostasy is ice, which is slightly less dense than seawater. Therefore, icebergs float but only about 10% of the volume sticks above water (Figure 3.14(d)).

(a) Pratt Isostasy

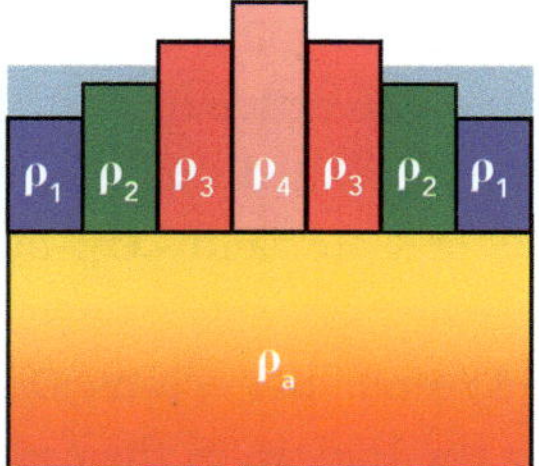

(b) Airy Isostasy

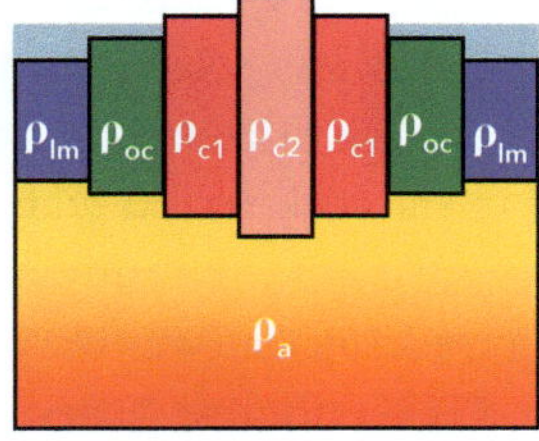

(c) Continental example

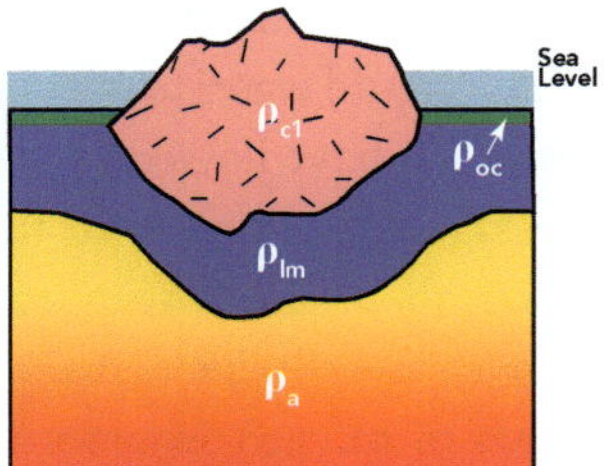

(d) Iceberg example

Figure 3.14 (a) Pratt and (b) Airy models for isostasy. (c) Schematic example of changes in the thickness of the lithosphere correlated to elevation. (d) The ice in an iceberg is less dense than seawater and thus is able to float. As the iceberg melts it isostatically adjusts such that a small part is always above water.

The 90% below water displaces a volume of denser seawater equivalent to the mass of the entire iceberg, but because the ice is less dense, the rest rises above water. As an iceberg melts from the top, removing ice, the mass decreases and the iceberg readjusts its position so that the 90% of the now lower volume remains below water. The readjustment will involve an actual rise in the top to compensate for the volume lost. Regardless of how much melting occurs, removing ice from the top of an iceberg requires some ice to stick above water. About 10% of the volume of the iceberg will always be above sea level, whether the surface area is several thousand km^2 or just $1\,cm^2$, because the iceberg will continuously isostatically adjust its position to accommodate the removal of volume. As melting proceeds, parts of the iceberg that were originally under water will eventually be above water.

The crust of Earth behaves in much the same way, albeit over longer time frames than melting icebergs. Erosion of thick crust causes isostatic adjustment and over time, rocks that may have been buried deeper in the crust, such as plutonic or metamorphic rocks, can be exposed at the surface and weathered to produce sediment.

We will discuss the importance of isostasy to mountain building in subsequent chapters, but the reason this is important in considering the early evolution of Earth is that without low-density continental crust, very little land would be above sea level. The exception is volcanoes formed in areas where mantle upwells, such as seen today in Hawaii, and mantle hot spots were likely a key feature of the Hadean Earth. As we will see in Chapter 7 this may have been the key environment in which the earliest life appeared. On the Hadean Earth, clastic sedimentation would be limited to volcanic islands above mantle hot spots. Importantly, these would represent areas where new minerals would first appear, such as the clays and evaporites that may have played an important role in the emergence of life, as covered in Chapter 7. Permanent land ice would also be very rare, as any ice that might form would soon mix with the rest of the warmer global ocean. Perhaps most importantly, the presence of continents produces ecological niches not seen on a continent-free globe. This includes, of course, terrestrial environments, far from any ocean. Thus, understanding the evolution of continental crust on Earth has implications far beyond the understanding of the process of chemical and physical differentiation.

In Chapter 5 we will look more closely at oceanic lithosphere, which as it cools actually becomes denser than the underlying asthenosphere. As a consequence, large tears (i.e., fractures) can form and the denser oceanic lithosphere can sink. These tears are called subduction zones and are a key driver in making Earth a plate tectonic planet.

KEY POINT

The lower density of crust allows it to float on the asthenosphere by isostasy, forming terrestrial land.

3.5.6 Early Atmosphere

The early atmosphere of Earth was very different from today. Because of the proximity to the Sun, the early atmosphere would have had very little hydrogen (H) and helium (He), despite those gases making up most of the solar system. The early atmosphere would have been like the atmosphere of Jupiter if the He and H were removed. That is, it would have been rich in water (H_2O), carbon dioxide (CO_2), methane (CH_4), nitrogen (N_2), and ammonia (NH_4). Most of the early Earth atmosphere would have come from the degassing of these volatiles out of the mantle via volcanoes.

Because the interior of the planet was hotter at the beginning than it is now, it is not surprising that during early Earth evolution a lot of the *volatiles* (i.e., elements and compounds such as nitrogen, water, carbon dioxide, methane, and sulfur dioxide that have low boiling points) were extracted.

Of course, today's atmosphere is very different from early Earth (see Chapter 7). During the early Archean, collisions of meteors and comets with Earth were not big enough to break up the planet but every few tens of millions of years one big enough to reshape the surface and vaporize the hydrosphere could have contacted Earth. So, it may not have been until about 3.9 Ga that Earth began to have a stable atmosphere, made up of mostly H_2O, CO_2, and N_2. Initially, Venus, Mars, and Earth were thought to have had similar initial atmospheres, but there is newer evidence, from those 4.4 Ga zircons discussed above, that Earth's atmosphere was less rich in reducing components such as CH_4 and NH_4. Analysis of cerium (Ce), a trace metal that is found in zircon, provides evidence for more oxygenated magmas that are interpreted to reflect a more oxygen-rich atmosphere. Like many metals, cerium occurs in two states, Ce^{4+} and Ce^{3+}. Ce^{4+} is favored in more oxygen-rich conditions, whereas Ce^{3+} is formed in more reducing conditions. This is the same as iron, which occurs as either ferrous iron (Fe^{2+}) or ferric iron (Fe^{3+}). Ferric iron is more oxidized and forms the mineral hematite (Fe_2O_3), which is found as rust when metal is oxidized. In contrast, ferrous iron is less oxidized and occurs in a reducing environment, such as is found in anoxic bogs and deep ocean sediments, forming the mineral pyrite (FeS_2), which is diagnostic of reducing conditions. In the same way that Fe^{3+} preferably reacts with oxygen to produce hematite, the more oxic Ce^{4+} is more easily incorporated into zircon. Measurements of the Ce^{4+}/Ce^{3+} ratios showed that the zircons were enriched on Ce^{4+} and indicate that by 4.4 Ga, Earth's atmosphere was richer in H_2O, CO_2, and N_2, versus CH_4 and NH_4, suggesting that it was not strongly reducing. This also has implications for the origin of life, which we will discuss in Chapter 7.

A big question is why Earth's atmosphere today is so different from its early composition. Mars and Venus are still rich in CO_2 because neither has abundant liquid water at their surface. Venus is too hot, and all water exists as vapor, and Mars is too cold. Remote sensing as well as detailed surveying from the various recent Mars missions shows clear evidence that there

was a time early on in Mars history when there was flowing water on the surface (Figure 3.15). Most of this water either froze or escaped. On Earth, stronger gravity allowed the water to be retained, and plate tectonics, weathering, and the development of life on Earth had profound consequences on our atmospheric evolution, particularly the rise of oxygen. We will return to these stories in subsequent chapters.

KEY POINT

Earth's early atmosphere was rich in water, carbon dioxide, and nitrogen but lower in methane and ammonia than other planets, reflecting less reducing conditions.

3.5.7 Evolution of the Oceans

Sedimentary rocks are present in the rock record going back to at least 3.6 Ga, so we know that flowing water was active on Earth's surface at least that long ago. It has been speculated that comets may have delivered significant amounts of water and CO_2, but analysis of cometary ice shows that it is enriched in *deuterium* (one of the isotopes of hydrogen) at a level not seen in our oceans. One issue with water stability is that during the Hadean in the early phase of the Sun's main sequence, it was about 30% cooler than today. This is the faint young Sun hypothesis that we referred to earlier. A fainter Sun would have resulted in less solar energy reaching the surface of Earth, perhaps not enough to keep H_2O in a liquid state. And yet, there is evidence in the rock record going back to at least 3.6 Ga for sedimentary rocks deposited by running water on Earth and even Mars as well (Figure 3.15). One way out of this problem reflects the contribution of **greenhouse gases** – carbon dioxide, methane, and water – emitted from the vigorous volcanism at the time that were at high enough concentrations to have kept the atmosphere warm enough for liquid water to flow on the surface. As it does today, that water must have flowed to low spots, and these would have been the first ocean basins. In the Hadean, these oceans would have been about 4 km deep and punctured by volcanoes formed above areas where mantle upwelling (hot spots) occurred. Analysis of oxygen isotopes in the oldest 4.4 Ga zircons are also instructive as to the state of the earliest oceans. If there is standing water (i.e., oceans), lighter O^{16} preferentially enters the liquid phase, whereas O^{18} is more likely to enter the mantle and lithosphere. High O^{18}/O^{16} ratios were measured in the 4.4 Ga zircons, suggesting that they formed from magmas that interacted with ocean water, versus derived from primary partial melting of primordial mantle, and this also suggests that stable oceans formed early in Earth history. The extensive early oceans may have had an important control in making Earth less reducing than other planets.

Depending on the frequency and severity of extraterrestrial impacts during the hypothesized Late Heavy Bombardment, these early oceans may have experienced temporary episodes of vaporization followed by the return of that water to the

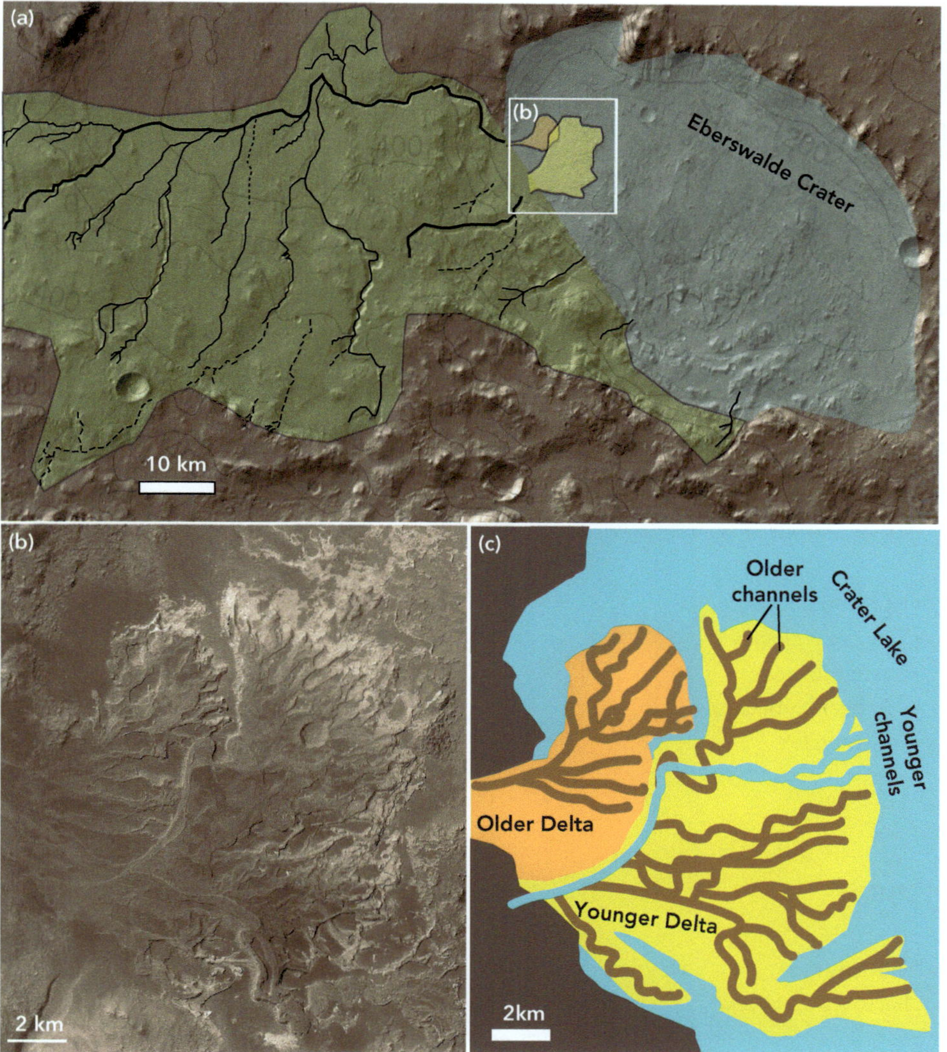

Figure 3.15 (a) A fan-like delta deposit in the Eberswalde Crater on Mars can be linked to upstream drainage channels (green area). (b) Close up of the delta and (c) interpretation of delta lobes and the associated channels that built into a crater lake. The delta was formed about 3.5 billion years ago when Mars had free-flowing water. Source: (a) NASA / JPL / Malin Space Science Systems, Public domain, via Wikimedia Commons; (b) Source: NASA / JPL / MSSS.

surface of the Earth via rainfall, and this also has implications for when Earth was habitable, a theme we return to in Chapter 7.

Of course, rainwater is fresh and marine waters are salty, so we also must explain the origin of the salt in the oceans. Simply, as fresh water runs over land, it acquires dissolved ions, such as Na^+, OH^-, Cl^-, SO_4^{2-}, and CO_3^{2-}, through weathering, and these are then delivered to the oceans via rivers. Ocean water is constantly undergoing evaporation, forming clouds, some of which rain over the land, returning that water to the sea by overland runoff, delivering more dissolved ions to the ocean. Over time, the dissolved ions built up and the ocean became saltier. Sometimes, oceans dry out, leaving *evaporite* deposits (see Chapter 16), removing some of the salt delivered to the oceans. In addition, after complex life evolved on Earth, marine organisms removed calcium carbonate ($CaCO_3$) and lesser amounts of magnesium and silicon to make skeletons that become limestone deposits, a type of sedimentary rock (see Chapter 7). Also, some seawater, along with its dissolved salts, is returned to the mantle

through the process of subduction of tectonic plates (see Chapter 5). Therefore, the world's oceans do not simply become progressively saltier with time (see Chapter 4) but appear to reach an equilibrium in which the long-term delivery of ions is balanced by their removal via the formation of evaporites, excretion of shells, and transfer into the mantle.

Like many Earth systems, whether it be the salt in the oceans, the various gases in the atmosphere, or the proportion of oceanic versus continental crust, over billions of years many of these may reach some sort of equilibrium. One of the most fascinating aspects of Earth history is deciphering when this equilibrium changes, whether gradually or sometimes suddenly, resulting in changes in the positions of land and ocean, changes in the composition of the atmosphere, and the ability of life to adapt to these changes.

KEY POINT

Most of Earth's ocean water likely evolved by outgassing from the mantle as the Earth cooled. Weathering allowed salts to build up.

3.6 Summary

- The Big Bang produced our universe about 13.8 billion years ago.
- Fusion in stars and recycling of material produced in supernovas created the elements.
- Our solar system was produced about 4.6 billion years ago by the coalescing of material created in earlier stars.
- The Moon was created by a giant impact with a Mars-sized body (Theia) that knocked the Earth, resulting in its tilted axis and faster degree of rotation.
- Chemical and physical differentiation in a spinning Earth resulted in a layered Earth with a metallic iron and nickel core, and a rocky mantle and lithosphere.
- Continental growth probably started very early on in Earth history.
- Continental lithosphere is less dense than the asthenosphere and floats by the process termed isostasy forming land masses that are higher than surrounding oceans.
- The oceans and atmosphere were primarily the result of volatile outgassing through volcanoes.
- The oceans became salty through chemical erosion, but local evaporation as well as removal of ions by precipitation of organisms prevents ocean water from simply becoming saltier with time and provide a good example of the equilibrium conditions that characterize many Earth systems.

Key Words

- Big Bang
- inflationary epoch
- Milky Way
- supernova explosions
- nebular hypothesis
- planetesimals
- protoplanets
- escape velocity
- asteroid belt
- impact hypothesis
- ultramafic
- isostasy
- core
- mantle
- crust
- greenhouse gases

Further Reading and References

Allegre, C. J., Poirier, J.-P., Humler, E., and Hofmann, A. W., 1995, The chemical composition of the Earth, *Earth and Planetary Science Letters*, 134, 515–525.

Asphaug, E., 2014, Impact origin of the Moon?, *Annual Reviews of Earth and Planetary Science,* 42, 551–578.

Chambers, J., and Mitton, J., 2017, *From Dust to Life: The Origin and Evolution of our Solar System*, Princeton University Press.

Haynes, W. M., 2016, Abundance of elements in the Earth's crust and in the sea, in *CRC Handbook of Chemistry and Physics*, 97th ed. Taylor and Francis.

Kring, D., 1997, Composition of Earth's continental crust as inferred from the compositions of impact melt sheets, in *Lunar and Planetary Science XXVIII*, https://adsabs.harvard .edu/pdf/1997LPI....28..763K.

Review Questions

1. When did the Big Bang happen?
2. How long was there between the Big Bang and the formation of our solar system?
3. How does turbulence and uneven mixing result in the formation of stars, solar systems, and galaxies?
4. Explain how it is that all the chemical elements that are found on Earth today were produced in stars.
5. How does the nebular hypothesis explain the relative size and composition of the inner rocky and the outer gas planets?
6. What evidence favors the giant-impact model for the formation of the Earth–Moon system?
7. Why was the interior of Earth much hotter in the Archean?
8. What produced the major chemical differentiation of Earth into its core, mantle, and crust?
9. Why is the evolution of continental growth so important to Earth history?
10. How does the principle of isostasy explain the heights of mountains and the depths of oceans?
11. When were most of the gases in today's atmosphere extracted from the mantle? What evidence supports this conclusion?
12. Why don't Earth's oceans simply become saltier and saltier over geologic time?

Geological timescale and life.
Source: Gary Hincks / Science Photo Library.

Chapter 4

The Age of Earth

Historical Versus Modern Dating Approaches

LEARNING OBJECTIVES

- Describe the wide variety of religion-based estimates of the age of Earth and the universe from a near infinite universe in Hindu cosmology to the few thousands of years in some Judeo-Christian theologies.

- Define some of the Enlightenment approaches to determining the age of Earth, such as the hypothesized 50–150 million years it would have taken to make the oceans salty from a hypothesized freshwater beginning, based on knowledge of the delivery of dissolved salt via rivers, or estimates based on measured sedimentation rates integrated over the thicknesses of ancient sedimentary rocks and finally based on heat flow and the time needed to cool Earth from a molten state.

- Discuss how the discovery of radioactive elements led to the ability to obtain absolute dates of rocks and minerals.

- Explain how the concept of closure temperature can be used in dating the thermal history of rocks and minerals, and why some of the first U/He dates were much younger than U/Pb of the same rock samples.

- Discuss how measurement of detrital zircons and ultimately meteorites led to our current understanding that Earth is about 4.6 billion years old.

Introduction

Perhaps no scientific question has been more widely considered across the span of human history as the age of Earth. The question, "How long has this planet been here?" dovetails quite reasonably into, "How long have we been here?" and so the problem of geochronology is essential to much more than just rocks. Whether we consider this problem from Earth or from outer space (Figure 4.1), the issue remains essential.

In Chapter 1 we discussed a few simple rules that allow the determination of the *relative age* of rocks, but these methods give no indication of how old in absolute terms any of the rocks in question are. Steno's principle of superposition is very helpful, but it only sets two strata apart in a relative sense. If a limestone sits on top of a sandstone, we know the limestone is younger, but we can say nothing of exactly when either rock was deposited in terms of years before today.

William "Strata" Smith's *principle of faunal succession*, introduced in Chapter 2, is an outgrowth of superposition, allowing fossils to inform us of the relative order of events without having to see two rocks in physical contact. Thus, if we find one rock that contains trilobites and another that contains *Tyrannosaurus rex*, we know the former is older and the latter is younger. However, on the question of how many years ago did the dinosaurs or the trilobites walk the Earth, the principle of faunal succession is mute.

Figure 4.1 How long has this been in existence? The view of Earth from Apollo 8. Source: NASA / Goddard Space Flight Center Scientific Visualization Studio.

In this chapter we will discuss the modern methods for determining the age of rocks, and indeed the age of the planet itself. First, however, because humans have only had a good grasp of the true span of geologic time for about the past 100 years, and we have only had our modern estimate for the age of Earth for the past 50 years, we will first review some of the important events in the history of thinking about this fundamental problem.

4.1 Historical Approaches to Estimating the Age of Earth

Because the question of the age of the Earth is so fundamental to all cultures, methods of assessing this quantity go back thousands of years and cover a wide range of methods. In this section, we discuss just a few of these methods, which precede our modern approach using geochemistry.

4.1.1 Religion-Based Approaches

Human religions vary widely in their understanding of the age of Earth. Among the world's religions, Hindu philosophy is probably most closely allied with our modern scientific understanding. In Hindu cosmology, time is considered both infinite and cyclic. Hindu theologies postulate multiple universes, which pass through cycles of birth from initial chaos, through ordered universes that die and revert to chaos, only to be reborn again. The duration of these cycles ranges from our ordinary day and night to a day and night of the god Brahma, which is 8.64 billion years long.

The age of the universe and Earth in the monotheistic Abrahamic religions, including Judaism, Christianity, and Islam, are largely described in the book of Genesis in which an omnipotent God creates the universe, Earth, and everything living upon it in six days. The book of Genesis contains two separate, and in places conflicting, accounts of this creation, and there is continued debate, especially among Christian groups, as to whether these six days should be taken as six literal Earth days or metaphorically longer periods, more along the lines of the Hindus.

The book of Genesis includes genealogies from the first humans (Adam and Eve) to Abraham, and numerous Christian scholars have used those genealogies, followed by estimates of the reigns of subsequent biblical kings (e.g., David, Solomon, etc.) to suggest an age for the time of creation. In 1738, Alphonse de Vignolles compiled in his *Chronologize de l'histoire sainte*, more than 200 computations of this sort, whose estimates for the date of creation ranged from 6984 BCE to 3483 BCE. Sir Isaac Newton, although much better known as the founder of physics and calculus, spent much of his life in studies of alchemy and biblical chronology and suggested an age of creation of 4000 BCE.

Most of the authors of these chronologies were prominent scholars and theologians of their day. Perhaps the two most well-known were John Lightfoot (1602–1675), a biblical and Greek scholar at Cambridge University, and James Ussher (1581–1656; Figure 4.2).

A theologian and scholar, Ussher rose to be an Anglican Archbishop in Ireland. In 1647, Lightfoot published the results of his analysis of the biblical chronologies, concluding that the date of creation was 3928 BCE on the autumnal equinox. In an earlier book (*A few, and New Observations Upon the Book of*

Figure 4.2 James Ussher. Source: After Peter Lely, Public domain, via Wikimedia Commons.

Genesis, the Most of Them Certain, the Rest Probable, all Harmless, Strange and Rarely Heard of Before), published in 1642, Lightfoot suggested the creation of humans occurred at nine o'clock in the morning. This mention of 9:00 AM is often mistakenly attributed to Ussher as his estimate for the time of creation (not the creation of humans). Ussher used a variety of astronomical, historical, and biblical sources published in two works (one in 1650, the other in 1654) to calculate the date of creation as October 22, 4004 BCE. Of all the religion-influenced estimates for the age of Earth, Ussher's is the most famous. This fame can be traced to the 1701 edition of the English bible, in which the results of his calculation were added as a marginal note.

KEY POINT

There have been a wide variety of religion-based estimates of the age of Earth from near infinity in the recycling concepts in Hindu cosmology to several thousand years based on biblical creation myths described in the book of Genesis and the chronology of the descendants of Adam and Eve.

4.1.2 Approaches Considering Observation and Modeling of the Natural World

The nineteenth century saw a shift in the approaches to this problem from the scriptural to the observational. The Enlightenment, also known as the Age of Reason, was the intellectual and philosophical revolution that dominated Europe during the eighteenth century. Enlightenment philosophers generally supported individual liberty, reliance on observation of the natural world, as opposed to fealty to revealed truth through religious texts, and separation of church and state. During the Enlightenment, the first scientific societies and scholarly journals were established, and we discuss several of the most influential ideas that stemmed from Enlightenment scientists below.

4.1.2.1 Salt in the Ocean

One approach for estimating the age of the Earth using gradualist thinking (see Chapter 2) was advanced in the late nineteenth century. This method assumes that the reason that the oceans are salty and the water on the continents is not that the oceans were originally fresh, just like the lakes and streams that fed water into the oceans, but that they have become saltier over time because of the erosion of soluble elements, such as sodium and calcium, from the continents. This approach recognizes that the fresh water on the continents is not pure H_2O as it has small amounts of dissolved ions. Thus, so the method proposes, if we can (1) measure the concentration of salt in the modern ocean (about 35 parts per thousand), (2) determine the salinity of the rivers delivering water to the ocean (typically less than 0.5 parts per thousand), (3) determine the rate at which water is

transferred from the continents to the ocean, and (4) calculate the volume of the oceans, we could calculate the time it took for the ocean to achieve its present level of saltiness. This method was first proposed by Sir Edmund Halley (who also discovered Halley's comet) in 1715 but this idea was given the greatest attention in the final decades of the nineteenth century. Depending on the values assigned to the several parameters that were used in the calculation, various naturalists derived estimates that ranged from around 50 to 150 million years.

4.1.2.2 Sediment Accumulation

A second way of applying knowledge of modern surface processes to the problem of the age of the Earth is using erosion rates of continents. Like the approach of salt in the ocean, this approach tracks the transfer of material from the continents to the ocean, but here we concentrate on solid material rather than dissolved salts. The approach is very similar. If we could know the rate of sedimentation in the ocean and if we could determine the thickness of sedimentary deposits on Earth, dividing the thickness by the rate of accumulation would simply provide the total time it took to deposit all the sediment. Despite many shortcomings, which we will discuss below, the sediment-thickness method was commonly used to estimate of the age of the Earth at the end of the nineteenth and beginning of the twentieth centuries. More than 30 estimates using this approach were published from 1860 to 1917, and these ranged from a few million to several billion years, with a median value of around 100 million years.

KEY POINT

Enlightenment approaches to determining the age of Earth included estimates based on the time it would have taken to make the oceans salty from a hypothesized freshwater beginning based on knowledge of the delivery of dissolved salt via rivers, and estimates based on measured sedimentation rates integrated over the thicknesses of ancient sedimentary rocks and ranged from a few million up to a several billion years.

4.1.2.3 Calculations Based on Heat Flow

The most famous of the nineteenth-century attempts to determine the age of the Earth was certainly that of William Thompson (Lord Kelvin), who estimated the age of Earth based on calculations of heat loss associated with the cooling of the Earth. Similar approaches of estimating the age of the Earth and Sun by reference to bulk cooling history were first attempted in the early eighteenth century.

It is hard to overestimate the intellectual status of William Thompson (1824–1907; Figure 4.3) in the middle of the nineteenth century. He wrote his first scientific paper at age 16 and became Professor of Natural Philosophy at Glasgow University in 1846 at age 22. At his retirement at the turn of the century he

Figure 4.3 Sir William Thompson, the first Baron Kelvin (1824–1907). Source: Russell & Sons, London. No restrictions, via Wikimedia Commons.

had authored more than 600 scientific papers and books and held 70 patents. More than a dozen physical phenomena are named after him (either with the name Thompson or Kelvin) as well as a temperature scale. In 1866 he was knighted by Queen Victoria, becoming Sir William Thompson, and in 1892 he became the first scientist to be honored with a peerage and took the title Baron Kelvin. Thompson is thus now generally referred to as Lord Kelvin.

It is important to appreciate Kelvin's scientific prestige to put his estimate of the age of the Earth in a historical and social context. In the last half of the nineteenth century, when Lord Kelvin spoke, people listened. Kelvin's interest in geochronology was an offshoot of his broader work in thermodynamics. Through the second law of thermodynamics, he understood that anything doing work will eventually run out of energy and stop. He applied this reasoning to the Earth: "Within a finite period of time Earth must have been, and within a finite period of time to come, Earth must again be unfit for the habitation of man."[1]

In other words, Kelvin reckoned there was a time in the past when it was too hot and there will come a time in the future when things will be too cold for humans to survive. With this in mind, Kelvin attempted to estimate the lengths of time in question. The first attempt he made was to estimate the age of the Sun based on estimates of the temperature of the surface of the Sun, the rate of cooling of the Sun, and several other parameters that were poorly known at the time, but the best-known effort along these lines are his estimates for the time of cooling of Earth.

In Kelvin's time it had long been established that Earth is roughly a sphere with a radius of about 6,300 km, so one quantity Kelvin needed for his calculations, namely the size of the Earth, was already well known. Next, Kelvin needed an estimate of the initial temperature of the Earth. Obviously, the hotter the initial temperature, the longer it will need to cool down. But what was the initial temperature of Earth? In Kelvin's time there was no way to know, so he simply assumed that the initial temperature was the melting temperature of iron. This had the advantage of being a temperature that could be measured using the technology of the time. The final bit of information he sought for his calculation was the gradient of heat at the surface of Earth. To do this, he took advantage of the presence of underground coal mines in Wales. The deepest mines available allowed the temperature of rocks to be measured about 1 km below the surface. Now armed with the requisite data, and some innovative mathematics, Kelvin was able to estimate how long a body the size of Earth, initially at the temperature of molten iron (1,510 °C or 2,750 °F), needed to cool to the point that the gradient of temperature over the outermost 1 km matched what was measured in the Welsh mines. Depending on various assumptions, including the thermal conductivity of rocks as well as the initial temperature, in his first publication on this topic in 1862, Kelvin calculated that Earth was in the range of 30 million to 400 million years old. In later work, he refined this estimate to a value at the lower end of this range.

4.1.3 Problems with Historical Methods

By the end of the nineteenth century, estimates for the age of Earth ranged from around 6,000 years (based on religious arguments) to several hundred million years (based on surface processes and heat-loss arguments). However, each of these methods is flawed in some important way.

Prior to the Enlightenment, the Church dominated European thinking, not just in matters theological but also secular and scientific. It is now generally (but not, it is worth noting, entirely) accepted that theology and science are complimentary but separate ways to investigate questions of our existence. Observation of the natural universe is considered to be the only way to quantify its physical and temporal limits; and is not considered to be a spiritual question, certainly in the scientific community.

The salt-in-the-ocean approach suffers from several serious flaws. First, it assumes that at some time the oceans existed but were free of dissolved salt. Although it is necessary to proceed with this assumption to make the calculation simple, there is no

---

1 "On a universal tendency in nature to the dissipation of mechanical energy," *Proceedings of the Royal Society of Edinburgh*, April 19, 1852.

observation of the natural world, either today or from the rocks formed in the ancient past, that requires this assumption. Second, although it may be possible to measure the modern flux of water and dissolved salt in rivers, it is not very likely that whatever values one might obtain for today would be representative of all the geologic past. But the third and most striking flaw of the salt-in-the-ocean method for calculating the age of the Earth is the assumption that the path of salt into the ocean is a one-way process. In fact, observations of salt deposits around the world are testament that salt has indeed been removed from the oceans throughout at least the past 600 million years (e.g., see Chapter 16). The pink Himalayan salt that can be found in supermarkets today formed about 550 million years ago in the Lower Cambrian Period. There is a very good chance that the last time you put salt on your food, that salt came from a salt deposit that was produced when a portion of the ocean evaporated enough to precipitate (i.e., remove) salt. If all the salt in all the salt beds around the world were accounted for in this method of age determination, the resulting calculation would be much older than the estimate of the age of the Earth of 150 million years discussed above.

For the matter of using sediment thickness to estimate the age of Earth, the first problem to be dealt with is what is the rate of sediment accumulation? One cannot go to the geologic record for this because that would require knowledge of timing, and timing is the very quantity in question. Therefore, one can either make broad assumptions about time, and use these estimates as bounding values, or one could visit modern sites of deposition, measure the rates of sedimentation, and calculate an average value. Even this has problems. First, the rate in question is so slow that even measurements made over several decades would probably not encompass enough time to determine a good long-term average. Second, even if the rate of deposition was measured accurately at one or even several locations, different places around the world are likely to have significantly different rates, owing to the differences in tectonic setting and climate. So, the determination of a single global sedimentation rate is not a straightforward matter.

But then comes the matter of thickness. What is the right value? To tackle this problem, one could go to some place of exceptionally thick sedimentary rocks, such as the Grand Canyon, and use the thickness observed there. However, most thick sedimentary sequences, including the Grand Canyon, have within them important unconformities (see, for example, Figure 4.4). Indeed, the contact between the basement rocks in the Grand Canyon and the overlying Cambrian Tapeats Sandstone represents more time (about a billion years) than it took to form all the rocks present in the canyon (see Chapter 1, Figure 1.19). Recognizing the unconformity problem, researchers at the end of the nineteenth century took the approach of looking for the thickest sections of each of the known geologic periods of the time; the thinking being that although any one place, such as the Grand Canyon, may not have a continuous record of

Figure 4.4 The problems of measuring time using sedimentary strata are illustrated in this outcrop. We can see two sequences of layered rocks: the first are the layers dipping from left to right in this image. Following the law of original horizontality, these strata started out flat and were subsequently tilted by tectonic forces. The second sequence of rocks lies horizontally on the top of this hill. Although it took thousands of years (or tens of thousands, perhaps) to deposit the two sequences shown here, the time to tilt the first sequence and then erode those rocks to produce the flat surface on which the second sequence was later deposited surely took much longer. In cases of angular unconformities such as this one (see Chapter 1), the rock record represents only a small amount of time of the history necessary to produce this outcrop.

sedimentation, by looking at each geologic period in turn (e.g., Cambrian, Ordovician, Devonian, etc.) separately where it was thickest, a better estimate of total thickness could be obtained.

Although knowing which is the thickest section of Cambrian rocks or Cretaceous rocks is relatively simple (because these ranges are defined by the fossils they contain), as we shall see in more detail in subsequent chapters, rocks older than the Cambrian have very little in the way of age-diagnostic preserved fossils. Therefore, any estimate of the thickness of Precambrian rocks is going to be hugely uncertain, and we now know that the Precambrian constitutes more than 88% of Earth history (see Figure 1.2). Therefore, attempts to calculate the age of Earth using sediment thickness are crippled by not really knowing what a good average rate of accumulation is and not really knowing what a good total thickness is to use in the calculation.

Because of the evident problems with the other methods and the very high esteem in which Lord Kelvin was held, his estimate, based on planetary cooling, held the most sway at the end of the nineteenth century. However, most geologists at the time thought that even the upper end of Kelvin's range was far too young based on the shear complexity of the geological record. Utilizing the doctrine of uniformitarianism, discussed in Chapter 2, geologists did not think that a few hundred million years was enough time to produce the variety of rocks and geologic features documented around the world at the slow rates they observed shaping the modern landscape. But geologists were handicapped in their arguments because they could offer no firm alternative to the estimate based on Kelvin's equations. Many biologists also had concerns over Kelvin's estimates. Although his first paper on the age of Earth was published only three years after the publication of Charles Darwin's *On the Origin of Species* (see Chapter 6), the idea of modification of species through natural selection had gained enough currency that many argued that the evidence from the fossil record was inconsistent with an Earth as young as Kelvin suggested. The biologists argued that Kelvin's estimates did not provide nearly enough time for life on Earth to develop the diversity and complexity observed.

In the preeminent textbook of the time, *Principles of Geology* by Charles Lyell, which Charles Darwin took with him on his famous trip on the HMS *Beagle* (see Chapter 6), Lyell championed the gradualist school of thought regarding the pace of geologic events (note that we discussed gradualism in Chapter 2). Lyell certainly did not think that all the variety of geologic features that had been described in the 1800s could have been achieved in just 30 million years. Similarly, the concept of speciation through natural selection led others to conclude that the Earth must be very old indeed because they did not think that the variety of life on Earth could have come to the modern complexity in just 30 million years.

Kelvin was not sympathetic to these arguments. He claimed that he practiced rigorous science, armed with equations and numbers, and he designated geologists as mere casual observers who walked around the countryside kicking at rocks with their boots. Kelvin opined, "When you measure what you are speaking about and express it in numbers, you know something about it, but when you cannot express it in numbers your knowledge about it is of a meagre and unsatisfactory kind."[2] Until the end of the nineteenth century, all geologists, including Lyell and Darwin, countered this attitude with the well-worn aphorism of Hutton, regarding the age of the Earth, that there was "no vestige of a beginning, no prospect of an end." That's certainly more *poetic* than "30 million" but it was not the kind of rigor that Kelvin demanded. As with many ideas in science, new discoveries in the late nineteenth century, and particularly radioactivity, forced a reconsideration of Kelvin's estimate and we will discuss this in the next section.

KEY POINT

Problems with Enlightenment methods for estimation of Earth's age included that salt is sometimes removed from the oceans, and sedimentary rocks include unconformities that commonly represent more time than the preserved rocks.

---

2 *Popular Lectures and Addresses*, vol. 1 (1889) 'Electrical units of measurement', delivered May 3, 1883.

4.2 Radioactivity and Geochronology

Along with an understanding of radioactivity, it is geochemistry that is the foundation of our modern understanding of the age of rocks and the significance of isotopic information obtained from minerals.

4.2.1 Discovery of Radioactivity and Kelvin's Dilemma

In 1895, 12 years before Kelvin died, Henri Becquerel discovered that uranium can spontaneously transform into another element through the process of **radioactive decay**. From the perspective of Kelvin's calculation, a key aspect of radioactivity is that during the decay from a radioactive **parent element**, such as uranium, to a **daughter element**, such as lead, energy in the form of heat is released. Indeed, it is for this property – heat generation – that radioactivity gets its name. Reference is made to the radiation of heat, in the name of the element *radium*, first recognized in 1898 by Marie Curie.

Kelvin was using the cooling of Earth's interior to estimate its age, but he could not appreciate the additional heat generated by radioactivity in the deep Earth. Without this knowledge, Kelvin's method would *underestimate* the age of Earth, because a body with internal heat generation will cool more slowly than one without. The inability to consider heat generation by radioactivity in Kelvin's calculations is much more important than any of the other parameters that he had to guess at. That is, any variation in the choice for the initial temperature, the average heat conductivity of rock, or the modern geothermal gradient makes a much smaller difference in the outcome than the addition of a term to account for heat we now understand to be produced by radioactivity.

Thus, we now know that Kelvin was wrong because, at the time he performed his calculations, he was unaware of radioactivity. Kelvin's approach was robust based on the knowledge that he had, but his estimates ultimately were proved incorrect because he could not account for additional heat generated by radioactive decay deep inside Earth after it began to cool from its original high temperature (whatever that was). This is a good example of significant revision of hypotheses as a result of new evidence, as discussed in Chapter 2 (Figure 2.10). Of course, now, one can redo Kelvin's calculation with an added term to account for the additional heat generated by radioactivity and this will yield values in the billions of years. However, now that we know that rocks contain clocks within themselves, it is a much better approach to just date the rocks directly rather than by estimates of cooling time.

KEY POINT

Although Lord Kelvin's heat-loss method seemed to estimate a time that was too short to account for evolutionary changes seen in the rock record, geologists had a hard time countering his argument until the discovery of radioactivity that provided a heretofore unknown heat source.

4.2.2 Basics of Radioactivity

4.2.2.1 Isotopes

Radioactivity is the decay of unstable, radioactive elements to stable elements. To understand radioactive decay, we must consider **isotopes**, not just elements. Except for some hydrogen atoms, the nucleus of all atoms consists of a combination of protons and neutrons (plus even-smaller sub-atomic particles) surrounded by negatively charged electrons. For chemical reactions, the protons and electrons play the major role, whereas the neutrons play only a minor or have no influence. For example, carbon has six positively charged protons; it is the protons that make carbon behave differently from nitrogen, which has seven protons, or oxygen, which has eight protons. Carbon always contains neutrons in its nucleus, but the numbers vary. Carbon atoms may have six, seven, or up to eight neutrons (Figure 4.5). These different types of carbon atoms are called its isotopes and are referred to by the total number of protons, which is always six, plus the number of neutrons, which vary from six to eight. This total is called the **atomic number** and is used to describe the various isotopes of carbon as ^{12}C, ^{13}C, and ^{14}C (carbon-12, carbon-13, and carbon-14, respectively). In most chemical reactions, the isotopes behave similarly, but ^{14}C, unlike the other two, is radioactive. Occasionally, an atom of ^{14}C will spontaneously decay to become ^{14}N.

Uranium has 92 protons. Two of the isotopes of U (^{235}U and ^{238}U) are shown in Figure 4.5; they vary only in the number of neutrons. Both isotopes are radioactive, and this decay and the radioactive decay of other isotopes (most notably ^{40}K) are what allow us to determine the absolute age of individual rocks and minerals (Table 4.1).

4.2.2.2 Parents and Daughters

One important aspect about radioactive isotopes is that it is impossible to tell by looking at a congregation of atoms which specific one will decay next. We do know, however, for a given amount of time, the probability that an atom will decay. Take the simple example of flipping a coin. We know that for any given flip, the chance of getting heads is 50/50. If we ask 1,000 people to flip a coin, the number of people who flipped heads would almost all the time be very near 500. And if we asked the

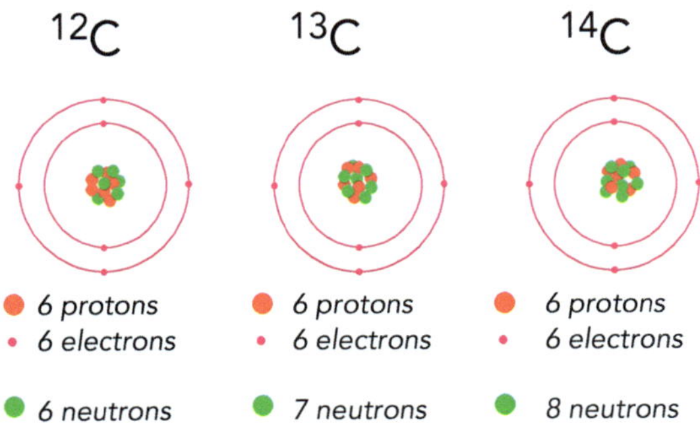

Figure 4.5 An illustration of carbon isotopes and uranium isotopes. ^{12}C and ^{13}C are stable but ^{14}C and both ^{235}U and ^{238}U are radioactive but with very different half-lives.

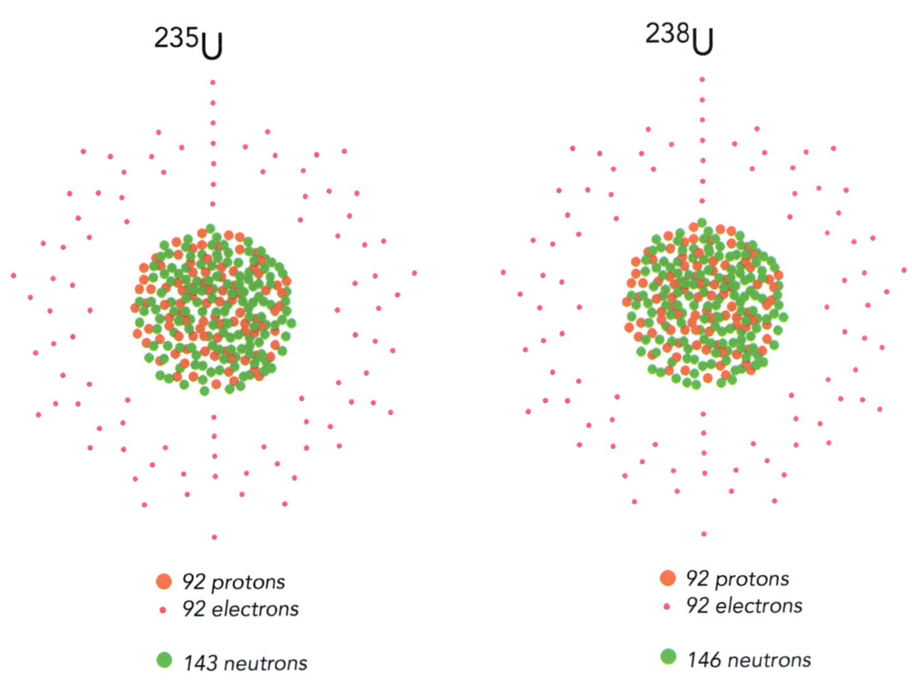

Table 4.1 Commonly used systems for isotopic dating in geology and their half-lives

^{238}U	$\rightarrow$	^{206}Pb	$t_{1/2} = 4.47 \times 10^9$ a
^{235}U	$\rightarrow$	^{207}Pb	$t_{1/2} = 0.704 \times 10^9$ a
^{232}Th	$\rightarrow$	^{208}Pb	$t_{1/2} = 14.0 \times 10^9$ a
^{40}K	$\rightarrow$	^{40}Ar	$t_{1/2} = 1.25 \times 10^9$ a
^{87}Rb	$\rightarrow$	^{87}Sr	$t_{1/2} = 48.9 \times 10^9$ a
^{147}Sm	$\rightarrow$	^{144}Nd	$t_{1/2} = 106 \times 10^9$ a
^{14}C	$\rightarrow$	^{14}N	$t_{1/2} = 5{,}730$ a

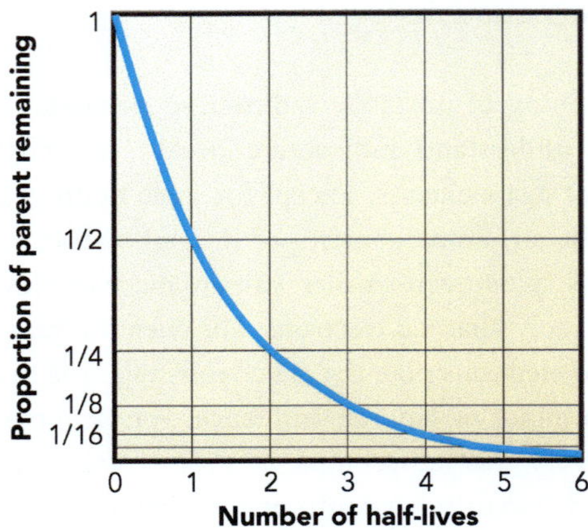

Figure 4.6 The decay of radioactive parents and growth of daughters as a function of half-life.

500 folks who got heads the first time to flip again, we would get about 250 heads, then about 125, 63, 32, and so on until there were only a few people left. (Of course, if we repeated the trial, it is very unlikely that the person at the end of the first run, who had flipped heads six or seven times in a row, would be the same at the end of a second run.) The fact that we have fewer heads each time than the time before does not mean that heads became less likely, just that we had fewer flips. In other words, the number of heads we turn up in any such exercise is proportional to the number of flips, but the chances of any single coin coming up heads is always 50%. The math is the same for radioactivity, although the likelihood of change is smaller (in a given year). And in this case, we will call the atom that will decay the *parent* and the atom it will become the *daughter*.

As an example, we know that each atom of ^{238}U has a chance of decay in a given year of about 1 in 10 billion. That may not seem very high, but it is key to consider that in 1 nanogram (i.e., one billionth of a gram) of ^{238}U there would be more than *two thousand billion* atoms, and some natural crystals, as small as 2 mm in diameter, contain 2 nanograms of ^{238}U. So even if the chances of the decay of an individual atom is only one in a billion, in a crystal with this amount of uranium, about 400 atoms would decay every year and so, after 50 million years, there would be about 20 billion daughter products (in this case ^{206}Pb) to measure. Modern labs can easily measure daughter products in these amounts.

The chance of decay in a given year is called the **decay constant**, but for many reasons it is more convenient to speak of a complementary term, the half-life. The half-life is the time it takes for the radioactive isotope in question to decrease by half. So, if we started with 1 microgram (1 μg) of ^{238}U, it would take one half-life to drop to 0.5 μg. It would also take one **half-life** to decay from 0.5 μg to 0.25 μg, and another half-life to decay from 0.25 μg to 0.125 μg. Note that as we move from one half-life to two to three the rate of decay decreases. This is because the number of atoms that decay is proportional to the number of parents (which is always decreasing).

Figure 4.6 shows the proportion of remaining parent atoms plotted against the number of half-lives. Here we see that the slope of the line is initially very steep and flattens out over time. In fact, after about four or five half-lives, the slope of the line becomes very shallow. After five half-lives, the ratio of parents to daughters is changing so slowly that it becomes very difficult to measure precisely, and thus it is difficult to know where we are on the curve. Similar problems occur when the system is very young, as there is a limit to how small a value any measuring device (typically a mass spectrometer) can reliably measure, although improvements in technology continue to push that value down over time.

The most well-known radioactive atom is ^{14}C, as it is frequently mentioned in the popular press. Carbon dating receives a lot of attention because it is widely used in archeological investigations, which are generally of great general interest. From a scientific perspective, ^{14}C is particularly well suited to the study of things that are hundreds to a few tens of thousands of years old because the half-life of ^{14}C is only 5,730 years. This makes it ideal to show that an artefact such as the Shroud of Turin, thought by some to have been the actual linen used to bury Christ, is actually about 700 years old, or the age of the trees used in Anasazi homes in Utah are about 1,000 years old, but because the half-life is so short, even the best, extremely specialized laboratories cannot date material older than about 50,000 years. Hold on, you might think. If ^{14}C only has a half-life of 5,730 years, how can there be any left? In this case, the main source of ^{14}C is incoming cosmic rays that convert ^{14}N to ^{14}C by knocking out a proton and adding a neutron. The resulting ^{14}C, however, is ultimately unstable and reverts back to ^{14}N by radioactive decay.

^{14}C is good for measuring material less that about 60,000 years old, but for measurement of events that are greater than many millions of years old, different chronometers that don't decay away so fast are required (Table 4.1). The next section considers this and other challenges of geochronology.

KEY POINT

Certain isotopes such as ^{14}C, ^{235}U and ^{238}U, and ^{40}K spontaneously decay with a known half-life and the ratio of parent and daughter produced can be used to determine the age of a sample.

4.3 Requirements of Radioisotope Dating

Once radioactivity was discovered, it became clear that three things needed to be true for isotopic dating to be a useful tool for geologic histories. These are (1) the decay rate must be known and it must be invariant over time, (2) we must be able to account for any daughters that were present at the time of mineral formation (i.e., those that are not produced by decay of the parent after mineral formation), and (3) the material being dated must have been a closed system, by which we mean that the only changes in the ratio of parents to daughters has been because of decay of parents into daughters without any additions or subtractions because of other processes. We discuss each of these in turn.

4.3.1 Knowing the Rate of Radioactive Decay

Clearly, knowing the decay constant is essential to this enterprise, but if the Earth really is millions or billions of years old we are going to need to employ isotopic systems with half-lives of millions of years or longer. But how can we be sure of something that takes place over timescales so different from a human lifetime? One approach has been to measure this *within* a human lifetime. The decay constant for ^{87}Rb has been measured directly. In one such measurement, researchers prepared $\sim$20 g samples of rubidium perchlorate ($RbClO_4$). They then measured the amount of the parent (^{87}Rb) and the daughter (^{87}Sr) to establish a baseline. Then they let the samples sit in a cabinet for 19 years. Then they measured the samples again and found that there was more ^{87}Sr and less ^{87}Rb than they started with. Because they knew the magnitude of these changes and the time over which they occurred, they could calculate the decay constant, a value that corresponds to a half-life of 48.9 billion years. Of course, 19 years is a very small portion of 48.9 billion but 20 g of $RbClO_4$ contains a *lot* of Rb atoms for these purposes. With that amount of parent, 19 years is enough time to build up a measurable amount of daughter, even though it has only been 0.00000004% of one half-life.

The half-lives of some other radioactive atoms have also been determined by direct measurement in similar experiments, but others have been determined by comparison. In certain situations, it is quite reasonable that a single rock should yield the same geochronologic result regardless of the method used to determine the age. In those cases, if we can date the rock with a system for which the decay constant has already been determined, for example by direct observation, then we can use the age obtained from one system to determine the decay constant from another system.

A related issue is, even if we can measure the decay of something today, how can we be sure that the modern decay is characteristic of decay from the beginning of Earth history? If it were the case that the half-lives of radioactive systems were shorter in the past but have for some reason become longer over time, we would have great difficulty calculating accurate ages based on laboratory measurements of parents and daughters.

One of the best ways to study this problem is to use rocks from the Moon. Lunar samples are ideal because the Moon lacks two things that can potentially complicate interpretation of isotopic data. First, the Moon has no atmosphere (see Chapter 3), so it never rains on the Moon, which means that we don't have to worry that the rocks were ever weathered by fluids passing through them.

Another big thing we worry about when interpreting isotopic data is the thermal history of the sample and we will elaborate on this later in the chapter. In this regard, volcanic rocks are simpler than plutonic rocks because volcanic rocks pass from very hot to very cold very fast, whereas plutonic rocks may take tens of millions of years (or more) to cool from their crystallization temperature to surface temperatures. The simpler history of volcanic rocks makes them easier to interpret. There are lots of volcanic rocks exposed on the surface of the Moon (basalts mostly) and they are easier to interpret than rocks on Earth because they have been sitting there without any disturbance, such as by weathering or metamorphism, for billions of years. This is the ideal laboratory for checking to see if we have the decay rates correct because we expect these rocks to give the same age no matter what system is applied to them.

Examination of Table 4.1 shows that the various radioactive systems used in geology have half-lives that vary by a factor of more than 150 (excluding ^{14}C). If we used incorrect values for the half-lives of these different isotopic systems, the large range in these values coupled with extreme age of the basalts on the Moon would make the calculated ages come out different. But in fact, we do not obtain different ages. Moon rocks dated by these different isotopic systems show the same age, regardless of which method is used. This indicates that we have a robust understanding of the half-lives of the systems being used.

4.3.2 Knowing How Many Daughters Were Present at the Beginning

To use radioactivity to date rocks, we measure the amount of parent and daughter in the sample; if we know the half-life of the parent, then we can then calculate the age. However, what if there were some extra daughter atoms incorporated into the mineral at the time it was first formed? In this case, the amount of daughter we measure would be the sum of whatever was there to begin with plus the daughters that were subsequently formed by radioactive decay of the parent. If we had no way to account for daughter atoms in the system at the beginning, we would overestimate the age of the system.

The best way to deal with this problem is to work with systems where we have a good reason to assume that there was no daughter present at the beginning. This can be achieved when the chemical characteristics of the parent and daughter are very different. The mineral **zircon** (zirconium silicate or $ZrSiO_4$) gives us a good example of this. As with most minerals, a grain of zircon is never exactly the composition described by its nominal chemical formula; there will always be small variations or substitutions. However, the substitutions are not random. They must follow the rules that the main components set up. In these situations, we must consider the size and charge of the ions that are being put together to build the mineral. For example, zirconium has a charge of 4+ and a radius of 0.092 nm and uranium also has a charge of 4+ with a radius of 0.108 nm. Therefore, because the charge is identical (and charge is more important than size in this instance) and the sizes are very similar, whenever a zircon is crystallizing from a magma, if there is uranium present, some U^{4+} will be incorporated into the site that would otherwise be occupied by a Zr^{4+}. The uranium ultimately decays to lead (Pb) but Pb is incompatible in a zircon at crystallization; it has a radius of 0.137 nm and a charge of +2. If Pb atoms are available in a cooling magma, they could never be incorporated into the growing zircon crystal because the atoms are far too big, and they have the wrong charge to incorporate into the zircon's crystal structure. Therefore, when we find Pb in a zircon, we can be confident that all of it is there because of the decay of U and therefore no subtraction of initial Pb is required.

It is ideal when we can be sure that there were no daughters to begin with but sometimes that is not always the case. However, this is not a fatal flaw; we have a way around this. Consider the isotope rubidium-87 (^{87}Rb), which we discussed earlier. It is radioactive and decays to strontium-87 (^{87}Sr) with a half-life of 48.9 billion years. Rb behaves a lot like the element potassium (K) and Sr behaves a lot like calcium (Ca). The problem from our dating perspective is that K and Ca (and therefore Rb and Sr) behave somewhat the same in many geologic systems, so it is very rare that a mineral will crystalize containing Rb but totally free of Sr. This means that even at the very moment of crystallization, in minerals that have K and Ca, the ratio of daughters (^{87}Sr) to parents (^{87}Rb) will not be zero. If we were to analyze such a mineral without assuming an initial ^{87}Rb (i.e., daughter) then it would appear to be older than it really is.

The way around this problem is to analyze more than one mineral from a single rock, but before we get to that we have one more bit of chemistry to deal with. There are four isotopes of Sr (^{84}Sr, ^{86}Sr, ^{87}Sr, and ^{88}Sr). All of these isotopes are stable but the amount of ^{87}Sr in a sample that contains ^{87}Rb always increases because the ^{87}Rb is radioactive and decays to ^{87}Sr, adding ^{87}Sr to the mineral. The machines we use to measure these isotopes are actually not very good at measuring absolute

amounts but quite good at measuring ratios. So, we tend to think of the increase in ^{87}Sr not so much as its absolute increase but in the increase in the $^{87}Sr/^{86}Sr$ ratio.

When a magma is crystallizing, minerals forming with Sr in them don't discriminate between the four isotopes of Sr. For most of these igneous chemical reactions, the extra neutrons don't matter so there will be no discrimination of Sr isotopes during crystallization. Therefore, at the time of crystallization, all minerals will have the same $^{87}Sr/^{86}Sr$ value as the magma began with (Figure 4.7). However, different minerals can have very different $^{87}Rb/^{86}Sr$ values because Rb and Sr behave differently enough that some minerals will favor more Rb, and others will favor more Sr. So, immediately after crystallization, different minerals will all have the same $^{87}Sr/^{86}Sr$ value but different $^{87}Rb/^{86}Sr$; this will appear as a horizontal line on Figure 4.7. But with time some of the ^{87}Rb will decay to ^{86}Sr, so $^{87}Rb/^{86}Sr$ values will decrease and $^{87}Sr/^{86}Sr$ will increase. However, they will not increase at the same rate because the number of decays from ^{87}Rb to ^{87}Sr will depend on the number of ^{87}Rb in each mineral (as discussed earlier). The effect of this will be to rotate the horizontal line in Figure 4.7 counterclockwise and the steepness of the line is proportional to the age. This line is called an isochron (equal time) and these kinds of

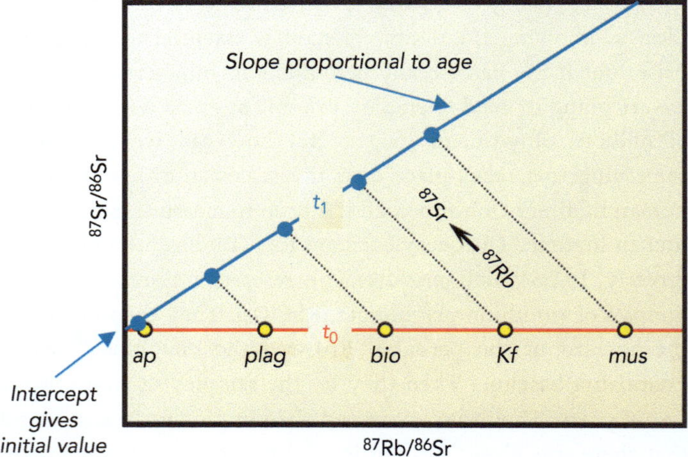

Figure 4.7 An example of an isochron diagram. In this case we are showing minerals that would be typical of a granite (ap = apatite, plag = plagioclase feldspar, bio = biotite, Kf = potassium feldspar, mus = muscovite) but not all granites will have all of these minerals. At the time of crystallization, all minerals form with the same $^{87}Sr/^{86}Sr$ value as the original magma, because different isotopes of Sr behave the same chemically. However, Rb and Sr do have different geochemical behavior and therefore different ratios of $^{87}Rb/^{86}Sr$. So, at the time of crystallization all minerals plot along the red line marked t_0. Over time, ^{87}Rb in these minerals decay to ^{87}Sr along the paths shown by the dotted lines. The amount of decay over time is proportional to the amount of ^{87}Rb initially present so the dotted line for muscovite is longer than the line for apatite. So, at any time after crystallization, t_1, we can measure several minerals from a single rock and determine the age by reference to the slope of the isochron line. This method allows us to determine the age of an isotopic system even when there are daughters present at t_0.

diagrams are called **isochron diagrams**. So, if we measure at least three minerals formed at the same time from an igneous rock we can get around the problem of there already being daughters present at $t = 0$. Geologists can usually determine the timing at which crystallization of specific minerals occur, by examining the microscopic texture.

The isochron approach is used in many other systems to solve the problem of having some daughter present at the time of formation. This, combined with the few examples of minerals that do not have any initial daughters, allows us to produce reliable isotopic ages for rocks and minerals.

KEY POINT

Determining ages require the initial proportion of daughter products to be known. In some cases, that daughter element could never have formed within an original mineral as it is chemically incompatible, but in other minerals, some proportion of daughter and parent products may be present at the time of crystallization. In these cases, comparison of several minerals with differing amounts of initial daughter and parents can be used to determine the age with the isochron approach.

4.4 Geochronology of the Oldest Rocks

The search for the oldest rocks has taken geologists to many continents as well as to investigations of meteorites fallen from outer space.

4.4.1 The First Attempts at Isotopic Dating

When radioactivity was discovered, geologists soon realized that this could allow a quantitative measure of the age of rocks, a measure that had been missing from all previous studies of Earth history. This was particularly important in light of the debates that geologists and biologists had with Lord Kelvin about his estimate for the age of the Earth based on simple heat-loss considerations.

One of the first radioactive schemes that came to be well understood was the decay of U to Pb by a series of steps, many of which produce an atom of helium (He) along the way. It was understood that either the modern measured ratio of U to Pb or U to He would be proportional to the age of the sample. Because of the several steps from ^{238}U to ^{206}Pb, eight ^{4}He atoms are produced for every ^{206}Pb atom produced by U decay. Because of this, the first attempts to date rocks used the U/He value because it was thought there would be eight times more He than the Pb, making it easier to measure. Applying Steno's principle of superposition (see Chapter 1), geologists looked for the oldest rocks available to them at the time, which were in East Africa. With these rocks in hand, they measured the U and He.

The results suggested an age of about 30 million years! This was quite concerning to the gradualists (both geologic and biologic, see Chapter 2). Did this mean that Kelvin was right?

If so, much revision of geologic history would be required. However, before too much panic set in among the gradualists, the analysts went back to the lab, this time to measure the U/Pb value. With these values, an age of around two billion years was obtained. This was a number that the gradualists were much more comfortable with but needed to be reconciled against the younger age obtained using the U/He system.

The approach of using the U/He values, which produced such a surprisingly young age, was initially not well understood but we now know that because it is a very small atom, He, can be driven out of a mineral if the temperature is high enough. Simply stated, He can vibrate out of the crystal above a certain temperature and is retained only at low temperatures; the temperature at which the He in these systems become locked in, or *closed*, is around 150 °C. The temperature at which daughter products become locked in a mineral is referred to as the **closure temperature**, which can be quite different for different minerals and radioisotope systems. So, when we date a mineral using the U/He method, we are determining how long it has been since the rock last experienced a closure temperature of about 150 °C, which can be much younger than the age of formation.

Because Pb is a very large atom, and much larger than He, it is retained at much higher temperatures, sometimes even higher that the temperature of high-grade metamorphism, so analysis of the U/Pb system is much better to estimate the formation age of an igneous granite or a metamorphic gneiss. In contrast, analysis of the U/He system may be used to learn when the rock later cooled to below 150 °C.

In general, this concept of closure temperature is applied to all isotopic dating. What we are dating is not necessarily the time a mineral crystallized, but rather the time the mineral was last at its closure temperature. Work in the lab has established the different closure temperatures for different systems and some of the more common ones are shown in Table 4.2. Some systems, such as Pb in zircon, are very hard to disturb with typical processes in the rock cycle such as burial and metamorphism. A zircon needs to be heated above 800 °C before the daughter Pb products can escape. In contrast some, such as He in the mineral apatite, are very easy to disturb, with heating to temperatures less than 100 °C. Temperatures of 100 °C are not common on the surface but may be encountered within a few kilometers of the surface. So, if a rock buried a few kilometers deep is exposed by later erosion, producing sediment, estimation of the U/He ages of apatite grains extracted from the sediment can tell how long ago the sediment was buried as part of the precursor rock. This can be useful for estimation of uplift and erosion rates in active mountain belts (see Chapter 15).

Although this property of mineral dating was originally considered a serious problem, leading to an Earth-age estimate of 30 million years when all the geologist and biologists were expecting something much older, we now understand this as a

Table 4.2 Closure temperature of different mineral dating systems

Daughter element	Closure	
	Mineral	Temperature ($^\circ$C)
Pb	zircon	~800
Pb	apatite	~500
Ar	hornblende	~500
Ar	muscovite	~400
Ar	biotite	~300
Ar	K-feldspar	~200
He	zircon	~175
He	apatite	~70

useful property that we can use to estimate the age at which a rock or mineral sample was last heated to a given closure temperature. We now know that when we date a mineral by the U/Pb system using the mineral zircon we are learning about when the rock was last at a very high temperature, and in this instance a temperature essentially the same as the crystallization temperature, but when we date the same mineral using the U/He technique we are learning about the last time the rock was at a temperature somewhere near 150 $^\circ$C. For rocks with complicated histories, those temperatures may have been reached at very different times. Applying different dating systems to the same rock allows us to understand the time it took for a rock to cool from one closure temperature to another, and this information can be very useful in determining the tectonic events that affected a region, because the rate at which a rock is changing temperature is related to the broad tectonics of the region at the time of cooling.

The fact that two different isotopic dates can be obtained from the same rock (e.g., U/Pb dating of a zircon and K/Ar dating of a muscovite) has been used by some as evidence for questioning the entire enterprise of isotopic dating. This comes from a mistaken expectation that these dates should be the same because they were obtained from the same material. This criticism is a bit like trying to call into question a person's personal papers by noting that the date on the birth certificate, high-school diploma, and marriage license are all different. "How can we trust any of these documents," the critic might exclaim, "if they don't agree on the timing!?" Well of course, these documents attest to different events and we therefore *expect* them to have different dates. A modern understanding of the theory of closure temperature, sketched out in Table 4.2, shows that we

should expect different — not similar — ages from different systems from all but the most rapidly cooled volcanic rocks. Now that geologists have a detailed understanding of how to interpret these data, isotopic dates provide a valuable tool in assessing the history of many rocks.

KEY POINT

Measurements of radioisotopes determine the time at which a rock or mineral reached its closure temperature. For some minerals, such as zircon, the closure temperature is very high and usually reflects the time at which the mineral crystallized from a magma. For other minerals, such as apatite, the daughter products can be released at relatively low temperatures and the parent to daughter ratio is a function of how much time has passed since a rock or mineral was last at the closure temperature.

4.4.2 The Problems of Geochronology of Older Rocks

So, what is the age of Earth?

In theory, determining the isotopic age of a mineral is relatively straightforward. Measure the ratio of parent to daughter and, with knowledge of the decay constant, closure temperature, and initial daughter if any, calculate the age. If one wants to know the age of Earth, one can simply use Steno's laws of superposition and cross-cutting relationships to find the oldest rocks, and then take the rock back to the lab and determine its age. However, there are some obvious and some less obvious problems with that approach.

First, if one is trying to find the oldest rocks on the planet, they may not be available. On a dynamic planet with weathering and plate tectonics, it is quite possible that the rocks that were a part of early Earth have long ago been destroyed, either worn away by erosion, melted by younger magmas, or destroyed by tectonic processes. Alternatively, old rocks may not be destroyed but transformed by metamorphism into a new form that makes their participation in the early days of Earth unrecognizable. Moreover, rocks need not be destroyed to be unavailable to the modern geologist. As some old rocks are eroded, the detritus from them is deposited elsewhere, becoming sedimentary rocks, covering older rocks. So, although finding rocks from the beginning of Earth history is rather unlikely, finding detritus in younger rocks that was eroded from the oldest rocks is much more likely.

But even for the rocks that we do have access to, there are other less obvious issues that must be considered when undertaking isotopic dating. A dynamic planet means water moving about the surface. Water interacts with rocks and minerals to produce chemical weathering at the surface. Before this process proceeds so far as to be obvious to the naked eye, it can still alter the composition of minerals, and can result in changing the apparent age of the mineral. For example, under some conditions, in some minerals, water can leach out the radioactive elements such as uranium and potassium. If this were to occur without a corresponding removal of the daughter products (in this case, Pb and Ar, respectively) the apparent age of the mineral would increase, because the daughter-to-parent ratio would increase. Fortunately, this kind of alteration is often easy to detect microscopically so, with proper care, such minerals can be excluded from analysis.

However, another complication to isotopic dating is generally not something that can be recognized by optical inspection (even with a microscope). This is because the complicating factor here is not water but heat, and short of the very high temperatures of metamorphism, warming up a rock may leave no visible alteration but, from the perspective of isotopic dating, nevertheless produce important changes (Table 4.2). So, if we are seeking to know the age of Earth, we should only use systems with very high closure temperatures and with minerals that are highly resistant to weathering or interactions with water. In this regard, zircon is particularly useful because it lacks cleavage (i.e., planes of internal crystal weakness that water can exploit and along which daughter parent elements may escape), is highly resistant to weathering and chemical alteration, and the U/Pb system resists escape of daughter lead, below temperatures of 800 °C. For these reasons analysis of the U/Pb system in the mineral zircon is most commonly used in the search for the age of the Earth.

4.4.3 The Oldest Terrestrial Evidence

The principle of superposition can lead us to locations where we might find old rocks. The central areas of the continents in South Africa, Greenland, and North America have all been shown to include rocks with formation ages more than 3.8 Ga, but the rock unit that currently has the oldest-known formation age is a gneiss found in the Northwest Territories in Canada.

These rocks, called the **Acasta Gneiss** (so named because they are located along the Acasta River), are a group of strongly metamorphosed granitic rocks (Figure 4.8). This region has experienced at least three different episodes of metamorphism and deformation, which make it difficult to see back to the original formation. However, as discussed above, the U/Pb system, when applied using the mineral zircon, retains the oldest age of formation and may be unaffected by the later stages of metamorphism. Zircons from some of the Acasta Gneiss have been analyzed in several labs and results suggest an original crystallization age of about 4.1 Ga; other techniques have been used to determine the ages of the later metamorphic episodes. No other *rocks* have been shown to be clearly older than 4.1 Ga, and it may be that some older rocks exist on Earth but have not yet been identified. However, we do have evidence of some older *minerals*.

The **Mount Narryer Quartzite** is exposed in the Jack Hills of Western Australia (Figure 4.9). This rock was originally a sandstone, deposited about 3 Ga and metamorphosed about

🔑 **KEY POINT**

Zircon is a favored mineral for determining the formation age of igneous rocks as it has one of the highest closure temperatures and is highly resistant to subsequent weathering.

Figure 4.8 An outcrop of the Acasta Gneiss in Northwest Territories, Canada. Photo from Acasta Gneiss Complex – four-billion-year-old rocks with folded layering. Source: Northwest Territories Geological Survey, Government of Northwest Territories, Yellowknife, Canada.

Figure 4.9 An outcrop of the Mount Narryer Quartzite in the Jack Hills of Western Australia. Source: Photo courtesy of Bruce Watson.

2.5 Ga. So, this rock in no way qualifies as the oldest rock on Earth, but by studying its constituents geologists have been able to see farther back in time with this rock than any other.

Sandstones are made of up bits and pieces of older rocks. The grains of the Mount Narryer Quartzite include mostly quartz but also a significant proportion of *detrital zircon*, so called because these zircons represent sedimentary (or detrital) sand grains that were eroded from an older, originally igneous, or metamorphic source rock. Because the U/Pb dating system is usually not reset by metamorphism, these zircons have been the subject of intense study; early work in the 1980s suggested the rock contained some very old zircons and since then literally hundreds of thousands of zircons have been analyzed from this geologic unit.

Of the thousands of grains analyzed, most of the zircons in the Mount Narryer Quartzite have the relatively unremarkable age of around 3.5 Ga. Although zircons of this age might be interesting from the perspective of the study of the source area of the Mount Narryer Quartzite, from the perspective of our interest in the age of our planet, lots of other rocks have been shown to be older than this, including the Acasta Gneiss. However, a small proportion of these detrital zircons (i.e., less than 5%) have been shown to be *older* than 4.0 Ga, with some as old as 4.4 Ga. So, although the sandstone these grains were obtained from was deposited "only" three billion ago, it contains the oldest-known minerals. Based on this information, Earth must be at least 4.4 billion years old. These zircons are also particularly important in understanding not just the age of the planet but its early history (see Chapter 3). Zircons are found almost exclusively in granites and granites are almost exclusively found in continents. Therefore, these zircons tell us not only that Earth is at least 4.4 Ga in age, but that some continents had begun to form by then.

KEY POINT

Analysis of detrital zircons in the 3 Ga Mount Narryer Quartzite shows a small proportion with ages as old as 4.4 Ga, which were most likely eroded from granitic crust of that age, and indicates that the Earth is older than 4.4 Ga.

4.4.4 Extraterrestrial Evidence

As with the Acasta Gneiss in Canada, it could be that older rocks, as yet unexamined, exist somewhere on Earth. So, how much older could Earth actually be? To understand the age of formation of our planet we have found it necessary to look elsewhere.

What makes understanding the very beginning of Earth so difficult is the dynamic nature of our planet. Studying the beginning of any planetary body with plate tectonics and weathering, such as ours, will be difficult, because the older the planet, the greater the likelihood that the oldest material will

have been destroyed. However, other planetary bodies in our solar system have had a very similar genesis but have not been complicated by weathering or plate tectonics. Analysis of these materials provides our best understanding of the beginning of Earth.

First, consider the Moon. From our discussion in Chapter 3, we understand the current favored model for the formation of the Moon was a giant impact of Theia, a Mars-sized planet, with Earth; the resulting debris coalesced into Earth and the Moon. Some rocks from the Moon have been dated at 4.51 Ga. With the Moon being even more poorly studied than Earth, it's quite possible that older rocks may exist on the Moon, but even so, this pushes the age of the Earth–Moon system slightly older than the oldest zircons found in the Mount Narryer Quartzite. But our model for the formation of the Moon requires Earth to have existed even before that. To understand Earth's ultimate origin, we must look even further out to the meteorites.

The nebular hypothesis for the formation of the solar system, discussed in Chapter 3, explains the formation of all the planets out of a single cloud of cosmic dust. From this cloud (or nebula), the individual planets coalesced, and it appears that in addition to the planets we have today, an additional planet might have formed in the space between Mars and Jupiter. However, the gravity of the giant planet Jupiter kept the planetesimals in this region from forming a planet such as Mars or Earth.

From the perspective of our understanding of the early solar system, this is good news. The asteroid belt is millions of pieces of rocky and metallic material that formed very early in the history of the solar system that never coalesced to form an actual planet. These fragments of our early solar system have not undergone significant modification and have been spinning in the frozen void of empty space since the solar system first formed. In other words, many of the small pieces now occupying the asteroid belt represent what Earth would have looked like at the time it first began to form.

Fortunately, on occasion, the gravity of the larger planets perturbs the orbit of some of the material in the pre-planetary asteroid belt that are sent on a collision course with Earth (not all meteorites originate from the asteroid belt, and some clearly come from Mars; these Mars meteorites are blasted off Mars in smaller collisions). Meteorites from the asteroid belt allow us to go back in time to sample materials that would have originated at the same time as the planetesimals that eventually formed Earth, but before the Moon-forming event.

Dating these rocks has shown there are a variety of ages, but from the perspective of the origin-of-Earth question, we are interested in the oldest of these results, and the evidence we now have shows that the material in the asteroid belt formed about 4.6 Ga. This is the case for many individual meteorites (such as Juvinas, Figure 4.10) as well as groups of meteorites of similar texture and composition.

This age of 4.6 Ga is consistent with data from the isotopic composition of some rocks on Earth. For example, in the 1950s,

(a)

(b)

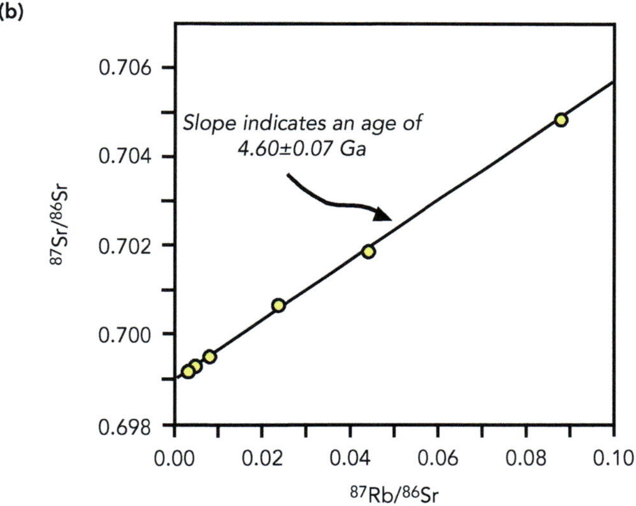

Figure 4.10 (a) Photograph of one of the fragments of the Juvinas meteorite. (b) Rb-Sr isochron diagram from Juvinas meteorite showing an age of 4.6 billion years (each point on the graph represents the isotopic composition of a different mineral in the meteorite). Source: (a) BUFFET Gaelle, CC BY-SA 4.0 <https://creativecommons.org/licenses/by-sa/4.0>, via Wikimedia Commons; (b) Data source Allegre et al. (1995).

Clair Patterson, a geochemist at the California Institute of Technology, studied the Pb isotopic composition of rocks of various ages. By comparing these values to some meteorites, he was able to calculate that Earth must have been evolving for 4.55 Ga.

KEY POINT

Dating of rocky meteorites that originated in the pre-planetary asteroid belt shows ages of 4.6 Ga and indicates that Earth began to form at about that time.

4.5 Conclusion

Our modern understanding of the age of our planet, which has matured during the past 100 years, is that the Earth and all the other planets were formed 4.6 billion years ago. The oldest rocks we have found on Earth are 4.1 Ga, but we now have individual minerals from Earth older than 4.4. Ga and meteorites, which are our best samples of material formed at the very beginning of the solar system, about 4.6 Ga. The evolution of our understanding of this problem involved learning more about what different isotopic systems are really telling us, as well as finding the best rocks to study. This improved understanding is used in essentially all geologic investigations, several of which we will describe in the following chapters.

It may be hard for some to fathom 4.6 billion years! For others, the notion that Earth and our universe have a distinct beginning, as well as a probable end for the Earth, albeit not for another five billion years or so, may be sobering (see Chapter 3). There are an enormous number of consequences of an Earth that is billions of years old, not the least of which is that it provides time for small changes to accumulate into large effects, as we explored in the concepts of uniformitarianism and gradualism in Chapter 2, but it also means that rare and improbable events such as meteorite or comet impacts, sometimes with drastic consequences such as the extinction of the dinosaurs (see Chapter 13), are likely to occur eventually. An Earth of great antiquity also provided Charles Darwin with the time needed for his theory of natural selection to work, resulting in the geological record that shows how life has changed through time according to evolutionary principles (see Chapter 7). The concept of deep time is one of the fundamental discoveries that geology has provided in answering the question of when our planet originated. Much of the rest of this book will explore the great events and stories of this 4.6-billion-year history, the processes that caused these events and changes, the observations, and in some cases the people who were involved.

4.6 Summary

- A wide variety of religion-based estimates of the age of Earth and the universe ranged from a near infinite universe in Hindu cosmology to the few thousands of years in some Judeo-Christian theologies.

- There were a variety of Enlightenment approaches to determining the age of Earth, including the time it would have taken to make the oceans salty from a hypothesized freshwater beginning based on knowledge of the delivery of dissolved salt via rivers, estimates based on measured sedimentation rates integrated over the thicknesses of ancient sedimentary rocks, and finally estimates based on heat flow and the time needed to cool Earth from a molten state. These estimates varied from a few tens of millions to a few billion years.

- These Enlightenment methods suffered from several flaws. As well as the problem that salt is sometimes removed from the oceans, sedimentary rocks include unconformities that commonly represent more time than the preserved rocks. Lord Kelvin's heat-loss method failed to account for the generation of new heat by radioactivity, which was unknown in his time.
- The discovery of radioactive elements led to the ability to obtain absolute dates of rocks and minerals by analyzing the ratio of parent and daughter products integrated with knowledge of the half-life.
- Radioisotope systems have different closure temperatures, and these can be used in dating the thermal history of rocks. Systems with low closure temperatures such as U/He may yield younger cooling or reheating dates, versus U/Pb in zircons which have much higher closure temperatures of 800 °C.
- The 3 Ga Mount Narryer Quartzite in Australia contained detrital zircon grains that have ages of 4.4 Ga, indicating there was continental crust in the Hadean Earth. Ultimately, dating of stony meteorites led to our current understanding that Earth is about 4.6 billion years old.

Key Words

- radioactive decay
- parent element
- daughter element
- isotopes
- atomic number
- decay constant
- half-life
- zircon
- closure temperature
- isochron diagram
- Acasta Gneiss
- Mount Narryer Quartzite

Further Reading and References

Allegre, C. J., Poirier, J.-P., Humler, E., and Hofmann, A. W., 1995, The chemical composition of the Earth, *Earth and Planetary Science Letters*, 134, 515–525.

Lewis, C., 2000, *The Dating Game: One Man's Search for the Age of the Earth*, Cambridge University Press.

Review Questions

1. Describe the wide variety of religion-based estimates of the age of Earth.
2. Describe some of the Enlightenment approaches to determining the age of Earth.
3. What were the problems with estimating Earth's age from the sedimentation rates and thicknesses of sedimentary rock?
4. Lord Kelvin suggested the Earth was about 30 million years old, based on heat-loss calculations. Why was his age respected and how were his calculations debunked?
5. Discuss how the discovery of radioactive elements led to the ability to obtain absolute dates of rocks and minerals.
6. Explain the concept of a radioactive half-life.
7. How are the ratios of parent versus daughter isotopes used to date rocks and minerals and what are the benefits of different systems such as ^{14}C versus U/Pb?
8. What is the closure temperature and how are the variations in closure temperatures of different systems used to determine the origin and thermal history of rocks and minerals?
9. In the early days of radioisotope chronometry, why were the first U/He dates much younger than U/Pb of the same rock sample?
10. Although we have found no rocks on Earth that are 4.4 Ga, how do we know rocks of this age existed?
11. What are detrital zircons and what do they tell us about the age and composition of Hadean crust?
12. What does analysis of meteorites tell us about the age of Earth's formation?
13. How old is planet Earth?

Indonesia, Java Island, Bromo (2,392 m) and Semeru (3,676 m) volcanoes, elevated view. Source: Tuul & Bruno Morandi / Getty Images.

Chapter 5

Plate Tectonics

Our Unifying Geological Concept

LEARNING OBJECTIVES

- Describe some of the key evidence that Wegener used in developing his continental drift hypothesis, such as similarities in fossils and rocks, as well as the apparent jigsaw fit.
- Explain how paleomagnetic studies helped understand continental drift.
- Recall how mapping of dipping deep earthquake centers (Wadati–Benioff zones) led to the discovery of subduction zones.
- Illustrate the main types of plate boundaries (divergent, convergent, and strike-slip) and their relative motion to each other.
- Describe the main layers of the Earth and how they are important in plate tectonics.
- Discuss the changes and controls on the density of oceanic lithosphere, especially densification of down-going slabs by metamorphism, and how that drives plate tectonics.
- Explain how tectonics and sedimentary basins are linked.
- Discuss some of the key diagnostic features of plate tectonics, and subduction zones, such as blueschists and eclogite and the implications for when plate tectonics began on Earth.

Introduction: Up and Down or Side to Side?

As the nineteenth century turned to the twentieth, the overwhelming majority of geologists thought that Earth's great geographic variety was primarily the consequence of bodies of rock moving up and down. Put simply, mountains were places that recently moved up and oceans were places that had recently moved down. Sometimes it was suggested that regions had gone from being high to being low in several episodes. The proposed driving mechanism for these changes were cooling of the Earth and gravitational instabilities. Cooling was used to explain contraction, compression, and formation of mountain belts.

Some Earth scientists began to challenge that viewpoint in the 1910s and slowly, over the next half a century, the idea that major geologic features were produced mostly by up-and-down motion was set aside, and by the 1960s most geologists had come to understand that Earth's great mountains and vast oceans are primarily the result of things moving side to side rather than up and down. The theory of up-and-down motion was known as the **geosynclinal theory** of continental deformation, and it was supplanted by what is now known as **plate tectonics**. The change from thinking geology is the way it is mostly because rocks move up and down to thinking the main driver is side-to-side motion is a story of fundamental scientific change, and the fact that this revolution in thinking only came into full swing about 50 years ago is one of the reasons that the study of geology is as fascinating as it is. Compared to some other fundamentals of science, the theory of plate tectonics is still in its early stages of development, and, with the aid of new technologies, we are continuing to learn new things about the way the Earth works every year. In this chapter we will begin with a brief history of this idea and then move on to the major components of the theory as it is now understood.

5.1 Continental Drift: Alfred Wegener and his Big Idea

Even a casual inspection of a globe or map of the world leads to the observation that the shorelines of eastern South America and western Africa are about the same shape (Figure 5.1). A more detailed inspection of these regions reveals other similarities, including the type of rocks, the age of rocks, the types of fossil flora and fauna, and the timing of deformation of many of the rocks (Figure 5.2(a)). In 1915, German scientist Alfred Wegener published a book, *The Origin of Continents and Oceans*, which noted these similarities and other similar correlations between other continents. In making these comparisons, Wegener proposed that several of the modern continents had at one time been connected into a single supercontinent (Figure 5.2(b)), which he called the **Urkontinent**, German for "primal continent." The term **Pangea**, Greek for "all lands" is now more commonly used (we will discuss the formation of Pangea in Chapter 10). In broad strokes, Wegener proposed that in the Carboniferous Period, all the continents were joined together (Figure 5.2(b)) and later broke off from the main body at various times in the ensuing 300 million years. The idea that the location of continents and oceans was not fixed over time came to be known as **continental drift**.

Contemporaries of Wegener's, including the Italian geologist Roberto Mantovani, and Americans Frank Taylor (in 1908 and 1910) and Howard Baker (in a series of publications from 1911 to 1928), independently outlined ideas regarding the mobility of the continents. Except for the presentation by Wegener, the most complete of these discussions came from Taylor, such that in the early part of the twentieth century references can be found to the "Taylor–Wegner hypothesis." Before 1900, the concept of moving continents was considered to be part of catastrophist thinking (see Chapter 2), but Taylor and Wegener put continental drift firmly on the philosophical foundation of uniformitarianism.

The treatment these ideas now receive, more than 100 years after the publication of the first edition of his book, stands in stark contrast to the reception throughout most of the first half of the twentieth century. Some of the initial resistance was founded in ignorance. Early supporters of Wegener and continental drift were mostly from Europe and Africa and the opponents mostly from North America; this may be because much of the geologic evidence used by the drift proponents came from the "Old World" in locations most North American geologists did not know well. However, Wegener was rightly criticized for proposing mechanisms for the movement of the continents that were not viable.

KEY POINT

Alfred Wegener proposed the continental drift hypothesis, the idea that the continents were mobile, originally joined together 300 million years ago, and later drifted apart.

We will next discuss the various lines of evidence Wegener and his contemporaries used to argue for continental drift and the critiques made against the idea.

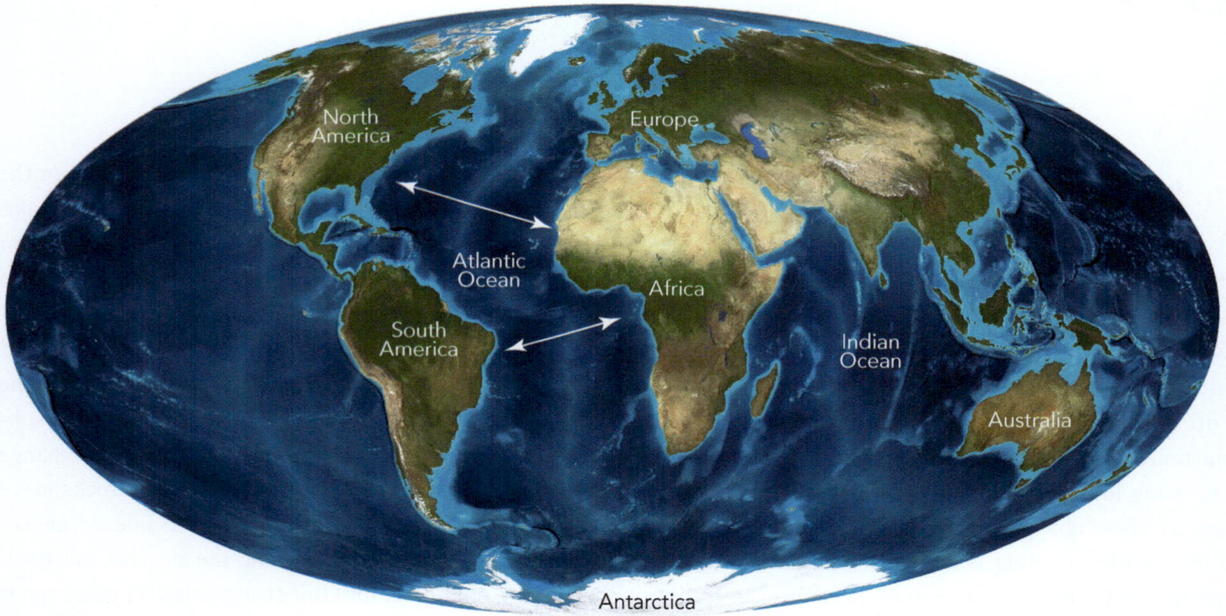

Figure 5.1 Map of Earth and fit of continents. Light blue around land masses are the continental shelves. Source: © DeepTimeMaps.

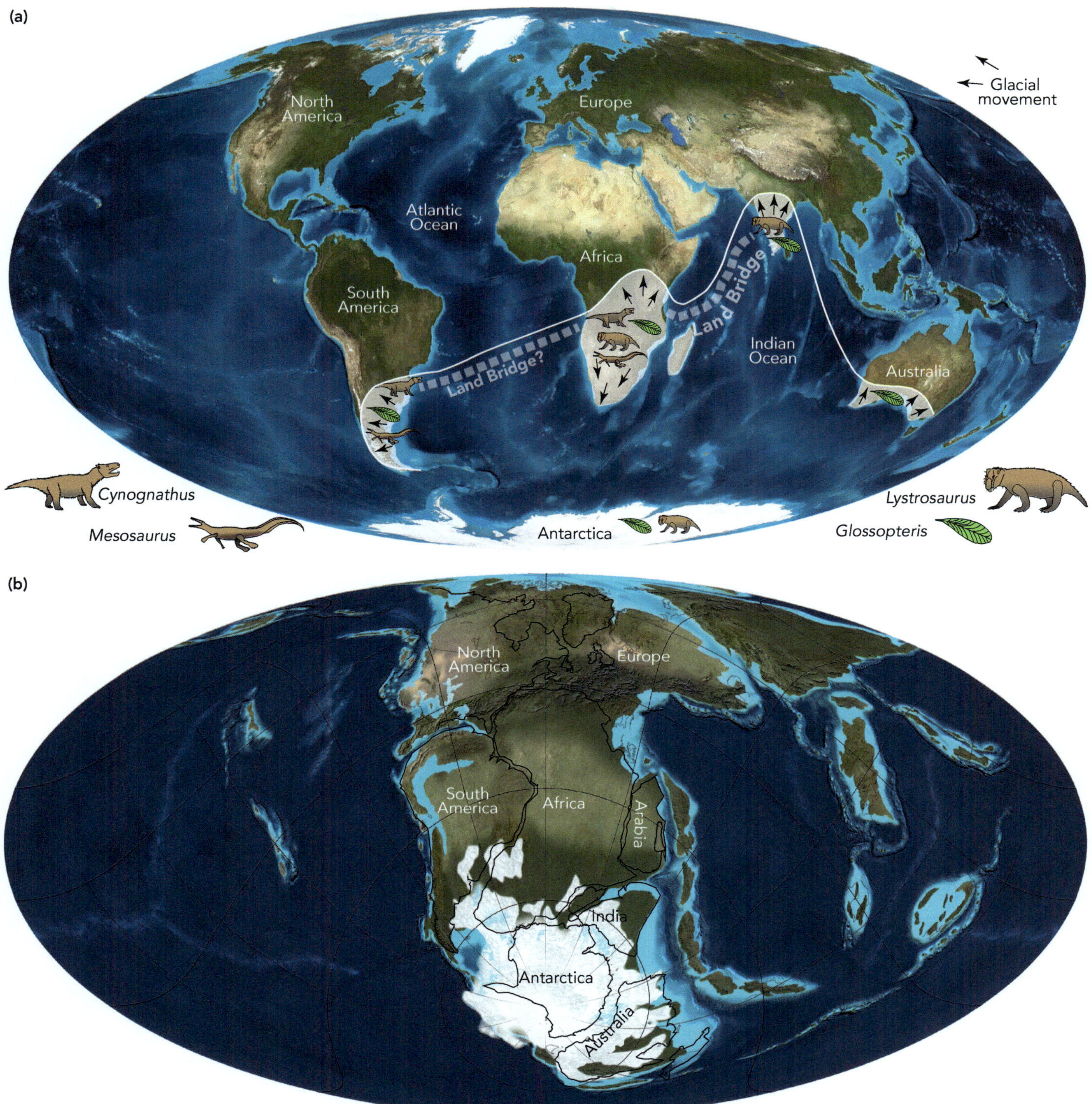

Figure 5.2 (a) Earth map showing fossil similarities in different continents, (b) Pangea, 280 Ma. Source: © DeepTimeMaps.

5.1.1 Reassembling a Supercontinent: The Fit of the Puzzle

It is not just the match between South America and Africa that lead so many to conclude that the continents were once together in a single grouping. Closing the Mediterranean Sea brings Europe in line with Africa. The eastern edges of North America and Greenland fit very well against northern Africa and Europe; and Madagascar, India, Antarctica, and Australia can all fit together on the eastern side of Africa (Figure 5.2(b)).

Moreover, it has been shown that the best fit is obtained not using the shapes of modern shorelines (which are dependent on how much water is partitioned between the oceans and in glaciers on land) but rather the trace of the transition from the continental shelf to the continental slope, around

200–900 m below modern sea level, extending 1–1,300 km from the present shoreline. (These are the lighter blue in Figures 5.1 and 5.2.)

KEY POINT

The continents, and especially the eastern Americas and western Africa and Europe, seem to fit together like a jigsaw, suggesting they were once joined together.

5.1.1.1 Geologic Evidence

Wegener pointed out that where continental outlines could be matched, many instances could be found where the types of rocks on either side of the suture were quite similar. Extending the jigsaw-puzzle analogy, this is like finding that the shapes of the pieces not only fit together but that the colors and patterns are a match. The similarities included stratigraphic sequences that matched in thickness, rock type, and age over significant thickness, such as in southern South America, southern Africa, and Antarctica; as well as the presence of mountain ranges in which the rocks had been deformed in similar styles at similar times. Similarities can be seen between the southern Appalachian Mountains in North America and the Atlas Mountains in Morocco, and rocks of the northern Appalachian Mountains are like the rocks in Ireland and Scotland. Also, the mountains of eastern Greenland can be matched to the Caledonian Mountains of Norway (see details in Chapter 10).

5.1.1.2 Fossil Evidence

In addition to geologic similarities, many instances could be found where the age of the rock and the fossils they contained were similar across the modern-day oceans, further reinforcing the idea of a common past. One key aspect of this line of reasoning is that similarities could be found not just for marine organisms, but also for non-marine plants and animals. Many organisms that live in the sea swim around so it is possible that similar fossils can be seen across oceans not because the land was once together but because the organisms were able to move from one side to the other. Even marine organisms that live on the seafloor could be swept across an ocean by strong currents. However, it is much harder to explain the distribution of non-marine flora and fauna because of swimming or ocean currents. In southern Africa and southern South America, for example, Wegener noted the presence in Permian rocks of *Mesosaurus*, a small freshwater reptile, as well as *Glossopteris*, a large fern with leaves up to half a meter long, and too big to be carried across the ocean by wind (Figure 5.3).

Acknowledging that such land plants and animals could not have made it across the southern Atlantic as we know it today, skeptics of continental drift invoked explanations that attempted to explain how *Mesosaurus*, for example, could have walked from South America to Africa without requiring the

Figure 5.3 (a) *Mesosaurus* and (b) *Glossopteris*, terrestrial species that were found in widely separated continents. Source: (a) Photo by JPB. With permission of Houston Museum of Natural History; (b) Photo by JPB. With permission of Perot Museum, Dallas.

continents to have changed positions. They suggested that a now-drowned Permian land bridge extended across the southern Atlantic, allowing mixing of flora and fauna between Africa and South America (Figure 5.2(a)). They further suggested that this land bridge was subsequently eroded away and might be found below the Atlantic Ocean. In the early twentieth century, bathymetric surveys of the oceans found no evidence of a buried or eroded land bridge, debunking this hypothesis.

KEY POINT

In the absence of mobile continents, land bridges seemed to be the only mechanism that would allow terrestrial plants and animals to migrate across the oceans to be found on different continents.

However, speculative land bridges, thousands of kilometers long, could not remove another strong line of evidence supporting the idea that once-joined continents had drifted apart. The fossil plants, including *Glossopteris*, found in Permian rocks of South America, Africa, and India, parts of which now lie in equatorial climates, are characteristic of a cold climate and these fossils are often found interbedded with glacial deposits. Distinct linear striations were observed on old, glaciated surfaces that yielded information as to the directions the glaciers moved (arrows on Figure 5.2(a); see also Figure 8.5 in Chapter 8 for photos of glacial striations). A simple explanation of the evidence for glacial conditions in all these areas is that they were together near the South Pole (Figure 5.2(b)). This is particularly true given that, during most times in geologic history, such cold climates have been restricted to areas near the poles. Some hypothesized that glacial conditions may have expanded to the equatorial regions over which South America, Africa, Antarctica, and India are now found (see Chapter 8 on the snowball Earth hypothesis), but had that been the case in the Permian many more areas should contain evidence for this widespread cold, and such evidence has not been found.

 **KEY POINT**

Paleozoic glacial deposits near the equator could be best explained only if these continents were originally at the South Pole.

5.1.2 Early Ideas of Mechanisms for Continental Drift

Despite several lines of evidence suggesting continents now thousands of kilometers apart were once adjacent, Wegener did not offer a good explanation for *how* the continents could have moved. He speculated that perhaps the continents plowed through the oceans, much like an icebreaker through sea ice, possibly driven by tidal or centrifugal forces. However, even in his time it was recognized that basaltic oceanic crust is stronger than granitic continental crust, and even if they were about the same strength, such motion would deform and distort the rocks of the continent in ways not observed. Moreover, tidal forces strong enough to move continents would cause the Earth to stop rotating very quickly. Until a more plausible mechanism could be found, the idea of continental drift remained on shaky ground.

It was not until 1929, just before Wegener's untimely death on a research expedition to Greenland, that English geologist Arthur Holmes proposed that Earth's mantle undergoes **convection** and hypothesized this as a mechanism for continental drift. We have all seen thermal convection in action when we boil a pot of water on the stove (Figure 5.4). Water at the bottom of the pot is hotter and therefore it expands and becomes less dense; this allows it to rise to the surface, being replaced by water at the top of the pot, which has cooled and become less buoyant. Holmes proposed that the same thing might occur in the mantle but far less vigorously than boiling water and on a far longer timescale. In the case of the pot of water, the energy comes from the burner on the stove; for the mantle, the energy comes largely from radioactive decay in Earth's interior (see Chapter 4). Holmes suggested that the convective motion of the underlying mantle (Figure 5.5) could transfer some energy

to the rocks at the surface and produce the motion of the continents as proposed by Wegener.

Figure 5.4 Convection in a boiling pot of water.

 **KEY POINT**

Wegener speculated that continental drift might be driven by tidal forces, but his mechanisms seemed untenable given what was then known about the rigidity of oceanic crust. Arthur Holmes suggested that mantle convection might play a role.

5.1.3 Paleomagnetism Confirms Seafloor Spreading

Even before Wegener proposed his continental drift idea, it was known that rocks contain within them iron-rich minerals that can be affected by Earth's magnetic field. When some rocks form, the direction of the magnetic field is locked into the magnetic minerals in the rock (Box 5.1, Figure 5.6). This information allows geologists to determine where Earth's magnetic pole was at the time the rocks formed. In the years following World War II, geologists in the UK and the USA made two significant observations of magnetic information from rocks that led to a widespread acceptance that the continents had been in different positions relative to each other in the past.

The first of these observations was that in rocks the measured magnetic signature indicated a **paleopole** very different from the modern magnetic pole (Figure 5.7). This divergence tended to increase with the age of the rocks. In particular, rocks from a given continent commonly showed a smooth variation of the apparent location of the paleopole over time, as observed by measuring the paleopole in rocks of different ages. Two reasons for these shifting magnet poles were considered possible. The first could be because the continents had moved around, as suggested by Wegener. Alternatively, the continents may have stayed put but the magnetic poles may have wandered about the globe over hundreds of millions of years. However, the apparent polar-wander paths for different continents showed different evolutions (Figure 5.7) and this is not consistent with the idea

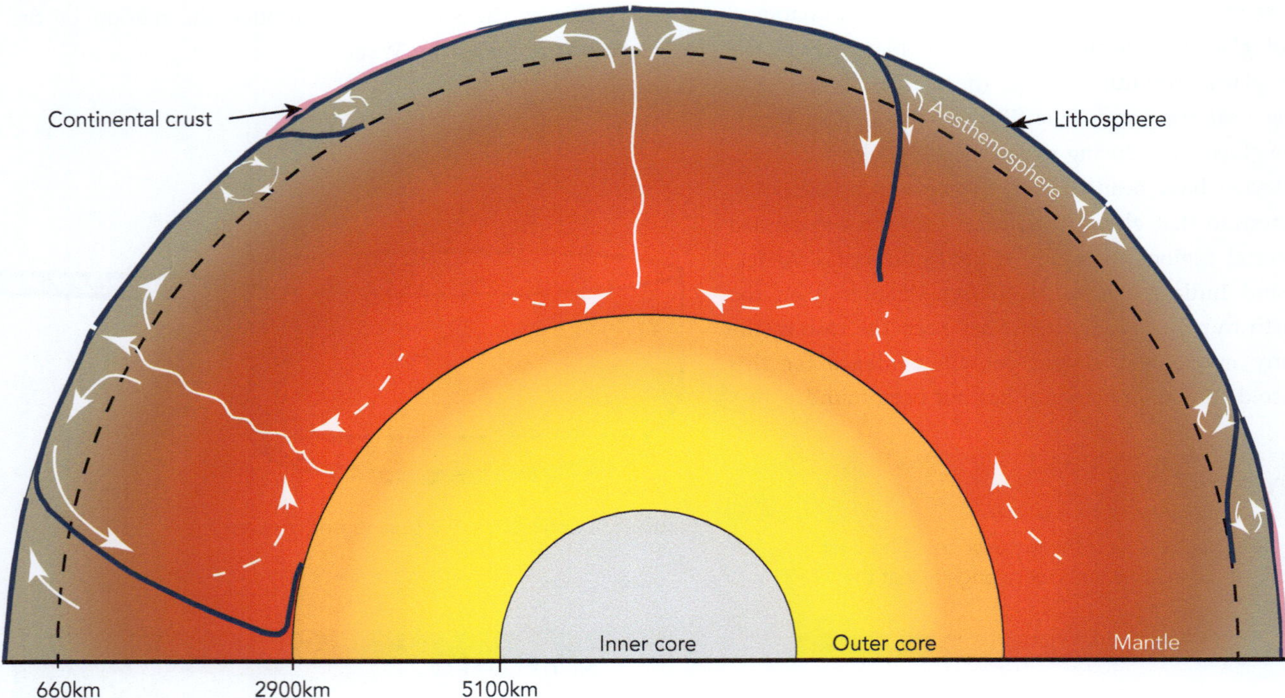

Figure 5.5 Convection in the mantle and associated lithospheric plates.

BOX 5.1 Magnetism in Rocks

Elements that have unpaired electrons (e.g., iron, manganese, chromium, and cobalt) are affected by a magnetic field. In 1900, Pierre Curie recognized that permanent magnetism is lost from magnetizable materials at temperatures from 500 to 700 °C (known as the Curie temperature), depending on the composition and structure of the minerals. If a mineral containing these elements cools below its Curie temperature in the presence of a magnetic field, the unpaired electrons become aligned with the external field, causing the mineral to exhibit magnetization. The strength of the magnetization will depend on the strength and direction of the external magnetic field, as well as the magnetic properties of the mineral itself. The Earth behaves as a magnet (Figure 5.6) whose poles are nearly coincident with the spin axis (i.e., the geographic poles). Magnetic lines of force emanate from the magnetic poles such that a freely suspended magnet is inclined upward in the southern hemisphere, horizontal at the equator, and downward

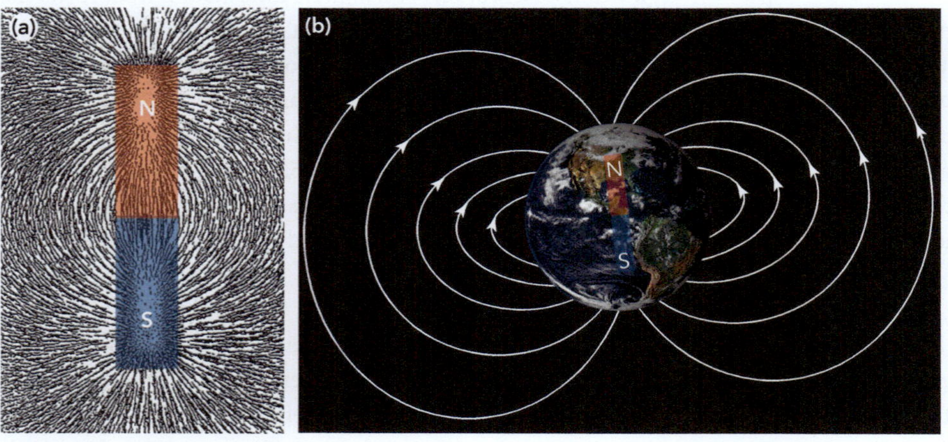

Figure 5.6 Magnetic lines of force: (a) bar magnet and (b) Earth. Source: (a) Adapted from Newton Henry Black, Public domain, via Wikimedia Commons; (b) Reto Stöckli, Nazmi El Saleous, and Marit Jentoft-Nilsen, NASA GSFC.

in the northern hemisphere. It was first thought that Earth's magnetic field was caused by a large, permanently magnetized material deep in the Earth's interior. Because the interior of Earth is hotter than the surface and increases at a rate of about 25 °C/km within the uppermost 100 km, nothing can be permanently magnetized at a depth greater than about 30 km (where rocks are hotter than 700 °C). Therefore, another explanation is needed. A dynamo produces electric current by moving a conductor in a magnetic field and, vice versa, an electric current in a conductor produces a magnetic field. It is believed that the outer core is in convective motion (because it is liquid and in a temperature gradient). A "stray" magnetic field (probably from the Sun) interacts with the moving iron in the core to produce an electric current that is moving about the Earth's spin axis, yielding a magnetic field – a self-exciting dynamo! This theory for Earth's magnetic field fits the available data, and predicts that the magnetic and geographic poles should be nearly coincident and that they would move slowly until they may quickly flip to the opposite orientation, with the magnetic pole moving from the geographic northern hemisphere to the south.

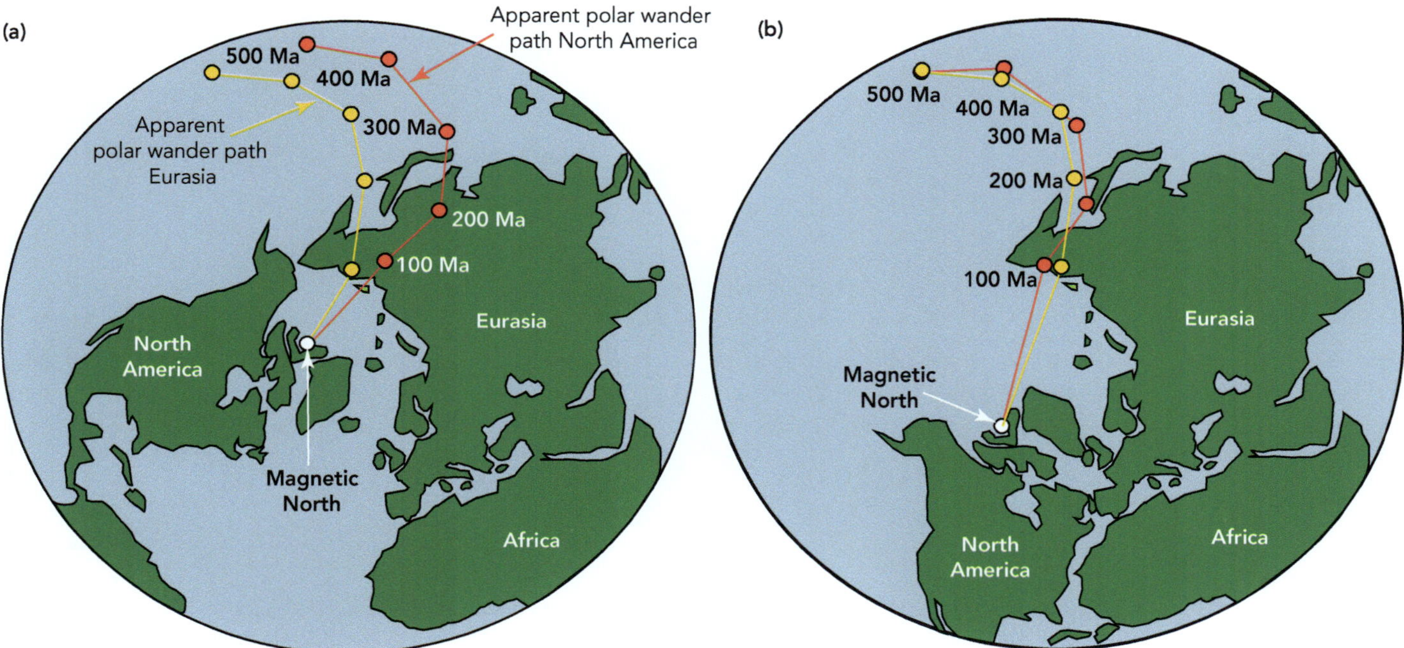

Figure 5.7 Polar wander: (a) poles between North America and Eurasia do not coincide unless (b) polar paths coincide if continental drift is assumed and the Atlantic Ocean is closed.

that the magnetic pole had moved (assuming just one magnetic pole), and is more consistent with the idea that the magnetic pole is relatively stable and that it is the continents that have moved relative to the pole and to each other.

KEY POINT

Paleomagnetics showed that paleopoles indicated either that the magnetic pole had wandered over geologic time or that continents had moved. We now know that magnetic poles are relatively stable compared to the movement of plates.

One of the important players in the development of Wegener's ideas was the American geophysicist and former Naval officer, Harry Hess, who played a key role in the second paleomagnetic observation that swayed minds towards continental drift. During his naval cruises during World War II, Hess ordered that the sounding equipment be in continuous use, thus greatly expanding the understanding of the nature of the seafloor. One key finding of this surveying was that the ocean basins are anything but flat and featureless – as they were often imagined before Wegener – and included a system of underwater mountain chains that we now know encircles the globe like the seams on a baseball (Figure 5.8).

After the war, Hess became a professor at Princeton, returning to the sea to continue surveying the rocks beneath the waves. The research group at Princeton, along with a similar group at Columbia University, led by Maurice Ewing, Bruce Heezen, and Marie Tharp, began investigating the magnetic signature of the ocean floor, as well as learning more details about the deep-sea bathymetry (Figure 5.8). Tharp and Heezen first discovered that the underwater mountain ranges seen in most ocean basins generally had a deep canyon in the middle of the otherwise high-standing feature. These underwater mountains came to be called **mid-ocean ridges**, although not all of them were actually near the middle of the ocean basin. The

ridges were covered with a thin layer of sediment and the rocks that made up the ridge were not sunken granitic continental rocks but rather oceanic basalts and their plutonic equivalents. Occasionally, these rocks are preserved on land in what are called ophiolite sequences (see Box 5.2 and Figures 5.10 and 5.11).

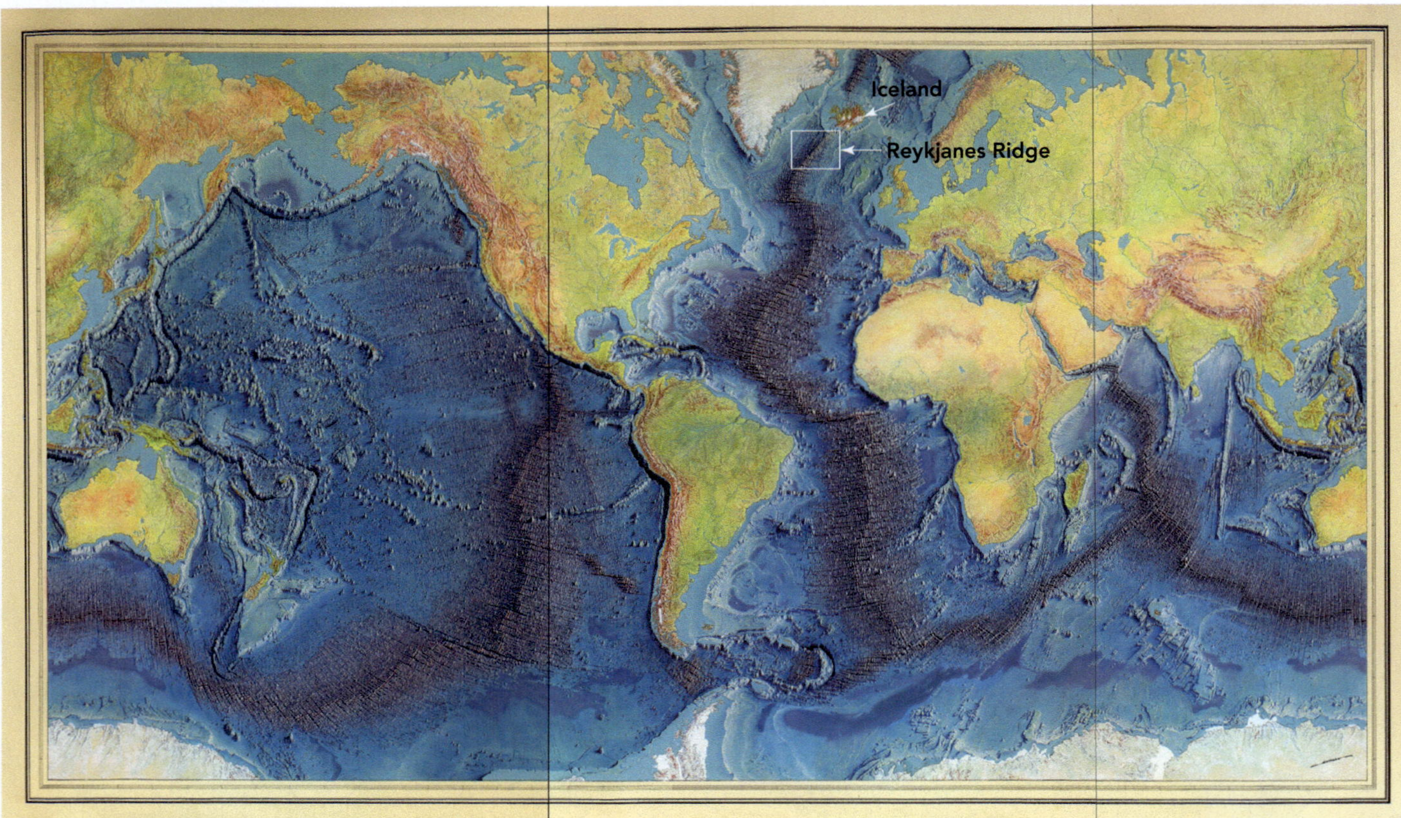

Figure 5.8 Map of the world's oceans showing elevated mid-ocean ridges. Box around Reykjanes Ridge below Iceland is highlighted in Figure 5.8. Painting by Heinrich Berann based on the scientific profiles of Marie Tharp and Bruce Heezen (1977). Source: Berann, Heinrich C., Heezen, Bruce C., Tharp, Marie., CC0, via Wikimedia Commons.

BOX 5.2 Ophiolites

Ophiolites consist of a vertical arrangement of rock types, each layer of which is interpreted to form part of the oceanic lithosphere initially formed at an ocean spreading center (Figure 5.9). At the base is a highly deformed metamorphic zone that likely represents the base of the lithospheric mantle, where it decouples from the less rigid asthenospheric mantle. This is overlain by an ultramafic serpentinite layer, which forms part of the rigid upper mantle and consists of altered **lherzolites** – ultramafic plutonic rocks made of mostly olivine and pyroxene.

Ophiolites were well known in the European Alps, in the Mediterranean, and other places, but until plate tectonic theory was developed their origin was controversial. The three main hypotheses as to their origin were that they were (1) intrusive

igneous complexes, formed by ultramafic magma, (2) upwelled **diapirs** of partly fused mantle material, or (3) slices of oceanic crust and mantle. Once plate tectonic theory became established, the third explanation became widely accepted.

As part of the process of forming oceanic crust, the underlying mantle initially experiences partial melting, which releases the magma at an oceanic spreading center that then cools to form the oceanic crust. In that sense, the first hypothesis, that ophiolites are formed by cooling of magmas, is not totally off the mark, as magmatic processes are certainly involved in their formation. In the process of partial melting, the remaining mantle is said to be depleted of elements that preferentially enter the magma phase. The mantle is made up of oxygen-poor

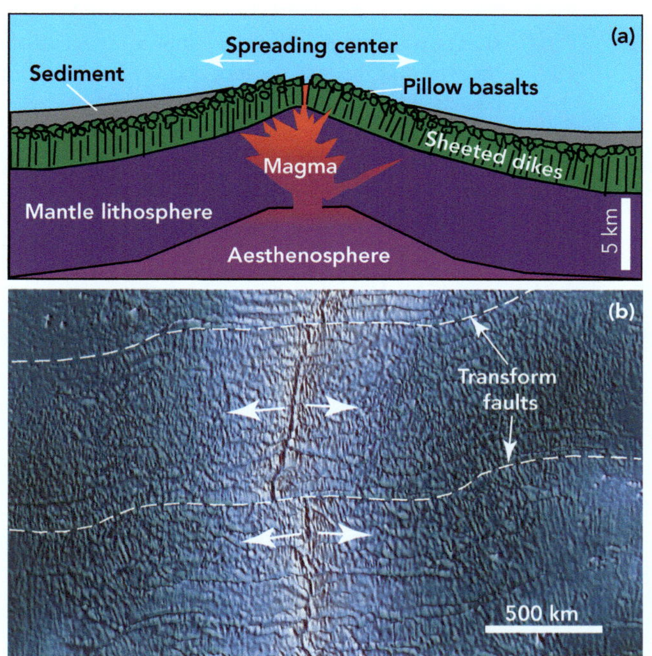

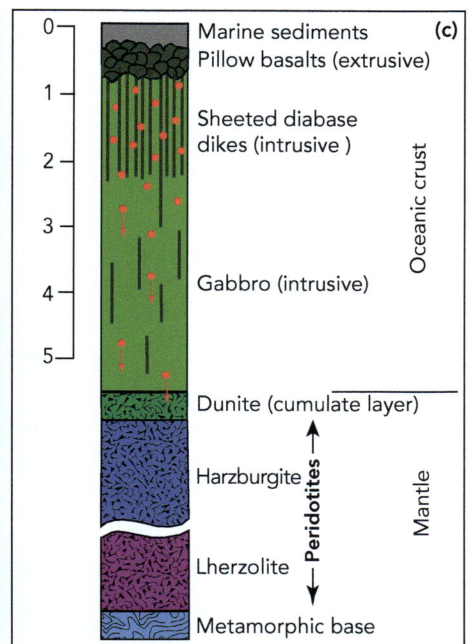

Figure 5.9 (a) Formation of oceanic crust at a spreading center. (b) Image of the mid-Atlantic Ridge shows the central valley spreading centers that are offset along transform faults. (c) General vertical section through an ophiolite sequence formed from oceanic lithosphere (crust and upper mantle). Heavier crystals (red hexagons), mostly olivine and chromite, formed in the gabbro layer sink and accumulated forming a cumulate layer at the mantle–crust boundary.

and iron- and magnesium-rich rocks such as peridotites (Figure 5.9). Peridotites are named after the most common mineral that they contain, which is olivine, the gem-quality variety of which is called **peridot**. The depleted layer forms **harzburgite** whereas the underlying lherzolite is less depleted. The overlying coarsely crystalline gabbro layer is formed by slow cooling of the magma deeper in the lithosphere to form a rigid crust (Figure 5.9). Dense minerals, including olivine and chromite ($FeCr_2O_4$), can sink through the cooling magma settling to form the so-named cumulate layers (Figures 5.9 and 1.13).

As the crust cools and becomes rigid it is constantly re-intruded by magma from below. The intrusion occurs along vertical cracks, forming dikes, and these dikes coalesce to form a layer that resembles a series of vertical sheets (Figures 5.9 and 5.10(a)). Because these form at shallower levels than the gabbro, they cool faster forming a more finely crystalline diabase. Some of the magma escapes to the surface and flows under the ocean, where it cools very quickly, forming distinctive basaltic lavas that look like pillows (Figure 5.10(b)). These distinctive **pillow lavas** are diagnostic of cooling under water and sometimes show a glassy texture characteristic of the virtually instantaneous cooling that happens when magma is quenched as it meets

water and cools too fast to allow crystals to form (see Chapter 1). The pillow lavas may be overlain by sediments typical of deep water environments, such as chert (Figure 5.10). The sequence thus consists of undepleted and depleted mantle overlain by gabbro, intruded by sheeted diabase dikes that feed basaltic pillow lavas that are in turn buried by deep water sediments. The difference in composition of the mantle versus crust is because the crust formed as a partial melt of the upper mantle. The decrease in crystal size from coarser-grained gabbro to the finest grained and glassy pillow basalts reflects the faster cooling of the shallower extruded lava under water, versus slower cooling at depth that allows larger crystals to be formed in the gabbro. The vertical layering and cross-cutting relationships (introduced in Chapter 1) allow us to infer how ophiolites form. Ophiolites are the remnants of oceanic lithosphere thrust onto or trapped between continents. Drilling and sample recovery of modern oceanic spreading centers has revealed similar rocks in similar stratification to that seen in many ophiolites around the world, confirming the oceanic origin of ophiolites. Recognition of ophiolites has important tectonic significance because such small silvers of mafic rock show where a sometimes-vast ocean was closed between two continents (see Chapters 10 and 15).

Figure 5.10 (a) Vertical dike cuts horizontal sheets, which show columnar jointing, reflective of the way that the lava cooled, and feeds pillow basalts. This is from the Troodos ophiolites in Cyprus. (b) Pillow basalts, offshore Hawaii. Source: (a) Public fomain, https://en.wikipedia.org/w/index.php?curid = 10947084; (b) OAR/National Undersea Research Program (NURP)).

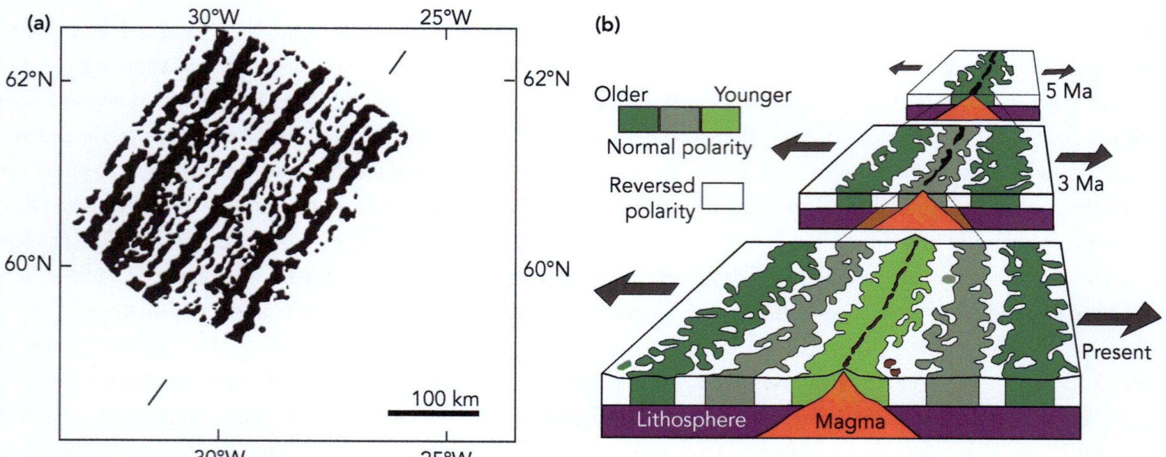

Figure 5.11 (a) Magnetic stripes on the seafloor over the Reykjanes Ridge, south of Iceland show a remarkable symmetry. Location in Figure 5.8, modified after Vine (1966). (b) Block diagram illustrating formation of stripes. As the new seafloor grows along the ridge axis, cooling magma locks in Earth's magnetic field. This has reversed over time and the striping records times of normal versus reversed polarity. Source: (a) Used with permission of Science, from Vine (1966), copyright 1966; permission conveyed through Copyright Clearance Center, Inc.; (b) Modified from Chmee2, Public domain, via Wikimedia Commons.

The research ships gathered data about the depth of the seafloor and also measured variations in the strength of the gravity and magnetic field as they moved about. The magnetic signature proved to be quite variable, with variations that at first were hard to decipher. The records showed **magnetic anomalies** in which areas with fairly constant magnetic intensity pass relatively quickly into areas with quite different intensity, then switch back (Figure 5.11(a)). As dozens of these lines were placed on a large map of the seafloor, the magnetic anomalies lined up, forming broad bands of high and low intensity that were symmetrical about a line going right through the mid-ocean ridge valleys (Figure 5.12).

It had been known since the 1930s that some rocks had magnetic signatures with a polarity opposite to the modern "normal" field, such that a compass would point to the south magnetic pole, suggesting that Earth's magnetic pole had changed or flipped in the past, and today we can use paleomagnetic measurements to distinguish times of normal and reversed polarity of Earth's magnetic field. The variation in the magnetic signature of the ocean basins was thought to be somehow related to the flipping of Earth's magnetic field, but a complete synthesis of all this information, along with a proposed mechanism for formation, would not appear until the 1960s.

In 1962, Hess published *History of Ocean Basins*, and in 1963, PhD student Fredrick Vine and his supervising professor Drummond Matthews, from Cambridge University, England, published a paper in *Nature* that began to shift the opinion regarding the motion of continents (see Box 5.3). Hess proposed that continents were moved about by "seafloor spreading," in which convection in the mantle caused rifts in the crust, pushing continents away from each other. As the rifts formed, magma welled up forming new oceanic lithosphere (broadly similar to rocks seen in ophiolites on land) that filled the cracks and erupted as pillow lavas at the seafloor (Figures 5.9 and 5.10 (b)). Vine and Matthews realized that if Hess was correct, the pattern of magnetic anomalies should be symmetric about the mid-ocean ridges. They realized that the magnetic stripes might record oceanic crust formed during times of normal and then reversed magnetic polarity. They suggested that this occurs as hot mantle material cools through its Curie temperature (see Box 5.1) as it is added to the crust at the mid-ocean ridge, taking on the magnetic field at the time of formation. This interpretation fully integrated the geologic, paleontological, and

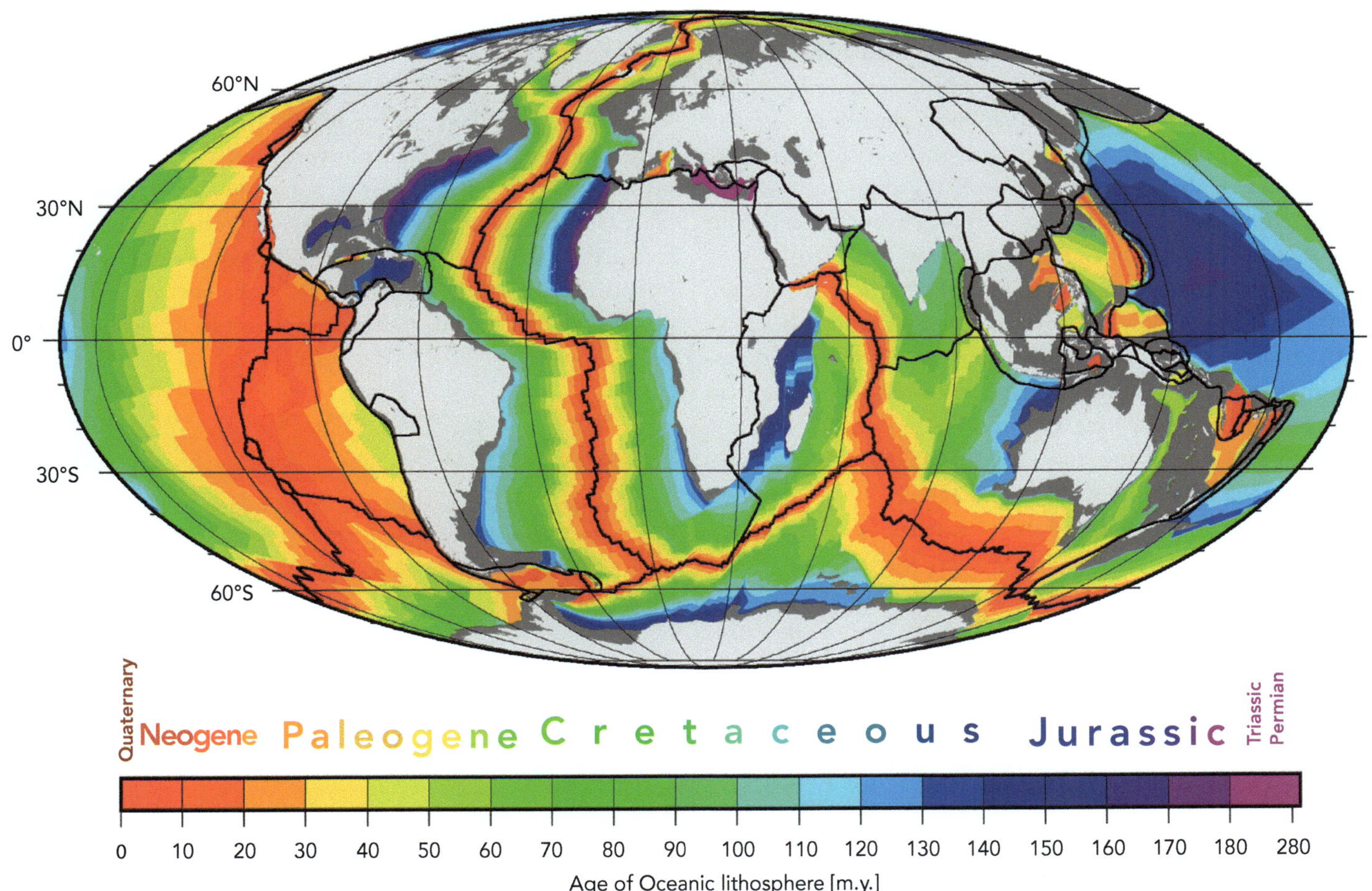

Figure 5.12 Age of oceanic crust. Source: Reproduced with permission from Müller et al. (2008). With permission by John Wiley and Sons.

BOX 5.3 Publication of the Idea of Seafloor Spreading

The publication in 1963 in *Nature*, the preeminent journal of science and engineering, by Vine and Matthews, two researchers at the prestigious Cambridge University in England, is rightly regarded as one of the most important papers in the history of plate tectonics, or even the history of geology in general (Vine and Matthews, 1963). It is so important because this marked the shifting of opinion from skepticism about continental drift to acceptance of what would eventually be called plate tectonics. But if not for the vagaries of scientific publishing, another name might be associated with this idea. In early 1963, Canadian geophysicist Lawrence Morely, submitted manuscripts to both *Nature* and the *Journal of Geophysical Research* on the topic of seafloor spreading and the significance of marine magnetic anomalies. However, both papers were rejected. At the time

Morely worked at the Canadian Geological Survey, a solid institution, but not one with perhaps the same stature in the minds of the editors of *Nature* (published in London) as Cambridge University.

This idea was controversial at the time and reviewers of innovative ideas often (perhaps too often) look more for reasons to discount a new idea rather than let it be published and debated. Later in the same year, Vine and Matthews' paper was accepted in *Nature*. As it became well known that Morely had used similar data and came to similar conclusions as Vine and Matthews, the idea that oceans grow larger by symmetrical spreading at the mid-ocean ridges came to be known as the Vine–Matthews–Morely hypothesis, although some 60 years later Morely is not often mentioned as we cannot cite a paper that was never published.

geophysical data from the continents with the geophysical data from the oceans, and provided independent evidence for Wegener's continental drift.

In addition to the magnetic data, the bathymetry of the seafloor highlighted the mid-ocean ridges (Figures 5.8 and 5.9), where the thickness of sediments overlying the basaltic oceanic crust became thicker farther away from the spreading centers, indicating that older ocean crust had more time to be overlain by marine sediment. It was also clear that earthquakes were concentrated at the ocean ridge where spreading occurs, indicating crustal movement. All these independent observations supported the seafloor-spreading hypothesis.

In this model, as the oceanic plate splits apart, the pressure on the rocks just below the surface decreases, and this allows partial melting of the mantle by a process called **decompression melting**. The resulting basaltic magma is less dense than the more mafic mantle (see Chapter 1), and rises to fill the newly formed cracks forming the sheeted dikes (Figures 5.9 and 5.10) that feed pillow lavas extruded on the seafloor. As the older crust moves away from the ridge, it cools, becoming denser, sinking down a bit as it rides over the underlying mantle. It can now be shown that the depth of the oceanic crust below sea level is proportional to the age of the crust, with the youngest rocks being at the highest points (compare Figures 5.8 and 5.12). The thickness and age of the sedimentary rocks that cover the ocean floor can also be used to measure the age of the oceanic crust that it covers, with older sediments over older crust and younger sediment on younger crust. Finally, the location of marine earthquakes closely coincides with the mid-ocean ridges, consistent with these places being areas of active **rifting**, compared to the regions on either side of the ridges.

Thus, geophysics provided a new sort of data after World War II, especially the bathymetric and **magnetic surveys**, which showed how ocean crust grows through time via seafloor spreading, reconciling the geologic evidence from the continents that had prompted the idea of continental drift in the first place.

However, this created a balance problem. If the oceans (or at least some of them) are getting bigger, then either Earth overall is getting bigger, or some other part of the planet must be getting smaller. During the early days of the development of plate tectonic theory, some scientists suggested that Earth may have expanded over time, and we discuss the expanding Earth hypothesis more fully in Chapter 10. In the next section we will discuss the evidence that showed how ocean lithosphere could be destroyed.

KEY POINT

Bathymetric observations of mid-ocean ridges, along with magnetic surveys that showed striping patterns, helped confirm the idea that oceans are drifting apart.

5.1.4 Subduction: Where Oceans are Destroyed – The Independent Observations of Wadati and Benioff

The discovery of the places where oceans were destroyed involved seismology, the branch of geophysics that studies the way energy passes through the various layer of Earth (see Box 3.2); mostly the energy is associated with earthquakes, but explosions and other artificial sources can be used. The instruments used to study this energy are called **seismometers** and can detect the magnitude and duration of ground motion caused by earthquakes. The first seismometers were produced in the last half of the nineteenth century in Europe and the first modern seismometer was installed in North America in 1897. The data obtained by these instruments allow researchers to know how far away an earthquake was from the recording station but not the direction. It was thus soon understood that to pinpoint the location of individual earthquakes, many seismometers were needed. This location can be pinpointed in three dimensions. That is, not only can we place this on a map, but we can also know the depth at which the earthquake initiated; this point of initial energy release is called the **focus** or **hypocenter** of the earthquake. A global network of seismometers was deployed to allow better mapping of earthquake centers.

In a series of publications from about 1928 to 1935, Japanese geophysicist Kiyoo Wadati noticed a pattern of earthquake foci in Japan in which earthquakes off the east coast of Japan had a shallow focus, whilst earthquakes farther west in China across the Sea of Japan had much deeper foci (Figure 5.13(a)). The recognition of deep earthquakes, with focal depths greater than a few hundred kilometers, was a fundamental observation for our understanding of tectonics as it would eventually come to be known, but this work remained unappreciated or indeed unknown for decades. The cause of this lack of acknowledgment was twofold. First, Wadati's work was published in Japanese and was inaccessible to most western scientists. The second reason was that during World War II, Wadati's research had no obvious utility.

After the war, Hugo Benioff, a seismologist at the California Institute of Technology, began a similar study of the distribution of earthquakes and their focal depths. Ironically, although the war initially kept researchers away from study of deep earthquakes, new military concerns, namely nuclear technology, which was born in 1945, suddenly made this a priority. Seismology could be used to pinpoint nuclear explosions and there was now a military need to distinguish the energy released by earthquakes versus energy released by atomic bomb tests.

Benioff compiled a catalog of earthquakes near the Andes in South America and the Tonga-Kermadec islands in the south Pacific, in which he also observed an inclined zone of hypocenters down to depths exceeding 600 km (Figure 5.13(b)). Thus, the same fundamental observation made by Wadati in Japan in the late 1920s and early 1930s was replicated by Benioff in California in the late 1940s and early 1950s. Until as recently as the 1980s, this relationship between depth and location earthquakes was referred to as a Benioff zone but today this relationship is known as the Wadati–Benioff zone in recognition of Wadati's earlier insights.

Once these zones were identified it was realized that they are the areas where oceanic lithosphere is lost by sinking into the asthenospheric mantle in what are now called **subduction zones**. This solved the balance problem. Oceanic lithosphere is created at mid-ocean ridges (via seafloor spreading) and destroyed along subduction zones. By and large, continents are carried along for the ride, but occasionally collide and stall the subduction process that we will explain below.

KEY POINT

Wadati and Benioff identified dipping earthquake centers associated with oceanic trenches and it was later realized that these were subduction zones, places where oceanic lithosphere was sinking and recycling back into the mantle, balancing the new ocean lithosphere created at spreading centers.

5.2 Plate Tectonics

Integrating the Tharp–Heezen seafloor maps (Figure 5.8) and the Wadati–Benioff zones it is now understood that Earth's lithosphere consists of about a dozen major plates and numerous smaller ones (Figure 5.14). A plate is defined as a sheet of lithosphere that moves in the same direction. In the next sections we discuss the main mechanisms that drive plate tectonics and then briefly review the different types of plate boundaries.

5.2.1 Control of Earth Structure on Plate Tectonic Mechanisms

We introduced the main chemical division of Earth's layers (core, mantle, and crust) in Chapter 3, but the broad understanding of plate tectonics requires understanding that the layers are controlled by changes in the physical versus chemical properties. In this consideration, the two main divisions are the **lithosphere** and the **asthenosphere** (Figures 3.11 and 5.5). The key differences are the strength, temperature, and density of the material and this affects the **rheology**, which refers to the way that solids or fluids deform and flow. The concept is related to the concept of viscosity, which is a measurement of flow resistance. Water, for example, has a low viscosity and flows easily, whereas molasses has a higher viscosity and flows less vigorously. Of course, if you heat molasses it flows more easily, so it is clear that temperature affects the viscosity. In Chapter 1, we introduced a related idea that rocks can deform in either a brittle (i.e., characterized by fractures and faults) or ductile manner (folding without breaking).

The crust and the upper part of the mantle, although compositionally different, behave broadly as a single mechanical layer that constitutes the rigid lithosphere that forms Earth's plates. The behavior of the lithosphere depends very much on its composition (Figure 5.15). Oceanic lithosphere comprises a crust (about $2.96\,\mathrm{g/cm^3}$) overlying a denser mantle lithosphere (about $3.23\,\mathrm{g/cm^3}$). As the mantle lithosphere moves away from spreading centers and cools, it becomes denser.

The asthenosphere is about 1% less dense than the mantle lithosphere due to thermal expansion. This is a common property of many materials. Maybe you have had difficulty getting a

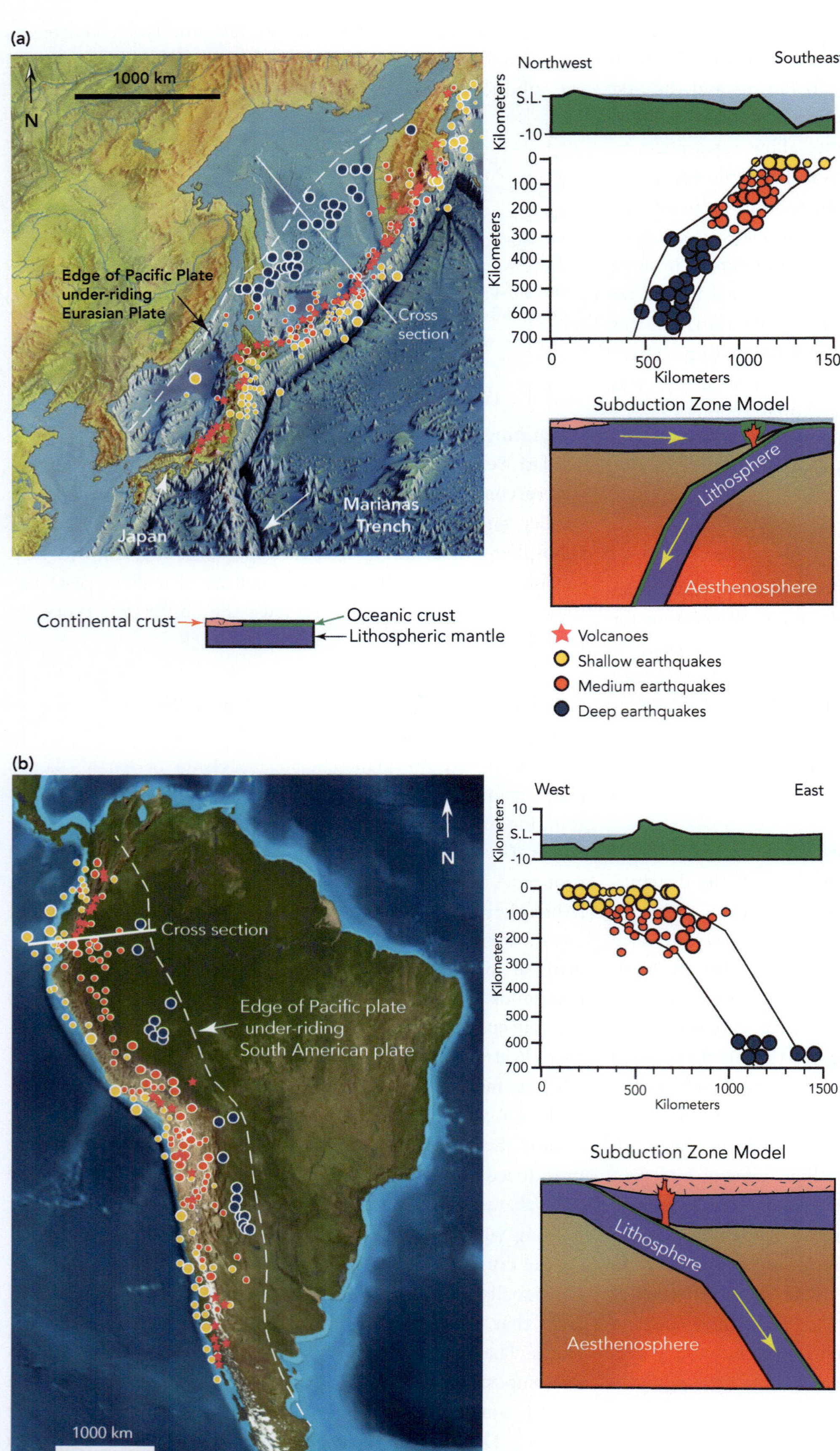

Figure 5.13 Wadati–Benioff zones defined by earthquake locations: (a) Japan; (b) Andes. Source: (a) Adapted after Figure 1 in Benioff (1964); (b) © DeepTimeMaps.

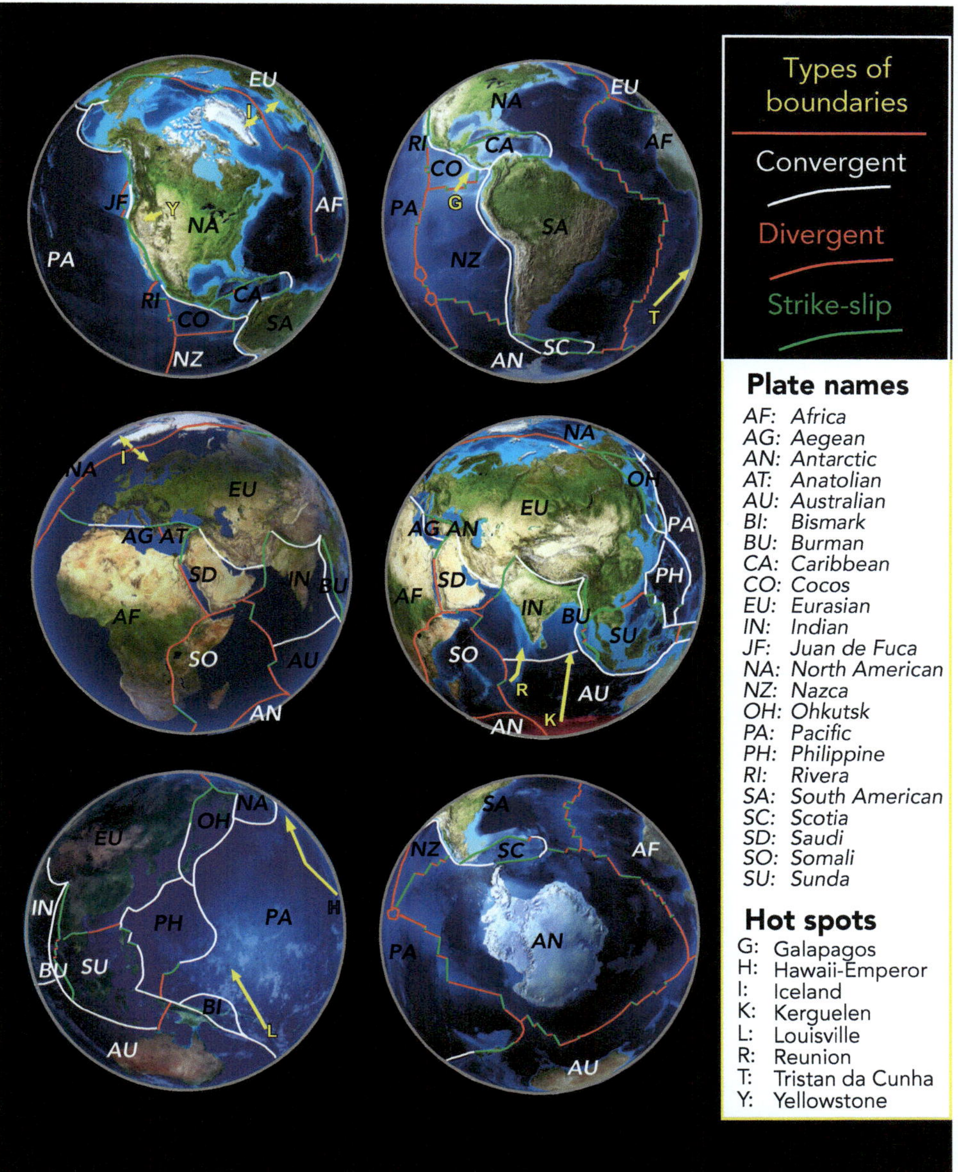

Figure 5.14 The Earth's major plates. Source: Top globes: Leonello Calvetti via Stocktrek Images / Getty Images. Middle left globe: Matthias Kulka via The Image Bank / Getty Images. Middle right globe and bottom right globe: Leonello Calvetti via Science Photo Library / Getty Images. Bottom left globe: Andrzej Wojcicki via Science Photo Library / Getty Images.

lid off a jar? An easy solution is to put the lid under hot water. The heat causes the metal lid to expand a bit (but not so much the glass jar), allowing easier removal. As we will see in Chapter 19, one of the causes of rising sea level is thermal expansion of the ocean water due to global warming.

So, the asthenosphere has a density of about $3.21\,\text{g/cm}^3$ (Figure 5.15). The effective density of the lithosphere reflects the combination of crust *plus* mantle lithosphere. As long as this is *greater* than $3.21\,\text{g/cm}^3$, the lithosphere will effectively float on the asthenosphere, but old, cold, thin-crusted oceanic lithosphere can become *denser* than the asthenosphere as it moves away from a spreading center, and it is this type of lithosphere that can sink along subduction zones. Gravity literally pulls the denser lithospheric slab down into a slightly less dense asthenospheric mantle.

A key feature of the colder lithosphere is that it is *much stronger* than the asthenospheric rocks below, which are hotter and therefore weaker. The lithosphere's greater strength allows the oceanic lithosphere to sink as a coherent slab, pulling the adjacent oceanic plate down with it (Figure 5.5). The pulling force of the sinking slab in subduction zones is transmitted to the other end of the plate that forms the mid-oceanic spreading centers so, in a sense, subduction at one end of a plate drives seafloor spreading at the other end.

Although the density of the oceanic lithosphere is initially only slightly less than the density of the asthenosphere, as the descending oceanic slab is pushed deeper it may be metamorphosed into a metamorphic rock called **eclogite** (Figure 5.16(a)), which has a density of 3.45 to $3.75\,\text{g/cm}^3$. This increase in density, combined with the strength of the lithosphere, which

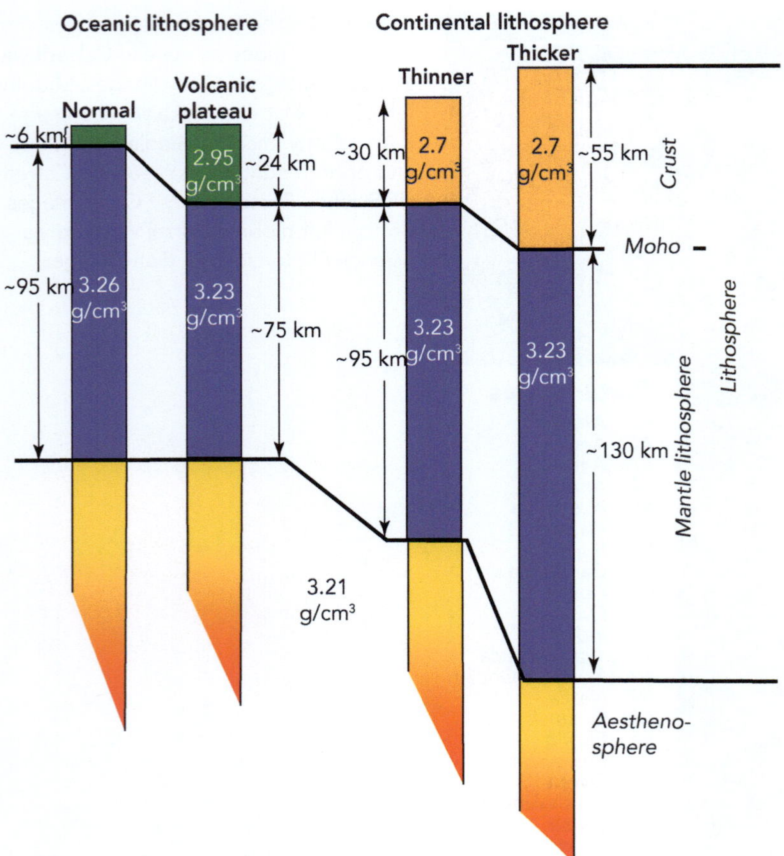

Figure 5.15 Density and thickness differences of Earth's lithosphere.

inhibits the slab from simply breaking off, causes descending slabs to sink even faster, pulling plates deeper into the mantle. The force produced by the densification of the subducting oceanic lithosphere is called **slab pull** and the force produced by upwelling magma at the spreading centers is called **ridge push**.

Although asthenospheric mantle is technically solid rock, it contains about 0.1% magma, as well as a temperature gradient from 1,500 °C near its boundary with the lithosphere, to 3,700 °C at the outer core, and this allows the asthenosphere to convect. Some of the convection is driven by rising heat, and some of it is driven by sinking plates at subduction zones. Although convection is not the main driver of plate tectonics, the ability of the asthenosphere to convect is required to allow plates to move and especially subduction zones. Ridge push and slab pull add to the effect of mantle convection.

Continental lithosphere has only a very minor role in subduction because it is much less dense (Figure 5.15), and for this reason it is practically impossible to subduct. Thus, when two plates move towards each other due to the consumption of an intervening ocean by subduction, the convergence between the two continents will either be accommodated by shortening within the continental crust (making mountains) or the locus of convergence will be transferred elsewhere (see Chapter 15).

KEY POINT

Rigid oceanic lithosphere moves over the less rigid asthenosphere, sinks at subduction zones, and pulls apart any ocean ridges, forming the Earth's plates.

5.2.2 Plate Boundaries, Plate Motion

We can now use GPS to actively measure plate motions. This shows that plates move at average rates of 1–20 centimeters per year on the timescale of our modern measurements. Based on geologic and geophysical evidence, this is also the rate at which mantle convection occurs over millions of years. As plates are restricted to form the outer shell of a sphere, there are only three ways in which two plates can interact: they can move away from each other, towards each other, or slide past each other. Also, as plates interact, they may cause changes in how other plates move. We will next discuss some of the details of each of these configurations.

5.2.2.1 Divergent Plate Boundaries

We have seen that subducting plates can induce splits in the trailing parts of the plates, producing elongate rifts that create a **divergent plate boundary**, where plates move in opposite

Figure 5.16 Common metamorphic rocks found in subduction zones. (a) Eclogite from Norway. The reddish pink mineral is garnet, the green mineral is pyroxene and the blue mineral is kyanite. (b) Blueschist, Mesozoic, California USA. The dark red mineral is garnet and the blue mineral is amphibole. Source: (a) No machine-readable author provided. Woudloper assumed (based on copyright claims), Public domain, via Wikimedia Commons; (b) James St. John; https://www.flickr.com/photos/jsjgeology/50514274328/in/photostream/

directions. An example is the mid-oceanic ridge that runs down the middle of the Atlantic Ocean, which separates the North American and African plates (Figure 5.17) and along which new oceanic lithosphere and crust are created. In some cases, rifting may start in the center of a continent, such as is seen in East Africa (Figure 5.18), but if this continues long enough, a new ocean basin may be formed with a mid-ocean ridge in the middle. An example of a young ocean is the Red Sea (Figure 5.18). When rifting starts in a continent, a valley is formed into which sandstones and conglomerates can be deposited, and in some cases rift

lakes are formed, such as lakes Malawi, Tanganyika, and Victoria in the East African rift system (Figure 5.18). In early rifting stages it is common for several rifts to form, and in this case in East Africa this is called a **triple junction**, reflecting three rifts. Two of them are underlain by new ocean crust, including the Red Sea and Gulf of Aden. In early stages of continental rifts, basalts may also be interbedded with or may form dikes in the sediments. Recall in the Grand Canyon that the Upper Unkar Group is capped by the Cardenas basalt, and the overlying Chuar Group was deposited in an extensional rift associated with the breakup of Rodinia, and there are also abundant dikes (Figure 1.18).

Other examples of triple junctions can be seen in the Pacific and Indian Oceans (Figure 5.14). Such configurations occur when lithosphere is stressed from forces below at a point. The angle between the rifts is usually close to 120° because this is the energetically simplest way to tear apart from a point.

Earthquakes in divergent plate boundaries are almost always shallow and metamorphic rocks are rare. Deformation of rocks is limited mostly to normal faulting in a zone a few tens of kilometers wide that occurs during active rifting.

KEY POINT

At extensional plate margins, where plates split apart, they form rifts such as the East Africa Rift Valley. These can evolve onto new oceans.

5.2.2.2 Convergent Plate Boundaries

Convergent plate boundaries can be ocean–ocean, ocean–continent, and continent–continent (Figure 5.19). When two oceanic plates converge, one of them will be *subducted* and descend into the asthenosphere and one will ride over the subducting plate. Which one stays up, and which one goes down is dependent on their densities (Figure 5.15).

In addition to eclogite (see above), distinctive **blueschists** are also a common metamorphic rock associated with sinking oceanic slabs (Figure 5.16(b)). Blueschists are named after the common blue metamorphic minerals found within them (e.g., lawsonite, glaucophane). These minerals are important because they are formed under the unusual conditions of high pressures but quite low temperatures. As a cold oceanic slab sinks, it experiences changes in pressure with depth instantaneously, but heats up slowly. Where plate convergence is rapid, this can result in rocks at great depths (i.e., pressure) but not the temperature we would expect at this depth in other tectonic settings. In looking for evidence of subduction in older rocks, finding blueschists can be diagnostic of the low-temperature, high-pressure metamorphic conditions that are only found in subduction zones.

As the subducting slab descends, these metamorphic reactions also dehydrate the oceanic crust. Recall that oceanic crust can become quite hydrated due to formation of serpentinite via hydrothermal metamorphism at ocean spreading centers. During

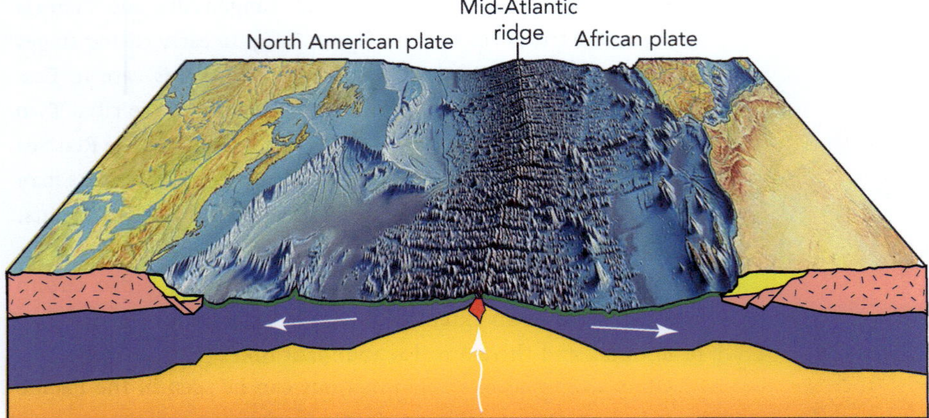

Figure 5.17 The mid-Atlantic Ridge marks a divergent boundary that separates the North American and African/Eurasian plates.

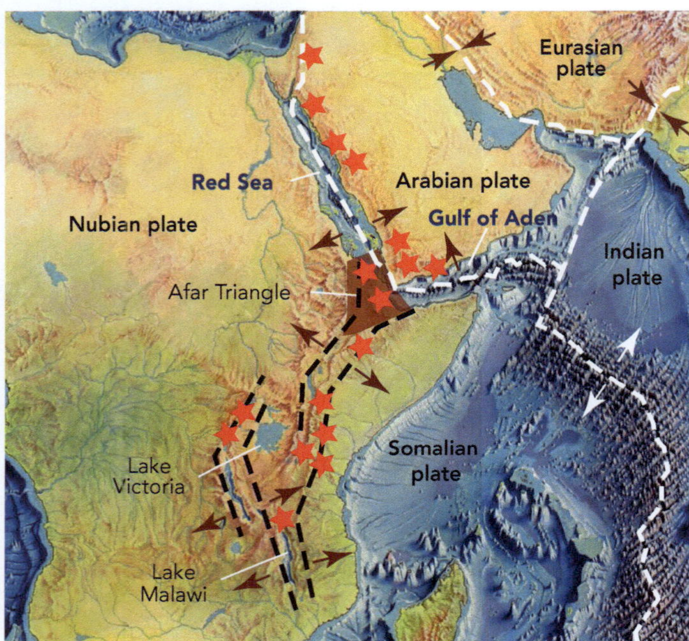

Figure 5.18 Africa consists of two major plates, the Nubian and Somalian, that are splitting apart and forming a series of rift lakes, such as Victoria and Malawi. Volcanoes (red stars) are formed along the rift faults. The Afar Triangle marks a triple-point where three plates (Arabian, Somalian, and Nubian) are all splitting apart, forming what is called a triple junction.

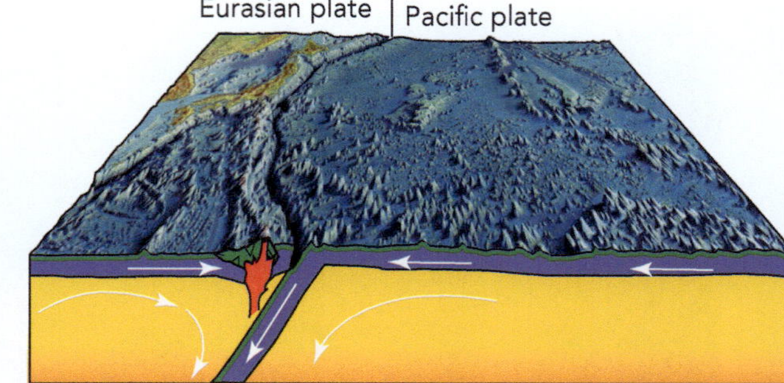

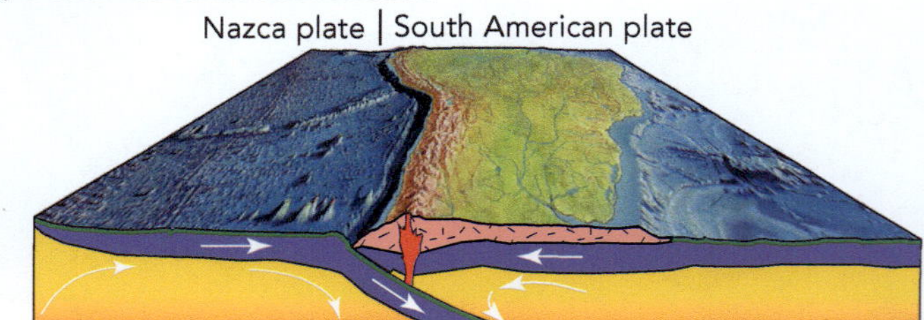

Figure 5.19 Convergent plate boundaries. (a) The subduction of the older Pacific oceanic plate beneath the Philippine oceanic plate. (b) The subduction of the Nazca oceanic plate beneath the South American continental plate. (c) The collision of the Indian and Asian continental plates.

subduction, water from serpentinized basalts is released and rises into the overriding plate. As discussed in Chapter 1, this addition of water into already hot rocks causes melting and these melts can rise to produce the arcs of volcanoes that are common on the upper plates of subduction zones. These island arcs commonly have an andesitic composition and produce distinctive cone-shaped **strato-volcanoes**, such as form much of Indonesia. The eruptions from these stratovolcanoes can be quite devastating, as exemplified by the eruption of Krakatoa in 1883, generally considered to be one of the largest in recorded history (Figure 5.20).

Modern examples of ocean–ocean collisional settings include the Marianas Arc in the western Pacific, where the Pacific plate

Figure 5.20 The 1883 eruption of Krakatoa. Report of the Krakatoa Committee of the Royal Society (London, Trubner & Co., 1888). Source: Lithograph: Parker & Coward, Britain, Public domain, via Wikimedia Commons.

is being subducted beneath the Philippine plate (Figures 5.13 and 5.19(a)), and the Tonga Arc in the southwest Pacific, where the Pacific plate is being subducted beneath the Indo-Australian plate. The oceanic crust being subducted in the western Pacific was formed at the spreading center known as the East Pacific Rise, currently over 12,000 km from the convergent boundary in the west. This crust is therefore very dense because it has had a long time (about 180 million years) to cool from its initial hot state. The Marianas Trench is the deepest point in the world's oceans because the very dense crust subducted here promotes a very steep angle of subduction, pulling the ocean floor down.

KEY POINT

Ocean–ocean boundaries are characterized by subduction zones and volcanic island arcs. Eclogite and blueschist are diagnostic metamorphic rocks.

When an oceanic plate and a continental plate converge (Figure 5.19(b)), the oceanic plate is always the one that is subducted, as continental lithosphere is too light to subduct.

Modern examples of ocean–continent convergence include the Andes (Figures 5.13(b), 5.14, and 5.19(b)), where the Nazca oceanic plate is being subducted beneath western South America, and the Cascades where the small oceanic Juan de Fuca plate is being subducted beneath northern California, Oregon, and Washington (see Chapter 14).

KEY POINT

Dangerous stratovolcanoes and a wider variety of volcanoes are associated with ocean–continent collisions.

The third type of convergent plate boundary is where two continental plates collide (Figure 5.19(c)). In this case, subduction does not occur because continental plates are too buoyant for either to be pushed down into the asthenosphere. The best modern example of continent–continent collision is the Himalaya and Tibetan plateau (Figure 5.19(c)), which we will discuss in more detail in Chapter 15.

KEY POINT

Continent–continent collisions lack subduction and volcanoes, have intense earthquakes, and result in extremely high mountains, such as the Himalaya.

5.2.2.3 Strike-Slip Plate Boundaries

Lastly, plates can slide past one another without getting closer or farther away from each other (Figure 5.21). This tectonic setting is often called a **transform** or **strike-slip** boundary for the kind of fault that marks the boundary between the two plates. Such a boundary can be between any two types of lithosphere (i.e., ocean–ocean, ocean–continent, or continent–continent). This setting is essentially a conservative boundary in that no new crust is formed nor is any old crust destroyed.

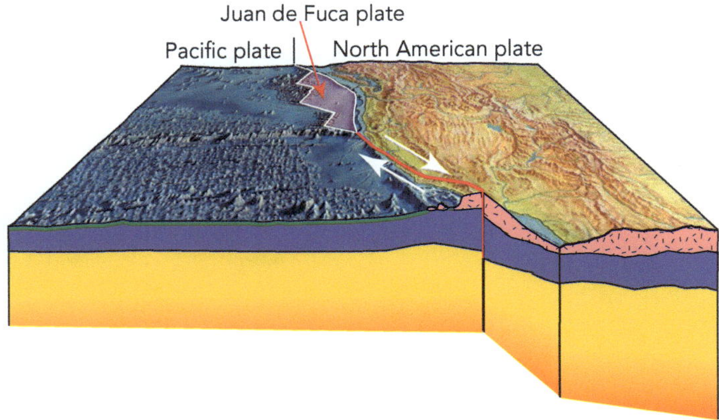

Figure 5.21 California includes the Pacific plate sliding north against the North American plate along the San Andreas Fault. See Chapter 14 for more details.

When the plate boundary is a straight line, deformation is restricted to a zone very near the boundary, but when the boundary has a bend in it, zones of *transtension* or *transpression* can develop, depending on the geometry of the bend. The San Gabriel Mountains, north of Los Angeles, are an example of deformation producing a mountain due to compressional forces in a **restraining bend** of the San Andreas fault system and the Salton Sea is an example of a basin being formed due to extension associated with a **releasing bend**.

Modern examples of major strike-slip plate boundaries include the San Andreas fault zone in California (Figure 5.21), which we review in Chapter 14, and the Alpine fault in New Zealand, which marks the boundary between the Pacific plate and the Indo-Australian plate.

KEY POINT

Where plates slide past each other, shallow strike-slip faults are common such as the San Andreas in California.

5.2.3 Hot Spots

Hot spots are points, or areas, where mantle upwelling occurs. Upwelling is the surface expression of convection in the mantle and can reflect plumes that originate at the mantle core boundary or at higher levels in the asthenosphere (see Figures 5.5 and 5.14). As plates move over a hot spot, it commonly forms a line of volcanoes that are progressively older away from the active hot spot.

Hawaii is one of the best-known modern examples of a hot spot (Figures 5.14 and 5.22). Today, the hot spot lies under the Island of Hawaii (also called the big island), but Hawaii is the only one of the Hawaiian island chain that contains an active volcano. The other islands, such as Maui and Oʻahu, went extinct as the plate moved past the hot spot. Figure 5.22 also shows an older chain, the Emperor Seamounts, of inactive and now submerged volcanoes that formed 85 to 50 million years

ago. You will notice a sharp kink, or dogleg, that marks the transition from the older Emperor to the modern Hawaiian island chain, and this is understood to mark a change in the orientation of the movement of the Pacific plate from north to northwest, although there is some evidence that the hot spot itself may have wobbled around a bit. In 1963, Canadian geophysicist J. Tuzo Wilson, who we introduce in Chapter 10, hypothesized that the Pacific Ocean moved over a hot spot produced by mantle convection and rejected the earlier "fixist" views that Earth's surface was relatively immobile.

One model for hot spots suggests that magma is formed deep within the mantle, perhaps as deep as the core–mantle boundary. This hot magma can rise through the mantle in plumes to the base of the lithosphere, and sometimes breaks through the crust to produce the surface expression of a hot spot, long-lived volcanic centers. Because these plumes are formed at such depth, they seem to act independent of the motion of the lithospheric plates. The interaction of hot spots with the plates suggests the hot spots may be relatively stationary compared to the ever-changing geometry of the plates, but recent analysis suggests they may not be as fixed in space as originally thought.

Depending on the definition of a hot spot, there are several dozen hot spots active today beneath both oceanic and continental lithosphere. Major hot spots include Iceland and the Galapagos, focused at or near a divergent plate boundary, Hawaii, Kerguelen, Louisville, Reunion, and Tristan de Cunha, all under oceanic lithosphere, and Yellowstone, beneath western North America (Figure 5.14).

5.2.4 Tectonics, Sedimentation, and the Rock Cycle

In Chapter 1 we introduced the different rock types and the idea that sediments are deposited in sedimentary basins (Figure 5.23). One of the key features of Earth is that it is a water- and ice-rich plate tectonic planet with a vigorous atmosphere. These spheres (lithosphere, cryosphere, and atmosphere, and to a degree the biosphere) are tightly linked and result in constant interaction and recycling. Sedimentary basins form as low areas on Earth's surface and are commonly covered by water, the oceans being the most common type of basin. Most of the deposition in oceans occurs along continental margins and can be siliciclastic (i.e., gravels, sands, and muds eroded from elevated land masses) or carbonates, especially in tropical areas.

Rift basins occur where continental crust is being extended and thinned (Figure 5.23) and can be filled with coarse material eroded off the flanks of fault scarps, as well as volcanic rocks fed from magmas extruded above faults. If a rift connects to an ocean, it may also contain marine sediments and as rifting proceeds into continental drift a rift basin can evolve into a **passive continental margin**, that is, where continental lithosphere is attached to ocean lithosphere, such as along the margins of the Atlantic Ocean (Figure 5.8).

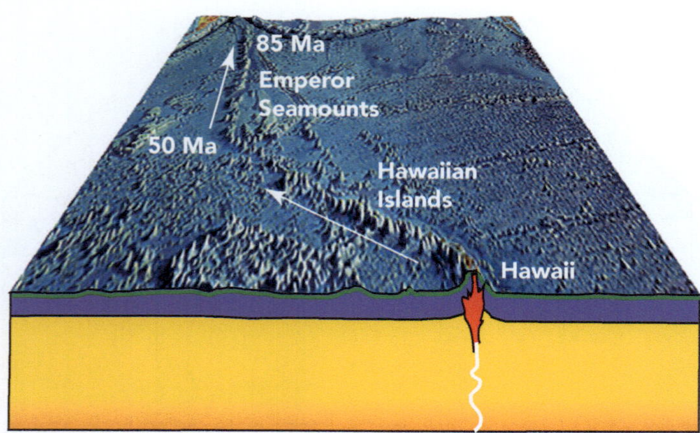

Figure 5.22 Hawaiian hot spot.

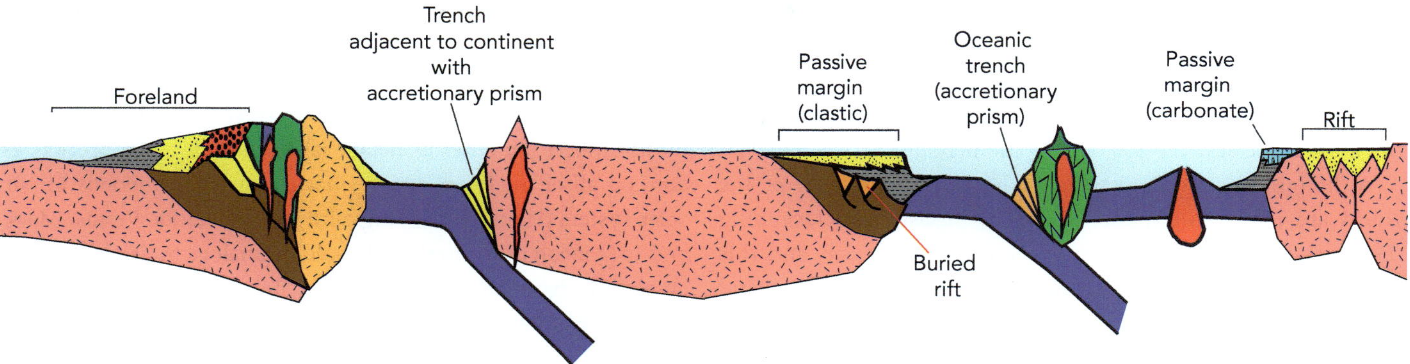

Figure 5.23 Sedimentary basins and tectonic setting. The sequence of sedimentary strata and the character of the sediment within the layers varies depending of tectonic setting. Geologists can recognize these distinctive sequences in ancient rocks when reconstructing past Wilson Cycles.

In compressional tectonic settings, deep trenches associated with subduction zones can be filled with volcanic material eroded from adjacent island arcs, and much of this sediment will eventually be incorporated into the compressional zone, forming a deformed wedge called an **accretionary prism** (Figure 5.23). On the landward side of a compressional mountain belt, thickening of the crust can place an excessive load on the underlying lithosphere, causing it to subside by isostasy forming a **foreland basin** (Figure 5.23), and we will discuss several of these in Chapters 10, 14, and 15. The process of bending the lithosphere is very similar to a diver on a diving board. As the diver stands on the edge, they bend the board down and once they dive off the board, it bounces back. In foreland basins, an episode of thrusting can thicken the crust, causing the adjacent area to subside, like the diver, but later erosion reduces the thickness and can allow the basin to rebound. Foreland basins may or may not be connected to an ocean and can fill with marine or non-marine sediments.

As we will see in later chapters, analysis of the sediments eroded from mountain belts can provide important clues about how orogenesis (i.e., mountain building) occurs. As an example, as erosion exposes deeper roots of mountain belts, younger sediments may contain more metamorphic and igneous material versus older sediments that may record erosion of the shallower parts of a mountain belt that may contain more sedimentary rocks.

In many places, the only record of an orogenic event may be the eroded sediments and analysis of the sediment composition can help deduce the tectonic setting. As we saw above, Wegener used the evidence from sedimentary rocks, such as the similarity of fossils and sediment, to deduce that now separated land masses were once together.

The idea that igneous and metamorphic rocks can be uplifted, converted by weathering to sediment, and later buried to form sedimentary rock, and perhaps even later metamorphosed or partially melted to form an igneous rock, is termed the **rock cycle**. Tectonics enables surface material, such as water, to be recycled into the mantle in subduction zones, and then recycled back to the surface through island arc volcanoes. Weathering, both physical and chemical, has resulted in salty oceans (as discussed in Chapters 2, 3, 4, and 16) but also provides key nutrients that in turn feed the biosphere and drive evolution. Splitting and joining of plates changes the configuration and connectivity of lands and seas, and this plays a major role in the evolution of life on Earth.

5.2.5 Super-Volcanoes and Other Plate Tectonic Hazards

The largest volcanic eruption in recorded human history was in 1815 in Tambora, Indonesia, which forms part of the western Pacific "Ring of Fire." About 10,000 people died from the direct eruption and a further 50,000–90,000 are estimated to have perished as a consequence of post-eruption disease and famine. In Chapter 17 we will review the history of human evolution and given that *Homo sapiens* did not appear until about 250,000 years ago we are not sure than any human has ever experienced a super-volcanic eruption. In Chapter 11 we will discuss the idea that the eruption of the Siberian Traps disrupted Earth's climate and oceans, triggering the greatest mass extinction of all time at the end of the Permian Period. Volcanoes, whether they are associated with mid-ocean spreading centers, subduction-zones, or hot spots, represent one of the potentially deadliest of **natural disasters**.

Earthquakes associated with plate boundaries can be locally devastating, but there is no evidence that they match super-volcanoes in their potential for destruction. Cities on the American and Canadian Pacific Rim, such as Anchorage, Alaska; Vancouver, British Columbia; Seattle, Washington; and San Francisco and Los Angeles, California, are all at risk of natural disasters related to plate tectonic processes, including volcanoes, earthquakes, and also the tsunamis that can be generated by offshore earthquakes. The USA and Canada are extremely wealthy countries with rigorous building standards as well as early warning systems that are able to mitigate some of the worst effects. As an example, USGS geologists noted bulging of Mount St. Helens in the weeks before it erupted and were able to evacuate most of the residents who were in the path of likely destruction, but this is not true on the other side of the Pacific. Poorer countries such as Indonesia and Thailand are far more vulnerable. During the 2004 Boxing Day earthquake under the Indian Ocean, over 200,000 people perished.

Although most geologists can tell you which areas of the world are likely to experience natural disasters, determining the precise details of when and where are far harder, and this reflects the complexity of plate margins. It is easy to understand that the Yellowstone hot spot will continue to move eastward as the North American plate moves westward, and that within the next million years or so another super-eruption is likely, but predicting the precise timing remains elusive.

5.2.6 When did Plate Tectonics Begin?

It is now well established that modern-style plate tectonics produces distinctive features in the geologic record. These include ophiolites, island arc volcanic rocks, and unique styles of metamorphism, such as blueschists and eclogite (Figure 5.16), that reflect conditions of high pressure and low temperature, as well as distinct structural styles, such as fold and thrust belts formed by lateral compression. These were first largely identified through observation of active processes using mostly geophysical observations, but similar geological features can be identified in ancient rocks using the principles of uniformitarianism discussed in Chapter 2. As with almost every problem of the deep past, the question of how plate tectonics has operated over Earth history becomes more difficult the farther back in time we travel. One of the big questions currently under debate is when did plate tectonics begin?

The fact that plate tectonics destroys important evidence of its existence through subduction and recycling is fundamental to this problem. The oldest oceanic crust in the world's oceans is found in the Mediterranean and is no more than 340 million years old (see also Chapters 10 and 16), and most oceanic crust is younger than 180 Ma (see Chapter 10). This may partly explain why, during the 1980s, after the idea of plate tectonics had become the dominant theory for modern-day observations, there was still significant opposition to applying the ideas of plate tectonics to rocks older than Mesozoic. Today that notion is no longer viewed as appropriate, but contemporary estimates for when modern-style plate tectonics began on Earth range from as early at 4.4 Ga to around 1 Ga or less.

The advocates for modern-style plate tectonics on Earth only since 1 Ga take note of the lack of certain geologic assemblages diagnostic of subduction, such as the blueschist metamorphic rocks that are formed in the very high-pressure and relatively low-temperature regimes characteristic of subduction zones, ophiolites, remnants of oceanic lithosphere, and island arc andesites, the volcanic rocks intermediate between granites and basalts in composition formed today exclusively above subduction zones. They claim that absence of this diagnostic evidence is evidence of absence of subduction. Those who think something like modern plate tectonics began much earlier argue

that this evidence has simply not been preserved due to erosion and later magmatism and metamorphism. They argue that absence of evidence is not evidence of absence.

In the absence of plate tectonics, Earth would have a single lid of lithosphere (an outer planetary layer without major breaks), as seen on all of the other active silicate planets and moons in our solar system (Figure 5.24). Single-lid bodies include Venus, Mars, and Io (a moon of Jupiter), and could be analogs to Earth during a single-lid phase (Figure 5.25). Venus, Io, and Mars lack discrete and regional plates. All three bodies have a single-lid lithosphere, a mantle, and numerous large volcanoes, demonstrating that convection is taking place in the interior. Io (Figure 5.25(a)), similar in size to Earth's moon, has over 400 active volcanoes and may be analogous to the very early Hadean Earth characterized as a heat-pipe lid (Figure 5.24), reflecting a thin crust punctured by numerous pipes that feed volcanoes above. Venus (Figure 5.25(b)) is characterized as a vigorous single-lid planet (Figure 5.24) that shows evidence for volcanoes, bulges, and local tears in its lithosphere and some local compression and folding, mostly on the margins of its numerous large active volcanoes. These are thought to form above plumes associated with vigorous large-scale mantle convection cells. Venus also shows circular features on the order of a few hundred kilometers in diameter called coronea, which are interpreted to mark where mantle plumes impact the base of the lithosphere, locally causing subduction zones to form, with a few hundred kilometers of lateral movement of the associated lithosphere. However, there is no global plate mosaic such as defines Earth-type modern plate tectonics, where plates can move thousands of kilometers. Venus has vertical tectonics and some evidence of local lateral motion and has what is termed a "squishy plutonic lid." Venus also has a dominantly CO_2-rich atmosphere (a powerful greenhouse gas) that is 100 times denser than Earth, resulting in a scorching surface temperature exceeding 450 °C, disallowing formation of cool dense lithosphere that could sink into its mantle. Although Venus has a lithosphere and convecting mantle and perhaps limited subduction, it shows no evidence of the development of elongate trenches, island arcs, extensive fold and thrust belts, nor regional spreading centers that would define an Earth-like global plate network.

Mars (Figure 5.25(c)) has some of the largest volcanoes in the solar system as well as numerous local examples of both compressional and extensional faults, again indicating mostly vertical and minor lateral motion, thought to reflect underlying convection cells, but again, there is no evidence for development of extensive tears or formation of discrete plates, subduction zones, or spreading centers. That leaves Earth as the only body in the solar system with full-fledged planet-wide plates driven by subducting lithosphere.

Geologists who favor a much older age of initiation of plate tectonics consider the age of the oldest subduction indicators to be only the upper bound on the length of time plate tectonics has been active because erosion, metamorphism, and deformation are likely to have obscured much or most of the older

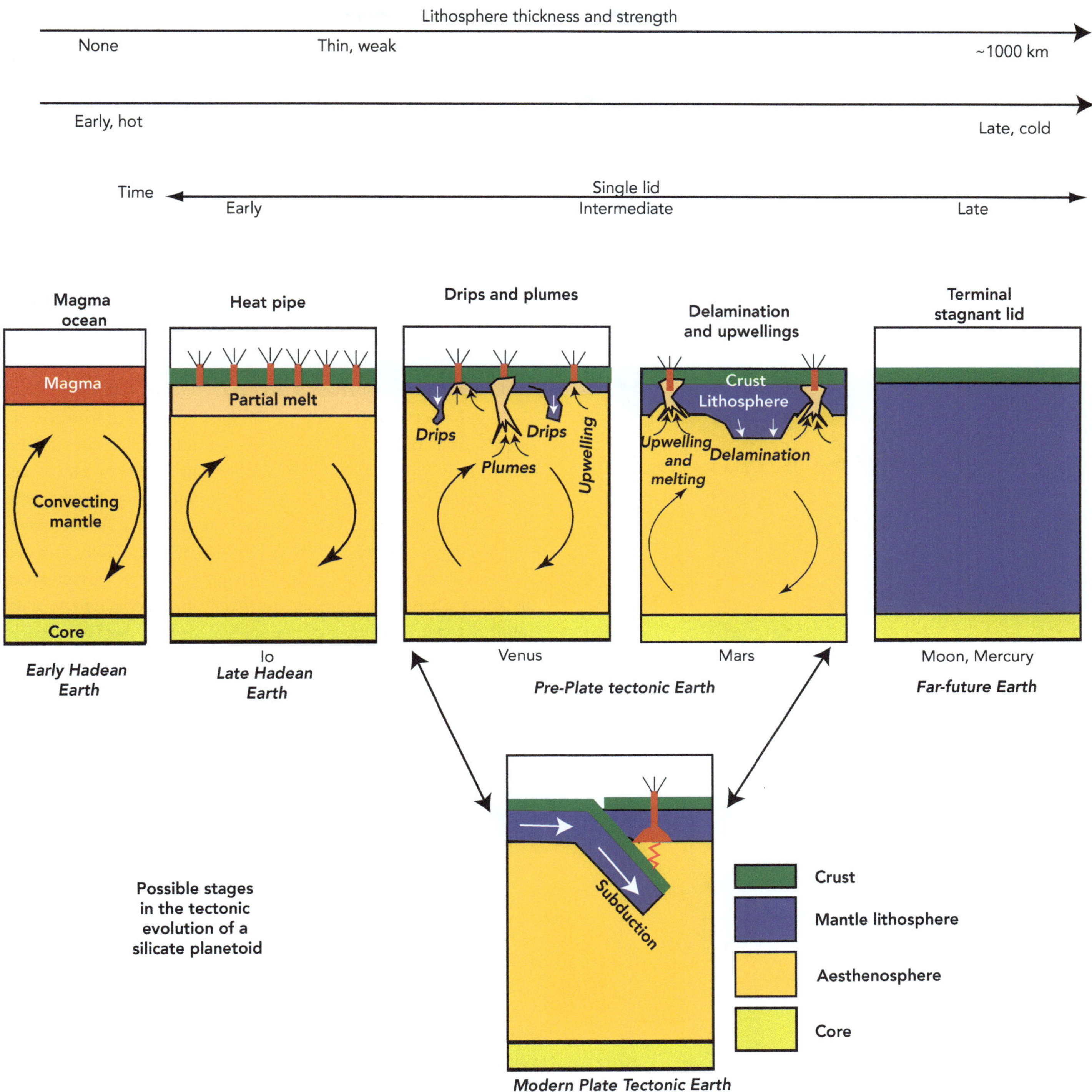

Figure 5.24 Models for different types of lithospheric plates in the solar system, and implication for initiation of plate tectonics. Source: Modified after Stern (2018).

evidence. For example, rocks subducted to great depth will warm over time and if not returned to near the surface quickly, they warm and will be transformed into other metamorphic rocks called greenschist. So, the paucity of very old blueschists may only be a preservation issue and not indicative of the absence of subduction before 1 Ga.

Zircons today are found almost exclusively in rocks of the continental crust. The presence of zircons of 4.0–4.4 Ga (found in western Australia and northern Canada) suggests that continental crust formed quite early in Earth history. Details of the composition of many of these oldest zircons further points to their continental origin. This leads advocates of early tectonics to strongly argue that some kind of tectonics must have been active since very soon after the collision event that formed the Moon.

Advocates for the late initiation of plate tectonics argue that episodes of partial melting due to convection alone could produce isolated granitic magmas without full-fledged plate tectonics. Modeling of early Earth also suggests that it was much

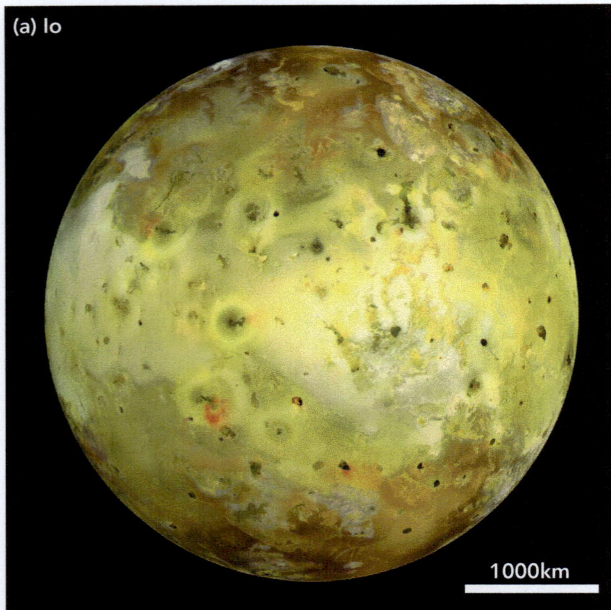

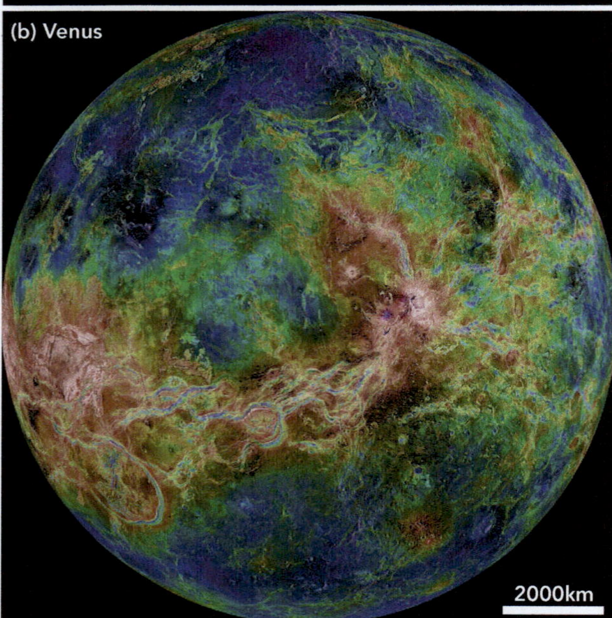

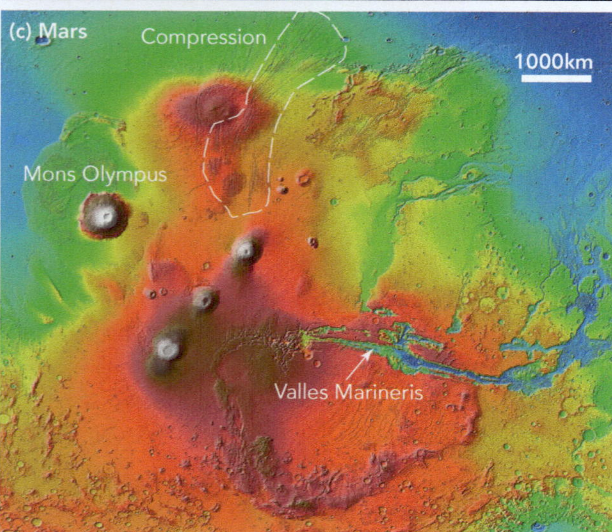

Figure 5.25 (a) Io, a moon of Jupiter, is a single-lid silicate moon with over 400 volcanoes, and is characteristic of a single-lid, hot-pipe body.

hotter, resulting in greater partial melting of the upper mantle, creating a thinner, hotter, lithosphere that was buoyant and weaker than today. Depending on the assumed initial mantle temperature, at hotter temperatures this buoyant weak lithosphere cannot subduct. Vertical tectonics, with areas elevated where upwelling occurs, such as associated with mantle plumes, may have allowed local tears at the surface and lithospheric drips and delamination below the plate (Figure 5.24). Uplift, analogous to Venusian coronae, may also allow local sliding and migration of continents, but driven by convection and overturn of the mantle rather than sinking of plates. Local compression may have allowed subcretion (under thrusting), forming folds, thrusts, and the greenstone belts. However, this same evidence has been cited as unequivocal evidence of modern subduction-driven tectonics.

Perhaps better resolution of the movement of Archean continents from paleomagnetic studies would help if it could be shown they had moved significantly prior to 2.5 Ga. However, heating and chemical alteration will degrade the paleomagnetic signal and most rocks that old have been heated at least to some degree. Furthermore, paleomagnetic studies can only resolve motion relative to the equator, so if continents moved great distances across the globe in the Archean, paleomagnetic data could not tell us if most of the motion had an east–west component. Studies of pressure–temperature histories of metamorphic rocks could also be helpful but such data are also subject to resetting by younger metamorphism and deformation. Add this to the likelihood of old rocks being destroyed by erosion or covered by tectonic or sedimentary processes, it is easy to see why study of Precambrian rocks (where very few fossils are present to aid us) will remain difficult.

Indeed, one of the greater challenges in Earth history is understanding events that occurred more than three billion years ago. In Chapter 2 we introduced the idea of actualism, the concept that Earth conditions and processes may have been quite different in the past. This is the basis for the advocate arguments against assuming that plate tectonics has been operating, much the same, since the Archean versus the idea that Archean, and perhaps much of the Proterozoic, were different enough to prevent modern-style plate tectonics. We argued in Chapter 2 that catastrophism is a poor doctrine to understand Earth history, but it is also not appropriate to apply

Figure 5.25 (*cont.*) (b) Venus is characterized as a single-lid planet with vigorous convection marked by large volcanoes and areas where the lithosphere has broken, sagged, and deformed. (c) Map of the Tharsis plateau on Mars caused by mantle upwelling, with several large volcanoes. The linear ridges in the area outlined with white dashed outline are compressional. The Valles Marineris is a giant outflow canyon that may be associated with an underlying strike-slip fault, indicating a local lithospheric tear. Source (a) NASA / JPL / University of Arizona; (b) NASA / JPL / USGS, Public domain, via Wikimedia Commons; (c) NASA / JPL-Caltech / Arizona State University, Public domain, via Wikimedia Commons.

uniformitarianism on a daily or even annual timescale (the meteor that probably killed the dinosaurs was a very different day but not one in which any laws of physics or chemistry need be ignored). The challenge of the study of the earliest of Earth history is to understand if there was ever a time in which uniformitarianism can't even be applied on timescales of tens of millions of years or if the large-scale processes we understand today have been operating from the beginning. One variant of actualism is to start with the observation that plate tectonics is clearly the dominant large-scale driver of geologic evolution today and work back to a time when it was not. As the focus is directed to older and older rocks, the evidence for plate tectonics moves from unquestioned to equivocal. The time of the switch depends on who is interpreting the rock record but varies from 1.0 to 4.0 billion years ago. Among scientists today there is no time for which there is evidence to support even a consensus for an interpretation of a time in which some kind of plate tectonics did not play a role in Earth history.

It is clear that Earth is the only planet in our solar system with life and it is also the only planet with subduction-driven plate tectonics. It may be that the development of complex life, and particularly metazoans, which we will review in Chapter 9, requires plate tectonics, with its consequent control on potential habitats. Understanding the interrelationship between plate tectonics and the origin and diversification of life on Earth may also be critical in answering questions about the possibility of life on other planets in the galaxy. At the time of writing, we know of no terrestrial rocks that were formed prior to 4.1 Ga and essentially all rocks we know of that were formed before 3.5 Ga carry with them the problems of metamorphism and deformation (and lack of fossils) discussed earlier, so a full understanding of what were the main drivers of the geodynamics of planet Earth at its inception may only be included in a future edition of this book.

The history of plate tectonics includes many examples of explanations for geologic observations that were changed substantially due to new information (i.e., from up and down to side to side), illustrating the iterative nature of the scientific method (see Figure 2.10).

There is debate as to when plate tectonics began. The absence of ophiolites, blueschists, and island arc volcanic rocks has led some to suggest plate tectonics only began in the Neoproterozoic, whereas others believe it started in the Archean Eon.

5.3 Summary

- Wegener used observation of the apparent jigsaw pattern of Atlantic continental margins as well as similarities in fossils, rock types, and glacial deposits to suggest that around 300 Ma North and South America were joined to Europe and Africa, and that South America, Southern Africa, India, and Australia were connected as Pangea and lay close to the South Pole.
- Wegener was vague in suggesting a mechanism, hypothesizing that tidal forces might be important, but physicists of the time understood that these mechanisms were not plausible.
- Without a viable mechanism to move plates, there was the enthusiasm for the idea of land bridges that might have allowed migration of terrestrial faunas found as fossils in now separate continents, but mapping of the ocean floors showed no evidence for these.
- Paleomagnetic studies showed that apparent polar wander curves only made sense if continents were allowed to move.
- Mapping of oceans, as well as paleomagnetic surveys, showed mid-oceanic ridges that had matching patterns of magnetic stripes. The striping patterns were produced as new ocean crust was formed continuously and Earth's magnetic field switched polarity episodically and migrated away from the ocean ridge center, confirming the idea of seafloor spreading.
- Mapping of dipping and unusually deep earthquake centers (Wadati–Benioff zones) led to the discovery of subduction zones: areas where ocean floor sinks back into the asthenosphere. This solved the problem of how space could be created for new oceans at spreading centers.
- Plate boundaries can be extensional (divergent) compressional (convergent) or strike-slip.
- The main layers in Earth's outer mantle include rigid oceanic and continental lithosphere and the underlying mobile asthenosphere. Convection in the asthenosphere enables recycling of sinking oceanic slabs in subduction zones and upwelling allows new ocean lithosphere to be formed at spreading centers, whereas continental lithosphere cannot be subducted, and collisions result in the formation of major mountain belts and additional continental crust.
- The main force that drives plate tectonics is slab pull, primarily caused by higher density of cold ocean lithosphere. Slab pull is enhanced by the metamorphism of sinking slabs to eclogite. Elevation of ocean lithosphere at mid-oceanic ridges also adds a component of ridge-push forces and without convection and a mobile asthenosphere the plate would not be able to move.
- Blueschist is metamorphic rock characteristic of the low-temperature/high-pressure regime thought to be diagnostic of subduction zones. A lack of Archean blueschists has been cited as evidence that early Earth did not have plate tectonics and that it may have developed later in the Proterozoic Eon, but this is countered by observations of Archean zircons that may imply early tectonic processes required to produce continental crust.

Key Words

- geosynclinal theory
- plate tectonics
- Urkontinent
- Pangea
- continental drift
- convection
- paleopole
- mid-ocean ridges
- lherzolites
- diapirs
- peridot
- harzburgite
- cumulate
- pillow lavas
- magnetic anomalies
- decompression melting
- rifting
- magnetic surveys
- seismometers
- focus
- hypocenter
- subduction zones
- lithosphere
- asthenosphere
- rheology
- eclogite
- slab pull
- ridge push
- divergent plate boundary
- blueschists
- stratovolcanoes
- transform boundary
- strike-slip
- restraining bend
- releasing bend
- passive continental margin
- accretionary prism
- foreland basin
- rock cycle
- natural disaster

Further Reading and References

Benioff, H., 1964, Orogenesis and deep crustal structure - additional evidence from seismology, *Bulletin of the Geological Society of America*, 85, 385–400.

Cox, A., and Hart, R. B., 1986, *Plate Tectonics: How it Works*, Wiley-Blackwell.

Müller, R. D., Sdrolias, M., Gaina, C., and Roest, W. R., 2008, Age, spreading rates, and spreading asymmetry of the world's ocean crust, *Geochemistry, Geophysics, Geosystems*, 9 (4), Q04006, https://doi:10.1029/2007GC001743.

Oreskes, N., 2018, *Plate Tectonics: An Insider's History of the Modern Theory of the Earth*, CRC Press.

Stern, R. J., 2018, The evolution of plate tectonics, *Philosophical Transactions of the Royal Society*, A376, 20170406, http://dx.doi.org/10.1098/rsta.2017.0406.

Vine, F. J., 1966, Spreading of the ocean floor: new evidence, *Science*, 154, 1405–1415.

Vine, F. J., and Matthews, D. H., 1963, Magnetic anomalies over oceanic ridges, *Nature*, 199, 947–949.

Review Questions

1. What was the key evidence that Wegener used in developing his continental drift hypothesis?
2. What were the shortfalls in early ideas about continental drift?
3. How did paleomagnetic studies help develop plate tectonic theory?
4. What is a land bridge and why were they a popular hypothesis before plate tectonics was widely accepted?
5. What insights in mapping and surveying the world's ocean helped to develop plate tectonic theory?
6. How did magnetic surveys across oceanic ridges support the idea of continental drift?
7. How did earthquake surveys help in the discovery of subduction zones?
8. What are the three main types of plate boundaries?
9. Describe the main layers in Earth and how their differences in properties contribute to the development of plate tectonics.
10. How are volcanoes and plate tectonics related?
11. What are the main forces that drive plate tectonics?
12. How does oceanic lithosphere change from the time it is produced at an oceanic spreading center, to subducting into the asthenosphere?
13. What are some of the key metamorphic processes and rocks associated with subducting oceanic lithosphere and what is their role in driving plate tectonics?
14. Describe some of the main types of sedimentary basins and their link to plate tectonics.
15. What is some of the evidence that helps determine when plate tectonics began?

Ammonite types of fossil, showing now extinct marine mollusks called Ammonoids. Source: Raj Kamal / Getty Images.

Chapter 6
Evolution

Natural Selection and the Organization of Life

LEARNING OBJECTIVES

- Outline the main components of evolutionary theory and their relationship to the geological evidence.
- Explain how artificial selection provides a basis for understanding natural selection.
- Explain how evolution is a product of natural selection that takes advantage of genetic variation within a population.
- Outline some of the key anatomical and genetic evidence for the relatedness of species that indicates inheritance by evolution, such as homology, vestigial structures, and imperfect adaptations.
- Describe the concept of a clade and the basic organization of life.
- Explain the genetic evidence that shows how all life is related.
- Explain the idea of a molecular genetic clock.
- Explain some of the mechanisms and modes by which evolution occurs.
- Explain how evolution is driven by limiting factors and global change.

Introduction: The Observation of Evolution

William "Strata" Smith's "principle of faunal succession," introduced in Chapters 1 and 2, codified one of the key observations in geology: strata of different ages contain unique, age-diagnostic fossils. Today anyone can repeat these observations, either in northern England, where Smith worked, or with any other fossiliferous sequence of rocks. The rock record is clear, life on Earth has changed over time. As one species becomes extinct it is often succeeded by a new species. This is **evolution**, a simple and straightforward observation. The ability of life to diversify was made possible by the immensity of geological time and was largely driven by the changes in local or global environments. The only theoretical framework needed to start this work is Steno's principle of superposition, which we explained in Chapter 1. Applying this idea with observations of fossils in different strata clearly show that life has evolved. Observations from other fields of science, including anatomy and physiology, molecular biology, and genetics, are consistent with the data from paleontology and stratigraphy and show that all life is related. The idea that best explains the observed evolution of life over time, as well as how species are related, is the theory of natural selection and is the focus of this chapter.

KEY POINT

That life has evolved on Earth over billions of years is an observation (not a theory) shown clearly using only the principle of superposition and the principle of faunal succession. Natural selection is a theory that explains why evolution is likely to occur. Natural selection favors traits in organisms that increase the likelihood of the production of healthy offspring to transfer genetic material into the next generation.

6.1 Darwinian Versus Lamarckian Ideas of Evolution

In 1859, Charles Robert Darwin (Figure 6.1) published his landmark book, *On the Origin of Species by Means of Natural Selection, or the Preservation of Favoured Races in the Struggle for Life*, which eclipsed previous ideas about evolution. Darwin was well-steeped in the geological sciences and had a copy of Charles Lyell's *Principles of Geology* on his five-year voyage on HMS *Beagle* in which Smith's principle of faunal succession was prominent. In Darwin's time, extinction was widely accepted, and led to the earlier ideas of Georges Cuvier that Earth had experienced 'revolutions', which some attributed to divine catastrophes as we discussed in Chapter 2. Cuvier was no catastrophist, but in his time, it was not understood how one species could transform into another.

Figure 6.1 Charles Darwin. Founder of the theory of natural selection by modification with common descent. Source: Charles_Darwin_seated.jpg: Henry Maull (1829–1914) and John Fox (1832–1907) (Maull & Fox). Derivative work: Beao, Public domain, via Wikimedia Commons.

Jean-Baptiste Lamarck published a long entrenched early theory of evolution, in which he suggested that a "complexifying force" drove life towards more progressive forms through time and that organisms also adapt to their environment. He suggested that this adaptation results from two laws: (1) the law of use and disuse – the more a specific trait is used, the more enhanced it becomes and the more useless it is, the greater is the likelihood that trait will disappear; and (2) the inheritance of acquired characteristics – the idea that traits acquired during a lifetime are passed directly on to successive generations.

Darwin and Alfred Russel Wallace challenged Lamarck's ideas in their development of the theory of natural selection. Neither Darwin nor Wallace coined the term evolution but rather explained how evolution occurs via the process of natural selection. Darwin eventually eclipsed Wallace in fame and reputation following publication of his landmark 1859 publication.

In presenting his argument, Darwin began by describing the process of **artificial selection**. Artificial selection was routinely practiced by humans, both in the development of favored breeds of domestic animals, flowers of favored styles or colors, and of course the agricultural manipulations that improve produce (Figure 6.2). To produce a rare, recessive variety, like a black tulip, one must cull all lighter tulips, retaining only the more generally dark tulips, and only cross breeding the very darkest until a black tulip is achieved. Breeds of dogs, or any other favored animal breed, like racehorses, can only be maintained by preventing interbreeding with the general population of dogs or horses (Figure 6.2(b)). If special breeds of dogs are allowed to interbreed, such as occurs in the wild, mutts and mongrels quickly homogenize the population. Artificial selection of wild corn led to modern corn with more and larger kernels (Figure 6.2(a)). A key aspect to artificial selection is that it requires some degree of variability within a population and can be achieved only by allowing for interbreeding of individuals with favored traits that can become dominant. Darwin thus understood that variations could be passed on or inherited by offspring and that they could be selected for. His insight was in deciphering that this process also operates under natural conditions, where individuals and species compete for resources and reproductive success.

Charles Darwin was influenced by reading about the artificial selection of livestock through breeding techniques, and was himself a breeder of pigeons. Darwin realized that a similar process must be at work in nature but driven by environmental change, rather than choices of the livestock managers, and at a much slower pace, which he called **natural selection**.

Darwin harnessed three observations to lay the foundations for his explanation of evolution by natural selection of favored characteristics in the struggle for existence. The first is that within any population there exists a range of variation within species that he referred to as *varieties*. We now know that these varieties mostly come about from variations in an organism's genes, resulting in a variable gene pool. **Mutations** of genes that produce traits that are disadvantageous (e.g., a mutation that

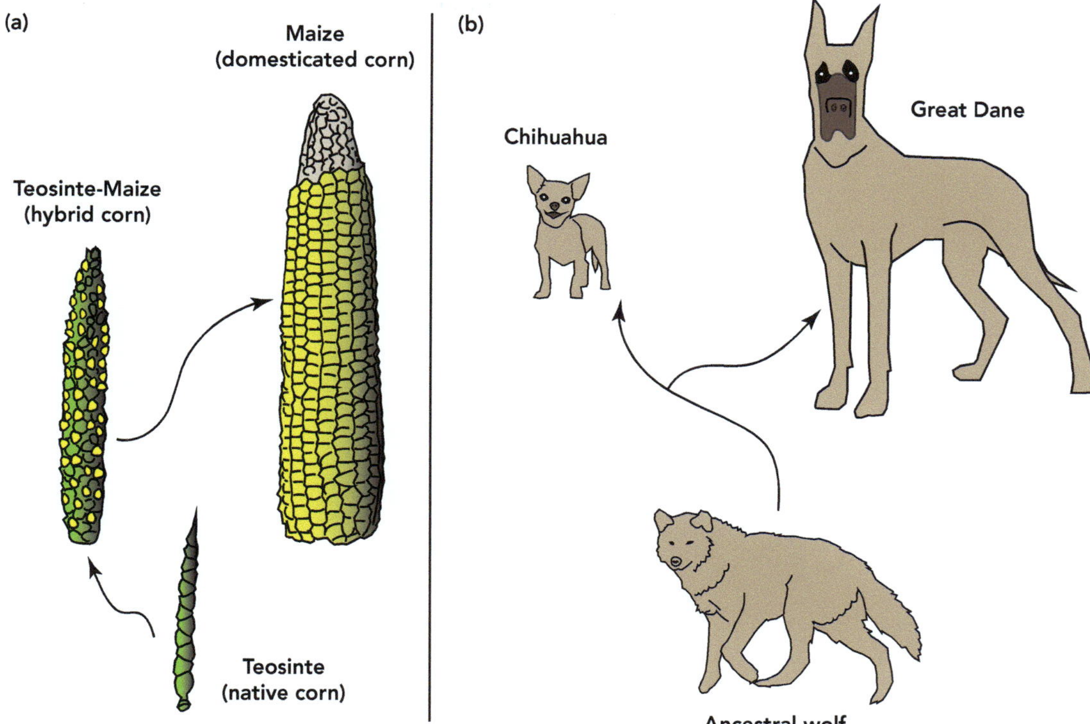

Figure 6.2 Examples of artificial selection: (a) native variety of corn, Teosinte, is eventually bred into modern maize with larger and more abundant kernels; (b) all of breeds of dog, such as Great Danes and miniature Chihuahuas, originated from domestication of a wolf-like ancestor.

produced blindness) are quickly removed from the population, as these disadvantages make it less likely that the individuals will produce offspring that also have these traits. Some mutations are beneficial, but most mutations are benign, conferring neither benefit nor disadvantage. Benign mutations will remain in the population unless the environment changes to a degree, such that what was once benign becomes disadvantageous or advantageous. Blindness may be a liability for an animal that lives in daylight but may be an advantage for deep-cave dwellers, who conserve energy by evolving a loss of eyesight.

The second of Darwin's key observations was that most organisms produce far more offspring than can possibly survive. For example, female mosquitoes typically lay around 300 eggs during their lifetime of 40–50 days. This is a strategy that assures the continuity of the species by overwhelming the very small odds of an individual surviving with very large numbers of individuals. Clearly, not all the 300 eggs survive to produce offspring of their own. If that was the case, in a few years, a pair of mosquitoes would produce a number of mosquitoes greater than the number of atoms in the planet Earth!

The third observation that provides the foundation for Darwin's explanation for biologic evolution is that competition for food and living space results in the elimination of those least well adapted to their environment. Two things are important to note about this. First, it is often said that Darwin argued for the **"survival of the fittest,"** but this was a later term, coined in 1864 by English philosopher Herbert Spencer after he had read

Darwin's *On the Origin of Species*. It is important to understand that "fittest" here means all those organisms that meet some minimum requirements to survive in a given environment, keeping in mind that environments can change such that organisms, once well adapted to their environment, may find themselves disadvantaged to new conditions, even to the point of extinction if the environment changes beyond their ability to survive or adapt. In this case, the organisms that survive may or may not have been the fittest for conditions prior to an extinction, but find themselves to be the lucky survivors after an extinction event, and this has sometimes been called "survival of the luckiest." We will examine this concept further in Chapters 11 and 13 when we look at some of the mass extinction events that punctuate the geological timescale.

KEY POINT

Three key factors promote biologic evolution by natural selection: (1) variation within populations; (2) more offspring are produced than can survive; and (3) there is a competition for food and living space.

6.2 Evidence of our Shared Ancestry

The main approaches that are used to infer ancestry for all life on Earth are anatomical and genetic analysis, which we will examine in more detail below.

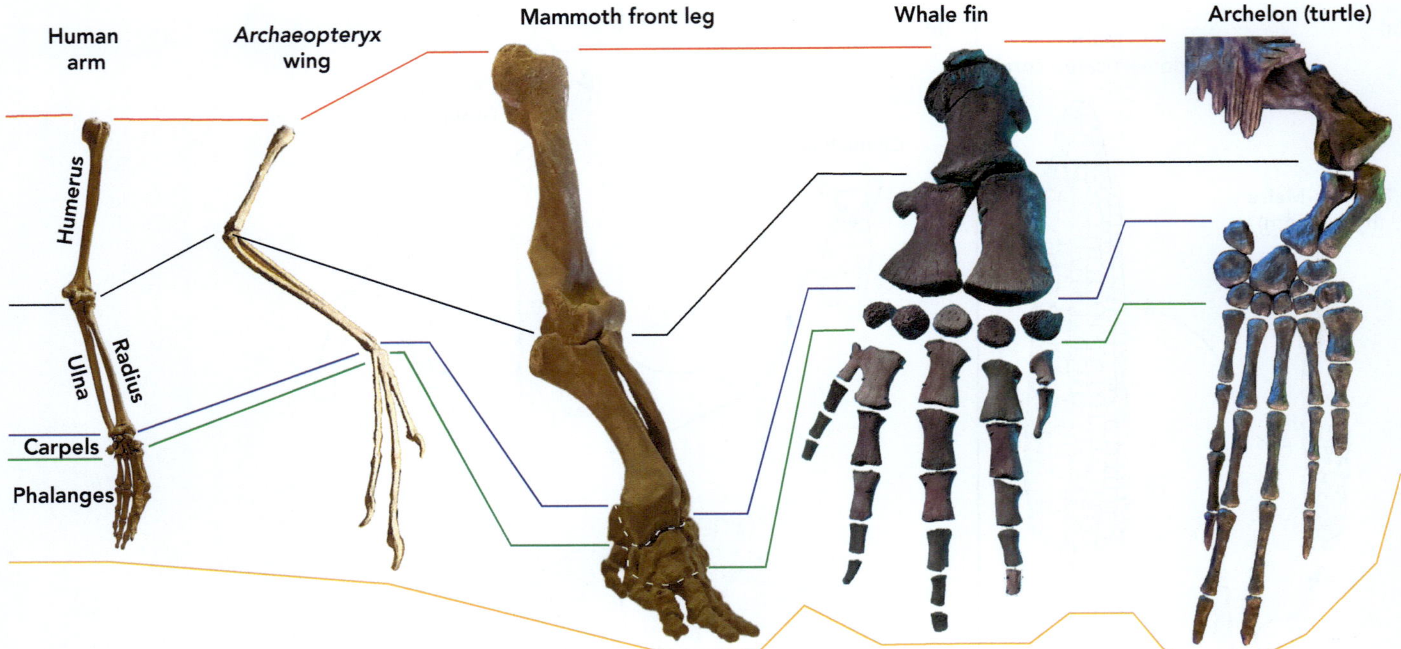

Figure 6.3 Comparison of the bones between humans, *Archaeopteryx* (an ancient bird), mammoth, whale, and giant sea turtle (*Archelon*, an extinct Cretaceous turtle thought to be the largest that ever lived) shows that the position and number are invariant and indicates that all these different animals were descended from a common four-legged ancestor. Similar features indicating shared ancestry are called homologous. Source: (Human arm) Photo by JPB. With permission of Museum of Natural History, Kolkata; (*Archaeopteryx* wing) Photo by JPB. With permission of Perot Museum, Dallas; (Mammoth front leg) Photo by JPB. With permission from Houston Museum of Natural Science (HMNS); (Whale fin and *Archelon*) Photos by JPB. With permission of ROM (Royal Ontario Museum), Toronto, Canada.

6.2.1 Homology and Analogy

One of the most compelling pieces of evidence for evolution is the remarkable similarity of life forms. **Homology** refers to similar anatomical or genetic structures attributable to a common origin. For example, at the anatomical level, terrestrial vertebrate limbs have a remarkably similar patten (Figure 6.3). A single long bone is connected to a pair of bones that are in turn connected to a larger group of circular wrist and ankle bones and then the elongate bones that make up fingers and toes. This same basic pattern can be observed such that you can readily match the bones on your own arm with limb bones in the wings of an *Archaeopteryx* (an ancient bird), a mammoth leg, and the flippers of whales and marine turtles (Figure 6.3). These are interpreted to be the consequence of a shared ancestry in a common, four-limbed ancestor (technically a *tetrapod*, literally "four footed"), which is discussed in more detail in Chapter 11. Subsequent evolution has driven the differentiation of the ancestral tetrapod limbs into the variations shown in Figure 6.3. Even though these bones are used for flying by birds (and *Archaeopteryx*), walking by mammoths, swimming by whales and sea turtles, and the various things that humans use their arms and legs for, the underlying anatomical structures are the same, indicating a shared ancestry.

KEY POINT

Homologous anatomical or genetic structures show similarity between species indicating shared ancestry, even if the structures serve different purpose such as human arms, dogs' legs, and whale flippers.

These contrast with *analogous* structures, which are features adapted for the same purpose, but which do not have a common origin (Figure 6.4). Birds and crabs both have eyes and legs, and both use these for much the same purpose, but birds and crabs are not closely related, and these features evolved independently. Birds and butterflies also both have wings that they use to fly, but again these structures are analogous and evolved independently. Analogous structures present good examples of *convergent evolution*, in which different organisms develop similar structures in adapting to similar environmental conditions. However, even in the case of analogous structures, it is now recognized that at the genetic level, key genes, such as the *Hox genes*, control very fundamental aspects of how an animal develops from the embryonic to adult stage (Figure 6.5). This includes differentiation into a head and tail, formation of the

Figure 6.4 Birds (a), crabs (b), and butterflies (c) all have legs and eyes and use them for much the same purpose. Butterflies (c) and birds (a) also have wings and fly. However, crabs, butterflies, and birds are not closely related. These features evolved independently and are analogous structures.

body cavity and gut, as well as controlling what appendages grow along the length of an organism. Thus, the appendages of an arthropod and limbs of vertebrates are controlled by homologous Hox genes (Figure 6.5).

Embryology provides critical insight into how evolution operates. All organisms that reproduce through sexual reproduction start by a sperm fertilizing an egg, producing a single cell that shares half of its genes with the "father" (sperm donor) and half with its "mother" (egg donor). During early embryonic development, the cell divides into two, then four, and so on, forming a hollow ball that begins to dimple and fold, eventually differentiating to form tissues, body layers, and organs. The instructions for all this complex differentiation are contained within the **DNA** (deoxyribonucleic acid). These instructions are somewhat akin to a computer code, and mostly involve quite local instructions among or between groups of cells that ultimately built our complex bodies. Sometimes an instruction can be delayed and turned off or on. Small changes in a local instruction (i.e., gene) during development of an embryo can cause major changes in the outcome of the adult animal, and we shall see how this may have driven human evolution in Chapter 17. If these changes are favorable, whether small or not, they may find themselves selected for.

KEY POINT

Convergent evolution in unrelated lines of organisms produces analogous structures that serve the same purpose (e.g., insect versus bird wings) but have a different evolutionary origin. Despite this, the development of many of these analogous structures are ultimately controlled by Hox genes, which reflect the deep homological genetic structures that indicate the shared ancestry of all life.

6.2.2 Clades

We commonly use the metaphor of a family tree to group related organisms together. A **clade** is a statistically defined grouping used to classify organisms that have a shared ancestor, as exhibited by shared anatomical or genetic similarities, and the study of clades is called **cladistics**. Ancestral organisms that lie at the base of a clade are commonly referred to as **stem** organisms, emphasizing their position at the base or stem of a clade, whilst groups at the top of the tree are referred to as **crown organisms**. In our dog example (Figure 6.2(b)), the wolf ancestor would be at the stem and Great Danes and Chihuahuas at the crown. A **cladogram** shows how organisms are organized according to their inferred **phylogenetic** history and the order in which they are related, as determined by the addition of newly derived traits (Figures 6.6 and 6.7).

One of the earliest cladograms was produced by Charles Darwin in his observation of different species of finches in the Galapagos Islands (Figure 6.7). Darwin hypothesized that an original "stem" finch from the South American mainland had colonized the Galapagos, but over time, due to different environments and food sources on the different islands, new species of finch had evolved from the common mainland ancestor and later adapted to the new environments by the process of natural selection (Figure 6.7). Isolation on different islands with different resources prevented inter-island breeding and allowed for evolution of finches unique to each of the islands. This process is referred as **allopatric speciation**, where geographic isolation of a population prevents the flow of genes that would normally occur through interbreeding.

The cladogram of the animal kingdom, also called Metazoa (literally "multi-celled animals"), shows how the major groups of animals, evolved through the addition of specific traits (Figure 6.8). The phylum *Porifera* (sponges) are the simplest

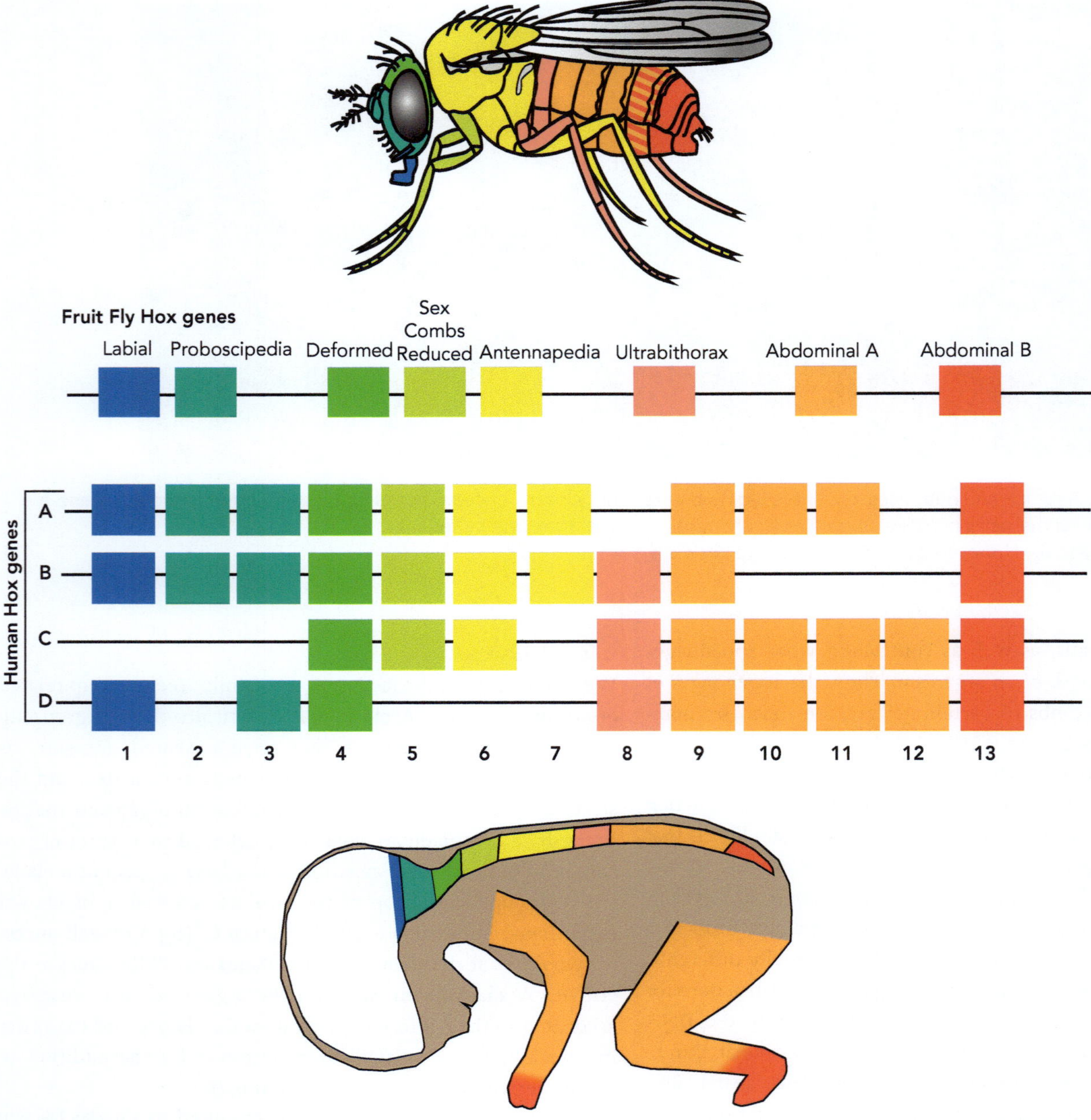

Figure 6.5 Comparison of Hox genes in fruit flies and humans demonstrates remarkable similarities. Hox genes control the differentiation into the front and back of organisms as well as development of appendages and segments. Source: Adapted from Mark et al. (1997).

and most primitive metazoans. They are multicellular but lack tissues or symmetry. The phylum *Cnidaria*, which includes jellyfish, sea anemones and corals, are the next in complexity. They show a radial symmetry and have tissues organized into two main layers (*diploblast*). The *bilaterians* are animals in which one side of the animal looks the same as the other. Simple bilaterians like the *Platyhelminthes* (flatworms) have tissues and an inner and outer layer but lack a *coelom* (body cavity). More complex bilaterian animal phyla, such as the *annelids* (segmented worms), have three layers (*triploblasts*) allowing formation of a coelom, distinct segments, and a more complex nervous system.

The *arthropods* (Figure 6.8) are the largest animal phyla and include familiar forms such as lobsters, shrimps, crabs, spiders, mites, insects, centipedes, and millipedes as well as the extinct trilobites. They have highly differentiated segments with the addition of complex jointed appendages, such as antennae, mandibles, and legs, as well as compound eyes. *Mollusca* are another large invertebrate phylum that includes clams, mussels, oysters, and scallops (forming the class *Bivalvia*), snails and slugs (forming the *Gastropoda*), and the squids and octopus, shelly nautilus, and extinct ammonites that form the *Cephalapoda*, and which have an excellent fossil record, as most make shells.

Our own group, the phylum *Chordata* (Figure 6.9), refers to organisms with an elongate cord down the axis of their body. In the earliest chordates, such as seen in the modern lancelet, this is a cartilaginous stiffening rod called a *notochord*, whereas in the vertebrates it includes the vertebral column and the spinal cord. The Chordata are now understood to belong to a broader group (**deuterostomes**) more closely related to the echinoderms (i.e., starfish and sea urchins), versus other invertebrates (**protostomes**), such as arthropods or mollusks, as shown in the animal cladogram (Figure 6.8).

The Chordata includes all organisms with a spinal cord (Figure 6.9). The *Cephalochordata* (e.g., lancets) have a notochord but lack bones or vertebrae. Hagfishes and lampreys are more advanced in that they added a head and vertebral column (backbone) respectively, but they lack jaws. Sharks and their kin (Chondrichthyes) and the bony fishes (Actinopterygii) show modification of the frontal gills into jaws, indicating that they evolved later (see Chapter 11). Coelacanths and lungfishes added lungs, amphibians and reptiles added four limbs (forming the tetrapods), with the single-bone, double-bone, clump-of-bones, and digit pattern described above (Figure 6.3). Mammals and reptiles developed an **amniotic** egg that allowed them to reproduce on dry land, and mammals added mammary glands that produce milk, as well as fur and hair. As we shall see in subsequent chapters, these additions of attributes typically occur along a distinct timeline, such that additional traits are added later in Earth history.

Homologies in anatomical structure, as well as genomic analysis, show how organisms in the clade are related, and the successive addition of new traits usually indicates the order in which organisms appeared. Of course, there are examples of "de-evolution," where more complex traits may be lost, although commonly some vestige may remain that provides evidence of the former ancestral origin (see Section 6.2.6). The observation of homological features and the determination of branching relationships can be tested against observations from the fossil record, using the "principle of faunal succession." For example, rocks of the Cambrian Period contain chordates, but no fishes. Fishes appear later in geological time in the Devonian

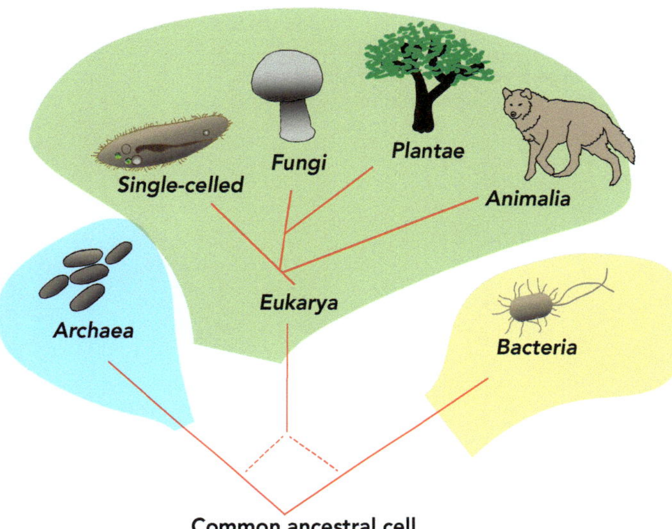

Figure 6.6 The "tree" of life showing the three major domains.

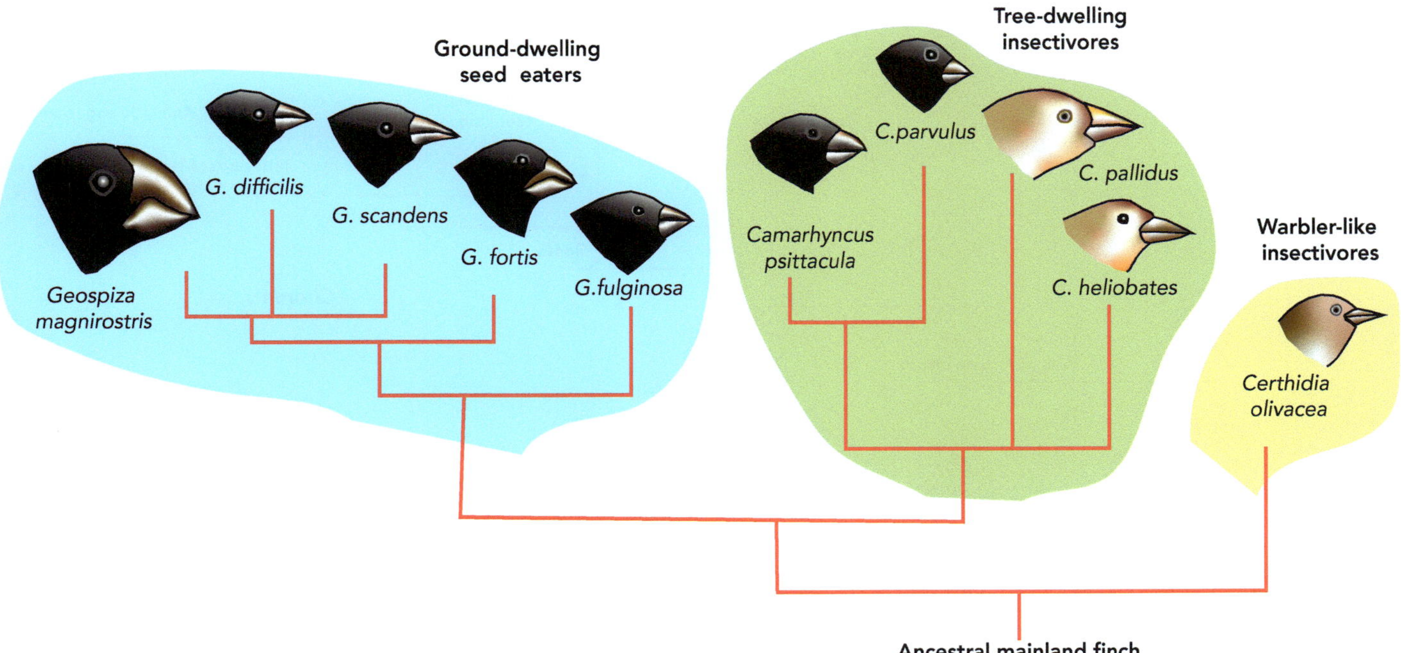

Figure 6.7 Cladogram showing how Darwin's finches are related. The diverse finches adapted to local conditions in the Galapagos Islands and evolved from a mainland ancestral finch.

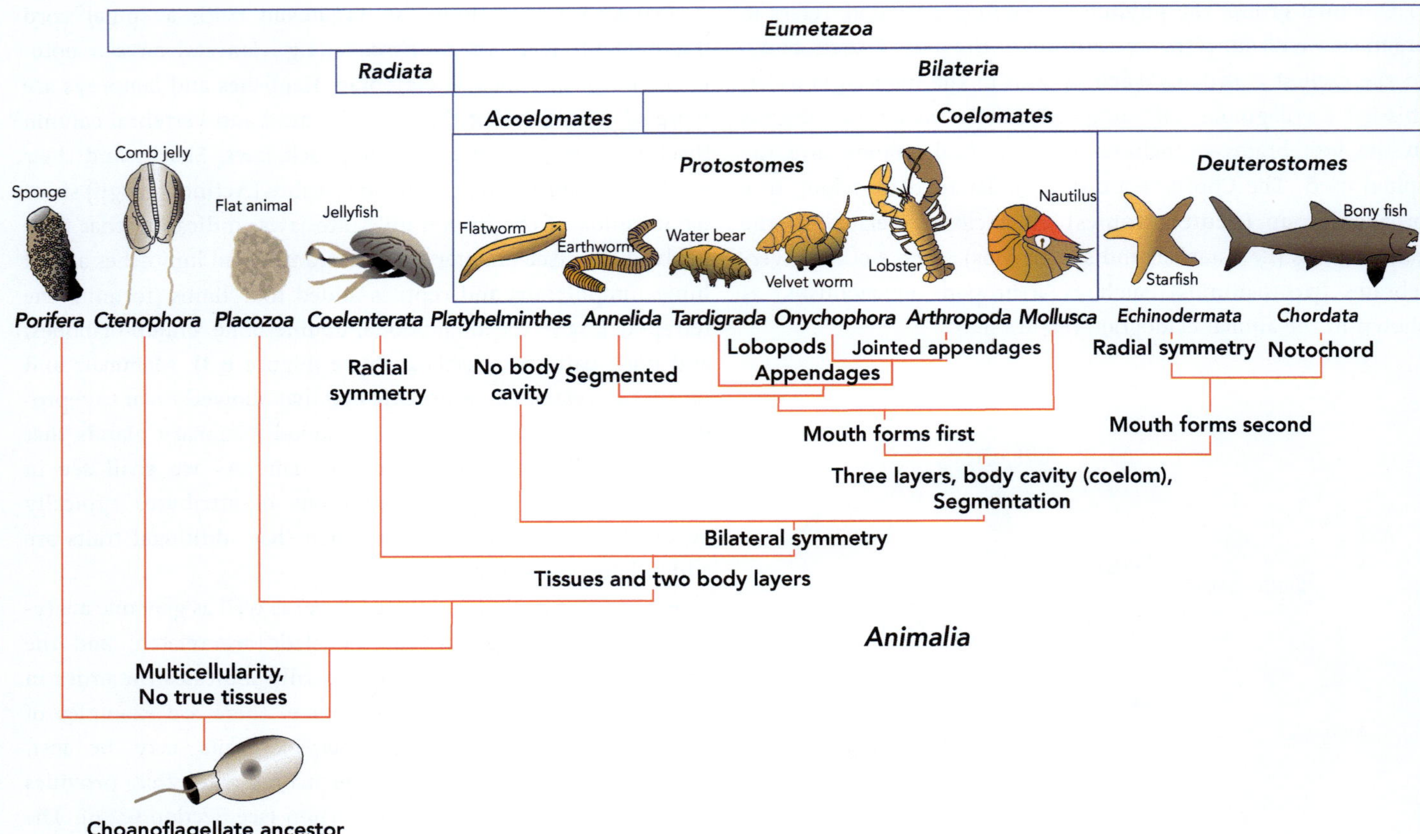

Figure 6.8 Animal cladogram.

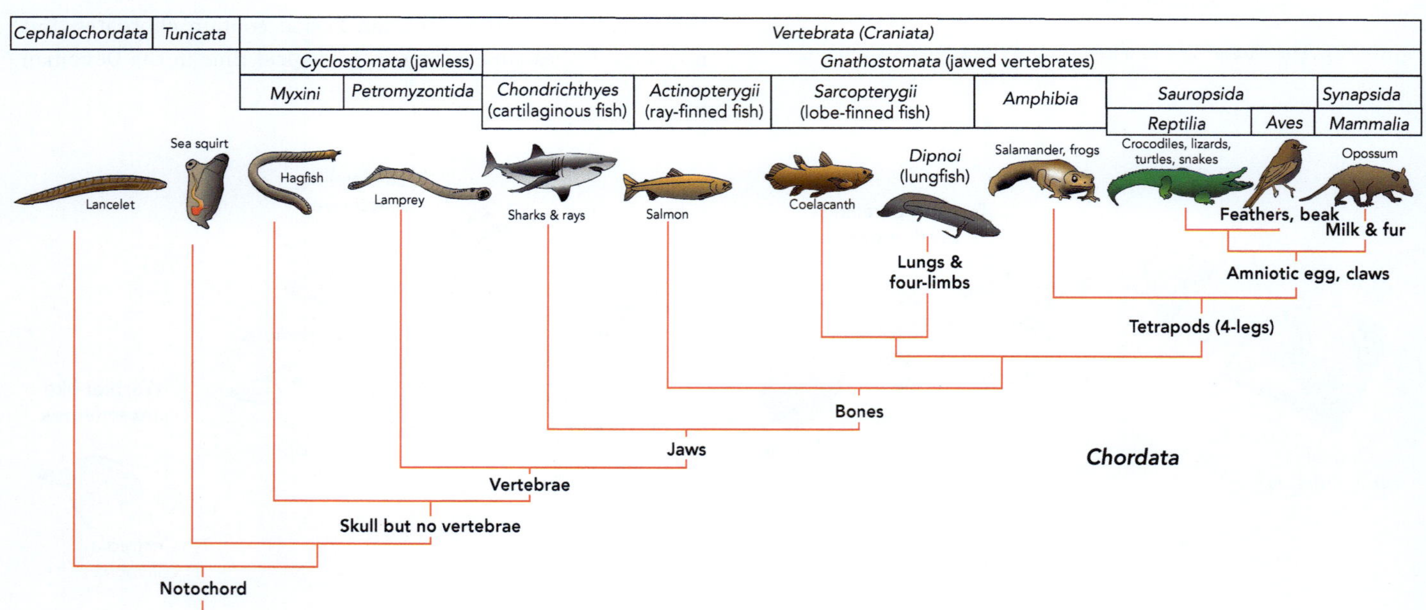

Figure 6.9 Chordate cladogram.

Period, but fully terrestrial amniotic vertebrates with four limbs (i.e., reptiles) don't appear until the Carboniferous Period (about 310 Ma, see Chapter 11). Unequivocal mammal fossils don't appear on Earth until after 250 Ma in the Mesozoic Era, and humans don't appear until about 250,000 years ago (see Chapter 17). Similar trends can be found in plants, with the earliest terrestrial vegetation appearing in the Silurian Period, expansion, and development of seed plants with vascular systems in the Devonian and Carboniferous periods during the so-called greening of the earth that we cover in Chapter 11, and the appearance of flowering plants and grasses quite late in Earth history in the Cretaceous Period of the Mesozoic Era.

The timing of the appearance of the fossil record of plants and animals matches the inferred history of the evolution of and order in which new traits appeared that define these groups.

KEY POINT

Cladistics use commonalities in anatomy and genes to infer the branching patterns of life from stem ancestors at the base of the tree of life to crown species at the top.

6.2.3 Organization of Life and Genetic Homology

Life is organized into three major **domains**: *Archaea*, *Bacteria*, and *Eukarya* (Figure 6.6 and Box 6.1). The Archaea have only been recognized as a separate domain within the prokaryotes since 1977, based on the pioneering analysis of ribosomal RNA by American biologists Carl Woese and George E. Fox. All these domains consist of organisms with cells, and most are single-celled (especially the Archaea and Bacteria). All cells contain DNA, and the similarity of DNA in all life provides some of the strongest homological evidence that all life is genetically related.

With enormous advances in **genomics**, it is now easier to examine the relatedness of organisms by directly comparing genomes versus relying on gross anatomy. Although this can't be done with most fossils, as the DNA is usually too degraded, this can be done with living organisms. Prior to genomics, organisms were classified and grouped based on anatomical homology. For example, animals with jointed legs and an exoskeleton form the arthropods, whereas all animals with a backbone are vertebrates. Life was also classified into kingdoms, of which the *Plantae*, *Animalia*, and *Fungi* (all eukaryotes) are the most well-known. However, cladistics and genomics are forcing a reconsideration of many of these classifications as we learn more about genetic homology. Although the earlier classifications based on broad body plans, skeletal features, and other macroscale features correctly reflect the phylogenetic history of many clades, genomic analysis has resulted in some significant changes to the details of our understanding of the structure and organization of the tree of life, especially single-celled life, and given how new the field of genomics is, more revisions are likely in the future.

Genomics can also be used to compare organisms and to deduce when they shared a common ancestor. Based on their anatomical similarity, Darwin suggested that our closest living relatives were the great apes, including chimpanzees and gorillas, all of which live in Africa (see Chapter 17). We now have genomic analysis that shows that human DNA is about 99% similar to chimps and a bit less to gorilla DNA (Figure 6.10). Nuclear DNA is housed in the **chromosomes**; humans have 23 chromosome pairs whereas chimps and great apes have 24 pairs. An explanation for this is that humans lost a chromosome by fusion of two into one. The ends of chromosomes are marked by a distinctive sequence of DNA molecules called **telomeres**. Examination of the second of the 23 human chromosomes (Figure 6.10), which is also the second longest, shows telomeres in the middle, indicating that it most likely formed by fusion of two separate chromosomes, and these can be matched to the unfused chromosomes in the great apes, supporting Darwin's idea that chimps and humans share a common ancestry. As we continue down the eukaryotic clade, we find that about 80% of our DNA is similar to fish, about 67% to corn (a eukaryote) and about 57% with a single-celled euglena.

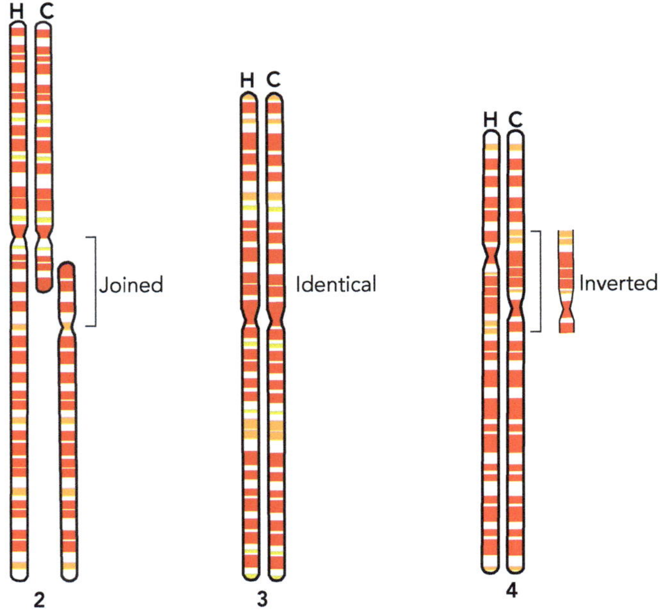

Figure 6.10 Human (H) versus chimp (C) chromosomes. The third chromosomes are identical. Chimps have an extra chromosome that is joined in human's second chromosome. Chromosome 4 is similar but has an inverted section.

BOX 6.1 Organizational Hierarchy of Life

Life is classified according to an organizational hierarchy. The traditional classification is domain, kingdom, phylum, class, order, family, genus, and species.

According to this, humans are: domain – Eukarya, kingdom – Animalia; phylum – Chordata, class – Mammalia, order – Primates, family – Hominidae, genus – *Homo*, and species – *Homo sapiens*.

In practice, many other groupings are used, such as super-, sub-, or infra-. In our case, humans are also: order – Primates, suborder – Haplorhini, superfamily – Hominoidea, family – Hominidae, subfamily – Homininae, tribe – Hominini, genus – *Homo*, and species – *Homo sapiens*.

6.2.4 Molecular Clocks

The structure of DNA was discovered by researchers from the University of Cambridge, James Watson, and Francis Crick, in 1953, based on X-ray diffraction analysis conducted by Rosalind Franklin at King's College London. Following their discovery, American biochemist Linus Pauling and his colleague Emile Zuckerkandl noted that differences in amino acids in the hemoglobin of different animal lineages seemed to change in a regular, linear fashion, especially when compared to the fossil evidence that showed when specific lineages appeared. They speculated that the rate of change of proteins in the DNA was likely to be approximately constant over time, even in different lineages. This was subsequently calibrated by taking the genetic distance between various organisms, based on analysis of the genome of modern living groups, and comparing the time of divergence of the groups in question based on estimates from the fossil record. This work demonstrated that the accumulation of genetic variation has been relatively constant over time, particularly among segments of DNA that do not perform essential regulatory functions. With this calibration in hand, we can now estimate the time at which two groups last shared a common ancestor by dividing their genetic distance by the general rate of genetic change. This approach is referred to as reading the **molecular clock**.

Based on this approach, it is thought that humans and great apes shared a common ancestor about 5 million years ago (see Chapter 17), whereas the split between single-celled *eukaryotes* (i.e., complex cells with a nucleus) and the first *metazoans* (multi-celled animals) occurred about 800 million years ago (see Chapter 7). Therefore, the genetic differences between modern metazoans, including humans, versus single-celled life are much larger than between organisms with a more recent divergence. The time of these inferred splits also matches the fossil record, which shows that the ape–human split was much more recent than the split of the branches between *prokaryotes*, simple single-celled organisms, and the larger eukaryotes, which we belong to (see Chapters 7 and 17).

KEY POINT

Genomics can be used to infer the timing of divergence between groups using molecular clocks that can then be compared with the geological record. The longer ago any two organisms share a common ancestor the greater the difference in genomes.

6.2.5 Imperfect Adaptations

Traits in some organisms can be traced back to ancestral organisms in which the traits played a different role. Darwin observed that gradual changes in characteristics are often accompanied by a change of function, recognizing that evolution occurs because of modification of pre-existing traits inherited through descent. The late Harvard paleontologist Stephen Jay Gould coined the term *exaptation*, as opposed to adaptation, to refer to the co-option of traits originally evolved for one purpose and later co-opted for another. This explains why homologous structures are so prevalent.

The evolution of whales provides a good example of this concept. Recent fossil finds have demonstrated that whales evolved from a land-based, four-legged, hoofed mammal (Figure 6.11). For whales to swim, natural selection is only able to work on the structures present in the ancestors. Therefore, limbs were modified into flippers, but retain the ratio of bones in the original that link them to their terrestrial ancestors (Figures 6.3 and 6.11).

Whales are mammals; they have lungs, they breathe air, their young suckle their mother's milk. Fish and most other aquatic animals have gills, allowing them to live underwater constantly, but although whales can hold their breath underwater for between 15 and 90 minutes they must ultimately come to the surface for air. This reflects their origin as land-dwelling air-breathing mammals. If they are unable to surface for air, they drown. Whales first appeared about 50 million years ago and became top predators. However, when humans in boats appeared they faced a new and unforeseen threat. Whales became highly valued for their oil, which was primarily used in lamps. The fact that they must return to the surface about every hour made them an easy target for hunting by surface-dwelling humans, versus sea creatures that live perpetually in the deep. The whales' **imperfect adaptation** to living in the sea has allowed humans to hunt several species of whales, including sperm, wright, humpback and blue, almost to extinction. In contrast the giant squids, which sperm whales prey upon, have never even been caught alive by humans, as they are adapted to living exclusively in deep water. Because sperm whales' respiratory systems developed from animals that lived in the air, they can be considered less perfectly adapted to life in the sea versus the giant squids that they hunt.

The 1846 invention of the kerosene lamp, by Canadian physician and geologist Abraham Gesner, heralded the beginning of the hydrocarbon industry that eventually lessened the reliance on whale oil, and strict rules imposed to limit whaling have allowed some of these endangered species to rebound. This provides an example of how selective pressures (in this case, hunting by humans) can devastate a population and push it towards extinction.

If we go even further back, all land vertebrates evolved from a fish ancestor, and we discuss this more extensively in Chapter 11. Gills are of no use on land, but lungs are useful. The earliest lungs developed from swim bladders, originally evolved to help maintain buoyancy in the bony fishes and *exapted* or co-opted as lungs in the earliest lungfish. As we will discuss further in Chapter 12, feathers appeared in dinosaurs long before they were used for flight.

The evolutionary concept exemplified by these imperfect anatomical adaptations provides clear evidence that life evolves

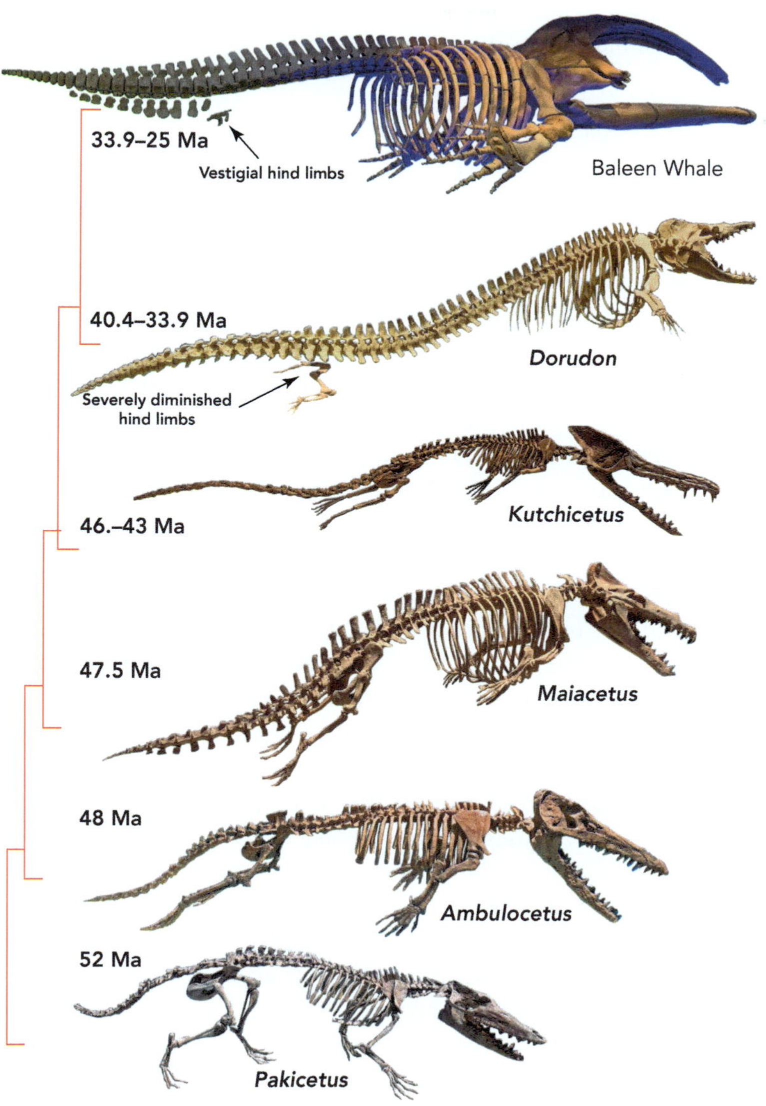

Figure 6.11 Fossils show the evolution of whales from a quadrupedal, terrestrial ungulate (*Pakicetus*) about 52 million years ago to modern whales. Note the reduction in hind limbs. Source: Photos by JPB. With permission of ROM (Royal Ontario Museum), Toronto, Canada.

because new evolutionary adaptations and modification can only be applied to the ancestral organism at hand with the body plans and anatomical features that are available, versus a separate "perfect" creation from scratch. The concept of homology is closely related (Figure 6.3). Whale flippers and mammal legs have the same bones: a single bone connected to a pair of bones, connected to a cluster of wrist bones and finally the digits, all in the same relative positions, again indicating that whale flippers evolved by modification of legs and hooves, versus a new design from scratch by an intelligent bioengineer. In fact, nearly every bone in any tetrapod, from skulls to jaws to limbs, can be matched, regardless of how they are used. Reptiles, for example have several bones that comprise the lower jaw, whereas in mammals the back part of the jaw bones have been co-opted and evolved to form the bones of the inner ear, leaving a single bone that builds mammal jaws (see Chapter 11), and before that, the foremost gill arches in the jawless fishes were modified to the form jaws in the jawed-fishes (see Chapter 12).

6.2.6 Vestigial Structures

Many organisms also show examples of a sort of reverse evolution, such as the loss of hind limbs in whales and snakes from former tetrapod ancestors. But evolution is commonly not capable of fully erasing the evidence of these former structures and they may be retained in a vestigial or shrunken fashion. Our appendix and tailbones (coccyx) are examples of such a **vestigial anatomical structure**. The coccyx is the highly reduced vestige that records our descent from tail-bearing ancestors (Figure 6.12). The appendix in herbivorous mammals, like horses or koala bears, is large and well-developed and is used to host bacteria that are helpful in digesting plants. Humans have relatively simpler digestive systems, compared to herbivores, and have thus lost the need for a larger appendix, but evolutionary processes have not erased the history of our descent from mammals with a larger appendix because the presence of our appendix is not a significant disadvantage from an evolutionary standpoint. That is, it doesn't make the production of offspring less likely. Therefore, we retain a vestige of this no longer very useful anatomical structure.

If we return to our discussion of whales, despite losing their hind limbs, whales also retain a pair of vestigial bones, unattached to the spinal vertebral column, and totally enclosed in the body of the animal (Figure 6.13). These bones represent shrunken vestiges of hind limbs and are another indication that whales evolved from ancestors with hind limbs. The fossil record of whales shows many intermediary forms that show progressive diminution of the hind limbs (Figures 6.11 and 6.13). Even though there is no external expression of hind limbs in modern whales, the adaptation of whales to an oceanic lifestyle and the resulting anatomical modifications by evolutionary processes were unable to completely erase the evidence

of their origin from an originally four-legged tetrapod land-dwelling air-breathing ancestor.

6.2.7 Vestigial Genes

It is even harder to erase DNA and there is ample evidence that modern DNA retains a record of its history. Evolution operates not only at the level of species, populations, and individuals but also at a more fundamental genetic level. Genes are constantly mutating, and this can occur by transposition of the so-called "jumping genes," discovered by American biologist Barbara McClintock (1902–1992). In her time women were generally discouraged from studying science. Barred from entering a biology program, McClintock ended up in the College of Agriculture at Cornell University in 1919, where she took her first course in genetics. Eventually she discovered that genes could be controlled and transposed by other genes along the chromosomes. Her work was largely disregarded until corroborated by French scientists in the 1960s. In 1983 McClintock was the first woman to win the Nobel Prize in Medicine.

Only a small amount of the DNA in our genes is used for building proteins. Some parts of DNA function as switches to other segments. DNA also replicates and competes within itself. Quite a lot of this replication and mutation is benign, having no measurable effect on an organism. Sometimes mutations have a negative effect, and these are typically weeded out of the gene pool. In contrast, mutations that provide a benefit will be passed on to offspring and preserved and enhanced via natural selection. The complex processes by which DNA molecules grow and change over time also leaves a record. The parts of DNA that are not involved in coding or switching are sometimes referred to as "junk" or non-coding DNA and represent vestiges that may have been of more use in the past but are no longer useful. We shall discus in more detail in Chapter 7 how symbiosis between prokaryotes and larger eukaryotes resulted in coding versus "junk" or vestigial DNA that has not been "erased" by evolutionary processes.

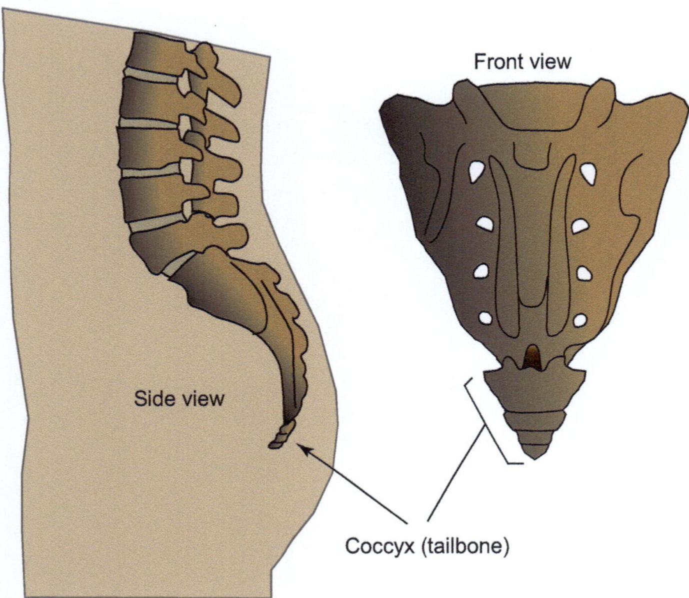

Figure 6.12 Your coccyx, also called the tailbone, represents the vestige of the longer tails that our ancestors had. Over time, the tail was reduced.

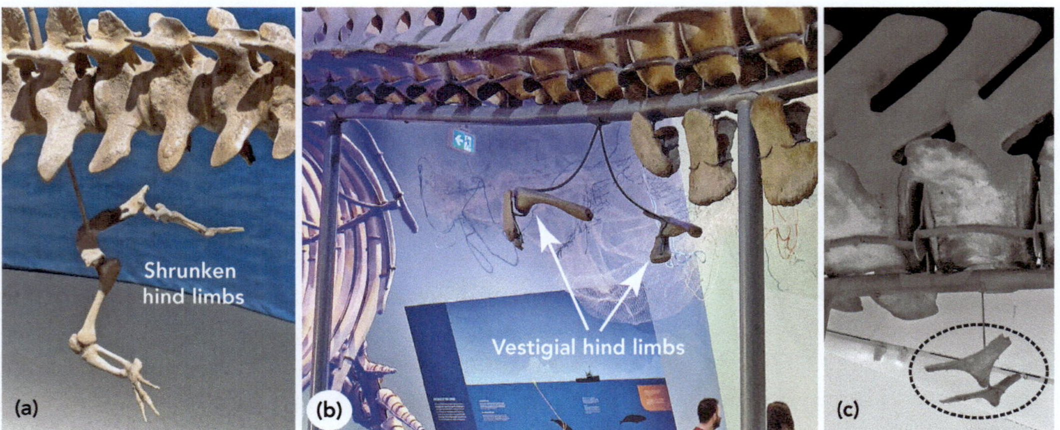

Figure 6.13 (a) Hind limbs of Dorudon. Vestigial hind limbs of (b) Wright whale and (c) blue whale. Source: Photos by JPB by permission of ROM (Royal Ontario Museum), Toronto, Canada.

KEY POINT

Evolution is not always very effective at erasing the history of previous ancestors. Vestigial structures, such as junk DNA, our appendix, and shrunken hind-limb bones buried in the flesh of whales, record evolution from previous ancestors in which these features were of more use.

6.2.8 What Use is Half an Eye?

An early criticism of Darwin's theory of natural selection wondered what use intermediate versions could possibly serve in evolving complex organs such as eyes. In later editions of Darwin's *Origin of Species*, he addresses this head on by pointing out that the gradation of characters along an evolutionary trend is often accompanied by a change in function. For example, swim bladders prevent fish from sinking but they were later co-opted to allow them to breathe air directly versus through gills alone. Fish record a long history of species that simultaneously used both gills and lungs. Such organisms were the ancestors of amphibians that today use both methods of respiration. As another example that we will discuss in Chapter 12, analysis of feathers showed they first appeared in the non-avian dinosaurs, perhaps for protection, insulation, or sexual display to attract mates and were only later co-opted for flight in their descendants, the birds (Figure 6.9).

It has been argued that the eye is so complex it could not possibly have evolved. This notion is championed by the rather strident statement – What use is half an eye? How could gradations in "eyeness" possibly be of any use? The short answer, as we will show next, is very useful indeed! Examination of modern organisms shows a complete gradation in complexity of eyes (Figures 6.14 and 6.15). Many single-celled free-swimming eukaryotes have a light-sensitive spot that allows navigation. Starfish have simple light-detecting clumps of cells, or eyespots, at the top end of their appendages. They have no focusing mechanism and no color vision, but they are able to detect light and shade and are able to use their crude eyespots for navigation. Starfish crawl quite slowly, and their simple eye spots have worked well enough for the past few hundreds of millions of years. The phylum Mollusca shows a wide variation

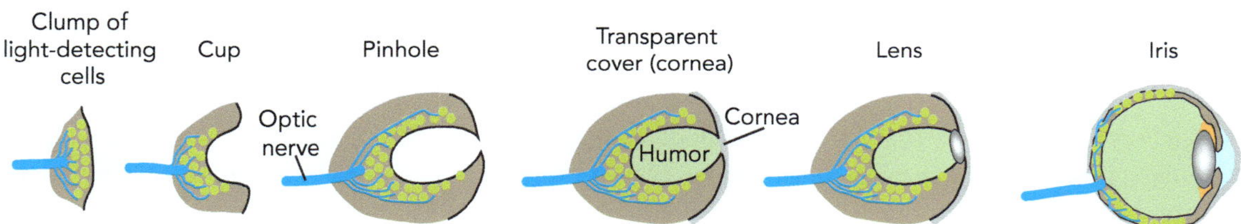

Figure 6.14 Evolution of eyes from simple light-sensitive cells to a camera eye.

Figure 6.15 (a) Nautilus has a simple pinhole eye; (b) a squid has an eye with a lens and iris; (c) horse fly eyes; and (d) wolf spider eyes. Source: (b) Gerald Corsi via iStock / Getty Images Plus; (c) WildPictures / Alamy Stock Photo; (d) Bryan Reynolds / Alamy Stock Photo.

in eye types. Snail eyes range from simple light-detecting eye spots to clumps of cells that lie in a simple depression or pit. A pit allows for improved focusing versus a clump. Some snails have eyes with lenses allowing for better focusing. The shelled nautilus has a simple pinhole eye (Figure 6.15(a)), allowing for even better focusing. Modern squids and octopuses have full-fledged camera-eyes (Figure 6.15b) with lenses, corneas, and irises and a complex musculature that allows focusing and control of the light proportion. Squids and octopuses are fast-moving predators and are by far the most intelligent of the invertebrates.

Arthropod eyes followed a rather different evolutionary path and include the compound eyes that are shared by insects and trilobites and the multiple eyes of arachnids, which include the spiders and scorpions (Figure 6.15(c) and (d)). The observation of these intermediate gradations, from simple clumps of light-sensitive cells to full-fledged camera eyes, demonstrates that intermediary forms are indeed effective. Such intermediary eyes are effectively used by many organisms today. "Half an eye," in this sense, is highly useful, when compared to blindness – so useful that a variety of eyes have independently evolved and are still used today, providing excellent examples of convergent evolution in many different clades including mollusks, vertebrates, and insects.

6.3 Models of Evolutionary Change

Although evolution is a well-established observation of biology and geology, our understanding of the mechanisms and patterns are constantly being revised and improved. This section examines some recent examples of how natural selection works.

6.3.1 Natural Selection in Action

Natural selection is the mechanism by which evolution of biologic systems occurs. The selection that takes place favors traits in individuals that are more likely to produce offspring. In other words, something is considered an evolutionary advantage only if it makes reproduction more likely than some other trait. However, a trait cannot be evaluated for its evolutionary advantage without an environmental context. A trait that might be highly advantageous in one environment, for example, white fur for polar bears in the Arctic, could be a distinct disadvantage in another. Also, there is a constant tradeoff. More attractive coloration may make a mate more interested, enhancing reproduction, but flashy colors may make it easier to be spotted by predators. Energy put into making a thicker or spikier shell may provide more protection from predators but takes energy away that could be used for reproduction. An environment can be influenced by climatic conditions such as temperature and rainfall that can drive natural selection, but natural selection can also be driven by competition with other organisms, either for resources or as predators or prey, and of course Darwin was

exquisitely aware of evolutionary changes driven by sexual competition for mates. Over the course of this book, we will discuss in detail some examples from the rock record where a changing environment played a major role in biotic evolution over geologic time, but we will now discuss some examples of natural selection that have been observed within human timescales.

Peppered moths were a common sight in England in the first part of the nineteenth century. Their name comes from their mottled coloration (Figure 6.16) with mostly light-colored bodies and wings decorated with darker spots. Before 1811 this was the only form of peppered moth that had been observed, but in the 1820s dark-colored peppered moths began to be noticed. An 1895 survey of moths in the city of Manchester found 98% of the peppered moths were of the dark variety.

This dramatic change can be explained by genetic variation and environmental change. A part of the moth's DNA has now been observed that seems to be responsible for the variation in coloration. The environmental variation was associated with the industrial revolution that was fueled by the burning of coal, which polluted the environment with dark soot. Peppered moths are nocturnal and spend their days resting on trees or walls. Prior to the industrial revolution and the pollution it produced, these surfaces were mostly light colored, so the peppered coloration of the moths provided good camouflage against being eaten by birds and bats. However, the rapid increase in pollution associated with the increase of coal usage in the eighteenth century during the industrial revolution, gave the heretofore rare, dark-colored moths a much better chance of hiding from hungry birds. Surveys at the end of the century were overwhelmingly dominated by the dark variety.

When the atmosphere in London became filled with soot, the white trees became darker and light-colored moths were eaten

Figure 6.16 Peppered moths. Source: Ian Redding / Alamy Stock Photo

by birds more readily. Within months, darker moths became more common and lighter moths became rare. In the twentieth century, when alternative energy sources were used and coal was phased out, and when laws designed to curb pollution were enacted, tree trunks were no longer covered in soot and light-colored moths were again found in greater numbers.

This is a clear example of natural selection in action. Pollution didn't make the moths black, it was a genetic mutation within the general peppered moth population that resulted in a small number of moths having a darker color, but because these black moths were easily spotted prior to the industrial revolution their numbers were always low. But once the environment changed, this variation within the gene pool provided a major advantage to black moths, which then became selected for survival as birds began preying on white peppered moths, missing the dark variety that subsequently became dominant during the time of pollution. When the pollution disappeared, so did the dark moths' advantage and the lighter moths resumed their dominance.

Since then, careful field experiments have been done to specifically test how birds select peppered moths and these have confirmed this example. There have now been several experiments on various living species to test how natural selection works. Work on Trinidadian guppies (a common aquarium fish) noted variations in the degree of coloration of males. Guppies live in the steep rocky streams of Trinidad. In some stream reaches there are few to no predators and male guppies there are brightly colored. In this case, brighter-colored guppies have more success in attracting females enhancing their reproductive success. In stream reaches with predators, the bright colors make them easy targets for predators, so they show less coloration in this environment. Where there are predators, bright coloration is a liability. Experiments have now been done that take the drab-colored males and put them in areas with no predators and conversely place colored males in areas with predators. Within 30–60 generations the experiments showed that the males adapt to the new condition. Populations with brightly colored males with predators became drab colored and vice versa. Even more compelling, these experiments were done with sand or gravel river sediments. In sandy areas with no predators, guppies developed larger gravel-sized spots, more attractive to females, but where there were predators, the mottling tended to be smaller and more sand-size, providing camouflage. These and other experiments show that natural selection can cause variation over several tens of generations, which is quite fast (much faster than can be recorded in the geologic record).

The environment of many hospitals also shows examples of natural selection active today. Strong use of antibiotics mostly does what it is supposed to do – kill bacteria – but bacteria reproduce extremely quickly, allowing the population to adapt to external stresses, such as antibiotics. In any bacterial population, even if only one in a billion bacteria has a mutation that is resistant to the antibiotic, that strain will quickly become dominant if it is exposed to antibiotics. It is now well known that some of the most infectious bacteria often occur in hospitals because of ubiquitous use of antibiotics. This has driven the population of bacteria in hospitals to be dominated by strains for which the antibiotics are not effective. In a most recent example, the more virulent mutations of the Covid-19 virus quickly became dominant as vaccines produced an evolutionary pressure against the less virulent varieties.

KEY POINT

Changes in environments are the main driver in changes in life on Earth. Experiments and observations on living populations demonstrate how selection occurs.

6.3.2 Extinction and Speciation

One of the hallmarks of the rock record is the sudden disappearance of old species and the appearance of new species. The disappearance of an organism marks its extinction, and the appearance of new organisms marks a speciation event. At any given time, some organisms are going extinct and new organisms are appearing. If the rate of speciation matches the rate of extinction, then the diversity of life on Earth should be relatively constant, but at certain times the rate of speciation has been greater than extinction. These bio-diversification events have happened a few times in Earth history such as during the Cambrian Explosion and the Great Ordovician Biodiversification Event, which we will review in Chapter 9. At other times in Earth history the rate of extinction has greatly exceeded the rate of speciation, especially during mass extinctions, such as characterized the end of the Permian Period and the end of the Cretaceous Period, which we will discuss in Chapters 11 and 13.

6.3.2.1 Phyletic Gradualism

The evolutionary change of a type of organism over time is called *phylogeny*. Since geologists first began examining the rock record, it was observed that any given formation might contain an assemblage of fossils with very little variation within the formation but with rather different looking fossils in older and younger layers. Across some boundaries, the change in fossil types was modest, but across others it is profound (see Chapters 11 and 13). For example, across the Permian–Triassic boundary entire groups of fossils, such as trilobites completely disappeared. We now attribute these sudden and widespread extinctions to major environmental change, but Darwin was quite clear that he thought his theory of natural selection involved a slow process of change of species by selection of what might seem trivial differences that accumulate into large changes over geological time periods across multiple

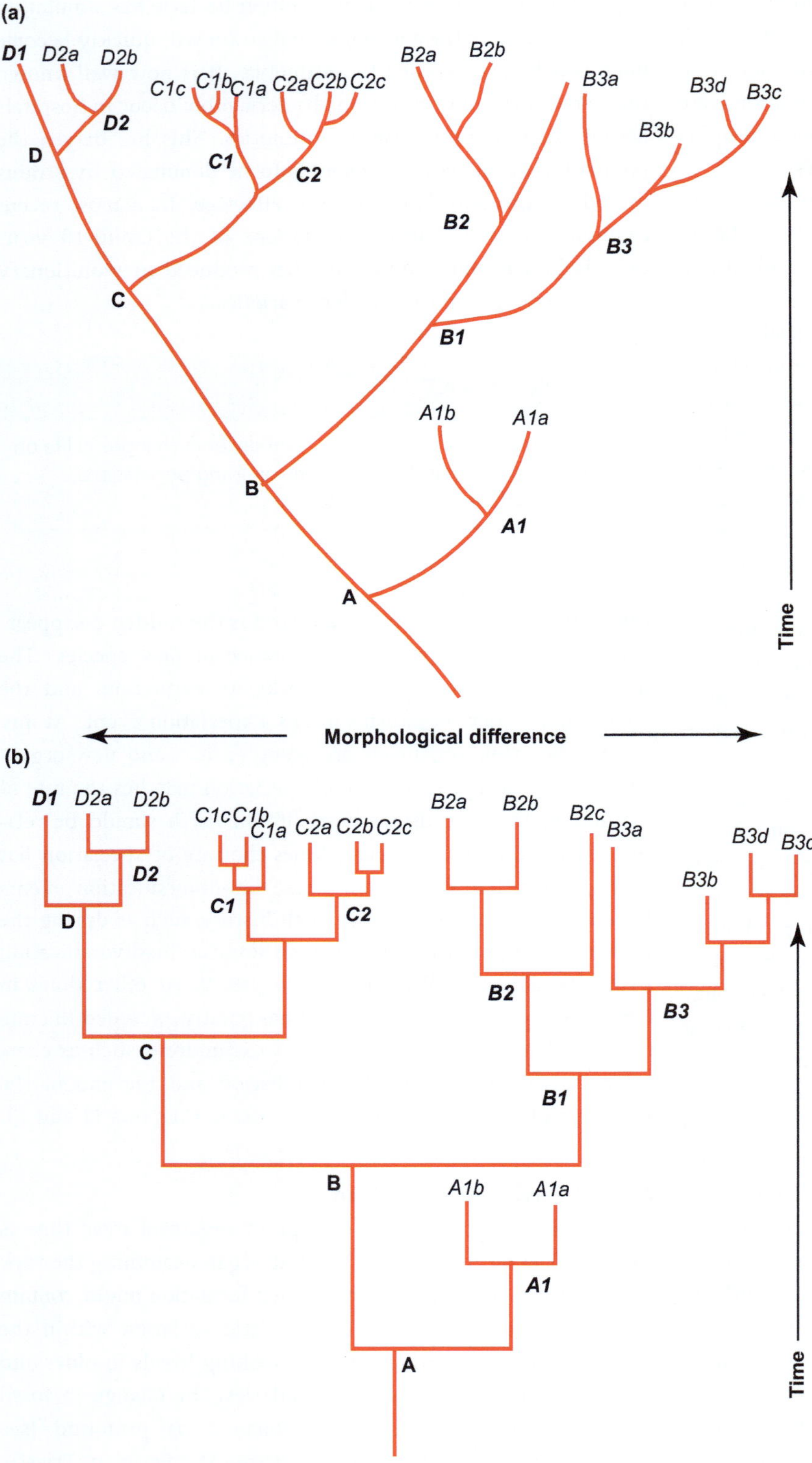

Figure 6.17 Models for evolution: (a) cone-shaped phyletic gradualism, indicating more gradual diversifications versus (b) step-like punctuated equilibrium. The letters represent hypothetical phyletic groups with species at the tops of the cladograms. Note that the cladograms show the same number of branches and species.

generations. Darwin attributed the abruptness of the fossil record to its highly incomplete nature, reflecting the difficulty in fossilizing life. This led to the model of **phyletic gradualism**, in which species change gradually (Figure 6.17(a)). This model assumes that change is always slow, and therefore, any abrupt changes in the fossil record will be the result of poor preservation of the intermediate species.

Phyletic gradualism also had a philosophical underpinning that guided the theory. The view of many nineteenth-century scientists was that humans were the pinnacle of evolution and that the purpose of evolution was an ever-improving progression that culminated in human beings. This suggests that evolution was progressing towards an ultimate goal (humans) and that all other organisms, however well suited for the environment they were living in, were simply links in the "great chain of being" that unrelentingly and gradually strove towards the ultimate goal, which is us.

6.3.2.2 Punctuated Equilibrium

An alternate view to phyletic gradualism is the idea that, for most populations and species, **stasis** (or equilibrium) is the long-term nature of things, and that evolution of organisms occurs in brief spurts, or punctuations, when stresses in the environment reach a tipping point beyond which organisms cannot adapt and survive. Steven Jay Gould and his colleague Niles Eldridge, curator at the American Museum of Natural History, used this line of reasoning to formulate the idea that evolution follows a **punctuated equilibrium** model (Figure 6.17(b)) and argued that this better explains the abruptness of changes seen in the fossil record. Punctuated equilibrium emphasizes the importance of environment and explains abrupt changes in the fossil record as the consequence of abrupt changes in the environment.

The idea is that in environments that are stable, organisms will typically become more diverse and more specialized, with highly optimized adaptations to sometimes peculiar ecological niches. Extreme specialization can be very good in the short term, but a serious liability for survival when environments change.

For the most part, paleontology supports the model of punctuated equilibrium, as there are significant stretches of the stratigraphic record marked by only modest change followed by brief intervals over which major change is observed. This model, then, assumes that fossils changes are faithfully recorded in the rock record as opposed to the assumption of phyletic gradualism, in which the fossil record is considered to be too flawed to illustrate the gradual change that was always occurring. However, some portions of the fossil record are consistent with phyletic gradualism, especially within-clade evolution that occurs between major extinction events. At any given time, there is always some inter-species and intra-species competition. In the dinosaurs, for example, such competition selected for larger sizes and allowed dinosaurs to evolve into giants, whether it was between predators and their prey or fighting within a species for the right to reproduce (see Chapter 12 for more details on dinosaurs).

6.3.3 Genes are the Agent of Evolution

There have been debates as to the fundamental level at which natural selection works. Does it primarily work at the level of species, a population or social group, or at the level of individuals, or at the genetic level? Richard Dawkins, professor of biology at Oxford University, has promoted the idea that selection primarily occurs at the genetic level. Dawkins was particularly interested in explaining the evolution of behaviors such as altruism, where one individual may make a sacrifice beneficial to another, and seemingly harmful to the giver. At face value, this would seem to be at odds with the concept of evolution as being the "survival of the fittest in the struggle for existence." Dawkins developed statistical models to show that many altruistic behaviors actually benefit the giver when measured by the longer-term success of genes of the giver and their offspring, and that altruistic behavior results in better survival of the genes of both giver and taker. This supports the idea that evolutionary processes operate at the level of genes. In his models, altruistic or "selfless" behaviors were demonstrated to be beneficial to the genes of the giver and their offspring, even if the giver died in the process of reproducing.

6.4 Limiting Factors

Earth and its environment have changed over the vastness of geological time, with a variety of effects on the biosphere and the history of life on Earth. Environmental changes include physical changes in the position and elevation of continents and oceans due to plate tectonics (discussed in Chapters 5, 10, 14, and 15), as well as climate and sea level (Chapters 8, 11, 16, 18, and 19). In addition, atmospheric and oceanic chemistry have changed throughout Earth history with sometimes severe consequences for the biosphere (see Chapters 3, 7, 8, and 11). Some of these changes, such as the position and elevation of plates, occur quite slowly, over millions to tens of millions of years, whereas others, such as climate, may change over tens of thousands of years. Rare extreme events, such as massive volcanism or extraterrestrial impacts, have caused almost instantaneous change, a few of which resulted in catastrophic mass extinctions (see Chapters 11 and 13).

Any given species on Earth is typically adapted to a range of **limiting factors**, such as temperature, salinity, pH, oxygen levels, food resources, and so on. If these limits are exceeded, organisms must either adapt, move, or die. Earth processes also commonly separate and isolate populations, which favors allopatric speciation. It's hard to produce new species if too much interbreeding occurs. The isolation of populations of finches to different islands of the Galapagos with differing climates and resources, limited interbreeding between islands and allowed different species to evolve adapted to the unique conditions on each island (Figure 6.7). Over longer periods, as plates drift, environments can change, forcing organisms to either adapt or migrate to more hospitable areas. Limiting factors play a large role in determining what drives extinction. For example, as the last ice sheets receded, about 11,000 years ago, animals that were adapted to the previously colder climates, such as penguins, polar bears, and mountain goats, either migrated towards the poles or became confined to higher elevations. Some organisms, such as mastodons and woolly mammoths, went extinct, but as we shall see in Chapter 17, hunting by humans may have played a role in their extinction.

Deciduous plants were able to recolonize areas that were formerly covered by ice, and similar patterns of population migration occurred in the oceans. As a result of global warming and increases in ocean acidity, which we discuss in detail in Chapter 19, reef-building organisms are under severe stress, and about 30% of the world's reefs have been destroyed. Global warming also promotes migration of equatorial species, such as insects like mosquitos, to higher altitudes and latitudes, bringing diseases such as the *Zika virus* with them.

Throughout the world's history, life on Earth has had to find a balance between adapting to the constantly changing environment versus extreme specialization that can make organisms hugely successful over shorter time periods but places them at risk when changes in limiting conditions occur. Evolution and extinction are best understood by studying the entire Earth system. Although mutations are required for genetic variability, adaptations and evolution of new species to changes in limiting factors are commonly an important driver of natural selective pressures that select genetic mutations that are favorable in the struggle for existence.

6.5 What Darwin Didn't Know

Darwin's theory of natural selection included several conditions:
- it required a mechanism to replicate and transfer genetic information, and a mechanism for that genetic information to vary;
- it required that Earth have a very long history;
- it required that environmental conditions change, thus creating natural selective pressures; and
- it assumed that there were transitional forms between species.

It is worth noting that in his time very few of these conditions were quantified. The age of the Earth could only be guessed at, and it wasn't until the 1950s that we had a robust quantitative measure that the Earth was 4.5 billion years old (see Chapter 4). Darwin would have been thrilled at this result with more than enough time for his process to work its natural wonders.

Gregor Mendel discovered the genetic basis of inheritance in 1866, but his work was largely ignored in Darwin's time. It was only in 1953 that Watson and Crick broke the genetic code and demonstrated the nature of DNA as a coiled molecule that can split and join. Genomics is now a robust field of biological research, and we now know how DNA molecules split apart and recombine during sexual reproduction, allowing the propagation of genetic variations as well as showing how organisms are related based on genetic homology. We also know how DNA in genes controls the growth and development of organisms from cells to embryos to adults and how small changes in instructions in the early stages of embryological development can cause significant changes in the resulting adult, some of which may be favorable. Darwin would likely have rejoiced at these discoveries, but in his own time he could only guess at hereditary mechanisms.

With the discovery of rare strata that preserve soft bodies, such as the Cambrian Burgess Shale in Canada (see Chapter 9), many transitional forms of countless related species have now been discovered, such that the fossil record is much more complete than it was in Darwin's day. Darwin would almost certainly be delighted that his theory of natural selection has stood the test of time and scrutiny by many generations of scientists.

6.6 Summary

- The idea that biologic systems have changed (i.e., evolved) over geological time is not a theory, but rather an observation as recorded in the fossil record.
- Charles Darwin and Alfred Wallace introduced the theory of natural selection to explain the observation of evolution. Organisms are selected for survival by the preservation of favored traits that enable them to adapt to changing environmental conditions.
- Genetic and anatomical homologies are used to reconstruct evolutionary pathways and to determine how life on Earth is organized. Cladistics is a way of organizing, grouping, and classifying organisms according to evolutionary relationships.
- Anatomically imperfect adaptations and vestigial anatomical and genetic structures can be used to decipher evolutionary relationships.
- Molecular genetic clocks, calibrated against the fossil record can be used to determine when organisms shared a common ancestor.
- Genes and mutations provide the biological underpinning that allows evolutionary change, but much of the pressure that drives natural selection are external changes in limiting factors that are in turn caused by changes in the position and elevation of plates, as well as changes in climate and ocean and atmospheric chemistry over a variety of timescales.
- Many of the factors that are critical in explaining the diversity of the fossil record, which requires a very old Earth, as well as the genetic basis for speciation were unknown in Darwin's time. Subsequent discoveries in geology and biology have been consistent with this and allowed elaboration of Darwin's theory of natural selection.

Key Words

- evolution
- artificial selection
- natural selection
- mutations
- survival of the fittest
- homology
- DNA
- clade, cladistics, cladogram
- phylogenetic
- stem organisms
- crown organisms
- allopatric speciation
- deuterostomes
- protostomes
- amniotic
- domains: Archaea, Bacteria, Eukarya
- genomics
- chromosome
- telomere
- molecular clocks
- imperfect adaptations
- vestigial anatomical structures
- phyletic gradualism
- stasis
- punctuated equilibrium
- limiting factors

Further Reading and References

Darwin, C., 1859, *On the Origin of Species*. Murray.

Dawkins, R., 1986, *The Blind Watchmaker*, Norton.

Dawkins, R., 2009, *The Greatest Show on Earth*, Bantam Press.

Dawkins, R., 2016, *The Selfish Gene*, 4th ed., Oxford University Press.

Gould, S. J., 1980, *The Panda's Thumb*, Norton.

Mark, M., Rijli, F., and Chambon, P., 1997, Homeobox genes in embryogenesis and pathogenesis, *Pediatric Research*, 42, 421–429, https://doi.org/10.1203/00006450-199710000-00001.

Schreiber, A., and Gimbel, S., 2010, Evolution and the second law of thermodynamics: Effectively communicating to non-technicians, *Evolution: Education and Outreach*, 3, 99–106. https://doi.org/10.1007/s12052-009-0195-3

Shubin, N., 2020, *Some Assembly Required*, Pantheon Books.

Review Questions

1. What is artificial selection and how did it provide Charles Darwin with the idea of natural selection?
2. What were some of the main observations that Charles Darwin used in developing his theory of natural selection?
3. What is some of the main anatomical and genetic evidence that evolution occurs by modification during descent from common ancestors?
4. What are Hox genes and what role do they play in evolution?
5. What is a clade? How are cladograms defined and constructed and how do cladistics help understand the organization of life from an evolutionary perspective?
6. In cladistics, what is the difference between a stem and a crown group?
7. What is convergent evolution and why do different organisms commonly show similar adaptations, such as flight in insects, bats, and birds?
8. Explain the difference between sympatric and allopatric speciation. Which type of speciation is likely to leave a fossil record?
9. What is homology? Give examples of anatomical and genetic homologies.
10. What are molecular (genetic) clocks?
11. How do imperfect adaptations (such as whales need to breathe air) and vestigial anatomical structures (such as buried rear limb bones in whales) provide evidence of evolution?
12. Evolution deniers commonly provide critiques such as "What use is half an eye?". What is wrong with this statement?
13. Contrast phyletic gradualism and punctuated equilibrium in explaining how evolution occurs.
14. What kind of Earth changes affect limiting factors for life?
15. What new discoveries, unknown in Darwin's time, have helped explain and bolster his theories?

Artistic reconstruction of the Hadean Earth four billion years ago. Source: Bruce Damer and Ryan Norkus, used with permission.

Chapter 7

The Origin of Life

Our Early Atmosphere and the Rise of Early Life

LEARNING OBJECTIVES

- Define life and its susceptibility to evolution.
- Describe the conditions required for habitability and the key fossil evidence for the earliest life.
- Explain the heterotrophic theory for the origin of life and the steps required for the formation of the first living cell.
- Explain why amphiphilic polymers are critical in forming protocells.
- Describe some experiments designed to replicate some of the processes required to build a cell.
- Contrast the shallow water hydrothermal field versus deep water hydrothermal vent hypotheses for the formation of the first cells.
- Compare top-down versus bottom-up approaches to understanding the origin of life.
- Explain the evolution of photosynthesis and the role of photosynthetic life in causing the Great Oxygenation Event.
- Explain how eukaryotes evolved from endosymbiosis with prokaryotic ancestors.

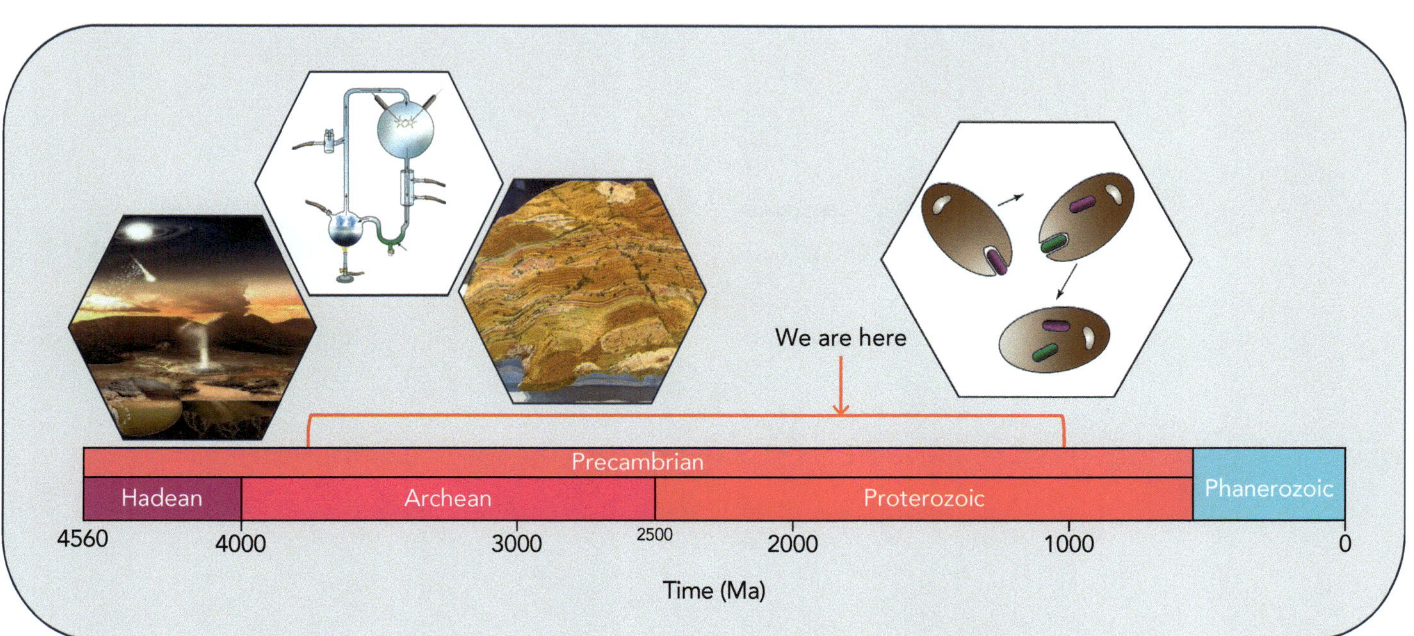

Introduction

NASA defines life as a self-sustaining chemical system capable of Darwinian evolution. Attributes of life typically include an organized internal structure surrounded by a membrane, the ability to regulate the internal environment of the organism (or cell), metabolism that allows growth and reproduction, and the ability to adapt to external change over time. The last point is key to the process of evolution that has driven the development of life on Earth, as introduced in Chapter 6. Even the most primitive prokaryotic bacterial cells are exceedingly complicated and include an external membrane and complex molecules, such as DNA and ribosomes that manufacture proteins.

How life originated from non-living matter on early Earth is one of the biggest questions in science. Fossil evidence tells us that life appeared relatively early in the history of our planet, and there are several hypotheses about how and where life appeared, including Darwin's idea that life may have begun in the "**primordial soup**," in a *shallow freshwater pond*, as well as the competing idea that life may have first appeared in association with *deep water oceanic hydrothermal vents* (Figure 7.1). This chapter reviews these ideas of how life may have evolved. This chapter also discusses some of the main evolutionary steps of the first 2.4 billion years of life, such as the appearance of more complex eukaryotes from smaller prokaryote cells (defined in Chapter 6) and the appearance of photosynthetic life, which fundamentally changed our atmosphere during the Great Oxygenation Event.

KEY POINT

The first cells required an internal structure within a membrane that allowed regulation of metabolism and reproduction, and the ability to adapt to external change over time, which is a fundamental driver of evolutionary processes.

7.1 Darwin's Idea: A Warm Little Pond

In his 1859 book, *On the Origin of Species,* Charles Darwin hypothesized that "all the organic beings which have ever lived on this Earth have descended from some one primordial form, into which life was first breathed." At the time of writing, Darwin shied away from further explanation, but by 1871 he wrote:

> it is often said that all the conditions for the first production of a living being are now present, which could ever have been present. But if (and oh what a big if) we could conceive in some warm little pond with all sort of ammonia and phosphoric salts, – light, heat, electricity present, that a protein compound was chemically formed, ready to undergo still more complex changes, at the present such matter would be instantly devoured, or absorbed, which would not have been the case before living creatures were formed.

Here Darwin outlined the basics for a laboratory experiment to investigate how life could have appeared in the conditions that characterized early Earth, a warm pond containing ammonia,

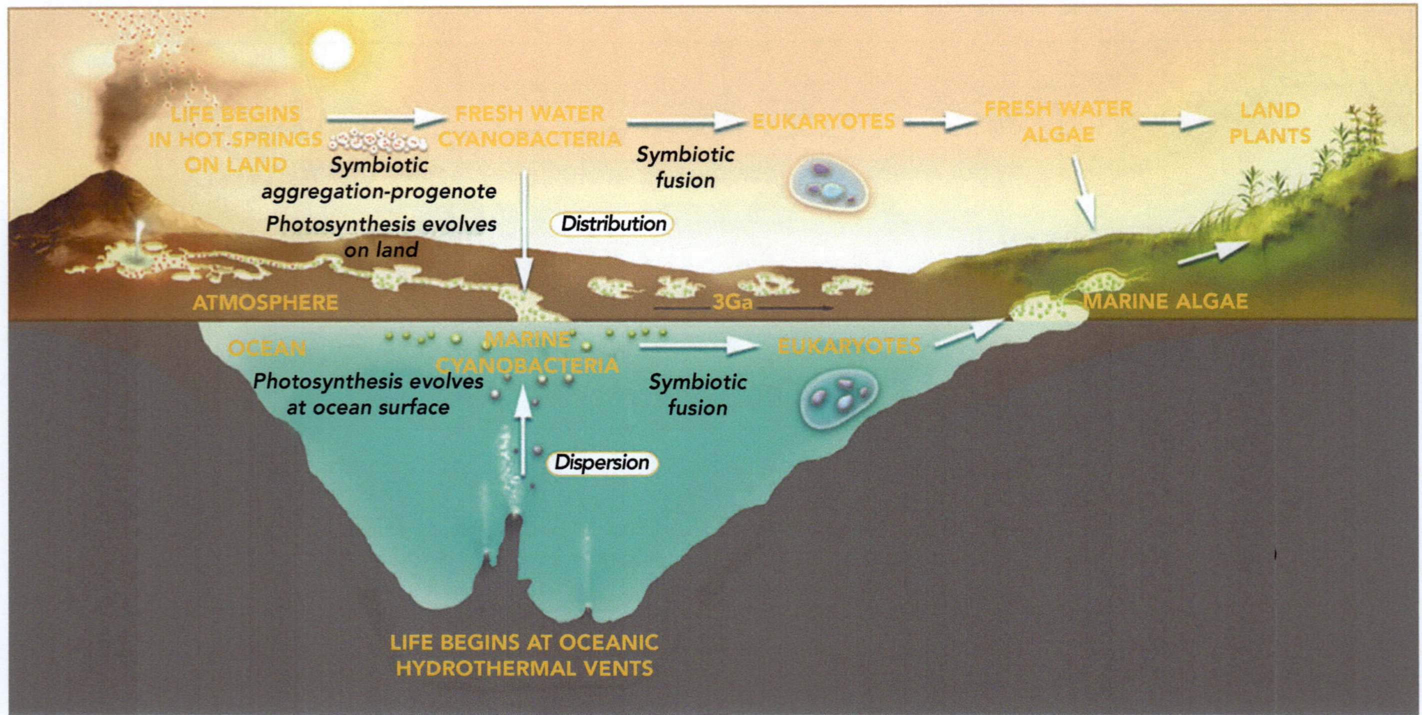

Figure 7.1 Alternative scenarios for an origin of life and adaptive pathways from freshwater hydrothermal field pools or saltwater hydrothermal vents to eukaryotes and land plants. Source: Bruce Damer and Ryan Norkus, used with permission.

phosphoric salt, light, heat, and electricity to drive a reaction that could create the first proteins. Darwin pointed out that if such an experiment were performed today, any "protein" compound formed in the experiment, would be "be instantly devoured, or absorbed" because we live on an Earth teaming with life. That is in stark contrast to an early Earth without life, where such protein compounds would not instantly be devoured but rather might undergo more complex changes that could have led to the first cell.

7.2 When Was the Earth Habitable?

One of the challenges in deciphering the origin of life is that Earth has a long history, and because of plate tectonics, the record of early Earth is rather poor. Most of the oldest rocks on Earth are highly metamorphosed, and metamorphism tends to destroy fossils that contain the evidence of earliest life. Before we can discuss the processes and environments associated with formation of the first cell, we must first ask, "When was Earth habitable?" As discussed in Chapter 3, the first crust and hydrosphere probably formed around 4.52–4.42 Ga, soon after cooling of the magma ocean generated by the impact with Theia that formed the Earth–Moon system (Figure 3.8). The Hadean Earth would have been covered by a deep, salty ocean punctured by large volcanic islands (Figure 7.2). Evaporation of ocean water would have resulted in significant rainfall and some of this would have collected as freshwater rivers and ponds on the emergent volcanic islands. Hydrothermal alteration of the porous volcanic rock would also have created secondary minerals, including clays, that would have plugged some of the pore space, preventing all of the fresh water from infiltrating the rock and allowing ponds to become stable. Hydrothermal emissions would have provided some chemicals as well as dissolved minerals, such as apatite ($CaPO_4OH$), which would have provided phosphate – a critical component of many organic molecules required for life, such as ADP (adenosine diphosphate). In this scenario we have the components that Darwin considered crucial in forming the first proteins.

A complicating factor is the hypothesized period of **Late Heavy Bombardment** (LHB), which may have lasted from 4.1 to about 3.9 billion years ago. The LHB hypothesis was largely developed based on dating of moon rocks collected during the Apollo missions, that showed a dominant age of 3.9 Ga, suggesting that the surface of the Moon was largely molten up to that point as a consequence of these late-stage impacts, although this is currently a topic of some debate. Life could have evolved as early as 4.5 Ga in the Hadean Eon when the first hydrosphere and crust was available, or as late as 3.9 Ga after the LHB. If life formed in the Hadean Eon there are questions as to how it could have survived the LHB. Given the abundance of life around deep water hydrothermal vents, as well **endolithic** (within rock) communities found kilometers below the surface in deep mine shafts, it is possible that life survived the LHB by taking refuge deeper in the lithosphere while "hell" was breaking loose at the surface. The observations of these hydrothermal vent communities in the deep oceans also led to the **autotrophic**

Figure 7.2 Artistic reconstruction of the Hadean Earth four billion years ago. Source: Bruce Damer and Ryan Norkus, used with permission.

hypothesis that life may have actually first appeared in these hot, deep water environments, driven by hydrothermal energy and the associated minerals such as pyrite in a reducing environment, and only later exploited Earth's surface.

KEY POINT

Earth had standing water soon after its collision with Theia that formed the Moon, but it probably wasn't fully habitable until after the period of Late Heavy Bombardment, which ended about 3.9 Ga.

7.3 Evidence for the Oldest Life

Some of the oldest generally accepted fossil evidence for the earliest prokaryotic cells is found in Archean rocks of the Pilbara Shield in northern Australia (Figure 7.3). The clearest example lies within cherts of the 3.43 Ga Strelley Pool Formation. Microscopic examination revealed clusters of cells with distinct membranes (Figure 7.3(b), (c), and (d)). The fossilized cells lie within finely laminated structures called **stromatolites** (Figures 7.3 and 7.4).

Figure 7.3 (a) Stromatolites from the 3.43 Ga Strelley Pool Chert, Pilbara Shield; (b), (c), and (d) show photomicrographs of the bacterial cells contained within, including (b) chains and (c) triplets; (d) shows the hollow nature and distinct cell wall indicating a membrane. Source: (a) Photo by JPB. With permission of ROM (Royal Ontario Museum), Toronto, Canada; (b), (c), and (d) Reprinted from Sugitani et al. (2015), with permission from John Wiley and Sons.

Stromatolites typically form in shallow water by the growth of **filamentous bacteria** that trap and bind fine particles of sediment (Figure 7.4). The bacteria form a sticky sheet, called a *matground*, which traps and binds sediment. The bacteria then rise through the sediment layer, forming another matground, and the process continues, eventually building a laminated deposit (Figures 7.3(a) and 7.4(a) and (b)). Somewhat older cherts of the Dresser Formation lie below the Strelley Pool Formation. Dated at 3.48 Ga, they also contain stromatolites, although they don't display clear cell structures. The Dresser Formation indicates deposition in hydrothermal pools on the flanks of an ancient volcano, and indicates formation on land versus the sea. Even older possible stromatolites have been found in the 3.7 Ga Isua Belt in Greenland, but the interpretation of these as biogenic remains controversial.

Given the youngest age that the Earth is considered to have been habitable is around 3.9 Ga, following the hypothesized LHB, gives about 200 million years until the first possible fossil evidence for life. Alternatively, if life appeared earlier, in the Hadean (4.5 Ga), that allows 800 million years of evolution before the first fossils appear.

The chemical and isotopic composition of some rocks and minerals (so-called "chemical fossils") can also be used to diagnose biotic processes and may actually pre-date the age of the oldest recognized fossils. For example, organic processes tend to incorporate the lighter isotope of carbon, carbon-12, versus heavier carbon-13 in their cells, so organic matter that is organic in origin will commonly have a higher proportion of carbon-12 (see Box 7.1). Measurement of the carbon isotope ratios of carbon associated with Strelley, Dresser, and Isua stromatolites all show depletion suggesting an organic origin.

In addition, as discussed in Chapter 6, it is now possible to calibrate the genetic code of living organisms to determine how closely related any given group of organisms are. In addition, assuming the rate of mutation is relatively constant, the diversity of genomes can be used as molecular clocks to determine how long ago groups of different organisms most likely shared a common ancestor. Genomic analysis suggest that the Bacteria are likely the earliest domain of life and that the Archaea split off later (Figure 7.5). These clocks suggest that the last common universal ancestor to all life may have originated around 4.2 Ga, which pre-dates the end of the Late Heavy Bombardment period. The split between the Archaea and Bacteria likely occurred in the early Archean (Figure 7.5), whereas the first eukaryotes probably appeared around 1.65 Ga in the Proterozoic Eon.

KEY POINT

The oldest unequivocal fossils are 3.43 Ga stromatolites from Australia, although there is some evidence of 3.7 Ga fossils. Molecular clocks indicate that life could have first appeared around 4.2 Ga before the end of the period of Late Heavy Bombardment.

7.4 Assembling the First Cell: Theory and Experiments

The fossil and geological evidence provide some constraints on when life could have appeared on Earth but still doesn't tell us much about how it occurred. Assembling the first cell requires a combination of physical and chemical processes and these have

Figure 7.4 Stromatolites: (a) Neoproterozoic, Miraflores Formations, Bolivia; (b) *Collenia symmetrica*, 1.1 Ga, Montana. Note that the curving laminated stromatolites overlie less curved layers, indicating that the curved layers are biological not the result of structural folding. (c) Modern stromatolites, Shark Bay, Australia. Source: (a) Photo by JPB. With permission from Houston Museum of Natural Science (HMNS); (b) Photo by JPB. With permission of ROM (Royal Ontario Museum), Toronto, Canada; (c) chameleonseye via iStock / Getty Images Plus.

BOX 7.1 Carbon Isotopes

Carbon isotopes are widely used in geochemical analysis. Carbon has three main isotopes, carbon-12, carbon-13, and carbon-14. Carbon-14 is radioactive and used for radiometric dating. The most common isotope is carbon-12 (99%) and about 1% is carbon-13. Most organisms preferentially incorporate lighter carbon-12 in their bodies, and photosynthetic organisms also preferentially incorporate carbon-12. Carbon isotopes are typically measured as a ratio that is compared to a global standard sample:

$$\delta^{13}C = \left(\frac{\left(^{13}C/^{12}C\right)\text{sample}}{\left(^{13}C/^{12}C\right)\text{standard}} - 1 \right) \times 1,000\%$$

The global standard has a high $^{13}C/^{12}C$ ratio and most natural materials yield a negative value. Samples depleted in ^{12}C give increasingly negative values.

---

been tested through experiments and modeling, within the constraints of our understanding of the nature of the Hadean and early Archean conditions on Earth described above.

7.4.1 Heterotrophic Theory for the Origin of Life

In the early twentieth century, Russian scientist Alexander Oparin and English scientist J. B. S. Haldane independently formalized Darwin's idea that life evolved in a warm little pond as the **heterotrophic origin of life theory**. The term *heterotrophic* refers to metabolic processes that use organic compounds of nitrogen and carbon, which today are obtained from plant or animal matter. Heterotrophs obtain their food from the external environment versus *autotrophs* that make their own food, like modern plants.

The theory assumes that early **pre-biotic** Earth had a highly reducing atmosphere, rich in carbon dioxide (CO_2), methane (CH_4), ammonia (NH_3), and hydrogen (H_2). The first step would involve mixing of these compounds with energy and water (H_2O) to form simple organic compounds called **monomers** (simple molecules consisting of a single part), and particularly amino acids that are essential to build the more complex macromolecules, called **polymers** (molecules composed of many parts), such as phospholipids (Figure 7.6), ribonucleic acid (RNA), deoxyribonucleic acid (DNA), as well as the proteins that are essential for life. The formation of polymers is thought to require concentration of monomers probably in shallow ponds where wetting and drying can occur, forming a "pre-biotic soup." Energy, either from the Sun, lightning, or other sources, results

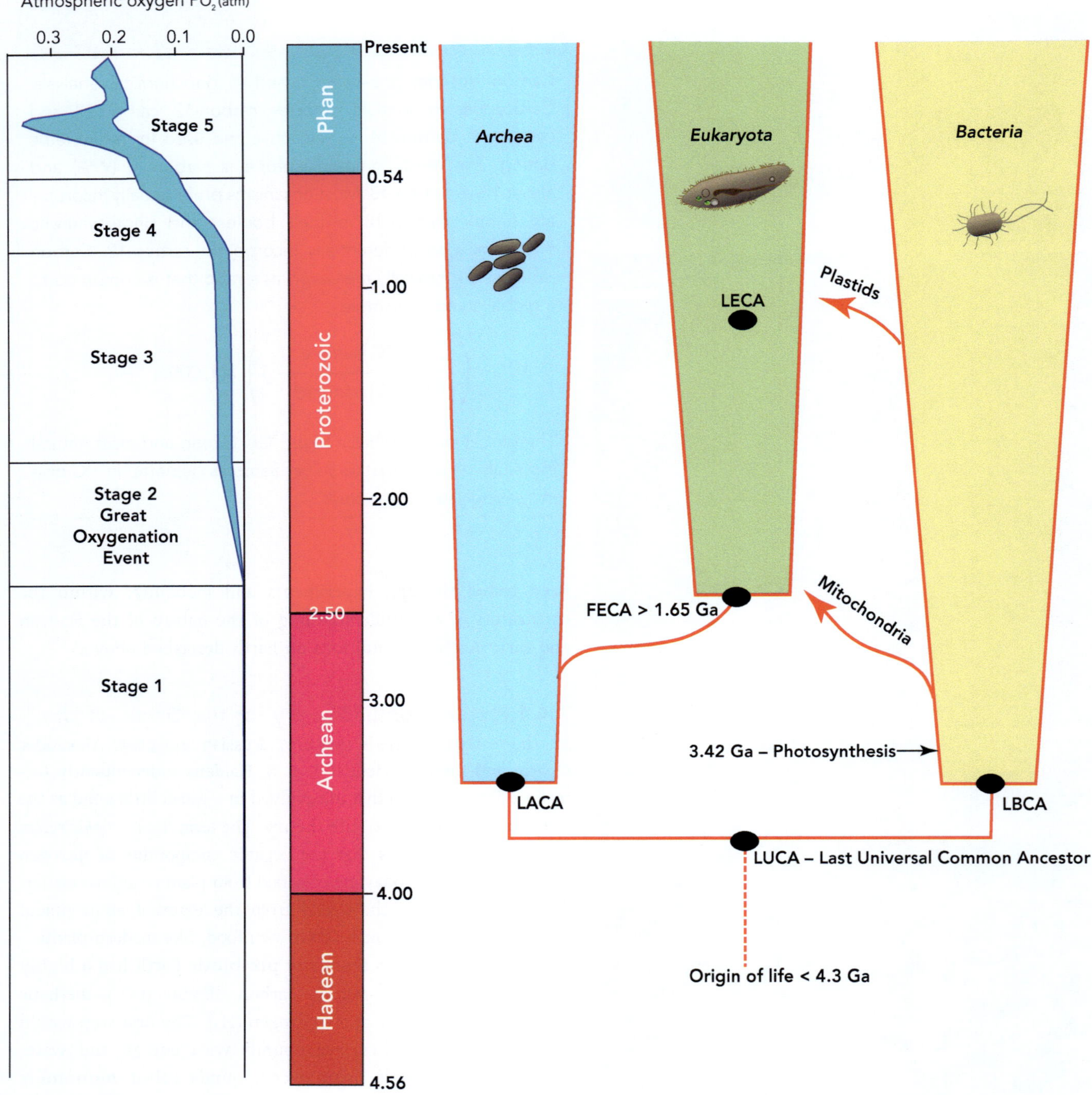

Figure 7.5 Timeline for the origin of the three main domains of life and their genetic relationships. LACA – last archaeal common ancestor, LBCA – last bacterial common ancestor, FECA – first eukaryotic common ancestor, LECA – last eukaryotic common ancestor.

in an **open system** that can help drive chemical reactions. Some long-chain molecules, such *lipids*, which include fatty acids and oils, are *amphiphilic*, that is, one end of the molecule is *hydrophilic*, preferring water, and the other is *lipophilic*, preferentially sticking to oils or fats (Figure 7.6). Detergents and soaps are common amphiphilic substances. One side of the soap molecules sticks to the grease on your fingers and pulls it off while the other side attaches to the water that you wash your hands with, carrying the grease away. In natural environments, amphiphilic chain molecules may be concentrated to form layers or balls with the hydrophilic end on the outside of the layer and the lipophilic side on the inside (Figure 7.6). In some cases, the two molecules will form a bi-layered ball (a *liposome* or *vesicle*) that can serve as a type of proto-cell container for chemical reactions to occur. Eventually, selection of protocells with a higher ability to contain reactions that could produce metabolic processes and eventually the ability to replicate are thought to have led to the first bona fide cells. Many of these steps have been reproduced in lab experiments, a few of which are described below, giving confidence that the same may have occurred in a natural environment.

KEY POINT

Amphiphilic organic polymers are able to form cell-like vesicles that are essential in forming cells and are relatively easy to reproduce in lab experiments.

7.4.2 Experiments on the Origin of Life

In 1952 Stanley Miller, assisted by Harold Urey at the University of Chicago, made the first experiment to test the heterotrophic, shallow water hypothesis (Figure 7.7). They assumed that the early Earth had a highly reducing atmosphere, and placed methane (CH_4), ammonia (NH_3), and hydrogen (H_2) in a sealed flask. Then they injected water vapor into the flask with the gases and next added energy by exposing the mixture to electrical sparks (Figure 7.7). The vapor was then cooled, and the condensed fluid was collected and analyzed. Within a day, the condensed liquid showed a pink color, indicating that some sort of chemical reaction had occurred. After a week, the fluid was extracted and sterilized to prevent any contamination with bacteria that might be present in the lab. This step addressed Darwin's concerns that any chemically formed "proteins" formed today would "be instantly devoured or absorbed." Analysis of the extract revealed that five **amino acids**, molecules essential for life, had been synthesized, including glycine (NH_2-CH_2-COOH) and alanine ($C_3H_7NO_2$). Following Miller's

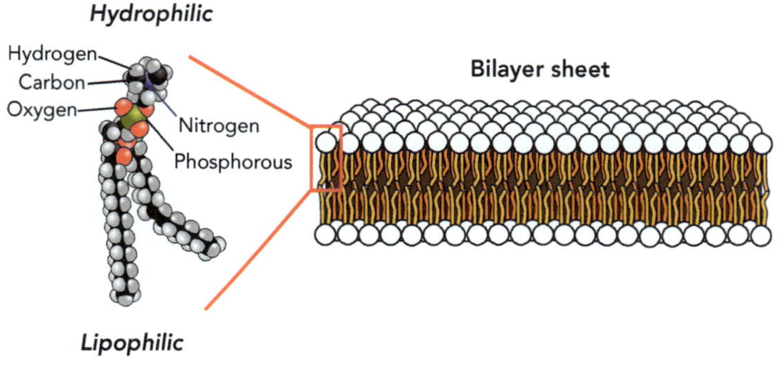

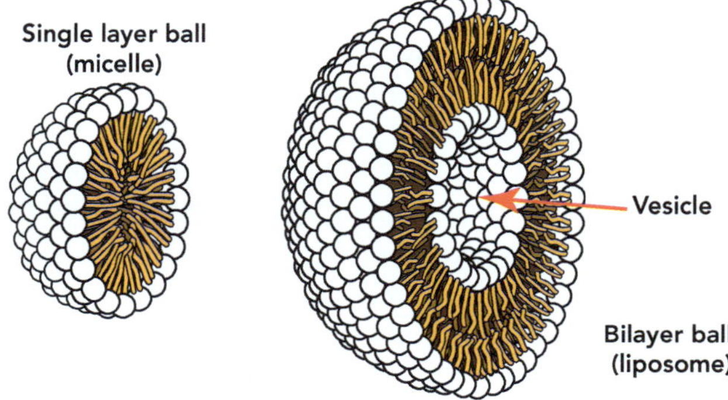

Figure 7.6 Amphiphilic polymer molecule consists of a hydrophilic (water-loving) phosphate molecule, shown as a circle, attached to a pair of fatty acid chains (lipids) that avoid water but are attracted to other fats. They can combine to form a sheet or a ball. Bilayer balls have a hollow interior that can contain reagents and may have been the precursors to the first cells. Source: First image on the left is from Cell_membrane_ detailed_diagram_4.svg: *derivative work: Dhatfield (talk) Cell_membrane_detailed_diagram_3.svg: *derivative work: Dhatfield (talk)Cell_membrane_detailed_diagram.svg: LadyofHats Mariana Ruiz derivative work: Mihaf, CC BY-SA 3.0 <https://creativecommons.org/ licenses/by-sa/3.0>, via Wikimedia Commons; all other images are from LadyofHats, Public domain, via Wikimedia Commons.

(a)

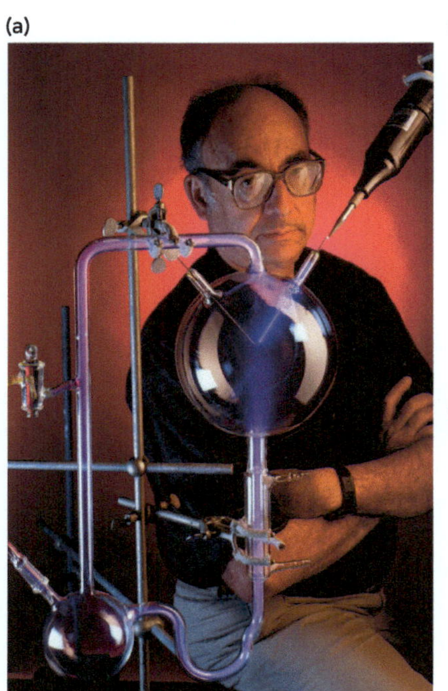

(b)

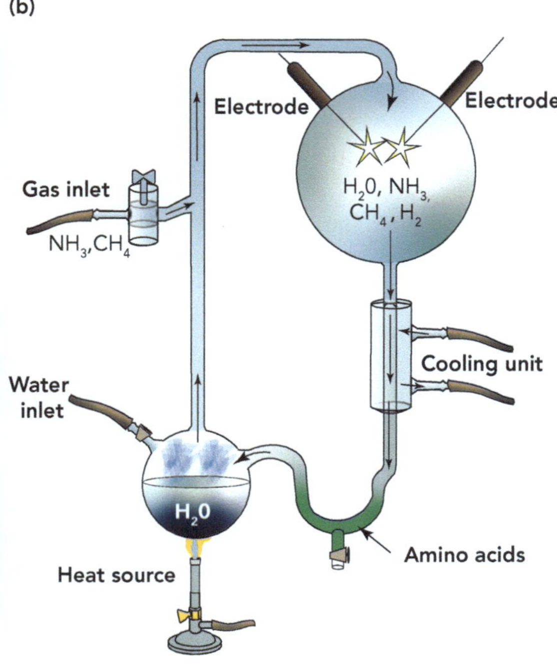

Figure 7.7 The Miller–Urey experiment recreated the assumed conditions on early Earth in an attempt to see if organic molecules could be made on a pre-biotic Earth. Source: (a) Jim Sugar / Collection: The Image Bank Unreleased / Getty Images.

death in 2007, further analysis of his samples revealed that 20 amino acids had in fact been synthesized.

However, as discussed in Chapter 3, we now know, from analysis of oxygen isotopes and cerium concentrations in the oldest 4.4 Ga zircons, that Earth's atmosphere was not strongly reducing. Additionally, there is abundant evidence that amino acids and other organic compounds were quite common throughout the solar system. In particular, *carbonaceous chondrites* are stony rather than metallic meteorites formed very early in the solar system that have never been heated to their melting point. Although they represent only about 5% of chondritic meteorites that fall on Earth, they are particularly important for origin of life theories because they contain hundreds of organic compounds including hydrocarbons, amino acids, aldehydes, phosphonic acids, and many others. In addition, other key organic compounds, such as formaldehyde, are made abiotically in the atmosphere. Experiments show that many of these organic compounds were likely made by photochemical reactions of ultraviolet radiation from the Sun that interacted with icy coatings of interstellar dust particles that contained simpler compounds such as carbon monoxide (CO), carbon dioxide (CO_2), methanol (CH_3OH), and ammonia (NH_3). The well-studied Murchison meteorite, which fell to Earth in 1969 near the town of Murchison in southeast Australia, contains amino acids that are found in DNA such as glycine and alanine. As a consequence, it is now thought that most of the early organic compounds necessary to build a cell were likely delivered to Earth by meteorites, rather than created in a shallow pond.

7.4.3 Protocells and the Shallow Water Hypothesis

It seems clear that early Earth had all of the basic organic ingredients required to form life, but a fundamental question is: how did the first cell arise? Cells require a membrane that provides a barrier for exchange of chemicals between the cell and the external environment. Energy-capturing chemical reactions must occur within the cell, and of course cells require a biochemical system that allows cells to reproduce, such as DNA. We have already introduced the idea that amphiphilic compounds can spontaneously form into hollow spheres that resemble cell walls. In 1964, Sydney Fox and Kaoru Harada synthesized more complex **proteinoids** from amino acid mixtures. They subjected their samples to episodes of wetting and drying and their proteinoids formed into microspheres with distinct membranes (Figure 7.8). Fox and Harada interpreted these to be protocells. Chemical manipulations, such as changing the acidity of the water in which the microspheres were formed, also showed that these proteinoid microspheres could replicate by a variety of processes, including budding, fission (i.e., splitting) and generation of spore-like particles (Figure 7.8).

These early experiments lent support to the idea that the formation of the first cells may have occurred in a "primordial soup," a shallow water environment in which wetting and

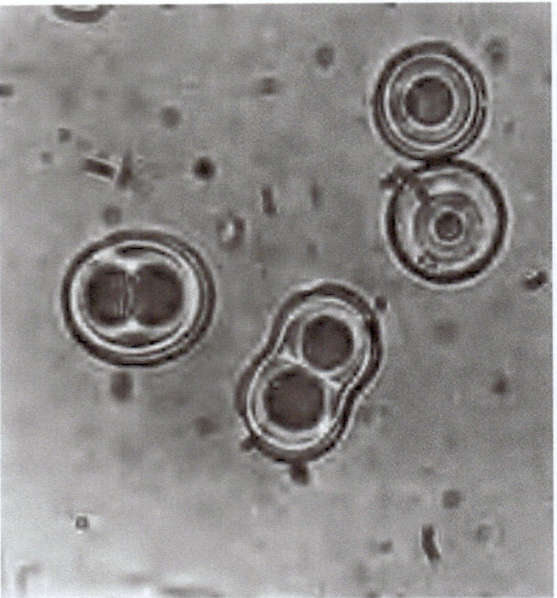

Figure 7.8 Proteinoid microspheres undergoing binary fission (from Fox, 1973). Photo is about 15 microns wide. Source: Reprinted with permission from The Walter de Gruyter publishing group.

drying occurred (Figure 7.9). Wetting and drying allows dissolved chemicals to become concentrated, promoting chemical reactions, and may also have been essential in forming the first cell membranes. As water dries out, amphiphilic layers can coat the bottom of a pond and trap concentrated chemicals (Figure 7.9). Re-wetting can cause the amphiphilic layers to bud off, forming spheres that can also trap chemicals within them. Of the many thousands or millions of such budded spheres, all that is needed is one or a few to have developed the ability to metabolize or replicate, enhancing their chance of survival. Consequently, it seems clear that Darwinian selection that favored replicating molecules likely also played an important role in evolving the first replicating cells.

KEY POINT

Early lab experiments produced the amino acids and replicating protocells that are required for building a cell.

7.5 Deep Water Hydrothermal Vents Versus Shallow Pond Hypotheses

One of the problems with the shallow water origin model is that early Earth had no free atmospheric oxygen and thus no *ozone* (O_3). Ozone is critical in providing protection against harmful ultraviolet (UV) radiation from the Sun. Without this layer of protection, the surface of early Earth would have been much more susceptible to irradiation by strong UV rays, making it more difficult to stabilize complex organic macromolecules. This led to the hypothesis that life may have originated in deeper water, away from the damaging effects of UV radiation.

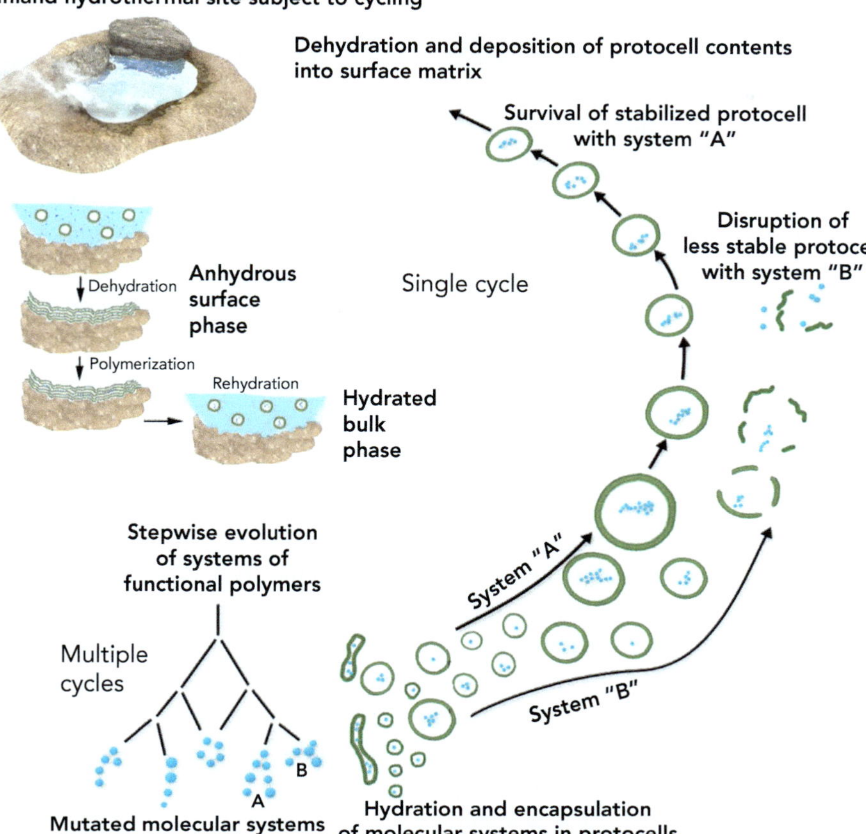

Figure 7.9 Model for formation of first cells by wetting and drying of organic-rich ponds in hydrothermal fields. Multiple cycles allow for concentration of organic reagents, such as amino acids and longer-chain proteins and fatty acids, that form amphiphilic polymers. Rehydration may form liposomic vesicles (protocells) that can contain replicating and metabolizing molecular systems. Most will be disrupted but successfully replicating protocells will be selected for and eventually dominate, forming the first cells. Source: Bruce Damer and Ryan Norkus, used with permission.

7.5.1 Hydrothermal Deep Water Life

The archaea show adaptations that suggest they may have originated in deep water, far away from the damaging UV radiation of the Sun. Many species of archaea show a tolerance for extreme environmental conditions and are referred to as **extremophiles** (literally, "organisms that love extreme environments"). Some of the first modern archaea were detected in areas of high heat, such as geothermal hot springs (the heat-loving **thermophiles**), as well as areas of high salt-content (salt-loving **halophiles**). The archaea also show metabolic adaptations to **anaerobic** reducing conditions including **methanogenesis** (see Eq. 7.1), which is the reduction of carbon dioxide that produces energy and methane as a byproduct

$$CO_2 + 4H_2 \Leftrightarrow CH_4 + 2H_2O + \text{energy}. \qquad (7.1)$$

This contrasts with **photosynthesis** (Eq. 7.2), which oxidizes CO_2, producing O_2 as a byproduct:

$$CO_2 + H_2O + \text{photons} \Leftrightarrow [CH_2O]_n + O_2. \qquad (7.2)$$

Chemosynthetic archaea also rely on sulfur reduction (see Eq. 7.3):

$$S + H_2 \Leftrightarrow H_2S + \text{energy}. \qquad (7.3)$$

As discussed above, in order to survive the Late Heavy Bombardment, taking refuge in the lithosphere using a metabolism based on reduction versus oxidation would be required.

Figure 7.10 A hydrothermal vent on the Niua underwater volcano in the Lau Basin, southwest Pacific Ocean. Source: Image by Coastal and Marine Hazards and Resources Program: https://www.usgs.gov/media/images/mineral-laden-water-emerging-a-hydrothermal-vent. Public domain, via USGS.

Studies of the life forms associated with deep water **hydrothermal vents** (Figure 7.10) show that these environments do not depend on photosynthesis to provide energy to the base of the food chain, but rather rely on chemosynthetic archaea and bacteria. Hydrothermal vents would have been common under the oceans of early Earth and would have provided chemical and thermal energy and an abundance of mineral and chemical compounds, such as sulfur and CO_2, both of which can release energy by reduction by the reactions shown above.

Equation (7.1) shows that reduction of carbon dioxide produces methane and Eq. (7.3) shows reduction of sulfur produces hydrogen sulfide. These are common byproducts of **chemosynthesis** that can be found in low oxygen-reducing environments, such as stagnant swamps, that release marsh gas (CH_4 and H_2S). Early life, prior to the availability of free oxygen, may thus have developed a metabolism based on reducing rather than oxidizing chemical reactions. The organic chemical reactions that led to development of life may also have been aided or catalyzed by the mineral surfaces that are common in hydrothermal vent horizons and in the lithosphere.

7.5.2 Hydrothermal Vent Versus Shallow Hydrothermal Field Hypotheses

As appealing as the deep water hydrothermal vent hypothesis is, there are some serious barriers to the idea. The biggest hurdle is how to concentrate organic molecules in a dilute oceanic environment. An abundance of water would promote *hydrolysis,* the breakdown of polymers by hydration, which is the opposite of condensation that allows polymers to form. In addition, salt, which is ubiquitous in seawater, would likely have inhibited the formation of polymers. Proponents of the deep water vent hypotheses suggest that containment within minerals may have been key for catalyzing and promoting reactions, but it is hard to see how it could have been possible to achieve the high concentrations required to drive chemical reactions.

Proponents of the shallow water hypothesis point out that UV radiation is strongly attenuated in a few meters of water and even more so if there is a layer of sediment that can protect organic polymers from disassociation during a dry phase. The big advantage of terrestrial ponds is the ability to concentrate reactants by wetting and drying cycles, and the freshwater nature removes concerns about the effects of salt. The proponents of the shallow water hypothesis suggest that ponds adjacent to terrestrial hydrothermal vents, such as are common around volcanoes, would have provided hydrothermal minerals such as clays that could serve as a catalytic substrate that promoted reactions required to build organic polymers. A key requirement is that the rate of production of polymers must exceed the rate at which they are removed by UV disassociation, hydrolysis, or other processes. Thus, one of the favored hypotheses is that life may have formed in shallow ponds associated with hydrothermal fields on the flanks of some of the earliest volcanoes. Although there may not have been much continental crust, oceanic volcanoes would have poked their heads above the sea and been exposed to terrestrial processes such as rain, erosion, and weathering. Much of the organic material is thought to have been extraterrestrial in origin, derived from organic material in chondritic meteorites. Certainly, there would have been no lack of extraterrestrial bombardment during the period of Late Heavy Bombardment. Wetting and drying in

shallow ponds would have concentrated these organic compounds, and higher concentrations, as well as catalytic reactions with clays and other minerals, are hypothesized to have promoted formation of amphiphilic polymers (Figure 7.6). Drying would promote formation of organic-rich amphiphilic layers that on wetting would form spheres that could contain self-sustaining reactions and eventually the first cells (Figure 7.9).

KEY POINT

Harmful UV radiation may have made it difficult to sustain a protocell in shallow ponds on early Earth without ozone, and this led to the hypothesis that the earliest life evolved around deep water vents.

7.6 From the Top Down: The RNA World

Although amino acids are not living, they form the basis for construction of RNA (single-strand ribonucleic acid) and DNA (double-strand deoxyribonucleic acid), which, along with lipids, carbohydrates, and proteins, are the essential macromolecules that form the basis for building a cell. RNA and DNA are essential in building genes and are the key in the ability of life to replicate itself. Although genes are built of DNA, RNA is a simpler macromolecule as it only has a single strand; it is now thought that the earliest life on Earth was initially RNA-based and later evolved into DNA.

So far, we have discussed a bottom-up approach to understanding how life evolved, but there are experiments and models that take a top-down approach. These include construction of synthetic cells to show how cells are built, as well as experiments to determine how the RNA world might have worked. Some of the recent critical breakthroughs have been the synthesis of catalyzing RNA molecules, called *ribozymes*, which demonstrate their ability to foster and drive biochemical reactions that would be required in an RNA world. In order to sustain a cell, metabolism is required, but to sustain multiple generations reproduction must also occur. It is not yet understood if these evolved simultaneously or separately.

KEY POINT

DNA is too complicated to have evolved first and it is thought that the earliest life was based on RNA. RNA is known to play a role in catalyzing reactions, supporting the idea of an early RNA world.

7.7 Origin of Photosynthesis

Except for a few hydrothermal vent and deep lithosphere communities, nearly all of modern life is driven by photosynthesis. An essential reason why we have life on Earth is that Earth is an open system and the energy supplied daily by the Sun to Earth's surface sustains the complex biochemical reactions and processes that allow life to thrive and evolve, as we introduced in Chapter 6. Photosynthetic organisms use sunlight to either incorporate or to manufacture organic molecules to be used as food for a cell (Eq. 7.2).

In Eq. (7.2) we showed that in the photosynthesis reaction, carbon dioxide (CO_2) and water (H_2O) are combined with the energy from the Sun to make carbohydrates ($CH_2O)_n$ and oxygen (O_2). Organisms that use sunlight to digest food are said to be **photoheterotrophic**, whereas those that use it to manufacture food are termed **photoautotrophic**. One of the diagnostic features made by photosynthetic organisms are stromatolites (Figures 7.3 and 7.4), which are bacterial in origin and, as discussed above, have been found in rocks as old as 3.7 Ga. There was no free oxygen on early Earth, so the earliest photosynthetic cells were almost certainly anerobic and likely used photosynthesis of compounds that were more common in a reducing Earth, such as arsenic:

$$CO_2 + AsO_3 + photons \Leftrightarrow AsO_4 + CO. \qquad (7.4)$$

In Eq. (7.4), carbon dioxide (CO_2) and arsenite (AsO_3) are combined with photons to make arsenate (AsO_4) and carbon monoxide (CO). Photosynthesis is a complex biochemical process that requires several components, which include (1) pigments (such as **chlorophyll**), (2) **reaction centers** (RC), which are the molecular structures that actually use the sunlight in metabolic processes, (3) light-harvesting antenna systems, and (4) a chemical pathway for incorporating carbon into the cell.

The earliest pigments in bacteria were likely associated with **hemes**, which are the components of molecules that include iron (Fe) and form part of the **hemoglobin** that gives blood its red color. Organisms that live in a reducing environment commonly metabolize iron, sulfur, and arsenic, and at some time, used sunlight in this process and would have developed hemepigments. Chlorophyll evolved later by using a magnesium (Mg) versus iron (Fe) atom at its center, giving it a green rather than red color. Chlorophyll is the pigment present in all cyanobacteria and plants. The chemical pathway required to synthesize hemes and chlorophyll are very similar, and this supports the idea that hemes appeared first, in an oxygen-poor environment, with later substitution of Fe with Mg and incorporation of oxygen that caused hemes to evolve into chlorophyll.

There are two types of reaction centers. Anerobic photosynthetic bacteria, which appeared earlier than eukaryotes (Figure 7.5), have one or the other, whereas photosynthesizers that use oxygen, rather than CO_2, have one of each and consequently it is thought they must have shared a common ancestor. There are indications that light-harvesting antennae and carbon reaction pathways evolved multiple times. Chlorophyll first appeared in the photoautotrophic oxygenic **cyanobacteria** (also called blue-green algae) around 2.1 Ga. Molecular clocks and genomic analysis indicate that chloroplast ancestors likely first evolved in fresh water and later migrated to marine habitats, and this indicates that prokaryotic life on land occurred

relatively early in Earth history, supporting the warm shallow pond hydrothermal field hypothesis. These organisms with chlorophyll were the first to produce oxygen and, as we will discuss in Section 7.9, caused a radical change in Earth's atmosphere.

KEY POINT

Photosynthesis uses light energy to drive reactions, and the earliest forms were anerobic and only later evolved the ability to produce oxygen.

7.8 From Prokaryotes to Eukaryotes

This section discusses the evolution of the larger and more complex eukaryotes from simpler prokaryotes (Figure 7.11). Gene-sharing between the prokaryotic archaea and bacteria, and the development of symbiotic relationships between prokaryotes, caused them to evolve into more complex eukaryotes.

7.8.1 Evolutionary History of Eukaryotes

Molecular clocks suggest that the eukaryotes first appeared in the Proterozoic Eon, likely after 1.84 Ga and possibly around 1.65 Ga (Figure 7.5). One of the main questions in eukaryotic evolution is how these complex structures and especially **organelles** such as **mitochondria** came to be (Figures 7.11 and 7.12). Mitochondria are of particular interest as they have their own DNA. Mitochondria are commonly referred to as the energy centers of cells as they convert glucose to **adenosine triphosphate (ATP)**, which provides the energy that drives many cell processes. Mitochondrial DNA, however, is not used as part of the cell reproductive process in contrast to nuclear DNA, which splits and recombines as part of the reproductive process.

Although there are a variety of hypotheses on the usefulness of mitochondrial DNA to the cell, from an evolutionary perspective it can be considered as a vestigial evolutionary structure (see Chapter 6). The best theory is that mitochondria originated as separate prokaryotic cells (Figure 7.12). This idea was first proposed (and initially rejected) in 1970 by American biologist Lynn Margulis and only confirmed much later by genomic analysis. Genomic analysis shows that the mitochondrial ancestor was a member of the Alphaproteobacteria, a class of bacteria that includes *Escheria* (i.e., *E. coli*) and *Salmonella*. Primitive heterotrophic eukaryotes (i.e., eukaryotes that *could not* make their own food by photosynthesis) likely entered a symbiotic relationship with an *autotrophic* alphaproteobacteria, which *could* make its own food. This process is called **endosymbiosis** (Figure 7.12), the process by which one organism, in this case the proto-mitochondrial archeobacteria, lives within the cell of another organism, in this case a primitive eukaryote. **Plastids**

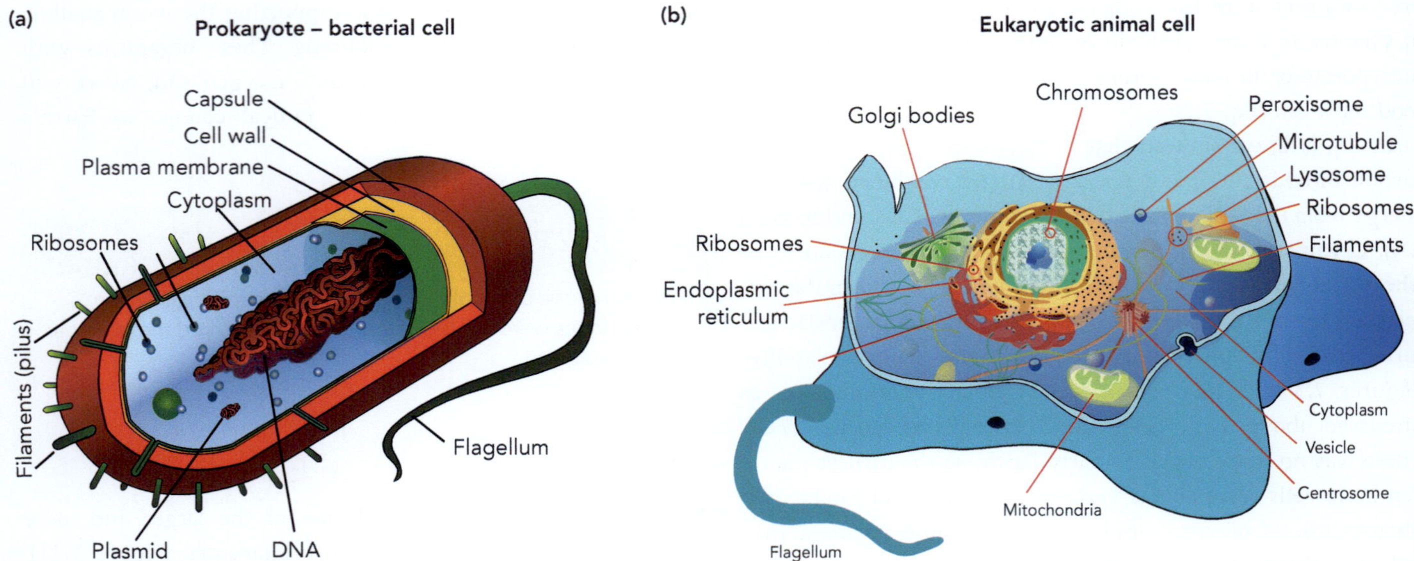

Figure 7.11 (a) Prokaryote cell and (b) eukaryotic animal cell. Eukaryotic cells can be tens to thousands of times larger than prokaryotic cells. Eukaryotes have a nucleus that houses the recombinant DNA, as well as a number of organelles, such as mitochondria and Golgi apparatus. Prokaryotic cells are smaller, and the DNA is not enclosed in a separate nucleus. Source: (a) Image modified from Ali Zifan, CC BY-SA 4.0 <https://creativecommons.org/licenses/by-sa/4.0>, via Wikimedia Commons; (b) LadyofHats, Public domain, via Wikimedia Commons.

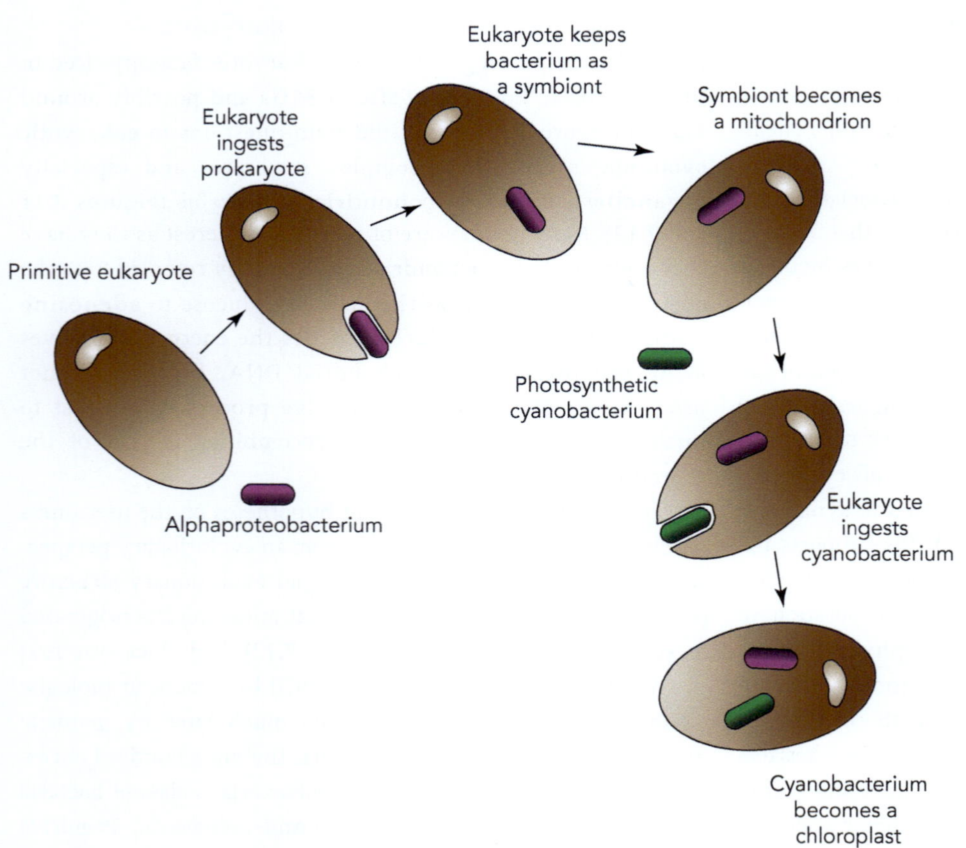

Figure 7.12 Illustration of the evolution of eukaryotes from prokaryotes by endosymbiosis.

are another type of membrane-bound organelle found in all plants. Like mitochondria, plastids have their own DNA and are also considered to have evolved from endosymbiotic cyanobacteria. *Chloroplasts* are probably the best-known plastid and contain chlorophyll, the key element in cyanobacterial aerobic photosynthesis. Other common plastids include *leucoplasts*, which are used by modern conifers to produce sticky saps, and **chromoplasts**, which manufacture and store pigments.

Genomic analysis shows that the eukaryotes are more closely related to the archaea, but also include endosymbiotic organelles derived from bacteria (Figure 7.5). This likely results from the ability of prokaryotes and some eukaryotes to transfer genes directly between cells by a variety of processes. These processes include **transduction**, whereby new genes are introduced by a virus, **conjugation**, which is the direct transfer between cells and can be achieved by connecting cells through a *pilus* (these are the filamentous or hair-like appendage shown in Figure 7.11), or **transformation**, in which genetic material is directly absorbed by a cell and incorporated into its own DNA (Figure 7.12). Such gene-sharing processes were likely common in early eukaryotes and provided an important mechanism for evolution to occur in these single-celled organisms. This allowed the eukaryotes to include genetic components and organelles descended from completely different domains of life, the archaea and bacteria. By and large, multi-celled organisms cannot do this. For example, you can't change your genetic structure by the simple act of eating. We don't gain chloroplasts by eating a green salad, and even when we get a bacterial or viral infection, our genome is not altered. This is in stark contrast to the eukaryotes, who were able to alter their genetic makeup by these gene-transfer mechanisms with prokaryotes.

KEY POINT

Unlike humans, the archaea, bacteria, and early eukaryotes were able to share genes. Many of the structures in eukaryotes (such as mitochondria and chloroplasts) likely evolved by endosymbiosis of eukaryotes with prokaryotes.

7.8.2 Fossil Record of Early Eukaryotes

Fossils of curved, tubular **Grypania** (Figure 7.13) have been found in rocks as old as 2.1 Ga and have been suggested to be the oldest eukaryotes. *Grypania* form narrow ribbons that represent compressed cylinders composed of a colony of cells that reached lengths of 13 mm and average 1.5 mm in diameter. The coiling is thought to indicate that these organisms formed as a helicoidal chain attached to the seafloor. Segmentation features (called *annulations*) and possible internal septa (Figure 7.13(b)) have been interpreted as individual cells. The size of the fossils, the colonial nature, and the evenness of the coil diameters have been used to suggest that they are eukaryotes, but this interpretation is not fully accepted, and others interpret *Grypania* as prokaryotic bacterial colonies.

The 1.492 Ga Roper Group in Northern Australia contains shallow water sandstones and shales that are richly fossiliferous. Large cells of *Tappania plana* (Figure 7.14(a)) and other organisms show evidence of an external **cytoskeleton**, with numerous filamentous protrusions, and a cell membrane system that enabled cells to change shape. These features are not usually found in prokaryotes, hence their interpretation

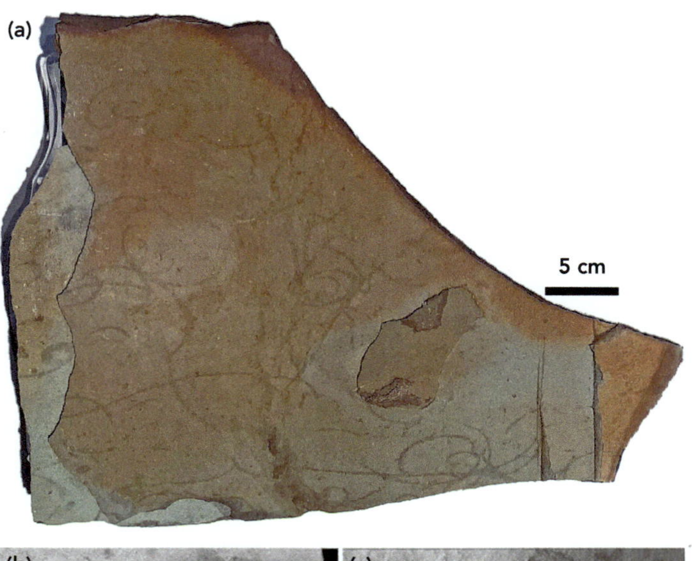

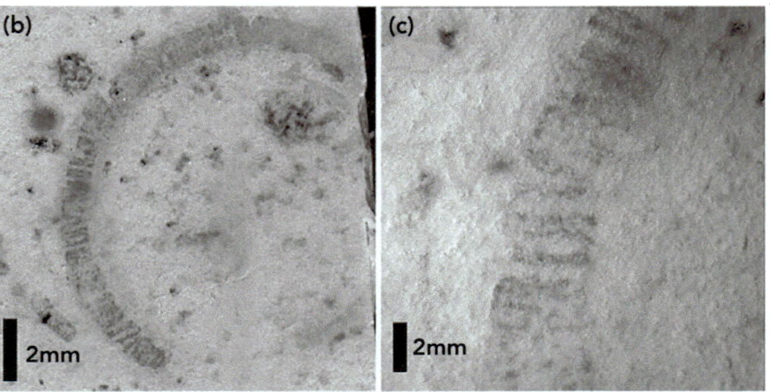

Figure 7.13 (a) *Grypania* fossils dated at 2.1 Ga; (b) and (c) closeups showing annulations and possible septa indicative of the cellular structure. Source: (a) Photo by JPB. With permission of ROM (Royal Ontario Museum), Toronto, Canada; (b) and (c) Reproduced from Henderson (2010).

as eukaryotes. Examination of the variety of cell morphologies suggest complex life cycles that show evidence of resting cysts and reproduction by budding and binary division. Similar fossils have been found in younger Neoproterozoic rocks in Arctic Canada and Siberia (Figure 7.14(b) and (d)). *Bonniea dachruchers* is an early example of a eukaryotic protist that secreted a small shell (technically termed a *test*, which means bowl- or bottle-shaped), which is related to the large group of single-celled *foraminifera*, common in the Phanerozoic seas. The enigmatic fossil *Diskagma* (Figure 7.15), found in 2.2 Ga **paleosols** (ancient soil deposits), is similarly bowl-shaped and may be the oldest non-marine eukaryotic fossil ever found. Its size and complexity have been cited as evidence to suggest it is a eukaryote, but if so, it would be older than predicted by molecular clocks.

7.9 Evolution of Earth's Atmosphere and the Great Oxygenation Event

As discussed above, prior to the appearance of chlorophyllic life, Earth's atmosphere was primarily nitrogen, CO_2, minor SO_2, and other volatiles (e.g., ammonia (NH_3), methane (CH_4), nitrate

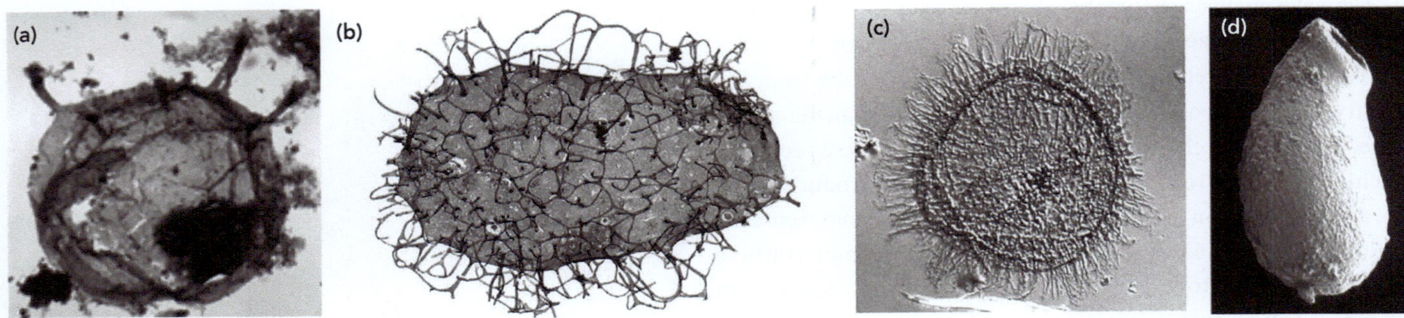

Figure 7.14 Examples of eukaryotes: (a) *Tappania plana*, from the 1.492 Ga Roper Group, Australia; (b) Neoproterozoic possible "*Tappania plana*," from Arctic Canada; (c) Neoproterozoic acritarch, *Appendisphaera grandis*, Miroyedikha Formation, Siberia; and (d) *Bonniea dachruchers*, a vase-shaped protist test from the Neoproterozoic Kwagunt Formation, Grand Canyon. Source: Reproduced from Knoll et al. (2006).

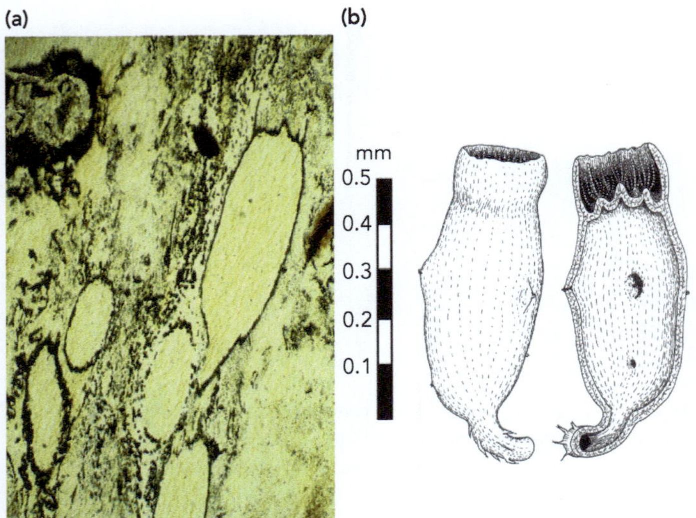

Figure 7.15 *Diskagma buttoni* fossils dated at 2.2 Ga from terrestrial paleosols that caps the Hekport basalt in South Africa. Source: (a) Retallack, CC BY-SA 3.0 <https://creativecommons.org/licenses/by-sa/3.0>, via Wikimedia Commons; (b) Reproduced from Retallack et al. (2013). Copyright (2013), with permission from Elsevier.

Figure 7.16 Precambrian banded iron formation. Source: Photo by JPB. With permission of ROM (Royal Ontario Museum), Toronto, Canada.

(NO_3)), as well as inert gases. Prior to chlorophyll, the world's oxygen was largely tied up in minerals, water, and gases. There is evidence that prior to the availability of free oxygen, the world's oceans would have had significant amounts of dissolved ferrous iron (i.e., Fe^{2+} in its reduced state). Once chlorophyllic cyanobacteria appeared, around 2.1 Ga, the free oxygen gas they produced would have quickly oxidized the ferrous iron in sea-water to produce a ferric (Fe^{3+}) iron oxide (i.e., either hematite, Fe_2O_3, or magnetite, Fe_3O_4), which is much less soluble in sea-water. Precambrian rocks include banded iron formations, rare in Phanerozoic rocks, that consist of alternating thin layers of iron-poor shale or chert and layers enriched in iron minerals (Figure 7.16). Banded iron formations have been found in rocks as old as 3.7 Ga, but are much more common after 2.4 Ga, which coincides with the appearance of the cyanobacteria.

The appearance of cyanobacteria eventually led to the **Great Oxygenation Event** (Figure 7.5). Oxygen-producing cyanobacteria may have evolved prior to 2.1 Ga, but any oxygen produced from about 3.85 to 2.45 Ga (Stage 1 in Figure 7.5) would have been largely consumed in the process of oxidizing iron, and it is thought that the world's oceans would have been largely anoxic throughout that time. During Stage 2 (2.45–1.85 Ga, Figure 7.5), O_2 began to increase but was largely absorbed in oceans and rocks on the seabed, including sequestration in banded iron formations. Atmospheric oxygen initially would have also reacted with methane (CH_4) to produce carbon dioxide (CO_2) and water (H_2O):

$$2C_2 + CH_4 \Leftrightarrow CO_2 + 2H_2O. \tag{7.5}$$

Both methane (CH_4) and carbon dioxide (CO_2) are greenhouse gases, but methane has a much stronger effect in this regard.

Figure 7.17 Glacial diamict Huronian Supergroup, Ontario, Canada.

Conversion of methane to carbon dioxide would have resulted in a cooling of Earth. There is widespread evidence that this process of conversion of methane to carbon dioxide drove the Proterozoic world into its first icehouse period, marked by the **Huronian Glaciation**, which lasted from 2.4 to 2.1 Ga, over a period of about 300 million years. The Huronian is characterized by glacial deposits worldwide (Figure 7.17), and although there is significant debate as to how persistent these glacial episodes were in time and space, it is clear that the world suddenly became cooler.

Between 1.85 and 0.86 Ga (Stage 3, Figure 7.5) oxygen began to build up in the oceans. The iron sink became depleted and banded iron formations are absent during this period. The loss of iron as a sink for O_2 ultimately allowed free oxygen to enter the atmosphere, however, atmospheric oxygen was still quite low throughout this period. The ability of oxygen to build up was likely inhibited by land weathering (such as oxidation of clays and iron minerals) and atmospheric buffering by methane and other gases.

Stage 4 (Figure 7.5), 850 Ma to 540 Ma, shows a massive build-up of free oxygen, almost to present-day values of about 0.2 atmospheres (i.e., about 20% oxygen), which reflects the depletion of oxygen reservoirs, such as iron on the oceans and land, and methane (CH_4), hydrogen (H_2), and hydrogen sulfide (H_2S) in the atmosphere. During this stage, the oceans would have become fully oxygenated. Stage 5 (540 Ma to present, Figure 7.5) shows significant variation, with a peak of about 0.35 atmospheres (i.e., 35% oxygen) around 300 million years ago in the Carboniferous Period, and we will discuss this oxygen peak in Chapter 11.

Clearly an oxygenated planet was required to allow the higher metabolisms that drove the next major step in evolution, mobile animals and complex plants, but all of that oxygen also caused a massive decrease in CO_2 and eventually drove the Earth into a deep freeze, and is the topic of our next story.

 **KEY POINT**

Once oxygen photosynthesis appeared, Earth's atmosphere became oxygenated during the Great Oxidation Event, paving the way for air-breathing life.

7.10 Summary

- Key hypotheses about how and where life appeared started with Darwin's idea that life may have begun in the "primordial soup" of some warm little pond.
- Early pre-biotic Earth had very different conditions than today. Origin-of-life experiments designed to replicate these conditions produced amino acids, demonstrated how a cell with membranes could arise, how reproductive processes may have arisen, and how informational macromolecules may have developed, all of which are critical steps in building the first cells.
- There is also evidence that life may have originated in deep oceans around hydrothermal vents, in the absence of oxygen.
- Single-celled prokaryotes dominated the first 2.5 billion years of Earth history.
- Endosymbiosis and gene-transfer between early simple eukaryotes and prokaryotes led to more complex and much larger eukaryotic cells, which are the basis for more complex multicellular life forms.
- The development of photosynthesis in the cyanobacteria transformed the atmosphere and oxygenated Earth during the Great Oxygenation Event, which commenced about 2.5 Ga.

Key Words

- primordial soup
- Late Heavy Bombardment (LHB)
- endolithic
- autotrophic
- stromatolites
- filamentous bacteria
- heterotrophic origin of life theory
- pre-biotic
- monomers
- polymers
- open system
- amino acids
- proteinoids
- extremophiles
- thermophiles
- halophiles
- anerobic
- methanogenesis

- photosynthesis
- hydrothermal vents
- chemosynthesis
- photoheterotrophic
- photoautotrophic
- chlorophyll
- reaction centers
- hemes
- hemoglobin
- cyanobacteria
- organelles
- mitochondria
- adenosine triphosphate (ATP)
- endosymbiosis
- plastids
- chromoplasts
- transduction
- conjugation
- transformation
- *Grypania*
- cytoskeleton
- paleosols
- Huronian Glaciation
- Great Oxygenation Event (GOE)

Further Reading and References

Deamer., D., 2019, *Assembling Life*, Oxford University Press.

Fox, S. W., 1973, Molecular evolution of the first cells, *Pure and Applied Chemistry*, 34(3–4), 641–670, https://doi.org/10.1351/pac197334030641.

Gradstein, F. M., Ogg, J. G., Schmitz, M. D., et al., 2012, *The Geologic Time Scale 2012*, Elsevier.

Henderson, M. A., 2010, A morphological and geochemical investigation of *Grypania spiralis*: Implications for early Earth evolution, Master's Thesis, University of Tennessee.

Holland, H., 2006, The oxygenation of the atmosphere and oceans, *Philosophical Transactions of the Royal Society B*, 361, 903–915.

Knoll, A. H., 2003, *Life on a Young Planet: The First Three Billion Years of Evolution on Earth*, Princeton University Press.

Knoll, A. H., Bergmann, K. D., and Strauss, J. V., 2016, Life: The first two billion years, *Philosophical Transactions of the Royal Society B*, 371, 20150493.

Knoll, A. H., Javaux, E. J., Hewitt, D., and Cohen, P., 2006, Eukaryotic organisms in Proterozoic oceans, *Philosophical Transactions of the Royal Society B*, 361, 1023–1038, http://doi.org/10.1098/rstb.2006.1843.

Retallack, G. J., Krull, E. S., Thackray, G. D., and Parkinson, D., 2013, Problematic urn-shaped fossils from a Paleoproterozoic (2.2 Ga) paleosol in South Africa, *Precambrian Research*, 235, 71–87.

Sugitani, K., Mimura, K., Takeuchi, M., et al. 2015, Early evolution of large micro-organisms with cytological complexity revealed by microanalyses of 3.4 Ga organic-walled microfossils, *Geobiology*, 13, 507–521, https://doi.org/10.1111/gbi.12148.

Review Questions

1. What is life and how is it susceptible to evolution?
2. What were the conditions on early Earth?
3. When may Earth have first been habitable? Could life have evolved during the period of Late Heavy Bombardment?
4. What are stromatolites and what other evidence is there for the earliest cellular life?
5. What are some of the key steps required to make a living cell?
6. What are amphiphilic polymers and why are they important in forming protocells?
7. What kind of experiments have been conducted to replicate the formation of the earliest organic compounds and protocells?
8. Explain the main arguments about the shallow water versus hydrothermal vent hypotheses for the formation of the first cells.
9. What is the difference between top-down versus bottom-up approaches to understanding the origin of life?
10. Why is it thought that the earliest life was based on RNA rather than DNA?
11. Explain how photosynthesis developed.
12. Explain how eukaryotes evolved and the concept of endosymbiosis.
13. How do mitochondria indicate they originated as separate prokaryotic cells?
14. When did photosynthesis arise?
15. How did the rise of oxygen photosynthesis change Earth's atmosphere and cause the Great Oxygenation Event?

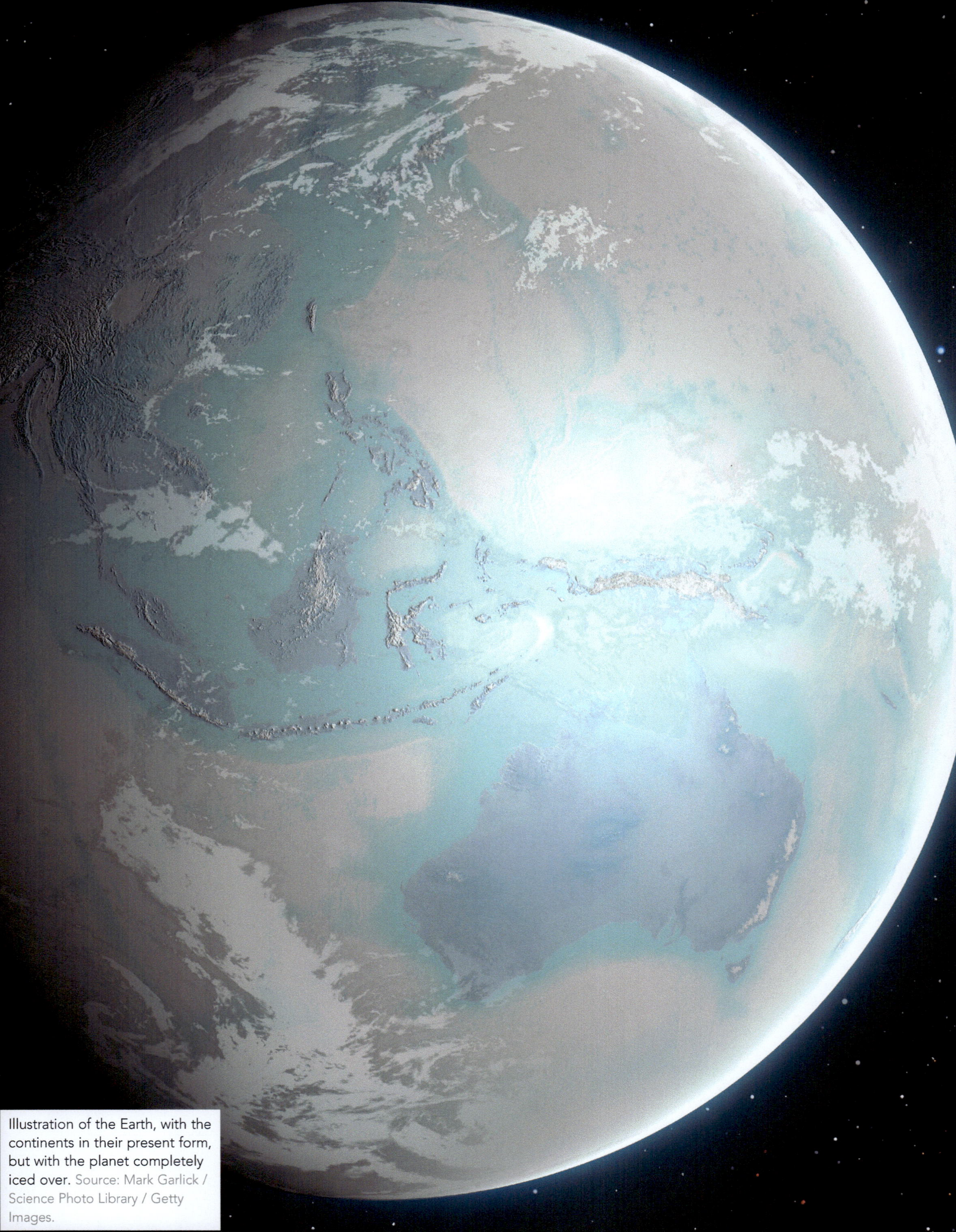

Illustration of the Earth, with the continents in their present form, but with the planet completely iced over. Source: Mark Garlick / Science Photo Library / Getty Images.

Chapter 8

Snowball Earth

A Neoproterozoic Frozen Planet

LEARNING OBJECTIVES

- Describe early proposals for a global freeze and how plate tectonics debunked the idea of a Paleozoic snowball Earth.

- List diagnostic criteria for recognizing glacial deposits and review evidence for synchronous, low-latitude glaciations in the Neoproterozoic Era.

- Explain factors contributing to snowball conditions, including the faint young Sun, greenhouse gases,

- runaway albedo, and the role of photosynthesis and oxygen.

- Explain what triggered the eventual collapse of glaciers and return to warmer conditions and why cap carbonates were deposited globally after the end of snowball conditions.

- Explain how life on Earth survived snowball conditions and why Earth has not experienced another snowball state.

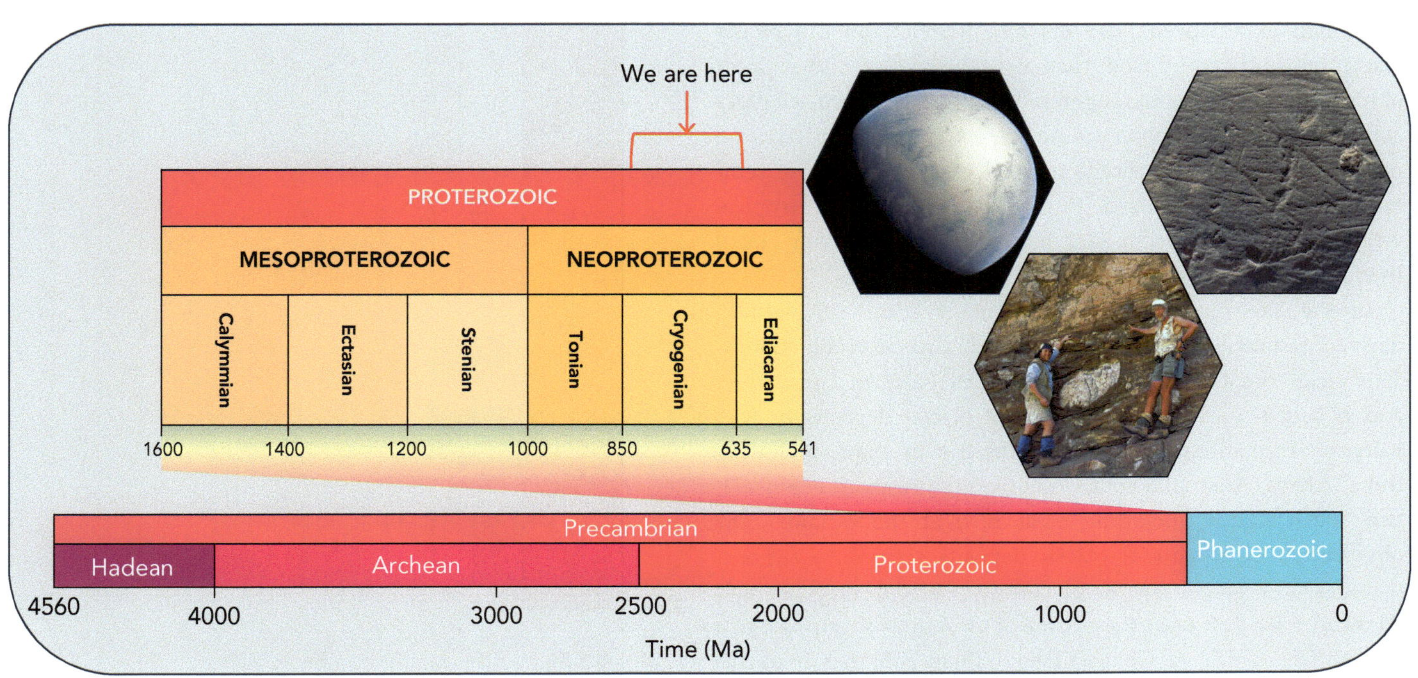

Introduction

Today, glaciers are found at high latitudes, closer to the pole, or at high altitudes, typically about 4,000 meters above sea level. However, the rock record tells us that during the past two billion years, Earth has experienced several episodes in which the distribution of ice across the globe was distinctly different than the modern arrangement. Earth has experienced about eight major glacial periods throughout its history, including the current stage (Figure 8.1). These glacial periods, also termed "**icehouses**," alternate with warmer so-called "**greenhouse**" periods when Earth was considerably warmer. Presently we are in an interglacial period of an overall icehouse and as recently as 18,000 years ago much of the Northern Hemisphere continents was covered with kilometers-thick ice sheets. The Late Cretaceous Period, in contrast, was a greenhouse and has been suggested to have been an ice-free time in Earth history.

In this chapter, we focus on a pair of global glacial episodes that occurred from 720 to 635 Ma, a time now known as the **Cryogenian** (or "cold birth") **Period** (Figure 8.1). Formally designated in 1990, it is one of the newest identified geological periods. During the Cryogenian Period it is thought that ice extended from high latitudes and elevations all the way to sea level in the tropics, covering most of Earth's surface, forming a "**snowball Earth**." This chapter will discuss the evidence and causes of the snowball Earth hypothesis, how life on Earth survived, and whether Earth could experience another deep freeze.

8.1 Did the Earth Ever Freeze?

One of the earliest suggestions of a recent global freezing was by Swiss-American scientist Louis Agassiz, who proposed that Earth had experienced "**die eiszeit**" (the ice-time), a recent glaciation, initiated by God, that covered the globe. In support of his hypothesis, Agassiz suggested that poorly sorted red clays containing boulders observed in the jungles of the Amazon provided evidence for recent tropical glaciations. These sediments were later realized to be the products of tropical weathering and Agassiz's idea of a recent God-driven global freeze fell out of favor.

In the late nineteenth century, observations of glacially derived sediments in older rocks were also receiving notice. This time, the focus of interest was on Paleozoic glaciation, and it began with the observation of glacial deposits in Late Paleozoic formations on all the southern continents, including India, Africa, Australia, South America, the Arabian Peninsula, and South Africa, including areas now at the equator. The observation of equatorial glacial deposits seemed to support the idea of a global freeze. In contrast, Alfred Wegener and Alexander Du Toit used these same observations in support of a very different idea. As we discussed in Chapter 5, they used this

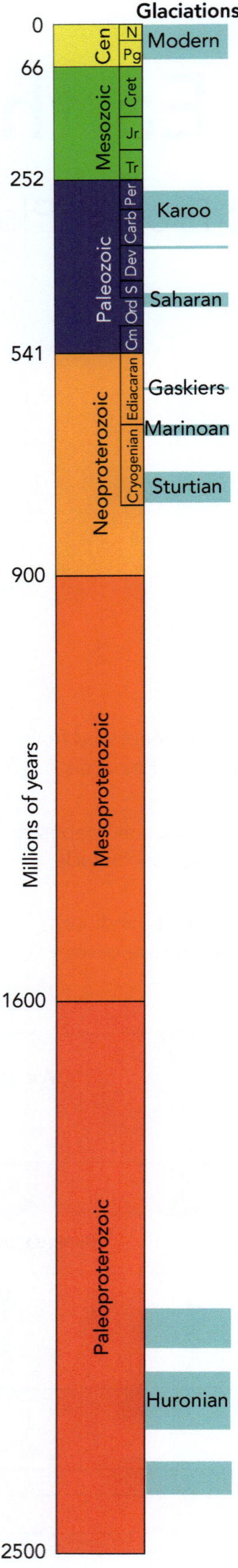

Figure 8.1 Major glacial periods in Earth history.

evidence to suggest that these Paleozoic glacial features formed when these continents had originally been joined together to form the supercontinent Gondwanaland and had lain at much higher latitudes, closer to the South Pole where the glaciers formed, and only later drifted into their present low-latitude equatorial position.

Many of the leading scientists of the time dismissed Wegener's idea of continental drift, assuming that continents were static. For them the more reasonable interpretation was that glaciers had extended to the tropics, indicating a global freeze. Eventually paleogeographic reconstructions based largely on paleomagnetic data demonstrated that the Late Paleozoic glaciations occurred when these continents lay close to the South Pole as part of Gondwanaland. It was only after the breakup that Australia, South America, South Africa, India, and the Arabian Peninsula migrated to a more equatorial position, corroborating Wegener's idea of continental drift. Plate tectonics provided an alternate mechanism to explain how the deposits of glaciers ended up at the equator, and debunked the idea that glaciers extended to the tropics.

KEY POINT

Previous ideas of a global freeze based on observations of equatorial glaciations were replaced by the idea that the glaciations occurred at polar latitudes and these continents later drifted into equatorial regions through the process of plate tectonics.

The most recent resurrection of the idea that the globe froze is focused on older rocks of the Neoproterozoic Era. This newest theory has been dubbed snowball Earth, and was introduced in 1992 by Joe Kirschvink, a professor at the California Institute of Technology, and further developed by Paul Hoffman and Dan Schrag at Harvard University. The snowball Earth hypothesis has been met with much interest and some skepticism, particularly given the fate of previous proposals regarding the idea of a permanently frozen Earth. In the following sections we outline some of the key evidence that has been used to test this hypothesis, as well as its consequences for life on Earth and whether Earth could experience such an event in the future.

BOX 8.1 Recognizing Glaciers in the Geologic Record

Glaciers represent a build-up of ice formed where accumulation exceeds melting. Once glaciers are thick enough, they flow from areas of net accumulation to areas where **ablation** or loss occurs by processes such as evaporation, sublimation, melting, and ice-calving, which is common where glaciers meet a lake or sea (Figure 8.2). Depending on the mass balance, which is largely controlled by the strength and duration of the summer melting season, glaciers can experience periods of advance and retreat.

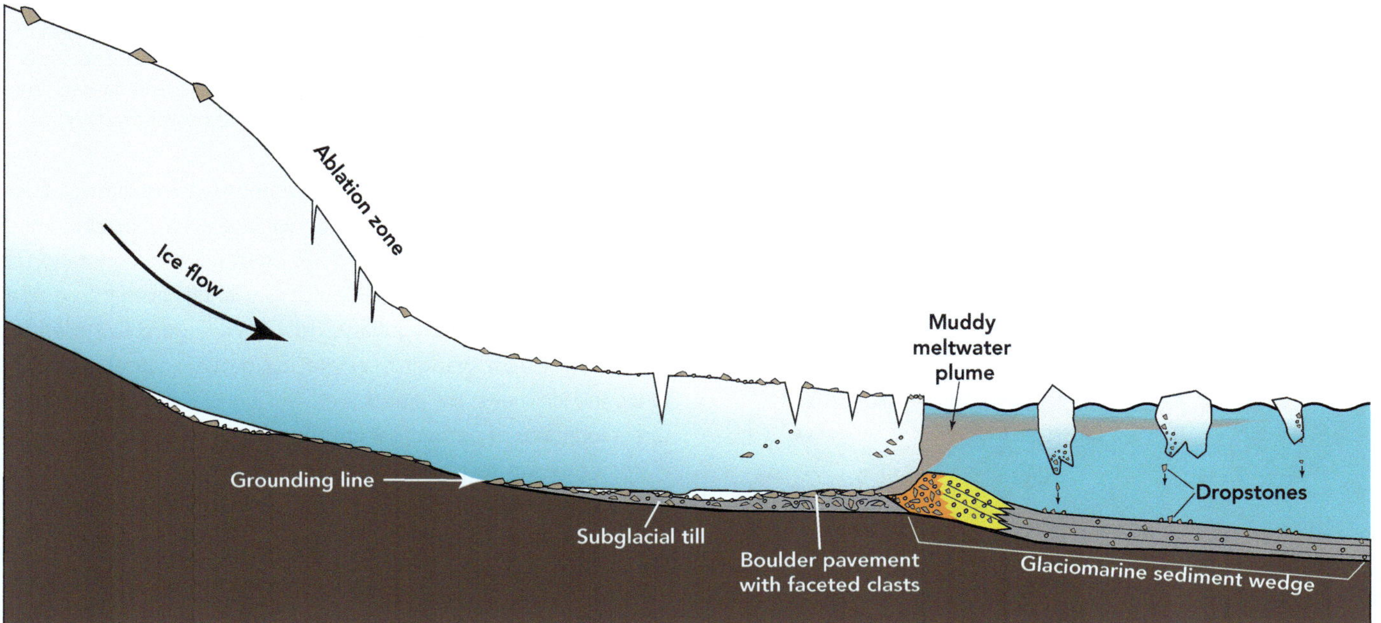

Figure 8.2 Cross section of a glacial ice sheet building into an ocean. The grounding line marks the area where the glacier sits on the seafloor and is marked by boulder pavements. Ablation by melting and ice calving forms icebergs at the terminal end and drives the down-slope flow of the ice sheet. A wedge of glaciomarine sediment forms that becomes muddier seaward and contains dropstones carried by icebergs and dropped to the seafloor as they melt. Source: Partly based on a figure by Antonio Valdisturlo, CC BY 4.0 <https://creativecommons.org/licenses/by/4.0>, via Wikimedia Commons.

When glaciers move, they may act as both a plow and a conveyer belt. A glacier can grind solid rock or sediment that it flows over, as well as carry sediment on its surface (Figure 8.3). When a glacier retreats and especially where it enters the sea, it leaves poorly sorted sediment consisting of mud, sand, gravel, and boulders referred to as a **diamict**, literally "two-mixtures" (Figures 8.4 and 8.5). Because diamicts can also be formed by a variety of processes unrelated to glaciers, such as landslides and debris flows, interpretation of an ancient sedimentary diamict as glacial in origin requires evidence of direct contact of the sediment with ice.

Specific geologic features that are uniquely formed by the direct contact with moving ice include glacial **striations** (Figure 8.5) and **chatter marks** (Figure 8.5(b)). Glacial striations are scratch marks made by pieces of rock stuck at the bottom of a glacier as they are dragged over a rocky surface. Chatter marks (Figure 8.5(b)) are curved, U-shaped cracks produced by chipping and fracturing of a bedrock surface, as a glacier flows over it. The orientation of striations is parallel to the ice flow direction and the chatter marks curve down flow. These marks can form on bedrock but can also be

Figure 8.4 Poorly sorted boulders, sand, and clay form a glacial diamict deposited from mountain glaciers in Montana. The dashed yellow line marks the top of a boulder pavement.

Figure 8.5 (a) Glacial striations formed by the Laurentide Ice Sheet carved into Devonian limestones exposed along the shore of Lake Erie. The arrow shows the direction of ice flow. (b) Glacial striations and chatter marks in Precambrian sandstones, Newfoundland. The arrow (with a pen for scale) shows the direction of ice flow. The dashed line outlines the chatter marks that are curved in the downflow direction.

seen in individual clasts, such as are common within a diamict. Such clasts can also show smooth surfaces, or facets, forming distinctive bullet-shaped clasts that record abrasion by ice. The orientation of clasts can also be used to distinguish glacial **tills** from non-glacial mass-flow deposits. Polygonal sand cracks (Figure 8.6) are also indicative of frost wedging, where ice growth causes cracking. Once the ice melts, the cracks are filled with sand and are a good direct indicator of permafrost environments.

It is common to preserve sediments that are deposited in a glaciomarine environment, where major glaciers lie next to or build into an ocean, such as on the modern Antarctic or Greenland shelves (Figure 8.2). Ice sheets may reach several kilometers in thickness, as is observed in Antarctic and Greenland. If an ice sheet of this thickness forms on the ocean or flows from land into the sea, it will sink into the sea. If the ice is thin compared to the water depth, then the ice will float, but over a continental shelf, where water depths are less than 200 m, the much thicker ice will ground itself on the seafloor. The

Figure 8.3 Crevasse on the Gorner Glacier, Zermatt, Switzerland. Large angular boulders and poorly sorted sediment, termed till, covers the top of the glacier. The photo was taken close to the glacial terminus, where debris tends to concentrate. Source: Adrian Pingstone, BY CC PD.

Figure 8.6 Ice-wedge polygons along the Mackenzie delta of the Arctic coast with a herd of caribou for scale. Source: Matti&Keti, CC BY-SA 4.0 <https://creativecommons.org/licenses/by-sa/4.0>, via Wikimedia Commons.

grounding line marks the area where submarine tills can be deposited (Figure 8.2). Distinctive **boulder pavements** can mark these grounding line areas where ice has come in direct contact with the seafloor (Figures 8.2 and 8.4) and later melted, leaving behind a layer of boulders. These boulders may also show the characteristic features associate with deposition by ice, such as glacial striations and faceted bullet-shaped clasts discussed above.

Another common features in glaciomarine sediments are **dropstones** (Figure 8.7). These occur where larger clasts, such as pebbles, cobbles, or boulders, are rafted out to sea, typically on icebergs or larger ice shelves (i.e., ice sheets that float above the sea but are still more or less connected to the land ice). As the ice melts, the rafted debris falls to the seafloor (Figure 8.2) where they can be found encased in much finer sediment, such as mud and silt, characteristic of a deeper water environment. In the process of falling through the water, clasts typically orient themselves vertically and they may also show distinct puncture marks as they impact the sediments on the seafloor.

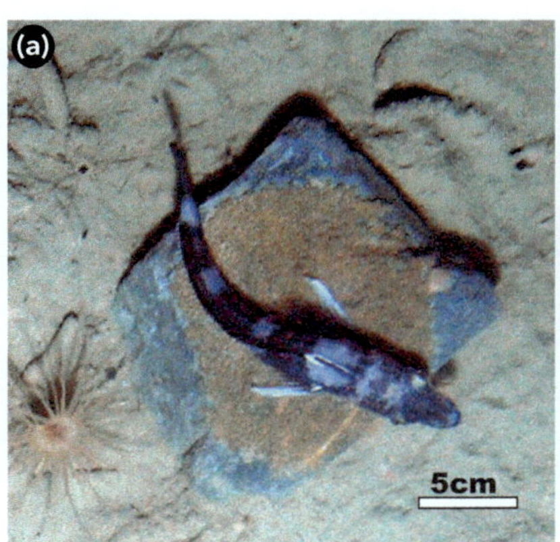

Figure 8.7 (a) Modern dropstone with fish, West Antarctica; (b) Glacial dropstone in lake deposits of the Late Paleozoic Itararé Subgroup in Brazil. Source: (a) Ziegler et al. (2017); (b) Photo by Eurico Zimbres, CC BY-SA 2.5 <https://creativecommons.org/licenses/by-sa/2.5>, via Wikimedia Commons.

8.2 Earth Freezes Over

Neoproterozoic glacial to **periglacial** (i.e., peripheral to a glacier) deposits have been recognized on all major continents except Antarctica, and include North and South America, Europe, Africa, Australia, and Asia (Figure 8.8). Key features include diamicts, poorly sorted mixtures of gravel and finer grained muds, that contain features diagnostic of abrasion by ice, such as striated and faceted clasts, and interpreted as glacial till deposits (Figure 8.9). Note that the term "till" refers to sediments deposited directly by glaciers, versus diamicts that may have other origins (see Box 8.1). Other features include polygonal sand-filled cracks (Figure 8.10), interpreted to be formed by ice wedges in areas of permafrost and in marine deposits, and dropstones (Figure 8.11), interpreted as large clasts rafted by ice and dropped to the seafloor as the ice ablates or melts, are also common (see Box 8.1 for explanations of these features).

Despite the ubiquity of these glacial deposits, the idea that they were synchronous and deposited at low latitudes has been questioned. Some scientists preferred a tectonic explanation, suggesting that the continents may have drifted into and away from polar positions such that continents were glaciated at different times as they moved into the polar zones, analogous to the migration of the Gondwanan continents that debunked the Paleozoic snowball Earth hypothesis. When first documented, the lack of a firm chronology and lack of paleomagnetic data allowed much speculation regarding the significance of these glacial deposits. Perhaps they formed at higher latitudes and drifted towards the equator like the Paleozoic example discussed above. Or perhaps they represent a number of short unrelated glacial episodes. It is only recently that we have been able to reliably date these deposits and it now seems clear that there were two glacial epochs. The older **Sturtian Epoch** lasted

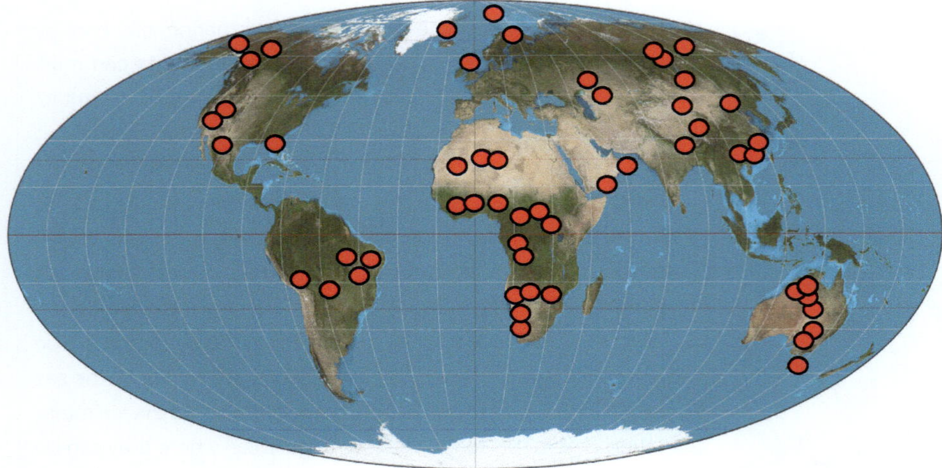

Figure 8.8 Modern distribution of Neoproterozoic Cryogenian tills.

Figure 8.9 (a) Glacial diamict overlies a striated quartzite. Arrow highlights direction of striations that show the direction of ice flow, Smalfjord Formation, Norway. (b) Striated and faceted boulder from the Jbélite tillite, Mauritania. Source: Photos by P.F. Hoffman.

Figure 8.10 Polygonal sand wedges in Scotland indicate subaerial exposure on the upper surface of a Sturtian glacial tillite (Port Askaig Formation) formed when glacial ice advanced across and later retreated. Source: Photo by P.F. Hoffman.

Figure 8.11 Vertically oriented dropstones show puncturing and deformation of laminated marine sediments, Ghaub Formation, Namibia. Source: Photo by P.F. Hoffman.

for about 59 million years (718–659 Ma), followed by an interglacial period of about 9 million years and the younger and shorter-lived **Marinoan Epoch** lasted for about 15 million years (650–635 Ma).

8.2.1 Evidence for Glaciers at the Equator

What makes this period of glaciation remarkable is that many of these sequences seem to have been deposited at or near the equator at low altitudes. Any Earth cold enough to have long-lasting ice at the equator would be cold enough to have ice everywhere, but proof of an equatorial origin rested largely on the paleomagnetic data. We introduced the concept of paleomagnetism in Chapter 5 (see Box 5.1). Paleomagnetic measurements that show low inclination of the magnetic field during deposition indicate formation nearer to the equator, whereas higher inclinations indicate formation at higher latitudes (Figure 5.6).

Analysis of the glacial deposits of the Sturtian and Marinoan epochs show low-inclination magnetic orientations, indicating that many of them were formed at or very near the equator. Initially there was skepticism regarding these paleomagnetic data. One argument was that the magnetic poles were reset by later tectonic processes. A more radical argument suggested that Earth's orbit was far more unstable during the Proterozoic, such that it experienced drastic changes in its **obliquity** – the degree to which Earth's rotational axis is tilted towards the Sun. The Earth is presently tilted at about 20° to the elliptical solar plane, but it was suggested that in the Proterozoic it may have been tilted up to 60° (Figure 8.12). In this argument, the poles would have faced the Sun and been much warmer, whereas the equatorial regions would have been much colder, potentially allowing for glaciers to grow there. However, astrophysical considerations of the Earth–Moon system show that high obliquity is virtually impossible, with only minor wobbles since 4 Ga. Improvements in paleomagnetic analysis have also confirmed a primary low-latitude setting in the Neoproterozoic and demonstrate that they were not reset during Paleozoic orogenesis.

The other compelling argument for equatorial glaciations was the observation that many of the glacial deposits are interbedded with carbonates. Carbonates are thought to be diagnostic of deposition in relatively shallow warm water. In addition, they commonly form in areas far away from major mountains, as mountainous areas typically deliver too much sediment to the oceans and impede the ability for carbonates to form. The ubiquitous carbonates are also difficult to reconcile with the high obliquity argument, as they would not likely form in the cold equatorial areas that high obliquity predicts.

KEY POINT

Paleomagnetic data and links to warm-water carbonates provide key evidence of low-latitude glacial deposits, supporting the idea of a global freeze.

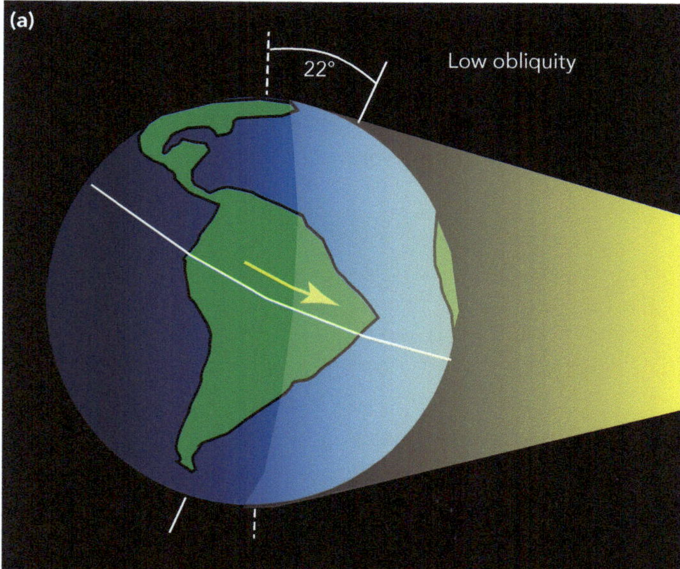

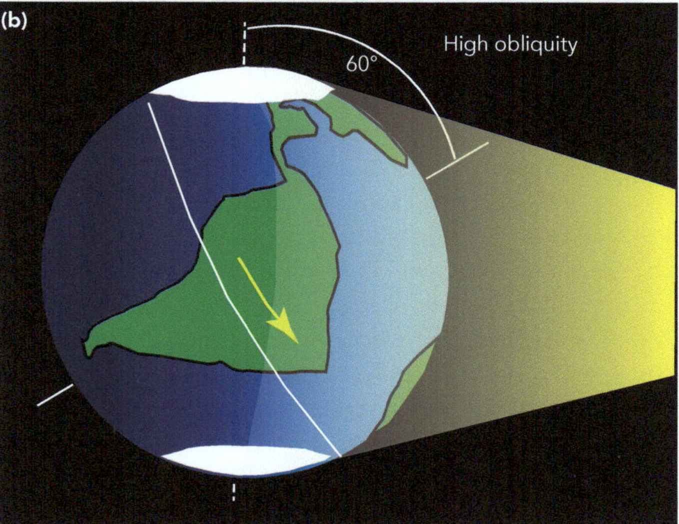

Figure 8.12 Large changes in obliquity were hypothesized to allow glaciers at the equator: (a) Earth orbit today at a 22° tilt; (b) Earth if it was tilted 60°.

8.2.2 Causes of Global Cooling

The observation of low-latitude, equatorial glacial deposits is quite remarkable, given that today, and through almost all of Earth history, glacial deposits are restricted to high latitudes or high altitudes. One must ask then, what made this time different?

8.2.2.1 The Faint Sun

As we saw in Chapter 3, the Sun was cooler in the past. In the Neoproterozoic, the Sun produced about 6% less energy than today, resulting in significantly less solar energy reaching Earth. In addition, changes in the incoming radiant energy from the Sun are primarily controlled by the distance of Earth from the Sun, which varies over 100,000- and 400,000-year cycles, and changes in the tilt (obliquity) of the Earth, which varies from about 22 to 24 degrees over 20,000 to 41,000-year cycles, rather than 22–60° as described in Figure 8.12(b). These orbital cycles cause climate to change and are an important driver of modern glacial cycles, discussed in more detail in Chapter 18. The surface temperature of Earth is primarily controlled by how much of the energy of the incoming radiation from the Sun is radiated back into space, and during the Neoproterozoic this would have been lower primarily because of the **faint young Sun**.

KEY POINT

The Sun was 6% cooler in the Neoproterozoic, and this helped drive Earth into a deep freeze.

8.2.2.2 Greenhouse Gases

The other major control on Earth's surface temperature are the **greenhouse gases**, including carbon dioxide (CO_2), water vapor (H_2O), and methane (CH_4). Of these three gases, CO_2 plays the largest role because it is abundant and has a long atmospheric residence time (from 50 to 200 years), allowing concentrations to build up, enhancing its heat-trapping effect and causing global warming. Water vapor, although abundant, has a very short residence time of only a few days and thus does not control long-term climate change. Clouds come and go over a few hours or days and if H_2O builds up too high it quickly falls back to Earth as rain or snow. Water vapor thus follows temperature changes rather than controlling it, but changes in water vapor can drive hydrological cycles and these can cause changes in the degree of weathering of silicate-rich rocks, such as granites and basalts (see Chapter 1), which can also consume CO_2.

An analog is the consumption of bread versus water. On a weekly basis, our weight may vary by a couple of pounds, but this mostly reflects retention of water. If you drink an excess of water usually it will result in more frequent trips to the bathroom with only a modest ability to retain the excess. Bread, in contrast, is stored much more readily as fat, and if one consumes an excess of calories this can result in weight gain that can only be mitigated by dieting or exercise. Atmospheric CO_2 behaves more like fat than water as it is more easily stored and can build up in the atmosphere due to its longer residence time. Methane molecules have a stronger heat-trapping ability than either CO_2 or H_2O, but the relatively low concentration of methane in the atmosphere and a short residence time of about 12 years generally make the long-term effects on global temperature subordinate to CO_2.

Carbon dioxide is primarily emitted to the atmosphere by volcanoes and metamorphism of carbonate rocks, and is consumed by two main processes, weathering of silicate rocks and by photosynthesis, the latter of which also produces oxygen. The rate at which CO_2 is produced or consumed plays a major role in determining the overall proportion of CO_2 in the atmosphere and consequently Earth's surface temperature.

Equation (8.1) is a simplification of the main components of silicate weathering:

$$2CO_2 + 2H_2O + CaSiO_3 \rightleftharpoons CaCO_3 + CO_2 + 2H_2O + SiO_2. \quad (8.1)$$

In the equation, CO_2 gas combines with liquid H_2O, forming weak carbonic acid (i.e., acid rain) that reacts with a calcium-bearing silicate mineral, $CaSiO_3$ in our example, and produces calcium carbonate ($CaCO_3$). The $CaCO_3$ typically precipitates in the ocean as limestone, and the silicon dioxide (SiO_2) precipitates as chert. These cherts and limestones become sediments, which are later compressed into sedimentary rocks, thus storing the CO_2 in rocks and decreasing CO_2 in the atmosphere. CO_2 is also consumed by photosynthesis, where it is stored in the organic material of the plants and algae that use it. However, for this CO_2 to be stored permanently the organic material must be buried in sediment versus oxidized and re-released at the surface back into the atmosphere. Limestones likely increased in abundance in the Neoproterozoic (Figure 8.13), reflecting the increase in oxygen photosynthesis and the development of calcification, which was able to store CO_2.

For most of Earth history the weathering and storage of CO_2 in sediment has been a highly constrained process that forms a **negative feedback loop** that disallows the catastrophic build-up of CO_2 that might otherwise cause runaway warming.

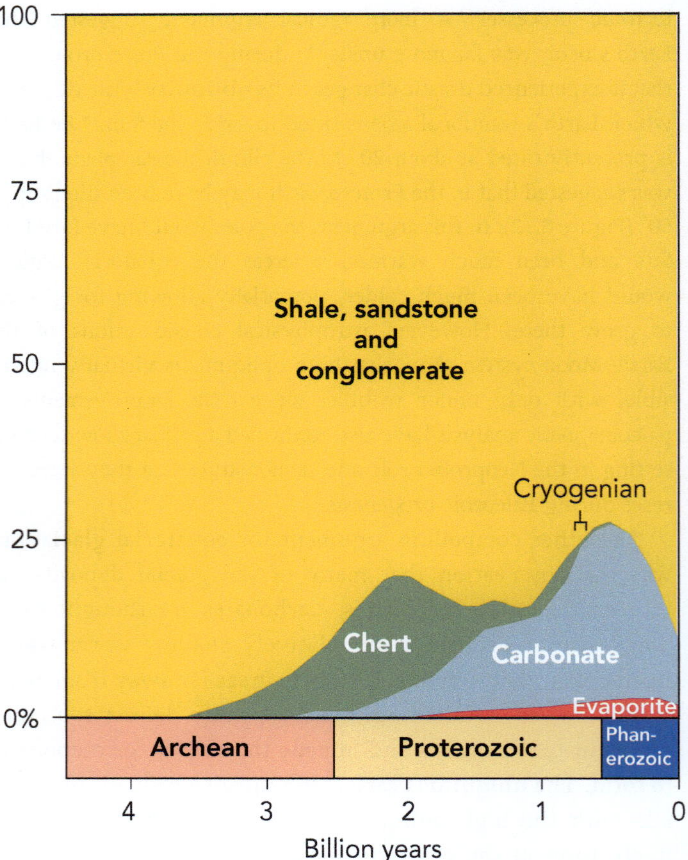

Figure 8.13 Sedimentary rocks through time. The sedimentary record is dominated by siliciclastic sediment-forming shales, sandstones, and conglomerates. Carbonate preservation coincides with the Great Oxygenation Event. Note the marked increase in carbonates in the late Proterozoic, coinciding with the Cryogenian. Although carbonates likely were deposited throughout the Archean they are rarely preserved. Source: Based on Ronov (1972).

As volcanoes and metamorphic processes emit CO_2, temperature increases. But this increase in temperature causes an increase in water vapor (recall H_2O follows temperature) and this extra water causes more vigorous **hydrological cycles** that in turn increase weathering rates. Increased weathering in turn increases sedimentation, allowing more storage of CO_2, which then causes a lowering of temperature, mitigating the effects of the excess CO_2 emitted by volcanoes and metamorphism in the first place. But as temperatures drop, because of increased weathering, H_2O drops, and weathering and plant growth slows down, thereby allowing CO_2 emitted by volcanoes and metamorphism to increase once again, which then increases temperature, and the loop continues. Because of this negative feedback loop, the average temperature of Earth's surface reflects the finely tuned balance of these competing processes. The snowball Earth hypothesis requires this finely tuned Earth system to have been seriously thrown off balance. By comparison, Venus has an atmosphere rich in CO_2 and CH_4 with very little H_2O and does not appear to have plate tectonics or a vigorous rock cycle. Venus has no life that we know of and consequently has a much warmer surface temperature than Earth as there is no mechanism to remove greenhouse gases.

8.2.2.3 Organization of the Continents, Weathering, Sedimentation, and CO_2 Sequestration

One key aspect that may have contributed to the freeze seems to have been the orientation of the continents. Between 1.1 and 0.9 Ga the supercontinent **Rodinia** was formed (Figure 8.14). Paleomagnetic study of rocks from this time indicates that this single continent spanned a region from near the South Pole to the equator and into the northern hemisphere to more than $60°N$ of the equator. Although Rodinia begun to break up around 720 Ma, the distribution of continents from pole to equator remained roughly the same and may have helped ice spread across the globe.

Recent tectonic modeling work of the breakup of the Rodinian supercontinent, by Adriana Dutkiewicz and her colleagues at the University of Sydney, Australia, suggest that volcanic emissions were lower than normal. In addition, the equatorial position of many continents in the Neoproterozoic likely resulted in more vigorous hydrological cycles. Weathering of volcanic rocks associated with the Franklin large igneous province in northern Canada may have provided an additional sink for CO_2. An increase in rain and running water are thought to have both enhanced weathering and increased sedimentation at their margins, both of which would have promoted CO_2 removal and enhanced cooling, and in combination with a faint young Sun and lower CO_2 emissions combined to trigger the prolonged deep freeze.

8.2.2.4 Runaway Albedo

In addition to greenhouse gases, the surface temperature of Earth is also partly controlled by the reflectivity of the surface of Earth, termed its **albedo**. Surfaces that reflect all the energy back into space have an albedo of 1 and surfaces that absorb all solar energy have an albedo of 0 (Figure 8.15). Albedo is largely controlled by the color and roughness of Earth's surface, the angle that the sunlight strikes the surface, and the insulating effects of the atmosphere, particularly greenhouse gases that aid in trapping some of the incoming radiant heat from the Sun versus simply reflecting it back into space. A significant proportion of the Sun's heat energy is also trapped in the world's oceans.

Fresh ice and snow on land have a high albedo of about 0.9–0.8; older snow, that is, snow that has lasted through at least one summer, is lower at 0.6–0.5; and sea ice ranges from 0.7 to 0.5. Some rocks and sediment have low albedos from 0.5 to 0.1, depending on their color and water content. Seawater has a low albedo of about 0.1 (Figure 8.15) and absorbs most of the solar energy it receives. Vegetation also reduces the albedo, but there were no plants in the Neoproterozoic, so in this case, the effect of plants can be ignored. Clouds have albedos between 0.8 and 0.4 and reflect a significant proportion of the Sun's energy, but as discussed above, clouds tend be quite ephemeral, so their effects are short-lived.

Once the poles began to freeze, the polar icecaps increased Earth's albedo, causing further cooling as more of the Sun's energy was reflected into space. Growth of marine ice shelves would have increased albedo (compared to the water the ice sheets flowed into), causing a **positive feedback loop**. This set up a **runaway albedo** event in which colder air produced more snow, which produced colder air leading to more ice in the oceans, leading to colder oceans, and so on. As the ice sheets advanced into the middle latitudes, meltwater and drift ice would have been deflected towards the equator by prevailing winds and **Coriolis forcing** (the deflection of wind and water flows due to Earth's rotation), enabling further advance of the ice sheets.

8.2.2.5 Peak Freeze

The peak freeze in the Neoproterozoic was driven by a number of factors. The cooler Sun and the development of photosynthesis following the Great Oxygenation Event (see Chapter 7) enabled removal of much of the early emitted CO_2. In addition, the increase in atmospheric oxygen would have caused removal of atmospheric methane (CH_4) by oxidation. The increase in weathering and sedimentation in the equatorial continents increased storage of CO_2, as seen by the marked increase in deposition of carbonates (Figure 8.13), and the modeling work that shows decreased CO_2 emissions from volcanoes. The colder Earth then experienced a runaway albedo that caused the amount of heat from the Sun reflected away from Earth's surface to drastically increase, and in combination with lower atmospheric CO_2 levels, which disabled the greenhouse effect, produced a snowball Earth, characterized by pole-to-equator ice

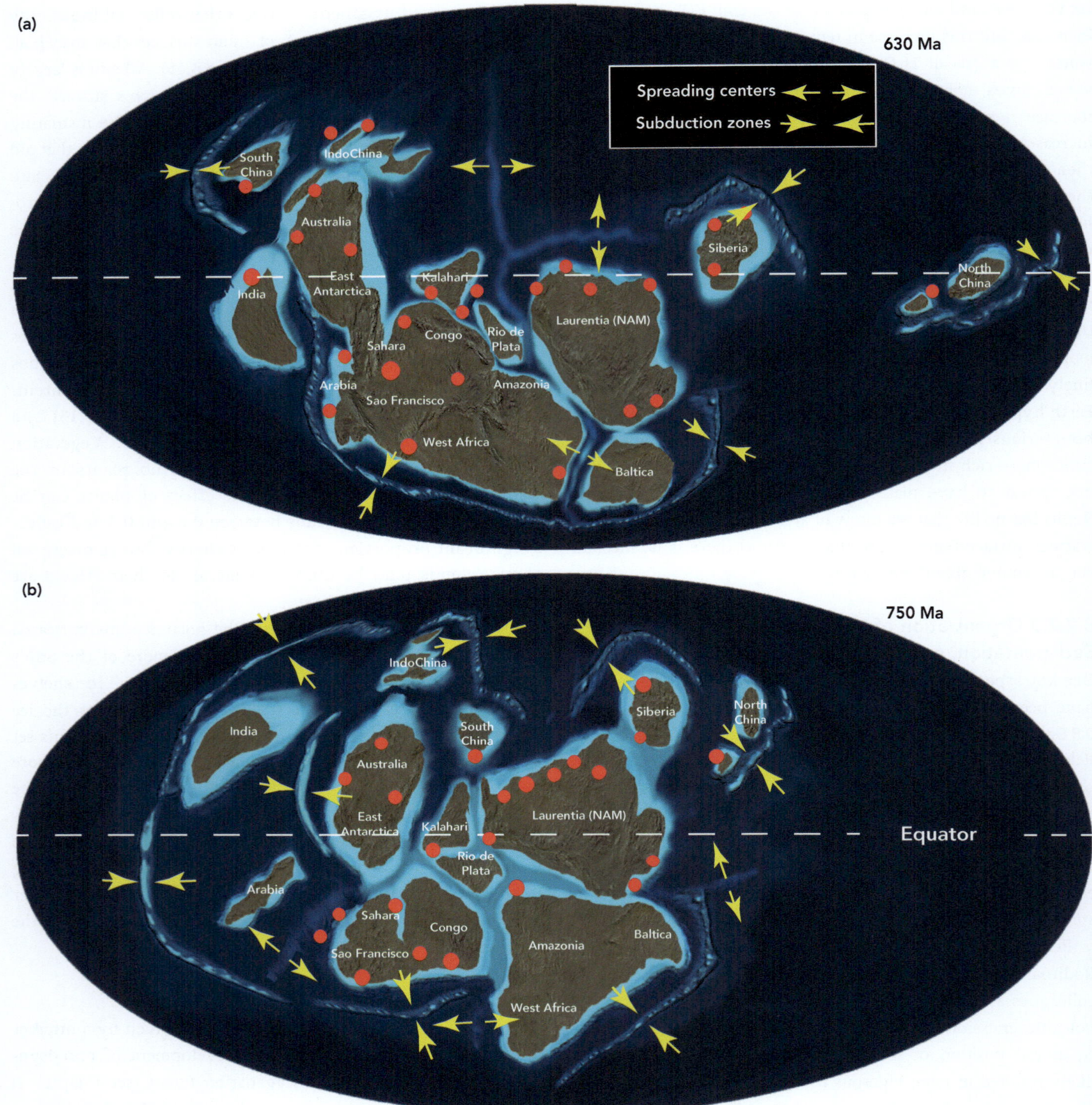

Figure 8.14 Neoproterozoic paleogeography and plate reconstructions at 750 (a) and 630 (b) million years ago. Red dots show the locations of glacial deposits and emphasize the ubiquity of development in low-latitude equatorial locations. The Rodinia supercontinent began breaking up at 750 Ma, prior to the older Sturtian glaciations. Source: Data from Hoffman et al. (2017); plotted on Blakey Deep Time maps (2 maps © DeepTimeMaps).

sheets. When marine water freezes, the surface ice is fresh, and the salt remains in the liquid phase. Therefore, as the tops of the oceans froze, the remaining seawater would have become increasingly salty, forming a dense brine. This would have created highly stratified oceans that likely experienced very little circulation and mixing. In addition, the excess salt would have lowered the freezing point of the oceans, allowing the briny water below the ice to remain in a liquid state.

Figure 8.15 Example of differences in albedo. The whiter sea ice covered with fresh snow reflects roughly 85% of the sunlight whereas the open Arctic Sea reflects about 7%. Source: Maxim Tupikov / Shutterstock.

KEY POINT

This is the first of several Earth's catastrophes linked to how changes in CO_2 change climate. The amount of CO_2 in the oceans and atmosphere is usually balanced, but in the Late Neoproterozoic, the rise of photosynthesis, methane reduction, and an increase in silicate weathering all conspired to cool Earth by removing CO_2 from the atmosphere. Once oceans began to freeze, a runaway albedo effect occurred that caused the snowball conditions.

With Earth covered with ice, the trend would be for the surface to maintain this condition. However, unless you are reading this at the South Pole or at the top of a very high mountain, you will notice that Earth is not covered by ice today. What then reversed the runaway albedo effect and brought the planet out of the deep freeze? Could this happen again? How did life survive during the deep freeze? In the next sections we will examine these questions to highlight the relationships between long-term climate changes, plate tectonics, and the negative feedback loops in the CO_2 cycle.

8.3 Out of the Deep Freeze

The runaway albedo affects the surface temperature of the Earth but has no effect on the heat *beneath* the surface that drives convection in the mantle, which is a critical component of plate tectonics, and for this story, the formation of volcanoes.

8.3.1 Volcanoes Change the Balance

While the surface of Earth froze over, radioactivity within the interior was still generating heat in the mantle, as discussed in Chapters 4 and 5. With plate tectonics continuing unabated (despite the covering of ice everywhere on the surface), volcanic activity would have continued at both divergent and convergent plate boundaries. This is important because, as discussed in Chapter 3, volcanoes are critical components of the chemical budget of the atmosphere, adding greenhouse gases such as methane and CO_2. Some of these volcanoes would have been underwater, such as along oceanic ridges, and this CO_2 would have been dissolved in seawater, increasing the acidity of the oceans. Other volcanoes would have out-gassed directly into the atmosphere. The gases released into the atmosphere would not have mixed with the oceans (as they do today) because most of the oceans were covered with a layer of ice. Therefore, over time, these gases would have built up to very high concentrations, producing a strong greenhouse effect in opposition to the runaway albedo effect. As we explained above, in the absence of snowball Earth conditions, weathering, photosynthesis, and storage of organic matter in sediments naturally buffers the warming effects from CO_2 release, preventing runaway warming. However, once the oceans were covered in ice, volcanism inevitably would have begun to cause a build-up of CO_2. However, in order to overcome the strong albedo of an ice-covered globe, climate modeling shows that a concentration of CO_2 in the atmosphere approximately 350 times greater than today would have been required to reverse the freeze.

Assuming Proterozoic volcanoes emitted CO_2 at rates comparable to modern volcanoes, it would have taken millions of years for them to emit enough CO_2 to reach concentrations sufficient to make the greenhouse effect greater than the effect of the runaway albedo. This is consistent with the durations over which snowball Earth conditions were in effect (59 million years for the Sturtian and 15 million years for the Marinoan). Snowball conditions would have shut down many of the primary sinks for volcanic CO_2 (i.e., photosynthesis and weathering) that are thought to have initially driven Earth into the snowball state in the first place. Increased ocean acidity (from the submarine volcanoes) may have caused an increase in weathering at the ocean floor, allowing some sequestering of the oceanic CO_2 that is thought to have lengthened the time of sustained freezing and may help to explain the long duration of the Sturtian glaciation.

After the tops of the oceans froze, Earth's surface temperature would have fallen to around $-50\,°C$. As CO_2 slowly accumulated in the oceans, global mean surface temperatures would not increase until a CO_2 threshold was reached and surface temperatures finally rose above freezing, ice would have begun to melt, and it is thought that this set up a runaway melting feedback loop opposite to what brought on snowball Earth in the first place. With a bit less ice, the albedo of the surface decreases, causing the surface to warm up, causing more ice to melt, and so on. The transition from snowball Earth to more normal ice conditions would have been quite rapid. It would have been so fast that the ice would melt faster than it takes for the CO_2 to be transferred from the atmosphere to the oceans. Without the balancing effect of an extremely low albedo, the extreme concentrations of CO_2 in the atmosphere would have pushed the surface temperature from icehouse to greenhouse very quickly, with average surface temperature rising to about $40\,°C$ in just a few thousand years (Figure 8.16).

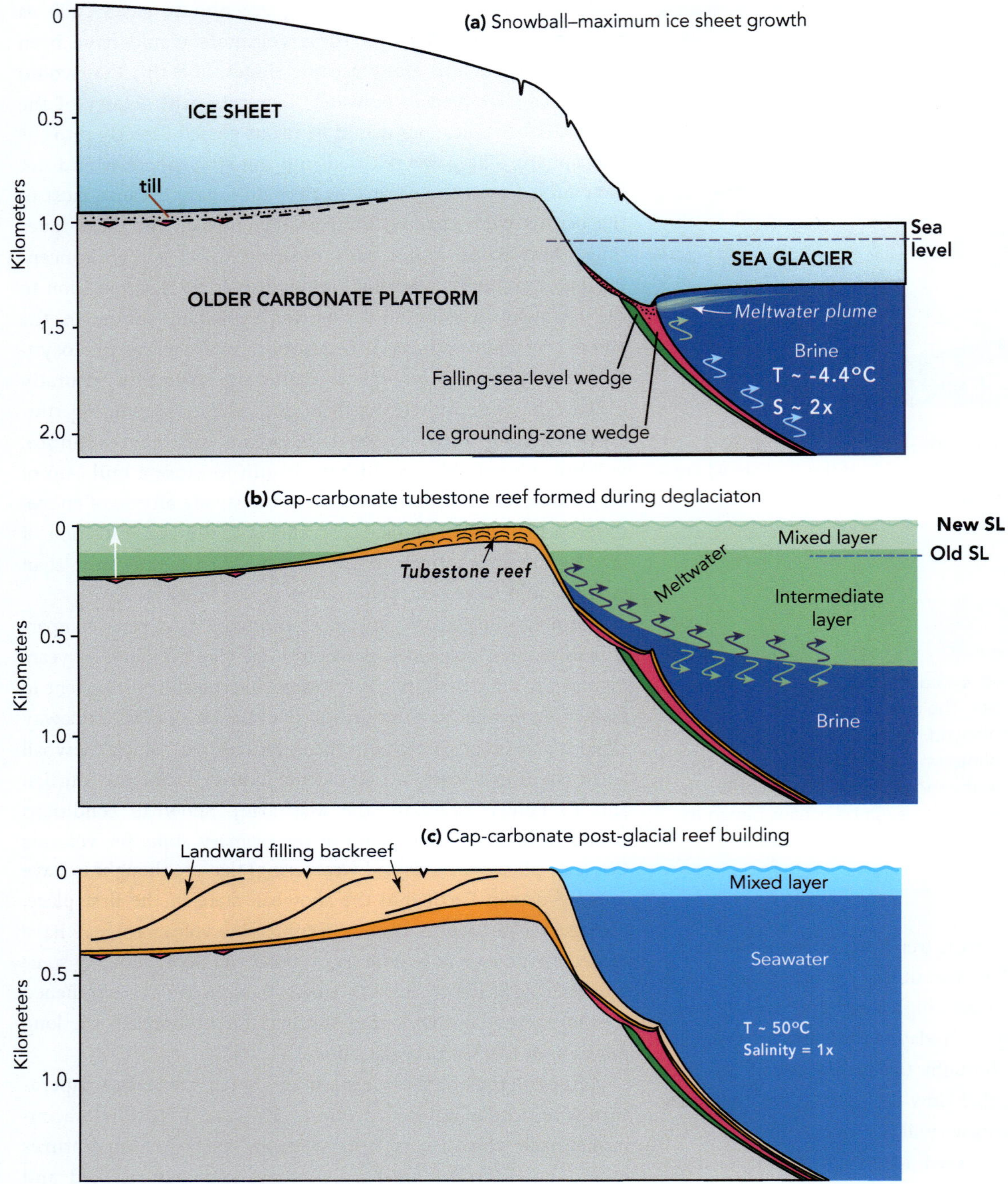

Figure 8.16 Evolution of the Late Neoproterozoic African continental margin and the formation of cap carbonates. (a) As ice sheets grow, sea level falls, producing a wedge of falling-stage sediments, ice grounding deposits, and glacial tills. Seawater becomes briny and prevents total freezing of the oceans. (b) During deglaciation, about 250 m of sea-level rise occurs and tubestone carbonate reefs form on top of the till deposits. A massive fresh to brackish water meltwater lid covers much of the ocean with a salty brine beneath. (c) Following deglaciation, oceans warm and sea level continues to rise. The flooded shelf, landward of the reef margin, fills with landward migrating back-reef limestones. Oceanic waters return to a warmer temperature and normal salinity. Source: Figure modified after Hoffman et al. (2017), Science Advances (AAAS).

8.3.2 Cap Carbonates

The evidence that a slowly accumulating but ultimately massive CO_2 build-up forced Earth out of snowball conditions is found in many rock sequences that lie on top of the Proterozoic glacial deposits. One of the key observations of Cryogenian stratigraphy is that most of the glacial deposits are overlain by a limestone or dolostone unit termed a **cap carbonate** (Figures 8.16 and 8.17). Recent isotopic dating shows that these carbonates appeared simultaneously at many locations around the world, within the precision limits of dating methods which range from 0.45 to 6 million years. The end-Sturtian cap carbonates are dated as occurring between 659.3 and 658.5 Ma and the Marinoan deposits are dated at between 636.0 and 634.7 Ma.

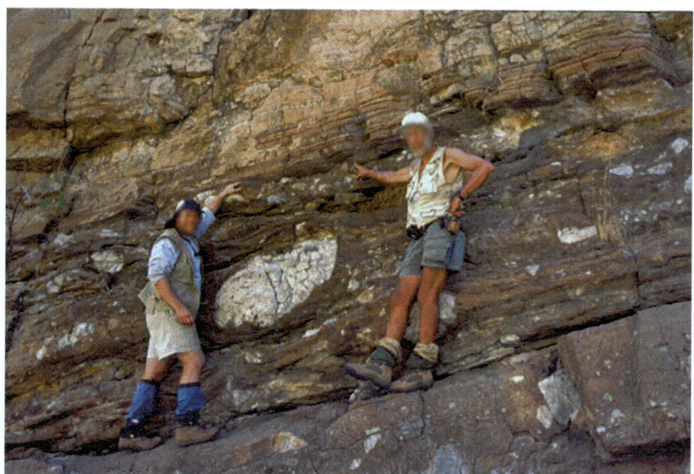

Figure 8.17 Daniel Schrag (left) and Paul Hoffman (right) stand next to glacial diamicts and point to the base of cap carbonates that have been correlated worldwide, indicating synchronous glacial melting and precipitation in relatively deep water. Source: Photo by Gabrielle Walker (Scientific American 2000).

Figure 8.18 Bladed crystals in Marinoan cap carbonates of the Hayhook Formation, Canada. The pink mineral is lime mud that was deposited between the crystals as they grew. Source: Photo by P.F. Hoffman.

During the snowball periods, the oceans became saturated with dissolved CO_2 and became increasingly acidic, marked by a lowering of pH, causing dissolution of carbonate rocks, as illustrated in Eq. (8.2):

$$CaCO_3 + CO_2 + H_2O \rightleftharpoons Ca^{2+} + 2HCO_3^-. \qquad (8.2)$$

In the equation, carbonate ($CaCO_3$) reacts with carbon dioxide (CO_2) and water (H_2O) to produce bicarbonate ($2HCO_3^-$) and dissolved calcium (Ca^{2+}). Although the oceans became more acidic, the increase was slowed (i.e., buffered) by carbonate sediment, mostly delivered to the oceans by glacial erosion of older widespread carbonate platforms. As glaciers reached the sea, the thickness of the ice sheet decreases, forming a steeply dipping transition zone (Figure 8.16(a)) where ablation (i.e., loss of ice) occurs. This ablation helps drive the glaciers to flow from land to sea, even in the absence of an open ocean. The dynamic flow of glaciers results in significant erosion of the underlying carbonate platforms and this likely promoted delivery of carbonate rock powder via subglacial plumes. The powder produced an effect identical to an antacid tablet, which you take to cure acid reflux. The carbonate in the tablets reacts with and neutralizes the effects of excess stomach acid. In the snowball oceans, the carbonate rock powder would be quickly dissolved by the acidic oceans, but this process consumes the acid and inhibits and slows the rate at which acidification occurs. All that excess dissolved carbonate also increases the alkalinity such that the oceans became supersaturated with dissolved calcium carbonate. The loss of seawater into ice also dramatically increased ocean salinity and prevented the oceans from freezing (Figure 8.16(a)).

Eventually CO_2 levels reached the critical threshold, and the greenhouse effect took over, causing a dramatic warming of Earth that triggered deglaciation. The meltwater would have formed a **meltwater lid** of relatively fresh water over 500 m in thickness (Figure 8.16(b)). Deacidification, combined with the extreme levels of dissolved carbonate in the highly alkaline oceans, as well as mixing with meltwater, provide the perfect conditions for mass precipitation of the carbonate that had been held in solution in the snowball oceans. The meltwater would have introduced fresh water over the briny and acidic deeper ocean water and the mixing of highly alkaline brines with fresh water would have created a zone of lowered acidity in an area that had an excess of carbonate in solution. Supersaturation would force carbonate to precipitate. Some of the cap carbonates are characterized by large radial crystals (Figure 8.18), indicative of **abiotic** precipitation in relatively deep water (i.e., hundreds of meters), and this contrasts with shallow water carbonates, which show features such as stromatolites, indicative of a photosynthetic biotic origin (see Chapter 7).

KEY POINT

Following the peak freeze, silicate weathering and photosynthesis largely shut down and the release of CO_2 by volcanoes gradually warmed the atmosphere enough for the globe-covering ice to melt. High acidity and high alkalinity as well as freshwater mixing from the meltwater caused widespread precipitation of cap carbonates as the Earth returned to a warmer state.

We know that, in the current glacial epoch (in which we are now in an interglacial interval), continental glaciers have advanced and retreated large distances many times over in episodes lasting a few thousand to a few tens of thousands of years. These fluctuations are due to variations in the orbits of Earth and other nearby planets. During the Cryogenian Period, Earth's orbit was surely also variable but the transition out of snowball conditions was driven solely by atmospheric chemistry over a period less than 10,000 years, during which the surface went from almost completely ice covered to almost completely devoid of surface ice.

8.4 How Many Glacial Episodes Were There?

There has been some debate as to whether the Sturtian and Marinoan represent a relatively stable and long-lived glacial episode or whether they represent a series of separate glaciations that occurred in different places over a prolonged time interval, because of tectonic uplift during Rodinian breakup. Although some of the early work suggested that the glacial deposits were synchronous, based on the ubiquity of the cap carbonate layer and a similar carbon isotopic signature, high-precision isotopic dates were lacking. In addition, a paucity of fossils hampered biostratigraphic correlation. However, since 2004, geochronology of rocks associated with the cap carbonates has shown they were deposited with remarkable synchroneity. It was also noted that many of the glacio-marine diamicts were interbedded with deep water deposits that lack evidence of glacial debris, indicating that the ice sheets were quite dynamic. Deep water deposits represent deposition by sediment-rich **turbidity currents**, which can be initiated by the gravitational collapse of sediments deposited at a continental margin or by major river floods. This sudden transport mixes sediment with seawater and the resultant flow moves down the continental slope as debris flows or turbidity currents. Of course, these types of currents do not require an open ocean, and it is possible that enough meltwater was generated at the base of the equatorial ice sheets to help generate these flows, but if these occurred associated with a floating ice sheet, more evidence of ice-carried debris, such as dropstones, might be expected throughout.

More difficult to explain are occurrences of wave-rippled sandstones interbedded with glacio-marine diamicts. Wave-ripple marks are formed in shallow water (usually less than 50 m) by the action of wind blowing over the water. Stronger winds, such as during a storm, can form wave ripples in deeper water than during non-storm times, and the longer the stretch of open water over which the wind blows (termed the **fetch**), the larger the waves. The presence of wave ripples suggest that there may have been periods of open water, but how much open water existed and for how long is still under debate. Another concern is how life might have survived a hard snowball, and this has led some to suggest a "**slushball**" Earth, which would have allowed low-latitude glaciations but with extended periods of open ocean that would have served as refugia for existing life on Earth.

Paul Hoffman, who has been at the forefront of the snowball Earth hypothesis, suggests an alternate explanation for the cycles of interbedded turbidites and diamictites. His research group suggests that equatorial continents likely were quite dry and, owing to a lack of vegetation, would have generated a large amount of dust. This dust would have accumulated on top of the ice sheets and over time become concentrated in the areas where ablation occurs, particularly at the point that they flow into the oceans. Ablation of the ice would increase the concentration of dust on top of the ice and over time this would form a dark crust termed **cryoconite** (Figure 8.19). Cryoconite

Figure 8.19 Dark cryoconite holes on the Canada Glacier, Antarctica. Warmer surface temperatures and high winds allow dust to accumulate on the surface as dark clumps, rich in organic matter, that sink to create the holes. Cyanobacteria as well as eukaryotic organisms and metazoans live in these holes and are considered as analogs to the refugia that may have harbored life on top of ice sheets during the Cryogenian. Source: Cryoconite holes on Canada Glacier (Taylor Valley, South Victoria Land, Antarctica), downloaded from the web. Original source unknown.

darkening decreases the albedo, which can cause partial melting of the ice surface, forming shallow ponds that may have served as refugia for photosynthetic organisms that would also cause further darkening of the ice surface. Darkening increases albedo and allows for partial melting, which may in turn have allowed for more dynamic ice sheets that experienced periods of advance and retreat, without requiring open oceans. This leaves the observation of wave-rippled beds unexplained, but, as mentioned above, the sparseness of wave ripples suggest that open water was a rare occurrence. In addition, climate modeling studies indicate that it would be quite difficult to sustain an open equatorial ocean and still allow for equatorial glaciations, as indicated by the ubiquity of observed glacial deposits.

8.5 Implications for Life on Earth

About 25 million years after the Cryogenian Period ended, Earth experienced a marked diversification of life, starting with the appearance of the first metazoans of the **Ediacaran fauna** and the later **Cambrian explosion**, which is discussed more fully in Chapter 9. There has been much speculation as to how life on Earth coped and survived during the Cryogenian glaciations, given that it appeared that marine photosynthesis effectively shut down. During global crises local populations may survive in

refugia (environments where survivors are able to take refuge). In the Cryogenian, these may have included deeper oceanic environments including hydrothermal vents. Paul Hoffman has suggested that photosynthetic organisms may also have taken refuge in cryoconite ponds on top of the Cryogenian ice sheets, whereas those that promote the slushball hypothesis would favor local open ocean refugia as areas where life survived. There is also evidence that photosynthetic life may have survived in freshwater lakes in areas that were ice-free; analogous to stratified lakes found in the dry valleys of Antarctica. Once deglaciations occurred, rising sea level would reconnect with these lakes and cryoconite ponds, reintroducing photosynthesizers to the oceans. Evidence from genomics, organic biomarkers, and molecular clocks shows that photosynthetic algae became the dominant photosynthesizers on Earth during the interglacial period between the Sturtian and Marinoan glacial episodes. This may have been driven by a massive increase in nutrient supply, and particularly phosphate, following the Sturtian deglaciation (see Chapter 6). As the food chain began anew, the next major step in the evolution of life on Earth occurred and will be the topic of the next chapter.

KEY POINT

Although photosynthesis almost shut down, life managed to survive in refugia, such as freshwater ponds on tips of ice sheets, around deep water hydrothermal vents, and possibly in freshwater lakes on land.

8.6 Why Hasn't There Been Another Snowball?

Following the Sturtian glaciations, the intervening warm period lasted about nine million years. During this time, resumption of photosynthesis, as well as the explosion of eukaryotic algae, and continental silicate weathering and sedimentation gradually reduced the CO_2 during the interglacial and once again allowed another snowball period to occur, because the continents retained their pole-to-equator north–south configuration, allowing enhanced hydrological cycles and enhanced silicate weathering, and the Sun was still about 6% colder than today. The ensuing Marinoan glacial period lasted about 15 million years. The shorter duration may reflect lower rates of submarine weathering that allowed CO_2 levels to build back up more quickly to the threshold required for deglaciation than during the prior Sturtian. Since then, Earth has experienced numerous other glaciations (Figure 8.1), but never to the extent of snowball Earth. A main reason is that the Sun has become warmer because of its natural evolution (see Chapter 3). As we shall see in later chapters, Earth has experienced other glaciations, such as the Late Paleozoic glaciations in **Gondwanaland**, which were also driven by a decrease in atmospheric CO_2 levels and will be discussed more fully in Chapter 11. Global climate models suggest that in the Paleozoic, CO_2 nearly dropped to the levels that caused the Cryogenian snowballs, but a warmer Sun and slightly higher CO_2 levels prevented another snowball from occurring.

KEY POINT

A warmer Sun is the main reason that Earth has not experienced another full-fledged snowball state, although variations in greenhouse gases, primarily CO_2, have caused cycles of glaciations (icehouse) and hotter "greenhouse" periods.

8.7 Summary

- The idea that the Earth was frozen was proposed and rejected a number of times but has received the most traction with the recent snowball Earth hypothesis, which proposes that Earth froze in the Neoproterozoic Cryogenian Period.
- Observations of low-latitude glacial deposits, using improved glacial depositional models integrated with paleomagnetic data, indicate that snowball conditions occurred in two pulses (Sturtian and Marinoan) during the Cryogenic Period (720–635 Ma) of the Neoproterozoic Era.
- The snowball was partly caused by a fainter Sun, generating 96% of the radiant heat compared to today.
- The breakup of the Rodinian supercontinent resulted in continents that extended from the South Pole, across the equator and up to 60° N latitude. An increase in the vigor of the hydrological cycled enhanced weathering and sedimentation and allowed more CO_2 storage, driving Earth into a cooler state. As ice began to accumulate, this resulted in a runaway albedo as glaciers at the poles expanded across the land surface all the way to the equator, eventually causing the ocean to freeze over.
- A gradual build-up of CO_2 following peak freeze conditions caused oceanic acidification that was buffered by delivery of carbonate from glacial meltwater which also increased ocean alkalinity. Shut down of photosynthesis and silicate weathering along with a steady release of CO_2 from volcanoes during the snowball period eventually caused a sudden global warming, as CO_2 levels reached a critical threshold that ended the first snowball period

and triggered the deposition of widespread cap carbonates.

- Resumption of photosynthesis and vigorous weathering in the ensuing interglacial once again allowed another snowball period to occur.

- Life on Earth may have survived in deep-sea refugia, in cryoconite ponds on the ice sheets and in dry-valley lakes.
- A warmer Sun may serve as the main explanation as to why Earth has never experienced another snowball period.

Key Words

- icehouse
- greenhouse
- Cryogenian Period
- snowball Earth
- die eiszeit
- ablation
- diamict
- striations
- chatter marks
- till
- grounding line
- boulder pavements
- dropstones

- periglacial
- Sturtian Epoch
- Marinoan Epoch
- obliquity
- faint young Sun
- greenhouse gases
- negative feedback loop
- hydrological cycles
- Rodinia
- albedo
- positive feedback loop
- runaway albedo
- Coriolis forcing

- cap carbonate
- meltwater lid
- abiotic
- turbidity currents
- fetch
- slushball
- cryoconite
- Ediacaran fauna
- Cambrian explosion
- refugia
- Gondwanaland

Further Reading and References

Dutkiewicz, A., Merdith, A. S., Collins, A. S., et al., 2024, Duration of Sturtian "Snowball Earth" glaciation linked to exceptionally low mid-ocean ridge outgassing, *Geology*, https://doi.org/10.1130/G51669.1

Hoffmann, P. F., and Schrag, D. P., 2000, Snowball Earth, *Scientific American*, 282(1), 68–75.

Hoffman, P. F., Abbot, D. S., Ashkenazy, Y., et al., 2017, Snowball Earth climate dynamics and Cryogenian geology-geobiology, *Science Advances*, 3(11), e1600983, https://doi.org/10.1126/sciadv.1600983.

Ronov, A. B., 1972, Evolution of rock composition and geochemical processes in the sedimentary shell of the Earth, *Sedimentology*, 19(3–4), 157–172.

Ziegler, A. F., Smith, C. R., Edwards, K. F., and Vernet, M., 2017, Glacial dropstones: islands enhancing seafloor species richness of benthic megafauna in West Antarctic Peninsula fjords, *Maritime Ecology Progress Series*, 583, 1–14, https://doi.org/10.3354/meps12363.

Review Questions

1. How did plate tectonic theory debunk the idea of a Paleozoic deep freeze?
2. What are some of the diagnostic features of glacial deposits?
3. What are some of the key lines of evidence that support synchronous, low-latitude, equatorial glaciations in the Neoproterozoic?
4. What are the main lines of evidence that Earth experienced a near total freeze in the Cryogenian Period?
5. What were the main causes of the snowball Earth?
6. Explain why CO_2 is a buffered system, in terms of its effects as a greenhouse gas.
7. What is a runaway albedo?
8. What triggered the end of the older Sturtian glaciations and what caused resumed glaciations in the Marinoan?
9. What is the significance of cap carbonates and how did they form?
10. How did life on Earth survive the snowball conditions?
11. Why has Earth not experienced another snowball?

View of Canadian Rockies from Burgess Shale, Yoho National Park, Alberta, Canada. Source: Matt Champlin / Getty Images.

Chapter 9

An Explosion of Life

Ediacaran Experimentation, the Cambrian Explosion, and Ordovician Biodiversity

LEARNING OBJECTIVES

- Describe how sea-level rise and the breakup of Rodinia led to the Sauk transgression and new habitats for metazoan evolution.
- Explain how molecular clocks and modern biology help infer the relationships of early metazoans and their single-celled ancestors.
- Describe the Ediacaran fauna and its unusual life strategies, including mobile feeders that may be precursors to modern animals.
- Explain the significance of the Cambrian explosion and the experimental body plans of its organisms.
- Describe the Great Ordovician Biodiversity Event and the modes of life diversification, including evolutionary experimentation and extinction.
- Explain the idea of a Cambrian, Paleozoic, and Modern Faunal overturn.
- Explain the concept of an evolutionary arms race.
- Describe the cause of the first mass extinction event at the end of the Ordovician Period.

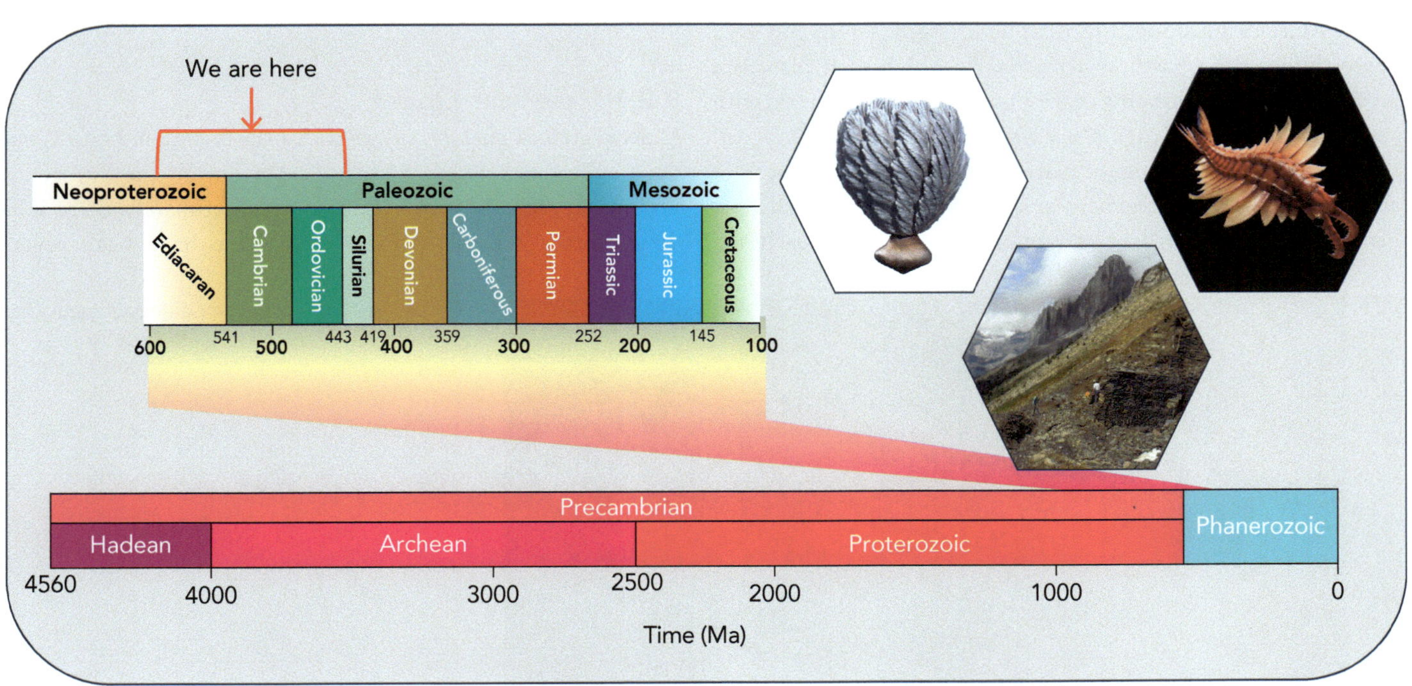

Introduction

In the nineteenth century, when geologists were developing the geological timescale, the newly defined Cambrian Period marked a profound change in rocks. Cambrian and younger formations were rich in fossils, whereas older Precambrian rocks appeared to be free of fossils and had historically been assigned to the **Azoic Eon** (literally "time without life"). As discussed in Chapter 7, we now know that the Precambrian has a rich history of single-celled colonial life. It is also clear that **metazoans**, or multi-celled animals, appeared in the Neoproterozoic Era as the Earth was coming out of the Cryogenian snowball Earth period, which was the focus of Chapter 8. This was followed by the **Cambrian explosion**, marked by the appearance of organisms with hard parts. Both the earlier Ediacaran and ensuing Cambrian systems contain fossils of several difficult-to-classify organisms. Are these enigmatic fossils failed evolutionary experiments in basic animal body plans in a less crowded Earth? Which of these organisms were the early ancestors of the more common forms of life that we see on Earth today? How explosive was the Cambrian explosion? This chapter examines the rise and subsequent diversification of metazoans, as well as the development of complex interactions between predators and prey leading to evolutionary arms races that illustrates some of the concepts of evolution introduced in Chapter 6.

9.1 Setting up the Cambrian Explosion: The Sauk Transgression

The Late Neoproterozoic was marked by the breakup of the supercontinent Rodinia to form Gondwanaland as well as several other continents, including Laurentia, which was ancestral North America introduced in Chapter 1 (Figures 1.32 and 9.1). If we think of the Earth as a puzzle, in which each plate is a piece, then supercontinents represent an assembled puzzle with shorelines around the edge. Once a supercontinent breaks apart, the increase in the number of plates increases the total length of available edges, yielding new areas of shoreline and shallow marine shelves, which are the habitats where much of early life

evolved. Separation of continents also allows more isolation of communities promoting allopatric speciation (see Chapter 6). In addition, the breakup was accompanied by creation of an extensive young and hot oceanic lithosphere, which isostatically rode high on the asthenosphere, due to its lower density (see Chapter 3 for explanation of isostasy), resulting in a worldwide decrease in the volume of ocean basins. In addition, the end of snowball conditions added significant volumes of water from glaciers back into the now-smaller oceans. Because the ocean basins were smaller, all that excess water had to go somewhere, resulting in extensive flooding of the newly formed shallow continental areas. Combined with the increase in shoreline areas resulting from the breakup of Pangea, vast new shallow marine habitats became available for metazoan diversification. This prolonged sea-level rise, referred to as the Sauk transgression (see Figure 1.21), is also marked by a worldwide increase in sediments deposited at the margins of these extensive new shorelines and shelves. These deposits contain a rich record of early multi-celled life that sheds light on the origin and diversification of metazoans.

KEY POINT

The end of snowball Earth conditions, the breakup of Rodinia, and the growth of new oceans caused a global rise in sea levels that provided vast new shallow water habitats, as well as isolation of areas that allowed evolution of the Ediacaran fauna and the later Cambrian explosion.

9.2 When did Metazoans Arise?

In this section we review the evidence from molecular clocks as well as fossils as to when the first metazoans appeared.

9.2.1 Molecular Clocks

Molecular clock analysis suggests that the earliest metazoans may have appeared between 800 to 750 Ma during the Cryogenian Period, or perhaps even earlier, but unfortunately there is no direct fossil evidence (Figure 9.2). There have been several

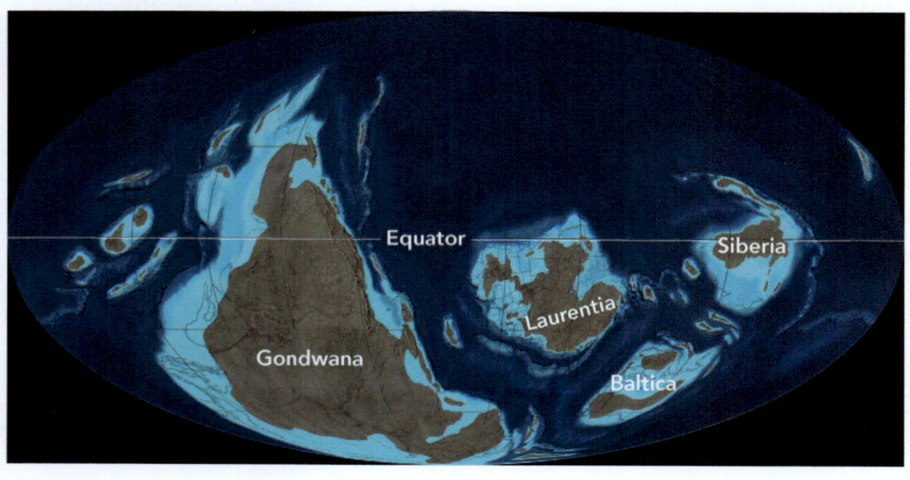

Figure 9.1 North America paleogeography in the Late Cambrian (500 Ma). Source: © DeepTimeMaps.

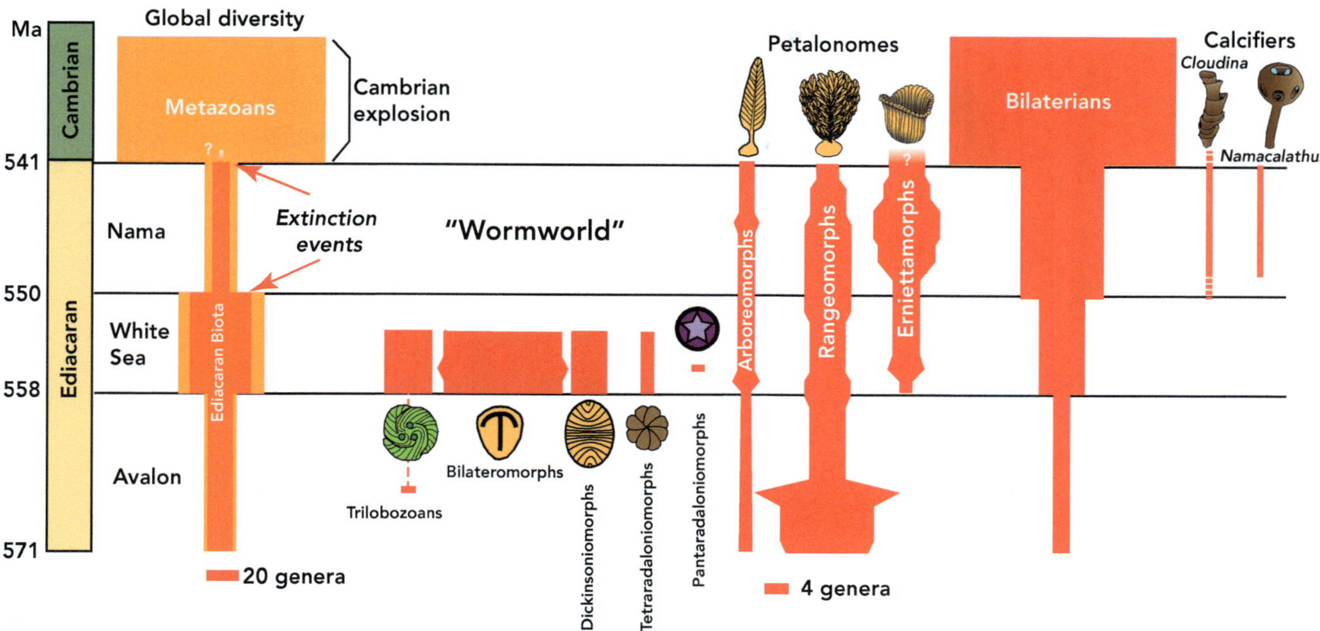

Figure 9.2 Diversification and extinctions associated with the Ediacaran and early Cambrian periods.

candidates for the oldest fossil metazoans, including multi-celled embryos from the Doushantuo Formation in China dating to around 600 Ma (Figure 9.3), but all have been controversial. The oldest unequivocal metazoans are found in the Upper Ediacaran Period of the Neoproterozoic Era, around 580 Ma.

The time gap from the origin of metazoans, as determined from molecular clock analysis, and the Doushantuo embryos is about 200 million years. Several explanations for this gap have been proposed, not the least of which is the difficulty of fossilizing soft-bodied organisms. Also, because these organisms evolved during the Cryogenian Period (see Chapter 8), life was likely confined to highly localized refugia, greatly reducing the likelihood of an extensively preserved fossil record. We also emphasize that the timings obtained using the molecular clock model have large uncertainties and are better considered as hypotheses with which to compare the stratigraphic record rather than inviolable observations.

9.2.2 Observations From Modern Biology

If we return to the animal cladogram, shown in Figure 6.8, there is wide agreement that the simplest metazoans are the **Porifera**, or sponges (Figure 9.4). Sponges are filter feeders that use cells with flagella to pump seawater into their bodies, extracting nutrients such as plankton, bacteria, and other organic detritus. The spent water is pumped out of the hole on top of the animal (Figure 9.4). Many sponges grow protective needles, called **spicules**, which are commonly fossilized, and some sponges developed calcareous skeletons. Unlike more complex animals, sponges have no distinct mouth or body cavity, like our own gut, and they show no symmetry. Interestingly, the cells that line the inside of sponges, called **choanocytes** (Figure 9.4(c)) are identical to single-celled **choanoflagellates** (Figure 9.4(d)), so-called because they have a collar around their flagellum.

Figure 9.3 Doushantuo embryos (scale bar is about 250 microns). Source: Image in the middle row, to the right: Reproduced from Gao et al. (2008). © 2008 Geological Society of China. All other images: Reproduced from Cunningham et al. (2017).

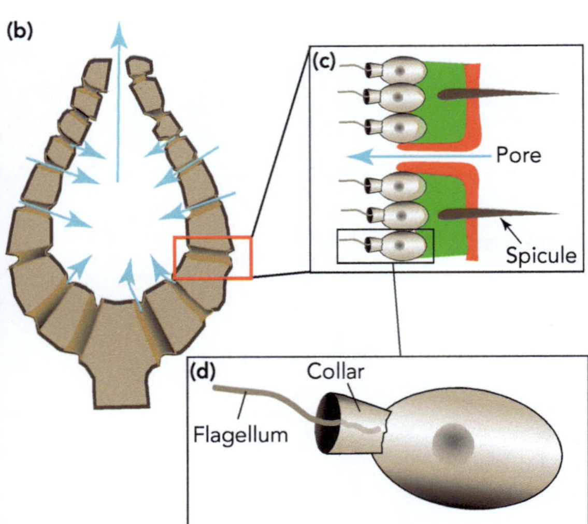

Figure 9.4 (a) Modern sponge, offshore California; (b) cross section of sponge shows water is drawn through pores and exhaled out of the top of the animal; (c) close-up of pores lined with choanocytes; (d) choanocyte is identical to a single-celled choanoflagellate and has a collar around its flagellum. Source: (a) NOAA/Monterey Bay Aquarium Research Institute, Public domain, via Wikimedia Commons.

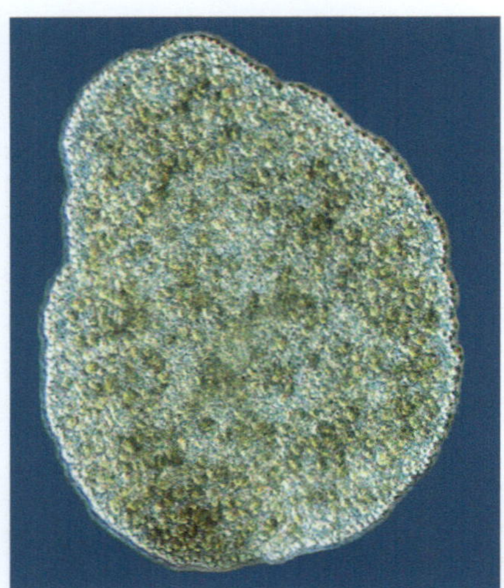

Figure 9.5 Placozoa, animal is about 0.5 mm across. Source: Bernd Schierwater, CC BY 4.0 <https://creativecommons.org/licenses/by/4.0>, via Wikimedia Commons.

Observation of modern choanoflagellates shows that they can form colonies by sticking together, and it is hypothesized that this type of behavior eventually allowed colonial choanoflagellates to evolve into sponges. The choanoflagellates are thus considered to be the closest single-celled ancestor to the earlier metazoans.

Given that the sponges are the simplest of animals, it is widely assumed that they were also the first metazoans. However, there are other animals at the base of the animal family tree, such as the **Placozoa**, or "flat animals," that are even more simple (Figure 9.5). Placozoans have no fossil record, but studies of modern animals show that individuals consist of a few thousand cells grouped into four different cell types. This contrasts to the 200 or so different cell types that humans have. The position of Placozoa on the animal cladogram is speculative, and there is molecular evidence that they may have appeared later than sponges, suggesting that

they descended from more complex animal ancestors. **Ctenophores**, more commonly known as comb jellies (Figure 9.6), are also thought to have appeared around the same time as sponges. The comb jellies have two layers, and a few species have a third middle layer, suggesting they may be primitive **triploblasts** (three-layered animals, see Chapter 6). They also show a radial symmetry, mouth, and anus as well as a pair of tentacles and they are covered with sticky cells that capture prey (Figure 9.6(a)). Like their sister group, **Cnidaria**, which includes jellyfish, anemones, and corals, they can also swim freely. Unlike placozoans, fossilized ctenophores have been found in Cambrian rocks (Figure 9.6(b)) so we know that they appeared at least 540 million years ago and probably earlier. It also seems clear that multicellularity evolved several times, probably stemming from colonial lifestyles in single-celled eukaryotes, as we saw with the choanoflagellates, that led to the differentiation of multi-celled organisms into the Fungi, Animalia, and Plantae kingdoms that are dominant today (Figure 6.6).

Part of our point here is that inferences about evolutionary relationships can be made by careful examination of the biology of modern organisms, integrating analysis of cells, genomics, embryological development, and anatomy, independent of the fossil record. This approach is particularly important in examining the relationships among the oldest phyla, especially those like placozoans that do not produce skeletons that can be readily fossilized. In the next section we focus on what is actually in the fossil record, which is another key archive that tells us the story of the origin of early metazoans.

KEY POINT

The evolutionary relationships of early metazoans can be inferred by careful examination of the biology and genomics of modern organisms, independent of the fossil record, suggesting that the Porifera were the earliest, and simplest, metazoans.

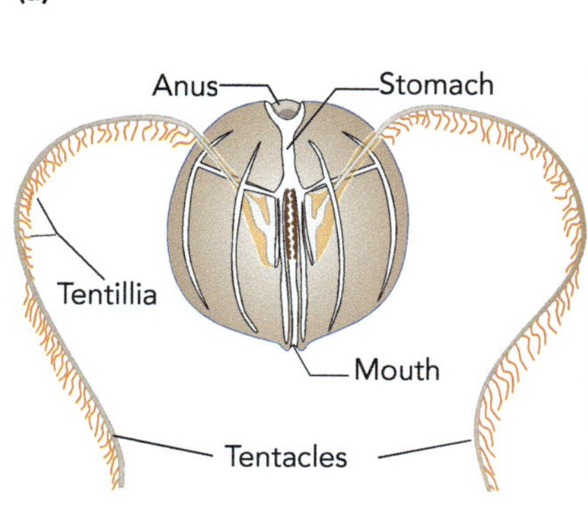

Figure 9.6 (a) Ctenophore body plan; (b) fossil ctenophore from the Cambrian, Burgess Shale, Canada. Source: (a) Adapted from Kaidor, CC BY-SA 4.0 <https://creativecommons.org/licenses/by-sa/4.0>, via Wikimedia Commons; (b) Photo by Jean-Bernard Caron. With permission of ROM (Royal Ontario Museum), Toronto, Canada.

9.3 Ediacaran Fauna

The first unequivocal metazoan fossils are found in rocks of the latter half of the **Ediacaran Period** (570–543 Ma and newly named in 2004). Discoveries in Namibia and Australia were initially assumed to be Cambrian in age, because it was thought that multi-celled life began in the Cambrian – a good example of circular reasoning – but we now know that these are in fact quite a bit older. Ediacaran fossils have now been discovered in Newfoundland Canada, the Ediacara Hills in the Flinders Ranges in Australia (after which the Ediacaran period was named), Namibia, China, and other sites around the world (Figure 9.7), and over 100 genera have been described. Enough fossils have been collected that a broad cladogram can be constructed that shows how the diversification and extinction of the Ediacaran fauna likely proceeded (Figure 9.2). Three major faunal assemblages have been recognized. The oldest Avalon Group, which is prominent in the Newfoundland fossils at Mistaken Point, recently designated as a UNESCO world heritage site, the middle **White Sea Group**, found in northeast Russia, and the youngest **Nama Group**, found in the desert outcrops of Namibia.

Some of these fossils appear to be the stem precursors of more common animal clades found today. Others may represent crown-group animals, but many seem to represent entirely extinct clades that are hard to link to modern phyla. Some scientists assume that the extinct clades are some forms of animal, while others have suggested that they may be more like plants or perhaps some form of lichen.

> **KEY POINT**
>
> The first metazoan fossils are found in the Ediacaran System and include the Avalon, White Sea, and Nama groups.

9.3.1 Soft Bodies and Osmotrophs

Most of the Ediacaran organisms were **sessile**, meaning that they lived attached on or partly buried in the sediment. One of the earliest fossils discovered is *Charnia masoni* (Figure 9.8 (a)), found in 1956 by English schoolchildren and eventually given to Trevor Ford, a geology professor at Leicester University. The fossils consisted of a leaf-shaped frond and stem. Later, a circular fossil called *Charniodiscus* was discovered in the same locality (Figure 9.8b). At first, these fossils were interpreted as separate organisms, with the circular holdfast interpreted as a possible jellyfish-type animal and the frond as a type of sea pen, both of which are cnidarians. It was then thought that *Charniodiscus* was the holdfast of *Charnia*, but later discoveries showed that *Charniodiscus* was the holdfast of a different but

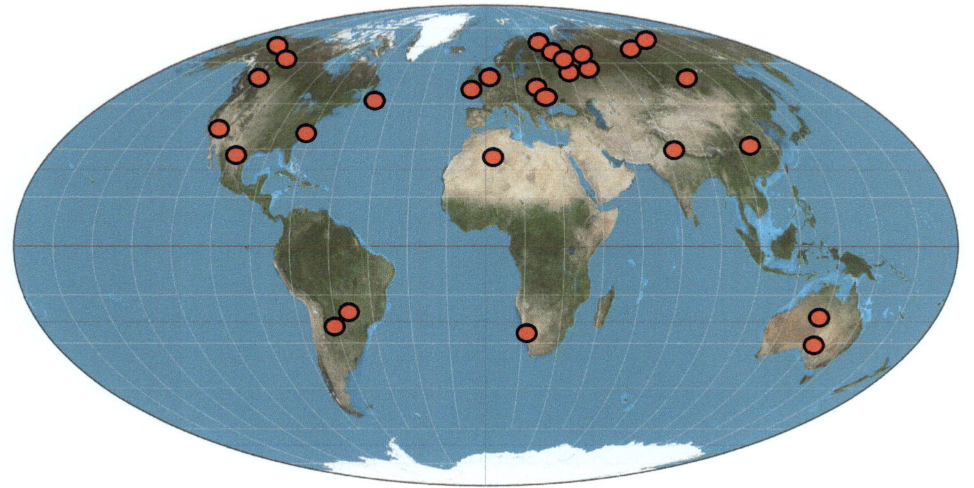

Figure 9.7 Locations where Late Neoproterozoic Ediacaran fossils are found.

Figure 9.8 (a) *Charnia masoni*, fossil reconstruction, organism is about 20 cm in length; (b) *Charniodiscus* fossil and reconstruction; (c) *Bradgatia* fossil and model; (d) *Fractofucus* fossils on bedding plane, Mistaken Point, Newfoundland, Canada and reconstruction. Source: (ai) Based on Charnia: https://alchetron.com/Charnia#charnia-af12c856-7bda-4017-9cc9–576e2343330-resize-750.jpeg; (aii), (b), and (c) Photos by Jean-Bernard Caron. With permission of ROM (Royal Ontario Museum), Toronto, Canada.

closely related species that belong to a diverse group of organisms that are now grouped into their own phylum, the Petalonamae (Figures 9.2 and 9.8), also called **frondomorphs** based on their petal- or frond-like forms. Petalonomae bodies consist of a series of alternating repeated elements, showing a fractal pattern in which smaller elements build progressively larger forms with the same basic shape, a bit like the branches and twigs of trees and bushes (Figure 9.8). In some of the organisms, such as *Charnia*, *Charniodiscus*, and *Bradgatia*, the upper part is attached to a bulbous or circular holdfast that sticks in the seafloor sediment. In others, like *Fractofucus*, the organism lies flat in the seafloor.

In all examples, the fractal growth pattern maximized the surface area of the organism and sheds light in understanding how they fed. The petalonomes show no evidence of a mouth, gut, anus, reproductive organs, body cavity (**coelom**), or other features of modern animals, such as cnidarians, to which they were initially thought to belong. The organisms are usually bilateral, but some show a trilateral symmetry. The petalonomes, found in Newfoundland, form part of the oldest Avalon Group and are found in sediments that show evidence of deposition in deep water, including alternating shale and sandstone beds that show grading and current ripples and a lack of wave-formed sedimentary structures, indicating they lived well below the *photic zone*, defined as the depth to which sunlight can penetrate, so it is unlikely that they were photosynthetic.

Younger rocks of the White Sea group contain *Ernietta* (Figure 9.9), another class of petalonomes, which consist of repeated tubes that are organized into offset, alternating pairs. *Ernietta* is found in sandstones that contain wave ripples, indicating a shallower environment.

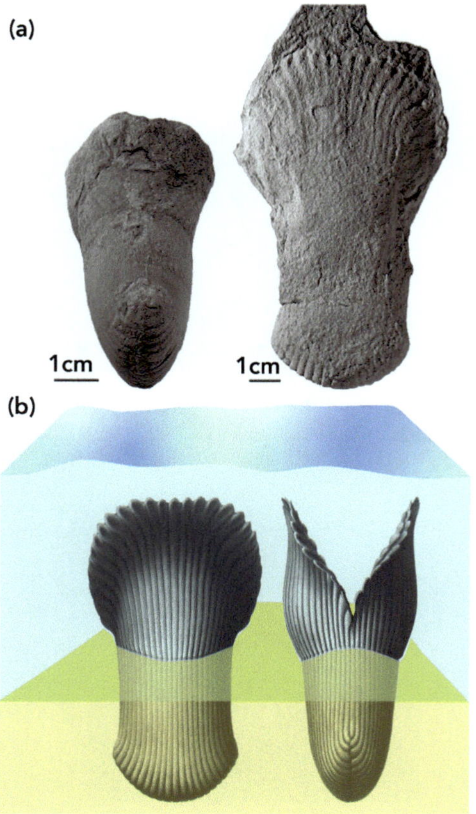

Figure 9.9 (a) *Ernietta* fossil from Namibia. Scale bar is 1 cm. (b) Reconstruction of *Ernietta*, suggesting it lived partly buried in the sediment. Source: (a) Reproduced from and (b) adapted from Ivantsov et al. (2016), © 2016 Lethaia Foundation.

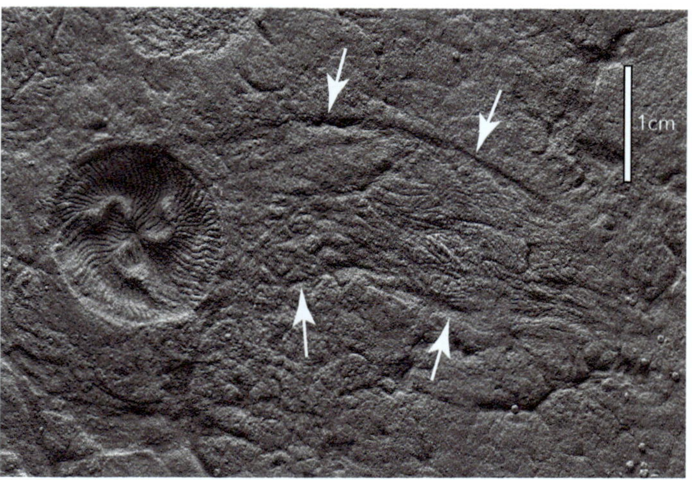

Figure 9.10 *Tribrachidium* fossil showing evidence of movement. White arrows outline the edge of burrow as the animal moved across the sand from right to left. Source: Reproduced from Iventsov and Zakrevskaya (2021). Springer Nature.

It is generally interpreted that the main way petalonomes exchanged food and oxygen with their environment was through **osmosis** – the passive transport of nutrients by diffusion across a membrane – hence they are termed **osmotrophic**. The large surface area of these membranes would have allowed the organisms to both feed on organic carbon and respire from oxygen in the seawater in which they lived. Our own lungs and circulatory systems are a good example of complex fractal branching patterns that allow the osmotic transfer of metabolic materials across a boundary, such as oxygen in the air sacs in our lungs that diffuses into arterial capillaries that is then used to oxygenate our blood. The petalonomes are thought to have used this type of feeding strategy exclusively.

Another oddball phylum is the Trilobozoa (no relation to trilobites), which show a three-fold, radial symmetry (Figure 9.10). Although they bear a superficial resemblance to coelenterates, they also have no distinct mouth parts, tentacles nor evidence of stinging cells. They are built of a series of radially oriented ridges and grooves, and they may have fed with the use of cilia to move organic material that passively settled on their bodies. There is convincing evidence that they were able to move (Figure 9.10), but their affinity to modern animals is unclear and they are another example of an early metazoan group that has been assigned to an extinct phylum.

9.3.2 Mobile Feeders: Precursors to Animals

In addition to these hard to classify organisms, there are a number of Ediacaran fossils that show a clearer affinity to modern animals (Figure 9.11). *Dickinsonia* and *Yorgia* (Figure 9.11(a) and (b)) both show bilateral symmetry, repeated elements, and some differentiation into a front and back. Geochemical studies have revealed organic **sterane** molecules in rocks that contain *Dickinsonia*. Steranes are hydrocarbon compounds diagnostic of the breakdown of cholesterol, an organic compound, found in animals but not fungi or plants. This evidence strongly suggests that *Dickinsonia* was an early animal. Recent studies demonstrate that *Dickinsonia* and *Yorgia* moved as they fed on bacterial mats (Figure 9.11(b)). In general, plants and fungi are not mobile, so the observation of movement, more complex feeding behavior, and steranes, all support an animal affinity.

Other Ediacaran bilaterians that are found in the middle White Sea group include *Parvancorina, Spriggina, and Kimberella* (Figure 9.11(c), (d), and (e)). All three show bilateral symmetry with more obvious differentiation of a head and tail than *Dickinsonia* and *Yorgia*. *Spriggina* (Figure 9.11(d)) shows repeated segments, traits that are common in the segmented worms or **annelids**, as well as a distinct head covering, or **carapace**, which resembles a trilobite, although it lacks evidence of distinctive eyes or appendages such as legs or antenna that the more complex arthropods have. The shield-like *Parvancorina* has a superficial resemblance to trilobites, and there is also evidence that it was mobile. *Kimberella* (Figure 9.11(e)) lacks segments but shows evidence of distinctive body layers, as well as a distinctive head and tail. *Kimberella* is also commonly found in association with distinctive scratch-like feeding marks indicating that it was quite mobile and grazed on the ubiquitous bacterial mats, rather than passive feeding by osmosis like the Petalanomae. *Kimberella* has been suggested to be an ancestor (i.e., stem organisms) of **mollusks**.

Figure 9.11 Ediacaran bilaterians. (a) *Dickinsonia* fossil. The "head" is on the bottom marked by larger elements; (b) *Yorgia* shows repeated elements. The dashed line indicates its former position as it moved over the substrate feeding on the bacterial mats; (c) *Parvancorina*; (d) *Spriggina*; and (e) *Kimberella*. Source: (a) DE Agostini Picture Library / Getty Images; (b) Aleksey Nagovitsyn (User:Alnagov), CC BY-SA 4.0 <https://creativecommons.org/licenses/by-sa/4.0>, via Wikimedia Commons; (c) Masahiro Miyasaka, CC BY-SA 4.0 <https://creativecommons.org/licenses/by-sa/4.0>, via Wikimedia Commons; (d) No machine-readable author provided. Merikanto~commonswiki assumed (based on copyright claims)., CC BY-SA 3.0 <http://creativecommons.org/licenses/by-sa/3.0/>, via Wikimedia Commons; (e) Aleksey Nagovitsyn (User:Alnagov), CC BY-SA 4.0 <https://creativecommons.org/licenses/by-sa/4.0>, via Wikimedia Commons.

The Ediacaran fossils include several stem lineages of modern metazoans but also a number of extinct clades based on a sessile osmotrophic lifestyle adapted to a world that lacked predators.

9.3.3 Animals with Hard Parts and Ediacaran Extinctions

Having a hard or spiky skeleton can provide protection from predators or can aid in being a predator. For example, squids and octopuses are soft bodied but both sport hard beaks, made of **chitin**, which is the same stiff material arthropods use to make their skeletons. The first organisms with *bona fide* hard skeletons are found in the latest Ediacaran forming part of the Nama Group, and include *Cloudina,* which forms a series of vertically stacked, millimeter-diameter calcitic cones, and *Namacalathus,* a small, stalked animal (Figure 9.12). *Cloudina*

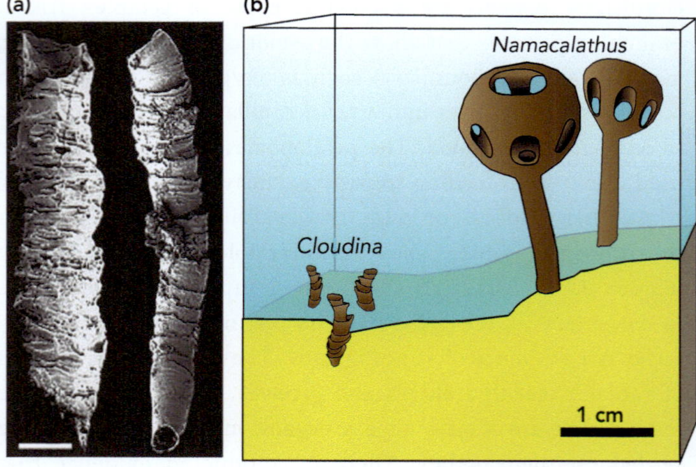

Figure 9.12 *Cloudina* fossils (a) and reconstruction (b) with associated *Namacalathus*. Like all shallow water Ediacaran biota, these lived on a substrate covered with algal and bacterial mats. Source: (a) Reproduced from Hua et al. (2003), © SEPM Society for Sedimentary Geology [2003]; (b) Reconstruction based on Seilacher et al. (2003).

Figure 9.13 Artistic reconstruction of Ediacaran fauna. Source: John Sibbick / Science Photo Library.

and *Namacalathus* were **vermiform** (wormlike) organisms that lived in shallow water, in close association with algal mats, forming small possible reef-like accumulations. Marks on *Cloudina* show that their shells have been bored into, but it is not clear if this is predatory behavior or other organisms using their shell as a refuge. Many modern large-shelled organisms are bored into by fungi, sponges, and other smaller organisms as a means of escape or as parasites. The Nama fauna have been designated as a "wormworld" (Figure 9.2) and have been hypothesized to be the precursor to the organisms that dominate in the Cambrian Explosion.

The appearance of skeletons indicates the beginning of an evolutionary arms race that was soon to result in an expansion of diversity in the upcoming Cambrian Period. The Ediacaran Period is characterized by two extinction events (Figure 9.2). The peak of Ediacaran diversity is marked by the White Sea fauna (Figure 9.13) and culminates in extinction of many of the organisms described earlier, including *Dickinsonia*, *Yorgia*, the Trilobozoa, and *Parvancorina*. The Petalonomae made it across this event but met their demise at the Ediacaran/Cambrian boundary. It is not yet known what drove these extinctions, whether it was interspecies competition causing biotic replacement, or whether it was caused by external environmental change. The disappearance of the osmotrophic petalonomes at the Ediacaran/Cambrian boundary may have been driven by the appearance of the new more mobile predators of the Nama "wormworld," but there is also geochemical evidence from carbon isotopes of a possible shift in the global carbon cycle, indicating a possible climatic event. An increase in competition for food may also have resulted in decreased availability of organic carbon in the ocean water. The sessile, soft-bodied osmotrophs likely had very slow metabolisms, and their unique style of living did not apparently survive the transition into the following Paleozoic Era. However, the subsequent dominance of trilobites and mollusks in the ensuing

Paleozoic Era may have been presaged by early mobile bilaterian ancestors, like *Kimberella*, whose life-strategy (e.g., grazing on algal mats) was ultimately more successful than relying on passive osmosis of nutrients wafting around in the ancient Proterozoic oceans.

KEY POINT

Extinction of the White Sea fauna may have been caused by external environmental change or internal interspecies competition and biotic replacement by the Nama "wormworld" fauna, which included the first organisms with hard parts, presaging the Cambrian explosion.

9.4 The Cambrian Explosion and the Burgess Shale

The late Ediacaran is characterized by the so-called "small shelly" or **Tommotian fauna** (Figure 9.14) and these persist into the early Cambrian, providing evidence that the "Cambrian" explosion started in the Late Ediacaran, as seen in the Nama Group fossils. Many common organisms can be recognized in the small shelly fauna, including **brachiopods**, which form a large group of two-shelled animals that dominated the Paleozoic Era, and look like bivalves but are unrelated. Gastropods, and **monoplacophorans**, both mollusks, are also found and primitive echinoderms, technically called **eocrinoids**, or "dawn crinoids," are also common. There are also fossils of unknown affinity that, analogous to squid beaks, probably represent hard parts, such as teeth, that were attached to larger soft-bodied organisms.

As we introduced in Chapter 6, biologists and paleontologists have identified 35 living animal phyla, most of which were well-established by the Cambrian Period. The distinction of these phyla is primarily based on their morphological characteristics, such as complexity of body plan, numbers of cell layers (i.e., two-layer **diploblasts** and three-layer **triploblasts**), types of appendages, and other diagnostic traits. Some phyla are extremely diverse, whereas others include a relatively small number of species. In addition to the 35 modern phyla, Cambrian rocks also contain fossils of rather strange animals whose ancestry and relation to modern phyla have long been controversial. One view is that these strange animals represent extinct phyla that are entirely beyond our modern experience, such as the osmotrophic petalonomes that dominated the preceding Ediacaran Period. These may reflect evolutionary experiments in a variety of basic body plans early in metazoan evolutionary history when Earth was still sparsely inhabited with many unfilled ecological niches available for natural selection to exploit. Alternatively, these organisms may represent ancestors (i.e., stem groups) that gave rise to the more common crown phyla at the tips of the bush of life that dominate Earth today. We will introduce

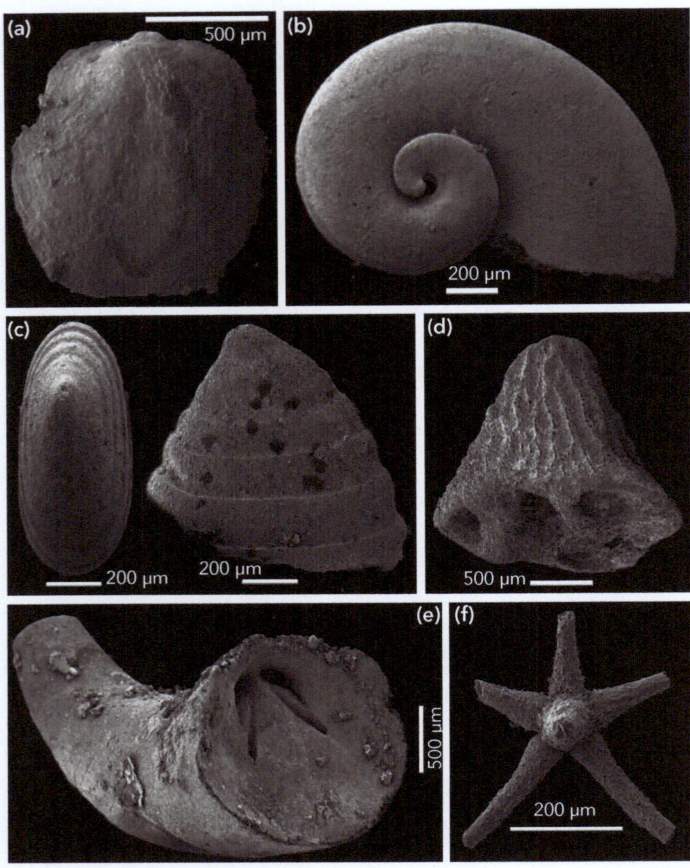

Figure 9.14 The small shelly fauna: (a) brachiopod; (b) gastropod; (c) top and side view of a monoplacophoran, an early mollusk; (d) echinoderm part; (e) *Hyolith*; (f) *Chancelloria*, which may be part of a sponge. Source: Reproduced from Skovsted and Peel (2007).

some of the weirder Cambrian animals below and then examine what they could mean in terms of the evolutionary history of metazoans.

9.4.1 The Burgess Shale

A ***Konservat lagerstätten*** is a sedimentary rock that preserves soft parts of animals. *Konservat lagerstätten* are of particular interest to paleontologists as they provide a more complete window into the diversity of life, given that most organisms are never fossilized and most of the fossil record is highly biased to preserving only the hard parts of organisms, such as shells. The Burgess Shale is a mid-Cambrian *Konservat lagerstätten* in which over 86% of fossils are soft bodied, exposed in the Walcott Quarry outside the small town of Field, British Columbia, close to Banff National Park (Figure 9.15). It was here, in 1909, that Charles Doolittle Walcott, former director of the US Geological Survey and then Secretary of the Smithsonian Institute, discovered the now-famous fossil site. The Burgess Shale is dated at about 508 Ma and provides a fossil record of the nature and diversity of marine life, 33 million years after the beginning of the Cambrian Period (541–485 Ma) and long after the appearance of the small shelly Tommotian fauna. The shales were deposited in about 160 m of water at the base of a submarine cliff of the Cathedral Formation, an ancient limestone reef deposit (Figure 9.15). Recall that in the Cambrian, North America lay south of the equator (Figure 9.1) and was rimmed by warm shallow water conducive to the development of carbonate reefs.

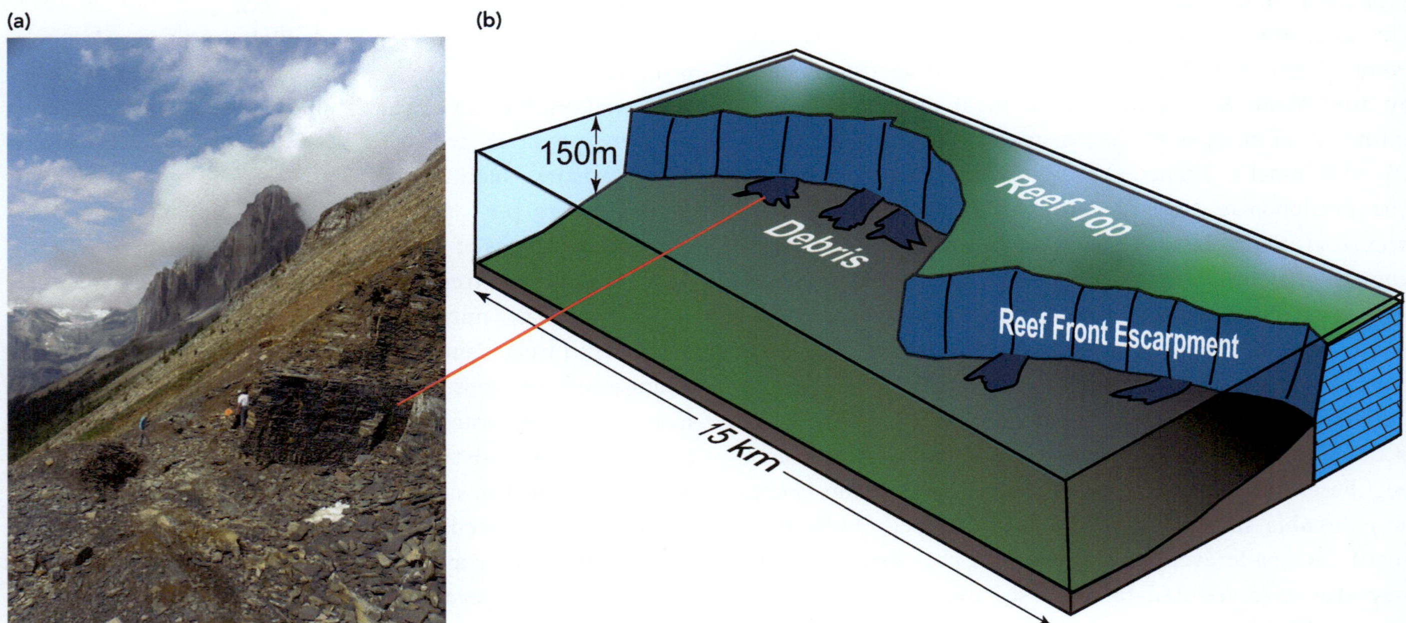

Figure 9.15 Photo of Burgess Shale (a) and block diagram (b) showing depositional setting. Shallow water communities that lived in the Cambrian Cathedral reef top were transported into the deeper water muddy environments at the base of the reef and preserved in the Burgess Shale. Source: (a) Mark A. Wilson (Wilson44691) (Department of Geology, The College of Wooster). Public domain, via Wikimedia Commons.

At the time of discovery, Walcott classified the Burgess fauna within the context of known animal phyla and did not appear to recognize the uniqueness of some of his finds. Although this was Walcott's discovery, he was never able to complete a thorough investigation owing to his administrative duties as director of the Smithsonian Institute in Washington DC.

Beginning in the 1960s, British paleontologist Harry B. Whittington, who taught at Yale and Harvard before returning to Cambridge University in England, initiated a major re-examination of these fossils, with two of his graduate students. Derek Briggs worked primarily on the arthropods and Simon Conway Morris was given the task of evaluating the soft-bodied "worms." Their analysis and reconstructions of the organisms were astounding. They found that there were many organisms that defied easy classification, several of which appeared to represent completely unknown, and presumed extinct, phyla.

Although the Burgess Shale preserves the soft parts of organisms, they are invariably squashed flat during the process of burial and lithification. Most of the animals preserved in the shale are thought to have lived in shallow water, at the edge of the Cathedral limestone reef, and were later carried into deeper water when parts of the reef collapsed (Figure 9.15). Therefore, the animals may display a variety of orientations, such that if an animal is buried on its side, the upper half of the body may cover the lower half. Whittington and his team used a traditional device called a *camera lucida*, in which the image of the fossil, seen under a microscope, is projected onto a piece of paper allowing the viewer to trace the image. They were able to literally peel away the organic films produced by the squished three-dimensional (3D) body, allowing them to visualize and reconstruct the animals in 3D as well as uncover lower body parts covered by upper body parts during burial. Reconstruction was made easier if there was more than one specimen with variable orientations.

In addition to the Burgess, there have been more recent *Konservat lagerstätten* discoveries, including the **Chengjiang biota**, of the Maotianshan Shale in China, which is about 10 million years older than the Burgess fossils, as well as sites in Greenland and elsewhere, and these have greatly expanded our knowledge of the evolution, diversity, and phyletic relationships of early metazoans.

9.4.2 Burgess Fauna from Weird to Normal

One of the odder animals, and one of the first to be described in detail by Whittington, is *Opabinia regalis*, a relatively small animal that ranged from about 4 to 7 centimeters long (Figure 9.16). It has bilateral symmetry and is well differentiated, with a distinct head, body, and tail region and it is built of a series of segments, sharing some similarities with *Spriggina*. In contrast to the Ediacaran animals, however, *Opabinia* has five well-developed eyes perched atop stalks. Details of the eyes are

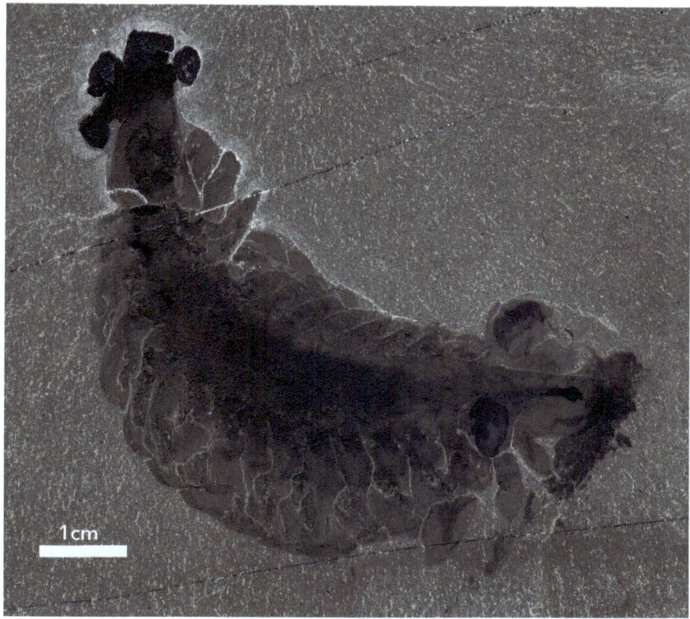

Figure 9.16 *Opabinia regalis*, fossil and reconstruction. Source: Photos by Jean-Bernard Caron. With permission of ROM (Royal Ontario Museum), Toronto, Canada.

hard to decipher, but it is assumed that they were compound eyes, such as found in insects today. *Opabinia* was an adept swimmer with two rows of flipper-like appendages along its sides, most of which were covered with gills, and a V-shaped tail that would have helped in balancing and maneuvering. *Opabinia*'s strangest feature is a single long flexible **proboscis** (i.e., mouth stalk) that it used to catch food with, which it would deliver to its mouth, located on the underside of its head. *Opabinia* shows a high level of anatomical complexity and was likely a capable predator.

The introduction of *Opabinia* to the paleontological community by Whittington in 1972 was met with disbelief and even some laughter because it was so very difficult to classify. It is clearly an animal, but what group does it belong to? It is certainly far more complex than a sponge or jellyfish. It has some aspects of annelids, such as segments, and it has its distinct head appendages like the stalked eyes and proboscis, but it lacks the diagnostic jointed legs that are required to call it an arthropod. In fact, it does not have legs at all! Whittington concluded that *Opabinia* is neither arthropod nor annelid, although he suggested that it might represent an earlier stem group of segmented animals from which both arthropods and annelids evolved.

Figure 9.17 *Anomalocaris canadensis* (a) fossil; (b) fossil mouth; and (c) reconstruction. (d) Reconstruction of *Aegirocassis benmoulae*, a 2 m-long Ordovician radiodont (copyright Marianne Collins ROM). Source: (a), (b), and (d) Photos by Jean-Bernard Caron. With permission of ROM (Royal Ontario Museum), Toronto, Canada; (c) Photo by JPB. With permission from Houston Museum of Natural Science (HMNS).

Another oddball, and the largest animals in the Burgess are the anomalocarids (literally "anomalous shrimp," Figure 9.17), which were up to a meter in length. Anomalocarids show many similarities to *Opabinia*, with a V-shaped tail, a row of "fins" along each side, a pair of eyes on stalks, and a mouth on the underside of the head, all indicating that they are related. But in contrast to *Opabinia*, anomalocarids have only two eyes versus five. Also, rather than a single proboscis, they have two large, curved, and clearly segmented appendages on their heads that are interpreted to have been used for capturing food. Like *Opabinia*, they also have a separate circular mouth on the underside of the animal's head (Figure 9.17(b)). Like *Opabinia*, *Anomalocaris* shows a high level of anatomical complexity and given its size was likely the *apex predator* of its day.

The history of how *Anomalocaris* was classified illustrates the challenges encountered in interpreting the Burgess Shale fauna. Initially, the different parts of *Anomalocaris* were interpreted as several different animals each assigned to completely different phyla. The jointed mouth appendages were initially interpreted as the tail of a shrimp-like crustacean (phylum Arthropoda), hence the name "anomalous shrimp." The circular mouth (Figure 9.17(b)) in contrast was interpreted as a jellyfish (phylum Cnidaria), and the body as a sponge (phylum Porifera). Later reconstructions of more complete fossils demonstrated that these bits-and-pieces in fact represented the disarticulated parts of a single, much larger soft-bodied organism. Like *Opabinia*, *Anomalocaris* was hard to classify and when Whittington and his student Derek Briggs presented the first reconstructions of the animal, they concluded that it did not belong to any known animal phylum.

Numerous other species of anomalocarids have now been documented worldwide in other Cambrian *lagerstätten*, and they have also been documented in younger rocks of the Ordovician and Devonian periods, with the youngest example found in the 407 Ma Hunsrück Slate in Germany. One Ordovician species, *Aegirocassis benmoulae* (Figure 9.17(d)), was the largest ever discovered, reaching over 2 m in length. *Aegirocassis* shows significant advances from the Burgess anomalocarids, in that the mouth appendages have been modified to form a series of fine sieve-like stems, analogous to the baleen filters in whales, and this has been interpreted to indicate it was a filter feeder. The younger anomalocarids also have an additional set of appendages down their sides. Such **biramous** (two-branch) limbs are common in modern crustaceans.

The anomalocarids are now assigned to the order **Radiodontia** (literally "radial teeth," named after their circular mouths), a diverse group of animals that share the common feature of a circular set of teeth on the underside of their bodies (Figure 9.17(b)). There now appears to be a consensus that they are a stem-arthropod, meaning that they evolved before the arthropods but share a common ancestor, indicating a close relationship. Their diversity, worldwide distribution, and continuation into the Devonian shows that they were a highly successful group of predators that lived alongside crown arthropods, such as the trilobites, for well over 100 million years.

One of the oddest organisms in the Burgess Shale is *Hallucigenia sparsa* (Figure 9.18(a) and (b)), a tubular animal, typically 0.5 to 5 centimeters in length, with two rows of puffy cylindrical legs, with small claws at the end, and a double row of sharp spikes on its back. Because of poor preservation, the earliest reconstructions did not recognize the two rows of legs and the animal was initially depicted upside down, with the spines dug into the sediment and a single row of tubes on its back. It was also initially unclear which end was the head and which the tail. *Hallucigenia* was initially assigned to the extinct phylum **Lobopodia**, worms with lobe feet, which includes other examples such as *Aysheaia* (Figure 9.18(c) and (d)). More recent evaluation revealed that the pair of tube feet and the double-row of protective spines clearly was an adaptation to escape predators. *Hallucigenia* is now thought to be closely related to the modern phylum **Onychophora** (velvet worms, see Figure 6.8) and the Burgess contains several species of this now-modest phylum. Newer cladistic and genomic analysis of the Onychophora suggest that *Hallucigenia* and its relatives, such as *Aysheaia*, belong to a superclade, called the **panarthropods** (Figure 9.19), that indicates an ancestral linkage between the radiodonts, lobopods, and crown arthropods. *Aysheaia* was also thought to have an affinity with the modern phylum of **Tardigrada**, also known as water bears, and may represent a stem ancestor to both the tardigrades and onychophorans (Figure 9.19).

Other weird Burgess organisms include *Wiwaxia*, which looks a bit like a walking pinecone, with two rows of blades sticking out of its back (Figure 9.20). The scale-like coverings (**sclerites**) and spines are hardened, showing that *Wiwaxia* was well-armored, indicating adaptation for protection against predators, perhaps *Anomalocaris*. *Wiwaxia* has mollusk-like rasping teeth that it likely used to graze on bacterial mats. *Wiwaxia* has also been hard to classify. When first discovered by Walcott in 1911, he classified it as a **polychaete**. Since then, it has been proposed to be a stem-group annelid or possibly a primitive mollusk, but its affinity remains controversial.

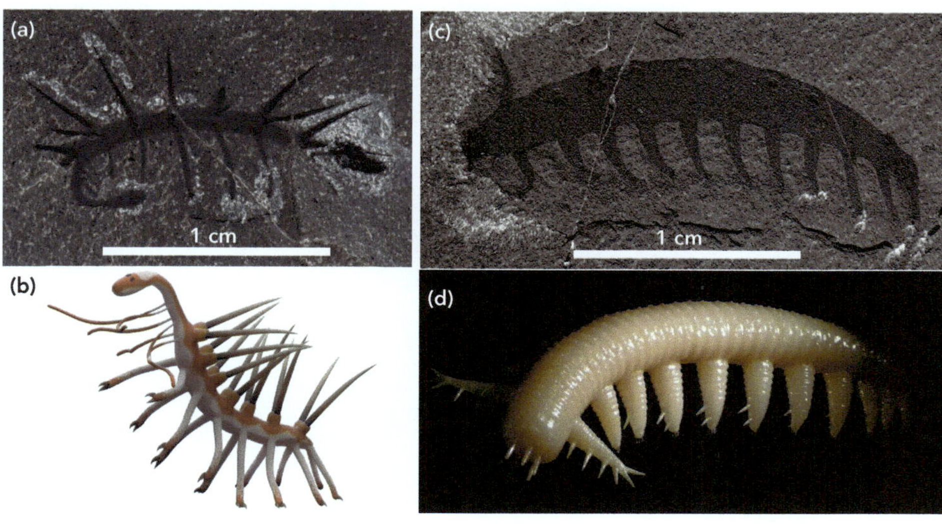

Figure 9.18 Lobopods: (a) *Hallucigenia sparsa*, fossil; (b) reconstruction; (c) *Aysheaia*; (d) reconstruction. Source: (a)–(c) Photos by Jean-Bernard Caron. With permission of ROM (Royal Ontario Museum), Toronto, Canada; (d) © Citron.

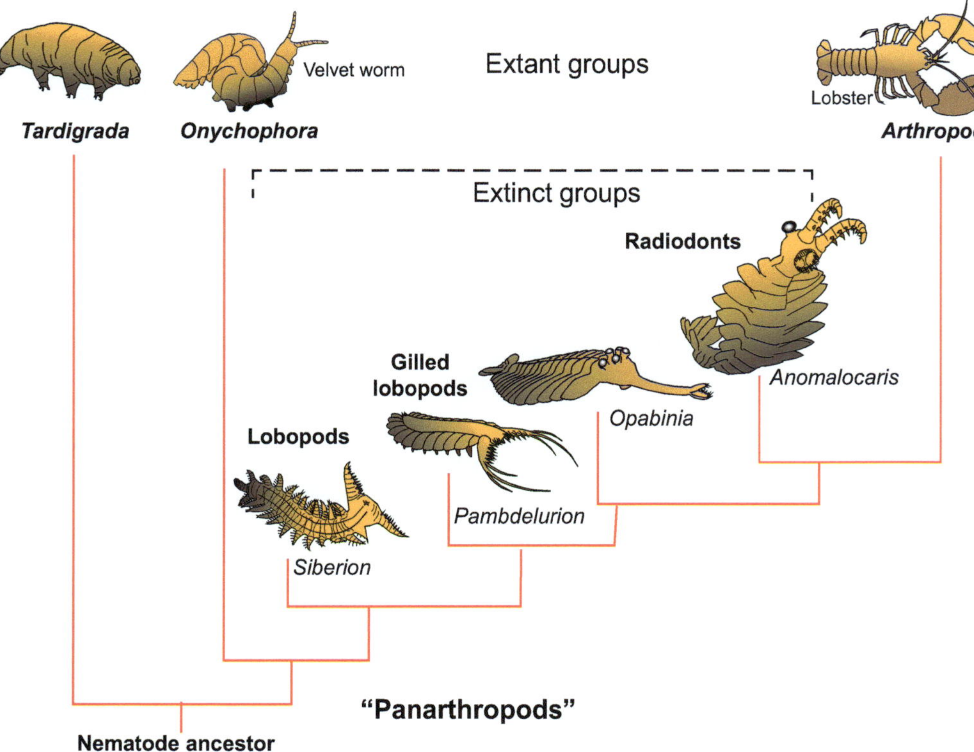

Figure 9.19 Family tree (clade) showing how the Burgess Shale fauna, including onychophorans, tardigrades, and arthropods, which are still around today, may have been related to the arthropod stem group, as well as extinct groups like the radiodonts and *Opabinia*.

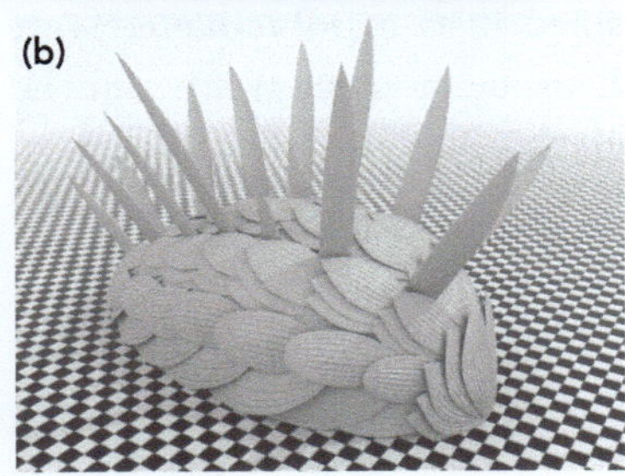

Figure 9.20 *Wiwaxia* fossil (a) and reconstruction (b). Source: (a) Photo by Jean-Bernard Caron. With permission of ROM (Royal Ontario Museum), Toronto, Canada; (b) Martin R. Smith, CC0, via Wikimedia Commons.

Figure 9.21 *Herpetogaster* fossil (a) and reconstruction (b). Source: (a) Photo by Jean-Bernard Caron. With permission of ROM (Royal Ontario Museum), Toronto, Canada; (b) Photo by Marianne Collins, 2010. With permission of ROM (Royal Ontario Museum), Toronto, Canada.

An equally strange animal is the recently named *Herpetogaster* (literally "creepy stomach"). *Herpetogaster* was a stalked animal with a large stomach and branched tentacles attached to its head (Figure 9.21). *Herpetogaster* is included in a group of animals called **cambroernids**, characterized by paired tentacles, a large stomach and a narrow intestine enclosed in a coiled sac. Like the animals described above their affinity is unclear. It has been suggested that *Herpetogaster* may represent a stem group to the deuterostomes (second mouth), a superphylum characterized by development of the first opening of the embryo becoming the anus and the second opening becoming the mouth, which includes the echinoderms and our own chordate phylum. This contrasts with the protostomes, which include most of the other animal phyla (see Figure 6.8).

In addition to the weirdos, and there are many other oddballs that we do not have time to mention, the Burgess also includes a wide variety of animals that can be more confidently assigned to one of the better-known animal phyla (Figures 9.22 and 9.23). For example, there are many Burgess arthropods, but even within these better-known phyla, some of the species are difficult to assign to any of the known classes of arthropod. The most common arthropod is *Marella splendens* (Figure 9.22(a)), which looks somewhat trilobite-like, but is no longer considered to be closely related. *Yohoia* is another odd arthropod with jointed legs and a pair of front appendages (Figure 9.22(b)). *Yohoia* belongs to an extinct class of arthropod called the **Megacheira** (great hands) and may be a stem ancestor to the **Chelicerates**, a group of arthropods that include modern organisms such as spiders and scorpions, as well as extinct groups like

the giant-clawed eurypterids, which are common in the Silurian Period. The Burgess also contains numerous species of trilobites (Figure 9.23), which of course are another extinct class of arthropod. In addition, a variety of brachiopods, several primitive echinoderms, as well as sponges are common. Recently discovered *Metaspriggina* (Figure 9.22(c) and (d)) and *Pikaia* are some of the earliest known chordates, which are the precursors to fish, the higher vertebrates, and our own group, the mammals.

KEY POINT

Despite their strangeness, many of the Cambrian oddballs are now thought to represent stem ancestral groups to modern phyla such as the arthropods. These stem ancestors also share traits with other basal groups such as the lobopodian onychophorans. Rather than evolutionary dead-ends, some groups, such as the Radiodonta, persisted well into the Devonian, as revealed by newly discovered lagerstätte.

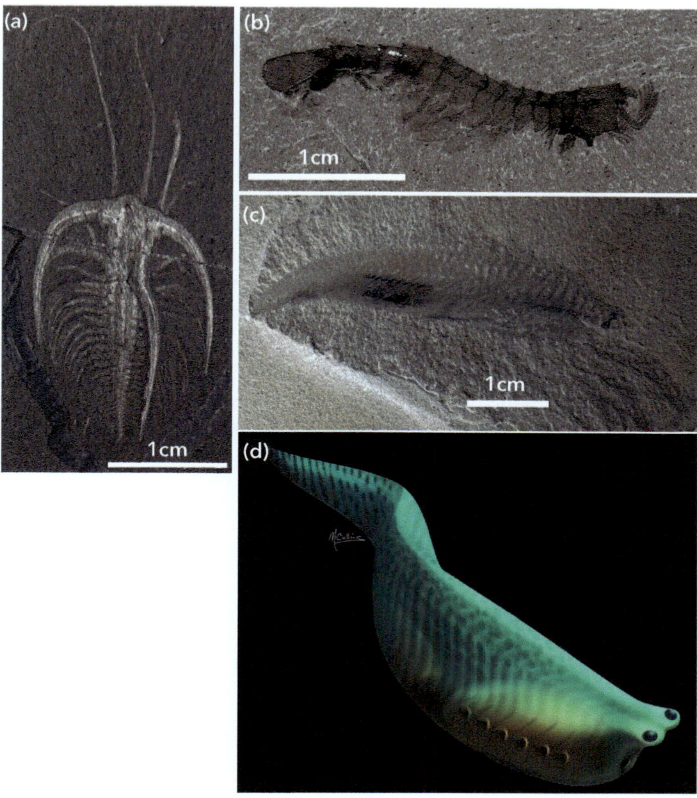

Figure 9.22 (a) *Marella*, an extinct arthropod, similar to a trilobite; (b) *Yohoia* has two front appendages; (c) and (d) *Metaspriggina* is an early fish-like animal. Source: (a)–(c) Photos by Jean-Bernard Caron; (d) Photo by Marianne Colllins with permission of ROM (Royal Ontario Museum), Toronto, Canada.

Figure 9.23 Artistic reconstruction of the Burgess Shale ecosystem. A large *Anomalocaris* swims over the substrate. An *Opabinia* hunts for prey above the sediment and several trilobites crawl over the surface. A school of *Pikaia* swim among sea anemones and stromatolite mounds. Source: Stocktrek Images / Getty Images.

9.5 The Great Ordovician Biodiversification Event

Following the Cambrian explosion (from 544 to about 524 Ma), life on Earth experienced a second stunning radiation and diversification referred to as the **Great Ordovician Biodiversification Event** (GOBE, Figure 9.24) from about 485 to 445 Ma, which saw an increase from a few hundred to around 1,500 families in about 50 million years. Major expansions occurred in the articulated brachiopods, corals (**Anthozoa**), starfish and related stalked **crinoids** (phylum **Echinodermata**), **ammonites**, and other cephalopods. There was also expansion of the **bryozoans**, small colonial filter feeding animals that look a bit like miniature corals, and the **ostracodes**, which are microscopic crustaceans that live in a hinged, clam-like shell.

This multifold increase in diversity was driven by persistently high sea levels, warm conditions, and an increase in predators, and may also have been reflective of increased oxygen (see Chapter 3). Predation forced prey to evolve survival strategies, such as heavier and more complex armor, as well as an increased ability to burrow deeper into sediment. Ordovician rocks typically contain much more complex trace fossils than Cambrian rocks, which reflects the evolution of more complex **infaunal** behavior (i.e., living in the sediment below the seafloor) as competition for food resources intensified and selective pressure of ever-increasing predations fueled an arms race.

The development and diversification of burrowing behavior is particularly important, as it represents a major transformation in bioengineering that can affect major Earth cycles such as the carbon cycle. For most of the Proterozoic and earlier, the dominant life forms were photosynthetic bacterial/algal mats, and we have already seen how oxygen photosynthesis fundamentally changed Earth's atmosphere during the Great Oxygenation Event, which ultimately allowed higher metabolic rate organisms and metazoans to evolve. However, in the older "matground world" the seafloor was a significant boundary wherein a sticky and probably relatively impermeable organic layer made it more difficult for exchange of oxygen and nutrients across the seafloor and into the sediment. This would tend to promote anoxia in the shallow sediment making it uninhabitable for oxygen users. Also, this would promote storage of organic matter in sediment, also affecting global atmospheric carbon. Disruption of this boundary by early burrowing organisms may have changed Earth's ability to cycle carbon as well as open new infaunal niches within the sediment. The ability for organisms to burrow into sediment, versus living on top of it, or simply attached by a stalks like the Ediacaran petalonomae, changed the ability for Earth cycling, as well as affecting the nature of the sediment surface as a habitat, and is an example of how new

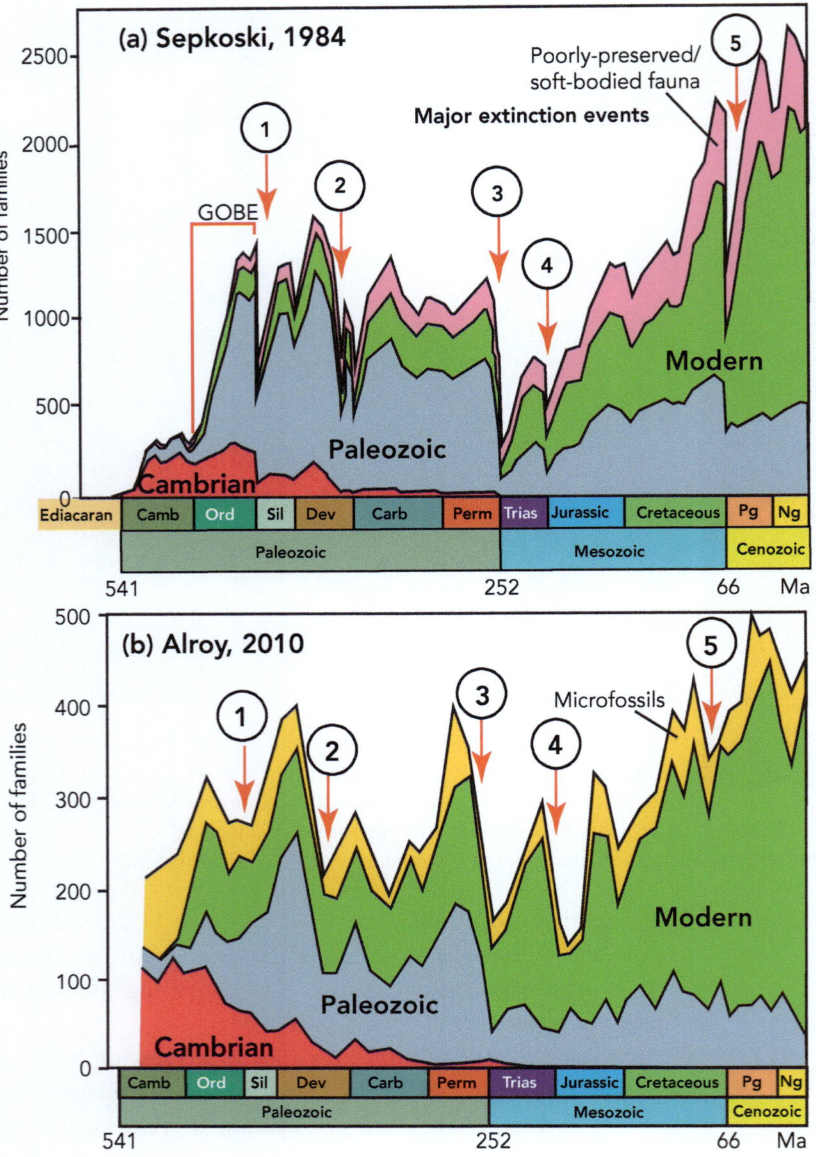

Figure 9.24 Phanerozoic diversity charts. (a) The Sepkoski Curve of diversity for oceanic animals during the Phanerozoic Eon. The Cambrian fauna (Cm) fade during the Paleozoic and are replaced by the Paleozoic fauna (Fz). Following the Permian extinction event (3) the Modern fauna (Md) dominate. Five major extinction events are shown, and there is an extinction of the Vendian (Ediacaran) fauna between the Proterozoic and Phanerozoic eons. (b) This figure suggests that Mesozoic diversification may not be as extreme as Sepkoski envisioned and is a result of bias in preservation of younger fossil groups. Source: (a) Modified after Sepkoski (1984); (b) Modified with permission of American Association for the Advancement of Science, from Alroy (2010); permission conveyed through Copyright Clearance Center, Inc.

life behaviors and bioengineering can drive evolutionary change.

KEY POINT

The Great Ordovician Biodiversity Event marks a continued diversification event with increases in burrowing behavior, where bioengineering took advantage of infaunal lifestyles as well as evolutionary arms races between predators and prey.

9.6 Modes of Biodiversification

One of the big questions in the evolution of life is how has life diversified? Are there predictable trends and how do extinctions control later diversification? We introduced this idea in Chapter 6 when we discussed the gradual versus punctuated model of evolution, proposed by Stephen Jay Gould and Niles Eldredge. There are also other questions, such as whether extinction is caused by interspecies competition or by more gradual biotic replacement, or driven by external environmental changes, which can be fast or slow. The Ediacaran turnover from the White Sea to the Nama "wormworld" is a critical event that presaged the Cambrian explosion, but it is not clear if the White Sea extinction was caused by gradual biotic replacement or a more sudden external environmental change. In evaluating the Ediacara and Burgess fauna in his 1989 book, *Wonderful Life*, Stephen Jay Gould suggested another mode of evolution marked by a high degree of experimentation of fundamental body plans at the level of kingdoms and phyla that characterized the Ediacaran and early Cambrian, as indicated by the significant number of unknown and extinct phyla found in those rocks that we discussed above, but once the losers and winners had been sorted out, life on Earth settled largely into diversification within the major clades that are dominant today (Figure 9.24).

9.6.1 The Sepkoski Curve: Patterns of Diversification and Mass Extinction

In 1953, the first edition of the *Treatise of Invertebrate Paleontology* was published, and, as of 2007, this had expanded to 50 volumes, authored by about 300 paleontologists, providing a thorough catalogue of all invertebrate taxa discovered across the stratigraphic record. This has led to comprehensive online datasets, with which it is now possible to go far beyond William "Strata" Smith, who discovered that diagnostic-age fossils could be used to make order of the stratigraphy of the units that he mapped in England 200 years ago (the principle of faunal succession, introduced in Chapter 2), and quantify how life on Earth has diversified and changed over geologic time.

The late University of Chicago paleontologist Jack Sepkoski took advantage of these publications and compiled all the available data that described the diversity of fossils throughout the Phanerozoic Era (Figure 9.24). Sepkoski identified three major groups, which he termed the "Cambrian," **"Paleozoic,"** and **"Modern"** fauna (Figures 9.24, 9.25, 9.26, and 9.27) and we could now add in the Ediacaran: Avalon, White Sea, and Nama groups. The Cambrian fauna include trilobites, the "small shelly fauna" (Figure 9.14), eocrinoids, and some of the earliest reef builders, including **archeocyathid** sponges.

Figure 9.24 shows that the Cambrian fauna (Figure 9.25) show a distinct pattern of increasing diversification during the Cambrian Period, which defined the Cambrian explosion, followed by a prolonged and rather gradual loss of diversity in the ensuing Ordovician, perhaps reflecting gradual biotic replacement by the more successful Paleozoic fauna (Figure 9.26). Figure 9.24 shows a more sudden drop in

Figure 9.25 Fossil examples of typical Cambrian fauna. Source: (a) Photo by JPB. With permission from Houston Museum of Natural Science (HMNS); (b) With permission of ROM; (c) Photo by JPB, McMaster collection, Canada; (d) Modified from Skye McDavid, CC BY-SA 4.0 <https://creativecommons.org/licenses/by-sa/4.0>, via Wikimedia Commons.

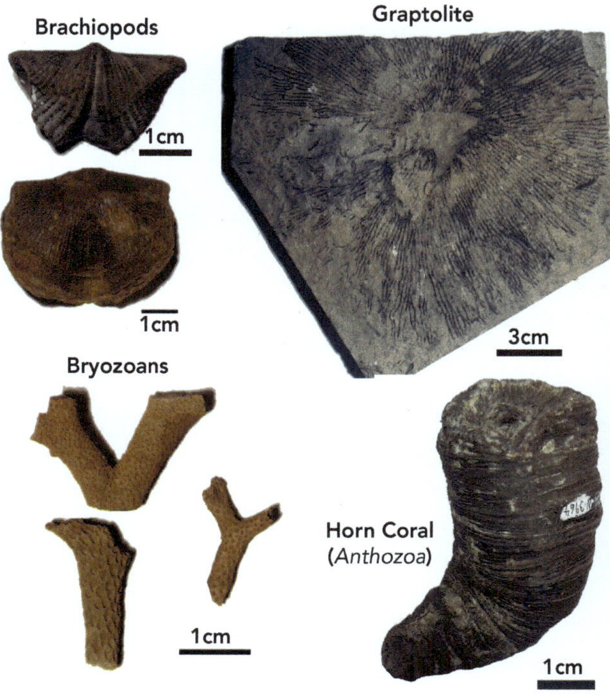

Figure 9.26 Fossil examples typical of the Paleozoic fauna. Source: (Crinoids) Photo by JPB. With permission from Houston Museum of Natural Science (HMNS); (all other photos) Photos by JPB, McMaster collection.

diversification at the end of the Ordovician, correlating to a mass extinction event. However, as we discussed above, discoveries of anomalocarids in younger Devonian *lagerstätte*, which had been assumed to have gone extinct long before, point out the difficulties of fully understanding patterns of life which are biased to the record of fossils with hard parts. Another sharp drop also occurs in the Late Devonian, which is also linked to another mass extinction event. A remnant of the Cambrian fauna lived throughout the remainder of the Paleozoic but ultimately completely disappeared at the mass extinction that ended this era (see Chapter 11). The pattern of diversification and decimation of the Cambrian fauna suggests that the Paleozoic fauna ultimately outcompeted and replaced the Cambrian fauna with more complex lifestyles, enhanced armor, and the consequent expansion during the GOBE.

The Permian extinction (see Chapter 11) opened new ecological niches and allowed the Modern fauna to become predominant. The Modern fauna (Figure 9.27) are dominated by gastropods (snails), bivalves, **Osteichthyes** (bony fishes), **malacostracan** crustaceans (e.g., lobsters, shrimps, and crabs), echinoids (e.g., sea urchins and sand dollars), **Chondrichthyes** (i.e., the cartilaginous fishes including sharks and rays), the **Gymnolaemata** bryozoans, and **demosponges** (Figures 9.24 and 9.27).

Sepkoski showed that although the Modern fauna appeared early in the Paleozoic, they showed a relatively static level of diversity throughout the Paleozoic and only experienced greater diversification in the Mesozoic and Cenozoic, following the decimation of the Paleozoic fauna because of the end

Permian mass extinction. However, more recent re-examination of Sepkoski's analysis (Figure 9.24(b)), using the far more extensive paleobiology database (https://paleobiodb.org/#/), suggests that the diversification of the Modern fauna may not have been as extreme as Sepkoski interpreted and may reflect sampling bias caused by the higher availability of fossil material in younger rocks.

Despite these caveats, Sepkoski's compilation shows that patterns of diversification, decimation, and extinction are quite variable. Figure 9.28 shows that some groups, such as the bivalves and gastropods, exhibit a more-or-less steady cone of increasing diversity, as modeled in Figure 9.29(a). Others, such as the articulate brachiopods and crinoids, show increasing and then decreasing diversity as shown in Figure 9.29(c). Some groups, such as starfish and **hexatinellids** (i.e., glass sponges), show periods of relative stasis with apparently sudden episodes of diversification that matches the punctuated equilibrium model of Eldredge and Gould (Figure 9.29(b), and discussed in Chapter 6), and some groups show very little change in diversity. The trilobites show initial diversification and a punctuated die-off, as diversity decreases with their eventual death knell at the end of the Paleozoic. In reality there is no single model for diversification or extinction and examination of the fossil record suggests a variety of patterns.

9.6.2 Five Mass Extinctions

Sepkoski's compilation also allowed him to quantify the magnitude of extinctions. He noted that there were five exceptionally large extinction events (Figure 9.24), and we will discuss two of

Figure 9.27 Fossils typical of the Modern fauna: bivalve, echinoid, gastropod, and demosponge, McMaster collection, Crustacean. Source: Photos taken by JPB. (Crab) Reproduced with permission of Perot Museum, Dallas; (Shark tooth) By permission of The Trustees of the Natural History Museum, London; (Fish) Reproduced with permission of Royal Ontario Museum; (all other photos) McMaster collection.

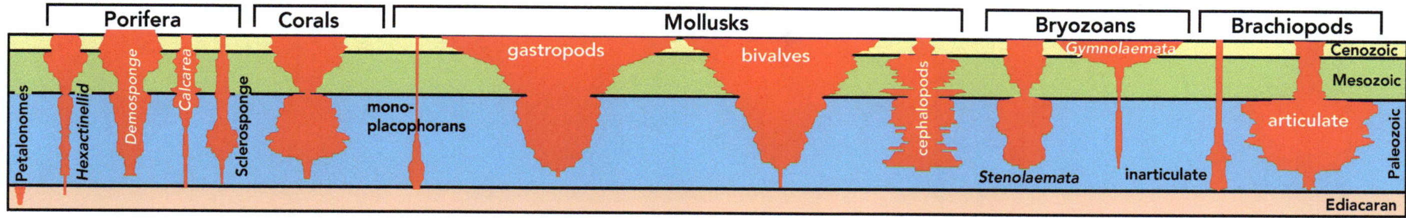

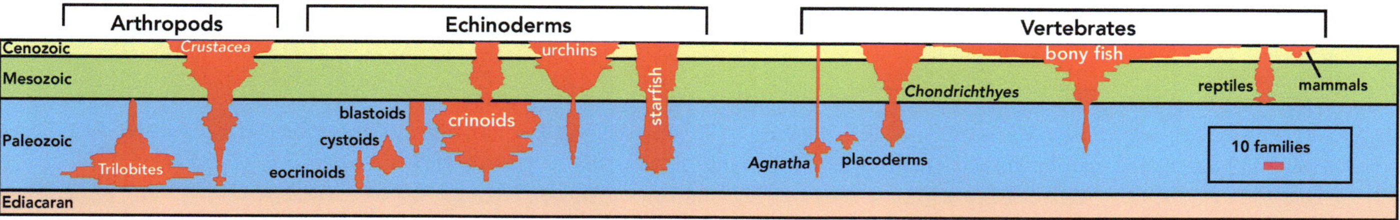

Figure 9.28 Spindle diagrams showing diversification, decimation, and extinction trends for a variety of animals. Source: Adapted from Sepkoski (1981). © The Paleontological Society, Cambridge University Press.

these mass extinctions events in more detail in Chapters 11 and 13.

Extinction is a routine aspect of evolution. During times when speciation outpaces extinction, we see an increase in diversity, such as the Cambrian Explosion and the GOBE. At other times, extinction exceeds speciation resulting in a loss of diversity. The GOBE marks an increase in diversity of the Paleozoic fauna, but this seems to be at the expense of the Cambrian fauna. There are many reasons for extinction, including biotic replacement where one species outcompetes another, changes in bioengineering, such as we saw in the development of burrowing behavior that changes the nature of infaunal habitats, but extinction can also be driven by external change, such as increases in volcanism, changes in climate, and, as we shall discuss in Chapter 13, extraterrestrial impacts. We are particularly interested in the causes of the five mass extinctions.

9.6.3 The Ordovician Mass Extinction

The end of the Ordovician marks the first mass extinction event. The Late Ordovician was marked by a severe global cooling, indicated by widespread glaciations in the southern hemisphere (Gondwanaland) followed by rapid warming and development of ocean anoxia. Major continental glaciations caused extensive lowering of sea levels, possibly up to 100 meters, and a corresponding loss of shallow marine habitats, ending the Sauk transgression. The extinction occurred in two pulses. The first decimated shallow water, and especially endemic species living in the vast shallow seas, technically referred to as **epeiric seas**, which characterized the Sauk transgression, as well as deep water fauna, and **nektonic-planktonic** fauna (i.e., free swimming or free-floating). A survival-fauna managed to re-establish itself, but these were decimated during the second extinction pulse. In the end an estimated 85% of all marine species were lost.

9.7 Evolution and the Fossil Record

The history of life, as recorded by the fossil record, shows how life has changed through time as well as providing insights as to the nature of extinct life forms. In combination with modern genomics, which tells us how living organisms are related, over the past few decades we have been able to greatly improve our understanding of how the diversity of life to Earth came to be.

The fossil record of the earliest Ediacaran metazoans suggests that there may indeed have been evolutionary experimentation in an uncrowded Neoproterozoic Era, as hypothesized by Gould (Figure 9.29(c)). This experimentation may reflect the lack of competitive and complex trophic relationships, which may have initiated with the earliest ecologies of the day but only became fully developed in the Paleozoic.

Quantitative analysis of diversity also allows us to observe how evolution proceeds, but it is worth realizing that the analysis of scientists such as Sepkoski rely on thousands of previous studies with careful cataloguing and classifying of fossil groups. These kinds of data were not available in Darwin's time. The fossil record is notorious for being incomplete, and this relates to the fact that not all environments are preserved, and most organisms are never preserved as fossils. However, organisms that have evolved shells and that live in marine environments have a rather better chance of fossilization than nearly any terrestrial organism and, as we have seen, the preservations of soft-bodied fossils in *Konservat lagerstätten* provide rare but critical insights about the broader diversity of life during the times that they represent. This was one of the reasons Sepkoski focused on the marine fossil record, and especially shelled organisms, because of their higher preservation potential. What is clear, especially in examining Figure 9.28, is that the evolution of life on Earth involved the appearance and disappearance of various groups at different times. In many cases, the extinction

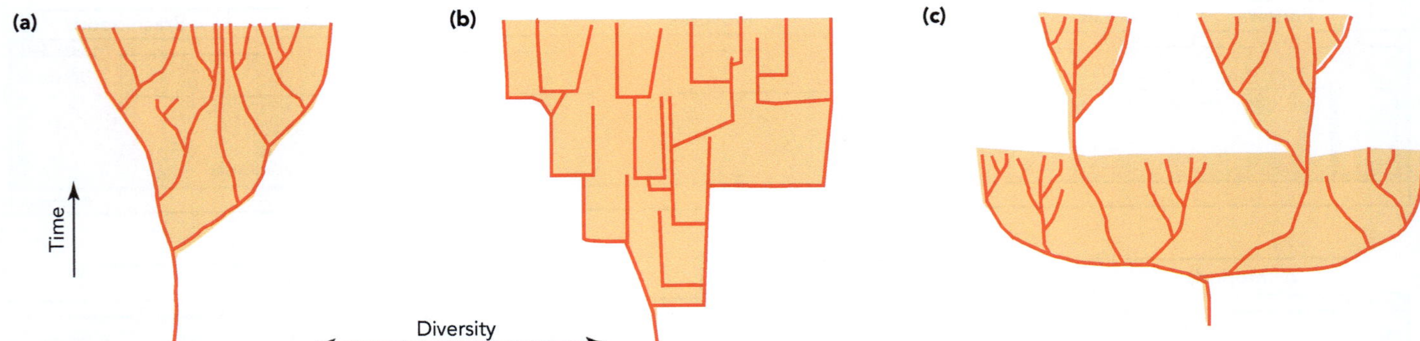

Figure 9.29 Models of evolution: (a) a cone of gradually increasing diversification; (b) punctuated equilibrium; and (c) early diversification and experimentation, followed by decimation and success and diversification of a smaller number of clades. Source: Based on Gould (1989).

of one group provides opportunities for others. In tandem with diagrams like Figures 9.24 and 9.28, the next step is to seek the relationships between major biological events, with other Earth events, such as plate tectonic changes, climate and sea level, ocean and atmospheric chemistry, and others. We have already seen how global warming and sea-level rise following the Cryogenian freeze led to a huge increase in habitats and consequent evolution and diversification of the metazoans. These examples illustrate that Earth can be viewed as a group of interacting systems through time. The answers to what caused life on Earth to change cannot be found by only examining modern biology but requires integration of life sciences and Earth sciences, and much of this record can only be read by picking up a rock.

9.8 Summary

- The end of the Cryogenian, breakup of Rodinia, and overall warming resulted in a global rise of sea level (Sauk transgression) that created vast new shallow water habitats that may have helped drive the evolution and diversification of the first metazoans.
- Molecular clocks suggest that the first metazoans appeared in the Cryogenian, but the first fossil evidence is found in rocks of the Ediacaran Period at the end of the Neoproterozoic Era.
- The multicellular kingdoms of life (i.e., animals, fungi, plantae) may have originated separately, but the simplest animals (Porifera) are thought to have evolved from single-celled choanoflagellates.
- Ediacaran fossils include the Avalon, White Sea and Nama groups. The White Sea Group shows the greatest diversity and includes the now extinct phylum Petalonomae, populated by strange organisms lacking most features of modern animals with fractal body plans, which lived a sessile lifestyle and fed by osmosis. The White Sea assemblage includes early bilaterians, such as *Dickinsonia*, *Yorgia*, *Parvancorina*, and *Kimberella*, which were mobile and employed more complex feeding strategies. Some of these may represent stem ancestors to the animal phyla that are more common today. The later Ediacaran Nama Group, referred to as the "wormworld," includes organisms with hard parts and may have been the precursor to fauna seen in the Cambrian Explosion.
- The Cambrian explosion marked an increase in organisms with hard parts, likely reflecting an increase in the evolution of predators partly driven by expansion of shallow marine habitats during the Sauk transgression. Most major animal phyla appeared in the Cambrian.
- Preservation of soft-bodied fossils in *lagerstätten* of the Ediacaran fauna, the Burgess Shale, and other formations worldwide reveal both unknown and stem phyla suggesting early evolutionary experimentation of metazoan body plans in a less crowded world. Subsequent diversification, and an expansion of organisms with hard parts as well as increases in burrowing, record an evolutionary arms race reflective of increasing competition as trophic webs grew more complex.
- Statistical compilations of biodiversity through time, based on analysis of the fossil record, reveal complex patterns of speciation, diversification, decimation, and extinction, and show that there have been three major faunal groups through time, the Cambrian, Paleozoic, and Modern.
- These compilations also allow the magnitude of extinctions to be measured and show that there have been five mass extinction events in the Phanerozoic Eon.
- The first mass extinction occurred at end of the Ordovician caused by massive climate change and the first ice age since the end of the Cryogenian Period.

Key Words

- Azoic Eon
- metazoan
- Cambrian explosion
- Porifera
- spicules
- choanocytes
- choanoflagellates
- Placozoa
- ctenophores
- **triploblasts**
- Cnidaria
- Ediacaran Period
- White Sea Group
- Nama Group
- sessile
- frondomorphs
- coelom
- osmosis
- osmotrophic
- sterane
- annelids
- carapace
- mollusks
- chitin
- vermiform
- Tommotian fauna
- brachiopods
- monoplacophorans
- eocrinoids
- diploblasts
- *Konservat lagerstätten*
- *camera lucida*
- Chengjiang biota
- proboscis
- biramous
- Radiodontia
- Lobopodia
- Onychophora
- panarthropods
- Tardigrada
- sclerites
- polychaete
- cambroernids
- Megacheira
- Chelicerates
- Great Ordovician Biodiversification Event
- Anthozoa
- crinoids
- Echinodermata
- ammonites
- bryozoans
- ostracodes
- infaunal
- Cambrian fauna
- Paleozoic fauna
- Modern fauna
- archeocyathid
- Osteichthyes
- malacostracan
- Chondrichthyes
- Gymnolaemata
- demosponges
- hexatinellids
- epeiric seas
- nektonic
- planktonic

Further Reading and References

Alroy, J., 2010, The shifting balance of diversity among major marine animal groups, *Science*, 329(5996), 1191–1194, https://doi.org/10.1126/science.1189910.

Conway Morris, S., 1998, *The Crucible of Creation, The Burgess Shale and the Rise of Animals*, Oxford University Press.

Cunningham, J. A., Vargas, K., Yin, Z., Bengston, S., and Donoghue, P. C. J., 2017, The Weng'an Biota (Doushantuo Formation): An Ediacaran window on soft-bodied and multi-cellular microorganisms, *Journal of the Geological Society*, 174 (5), 793–802, https://doi.org/10.1144/jgs2016-142.

Gao, L., Wang, Z., Liu, P., et al., 2008, *Octoradiate Spiral Organisms in the Ediacaran of South China*, Acta Geologica Sinica (English Edition), Vol. 81, John Wiley and Sons.

Gould, S. J., 1989, *Wonderful Life: The Burgess Shale, and the Nature of History*, W.W. Norton and Co.

Hua, H., Pratt, B. R., and Zhang, L.-Y., 2003, Borings in *Cloudina* shells: Complex predator-prey dynamics in the terminal Neoproterozoic, *PALAIOS*, 18(4–5), 454–459, https://doi.org/10.1669/0883-1351(2003)018<0454:BICSCP>2.0.CO;2.

Ivantsov, A. Yu., and Zakrevskaya, M. A., 2021, *Trilobozoa*, Precambrian tri-radial organisms, *Paleontological Journal*, 55(7), 727–741.

Ivantsov, A. Yu., Narbonne, G. M., Trusler, P. W., Greentree, C., and Vickers-Rich, P., 2016, Elucidating *Ernietta*: new insights from exceptional specimens in the Ediacaran of Namibia, *Lethaia*, 49(4), 540–554, https://doi.org/10.1111/let.12164.

Seilacher, A., Grazhdankin, D., and Legouta, A., 2003, Ediacaran biota: The dawn of animal life in the shadow of giant protists, *Paleontological Research*, 7(1), 43–54.

Sepkoski, J., 1981, A factor analytic description of the Phanerozoic marine fossil record, *Paleobiology*, 7(1), 36–53, https://doi.org/10.1017/S0094837300003778.

Sepkoski, J. J., 1984, A kinetic model of Phanerozoic taxonomic diversity. III: Post-Paleozoic families and mass extinctions, *Paleobiology*, 10, 246–267.

Skovsted, C., and Peel, J., 2007, Small shelly fossils from the argillaceous facies of the Lower Cambrian Forteau Formation of Western Newfoundland, *Acta Palaeontologica Polonica*, 52 (4), 729–748.

Review Questions

1. What was the Sauk transgression?
2. What conditions were favorable for the evolution and diversification of metazoans in the Ediacaran and Early Paleozoic that led to the Cambrian Explosion and Great Ordovician Biodiversity Event?
3. What do molecular clocks tell us about the appearance of metazoans and multi-celled life?
4. Why are choanoflagellates thought to be the single-celled ancestors to the earliest metazoans (Porifera)?
5. What are the main groups of organisms that characterize the Ediacaran Period? What is unusual about the Petalonomes and which organisms may have affinities with more modern animals?
6. Explain why it is that some of the organisms in the Burgess shale and many Ediacaran fossils are hard to assign to a modern group?
7. What do the fossils from the Burgess Shale and the several Ediacaran localities tell us about early metazoan life and evolutionary process?
8. What does the Sepkoski compilation tell us about how life on Earth has diversified, and particularly the idea that there are three great faunal groups, the Cambrian, Paleozoic and Modern?
9. How important are mass extinctions versus background extinctions in the history of life on Earth?
10. We now know that some of the soft-bodied oddballs found in the Cambrian *lagerstätte* persisted into the Devonian Period. What does this tell us about biases in the fossil record and the value of *lagerstätten*?
11. What kinds of patterns occur in the diversification, decimation, and extinction of clades? Are any of the proposed modes of evolution (e.g., gradualism, punctuated equilibrium) dominant?
12. How does the evolution of burrowing behavior drive evolutionary changes?
13. What are some of the proposed causes of the Ordovician mass extinction?

Water/snow melt flowing from the Tablelands in Gros Morne National Park, Newfoundland. Source: Deb Snelson / Getty Images.

Chapter 10

Iapetus and Pangea

The Lost Ocean and the Assembly of a Paleozoic Supercontinent

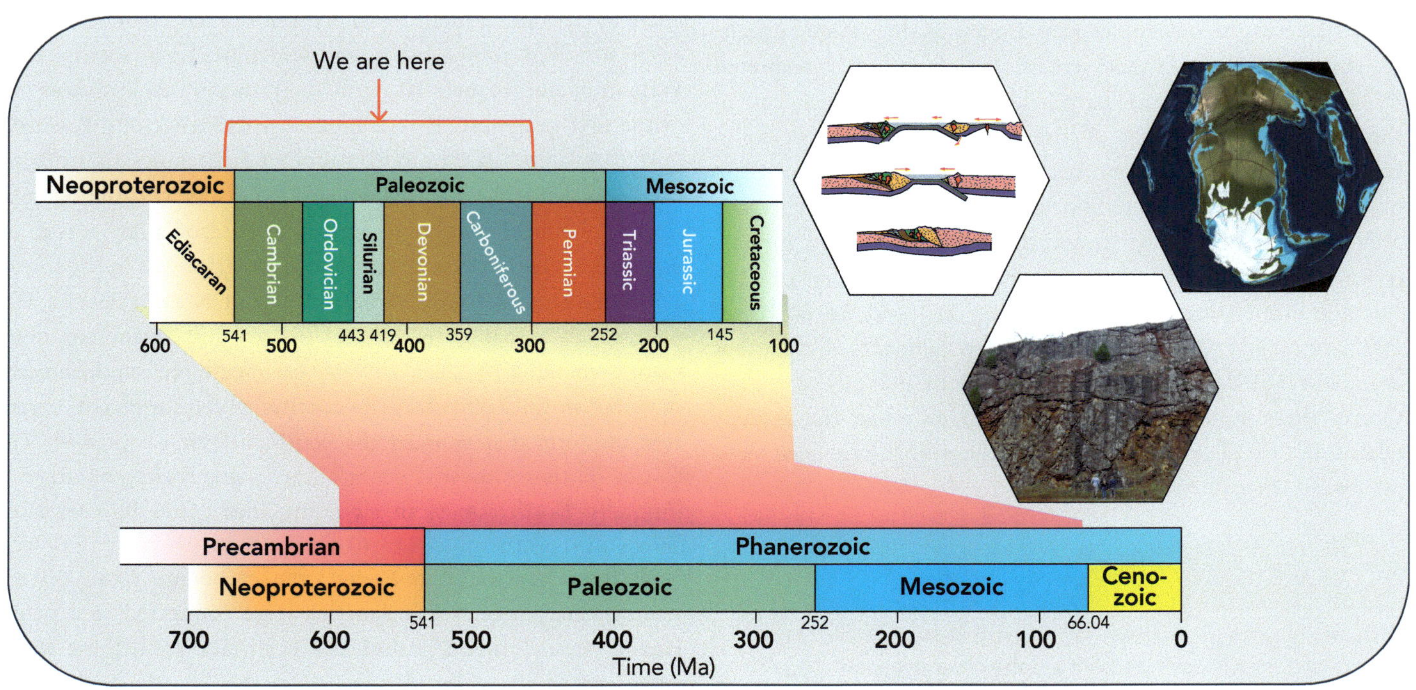

Introduction: The Search for Old Oceans?

In Chapter 5 we introduced the evidence that at the end of the Paleozoic Era most of the Earth's continents were joined together, forming the supercontinent **Pangea**, which in Latin translates as "whole mother-Earth land." The formation of the Atlantic and Indian oceans primarily records the subsequent breakup of Pangea and led to the hypothesis of continental drift. Although continental drift seemed like a viable mechanism to create new oceans, until we had a full understanding of the Wadati–Benioff subduction zones, where oceans were being destroyed (see Chapter 5), it was reasonable to consider alternate hypotheses. One such alternative was the **expanding Earth hypothesis**. Some scientists proposed that, in the Paleozoic, Pangea covered most of the Earth's surface, requiring a much smaller Earth that somehow became larger throughout the Mesozoic. Australian geologist Samuel Warren Carey embraced continental drift and was a chief proponent of the expanding Earth hypothesis, speculating that formation of low-density minerals as the Earth cooled may have caused the expansion. Carey's speculative attempt to address how the Earth could have increased in volume failed, but the full debunking required evidence for oceans older than the Mesozoic and led to a search for evidence of these older oceans. This chapter focuses on the search and discovery of ancient oceans, its implication for the emerging theory of plate tectonics, and the assembly of the supercontinent called Pangea.

10.1 The Age of the Modern Oceans

One of the first big questions is how old is the crust that floors the modern oceans. Answering required documentation of the ages of modern oceanic crust using isotopic dating and evaluation of the ages of fossils within sediment that covered this crust, as discussed in Chapter 4. In Chapter 5 (Figure 5.12) we showed a map of the age of crust in the world's oceans. The map shows a systematic pattern, where the crust becomes older away from spreading ridges, matching the magnetic stripes that were first discovered and interpreted by Harry Hess and Marie Tharp in the 1960s. The oldest crust in the Atlantic Oceanic lies at the outer edges (blue color in Figure 5.12), and is Early Jurassic in age, about 180 million years old, indicating that the Atlantic opened at around that time. The Mediterranean is the only ocean that has crust older than Jurassic (purple color in Figure 5.12). The map shows that most of the Mediterranean Ocean crust is between 280 and 220 Ma (Late Permian to Early Triassic), although a few remnants of crust that are about 340 Ma in age (mid-Carboniferous) have been found. Clearly there is no crust older than 340 Ma under the modern oceans, but for plate tectonics to work there must have been older oceans, so where are they?

KEY POINT

There are no modern oceans with a crust older than 340 Ma, so if plate tectonics is correct where are the older oceans?

One of the reasons for a lack of older oceanic crust is that over geologic time, most oceanic lithosphere is destroyed as it is recycled back into the mantle in subduction zones, as discussed in Chapter 5. Although oceanic crust may not be preserved, ancient continental margins that were attached to these lost oceans are commonly flanked by thick sequences of sedimentary rocks that have lower density and are thus more difficult to subduct, favoring their preservation over geologic time. These sedimentary rocks may be deformed and metamorphosed to form schists, slates, and marbles but their origin as sedimentary rocks are usually still apparent. Therefore, even if it is hard to find ancient oceanic crust, the sedimentary rocks, or their metamorphic equivalents, can used to identify ancient continental margins that must have lain on the flanks of now-vanished oceans. Also, as oceans close, the andesitic volcanic island arcs and associated accretionary prisms that formed in the overriding plate above the subducting slab (Figures 5.13 and 5.19) can also be difficult to subduct, as they have a lower density, and these can also be preserved at the sutures along which old oceans have closed. In the process of closing an ocean, these volcanic rocks may also be highly deformed into distinctive metamorphic rocks, forming **greenstone belts** (Figure 10.1).

KEY POINT

Sedimentary and metamorphic rocks can mark the position of now-vanished oceans.

10.2 Iapetus: An Early Ocean

The Canadian geophysicist J. Tuzo Wilson, was an early proponent of plate tectonic theory and suggested that Earth experiences cycles of continent assembly and breakup that occur over a few hundred million years, and which are now referred to as **Wilson cycles** (Figures 10.2 and 10.3). In 1966, Wilson was one of the first geoscientists to propose the existence of a Paleozoic "proto-Atlantic" ocean. Paleomagnetic data shows that during the Cambrian and early Ordovician periods, North America was rotated about 90° clockwise, such that Hudson's Bay was in an equatorial position and eastern North America was in the southern hemisphere (Figure 10.3(a)). Much of North America was covered by a shallow ocean and the margins of the continent were dominated by shallow water carbonate-reef environments recorded in thick limestone formations which formed an extensive platform that rimmed the continent (see Chapter 9). The fossils in these lower Paleozoic carbonate rocks contained a uniquely North American signature that could be correlated across the continent (Figure 10.4).

During the Cambrian Period, the carbonate rocks on the western margin of North America were connected to a paleo-Pacific Ocean, called **Panthalassa**. For this reason, these North American fossils were referred to as the "Pacific fauna" (see

Figure 10.1 (a) Metamorphosed Precambrian pillow basalt, typical of oceanic rocks. Sudbury Ontario. (b) Greenstone schist, Michigan. These types of rocks form serpentine, common in greenstone belts and commonly formed by metamorphism of oceanic rocks. Source: (b) Photo by James St. John, CC BY 2.0, https://www.flickr.com/photos/jsjgeology/8282176722.

Figure 10.4). Wilson also noted that these same "Pacific"-type faunas could be found on the eastern side of the modern Atlantic Ocean in Wales, Ireland, and Scandinavia. The boundary that separated these rocks appeared to be a major fault zone, or suture line. Wilson, invoking the newly developed plate tectonic theory, suggested that this suture line must have been separated by an ocean, older than the modern Atlantic, that disappeared down a subduction zone. This hypothetical ocean was named **Iapetus**, the father of Atlas, after whom the Atlantic Ocean is named. Critically, the suture line that marked the zone of closing of the Iapetus Ocean was not exploited when the new Atlantic Ocean opened, and thus several land masses that originally lay on the *eastern side* of old Iapetus were left on the *western side* of the newer Atlantic Ocean after it opened. One of

these sutures passed right through the island of Newfoundland in Atlantic Canada. The island of Newfoundland is commonly referred to by locals as "The Rock," owing to its extensive outcrops and rocky coastlines. Given the possibility of examining an ancient plate boundary, the island of Newfoundland became a focus of intense study to find the remnants of Iapetus.

> ### KEY POINT
>
> Fossils on either side of the modern Atlantic indicate that they were once sutured together along a line that marked the position of a long-lost ocean, named Iapetus.

10.2.1 Geology of "The Rock" (Newfoundland) and the Discovery of Iapetus

By the mid 1960s it was well understood that Precambrian rocks largely represented the deeply eroded roots of old continents and mountain chains. Canadian geologist Harold Williams, who had completed his graduate studies under Wilson, conducted extensive mapping of Newfoundland in the early 1960s to test Wilson's hypothesis. Williams showed that the island was marked by Precambrian basement rocks on the west and east coasts, and highly deformed younger Paleozoic rocks in the middle (Figure 10.5(a)). He further noted that the central area included distinctive volcanic rocks that were interpreted to have formed as oceanic island arcs associated with subduction zones (Figure 10.6).

Williams was one of the first to recognize that Newfoundland contained the remnants of a Paleozoic ocean that must have once separated two Precambrian continents. Paleontological analysis showed that the western basement was overlain by limestones containing the characteristic North America "Pacific-type" trilobite fauna, whereas the Cambrian trilobites in eastern Newfoundland showed the distinctly Atlantic affinity (Figure 10.4).

Williams mapped the major fault zones that marked the boundaries separating the western, eastern, and central zones (Figure 10.6(a)). Consistent with Wilson's idea, Williams suggested that the central area of Newfoundland represented the deformed remnants of the older, Iapetus Ocean (Figure 10.6). Eventually, Williams extended his geological maps of these zones throughout eastern North America across the entire **Appalachian Mountain** chain and interpreted all of it within the framework of the newly developed plate tectonic theory (Figure 10.5(b)). These zones were eventually also recognized and correlated into the United Kingdom, Ireland, and the **Caledonian** mountain chain in Scandinavia that lay on the opposite side of the modern Atlantic but were once sutured together along the line that Wilson had recognized (Figure 10.5(b)).

A series of major linear geological belts have now been identified. The ancient North American continent, which also

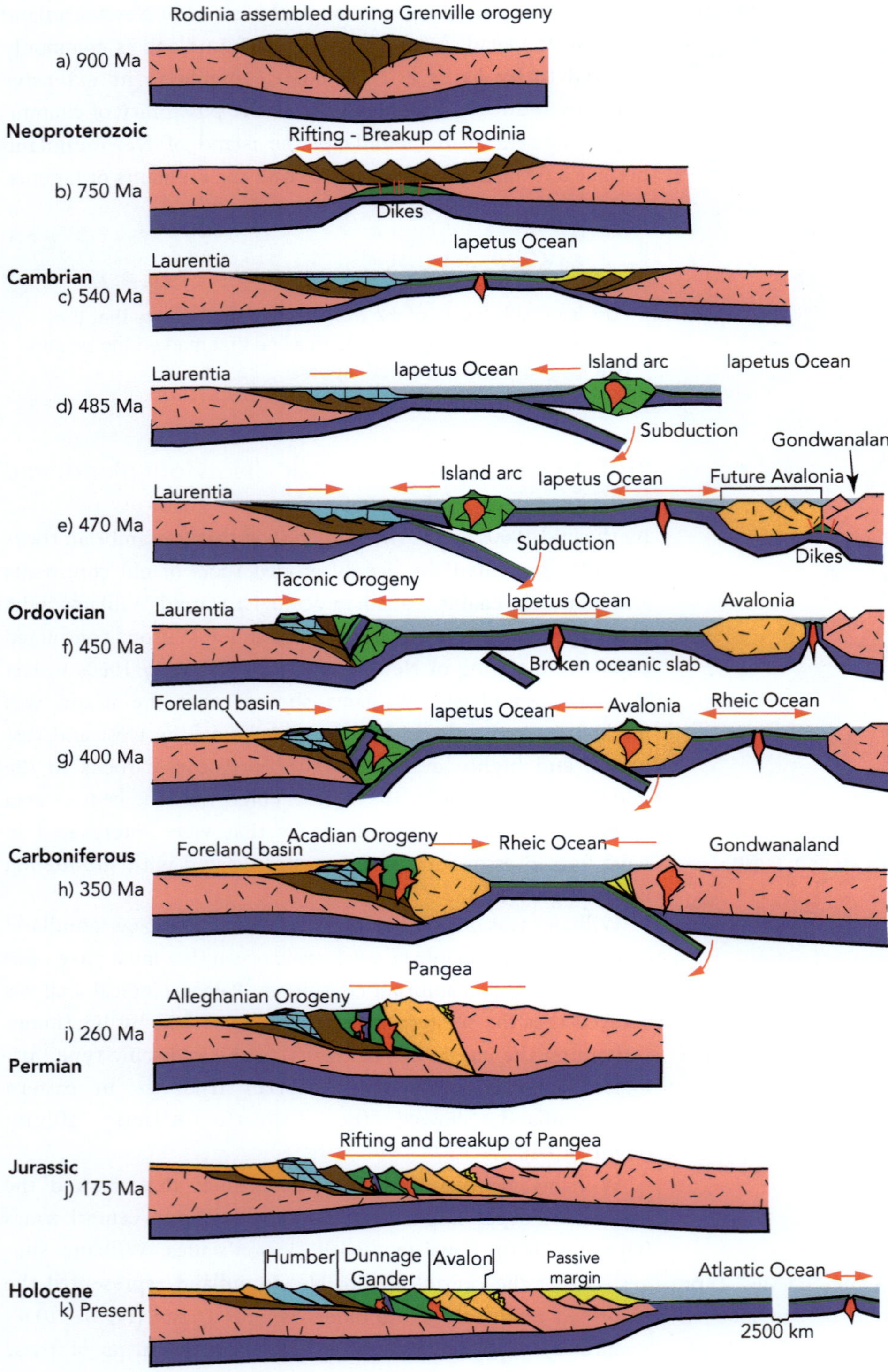

Figure 10.2 Cross sections showing the steps in the breakup of the Rodinian supercontinent and the assembly and breakup of Pangea.

includes Greenland, is referred to as Laurentia, as introduced in Chapter 1 (Figures 1.32 and 10.2), and includes granitic and metamorphic basement rocks, which form the craton (see also Figure 1.10). Overlying the craton is the platform of sedimentary rocks that lay at the margin of the continent, which faced Iapetus to the south and Panthalassa to the north. Because Laurentia has rotated (Figure 10.3), these margins are now on the eastern and western margin of North America respectively. These old equatorial carbonate platforms were deformed during the closing of Iapetus in the Paleozoic and Williams called this the **Humber Zone**. (Later destruction of the Panthalassa Ocean in the Mesozoic and Cenozoic Eras deformed the western margin and will be discussed in more detail in Chapter 14.)

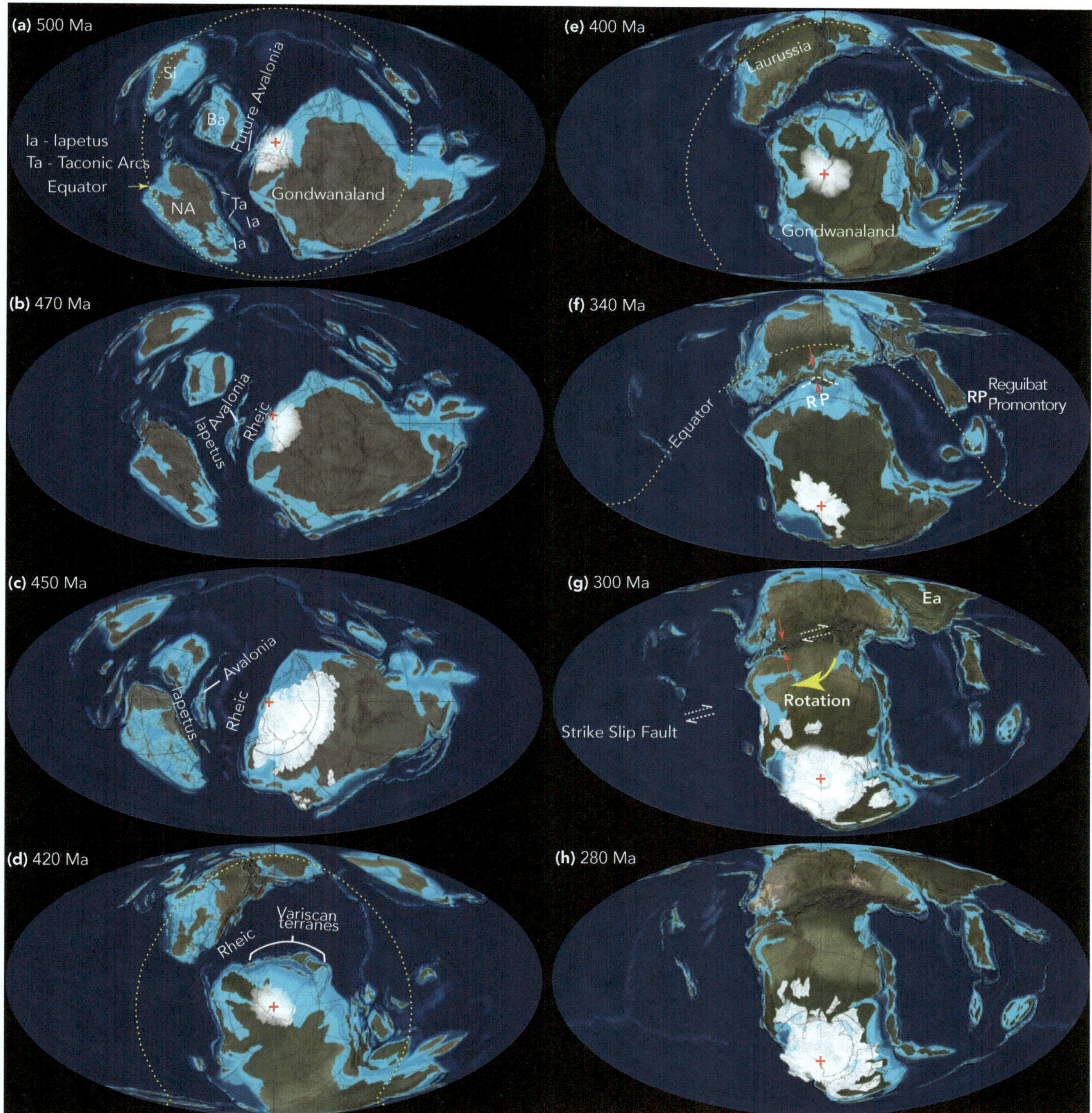

Figure 10.3 Paleogeographic evolution of Earth from 500 Ma to 280 Ma, showing the assembly of Pangea. The red cross marks the South Pole. The yellow dashed line marks the equator. Gondwanaland, which included Africa, South America, Australia, Antarctica, and India, lay close to or on the South Pole and was glaciated many times (see Chapter 11). As the Iapetus Ocean (IA) began to close, volcanic island arcs (Ta) formed above the subduction zones and the Taconic Orogeny began about 470 Ma. The formation of subduction zones in Iapetus allowed the Rheic Ocean to open up. Parts of Gondwanaland rifted to produce elongate microcontinents, including Avalonia and Ganderia. These eventually accreted to Laurentia (ancient North America, NA) as Iapetus closed. Laurentia (NA), Baltica (Ba), and Siberia (Si) were joined together by about 400 Ma. Eventually Africa and Laurentia collided, and the Rheic Ocean closed like a zipper, earlier in the north and later in the south. Source: © DeepTimeMaps.

During the closing of Iapetus, a series of volcanic island arcs formed above several subduction zones (Figures 10.2, 10.3, and 10.6) that were relatively close to Laurentia. These are referred to as the **Taconic arcs** (marked Ta in Figure 10.3(a)) and are more generally referred to as the **peri-Laurentian** arcs, because they formed above the subduction zones close to the periphery

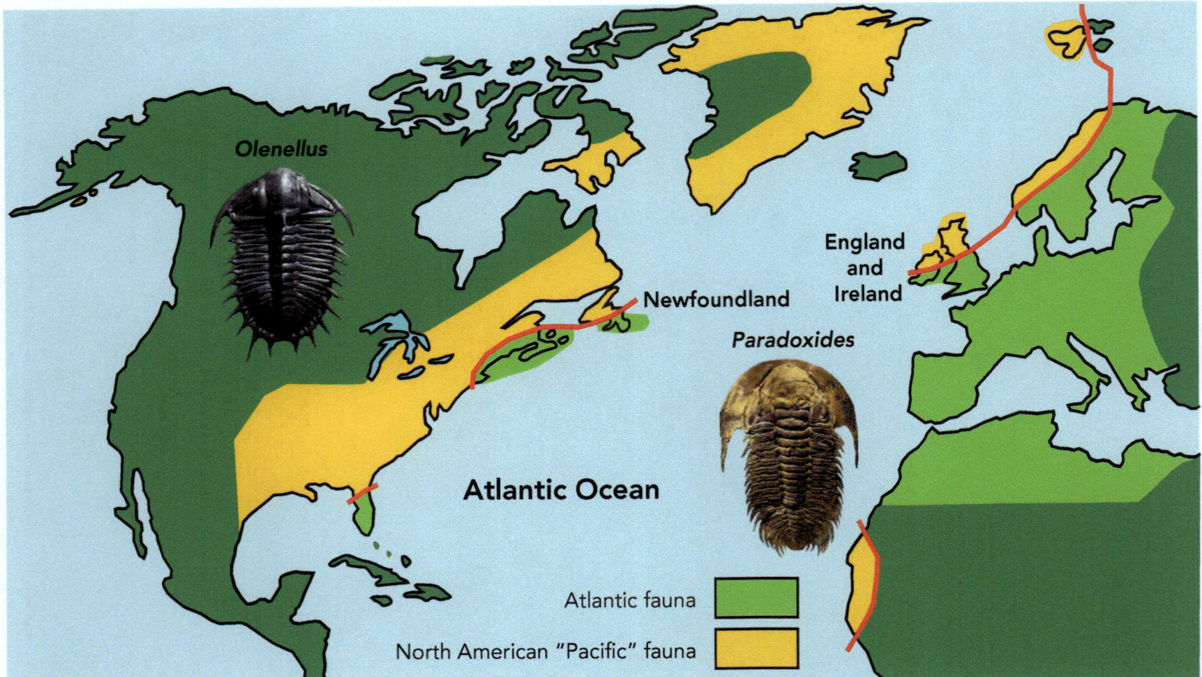

Figure 10.4 Map showing the distribution of fossils in Cambrian rocks. Most of North America is characterized by North American fauna that would have been linked to the paleo-Pacific Ocean to the west, long before the Rocky Mountains had formed. The North American fauna were thus termed the "Pacific fauna." Most of Europe and North Africa are marked by the Atlantic fauna. Eastern Canada, however, is marked by Atlantic-type fossils, such as *Paradoxides*, whereas northern England and Ireland and much of western Scandinavia contain the North American "Pacific" fauna, such as *Olenellus*. The suture between the two faunal provinces passes right through Newfoundland. Source: Modified from Wilson (1966), Springer Nature.

of Laurentia and down which much of the Iapetus lithosphere sank. The **Dunnage Zone** (Figure 10.5) represents the main deformed remnants of the Iapetus Ocean, and its associated island arcs. The **Gander** and **Avalon** zones (Figure 10.5) originally lay close to the margins of Gondwanaland (Figure 10.3(a) and (e)), and we will consider these in Section 10.3.

KEY POINT

The island of Newfoundland contains highly deformed central areas representing the compressed remains of the Iapetus Ocean. The western and eastern margins represent the land masses that lay at the margins of Iapetus.

10.2.2 Ophiolites: Remnants of Oceanic Lithosphere

The island arcs and the deformed and metamorphosed marginal sediments provided strong evidence of compression associated with subduction zones (see also Chapter 5), but was there any evidence of the actual oceanic crust?

10.2.2.1 The Bay of Islands Ophiolites

The search for oceanic crust associated with the closing of Iapetus led to closer examination of the Table Mountains in the Bay of Islands in western Newfoundland (Figure 10.7). These mountains show a distinctive light brown color and lack of vegetation that contrasts markedly with the surrounding greenery. Geological mapping shows that they are floored by thrust faults (see Figure 1.35 for explanation of thrust faults) that superimpose them on the platformal limestones that originally faced the open Iapetus Ocean. The Table Mountains are mostly made of metamorphosed igneous rocks consisting of serpentinite (see Chapter 5, Box 5.2), a hydrated ultramafic igneous rock made of the fibrous mineral **serpentine** (Figure 10.2), **diabase**, a fine-grained mafic intrusive igneous rock, and extrusive basalt. These are capped by sedimentary cherts, which are typically deposited in deep oceans. This distinct sequence of rocks is known as ophiolite (literally "snake rock," based on the preponderance of the snake-like serpentinite that is the common rock type). Similar sequences of more deformed ophiolites were also found farther inland in the Annieopsquotch Mountains.

It was long recognized that these ophiolites were "out of place," technically termed **allochthonous**, which refers to rocks or sediment formed in one place and later transported to another place, such as by thrusting. The antonym **autochthonous** means "formed in place." Plate tectonics, and the destruction of oceans by convergence, provided an explanation for how these allochthonous ophiolitic rock masses were initially formed in an ocean and later transported and thrust over the autochthonous sedimentary rocks that originally formed at the margin of that same ocean (Figure 10.6).

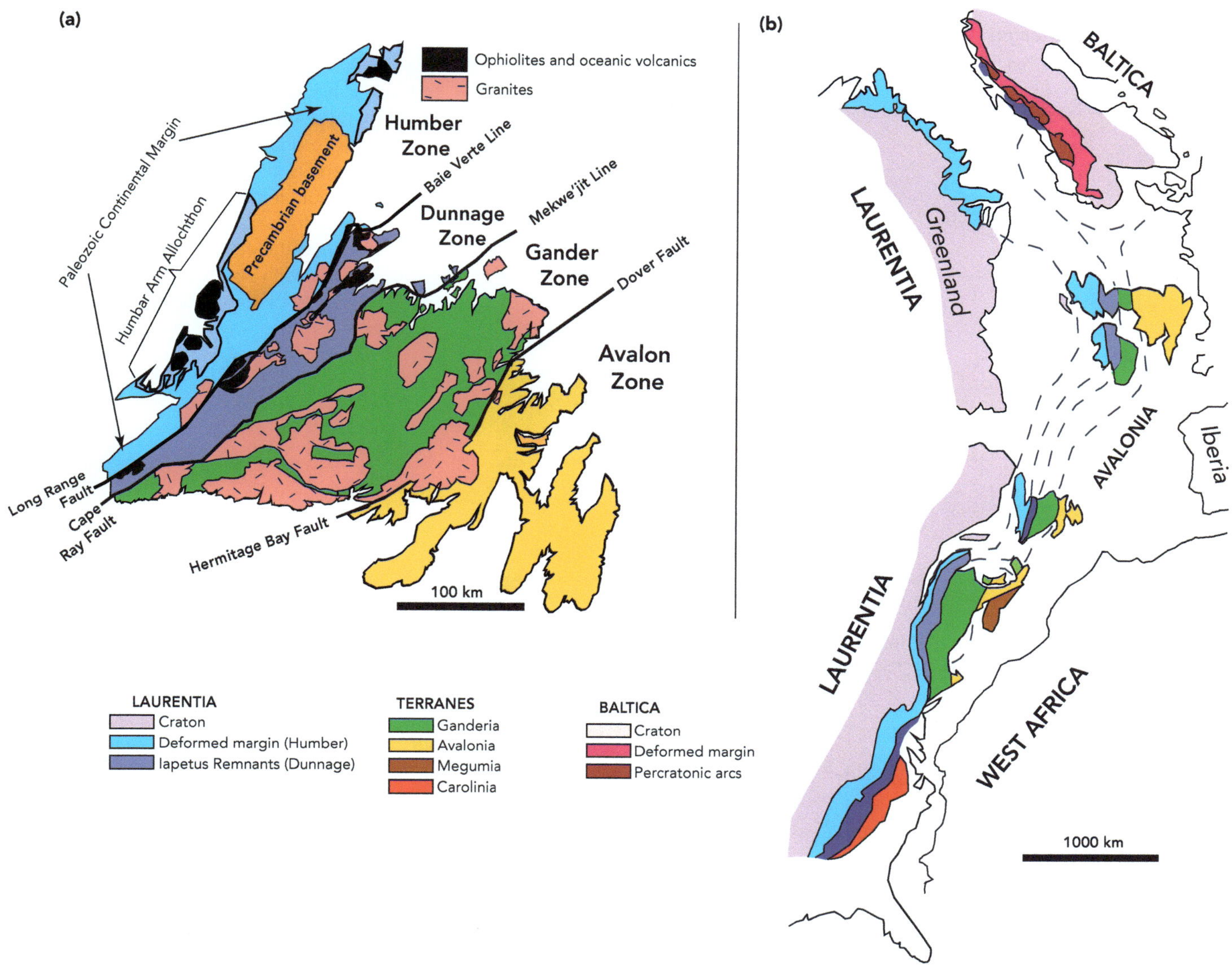

Figure 10.5 (a) Simplified map of Newfoundland, Canada highlighting major tectonic zones. The Humber Zone is the Paleozoic margin of Laurentia (ancient North America) that faced Iapetus. A large area of older Precambrian basement is exposed. The Humber Arm Allochthon includes over-thrust blocks of sediment (light blue) as well as ophiolites (black). The Dunnage Zone is the compressed remnants of Iapetus, including island arcs, ophiolites, and deep water sediments. The Gander and Avalon zones represent micro-continents carried on the Rheic Oceanic plate. Major fault zones mark the suture lines along which the various oceanic and continental plate were compressed together. As compression reached its maximum, metamorphism and melting resulted in the intrusion of extensive mostly Devonian-age granites. (b) Reconstruction of the Appalachian and Caledonian Orogens. Various tectonic zones and micro-continents can be traced between North America, the United Kingdom, Ireland, and Scandinavia (Baltica). Source: Figure modified after Van Staal et al. (1998); © The Geological Society 1998.

10.2.2.2 The Taconic Orogeny: Age and Emplacement of the Bay of Islands and Annieopsquotch Ophiolites

Dating of the Bay of Islands ophiolites showed that they were formed in an ocean in the Early Ordovician Period, about 485 Ma, and were later emplaced by thrusting about 470 Ma (Figure 10.2(d) and (e)) during a mountain-building event called the **Taconic Orogeny**, named after the Taconic Mountains in New York State. Unequivocal evidence of ancient oceanic crust had finally been found, but how did it end up being preserved as a mountain-top versus consumed by sinking into a subduction zone?

In addition to dates, detailed mapping and analysis of the structural geology of the ophiolites and their adjacent formation allowed the relative timing of deformation events and faults to be determined. Figures 10.2 and 10.6 show a series of cross sections illustrating how these ophiolites may have formed. The late Proterozoic breakup of Rodinia created a rift (Figure 10.2(b)) that opened and allowed formation of oceanic lithosphere at about 540 Ma (Figure 10.2(c)). Over time, the Iapetus Ocean developed a series of subduction zones (Figure 10.2(d)–(e)). The first dipped to the east (485 Ma, Figure 10.2(d)) and was overlain by island arcs (Ta in Figure 10.3(a)). As Iapetus subducted, these peri-Laurentian arcs

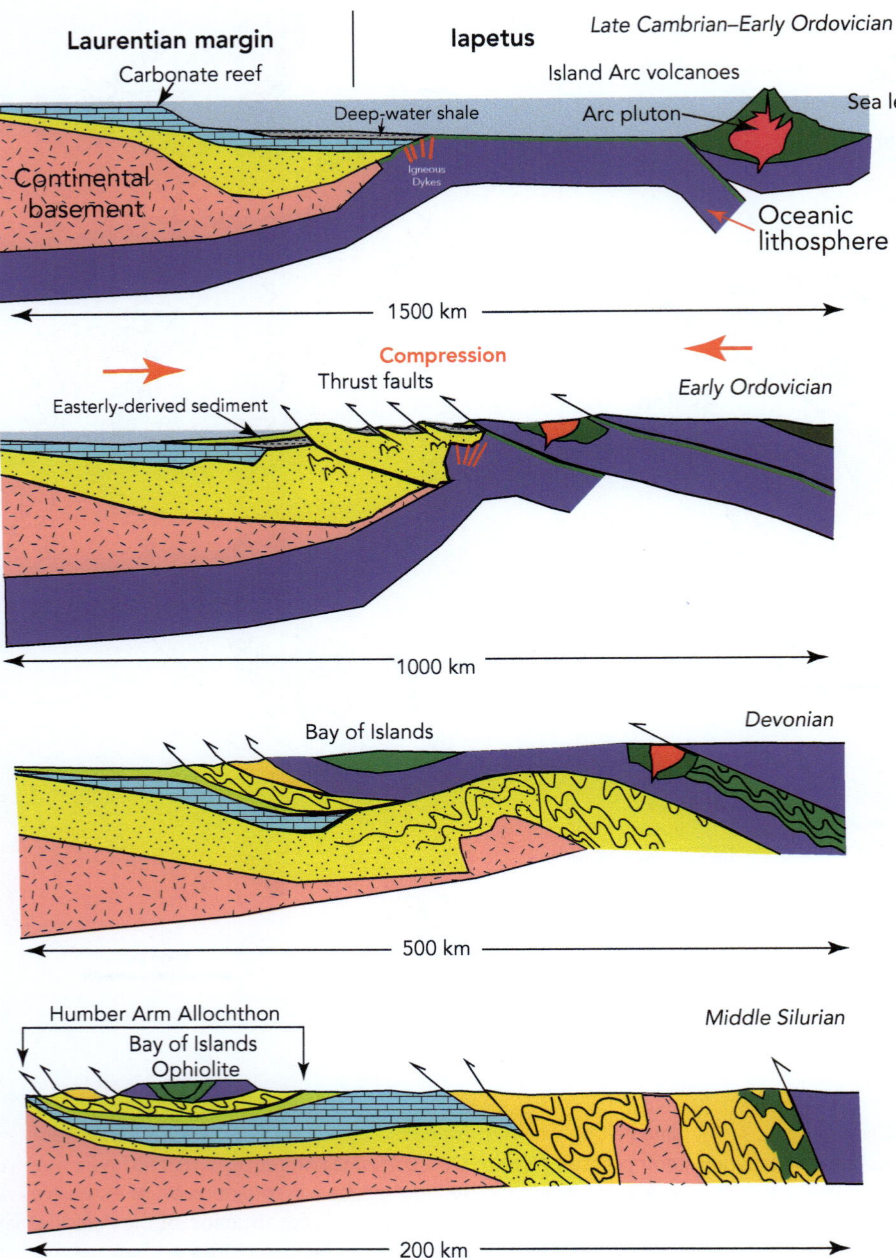

Figure 10.6 Geological evolution of western Newfoundland showing closing of the Iapetus Ocean and the accretion of oceanic volcanic island arcs to North America. The Humber Arm Allochthon includes metamorphosed remnants of the sediments formed on the old margin of Laurentia as well as ophiolites, remnants of oceanic crust that are thrust on top of the Laurentian margin as the ocean closed. Source: Figure modified with permission of Canadian Science Publishing, from Williams (1979); permission conveyed through Copyright Clearance Center, Inc.

accreted to Laurentia, causing compression and uplift associated with the Taconic Orogeny. Because of the eastern dip, there was no way to subduct the oceanic crust in the overriding plate and these rocks ended up being **obducted** (i.e., emplaced by thrusting) onto the Humber margin, as compression in the Taconic Orogeny reached its peak (Figure 10.2(e) and (f), and Figure 10.6). In like fashion, the Annieopsquotch ophiolites were also preserved, but their location caused them to be far more deformed and metamorphosed than the Bay of Islands ophiolites. Later compression and heating resulted in the area being further deformed and intruded by magmas, forming granitic plutons, which are extensive in the Dunnage and Gander zones (Figure 10.5(a)) as these areas experienced the brunt of the collisional forces.

KEY POINT

Discovery and dating of ophiolites provided the first direct observation of oceanic crust older than 340 Ma, which was thrust over the sediments that originally flanked Iapetus where the crust was formed.

10.3 The Assembly of Pangea

In Chapter 5 we described the three types of plate margins (divergent, convergent, and strike-slip), but many convergent margins are not formed in a simple accordion-style fashion, as Wilson originally envisaged, and may include both a convergent

Figure 10.7 The brown Table Mountains in Gros Morne National Park, western Newfoundland represent Ordovician oceanic lithosphere that has been thrust over Laurentian basement rocks. The brown color is typical of the weathering pattern of magnesium-rich silicate rocks that form in the upper mantle and oceanic crust and resist vegetation, compared to the less magnesium- and iron-rich rocks that form the underlying basement rocks. Source: Tango7174, CC BY-SA 4.0 <https://creativecommons.org/licenses/by-sa/4.0>, via Wikimedia Commons.

and strike-slip component. In addition, as oceans close, they can carry bits and pieces of lighter crust, island arcs, and microcontinents, which can be plastered onto an adjacent continent.

10.3.1 Suspect Terranes

In compiling his Appalachian map in the late 1970s, Williams noticed that there were several areas that formed discrete blocks, usually formed of more-felsic basement or sedimentary and metamorphic rocks and within which the geology seemed similar but were separated from the surrounding areas by major faults (Figure 10.5(a)). In some places, the fault zones were intruded by granites, or the fault zones were overlain by younger sedimentary rocks forming cross-cutting and overlap relationships that helped him understand the age of the faulting and of the blocks in between the faults. Soon after publishing his map, a group of geologists studying the west coast of North America suggested that much of that area was formed as a mosaic of **suspect terranes** (see Chapter 14), defined as relatively homogeneous geological provinces, that contain an internally consistent stratigraphy, deformational history, and paleomagnetic and magmatic signature and ages, but which contrasted sharply with those of nearby provinces. They interpreted these terranes as "suspect," in that they were not related to North America but rather were transported on adjacent plates associated with the Pacific Ocean and accreted onto the western margin by plate tectonic processes. Because western North America has a large strike-slip component, these terranes were interpreted to have first collided and then slid into their present position along strike-slip faults, such as the San Andreas (we will discuss this

in more detail in Chapter 14, Section 14.4.1). Williams wondered whether a similar mechanism had occurred in the Appalachians.

The earliest plate models for the evolution and destruction of Iapetus assumed a simple, accordion-like opening and closing, but what if the closing of Iapetus was more oblique and included a strike-slip component? Tuzo Wilson and Harold Williams initially assumed a simple collision of two major continents around a single symmetrical ocean, but the increasing recognition of geologically discrete blocks, or suspect terranes, that did not seem to have any geological relationship to each other suggested a more complex model. Also, it invited the possibility that there may have been several oceans that formed and were destroyed prior to the modern Atlantic rather than just Iapetus.

Inspired by these new ideas, Williams, and his USA colleague Robert Hatcher, then at the University of South Carolina, noted what they described as an "embarrassing number of terranes" along the eastern margin of the Appalachians that did not easily fit the simple symmetrical accordion model of Wilson. They suggested that these "embarrassing terranes" were better explained as suspect, with some emplacement along strike-slip faults. The terranes display a variety of shapes, sizes, and geological affinities. Some may originate as island arcs formed on oceanic lithosphere, such as the Aleutians in Alaska, and others may be formed by continental crust as small continents or islands, analogous to places such as Cuba in the Caribbean Ocean or Madagascar in the Indian Ocean.

Four major terranes containing continental crust were eventually identified based on Williams' tectonic map including **Carolinia, Ganderia, Avalonia,** and **Megumia** (Figure 10.5). A fifth **Suwanee terrane** was inferred and later identified based

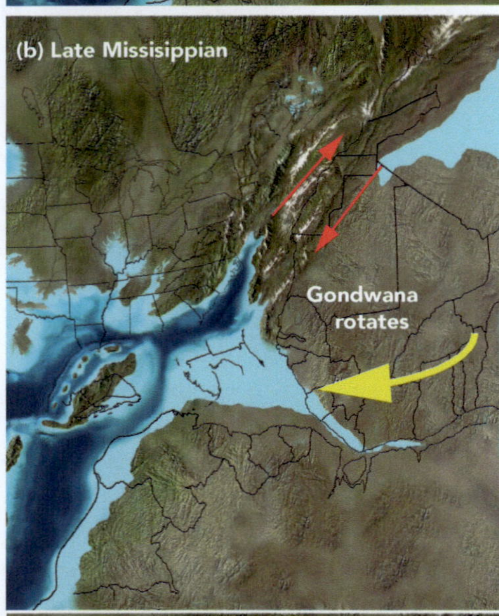

Figure 10.8 The north to south zipper-like closing of the Rheic Ocean begins (a) with collision of the Reguibat promontory. This is followed by (b) rotation of Gondwanaland, development of strike-slip faults in

on subsurface investigations of the geology underlying the younger coastal plain sedimentary rocks that covered it. Numerous other smaller terranes have also since been identified. Most of the larger continental terranes are thought to represent parts of the older Rodinian supercontinent (see Chapter 8) that were initially rifted away from Gondwanaland and later transported via oceanic plate motion to eventually collide with Laurentia (Figure 10.2(e) and Figure 10.3(a), (d), and (e)). The evidence for the original position of terranes includes their fossils as well as paleomagnetic data. Carolinia, Avalonia, and Megumia all contain the Atlantic-type fauna (Figure 10.4), indicating a probable origin from Gondwanaland. Also, the fossils on these terranes indicate organisms that lived at high latitudes and originated closer to the South Pole, where conditions were far colder in Gondwanaland versus the more equatorial environments in Laurentia during the early Paleozoic (Figure 10.3).

Paleomagnetic studies also confirm that Carolinia, Avalonia, and Megumia originated in the south and later migrated north and west, as shown in Figure 10.3. Analysis of the geochemistry of the volcanic rocks in these terranes, as well as the composition of the sedimentary rocks, also help us understand where these terranes originated. It appears that these terranes were rifted away from the western margin of Gondwanaland at various times during the Cambrian to Ordovician periods. This rifting opened a new ocean, called the **Rheic Ocean** (Figure 10.3(b) and (c)), which formed at about the time that Iapetus was closing (Figure 10.2(g)). As we have seen in Chapter 5, subduction of one oceanic plate commonly creates space for new oceans to open such that pull of the subducting slab on one end causes oceanic spreading at the other margin of the plate (see Chapter 5, Figure 5.5).

🔑 KEY POINT

Difficulty in matching geological blocks bounded by major faults indicated a more complex history marked by accretion of exotic "suspect" terranes, microcontinents, and island arcs that formed in the older Iapetus and Rheic oceans and later accreted to Laurentia as Pangea was assembled.

10.3.2 Orogenic Episodes and Zipper Tectonics

It has long been recognized that the formation of the Appalachians occurred in pulses. These pulses record periods of accretion of terranes to Laurentia, followed by periods during which the ocean was likely being subducted, but without significant accretion of terranes and consequently less compression. Mountain building, or **orogenesis**, is a combined response of terrane accretion, compression, and uplift. Orogenesis is also accompanied by deformation (i.e., folding and faulting), metamorphism, and partial melting, which in turn produces volcanic and plutonic igneous

Figure 10.8 (*cont.*) the central area, and finally (c) head-on collision in the south as the Rheic closes. Red arrows show the main directions of forces. Source: © DeepTimeMaps.

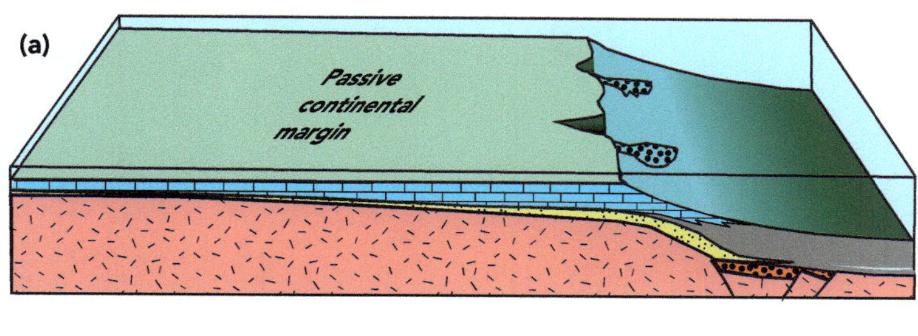

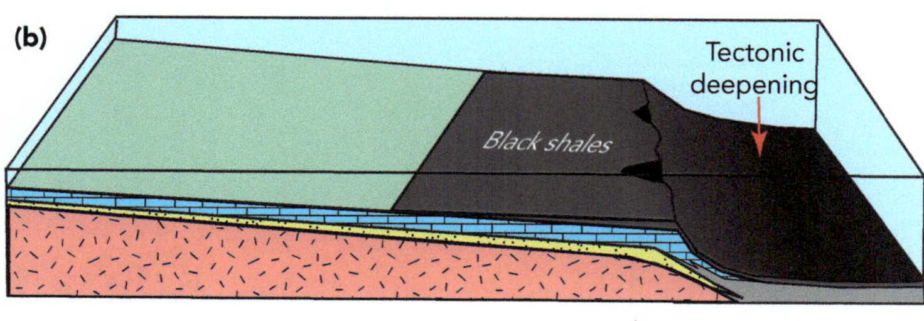

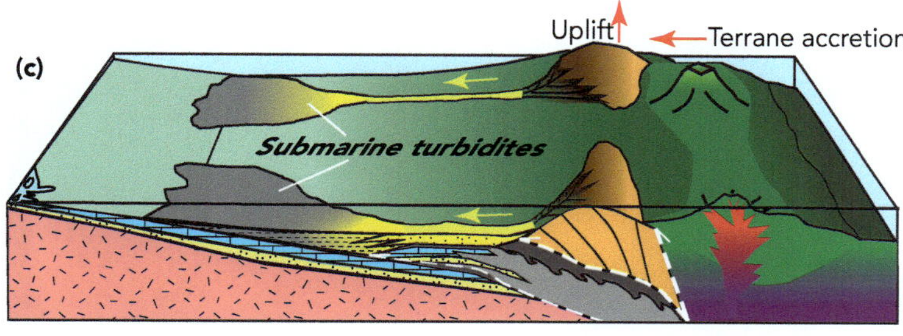

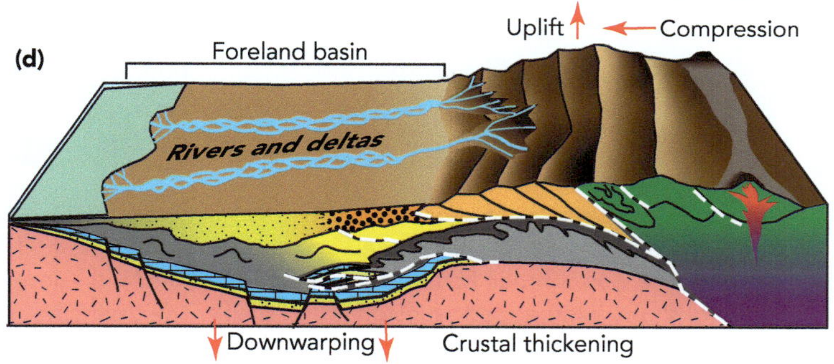

Figure 10.9 Evolution of passive to an active margin, typical of the Appalachian Orogeny. In (a) carbonate reefs form a shallow marine platform that faces the open Iapetus Ocean. In (b) sudden deepening is shown by the deposition of black shales indicating that the continent is being pulled down by an adjacent subduction zone (e.g., Figure 10.6(a)). In (c) accretion of an island arc allows deep water submarine turbidites to be deposited over originally shallow water carbonate platform. As the orogeny progresses (d), crustal thickening causes downwarping of the adjacent area forming a foreland basin that is largely filled with shallow water and non-marine sediments derived from the newly emerging land areas to the east where an ocean once lay.

rocks. The degree of metamorphism can provide insights as to the magnitudes of temperatures and pressures experienced, as well as how temperature and pressure change regionally and temporally. Analysis of the igneous rocks can also yield insights as to timing (i.e., the rocks can be dated), as well as the nature of melting.

The Appalachian orogenic pulses are usually named after a specific mountain belt. The major events include the oldest Ordovician Taconic Orogeny in northeast North America, followed by the Devonian **Acadian** and Carboniferous **Alleghanian** orogenies as shown in Figures 10.2 and 10.3. The Taconic Orogeny resulted in the obduction of the allochthonous ophiolites in the Bay of Islands discussed above (Figure 10.2(e) and (f), and Figure 10.6), which correlates to the closing of Iapetus. Accretion of the Carolinian terrane in the Late Ordovician to Early Silurian caused the **Cherokee Orogeny** in the south. The Devonian Acadian Orogeny is thought to reflect the docking of the Avalon and Meguma terranes, and the Carboniferous to Permian Alleghanian Orogeny correlates to the final assembly of Pangea, including the southern Suwanee terrane, with the ultimate collision

of Gondwana and Laurentia (Figure 10.2(g), (h) and (i), and Figure 10.3(f), (g), and (h)). At about 400 Ma (Figure 10.3(e)), Laurentia, Baltica, and Siberia had joined together, forming the greater land mass referred to as **Laurussia**.

It is thought that the closing of the Rheic Ocean is mostly associated with the final collision of Gondwanaland with Laurussia from about 340 to about 280 Ma (Figure 10.3(f)–(h)). The **Reguibat Promontory** was the first point along Gondwanaland to collide with Laurussia (Figures 10.3(f) and 10.8(a)). Once the Laurussian and Gondwana continents had collided, the plates locked up at the point of contact. Gondwanaland was then forced to rotate clockwise (Figures 10.3(g) and 10.8(b)), causing major strike-slip faulting in Canada and northeastern USA. In areas where the strike-slip faults defined a **releasing bend** (see Chapter 5), low areas (technically called **grabens**) formed and were filled with Late Paleozoic sediments. The rotating Gondwana block caused the Rheic Ocean to close from northeast to southwest, rather like a zipper. The final closing of the Rheic Ocean in the south culminated in a massive head-on collision in the southern USA, forming a major compressional fold and thrust belt during the final Alleghanian Orogeny that ended in the Early Permian Period at about 280 Ma (Figures 10.3(h) and 10.8(c)). The term **Appalachian Orogeny** collectively refers to the full suite of orogenic events including the Taconic, Acadian, Cherokee, and Alleghanian that culminated in the assembly of the Pangean supercontinent.

> ### 🗝 KEY POINT
>
> Once the continents Gondwana and Laurussia collided, subduction stopped at that point and the ocean to the south closed like a zipper, ending in a head-on collision in the Alleghanian Orogeny and the final assembly of Pangea.

10.4 The Stratigraphic Expression of Orogeny

Much of the geological record of orogenic events is recorded in the sediments eroded from uplifted areas, even if the areas of uplift

Figure 10.10 Cambrian–Ordovician sedimentary rocks of eastern North America. (a) Tilted gray shallow-marine tropical limestones of the Arbuckle Group, Oklahoma. The yellowish unit is dolomitic. (b) Close-up of shallow water stromatolites, indicating deposition in the warm waters of a hypersaline lagoon. (c) Coarser limestone consisting of crinoid skeletons, just below the tip of the pen.

Figure 10.11 Examples of deep water sediments deposited in the early stages of accretion. (a) Deformed interbedded sandstones and shales of the Cooks Brook Formation, western Newfoundland. Close up in (b) shows interbedded sandstones and shales. The distinctive sedimentary structures show a structureless (A) layer overlain by parallel laminated (B) layer, a ripple laminated (C) layer, and capped by shale layer (D). This distinctive sequence of sediments indicates deposition by submarine turbidity currents as they slow down. Contained fossils, as well as the lack of shallow water features, indicate deposition in deep water. The laminations in the rippled beds (marked with dashed lines), allow the direction of transport to be determined (see also Figure 1.31) and show that the sediments were derived from the east, an area that was previously open ocean, and indicated that the ocean was beginning to close.

are eroded or rifted away at some later time. Wilson's Pacific versus Atlantic faunas are contained within these types of sediments and are a good example of where the fossils, mostly diagnostic species of trilobites, provide critical clues to larger tectonic processes. Even Wegener used the similarity of fossils, such as the plant *Glossopteris* (Figure 5.3(b)), in the now-separated continents, to infer that they had been united as Pangea.

As we introduced in Chapters 1 and 5 (see Figure 5.23), mountain building causes thickening of the crust but areas adjacent to thickened crust may experience the opposite phenomenon of subsidence, forming sedimentary basins (Figures 10.9 and 5.23).

As the uplifted mountains erode, the produced sediment may be deposited into these adjacent basins, and the analysis of the stratigraphy of the basin fill can yield insights as to how the orogenic belt evolved. Orogeny can also generate angular unconformities (see Chapter 1) in the adjacent sedimentary sections.

Cross sections of the stratigraphy across the Appalachians (and other orogenic belts as we will see in Chapter 14), show a similar story (Figure 10.6). We have emphasized that the eastern margin of Laurentia was characterized by shallow marine sandstones overlain by limestones of the Sauk transgression (see Figure 1.20 in Chapter 1 and Figures 10.9 and 10.10). These Cambrian–Ordovician limestones can be found along the entire

Figure 10.13 (a) Pebbly sandstones, deposited in a river, erosionally overlie red floodplain mudstones of the Devonian Catskills Formation NY. (b) Fossil tree trunks, from Gilboa NY, further support a non-marine environment.

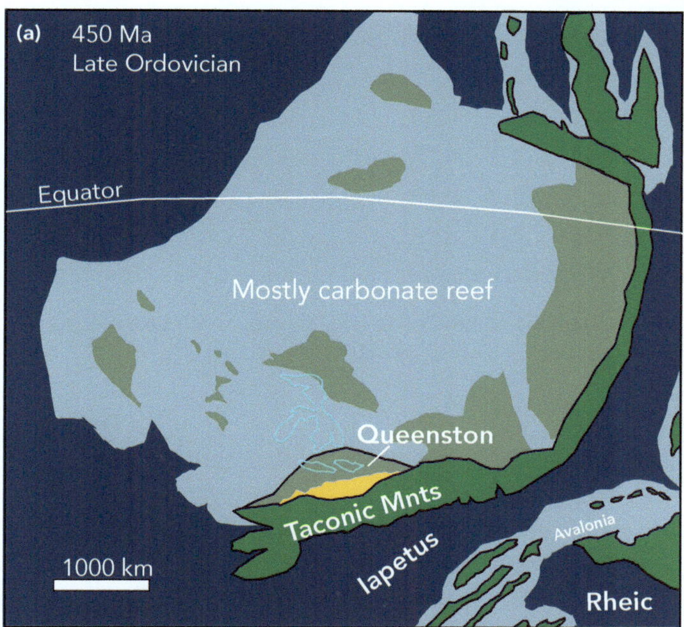

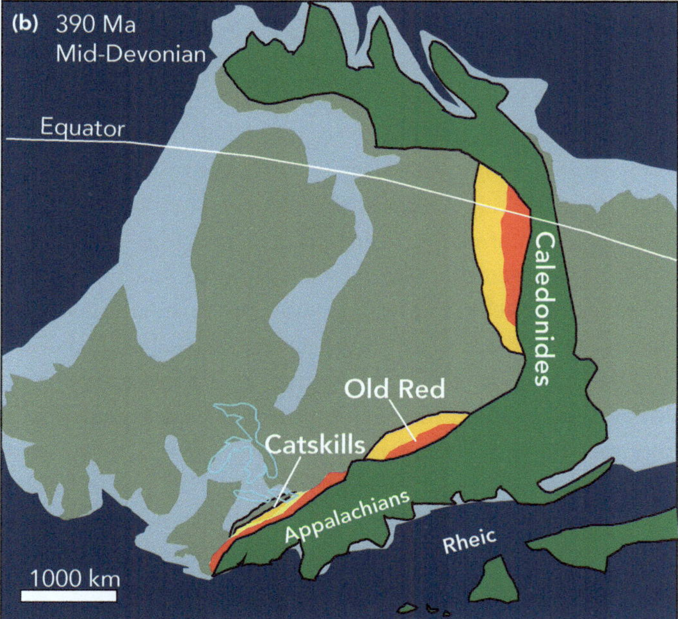

Figure 10.12 Paleogeographic maps of North America showing (a) Taconic and (b) Acadian clastic wedges formed in front of evolving mountain belts.

eastern margin of North America from Newfoundland to Texas and show the extent of the passive margin coast of Laurentia. As the Taconic Orogeny progressed, the old equatorial passive margin was first depressed by loading and thickening due to subduction farther east, resulting in deposition of deeper water black shales over the shallow water carbonate reefs (Figure 10.9 (b)). Island arcs migrated into the area that used to be the open ocean, and these eventually shed their sand and mud onto the deepened area, forming muddy sandstone-shale units several hundred meters to a few kilometers thick (Figures 10.9(c) and 10.11(a)). Improved developments in environmental analysis of these sedimentary rocks since the 1970s show that these sediments were deposited by turbidity currents in a deep water oceanic setting (Figure 10.11). Specific sedimentary structures such as the cross lamination produced by migrating ripples (Figure 10.11(b)), and scour marks at the base of beds allow the direction that the turbidity currents flowed to be determined. Analysis of these **paleocurrents** showed that the sediment was derived by erosion of Ganderian island arcs farther east (Figure 10.9(c)) from an area that used to be an open ocean

Figure 10.14 Alluvial-fan conglomerates of the Pennsylvanian Collings Ranch conglomerate overlie the tilted Paleozoic rocks of the Arbuckle Group Limestone, forming a classic angular unconformity marked with the white dashed line. Source: Photo taken along I35, near Ardmore Oklahoma by JPB.

(Figure 10.9(a) and (b)). This reversal of the direction from which sediment is derived is a common observation in margins that transition from a passive open ocean to an active margin characterized by subduction zones with associated areas that are uplifted.

As compression reached its maximum, and the ocean disappeared, mountains formed by the collision of terranes and eventually full-fledged continents caused the area of deep water deposition to become shallower. The younger wedges of sediments were formed by sediment deposited in alluvial fans, rivers, shorelines, and deltas rather than deep water (Figures 10.9(d) and 10.12). During the Taconic Orogeny, the interior part of North America, west of the Taconic Mountain belt, was covered by sandstones and shales of the shallow water Queenston Formation (Figure 10.12(a)) The red color of these shallow water to non-marine sedimentary rocks contrasts with the black shales and gray sandstones of the deeper water sediments. However, land plants had not yet developed in Ordovician time (see Chapter 11) and therefore there are no terrestrial fossils. During the Devonian Acadian Orogeny, however, similar wedges of sediment were deposited (Figure 10.12(b)), but this time terrestrial life was well-established (see Chapters 11 and 12), and the non-marine nature of the deposits is much clearer (Figure 10.13) as indicated by ubiquitous fossil plants. In the next chapter we will provide a more detailed explanation of the accompanying changes in biology and climate that allowed plants and animals to colonize the land, with significant consequences for changes in global climate.

One of the hallmarks of orogeny is deformation, uplift, and erosion, and these can be seen in the sedimentary record by the formation of angular unconformities that we discussed in the Grand Canyon (Figure 1.1). Limestones of the Ordovician Arbuckle Group in Oklahoma were originally deposited as horizontal layers in the warm shallow equatorial waters of the Cambrian–Ordovician seaway along the passive margin of Laurentia (Figure 10.9(a)), but are now tilted due to the later by compression during the Alleghanian Orogeny (Figures 10.10(a) and 10.14). Tilting was accompanied by erosion, forming the angular unconformity in Figure 10.14. The boulders in the overlying Pennsylvanian Collings Ranch conglomerate are composed of clasts derived from the underlying limestone formations and record deposition in alluvial fans associated with the uplift and erosion. Regionally, these conglomerates become finer grained to the northwest and passed into river and delta deposits that fed into the seaway that lay farther west, as shown in the schematic block diagram (Figure 10.9(d)). As with the earlier Taconic Orogeny in northeast North America, the Alleghanian causes similar effects farther south. The southern USA was underwater for longer than the north, despite the early accretion of the Carolinian terranes, but the orogenic processes form similar features in both areas, with the main difference being the time of formation, reflecting the complex history of terrane accretion and the zipper-like closing of the Rheic Ocean (Figure 10.8).

10.5 Conclusion

The closing of the Iapetus and Rheic oceans was complete by the end of the Paleozoic and resulted in final assembly of the Pangean supercontinent (Figures 10.2(i) and 10.3(h)). Pangean assembly culminated in the collision of the European and African continents with North and South America, and this continent–continent collision resulted in an extensive mountain chain that would have rivalled the Himalaya in its height and extent. Despite that fact that 250 million years have passed since Pangea began to break up, the remnants of this mountain chain remain as the modern-day Appalachians in North America and the Caledonides in Scandinavia and Scotland that can be mapped as the linear mountain belts shown in Figure 10.5(b). Following the assembly of Pangea, in the Mesozoic Era, the Atlantic Ocean formed by rifting of the supercontinent, initiating a new Wilson cycle (Figure 10.2(j) and (k)). As Pangea broke up, some of the old eastern margin terranes were left behind as well as bits and pieces of the now-vanished oceans, providing compelling geological evidence that plate tectonics had allowed entire oceans to be formed and ultimately destroyed.

10.6 Summary

- The observations of ocean crust not much older than about 180 Ma led to the expanding Earth hypothesis as a driver for continental drift. Debunking this hypothesis required finding evidence of older oceanic crust.
- J. Tuzo Wilson suggested that multiple cycles of continental assembly and breakup had occurred and that there may have been a pre-Atlantic Ocean, based on fossil evidence, with an ancient suture zone passing through the island of Newfoundland, Canada.
- Mapping of Newfoundland by Wilson's student, Harold Williams, revealed evidence for two land masses to the east and west with remnant of a Paleozoic ocean, called Iapetus, in between. These tectonic zones were later correlated and mapped across the entire eastern margin of North America and recognized in Europe and Scandinavia.
- Ophiolites, remnants of the Iapetus oceanic lithosphere, were discovered in western Newfoundland and provided direct evidence of an older ocean.
- These ophiolites were obducted onto the Laurentian continent because they formed on the overriding plate, as the rest of Iapetus was subducted eastward during the Taconic Orogeny.
- The closing of Iapetus resulted in accretion of suspect terranes (microcontinents) and each terrane typically marked an episode of orogenesis, deformation, and mountain building.
- Once the continents of Gondwanaland and Laurussia collided at the Reguibat promontory, further subduction was no longer possible at the collision point, and Gondwanaland began to rotate, closing the Rheic Ocean in a zipper-like fashion.
- This culminated in the Alleghanian Orogeny, when the final closure of the Rheic Ocean occurred, ultimately resulting in continent–continent collision and the final assembly of the Pangean supercontinent.
- Much of the evidence of these orogenic events is recorded in the sedimentary wedges and unconformities that developed on the landward side of the suture. Analysis of the environments of deposition and paleocurrents shows that the eastern margin of Laurentia started as a passive continental margin overlain by shallow equatorial reefs facing an ocean to the east. As subduction initiated, the shallow platform pulled the margin down and was covered with deeper water shales and sandstones derived from the rising volcanic islands to the east. Once the continents began to collide, even coarser sands and conglomerates marked the emergence of a Himalayan-scale mountain chain that formed the backbone of Pangea.

Key Words

- Pangea
- expanding Earth hypothesis
- greenstone belts
- Wilson cycles
- Panthalassa
- Iapetus
- Appalachian Mountains
- Caledonian
- Humber Zone
- Taconic arcs
- peri-Laurentian arcs
- Dunnage zone
- Gander zone
- Avalon zone
- serpentine
- diabase
- allochthonous
- autochthonous
- Taconic Orogeny
- obducted
- suspect terranes
- Carolina
- Ganderia
- Avalonia
- Meguma
- Suwanee terrane
- Rheic Ocean
- orogenesis
- Acadian Orogeny
- Alleghanian Orogeny
- Cherokee Orogeny
- Laurussia
- Reguibat Promontory
- releasing bend
- graben
- Appalachian Orogeny
- paleocurrents

Further Reading and References

Hatcher, R. D., 2010, The Appalachian orogen: A brief summary, in *From Rodinia to Pangea: The Lithotectonic Record of the Appalachian Region*, R. P. Tollo, M. J. Bartholomew, J. P. Hibbard, and P. M. Karabinos (eds.), Geological Society of America Memoir, 206, pp. 1–19.

Van Staal, C. R., Dewey, J. F., Mac Niocaill, C., & McKerrow, W. S., 1998, The Cambrian-Silurian tectonic evolution of the northern Appalachians and British Caledonides: History of a complex, west and southwest Pacific-type segment of Iapetus, *Geological Society, London, Special Publications*, 143, 197–242, https://doi.org/10.1144/GSL.SP.1998.143.01.17.

Williams, H., 1979, Appalachian orogen in Canada, *Canadian Journal of Earth Sciences*, 16(3), 792–807.

Williams, H., and Hatcher, R. D., 1982, Suspect terranes and accretionary history of the Appalachian orogen, *Geology*, 10 (10), 530–536.

Wilson, J. T., 1966, Did the Atlantic close and then re-open?, *Nature*, 21, 676–681.

Review Questions

1. Why are there no oceans with a crust older than 340 Ma?
2. What geological observations can be made to find evidence of vanished oceans?
3. How did we debunk the expanding Earth hypothesis?
4. What evidence did J. Tuzo Wilson use to hypothesize an older Iapetus Ocean?
5. What geological observations on the island of Newfoundland were made to test Wilson's hypothesis of an older Iapetus Ocean?
6. What is the significance of ophiolites in reconstructing plate tectonics over time?
7. How did the ophiolites in western Newfoundland end up where they are today?
8. How did the various linear geological belts (e.g., Humber, Dunnage, and Gander) and suspect terranes (e.g., Carolinia, Avalon, Meguma, and Suwanee) that make up the Appalachian Mountains form?
9. What is the Suspect Terrane concept and how does it relate to the Appalachian Orogeny?
10. Explain the idea that the Appalachian Orogeny occurred in a series of pulses including the Taconic, Acadian, and Alleghanian.
11. How did the final closing of the Rheic Ocean occur in the context of zipper tectonics?
12. How does stratigraphy record tectonics?
13. How does the stratigraphy of eastern North America record a change from a passive to an active margin associated with the assembly of Pangea?
14. Explain how angular unconformities mark orogenic events.

Artwork of giant dragonfly *Meganeura*. Source: Mark Garlick / Science Photo Library / Getty Images.

Chapter 11

Environmental Change in the Late Paleozoic

The Greening of Earth, Climate Change, and the Great Dying

LEARNING OBJECTIVES

- Recount some of the major steps in terrestrial evolution, including the development of vascular plants and the appearance of terrestrial insects and vertebrates.
- Explain how the expansion of plants led to changes in Earth's atmosphere, including higher oxygen levels that drove gigantism in insects, as well as reduced CO_2 that led to the Late Paleozoic Ice Age.
- Trace some of the evolutionary steps in land vertebrates from jawless fishes through early amphibians and the development of fully terrestrial amniotic reptiles.
- Provide evidence for the Late Paleozoic glaciations and explain the causes.
- Explain how extensive volcanism in Siberia caused the mass extinction at the end of the Paleozoic through global warming and oceanic acidification that selected out the low-metabolic calcifiers that dominated the Paleozoic fauna.

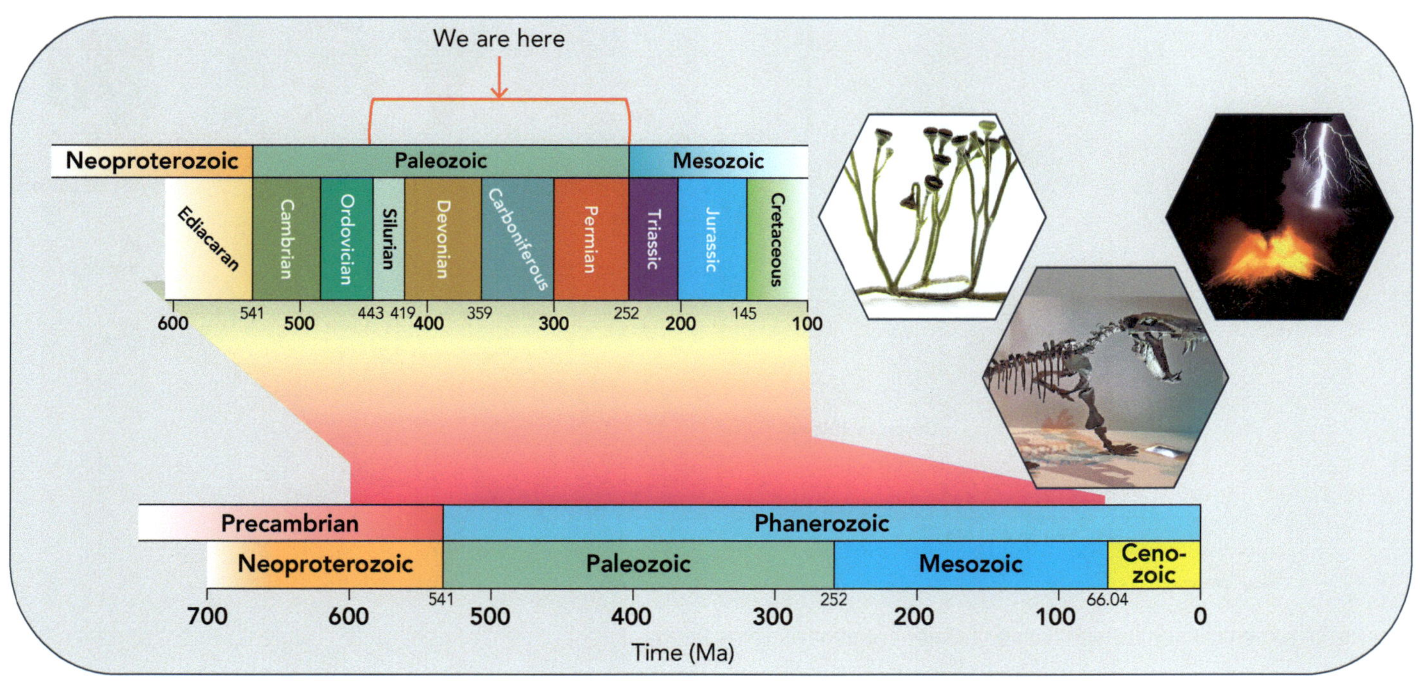

Introduction

As introduced in Chapter 9, from the beginning of the Paleozoic, life began to leave a much more tangible fossil record as well as experiencing major diversification during the Great Ordovician Biodiversity Event (Figure 9.24). Extinction events at the end of the Ordovician and the end of the Devonian decreased diversity by about 50%, but in each instance recovery to previous levels occurred within about five million years. The first part of this chapter reviews some of the major evolutionary events in the mid to later Paleozoic, with a particular focus on how the evolution of land plants paved the way for colonization by terrestrial animals and affected global climate, triggering a prolonged ice age. Massive volcanism reversed this trend and caused a hothouse at the end of the Paleozoic Era that initiated the most devastating of the five recorded mass extinctions known as the Great Dying and is the focus of the second part of the chapter.

11.1 Paleozoic Life: Invasion of the Land

As we saw in Chapter 7 (Figure 7.5), the first photosynthetic bacteria that used chlorophyll may have evolved about 2.1 Ga in fresh water, and photosynthetic eukaryotes may also have survived in freshwater environments during the Cryogenian snowball period, discussed in Chapter 8, but it took much longer for full-fledged plants to dominate terrestrial landscapes. Some of the major steps in plant evolution and their effect on climate are discussed below.

11.1.1 Plants Invade the Land

Genomic analysis shows that land plants, technically termed **embryophytes**, evolved from a single common freshwater algal ancestor. The first unequivocal fossil evidence of embryophytes (Figure 11.1) comprise microscopic reproductive spores in Ordovician rocks dated at about 470 million years ago. These fossil spores (Figure 11.2) were produced by **bryophytes**, simple **non-vascular plants** similar to modern mosses and liverworts (Figure 11.1). Adaptation to terrestrial conditions required the ability to avoid desiccation, as well as a system to regulate water. Bryophytes lack deep roots and would have been restricted to wetland environments. These primitive plants would also have consumed CO_2, and it has been hypothesized that enhanced depletion of CO_2 by these first primitive plants may have been enough to cause an ice age that ended the Ordovician Period and may have been the trigger for the first mass extinction. Gondwanaland lay close to the South Pole in the Ordovician (see Figure 10.3) and glacial deposits associated with these ice sheets are well documented. The mass extinction may have been caused by the loss of the extensive shallow water habitats of the Sauk transgression that had originally driven the Great Ordovician Biodiversity Event.

The first evidence of fossilized land plants with full-fledged **vascular systems** are the so-called **tracheophytes** (Figure 11.1) and are found in rocks of the Silurian Period (443.7 to 416.0 Ma). These include *Cooksonia* (Figure 11.3), a small, stalked plant with a spore-producing capsule on top. Vascular systems in plants are formed of **lignin**, a rigid organic molecule organized into pipe-like structures that are used to transport water and

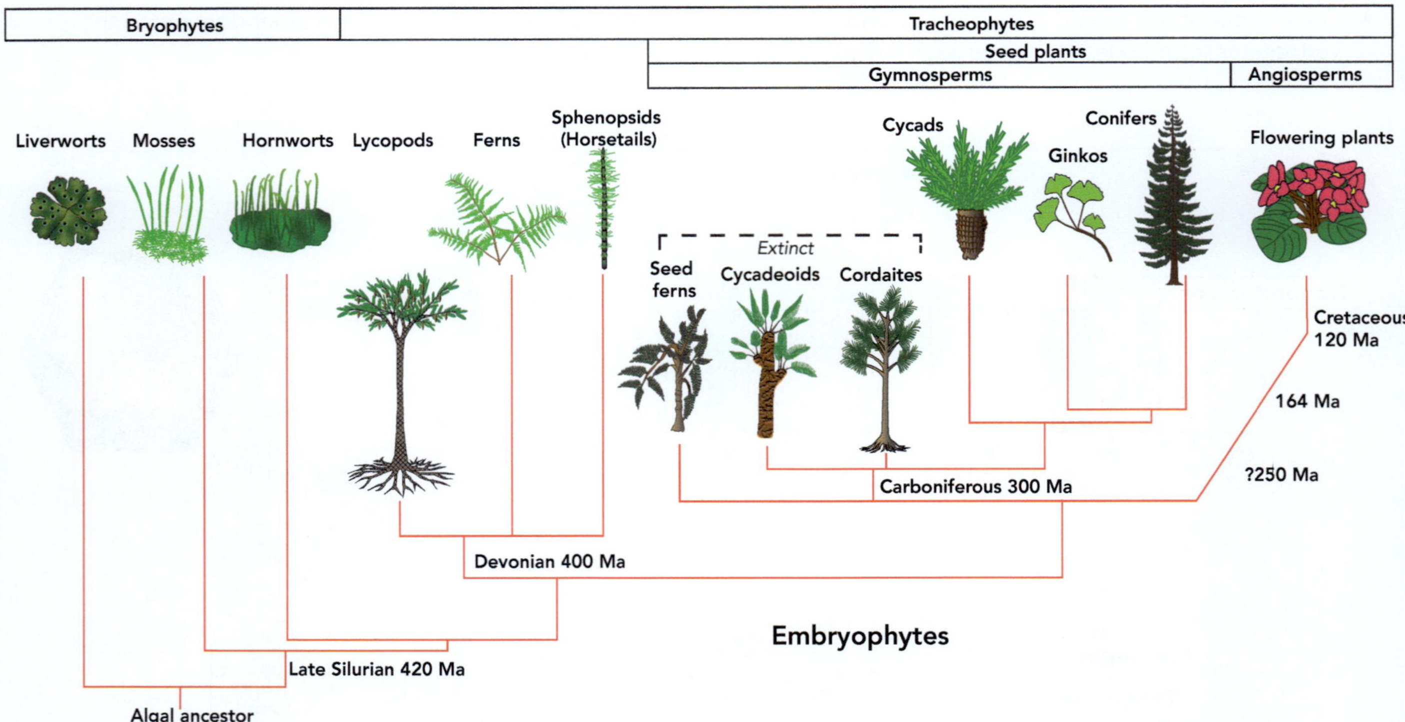

Figure 11.1 Plant cladogram with timing of major evolutionary steps.

minerals throughout the plant. The development of vascular systems was a key evolutionary adaptation that enabled plants to live farther away from wetland environments, allowing for extensive expansion across the world's land areas. The stiff coverings also prevented plants from desiccation. As plants invaded the land, animals from the sea followed.

The Devonian Period saw extensive diversification of land plants (see Figure 10.13(b)), resulting in the first extensive forests. This diversification also saw the appearance of **gymnosperms**, the first seed-bearing plants (Figure 11.1). The term gymnosperm means "naked seed," referring to the fact that their seeds do not have a covering. The gymnosperms include the **conifers** (i.e., evergreens), which are the main type of gymnosperm alive today, as well as **cycads**, **ginkos**, and **gnetophytes**. The flowering plants, or **angiosperms**, which mean "cased seed," did not arise until much later in the Cretaceous Period of the Mesozoic Era (Figure 11.1 and see Chapter 12).

The Carboniferous Period (358.9–298.9 Ma) is named after the vast coal deposits that characterize it, recording the enormous increase in global biomass resulting from the expansion of the seed-bearing vascular plants (Figures 11.1 and 11.4). This biomass formed a gigantic atmospheric carbon dioxide scrubber, removing CO_2 from the atmosphere. Because of extensive lowland areas that were either subsiding or flat, as well as generally

Figure 11.3 *Cooksonia* sp. (a) fossils and (b) reconstruction. Fossil is from the Late Silurian (420 Ma) Bertie Formation, NY from ROMIP49684–48992. Photo is about 7 cm high. Source: (a) With permission of ROM (Royal Ontario Museum), Toronto, Canada; (b) Reconstruction. Photo MUSE, CC BY-SA 3.0 <https:// creativecommons.org/licenses/by-sa/3.0>, via Wikimedia Commons.

humid conditions, much of this plant material was buried in wetlands as peat, eventually becoming transformed into coal as it was further buried and lithified. This resulted in permanent storage of the CO_2 consumed by plants and used to make their bodies. These plants also released massive amounts of O_2. The released oxygen gradually built up in the atmosphere to about 35%, compared to about 20% today (Figure 11.5), and these oxygen levels are thought to have been the highest ever in the history of Earth.

KEY POINT

The appearance and diversification of land plants, and storage as peat and coal, that characterizes the Carboniferous Period caused a decrease in atmospheric CO_2 and an increase in O_2. This is a good example of how the biosphere, lithosphere (which stores the plants as coal), and atmosphere are linked.

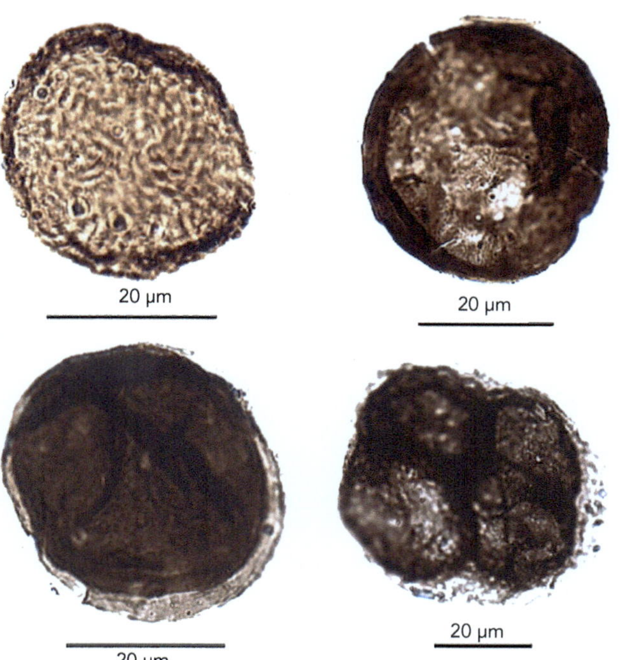

Figure 11.2 Fossil spores from the Ordovician Zanjón Formation, Argentina represent some of the earliest fossil records of land plants. Color plate of cryptospores from the upper part of the Zanjón Formation: (a) distal face of *Chomotriletes*? sp.; (b) *Gneudnaspora* (*Laevolancis*) *divellomedia* or *Laevolancis chibrikova*; (c) *Sphaerasacus glabellus*; and (d) new genus. The palynological slides are housed in the paleopalynological slide collection of the Unit of Paleopalynology, IANIGLA, CCT- CONICET Mendoza. Source: Steemand et al. (2010). With permission from John Wiley and Sons.

11.1.2 Evolution of Gigantic Insects and an Oxygen Spike

As we discussed in Chapter 9, arthropods were well established by the Cambrian Period, including the now extinct trilobites. Some of the sea-dwelling arthropods grew to large sizes, such as the group of scorpion-like **eurypterids** (Figure 11.6(a)), an extinct **chelicerate** predator that appeared during the late Ordovician Period, eventually going extinct at the end of the Permian. The first land-dwelling arthropods exploited the newly developed terrestrial habitats afforded by the expansion of plants. These included the **Hexapoda** (six-legged insects),

Figure 11.4 Fossil plants and insects from the Carboniferous: (a) seed fern (*Neuropteris*) (MNS); (b) *Calamites* stem, a common large plant; (c) *Lycopsid*, a common tree; and (d) *Meganeura monyi*, a giant dragonfly that had a wingspan of about 68 cm. Source: (a), (b), and (c): Photos by JPB. With permission from Houston Museum of Natural Science (HMNS); (d) Muséum de Toulouse, CC BY-SA 4.0 <https://creativecommons.org/licenses/by-sa/4.0>, via Wikimedia Commons.)

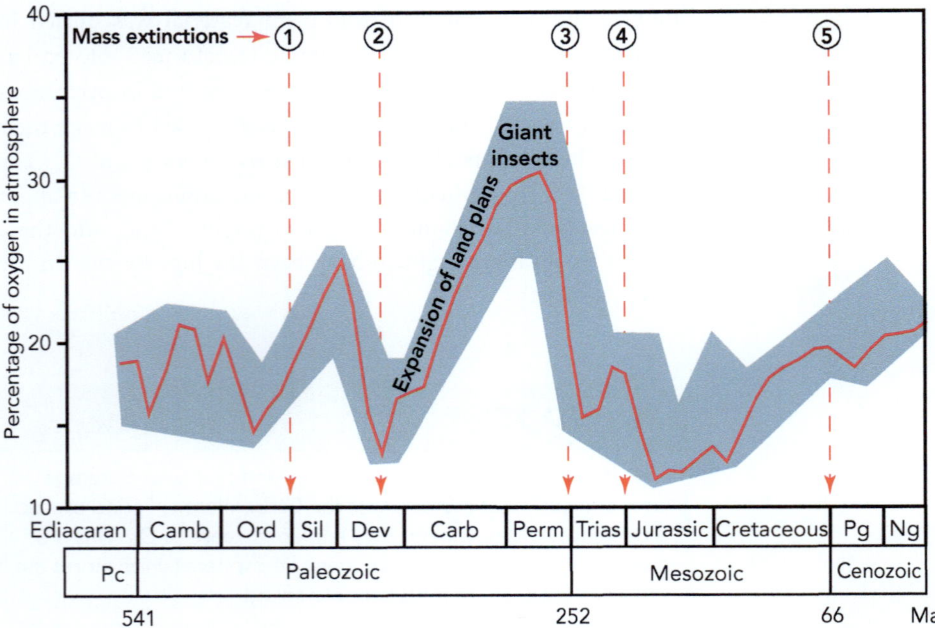

Figure 11.5 Oxygen levels through the Phanerozoic. The Late Paleozoic was characterized by high O_2, low CO_2, and low sea level, caused by glaciations. Source: Based on Berner et al. (2007).

Chelicerata (spiders and scorpions), and the **Myriapoda**, which include multi-legged centipedes and millipedes.

One of the oldest land-dwelling arthropod fossils is the ~400 Ma – early Devonian – ***Rhyniognatha hirsti*** (Figure 11.7). Originally thought to be the first winged insect, the tiny fossil was found encased in chert, but recent re-evaluation using 3D scanning microscopy suggests that it may have been a myriapod, more related to the millipede, rather than a winged hexapod.

Certainly, by the Carboniferous, land-dwelling arthropods were well established. Insects do not have lungs, but "breathe" through **spiracles**, openings along their sides, which connect to a system of tubes that convey oxygen through their bodies. If insects become too large, it becomes harder for oxygen to diffuse through their bodies, and this limits their growth. Also, arthropods have external skeletons and as they become bigger their skeletons become heavier and this also places limits on their growth. The increase in atmospheric concentration of oxygen in the Carboniferous and Early Permian (Figure 11.5), however, allowed insects to reach much larger sizes than are common today. Gigantism in land arthropods included *Arthropleura*, a 2 m-long millipede, which is also known from the distinctive trackways that it made (Figure 11.8), and **Meganeura**, a gigantic extinct dragonfly that had a wingspan of almost 70 cm (Figure 11.4(d)).

Figure 11.6 (a) Fossil of adult and infant eurypterids, from the Silurian period. (b) Aquatic scorpion, Middle Silurian. Source: (a) With permission from Houston Museum of Natural Science (HMNS); (b) Photo by JPB. With permission of ROM (Royal Ontario Museum), Toronto, Canada.

> **KEY POINT**
>
> Late Paleozoic increases in atmospheric O_2 allowed gigantism in insects.

11.1.3 The Earliest Tetrapods: A Fish Out of Water

The invasion of the land by plants and invertebrates provided food sources, and vertebrates were soon to take advantage of this. As we saw in Chapter 9, chordates such as *Pikaia* and *Metasprigginia* appeared early in the Cambrian and eventually evolved into vertebrates including the fishes. The first fishes are Late Ordovician jawless fishes, represented today in the class Agnatha, that include lampreys and hagfish. Fishes with a bony jaw evolved by modification of the front gill arches (Figure 11.9) and are common in Devonian rocks. Because of this, the Devonian Period is sometimes referred to as the "**Age of Fishes.**" Devonian fishes included the heavily armored *Placoderm* (Figure 11.10(a)). Apex predators included ***Dunkleosteus*** (Figure 11.10(b)), which grew to over 5 m long, as well as sharks and bony fishes, which were at the top of the food chain. The heavy armoring indicates the development of a complex predator–prey arms race, and this is also reflected by the increase in complexity and thickness of shells in invertebrates as well as increased burrowing – anything to stay clear of the apex predators of the time.

>  **KEY POINT**
>
> Gigantism and armoring in Devonian fishes indicate an evolutionary arms race in the first vertebrates.

The first land vertebrates evolved from the **lobe-finned fish** (Figures 11.11 and 11.12), the best known of which is the **Coelacanth** (Figure 11.11), originally thought to have been extinct since the Cretaceous Period, until its 1938 discovery off the coast of South Africa in the waters of the Indian Ocean. It is likely that the complex predator–prey relationships drove the lobe-finned fishes to shallow, coastal areas, while a complex land biota was developing and available for exploitation by any vertebrates that could adapt to a life out of water.

By the end of the Devonian, some vertebrates became amphibious and developed the capacity to spend some of their time on land. The characteristics that allowed amphibians to evolve include four well-developed limbs, and a moist, permeable skin. Amphibians can live on land but must return to the water to lay their eggs. Until recently, however, the fossil record of these early amphibians was very sparse and this is the story in our next section.

11.1.4 The First Amphibians: Filling Romer's Gap

During the middle part of the twentieth century (most notably the 1930s through the 1950s), Harvard professor Alfred Sherwood Romer was one of the leading American vertebrate paleontologists. Romer noted a lack of fossils that recorded the transition from fish to land vertebrates in the time spanning the latest Devonian to Early Carboniferous (i.e., from about 360 to 345 Ma). This 15-million-year period was subsequently referred to as **Romer's Gap**. In Chapters 6 and 9 we discussed modes of evolution and the contrasting views that the apparent lack of intermediary forms may be more consistent with punctuated equilibrium versus phyletic gradualism. Romer's Gap indicated that there may have been a short spurt of local allopatric speciation of lobe-finned fished into land-dwelling tetrapods.

In the 1980s, major field programs in Scotland, Nova Scotia, and Greenland discovered a plethora of transitional fossils that have effectively filled-in this gap. Much of this work was done under the leadership of Jennifer Clack, a leading British paleontologist at Cambridge University, starting in the 1980s and

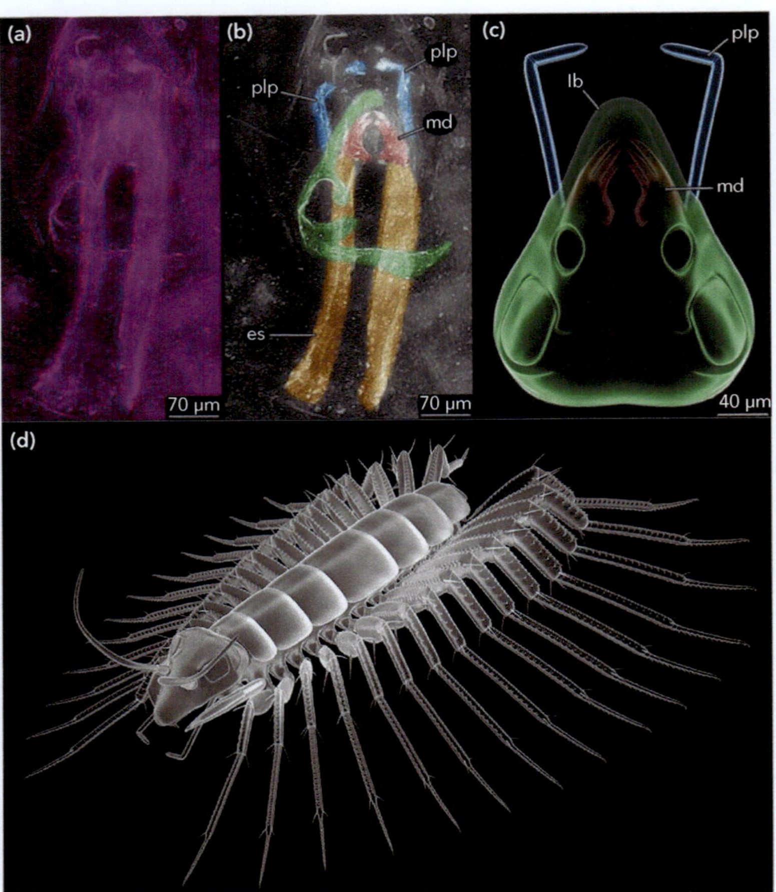

Figure 11.7 *Rhyniognatha hirsti*, originally thought to be the first winged fossil insect. (a) Photo of the fossil head; (b) outline of the key elements; (c) reconstruction of the head; and (d) speculative reconstruction of the animal. Plp – palps, md – mandible, es – elongate structures, lb – labrum or clypeo-labral complex. Source: Reproduced from Haug and Haug (2017).

continuing until just before her death in 2020. Much of Clack's work was focused on the **stegocephalians** (literally "armorheads"), which are **stem-tetrapods**. *Tetrapod* literally means "four footed," and stem-tetrapod refers to the early four-limbed vertebrates that were the ancestors of the four-legged **crown tetrapods** that we know today as the amphibians, reptiles, birds, and mammals. (Recall that in Chapter 6 we defined *stem* organisms as the ancestors that lie at the base or stem of a family tree (i.e., a clade), versus groups at the top of the tree, referred to as *crown* organisms.)

Most of the stegocephalians were relatively large, around 1 meter in length, and most were predators. As a group, the stegocephalians represent transitional forms between fish and tetrapods (informally referred to as "**fishapods**"), reflecting the fact that they enjoyed a largely aquatic lifestyle, but were able to breathe air and crawl out of the water in their search for food and new shallow water habitats. One of the most primitive of these animals is **Tiktaalik** (Figure 11.12), discovered in the Canadian Arctic in 2006 in rocks about 375 Ma in age in the Late Devonian Period. *Tiktaalik* is classified as a fish, as it has scales and four lobe fins, the fins have fish-like spines rather than bony digits, and it has gills, indicating an aquatic habitat. However, *Tiktaalik* lacks a dorsal fin and has a flattened crocodile-like head with a sturdy neck and shoulder bones, as well as robust wrist bones supporting its fins. A sister genus,

Epistostege (Figure 11.13), shows similar anatomical features. These anatomical adaptations would have allowed these "fishapods" to lift their heads out of very shallow water. Their skulls also have holes on top (spiracles), which may have indicated they had primitive lungs and could breathe air. These anatomical changes have been referred to as *exaptations*, since they represent modifications of features previously used for one purpose and co-opted for another purpose later in evolutionary history (see Box 11.1). It has been hypothesized that *Tiktaalik* and *Epistostege* lived in shallow and probably brackish water, and likely had far less competition for food resources versus the diverse but more competitive marine reef communities and their apex predators like *Dunkleosteus*.

Acanthostega and **Ichthyostega** (Figure 11.14) are found in rocks about 10 million years younger than those that contain *Tiktaalik* and *Epistostege*, and record the next steps in adaptation to a more terrestrial lifestyle. *Acanthostega* and *Ichthyostega* are more amphibian-like. They have bony digits at the ends of their strong forelimbs rather than spines (Figure 11.14), indicating an increased ability to lift their heads out of the water to breathe air, and they also developed nostrils, rather than simple spiracles, indicating fully developed lungs. However, they also sported gills, and studies of their limb bones indicate that they could not fully support themselves to walk on land. Their hind limbs were weaker than their forelimbs, and neither were capable of the rotary motion required for a truly *quadrupedal* four-legged gait that full-fledged walking crown tetrapods have. Consequently, they are thought to have lived in very shallow, mostly brackish to fresh water.

The newly developing marginal marine freshwater and terrestrial environments were evolving terrestrial invertebrates, such as insects, representing an exploitable food resource for a larger animal that could tolerate a life out of water. Developing the ability to breathe air may have allowed them to exploit very shallow water environments to escape predation by the larger marine fish of the day. Despite having four limbs, these stem-tetrapods were polydactyl, with more toes than the five digits that are the norm in today's land vertebrates (Figure 11.15). *Acanthostega* had eight digits and *Ichthyostega* had seven. To fully adapt to terrestrial life, however, vertebrates would have to develop limbs strong enough to support their weight without help from their buoyancy in water. Aquatic animals, such as fish, have a body density that is similar to water and are effectively weightless. Most of their muscle strength is used for propulsion through water and their swim bladders provide buoyancy. The extension of bones into the fins, in the lobe-finned fish, provided the basic body plan required to evolve

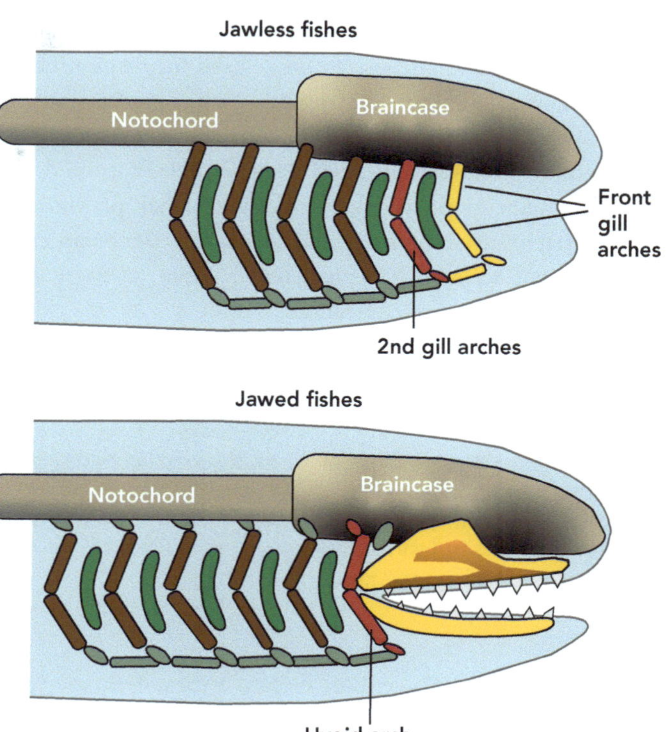

Figure 11.9 Evolution of jaws. The front gill arches in the jawless fishes (Agnatha) are modified to form the upper and lower jaws. The second gill arches become the hyoid arch in jawed fishes.

stronger limbs that were first used to lift their heads for breathing air and were later adapted for full-fledged walking on land.

Pederpes (Figure 11.12), discovered in 1971 and re-evaluated in 2002 by Clack, was an early Carboniferous tetrapod that lived between 348 and 347 Ma during the Late Mississippian, making it one of the earliest known animals capable of a partly terrestrial lifestyle. Its narrow skull indicates it breathed using muscles and its limb structure shows forward-facing feet and strong ankles, indicating that it was fully capable of a quadrupedal gait. However, small hindlimbs indicate that it retained an adaptation for swimming and its ear structure indicates that it could hear better underwater. Together, these anatomical attributes suggest that it was also adapted to aquatic hunting. Although we do not have fossil eggs, it is assumed that all the stegocephalians were amphibians and likely laid their eggs in water. They could leave their watery habitats only temporarily, disallowing full exploitation of terrestrial habitats.

The evolution of the tetrapods, and the filling of Romer's Gap, shows that the transition from a fully aquatic to partly terrestrial lifestyle occurred in a series of small steps, driven by evolutionary pressures that beckoned the vertebrates to exploit newly available terrestrial habitats. The abundance of transitional forms indicates that speciation of tetrapods was gradual, and that younger fossils consecutively show increasing adaptation to a terrestrial environment. There is also strong evidence that these changes were achieved by the exaptation or co-opting

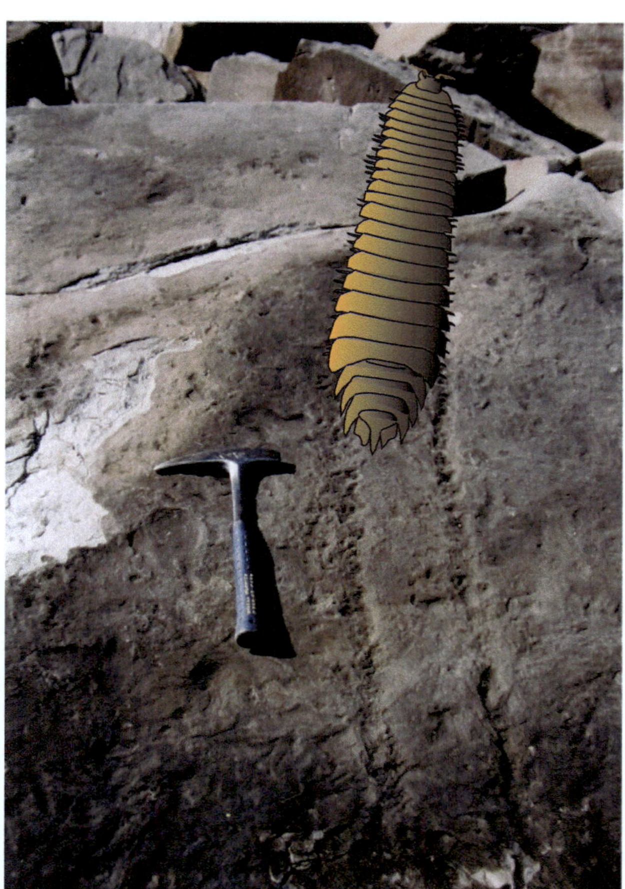

Figure 11.8 *Diplichnites* burrows made by the giant Late Carboniferous millipede, **Arthropleura** drawn on top. Source (main photo): Own work | Michael C. Rygel: https://www.wikiwand.com/en/Diplichnites Media/File:Diplichnites_mcr1.jpg.

of features used for one purpose, such as swim bladders for buoyancy, to a new purpose as lungs used for respiration, and fins for swimming into feet for walking. The Late Devonian was also marked by a mass extinction, but this chiefly decimated marine biota and especially reefs. It is possible that the decimation of marine populations caused additional pressure that forced vertebrates to invade the land, but the Devonian extinction does not seem to have had much effect on these newly developing terrestrial ecosystems.

> ### 🔑 KEY POINT
>
> Fossil evidence has filled Romer's Gap, showing how exaptation of features such as swim bladders and lobe fins were co-opted to lungs and feet in the earliest tetrapods. The abundance of intermediary fossils illustrates the many steps in the evolution from fish to land-dwelling tetrapods.

11.1.5 The First Reptiles: Freed From the Water

A key difficulty in fully adapting to land-dwelling, for both plants and animals, is the drying effect of air. To fully conquer

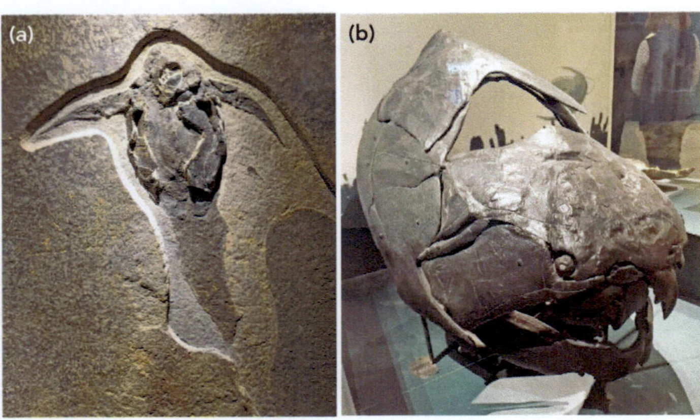

Figure 11.10 Devonian fishes: (a) armored *Placoderm*; (b) *Dunkleosteus*, a giant predator; and (c) a school of fossil lobe-finned *Holoptychius* sp. GSM1026. Source: (a) Photo by JPB. With permission from Houston Museum of Natural Science (HMNS); (b) Photo by JPB. With permission of ROM (Royal Ontario Museum), Toronto, Canada; (c) Photo by JPB. By permission of The Trustees of the Natural History Museum, London.

Figure 11.11 *Coelacanth*: (a) Triassic fossil and (b) reconstruction. Source: Photos by JPB. With permission from Houston Museum of Natural Science.

the land required the evolution of an encased egg (i.e., **amniotic**) that resists desiccation, as well as a watertight skin. This trait is shared among the amniotic tetrapods, including reptiles, birds, and mammals.

The fossil record shows that fully terrestrial vertebrates were established by the late Carboniferous Period. One of the oldest fossil reptiles include ***Hylonomus lyelli*** (Figure 11.16), discovered by Canadian geologist John William Dawson along the sea cliffs in northwest Nova Scotia in a hollowed-out stem of a fossil **Lycopsid** tree (Figure 11.4(c)), preserved in the Joggins Formation. Dated at 312 Ma (mid-Pennsylvanian), the fossil was named after Dawson's teacher, who was none other than Charles Lyell, one of the founders of modern geology (Chapter 2). Unlike *Perderpes* and *Ichthyostega* (Figures 11.12 and 11.14), *Hylonomus* has a more upright posture, both front and back legs face forward, and it has no evidence of gills, indicating a fully terrestrial habitat.

By the Permian, the amniotic vertebrates had experienced significant evolution and diversification, and had changed from tiny lizards like *Hylonomus*, that likely fed on insects, to much larger animals that exhibited a diverse variety of behaviors and adaptations. The earliest amphibians and reptiles, such as *Ichthyostega* (Figures 11.12 and 11.14) and *Hylomonus* (Figure 11.16), have only eyeholes and nostrils and form the group known as **anapsids**, literally "no arches," meaning that they had no other holes, or arches, in their skulls. As the vertebrates diversified, two main clades evolved including the **synapsids** (single arch, Figure 11.17) and the **diapsids** (double-arch). The synapsids represent the group that gave rise to the

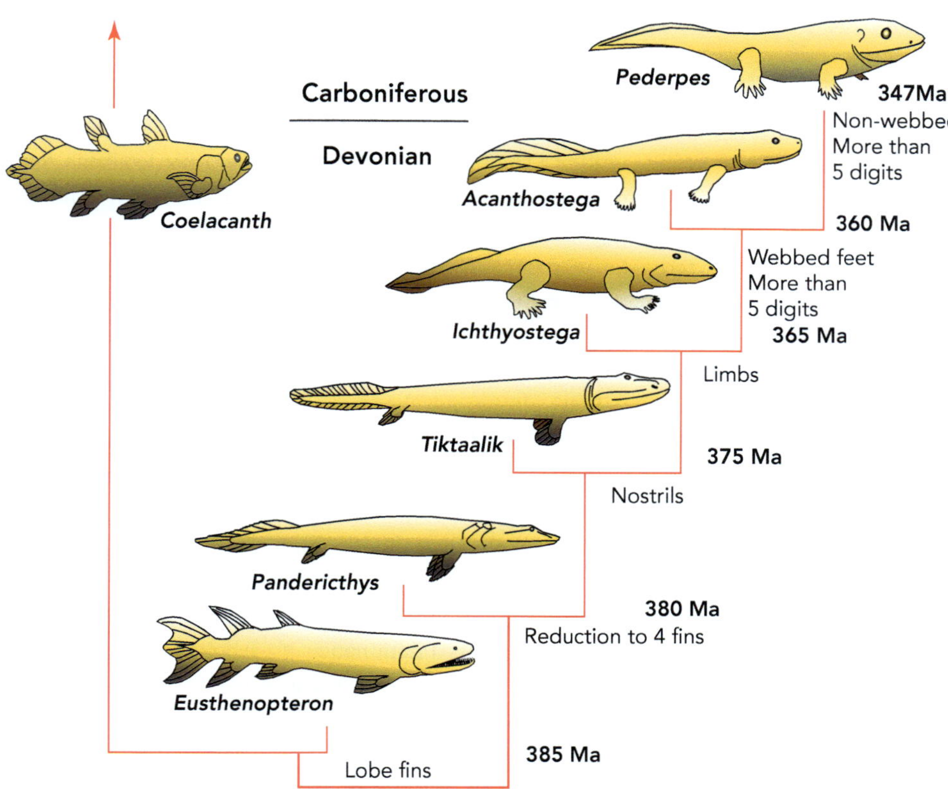

Figure 11.12 Evolution of amphibians. The lobe-finned fishes, including *Coelacanths* and **Eusthenopteron**, *Pandericthys*, and *Tiktaalik*, evolved into amphibious tetrapods, such as *Ichthyostega*, *Acanthostega*, and *Pederpes*.

(a)

(b)

Figure 11.13 *Epistostege* (a) fossil and (b) reconstruction. Note loss of top fin. Source: With permission of ROM (Royal Ontario Museum), Toronto, Canada.

mammals and have been historically referred to as the mammal-like reptiles.

Permian synapsids included the finbacks, such as **Edaphosaurus** (Figure 11.17(b)), an herbivore with peg-like teeth, and carnivorous **Dimetrodon** (Figure 11.17(a)). It was originally thought that the sail-like fins were used for thermoregulation, but a lack of **vascularization** (i.e., tubes for blood vessels) suggests that this is not the case, and it is now thought that their sails may have been used in courtship displays, more akin to a peacock tail. These animals were up to 3.5 m in length and over 300 kg in weight and indicate evolutionary gigantism.

Lystrosaurus (Figure 11.17(d)) is the most common of the Late Permian herbivorous synapsids and fossil sites commonly contain many individuals, suggesting complex herding behavior. About the size of a pig, *Lystrosaurus* was one of the few terrestrial vertebrates that survived the end-Permian mass extinction, and their ancestors were dominant in the following Triassic Period. *Scutosaurus* was one of the largest herbivorous synapsids, boasting a fully upright quadrupedal gait, and was heavily armored with scutes and spikes. The improved posture and armor indicative that a terrestrial arms race between herbivorous prey and their predators was well established by the Late

BOX 11.1 Exaptation

Lungs originally developed as swim bladders, used to control buoyancy, and were later adapted for respiration. In the same way, the front gill arches of the Agnatha were later adapted to jaws in the cartilaginous and bony fishes, and fins to legs from lobe-fined fishes to tetrapods. The term *exaptation* is used to refer to features that originated for one function and are later co-opted for another purpose, versus adaptation, which refers to novel features that are uniquely adapted to a specific function. Evolution is full of examples of exaptation, including feathers, now used by birds for flight but which originated in dinosaurs for other purposes, which we will discuss further in Chapter 12. This is a key point of evolutionary theory. Natural selection can only modify what previously exists. It is much easier to exapt or modify existing features to adapt to new functions versus creating entirely new features from scratch.

Permian. Large apex predators also evolved that likely fed on herds of lystrosaurs and scutosaurs. *Inostrancevia* (Figure 11.17 (c)), discovered in Russia, is an example of the largest of the saber-toothed animals, one of the **gorgonopsians**, reaching up to 3 m in length, also showing a trend to gigantism.

KEY POINT

Complex predator–prey relationships resulted in gigantism in Late Paleozoic synapsid reptiles.

11.2 Late Paleozoic Climate Change: From Icehouse Forests to Hothouse Deserts

As discussed in Chapter 10, Gondwanaland lay low in the southern hemisphere and Antarctica lay at the edge of the South Pole (Figures 10.3 and 11.18). A combination of uplift rates, caused by

the assembly of Pangea (see Chapter 10), and the expansion of plants with more vigorous root systems, may also have led to increased weathering. As we saw in Chapter 8, weathering of silicate rocks also consumes CO_2, further reducing its atmospheric concentration. These effects resulted in a negative feedback loop causing an overall loss of CO_2 stored in coal and via weathering that triggered prolonged glaciations in the southern hemisphere termed the *Late Paleozoic Ice Age* (Figure 11.19). It has been estimated that CO_2 levels almost dropped to the level that could have triggered another snowball epoch, but the by-then warmer Sun prevented a total global freeze (see Chapter 8).

The first glaciations began around 355 Ma in the late Mississippian Period and reached a peak in the Pennsylvanian Period. The icehouse lasted for about 70 million years, ending about 260 million years ago in the Late Permian (Figure 11.19). One of the reasons that the Carboniferous Period (as it was originally designated in Europe) is divided into an older Mississippian Period and younger Pennsylvanian Period in North America reflects a dominance of carbonate strata in rocks of Mississippian age recording warmer conditions, higher sea levels, and extensive warm shallow seas. As Earth cooled and the Pangean Orogeny proceeded (see Chapter 10), there was general uplift and an increase in sedimentation from rivers and deltas that covered over the old carbonate deposits. Thus, the global cooling and orogenic uplift are recorded by the general change from Mississippian age shallow-marine carbonate sediments to terrestrial clastic, coal-bearing rocks that characterized Pennsylvanian environments.

KEY POINT

Removal of atmospheric CO_2 in coal and via weathering triggered the late Paleozoic Ice Age, which lasted 70 million years. Because of a warmer Sun, Earth did not experience another snowball condition.

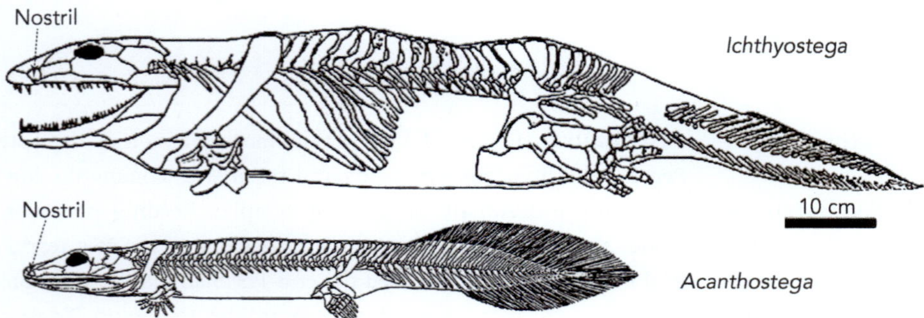

Figure 11.14 *Ichthyostega* and *Acanthostega*. *Ichthyostega* has a larger rib cage and more robust limbs and neck bones, enabling it to push up with its forelimbs to lift its head out of water. Small nostrils indicated that both could breathe air. However, the hind limbs of both animals face backward, showing an adaptation more suitable for swimming, indicating that neither could actually walk on land. Both are likely to have lived an essentially aquatic lifestyle. Source: Reproduced with permission by Springer Nature, from Ahlberg et al. (2005). Copyright, 2005.

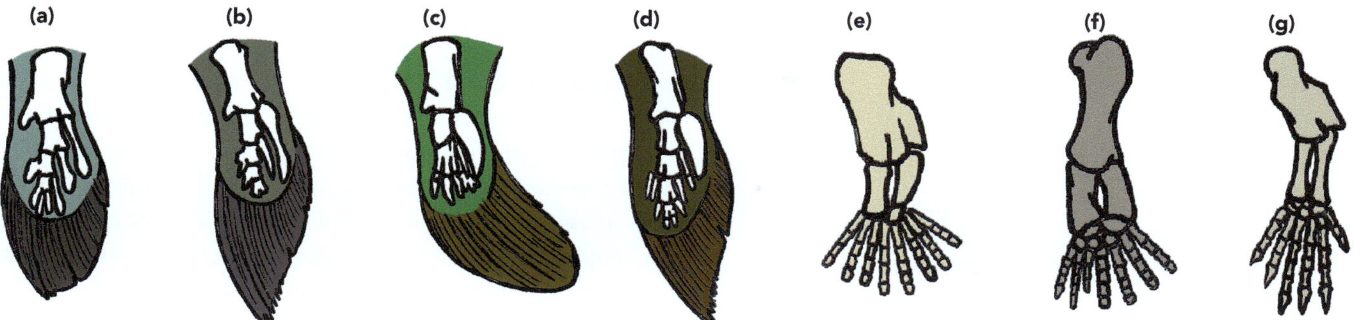

Figure 11.15 Evolution of limbs from fins. *Sarcopterygiin* fishes: (a) *Eusthanopteron*; (b) *Gogonasus*; (c) *Panderichthys*; and (d) *Tiktaalik*. Tetrapods: (e) *Acanthostega*; (f) *Ichthyostega*; and (g) *Tulerpeton*. Note the transition from fins with spines in (a)–(d), to limbs with digits in (e)–(g). Source: Conty, Public domain, via Wikimedia Commons.

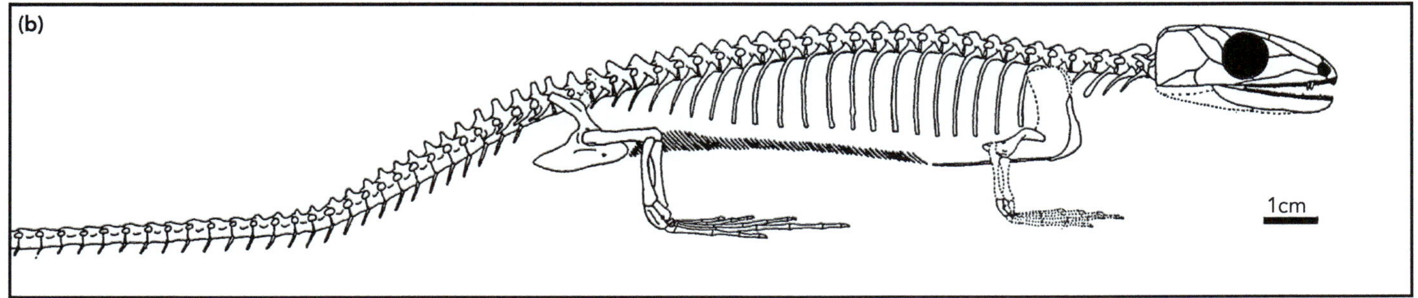

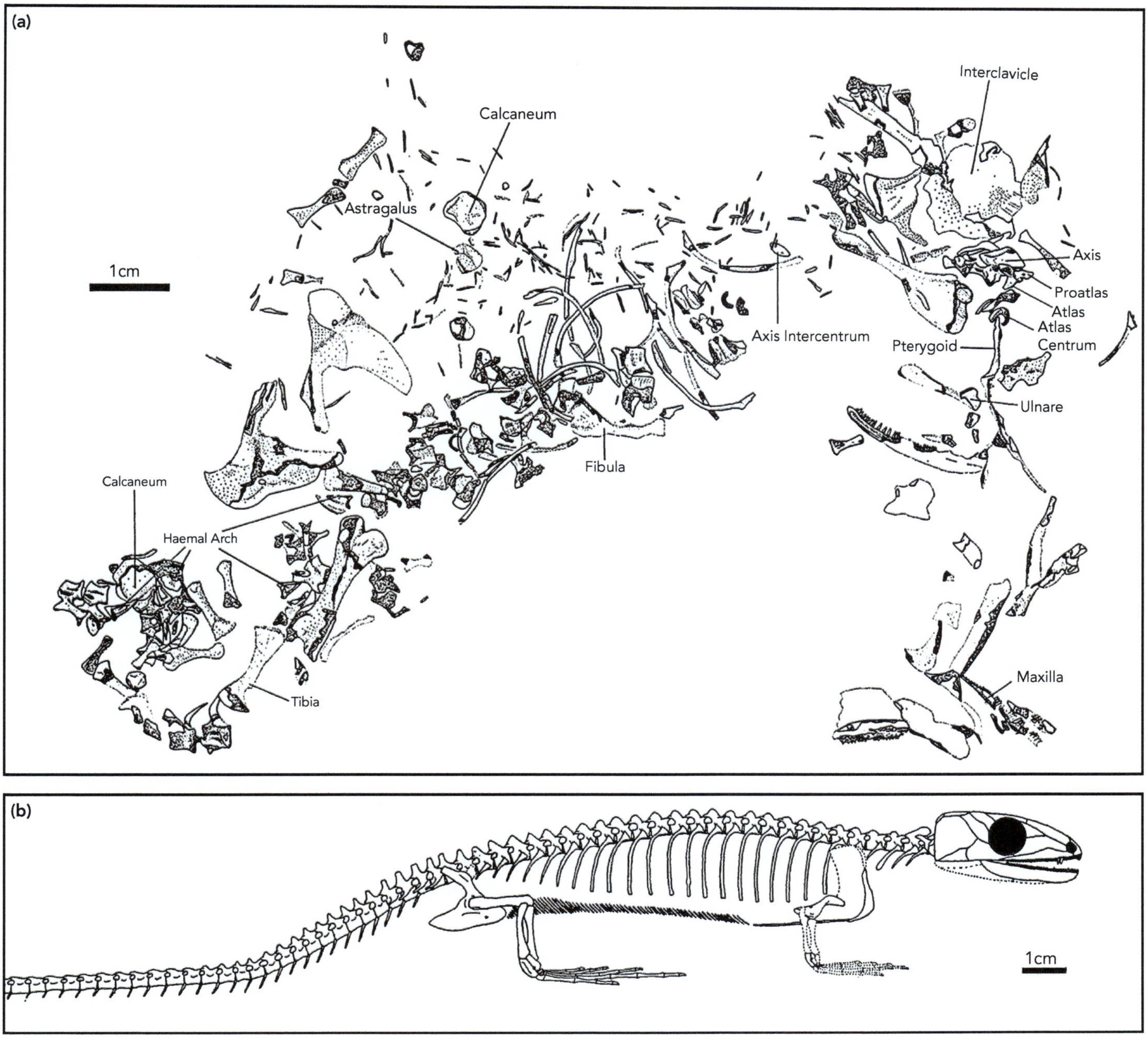

Figure 11.16 *Hylonomus lyelli*: (a) original fossil and (b) reconstructed skeleton, illustrating the complexity of fossil reconstruction. Source: Reproduced from Carroll (1964).

Figure 11.17 Permian synapsids: (a) *Dimetrodon*, a fin-back predator; (b) *Edaphosaurus*, a fin-back herbivore; (c) *Inostravencia*, a saber-toothed gorgonopsian; and (d) *Lystrosaurus*. All of the synapsids are characterized by temporal fenestrae, holes in the skull behind the eye socket. Source: (a) and (c) Photos by JPB. With permission from Houston Museum of Natural Science (HMNS); (b) Photo by Daderot,

As we saw in Chapter 5, Paleozoic glacial deposits have been found on all of the continents that made up Gondwanaland (Antarctica, Australia, New Zealand, India, South Africa, South America, and the Arabian Peninsula) and were important in deciphering the riddle of continental drift. The Late Paleozoic ice age experienced pulses of glaciations, much as have occurred in the more recent Pleistocene and younger ice ages that we are experiencing now (see Chapter 18). Once Earth was pushed into an ice age, high-frequency changes in climate, driven by regular perturbations of Earth's orbit, caused the surface of Earth to warm and cool, causing periodic fluctuations in the growth and decay of continental glaciers across Gondwanaland. These changes in climate are termed **Milankovitch cycles** and occur at frequencies of 20,000, 40,000, 100,000, and 400,000 years (see Chapter 18 for a fuller discussion). These rapid alternations produce alternating glacial and interglacial conditions that also cause rapid changes in sea level (Figure 11.19), with magnitudes up to 100 m. In equatorial areas, such as seen in the continental interior of the USA, the effects of these rapid sea-level changes can be clearly seen, based on observations of stratigraphic layering that shows rapid changes in depositional environments. These stratigraphic successions commonly show alternations of layers deposited in marine and non-marine environments and are termed **cyclothems** (Figure 11.20) due to the regularity of the layering. These can include alternations of non-marine coal-bearing layers with marine limestones and shales. Cyclothems were first identified by University of Illinois professor Harold Wanless in 1932 and were used as a basis for correlating and predicting the frequency and distribution of coal layers in the Illinois coal measures where Wanless worked. But in addition to the implications for managing coal resources, they also provided insights as to the rapid nature of global climate change and sea level that dominated the Late Paleozoic world and are similar to the glacial periods that characterize our more recent Pleistocene Epoch, which we will return to in Chapter 18.

As the Late Paleozoic climate became more erratic, the globe also became drier, and at around 305 Ma the extensive tropical *Lycopsid* rainforests that had dominated the Carboniferous Period collapsed. The forests were replaced with a greater proportion of tree ferns, and especially *Glossopteris* ferns, which prefer a cooler climate (see Chapter 5). The collapse of rainforests also led to a turnaround in CO_2 sequestration, resulting in an increase in atmospheric CO_2, which began a phase of global warming that eventually drove Earth into a hothouse. Although there was an overall warming from the Carboniferous into the Permian, cyclic changes in climate and sea level continued throughout the Late Paleozoic (Figure 11.19).

Figure 11.17 (*cont.*) CC0, via Wikimedia Commons; (d) Photo by Ghedoghedo, CC BY-SA 3.0 <https://creativecommons.org/licenses/by-sa/3.0>, via Wikimedia Commons.

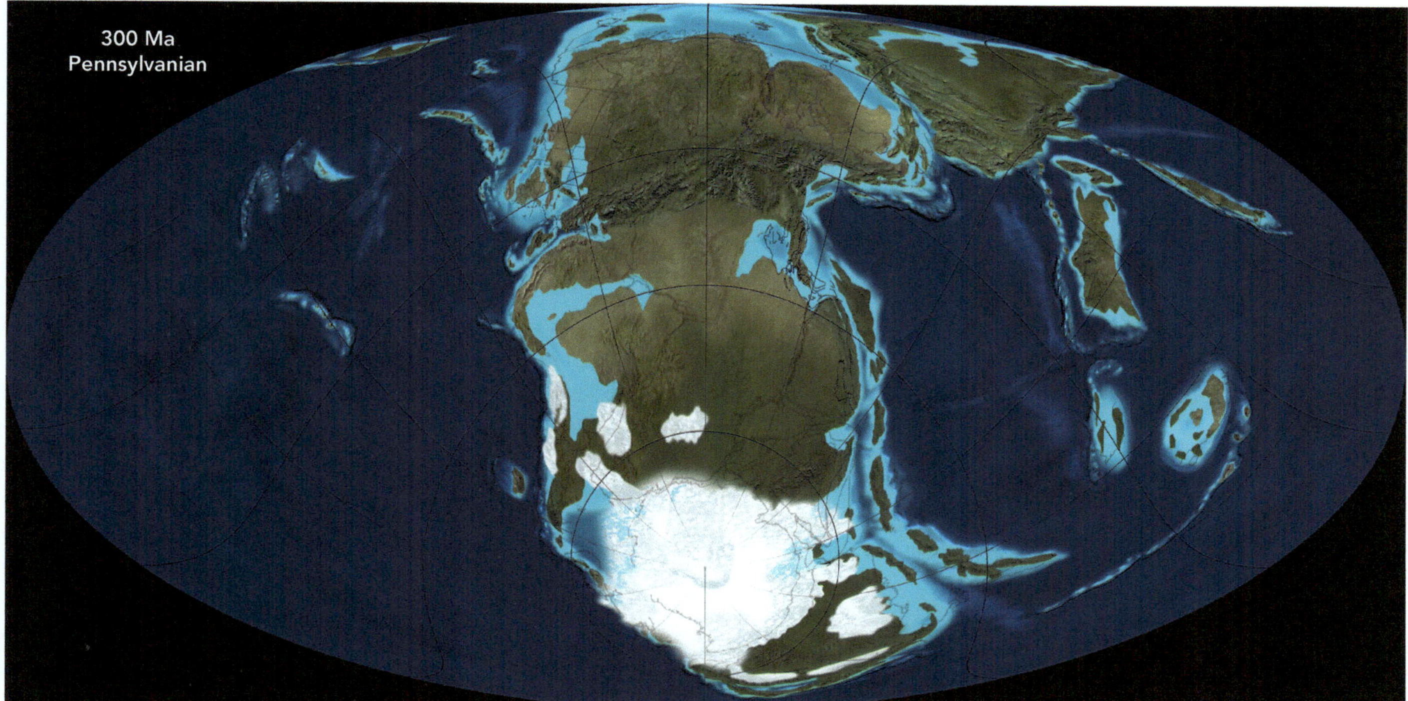

Figure 11.18 View of Earth from a south polar perspective in the Late Pennsylvanian (300 Ma). A major ice sheet covered Antarctica with smaller ice sheets over South America, Africa, India, and Australia. The greatest extent of these ice sheets occurred in the Latest Carboniferous to early Permian. Source: © DeepTimeMaps.

The Late Permian thus experienced a gradual increase in CO_2, an increase in aridity, and an increase in areas covered by deserts. The geological evidence for this includes globally widespread *eolian* (wind-blown) sandstones characterized by distinctive large-scale cross-beds diagnostic of formation in sandy deserts. Examples include the Coconino Sandstone near the top of the Grand Canyon (see Chapter 1, Figure 1.30(a)) and the Rotliegend Sandstone in Europe, attesting to the widespread nature of desert-ification. Part of this **aridification** may have been driven by the assembly of Pangea, which resulted in extremely extensive continuous and elevated land masses and the development of high mountain chains that increased the overall elevation of the continents. The Pangean mountains would have also created barriers for transfer of moisture on their landward sides (Figure 11.18), much like the Atacama Desert, which lies on the landward side of the Andes mountains and is one of the driest deserts in the world.

Despite the aridification of the Earth, the world's oceans nevertheless hosted diverse communities, such as seen in the Permian Reef complex in West Texas (Figure 11.21). This reef complex is well-exposed in the Guadalupe Mountains of West Texas and was built mostly of calcareous sponges (Figure 11.21 (b)), calcareous "phylloid" algae, bryozoan, and other encrusting organisms, as well as clusters of filter-feeding brachiopods, all typical of Sepkoski's Paleozoic fauna that we introduced in Chapter 9 (Figures 9.24 and 9.26).

The reef complex is overlain by extensive **evaporite** deposits, including gypsum ($CaSO_4$) and halite ($NaCl$) of the Castile and Salado formations (Figure 11.21(c)). These formed by the evaporation of seawater (see Chapter 16) and provide further evidence of the extreme aridity of that time. In Europe, eolian sandstones of the Rotliegend are overlain by the Zechstein Salt. The observation of eolianites and evaporites across the globe indicates that arid conditions extended over a very large area of Pangea in Late Permian time.

KEY POINT

Geological evidence such as eolian desert sandstones and evaporites, as well as the demise of rainforests and expansion of the cooler-climate *Glossopteris* ferns, indicates that Earth's climate changed to much drier conditions in the Permian as Pangea was fully assembled.

11.3 The Great Dying

The end of the Permian was marked by the greatest extinction event in Earth history (Figures 11.22 and 11.23). The causes of mass extinctions, in general, are obviously of great interest to us and a number of hypotheses have been proposed for the end-Permian extinction event, including:

1. massive volcanism, which released huge volumes of CO_2, causing a *hyperthermal event* (rapid global warming);
2. a runaway **greenhouse**, caused by methane release trapped in oceanic sediments, as well as climate changes that caused desertification of land areas;

3. oceanic **hypoxia** (loss of oxygen) and poisoning caused by upwelling of H_2S; and

4. a possible meteorite impact.

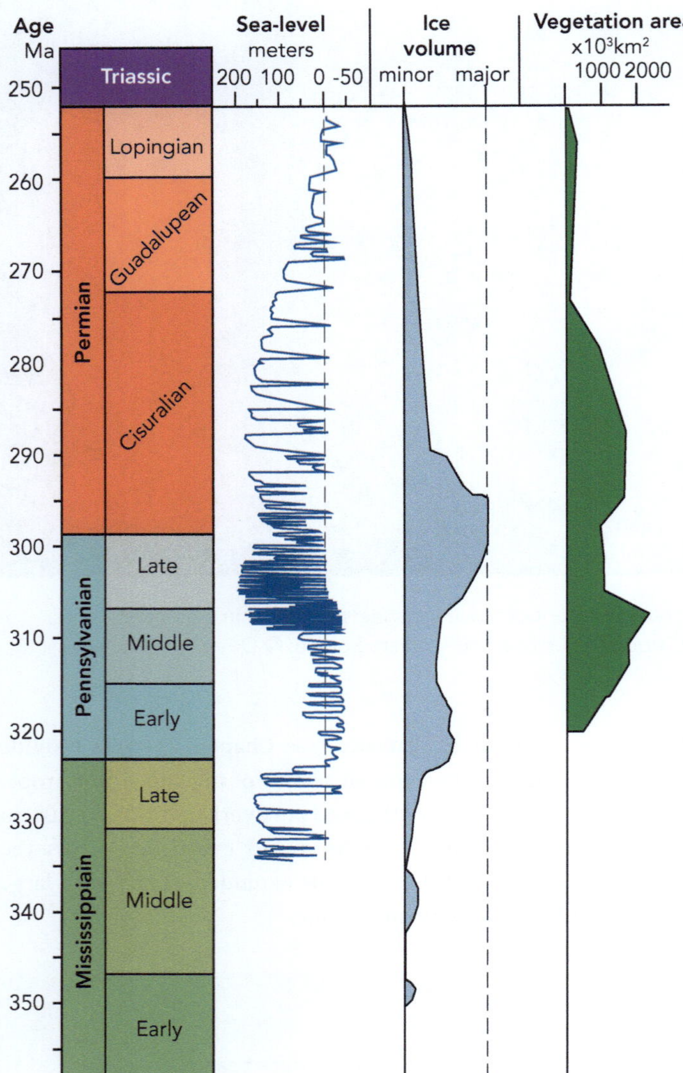

Figure 11.19 Major changes in climate, sea level, and vegetation extent over the Late Paleozoic. Source: Adapted from figure 2 in Montanez and Poulsen (2013).

11.3.1 The Siberian Traps

Volcanism is widely accepted as a major part of the story. The **Siberian Traps** is a massive volcanic deposit, formed of basaltic and rhyolitic lava (Figure 11.24). Such areas are referred to as **large igneous provinces (LIPs)** and are thought to form by upwelling from the mantle (see Chapter 5). The traps occur in Siberia, hence the name, and cover an area of 2 million km^2 (Figure 11.24(a)), but they are thought to have originally covered a significantly larger area of about 5 million km^2. They range up to about 3,000 m thick and have a volume of about 1 million cubic kilometers. This is the largest LIP that has been documented during the past 500 million years. Isotopic

dating shows that the eruptions initiated at 252.3 Ma, 300,000 years before the main extinction event, but it is estimated that two-thirds of the volume of these basalts were erupted over a 30,000-year interval, which coincides with the mass extinction event. Volcanism continued into the Triassic and the eruptions ceased at 251.4 Ma (Figure 11.25).

The magmas that fed the Siberian Traps were extruded through thick successions of carbonate and coal-bearing sedimentary strata of the Tunguska basin (recall all those Late Devonian and Carboniferous rainforests that we described above). It has been suggested that this magma would have vaporized much of the organic matter that was buried with these sediments, releasing a vast volume of CO_2 and other volatiles; up to 8.5×10^{19} g of CO_2 may have been released during the 30,000 years over which the eruptions occurred, at an average of 2.8×10^{15} g/year. Volcanoes today release about 2.7×10^{14} g (about 250 million tonnes) of CO_2 per year, or about 1/10th of the amount associated with the Siberian trap eruptions, but the CO_2 associated with these Siberian emissions may have been introduced into the atmosphere at rates similar to modern anthropogenic emissions of CO_2, which we will discuss further in Chapter 19.

In addition to the CO_2, the abundance of volcanic ash and presence of rhyolitic lavas would have resulted in very energetic volcanic explosions, releasing dust, ash, and other volatile gases, such as deadly hydrogen sulfide (H_2S,) and carbon monoxide (CO), as well as sulfur dioxide (SO_2). In addition, these eruptions resulted in large emissions of *halogens*, such as chlorine, bromine, and iodine, and these are known to act as a catalyst for atmospheric reactions that destroy the ozone layer, leading to increased ultraviolet radiation that would have been extremely stressful to terrestrial ecosystems. These releases of CO_2 and sulfur are also marked by major geochemical changes (Figure 11.25).

KEY POINT

Volcanic activity in Siberia formed the largest igneous province that has been discovered on Earth. This igneous activity released massive amounts of CO_2 stored in coal deposits, as well as SO_2 and halogens that triggered the end-Permian extinction.

11.3.2 A Selective Extinction

One of the peculiarities of the Permian extinction was that it was particularly selective against organisms that produced carbonate skeletons. The reef-forming tabulate and rugose corals (Figure 11.23(a) and (b)) went extinct, but soft-bodied sea anemones survived, and these anemones later evolved into the scleractinian corals that are the dominant reef builders today. Many more brachiopod species went extinct than bivalves. Giant calcareous *foraminifera* (single-celled organisms), including the **fusulinids**, were so abundant that they formed thick beds that look like layers of cooked rice (Figure 11.23(d)), but the

Figure 11.20 Outcrops of Pennsylvanian strata, near Taos, New Mexico, alternate between distal marine shales, shallow marine limestones, and non-marine incised rivers containing plant fossils. This records a cycle of regression and transgression of sea level, likely driven by alternating glacial and interglacial climate periods, and represents a classic Pennsylvanian cyclothem.

Figure 11.21 (a) Dipping beds of limestones of the Permian Reef complex in West Texas overlie flat-lying deep water sandstones and mudstones. (b) Massive fossil sponges form the core of the reef complex. (c) Finely laminated gypsum deposits of the Castile Evaporite overlie the reef complex.

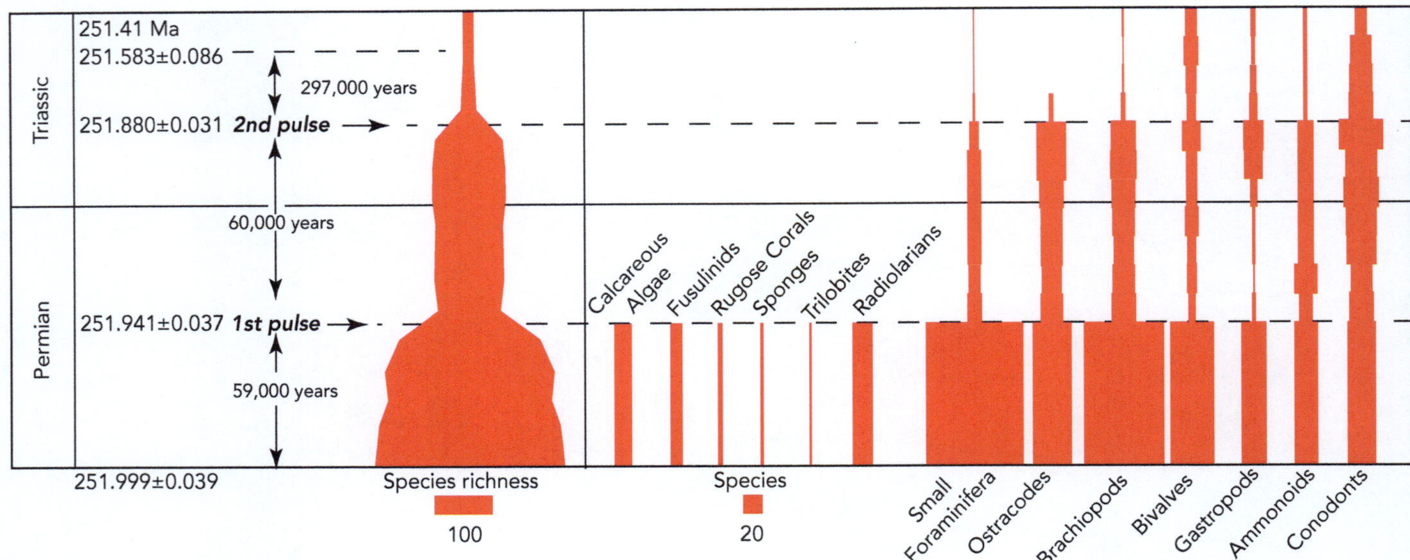

Figure 11.22 Time chart showing timing of pulses of extinction across the Permian–Triassic boundary. Widths of species columns show diversity. The first pulse of extinction occurs about 30,000 years before the end of the Permian, and a second pulse occurs about 60,000 years later, and 30,000 years into the Early Triassic. Source: Song et al. (2013). Modified by permission of Springer Nature (copyright 2013).

fusulinids experienced total annihilation. The fusulinids secreted a calcareous shell, but in contrast, the smaller **agglutinated foraminifera**, who literally constructed their shells by gluing together microscopic silt and clay-sized sediment particles, survived. This invites the question, why was the extinction event so selective? What was it about brachiopods that made it more difficult for them to survive the crisis versus the bivalves? Why were calcareous skeletons a liability?

11.3.3 Geochemical Changes and Duration of the Extinction Event

An enormous amount of geochemical evidence has also been collected across the boundary that shows several key changes (Figure 11.25). There is evidence for a sharp decrease in the ratio of $^{13}C/^{12}C$ isotopes extracted from sedimentary rocks, with an initial pulse at around 260 Ma, and several pulses after that. The decrease in $^{13}C/^{12}C$ likely indicates a large influx of organic carbon from the burning of Siberian coal deposits, which has a greater proportion of ^{12}C, compared to inorganic carbon, which is higher in ^{13}C. Analyses of limestone and evaporites that were precipitated from the oceans during the extinction event show sharp changes in the $^{87}Sr/^{86}Sr$ isotope ratios in seawater in the Late Permian (Figure 11.25). This shows a prolonged drop followed by a rapid rise after the extinction event. High ratios are thought to reflect times of enhanced weathering of continental land masses. The slow fall may reflect a slowing of weathering rates as the ice sheets receded and the world became hotter and drier, whereas the sudden increase may be the result of an increase in weathering and erosion due to the worldwide collapse of plants that stabilize land surfaces and decrease weathering rates.

Late Permian rocks also include abundant black shales, apparently deposited in relatively shallow waters. These black shales are rich in organic carbon and the mineral pyrite (FeS_2), both of which indicate reducing conditions, and suggests that shallow areas experienced anoxia. The evaporitic Castile Formation and Zechstein Salt also include gypsum ($CaSO_4$), a common sulfate-rich evaporite mineral. An increase in the ratio of sulfur isotopes, and specifically $^{34}S/^{36}S$, may be related to increased reducing conditions that extracted sulfur from seawater and fixed it in evaporites and pyrite. There was also a corresponding decrease in oxygen (Figure 11.5).

Dating of volcanic beds interleaved with sediments over the extinction event shows that the extinction occurred over a period of about 60,000 years, between 251.941 ± 0.037 and 251.880 ± 0.031 Ma (Figure 11.22). There is some indication that the extinction occurred in two pulses with the initial stage occurring during the 30,000-year period at the end of the **Changsingian Stage** of the Late Permian Period, followed by a second pulse about 30,000 years after the initial event in the earliest Triassic. Although these dates have a very high precision, some researchers have suggested that the first extinction pulse may have occurred during an even shorter period, but more precise dating is beyond the limits of current analytical techniques (see Chapter 4).

11.3.4 Effects of the Volcanism

The Siberian Traps caused havoc in the atmosphere and oceans, and this in turn had a profound effect on the biosphere, as we will discuss in the next three sections.

11.3.4.1 CO₂ and Oceanic Acidification

The enormous infusion of CO_2 led to an increase in atmospheric concentration that may have reached up to 2,000 ppm, which is seven times the concentration that characterized our atmosphere

Figure 11.23 Examples of groups of marine invertebrates that went extinct at the end of the Permian: (a) rugose corals; (b) tabulate corals; (c) trilobites and brachiopods; (d) fusulinids (giant single-celled foraminifera); and (e) blastoids, this is the top of the animal that would have been perched on a stalk. Source: (a), (b), and (e), Photos by JPB – McMaster University fossil collection; (c) Photo by JPB. With permission from Houston Museum of Natural Science (HMNS); (d) Guadalupe Mountains, Texas. Photo by JPB.

at the beginning of the industrial revolution (in Chapter 19 we note that CO_2 concentration was 280 ppm at the beginning of the industrial revolution and is currently over 400 ppm). Because of its effect as a greenhouse gas (see Chapters 8 and 19), this massive injection of CO_2 would have caused a hyperthermal event, and there are indications that global temperatures may have increased by as much as 10 °C. These temperature increases would cause significant warming of the oceans and a reduction in dissolved oxygen. These magnitudes of temperature change would have been extreme and likely occurred too fast to allow organisms to easily adapt to, which helps explain why so many species went extinct.

At the same time, the increase in atmospheric CO_2 would result in an increase in dissolved CO_2 in seawater. Recall that ocean water absorbs much of the global CO_2. The consequence of this additional dissolved CO_2 is an increase in **acidification** of seawater. Increasing the acidity of seawater would have made life very difficult for calcifying organisms, as increasing acidity increases the solubility of calcium carbonate, making it much more difficult to maintain a calcareous shell. This helps explain why calcifying organisms such as the tabulate and rugose corals went extinct, whereas the soft-bodied sea anemones did not. In addition, during colder icehouse periods, such as we are experiencing today, calcifying organisms, such as the scleractinian corals, mollusks, foraminifera, and algae, tend to construct

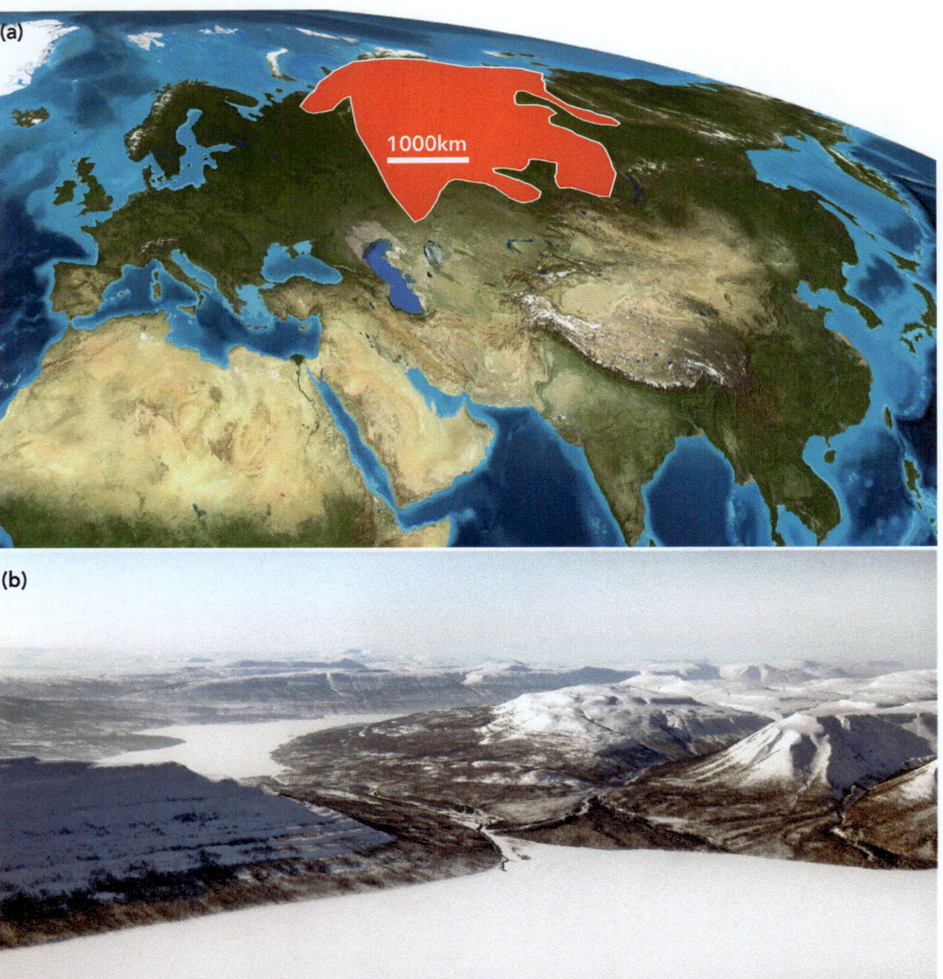

Figure 11.24 The Siberian Traps, a large igneous province that erupted at the end of the Permian and likely triggered a mass extinction. (a) The extent of the Siberian Traps and (b) image showing the staircase topography. Source: (a) © DeepTimeMaps; (b) OlgaChuma Ольга Чумаченко, CC BY-SA 3.0 <https://creativecommons.org/licenses/by-sa/3.0>, via Wikimedia Commons.

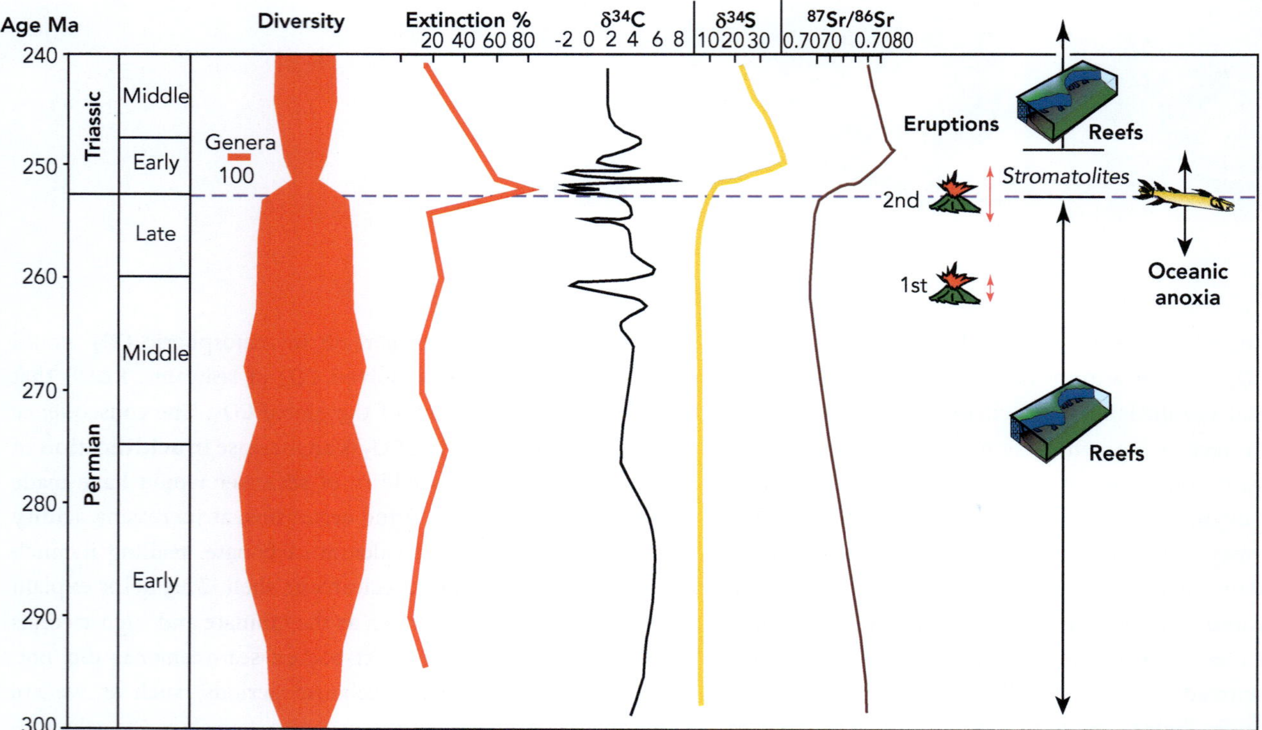

Figure 11.25 Geochemical trends versus biodiversity and major events.

their skeletons from the mineral aragonite, which has the same chemical composition as calcite ($CaCO_3$), but a different crystal structure and different solubility. Aragonite takes less energy to make than calcite but is more soluble. Greenhouse times, which also result in more acidic oceans, tend to be dominated by organisms that build calcite shells because calcite is more resistant to dissolution than aragonite. In the transition from the Late Paleozoic ice age to the end-Permian hothouse, the aragonitic organisms that dominated the Late Paleozoic seas were less able to cope with oceanic acidification compared to their calcitic compatriots and tended to extinction. Being well adapted to the colder Paleozoic ice age, they were unable to adapt to the warmer and more acidic oceanic conditions, allowing less cold-adapted organisms to be the "lucky survivors," a concept we introduced in Chapter 6.

The warming of oceans also tends to reduce the ability for ocean water to circulate, resulting in stratification (see Figures 14.9 and 18.12). Stratification disallows oxygenated water at the surface to reach the seafloor, and this can drive an ocean into an anoxic state. Anoxic conditions in deep water may result in the generation of methane (CH_4) and deadly hydrogen sulfide (H_2S) by methanogenic bacteria. This would also have created very inhospitable conditions for deep water benthic ecosystems, such as may have been exploited by the calcifying stalked echinoderms (**crinoids**, **cystoids**, and **blastoids**) that were also decimated.

11.3.4.2 Hypercapnia: Extinction Based on Metabolism (Physiology)

Although many calcifying organisms went extinct, some, like many mollusks, did not. It has been hypothesized that physiology and metabolism also played a role in the selectivity of the Permian extinction. Examination of metabolic rates show that organisms with low respiratory capacity and low metabolic rates, such as the filter-feeding brachiopods, were more susceptible to extinction than most mollusks, with much higher metabolisms. Swimming scallops or giant clams (i.e., bivalves) can close their shells quite quickly. Squids and octopuses (i.e., **cephalopods**) and their Permian counterparts the ammonoids, and even gastropods, all move much faster than slow-moving starfish and sea urchins (both echinoderms). The sessile, stalked echinoderms, such as blastoids, cystoids, and crinoids, also have low metabolic rates.

It has been hypothesized that many organisms experienced **hypercapnia**, or elevated CO_2 levels in their bloodstreams. Although not immediately toxic, and unlike carbon monoxide (CO) or hydrogen sulfide (H_2S), which are deadly to oxygen-loving aerobic organisms, hypercapnia slows metabolism, inhibiting the ability of organisms to reproduce and eventually causing death. Hypercapnia is especially toxic to organisms that already have low respiration rates. Also, it requires energy to build and maintain a calcareous shell. Calcifying organisms with slow metabolisms, such as brachiopods and stalked

echinoderms, had a harder time maintaining their skeletons as the oceans acidified and oxygen became less available.

It is also clear that benthic organisms (i.e., those that live on the seafloor) and especially sessile benthic organisms (i.e., those that are attached to the seafloor and lack motility) were more susceptible. Not being able to move was a clear liability. Organisms that were able to seek refuge in more hospitable environments included more motile animals, such as mollusks, fish, and other nektonic organisms, which survived the extinction event.

This example illustrates various features of evolution that we introduced in Chapter 6. There are advantages in having a low metabolism. Cold-blooded snakes can go for long periods without eating, and slow-moving or sessile echinoderms need far less food resources than faster moving mollusks. If food is scarce, organisms with high metabolic rates may find themselves starving to death. Conversely, high metabolic rates and higher motility may enhance the ability to escape fast-moving predators, but this also means more of the metabolism is used for escape versus reproduction. In the end, and especially during catastrophes, organisms suited for one set of conditions may find themselves ill-suited to the changed conditions. Such was the case for many brachiopods and echinoderms at the end of the Permian. Although they may have been superbly adapted to the pre-extinction conditions, they did not seem to have the metabolic capacity or ability to adapt to the catastrophic external environmental conditions that characterized the end of the Permian Period, and they were selected out of the biosphere. The mollusks and sea anemones ended up as the "lucky survivors."

Increased CO_2 led to oceanic acidification, which put stress on calcifiers and especially organisms that built skeletons of more soluble aragonite versus less soluble calcite. Slow-moving or sessile organisms, such as echinoderms and brachiopods with slower metabolisms, experienced hypercapnia, characterized by higher CO_2 levels in their blood and prolonged low oxygen stress that furthered their demise.

11.3.4.3 Land Extinction

The release of halogens disrupted the ozone layer, resulting in an increase in ultraviolet (UV) radiation, increase in temperature, and desertification, all of which combined to further the stress on plants that had already occurred with the earlier collapse of rainforests. Examination of spores and pollen at the extinction event shows high levels of mutation, consistent with higher UV levels. The Late Permian *Glossopteris* fern communities that lived in temperate latitudes in a contiguous belt around Gondwanaland (see Chapter 5, Figures 5.2 and 5.3(b)) disappeared very soon after the extinction event in the Early Triassic. The loss of terrestrial habitats further stressed terrestrial animals, and the lower oxygen levels caused the extinction of the giant insects and myriapods that had adapted to the much higher oxygen levels in the Carboniferous and early Permian (Figures 11.4(d) and 11.8). There were also major extinctions of vertebrates that paved the way for the rise of dinosaurs in the ensuing Mesozoic Era (see Chapter 12)

Loss of plant life on land may also have triggered an increase in weathering, allowing a greater proportion of phosphates to be delivered to the oceans. Algae and bacteria thrive on phosphates and create their own biogenic organic debris and waste products, which would sink to the seafloor and further rot by anerobic bacteria, producing toxic hydrogen sulfide (H_2S) and methane (CH_4), enhancing the greenhouse gas effect. Thus, there is some evidence that terrestrial and oceanic systems experienced a positive feedback loop that exacerbated the disaster.

This increase in weathering is also indicated in stratigraphic changes in paleoenvironments across the Permian–Triassic boundary worldwide. This includes a prominent coal-gap, indicating loss of forests, as well as sudden preponderance of coarse-grained, gravelly river deposits, reflecting the low-vegetation world, that overlie muddier pre-extinction marine deposits.

11.4 Other Hypotheses for Causes of the Mass Extinction?

A relatively small number of studies have presented evidence that an extraterrestrial impact (i.e., a large meteorite, see Chapter 13) triggered the extinction and several impact structures have been proposed as candidates. The meteorite hypothesis was largely driven by the overwhelming evidence of an impact that triggered the mass extinction that closed the Mesozoic Era and is the focus of Chapter 13. Evidence for the end-Cretaceous impact includes an increase in the element **iridium**, which is cosmogenic in origin and found globally in high concentrations at the boundary, abundant shocked quartz, and the deposits of **tsunamis** (massive tidal waves caused by the impact). We will review all of this and other evidence in Chapter 13, but tellingly, most of these observations, thought to be diagnostic of the large meteorite impact at the end of the Mesozoic, are notably lacking at the Permian–Triassic boundary.

Regarding the other hypotheses, oceanic anoxia certainly developed and may have led to the formation and release of hydrogen sulfide (H_2S). This H_2S would also have affected the ozone layer and is extremely toxic, and it is possible that it enhanced the effects of hypercapnia, accelerating the extinction. The idea of a gigantic methane release is also suggested by the decrease in ^{13}C. Methane, produced by bacterial breakdown of organic matter, is very common in subsurface sediments, and in some places, such as oceans and areas with permafrost, this methane can be trapped in a crystalline form (technically called a **gas hydrate** or *clathrate*). Changes in temperature, which might cause permafrost to melt or oceans to warm, can cause these clathrates to melt, releasing the gas. Recent studies, however, suggest that the amount of methane release needed is unrealistic.

11.5 A Slow Recovery

Recovery from the extinction was slow, and a complex robust ecological system may not have been in place until many millions and up to possibly 30 million years after the extinction event. The Early Triassic is marked by a lack of coal-bearing strata (called the coal-gap), indicating that it took almost 10 million years for coal-forming forests to return. This indicates that stressful conditions continued well into the Early Triassic.

Early Triassic marine environments saw an increase in algal stromatolites, showing an ecology more similar to the Proterozoic, which reflected the extinction of grazing animals, and a loss of reef-building organisms such as the **stromatoporoids** – calcareous sponges that had diversified in the Early Paleozoic. As we introduced in Chapter 9, the Sepkoski Chart (Figure 9.24) emphasizes that extinction of the Paleozoic faunal group made way for the eventual rise of the *Modern fauna*. The great dying preferentially selected calcifiers with high metabolic rates, such as mollusks, as well as mobile animals, such as fish, crustaceans, and echinoids that subsequently proliferated to form the Modern fauna. On land, one of the most important groups to appear were the dinosaurs, and their world will be the topic of our next major story in Chapters 12 and 13.

11.6 Summary

- Land plants (embryophytes) evolved from a single freshwater algal ancestor in the late Ordovician around 470 Ma.
- Fully terrestrial plants required resistance to desiccation and the ability to regulate water and atmospheric gases, which led to the development of seeds, vascular systems, and roots.
- A mass extinction at the end of the Ordovician may have been caused by expansion of oxygen photosynthesis that

used CO_2, initiating an ice age that caused retraction of the high sea levels of the Sauk transgression due to reduction of shallow marine habitats.
- The Devonian and Carboniferous Periods saw expansion and diversification of plants that were stored as coal. This reduced atmospheric CO_2 and increased oxygen levels, causing the Late Paleozoic Ice Age. The excess oxygen allowed for gigantism in terrestrial arthropods.

- Fishes were well developed by the Devonian Period and the lobe-finned fishes evolved into the first amphibious land vertebrates. Features such as swim bladders and lobe fins were co-opted to lungs and feet in the earliest tetrapods and the abundance of intermediary fossils illustrate the many steps in the evolution from fish to land-dwelling tetrapods. By the end of the Paleozoic, fully terrestrial vertebrates were common.
- Major glaciations occurred on Gondwanaland, and orbitally driven climate cycles caused rapid high-amplitude sea-level changes.
- As Pangea was assembled, Earth began to experience aridification and desertification as equatorial climates became dry and hot, marked by desert dune fields and evaporites.

- Massive volcanism in Siberia released enormous amounts of CO_2 stored in coal deposits during the earlier Devonian and Carboniferous periods. This caused a catastrophic hyperthermal event characterized by global warming and oceanic acidification that triggered the greatest of all mass extinctions at the end of the Permian Period.
- Oceanic acidification, stratification, and anoxia may have caused the release of H_2S and methane that exacerbated stress on Earth's biota. Release of halogens may have disrupted the ozone layer, causing stress on terrestrial biota.
- The excess CO_2 in the oceans selected against calcifying organisms with slower metabolic rates, such as brachiopods, stalked echinoderms, and corals.
- Recovery from the extinction was slow and hampered by renewed Siberian volcanism in the Triassic.

Key Words

- embryophytes
- bryophytes
- non-vascular plants
- cryptospores
- vascular systems
- tracheophytes
- *Cooksonia*
- lignin
- gymnosperms
- conifers
- cycads
- ginkos
- gnetophytes
- angiosperms
- eurypterids
- chelicerate
- Hexapoda
- Myriapoda
- *Rhyniognatha hirsti*
- spiracles
- *Meganeura*
- *Diplichnites*
- *Arthropleura*
- the Age of Fishes
- *Dunkleosteus*

- *Holoptychius*
- lobe-finned fish
- *Coelacanth*
- *Eusthenopteron*
- Romer's Gap
- stegocephalians
- stem-tetrapods
- crown tetrapods
- fishapods
- *Tiktaalik*
- *Acanthostega*
- *Ichthyostega*
- quadrupedal gait
- polydactyl
- *Pederpes*
- amniotic
- *Hylonomus lyelli*
- *Lycopsid*
- anapsid
- synapsids
- diapsids
- *Edaphosaurus*
- *Dimetrodon*
- vascularization
- *Lystrosaurus*

- *Inostrancevia*
- gorgonopsians
- Milankovitch cycles
- cyclothems
- *Glossopteris*
- aridification
- evaporite
- greenhouse
- hypoxia
- Siberian Traps
- large igneous provinces (LIPs)
- fusulinids
- agglutinated foraminifera
- Changsingian Stage
- acidification
- crinoids
- cystoids
- blastoids
- cephalopods
- hypercapnia
- iridium
- tsunamis
- gas hydrate
- stromatoporoid

Further Reading and References

Ahlberg, P. A., Clack, J. A., and Blom, H., 2005, The axial skeleton of the Devonian tetrapod Ichthyostega, *Nature*, 437, 137–140, https://doi.org/10.1038/nature03893.

Benton, M. J., 2023, *Extinctions: How Life Survived, Adapted and Evolved*, Thames & Hudson.

Berner, R. A., 1999, Atmospheric oxygen over Phanerozoic time, *Proceedings of the National Academy of Sciences*, 96 (20), 10955–10957.

Berner, R. A., Vandenbrooks, J. M., and Ward, P. D., 2007, Oxygen and evolution, *Science*, 316(5824), 557–558, https:/doi.org/10.1126/science.1140273.

Carroll, R. L., 1964, The earliest reptiles, *Zoological Journal of the Linnean Society*, 45(304), 61–83, https://doi.org/10.1111/j.1096-3642.1964.tb00488.x.

Clack, J. A., 2012, *Gaining Ground: The Origin and Evolution of Tetrapods*, Indiana University Press.

Haug, C., and Haug, J. T., 2017, The presumed oldest flying insect: More likely a myriapod?, *PeerJ*, 5, e3402, https://doi.org/10.7717/peerj.3402.

Knoll, A. H., Bambach, R. K., Payne, J. L., Pruss, S., and Fischer, W. W., 2007, Paleophysiology and end-Permian mass extinction, *Earth and Planetary Science Letters*, 256(3–4), 295–313.

Montañez, I. P., and Poulsen, C. J., 2013, The Late Paleozoic ice age: an evolving paradigm, *Annual Review of Earth and Planetary Sciences*, 41, 629–656.

Payne, J. L., and Clapham, M. E., 2012, End-Permian mass extinction in the oceans: an ancient analog for the twenty-first century?, *Annual Review of Earth and Planetary Sciences*, 40, 89–111.

Song, H., Wignall, P. B., and Yin, H., 2013, Two pulses of extinction during the Permian-Triassic crisis, *Nature Geoscience*, 6(1), 52–56.

Steemand, P., Astini, R. A., de la Puente, G. S., et al, 2010, Early Middle Ordovician evidence for land plants in Argentina (eastern Gondwana), *New Phytologist*, 188.

Review Questions

1. When did the "greening of the Earth" (i.e., evolution of the earliest plants) commence?
2. Describe the main steps in the evolution of plants.
3. Explain the main evolutionary steps in the evolution from aquatic to land vertebrates.
4. How did atmospheric oxygen change over the Phanerozoic? What effects did this have on the biosphere and how did it allow gigantism in insects?
5. How did the Pangean assembly affect global climate?
6. What led to the expansion and later collapse of rainforests in the Late Paleozoic?
7. What is the evidence for the Late Paleozoic icehouse and what were the probable causes?
8. What caused the transition from an icehouse into a greenhouse in the late Paleozoic?
9. What was the main cause of the end-Permian mass extinction?
10. Why was the Permian extinction selective against calcifying organisms and organisms with low metabolic and respiration rates?
11. How did emission of CO_2 and volatiles change atmospheric and oceanic conditions during the Permian–Triassic extinction?
12. Why was biotic recovery slow following the Permian–Triassic extinction?
13. How do the rates of anthropogenic CO_2 emissions compare with the rates hypothesized to have occurred during the Permian–Triassic extinction event?

Sauropod standing in the calm sea – an *Archaeopteryx* flying above it. Source: artpartner-images / Getty Images.

The Age of Dinosaurs

The Triassic, Jurassic, and Cretaceous World

- Describe the general conditions on Earth during the Mesozoic Era and the transition from arid-hothouse to tropical greenhouse.
- Discuss the difficulty in reconstructing dinosaurs and their first appearance and subsequent evolutionary history.
- Explain the evidence for upright walking and endothermy in dinosaurs.
- Discuss the evolution of feathers and the idea that they were not initially evolved for flight.
- Discuss the relationship between birds and dinosaurs.
- Explain the gigantism and diversity of dinosaurs.
- Discuss the lifestyles of mammals during the so-called "Age of Dinosaurs."

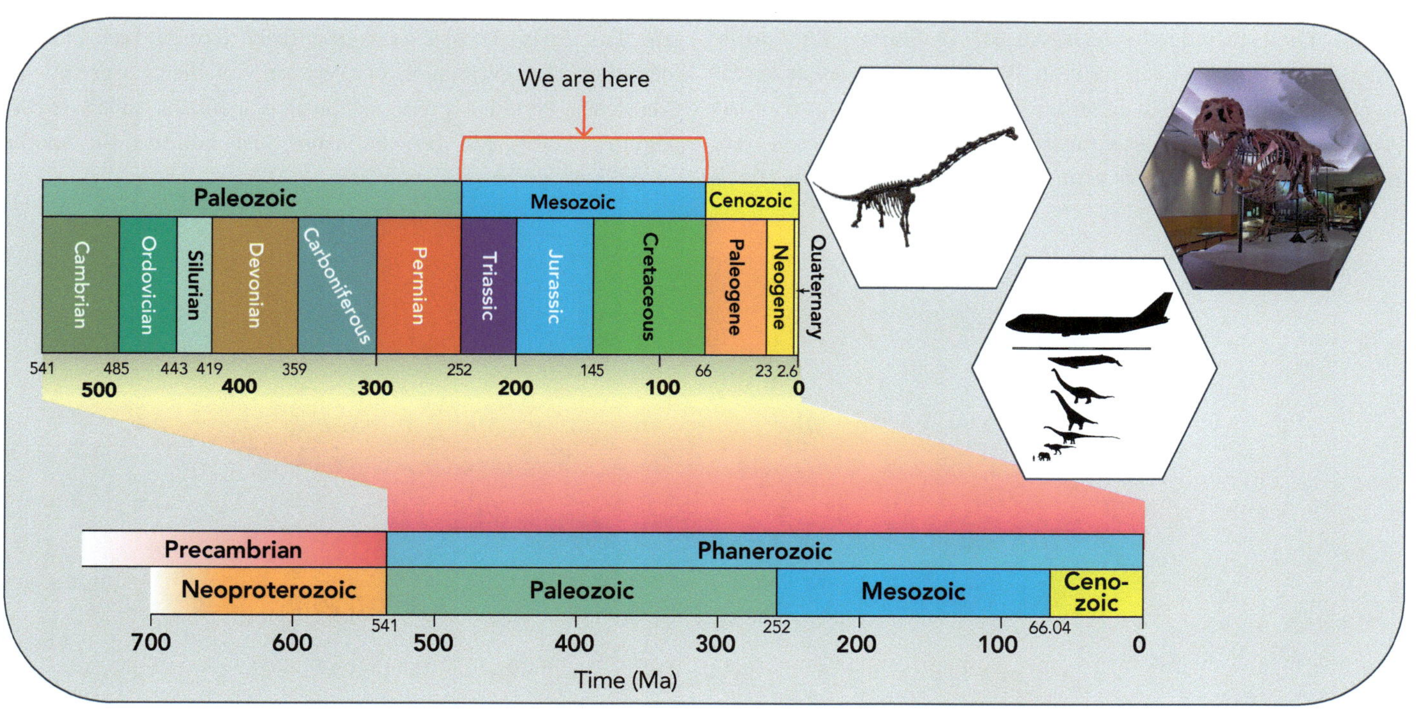

Introduction

It's hard these days to avoid dinosaurs. Kids' bedrooms are commonly full of toy dinosaurs, tourist spots like Niagara Falls feature dinosaur mini-golf (Figure 12.1), and the *Jurassic Park* movie franchise continues to delight many movie goers. Most readers of this book probably had both toys and books about dinosaurs growing up. These depictions, as well as more scientific documentaries such as the BBC series *Walking with Dinosaurs*, consistently show them as quick, smart, agile, and above all, dangerous, but this view of dinosaurs is actually quite recent. This chapter focuses on the evolution and lifestyles of dinosaurs that dominated the Mesozoic Era.

12.1 The Mesozoic

You will recall from Chapter 11 that the Late Permian was a hothouse period characterized by worldwide desertification, and these conditions persisted well into the Triassic and Early Jurassic periods. The deleterious effects of the Siberian supervolcanoes that triggered the end-Permian mass extinction continued into the earliest Triassic. The Mesozoic Era was marked by the breakup of Pangea with the initial opening of the Atlantic and Indian oceans starting in the Late Triassic to Early Jurassic (Figure 12.2). There is evidence that hot spots associated with these new oceans flared up at the end of the Triassic, triggering a fourth mass extinction.

These new oceans would have been very elongate (Figure 12.2), much like the Red Sea today, and at times they became cut off from the rest of the world's oceans. Because of the persistent dry conditions, many of the world's seas dried out leaving evaporite salt deposits behind (see Chapter 16 for more details about evaporites). These include the Zechstein Salt in Europe, the Castille Evaporites in the Permian basin in West Texas, as well as extensive salt formations that underlie the continental margins of the Atlantic Ocean. As the Atlantic opened, these salt deposits were overlain by a variety of younger sediments, including thick sections of muds and sands eroded from the adjacent mountains. Recall that Pangean assembly resulted in one of the highest mountain chains Earth has ever seen that would have rivalled today's Himalaya and Alps; once the Atlantic opened, erosion of these mountains would have delivered vast amounts of sediments to the continental margins of these new oceans. Salt has a lower density than other sedimentary rocks and eventually was able to flow upwards intruding the overlying rocks as *diapirs*, forming complex domes and ridges that formed traps for extensive oil and gas deposits, such as are found in the Gulf of Mexico, offshore eastern Canada, and many other places. The extensive wells, as well as seismic data collected for oil and gas exploration, have resulted in extensive knowledge about the geology of the world's continental margins and show how the modern Atlantic Ocean formed as a consequence of Pangean breakup.

The rate of seafloor spreading reached a high point in the Cretaceous Period and this caused an overall decrease in the volume of Earth's oceans that resulted in seawater being displaced into the continents. Although Earth remained in a greenhouse state, there was also a decrease in aridity in the Cretaceous Period, resulting in a much more humid Earth and an increase in vegetation. The combination of continued warming, smaller ocean volumes, and more humid conditions resulted in some of the highest sea levels since the Sauk transgression that characterized the Cambrian and Ordovician periods. The sedimentary sequences associated with these high sea levels were termed the **Zuni sequence** (see Figure 1.21) by Lawrence Sloss, a geology professor at Northwestern University, in 1963. The presence of vast shallow seas and global warming also coincided with an explosion in microscopic calcareous phytoplankton, which form extensive chalk deposits, after which the Cretaceous is named (Figure 12.3 (a)). The Early Triassic was essentially free of reef builders, following the extinction of Permian calcifiers, but by the Cretaceous Period a group of bivalve mollusks called **rudists** (Figure 12.3(b)) had become major reef builders, in another example of convergent evolution.

Figure 12.1 Dinosaur mini-golf, Niagara Falls, Canada. A *Tyrannosaurus rex* stands over its eggs flanked by a wide variety of dinosaurs.

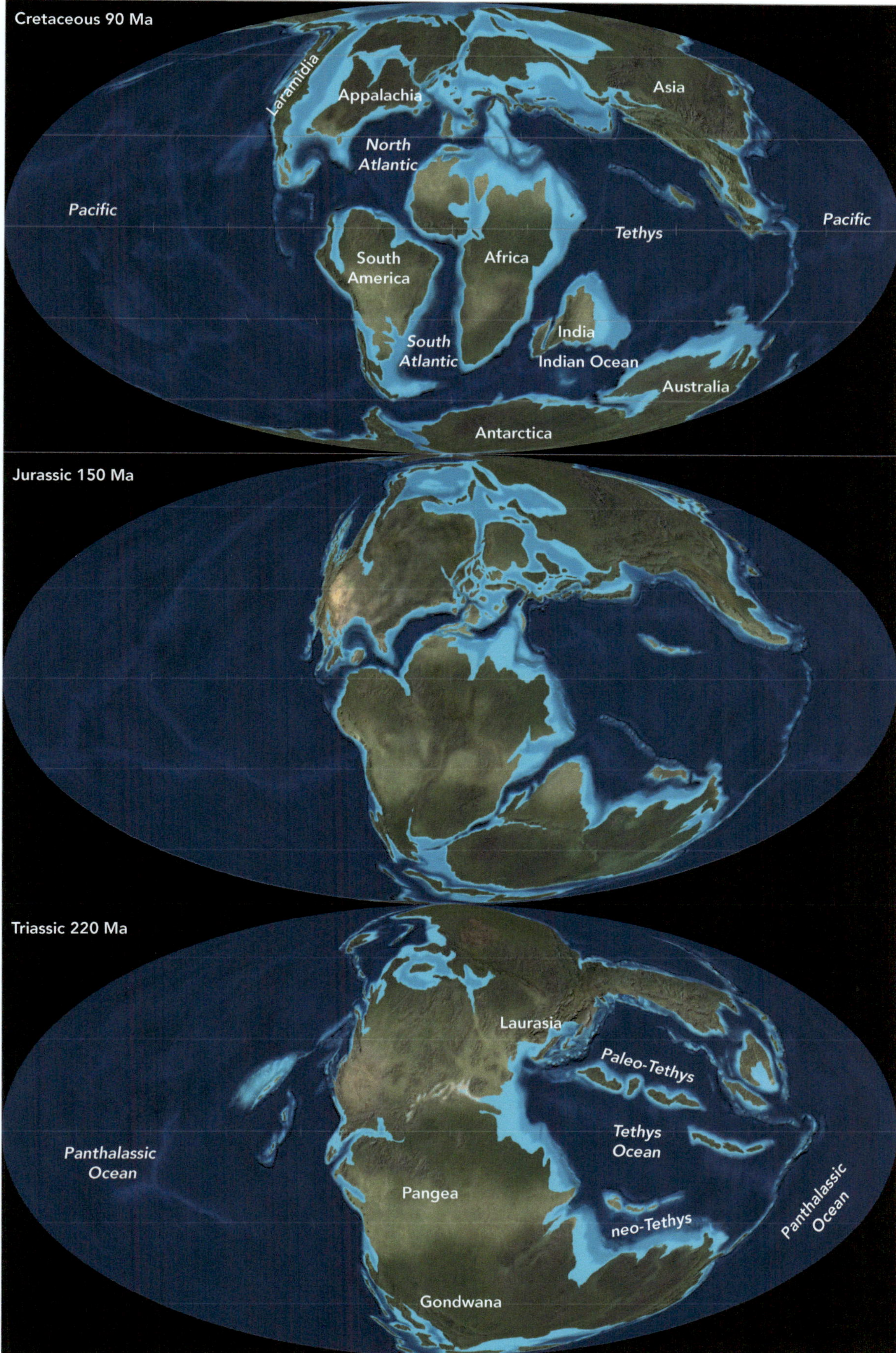

Figure 12.2 Triassic, Jurassic, and Cretaceous paleogeography. The Triassic is characterized by Pangea, and some of the earliest dinosaurs appeared in the jungles of southern Gondwanaland. The late Triassic and Jurassic shows the breakup of the Pangean supercontinent that formed the Atlantic Ocean. By 80 Ma, most continents are roughly in the positions we find them today. Due to high sea levels, North America was split into western (Laramidia), and eastern (Appalachia) lands and South-Central Europe was a series of islands. Source: © DeepTimeMaps.

Figure 12.3 (a) Chalk deposit; (b) rudists, giant bivalves that evolved into major reef builders in the Cretaceous Period. Source: (a) Gabriel Mello via Moment / Getty Images; (b) Photo by JPB. With permission of ROM (Royal Ontario Museum), Toronto, Canada.

Figure 12.4 Sir Richard Owen with a skeleton of a Giant Moa, an extinct bird. Note that the legs are fully tucked under the body, not splayed out like a reptile. Source: John van Voorst, Public domain, via Wikimedia Commons.

During the Cretaceous Period, sea level was hundreds of meters higher than it is today. This suggests that not much water was sequestered on land in glaciers, but there is some controversy as to whether or not the atmosphere was warm enough for the entire surface to be ice free. The Antarctic continent lay at the South Pole, as it does today, but was not fully disconnected from other land masses. Today, Antarctica is isolated and surrounded by the circumpolar sea, which keeps it very cold, enhancing the ability to grow massive ice sheets (see Chapter 18), but it was certainly much warmer in the Cretaceous Period. It is not clear what the elevation of Antarctica was in the Cretaceous, and this is important, as continental glaciation is favored by high latitude and higher elevation (reviewed in Chapter 18). Studies of Cretaceous stratigraphy show evidence for high-frequency, low-amplitude sea-level changes. The rapidity of sea-level changes indicate that they are almost certainly driven by rapid climate change (see also Chapters 11 and 18 for details on climate cycles) possibly driven by growth and decay of ice sheets, but the lower amplitudes suggest that these ice sheets would have been quite small. Dropstones have been found in circumpolar Cretaceous marine sedimentary rocks, indicating that both poles were cold and had local sea ice, and regardless of temperature, both poles would have been dark for half the year. Interestingly, dinosaurs have been found in the high Arctic and, although it was warmer than it is today, there is considerable speculation as to how they survived the darker and cooler winter months with periodic freezes.

KEY POINT

The Mesozoic Era was marked by the gradual breakup of Pangea and a transition from arid to humid conditions as sea level reached a maximum in the Cretaceous Period.

Figure 12.6 *Dryptosaurus* fighting, painted by Charles R. Knight in 1897. Source: Charles R. Knight, Public domain, via Wikimedia Commons.

12.2 Early Interpretations of Dinosaurs

In this section we review the naming of dinosaurs, the challenges in reconstructing dinosaurs, and the value of better fossils and trackways in developing a modern interpretation.

12.2.1 Reconstructing Dinosaurs from Bones

Dinosaurs, the "terrible lizards," were first named by Sir Richard Owen (Figure 12.4), in 1842, based on reconstructions of fossil remains (Figure 12.5(a)). Owen was trained as a surgeon and was especially gifted in his abilities in comparative anatomy, which led him to compare these fossils to modern-day lizards, and given their gigantic size gave them the name he felt was appropriate. However, early interpretations of dinosaurs were based on incomplete fossil remains. Many reconstructions depicted them to be slow, four-legged, lumbering "cold"-blooded gigantic lizards that dragged their tails on the ground, such as the **Iguanodon** shown in Figure 12.5(a), although the well-known American wildlife artist Charles R. Knight showed some of them to be rather more dynamic and even birdlike (Figure 12.6).

Reconstructing an ancient animal, and especially a vertebrate, is not always an easy task, and is dependent on the preservation and completeness of the fossils. This depends on how they become fossilized, the study of which is termed **taphonomy**. In Chapter 9 we discussed some of the challenges in reconstructing the compressed, flattened, and often contorted bodies of the Burgess Shale fauna. Similarly, most fossils of dinosaurs are isolated or jumbled bones, skulls are rare, and it is uncommon to find skeletons in which the bones are preserved in the original position that the animal lived or died. Even in relatively complete skeletons, parts may be missing, and the bones are usually contorted. In fact, many complete dinosaur skeletons show the head and necks twisted back, giving the appearance of a horrifically painful death (Figure 12.7), but this

Figure 12.5 (a) Model reconstructions of *Iguanodon* as a four-legged reptile, sculpted in 1854 for display at the Crystal Palace in London England. Models were sculpted by Benjamin Waterhouse Hawkins based on Owen's reconstructions. Note also that their forelimbs are shown tucked under the body, versus splayed out like a lizard, indicating a more mammalian gait. Fossil recovery was poor and the *Iguanodon* thumb was incorrectly thought to be a nose-horn. (b) Newer mount of the original *Iguanodon* fossils shows that it is bipedal, holding its tail off the ground, and the thumb-spike is correctly placed, versus stuck on the nose. (c) Older artistic reconstruction suggests a bipedal upright gait with tail dragging on the ground. (d) Newer artistic reconstruction suggests *Iguanodon* was actually quadrupedal. Source: (a) whitemay via iStock / Getty Images Plus; (b) Photo by JPB. By permission of The Trustees of the Natural History Museum, London; (c) Joseph Smit, Public domain, via Wikimedia Commons; (d) Sebastian Kaulitzki / Science Photo Library / Getty Images.

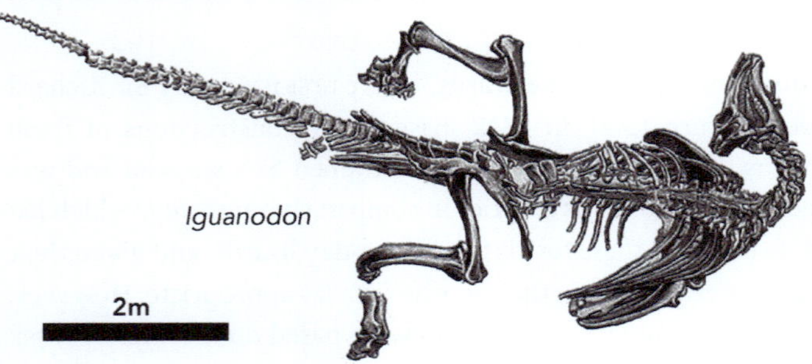

Figure 12.7 *Compsognathus* and *Iguanodon* skeletons show bent necks, originally thought to reflect agonizing death but later attributed to shrinking of neck cartilage. Cast of *Compsognathus* skeleton from the Bavarian State Institute for Paleontology and Historical Geology. Note that the head and neck are curved backwards. Source: (top) Nizar Ibrahim et al., The ornithologist Alfred Russel Wallace and the controversy surrounding the dinosaurian origin of birds, *Theory in Biosciences*, 2013, Springer Nature; (bottom) Gustave Lavalette, Public domain, via Wikimedia Commons.

was later interpreted as the result of contraction of the neck cartilage after death that folded the heads backwards, and, as such, this probably had nothing to do with the mode of death.

If you examine complete-looking skeletons on display in natural history museums, the fine-print will commonly show that only parts of the reconstructions are based on actual fossils, and it is not uncommon for a single mounted skeleton to include parts of several different animals. These days, most museums mount plastic or plaster casts rather than the original bones, and this allows mounts to be drilled through the

cast, as opposed to the heavy metal external mounts that must be used on real fossil bones (Figure 12.8). When the first ***Brontosaurus*** was mounted, the head was from a ***Camarasaurus***, a smaller and different species of **sauropod** with a distinctive stubby snout (Figure 12.9), until more complete specimens were found.

Another challenge in reconstructing dinosaurs specifically, is that some of their joints, and particularly the weight-bearing legs, had thick cartilage. If you look at dinosaur limb joints, there are commonly large spaces between bones, and it is not always clear how they fit together (Figure 12.8(a) and (b)). This contrasts with most mammals' limbs, which have considerably thinner cartilage, making it easier to determine how their bones fit together (Figure 12.8(c)). This caused some uncertainty in understanding how the four-legged dinosaurs (especially the horned **ceratopsians**, long-necked sauropods, and armored **ankylosaurs**) stood and walked. Some early depictions showed these animals with lizard-like postures, with their limbs splayed-out to the side, in a permanent push-up pose (Figure 12.10(b)), while others assumed they were **bipedal** (i.e., two-legged, Figure 12.5(c)). Given how unusual the dinosaurs were, it was not clear what modern animals might be comparable, and of course Owen's use of the term "lizard" invited comparison to the reptiles.

It was also originally thought that the gigantic sauropods were too big to have lived on land, and thus must have spent most of their time submerged in rivers or lakes (Figure 12.10(a)). The peg-like teeth of some of the largest sauropods, such as *Diplodocus* and *Patagotitan* (Figure 12.11), seemed unsuitable for chewing, so it was assumed they must have eaten soft marshy plants. Some of the duck-billed **hadrosaurs** were also depicted as swimmers, and their extensive head ornaments were hypothesized to have been used as some sort of aqualung, although they are now thought to have probably served as amplifiers for generating low loud sounds.

Taphonomic analysis now shows that sauropods (and hadrosaurs, ceratopsians, and ankylosaurs) lived on dry land. Many of the best-known sauropods are found in the Jurassic Morrison Formation, USA, which was formed by relatively small rivers that were far too shallow to have submerged a gigantic adult sauropod. They more likely spent time in the wetter **riparian** environments next to rivers where extensive vegetation provided a source of food, as well as a source of drinking water, in what is thought to have been an otherwise relatively dry area of North America. Rather than chewing, sauropods likely used their peg-like teeth to strip leaves and twigs as they browsed tall trees, relying on stomach stones (**gastroliths**) to "chew" their food, much like many birds, who lack teeth completely and many of which must ingest sand or gravel to break up hard-to-digest plant food in their gizzards. This was concluded from studies of dinosaur fossil feces (termed coprolites), as well as discoveries of gastroliths in the stomachs of several herbivorous dinosaurs.

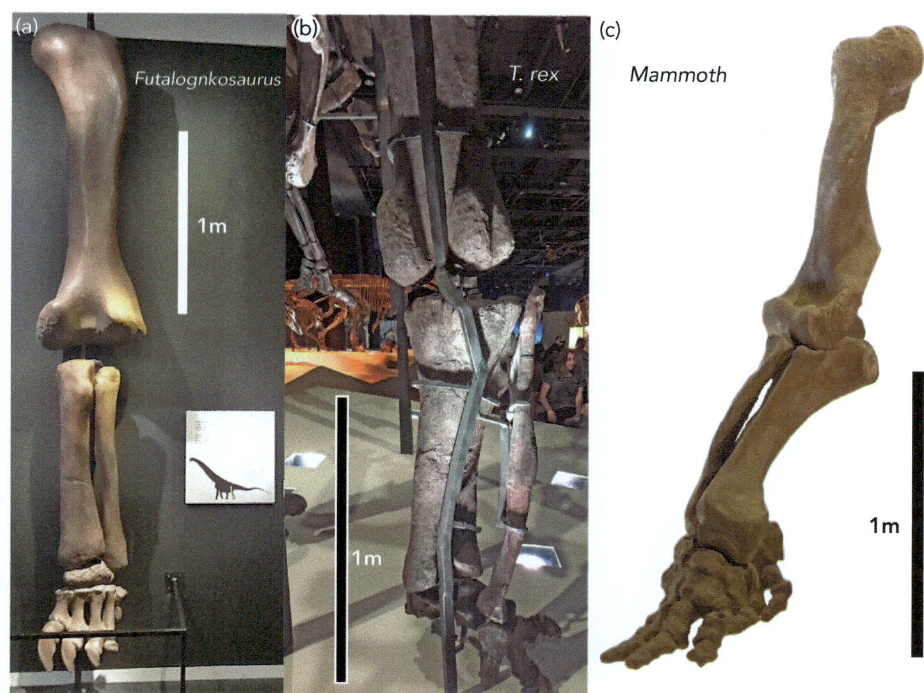

Figure 12.8 (a) *Futalognkosaurus*, (b) *Tyrannosaurus rex*, and (c) mammoth leg bones. Note large gaps where cartilage would have been in the *Futalognkosaurus* and *T. rex*. Metal mounting indicates that the *T. rex* is a real bone whereas the *Futalognkosaurus* is a cast. Mammoth bones show tight articulation of the humerus and fibula, allowing an accurate reconstruction of the skeleton. Source: (*Futalognkosaurus* leg) Photo by JPB, taken at Ultimate Dinosaurs touring exhibition, Science Museum of Minnesota; (*T. rex* and mammoth legs) JPB. With permission from Houston Museum of Natural Science (HMNS).

Figure 12.9 (a) 1905 mount of *Brontosaurus excelsus*, specimen AMNH 460, with sculpted *Camarasaurus* skull and tail dragging on the ground. (b) Modern reconstruction shows less curved neck and new skull. Source: (a) Dinosaurs, by William Diller Matthew, Public domain, via Wikimedia Commons; (b) Photo by JPB, TCU collection.

KEY POINT

Early reconstructions of dinosaurs were hampered by a poor fossil record, lack of modern animals to compare with, and uncertainty as to how bones may have articulated.

Figure 12.10 (a) Charles Knight's 1897 depiction of a half-submerged *Brontosaurus* in the foreground, with *Diplodocus* eating marsh plants in the background. In the further distance, a herd of submerged brontosaurs can be seen. (b) Henrich Harder's 1913 rendering of *Diplodocus*, showing splayed out legs with tail dragging on the ground. Source: (a) Charles R. Knight, Public domain, via Wikimedia Commons; (b) Heinrich Harder (1858–1935), Public domain, via Wikimedia Commons.

12.2.2 Evidence of Bipedalism and an Upright Gait

Owen initially assumed a more mammal-like posture for the dinosaurs, understanding that they were too heavy to hold their bodies in a lizard-like posture with their limbs splayed outwards. To that end, the standing **quadrupedal** *Iguanodon* in Figure 12.5(a) has its limbs underneath its body. Later reconstructions put the "nose" bone on the thumb (Figure 12.5(b) and (c)) and show *Iguanodon* as a bipedal plant eater, whereas even more recent reconstructions suggest that *Iguanodon* may have been quadrupedal (Figure 12.5(d)).

The discovery and analysis of dinosaur tracks (Figure 12.12) ultimately demonstrated that dinosaurs walked fully upright,

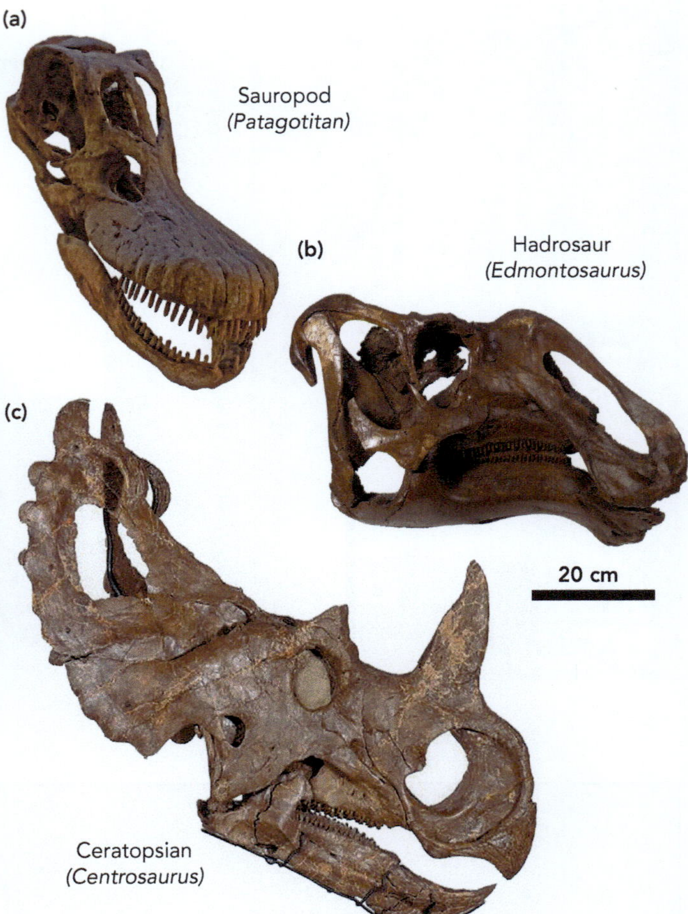

Figure 12.11 (a) The sauropod (*Patagotitan*) skull has small peg-like teeth, useful for stripping leaves, but not for chewing. (b) and (c) *Edmontosaurus* and *Centrosaurus* had banks of teeth adapted for chewing vegetation. Sauropods likely ingested pebbles, or gizzard-stones, to help process a plant-based diet. *Centrosaurus* also had a large bony beak, useful for chopping shrubs and woody plants. Source: (a) Photo by JPB. © 2024, Field Museum. All rights reserved; (b) With permission of ROM (Royal Ontario Museum), Toronto, Canada.

with their legs tucked in under their bodies as opposed to spayed to the side like lizards (compare Figures 12.12 and 12.13). The lack of tail drag marks also showed that they held their tails high, likely serving as a balance to their front ends. Many of the dinosaurs, and especially the carnivorous species such as *Tyrannosaurus rex*, were also bipedal (Figure 12.14). *Tyrannosaurus rex* has an enormous head, and it needed a large tail to ensure it was balanced.

Permian and older trackways show a dominance of a sprawling reptilian gait, such as shown in Figure 12.13, whereas Triassic trackways are dominated by an upright gait. Some of the earliest dinosaurs, such as **Herrerasaurus** (Figure 12.14), a small **theropod** (i.e., carnivore), were bipedal and show that bipedalism was an early adaptation and a major evolutionary change from their four-legged ancestors. Although the sauropods were quadrupedal, the earliest sauropod ancestors, including the prosauropod

Plateosaurus (Figure 12.14), were initially bipedal. Sauropod quadrupedalism is thus an example of reversal in which a quadrupedal reptilian ancestor evolved into a bipedal, herbivorous dinosaur (*Plateosaurus*), which eventually evolved back to a quadrupedal gait. Bipedalism also freed the forelimbs from being used for walking and were adapted for flying in the birds. The *pterosaurs* (literally "winged lizards") are a separate sister group that evolved alongside dinosaurs that also developed flight and provide a good example of convergent evolution. Getting off the ground is a useful strategy for avoiding predators and is also helpful in hunting small game that can be spotted below.

KEY POINT

A major anatomical change in dinosaurs was the appearance of an upright and bipedal gait, versus the sprawling gaits of the earlier tetrapods.

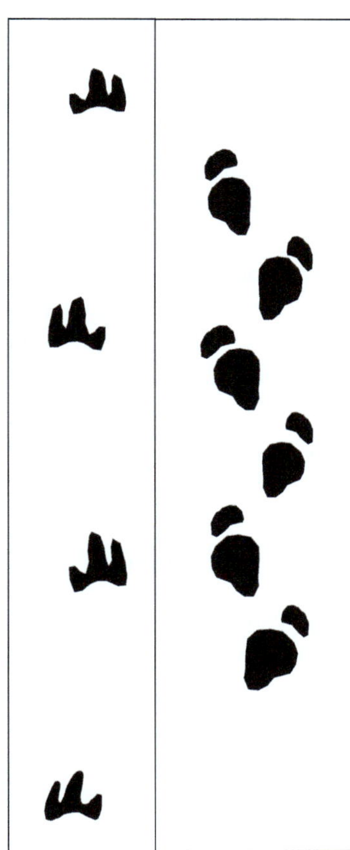

Figure 12.12 Schematic drawing of dinosaur trackways. (a) Three-toed theropod tracks show a single offset row of prints, much like modern bird tracks, and indicate that they were bipedal. (b) Sauropod tracks show alternating rows of left and right legs. The closeness of the tracks shows that the legs of both animals were tucked under their bodies and there is no evidence of drag marks from their tails, indicating that they were held off the ground, contrary to earlier reconstructions. Source: Adapted from Fig. 5.1 in Thulborn (1989).

Figure 12.13 Lizard tracks with tail marks. Lizards move side-to-side unlike dinosaurs and mammals who move in a vertical plane.

12.3 The First Dinosaurs

As we saw in Chapter 11, at the end of the Paleozoic, terrestrial vertebrates had divided into two major clades, the synapsids, which would eventually give rise to the mammals, and the diapsids. The dinosaurs first appeared in the Late Triassic Period, about 230 Ma, about 20 million years after the Permian extinction. Fully bipedal animals, such as **Marasuchus** (Figure 12.15), which is classified as a **dinosauromorph**, are almost indistinguishable from and are closely affiliated to dinosaurs. These dinosauromorphs have mostly been found in rocks about 235 Ma in southern South America, which would have lain in the sub-tropical belt, south of the equator and south of the deserts that characterized much of Pangea. These dinosauromorphs were all quite small, about the size of a chicken.

Eoraptor (Figure 12.15), at 231 Ma, was found in the Late Triassic Ischigualasto Formation in Argentina, South America, and is one of the earliest true dinosaurs. It was relatively small, about 1.5 m in length, shows an upright, bipedal gait, and likely had an omnivorous diet. *Herrerasaurus* (Figure 12.14), which was introduced above and was found in the same formation as *Eoraptor*, was an early carnivore but is quite a bit larger than *Eoraptor* – about the size of a pony or small horse. Other early dinosaurs are found mostly in rocks associated with southern Gondwana, and there are debates as to whether they represent the ancestors of the sauropods or theropod dinosaurs.

For most of the Triassic, dinosaurs were not the dominant theropods. The *pseudosuchians* (crocodile relatives) were larger and likely served as the top terrestrial predators, following the demise of the gorgonopsians mentioned in Chapter 11. Some of the larger synapsids, such as *Lystrosaurus*, made it across the Permian–Triassic boundary. The real so-called "Age of Dinosaurs" did not really begin until the Jurassic Period, following a mass extinction event that ended the Triassic Period.

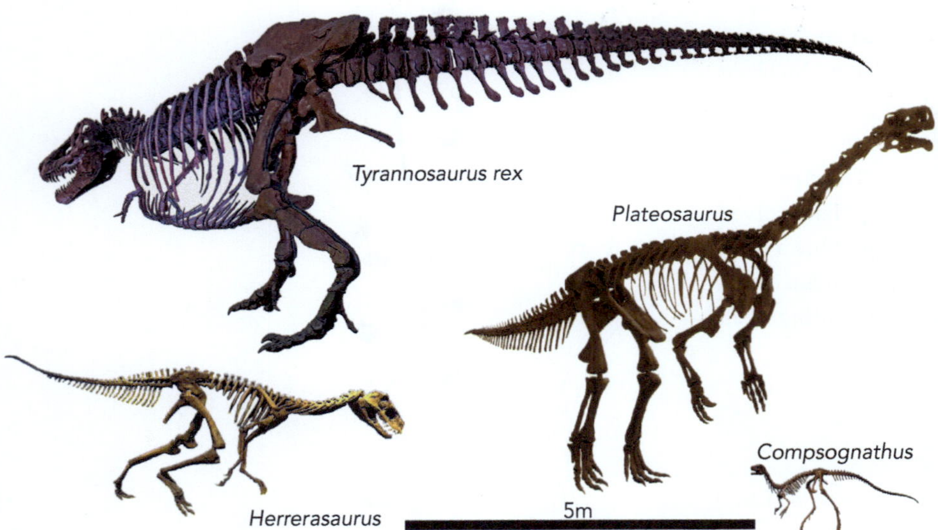

Figure 12.14 *Tyrannosaurus rex*, *Herrerasaurus*, *Plateosaurus*, and *Compsognathus* were all bipedal, using their tails for balance. *Herrerasaurus* was one of the earliest Triassic dinosaurs, whereas *T. rex* lived at the very end of the dinosaurs' reign in the Late Cretaceous. *Plateosaurus* was the bipedal ancestor to the quadrupedal sauropods. Source: (*T. rex*) Photo by JPB with permission of Field Museum. (*Herrerasaurus*) Photo by JPB, taken at Ultimate Dinosaurs touring exhibition, Science Museum of Minnesota; (*Plateosaurus*) breckeni / iStock / Getty Images; (*Compsognathus*) Photo by Jim, the Photographer from Springfield PA, United States of America. BYY CC 2.0, via Wikipedia Commons.

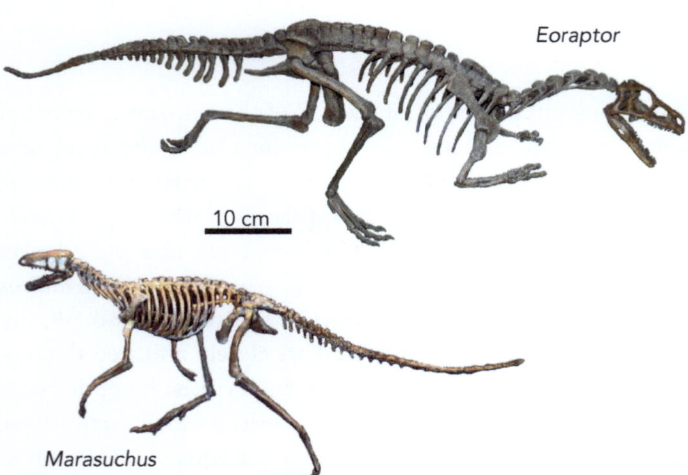

Figure 12.15 *Marasuchus* and *Eoraptor* fossils. Source: HoopoeBaijiKite, Public domain, via Wikimedia Commons.

KEY POINT

Although dinosaurs appeared in the Triassic Period, they did not dominate until after the end-Triassic mass extinction. During most of the Triassic, pseudosuchians were the top predators.

12.4 Dinosaur Adaptations and Behavior

Dinosaurs, of course, are vertebrates. The oldest vertebrates are the jawless fishes, which first appeared in the Late Cambrian Period in the Paleozoic Era and became dominant in the Devonian Period when they gave rise to the first terrestrial tetrapods (see Chapter 11). Fish swim using a side-to-side motion, requiring their vertebrae to flex laterally rather than vertically.

Amphibians and reptiles (including snakes) also move using a side-to-side motion (Figure 12.13), which they inherited from fish. One of the major adaptations of dinosaurs and mammals was a transition to a more vertical method of locomotion. It is very difficult to breathe using a side-to-side motion as one lung is being squeezed at the same time another is expanding. Lizards make poor long-distance runners. Dinosaurs (and mammals), in contrast, are able to breathe while running and this allows far faster mobility as well as long-distance running and was clearly a driver in their later success. This evolutionary change required a radical reorientation of limbs, from splayed outward to a more vertical position underneath the body, as well as changes in the vertebral column to allow it to flex in a vertical plane, rather than side-to-side. Whales and dolphins evolved from mammals and, unlike fish, also swim using a vertical motion, which is why their tail fins extend laterally rather than vertically. It is worth noting that the marine **ichthyosaurs**, which lived at the same time as dinosaurs and who evolved from land-dwelling reptiles, use a side-to side motion indicating that they are much more closely related to lizards versus dinosaurs. They also provide a good example of convergent evolution with whales and dolphins where a land-dwelling vertebrate readapts to an aquatic lifestyle.

KEY POINT

Dinosaurian bipedalism required a change from sideways to vertical motion of the spine.

The dinosaurs are also **amniotes** (Figure 12.16), that is, they belong to the group of vertebrates that have developed an amniotic sac required for living and reproducing on dry land versus life in water, such as amphibians and fish (see Chapter 11). The amniotes include reptiles, mammals, and birds (Figure 12.16). All

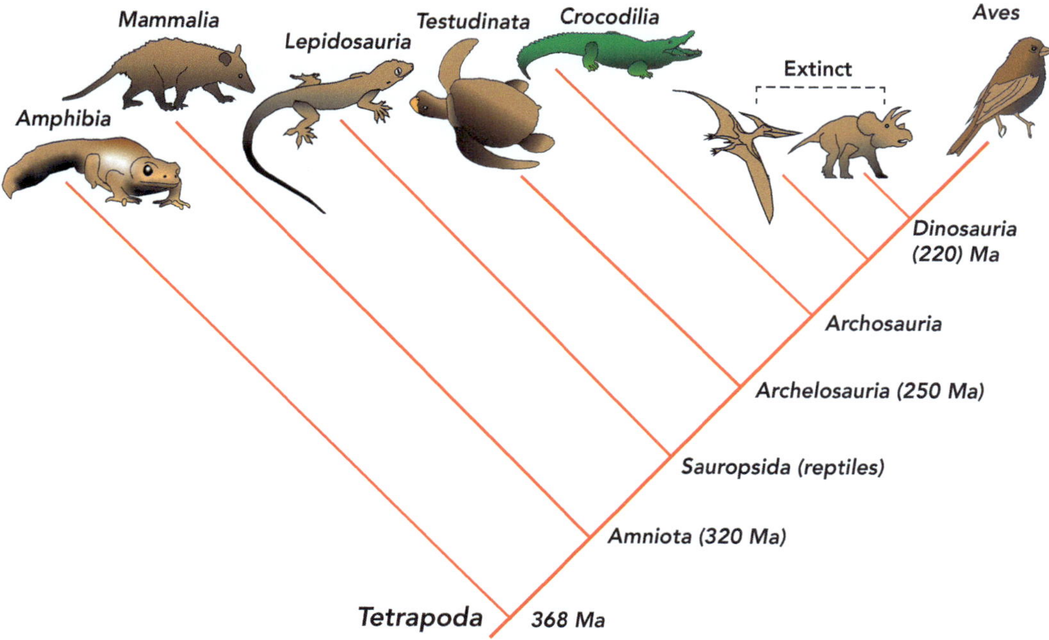

Figure 12.16 Tetrapod cladogram.

Figure 12.17 Poikilothermic marine iguanas on the Galapagos Islands, basking in the sun to heat their bodies. Their legs are splayed to the side. Also note the patches of shedding skin.

dinosaurs, reptiles, and birds lay eggs that contain an amniotic sac. In most modern mammals, the amniote is contained within the female body, and they birth their babies live, versus gestating an egg in a shell, although **monotremes** (like the duck-billed platypus) and many extinct early mammals were egg layers.

Once freed from an aquatic lifestyle, amniotes also require a watertight skin, but watertight skins are hard to stretch. To adapt to an unyielding skin, snakes regularly shed and replace their entire skin as they grow. Birds and reptiles also shed their skin, but in smaller patches. Mammals do not shed their skin wholesale, but anyone who has had a sunburn knows that even humans exfoliate their skin in patches or as particles (e.g., dandruff). We do not know if and how dinosaurs shed their skins, but it is likely that they did so in a more mammal-like patchwork fashion. Dinosaurs thus share both mammalian and reptilian traits. They have a watertight skin and lay eggs, like reptiles, but they walk fully upright like mammals. Later we will discuss the relationships of dinosaurs to birds and the evolution of feathers and flight.

12.5 Were the Dinosaurs Warm-Blooded?

Dinosaurs were originally thought to be "cold"-blooded, technically termed *ectothermic* (Figure 12.17), assuming that they were essentially giant lizards. The term *poikilothermic* is analogous and refers to animals that experience a wide range of body temperature. **Ectotherms** lack the ability to control their temperature by their own metabolic processes and they do not possess the complex heating and cooling abilities of **endotherms**, such as birds and mammals, which are able to regulate their temperatures by panting and sweating. Ectotherms include lizards, turtles, snakes, crocodiles, and all amphibians. Ectotherms are unable to engage in sustained high-energy activities like running long distances or powered flight and are unable to support the larger brains that characterize ectotherms. Endothermic metabolism favors "sit-and-wait" hunting strategies versus chasing prey, which is practiced by big cats, wolves, and hyenas. The advantage of ectothermy is that their metabolic and caloric requirements are far lower than endotherms, which must eat frequently using food to provide the energy to maintain their temperature. Ectotherms typically need only 5 to 10% of the energy of endotherms. For example, ectothermic marine iguanas spend much of their time basking in the sun, warming their bodies after gorging themselves on algae in the cool ocean waters (Figure 12.17).

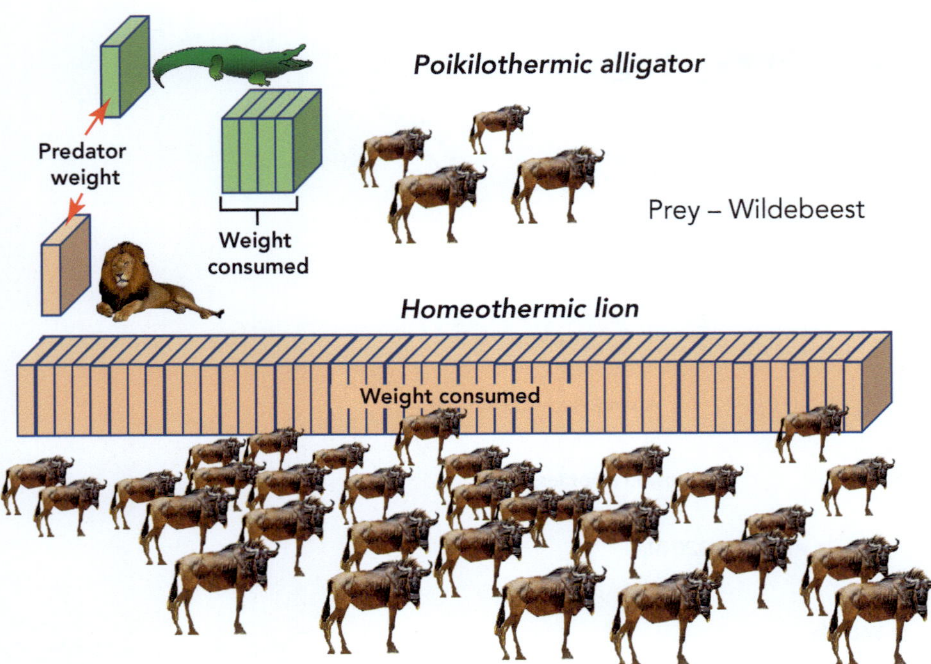

Figure 12.18 Predator–prey relationships for "cold"-blooded crocodile versus "warm"-blooded lion.

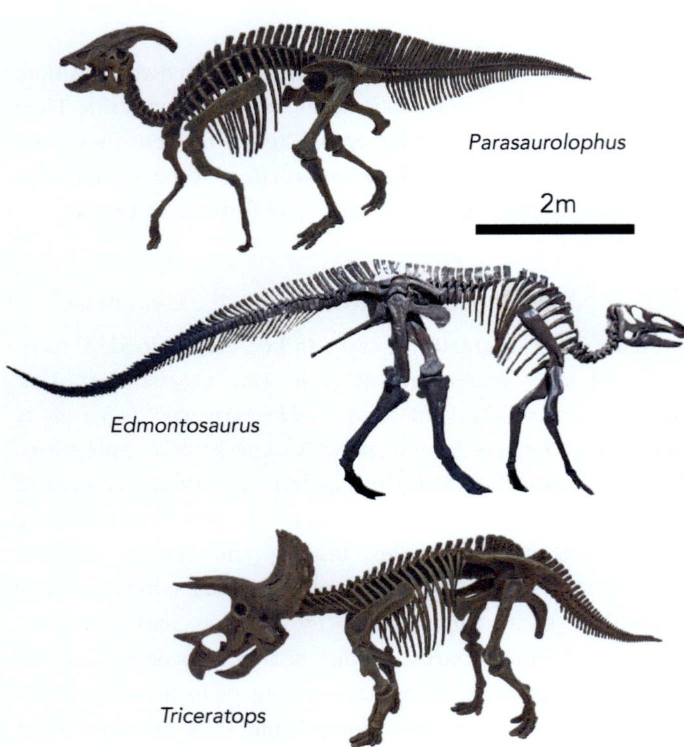

Figure 12.19 Late Cretaceous herding herbivorous dinosaurs included the duck-billed hadrosaurs, such as *Parasaurolophus* and *Edmontosaurus*, and the horned ceratopsians such as *Triceratops*. Source: (*Parasaurolophus* and *Triceratops*) Photos by JPB; (*Edmontosaurus*) Photo by JPB. With permission of ROM (Royal Ontario Museum), Toronto, Canada.

through regulation of their metabolic process, such as panting and sweating to cool down, and all of this requires food, lots of it and frequently. A consequence of greater caloric needs is that in endotherms the ratio of predators to prey in any given population is very low (Figure 12.18), and of course endothermic herbivores must eat constantly and require a large biomass of vegetation. In ectotherms, the ratio of predators to prey can be much larger. Snakes, for example, may survive for up to a year after a single meal, and in the American southwest there is a high proportion of predatory rattlesnakes versus the gophers that they prey on. Endothermic lions and tigers, in contrast, typically eat the equivalent of a deer-sized animal once a week and thus require a much larger proportion of prospective prey than ectothermic snakes or alligators (Figure 12.18). Even though the African savanna sports large herds of zebras, antelopes, and wildebeests, lions are rare by comparison, and the rarity of modern-day, large warm-blooded big cats is one reason that they are almost all considered endangered. In an endothermic population, if there are too many predators, they will quickly eat all the available prey and then starve themselves to death. Predator–prey ratios thus can be used to determine if the chief predators were endothermic or ectothermic (Figure 12.18). As we also see in Chapter 13, if external forces limit prey, the food chain may collapse causing extinction.

Although it is challenging to estimate predator–prey ratios based only on the fossil record of dinosaurs, there is evidence that many species of herbivorous dinosaurs moved in large herds (Figure 12.19). In Alberta, thousands of bones of the ceratopsian herbivore **Centrosaurus** (Figure 12.11), an earlier relative of **Triceratops**, have been found in extensive bone beds that contain

Endothermic birds and mammals, in contrast, must keep their body at a constant temperature, and this could be higher or lower than their surroundings. They accomplish this

the remains of thousands of individuals. These bone beds were likely formed when herds of centrosaurs drowned as they crossed rivers during catastrophic floods. Bone beds of the duck-billed hadrosaur **Edmontosaurus** (Figure 12.19) in the Lance Formation in Wyoming are estimated to contain the remains of up to 25,000 individuals, also indicating herding behavior.

In contrast, the fossil record of large predators, such as the tyrannosaurids, which lived at the same time as *Centrosaurus* and *Edmontosaurus*, are rather more limited. Only about 50 *T. rex* specimens are known worldwide, compared to thousands of herbivorous edmontosaurs and centrosaurs found in individual bone beds. This suggests that predator–prey ratios were low and support the hypothesis that the dinosaurs were indeed warm-blooded. Of course, this requires vast vegetation to support the herbivorous grazers and browsers at the bottom of the food chain.

In the 1960s, Yale paleontologist John Ostrom re-evaluated specimens of **Deinonychus antirrhopus**, a small Early Cretaceous bipedal carnivorous dinosaur with a huge retractable sickle-shaped claw on its foot and stiffening rods along its tail (Figure 12.20). It was clear that this was a highly agile predator that was able to use its hind legs for more than just running, as presaged by Charles Knight in 1897 (Figure 12.6). *Deinonychus* is a member of the **dromaeosaurids** (Figure 12.21), which are feathered theropod dinosaurs that include the **velociraptors**, featured in the Jurassic Park movies. The observation of feathers and the agility of these animals did not seem compatible with slow-moving, cold-blooded metabolisms and led to a renaissance in thinking about dinosaurs, and particularly the notion that they were warm-blooded.

Examination of dinosaur bone structures also revealed dense clusters of microscopic tubes called **Haversian canals**. These Haversian canals lie in the outermost region of bone and allow blood vessels and nerves to travel through them. Although they are found in both ectothermic and endothermic animals, the abundance and density in dinosaurs are more similar to endothermic animals and provide additional evidence that the dinosaurs were likely endothermic.

Figure 12.20 *Deinonychus antirrhopus* cast with artistic reconstruction by Emily Willoughby. Source: (top) Photo by JPB. With permission from Houston Museum of Natural Science (HMNS); (bottom) Emily Willoughby, (e.deinonychus@gmail.com, emilywilloughby.com), CC BY-SA 4.0 <https://creativecommons.org/licenses/by-sa/4.0>, via Wikimedia Commons.

KEY POINT

Dinosaurs were likely endothermic, as indicated by appearance of feathers, agile behavior, bone structure growth rates, and predator–prey ratios.

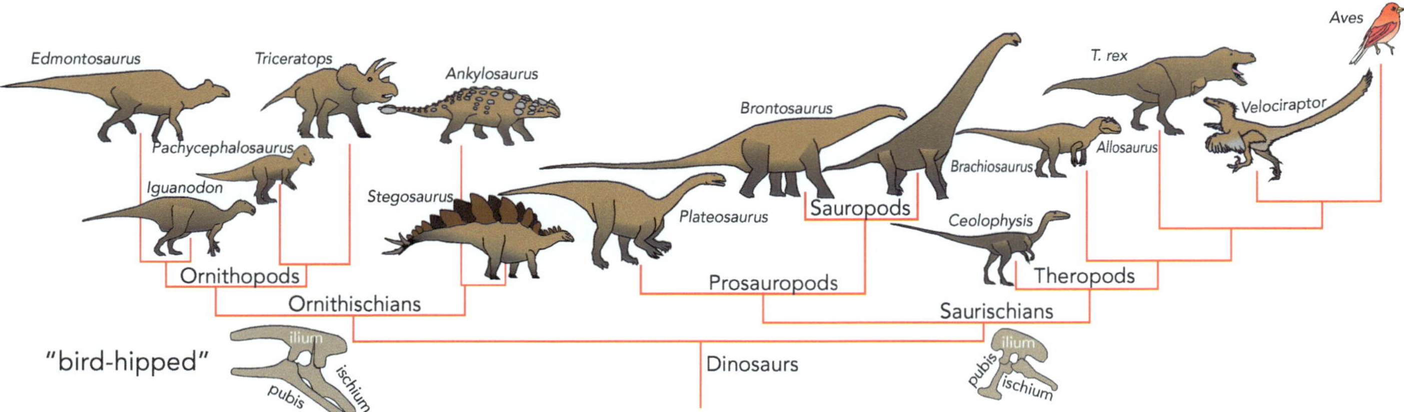

Figure 12.21 Dinosaur cladogram.

12.6 Dinosaur Classification, Are They Really Just Birds?

We now know that several dinosaurs had feathers, not all had wings, most did not fly, and they all had teeth rather than a beak. It is clear that dinosaurs are very closely related to birds. The discovery of feathered dinosaurs forced a major re-evaluation of the relationship between dinosaurs and birds (Figure 12.21). In Chapter 6 we introduced the idea of cladistics, which is the study of how organisms are organized according to their inferred phylogenetic history and the order in which they are related. The older eighteenth-century Linnean classification was based on broad body plans, skeletal, and other macroscale features, and in many cases did indeed reflect the broad phylogenetic history, but as we have seen in Chapter 9, cladistics has resulted in some rather significant changes in our understanding of what is related to what, and the history of dinosaur classification illustrates these changes quite clearly (Figure 12.21).

In the original Linnean classification, the sub-phylum Vertebrata was subdivided into several classes, including Mammalia, Aves (birds), and Reptilia, as well as five classes of fish. Imagine surveying the biology of our planet in the eighteenth century. It would be clear that these classes had some things in common, but it would be hard to imagine how mammals, reptiles, and birds were related, especially in terms of their phylogenetic relationships. It might be deduced that birds and mammals are "more advanced," as they have traits like fur and feathers that reptiles lack, and clearly birds' ability to fly seems unique. You might note that birds and reptiles lay eggs, suggesting that they are perhaps more closely related to each other than mammals, but reptiles walk on all four legs and birds only two, plus birds have feathers and beaks, whereas reptiles have teeth and a "scaly" skin, so they seem different after all. Recall that Linnean classification was developed nearly 100 years before Darwin's breakthrough in explaining the mechanism of evolution and long before we could compare genomes. But the observations that feathers evolved before flight, and the similarity of birds and theropods (Figure 12.22), has led to a rethinking of the classification of dinosaurs, enabling scientists to ultimately conclude that birds evolved from dinosaurs (Figures 12.16 and 12.21).

Richard Owen originally classified the dinosaurs as a sub-order of the Saurian Reptiles, which would have included the lizards, crocodiles, and turtles. The class Aves (i.e., birds) was thought to represent a different phyletic group. The classification of vertebrates, especially lizards and birds, has undergone a revolution in understanding over the last few decades (Figures 12.16 and 12.21) reflecting breakthroughs in genetic coding, and revised cladistics, and the discovery of

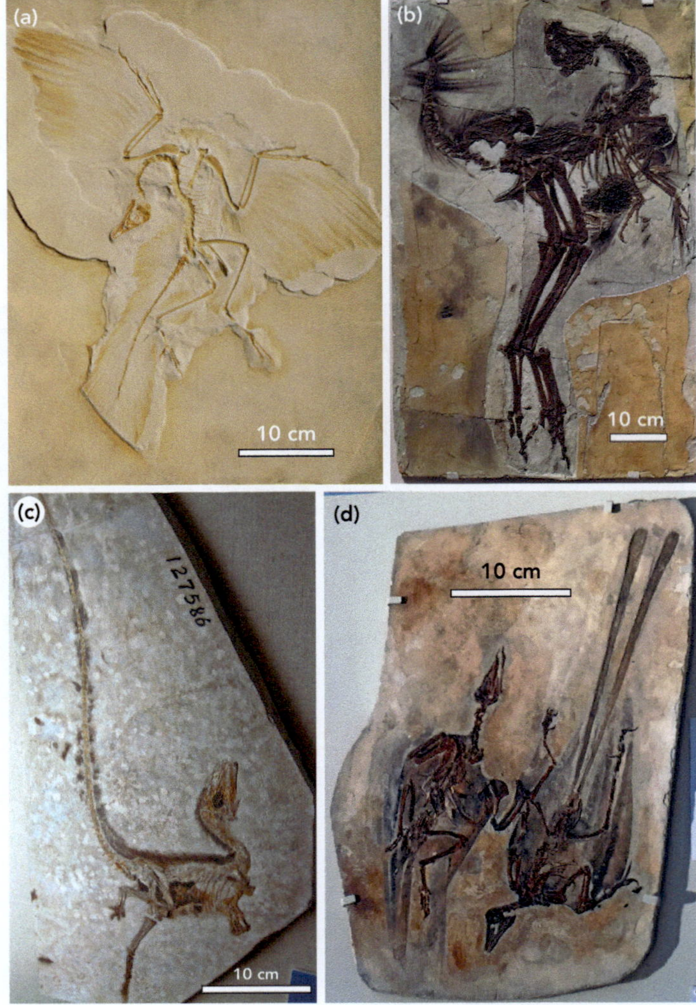

Figure 12.22 Feathered dinosaurs: (a) Late Jurassic *Archaeopteryx*; (b) *Caudipteryx*, Albian; (c) *Sinosauroteryx*; and (d) *Confuciusornis dui*, fossil casts at the ROM show that the male (right) has longer tail feathers than the female (left). Source: (a) Martin Shields / Alamy Stock Photo; (b) and (d) Photos by JPB. With permission of ROM (Royal Ontario Museum), Toronto, Canada; (c) Sam / Olai Ose / Skjaervoy from Zhangjiagang, China, CC BY-SA 2.0 <https://creativecommons.org/licenses/by-sa/2.0>, via Wikimedia Commons.

fossilized feathered dinosaurs has helped understand where dinosaurs fit.

Dinosaurs are now classified as belonging to the larger **archosaur** group, which excludes lizards and snakes but includes crocodilians and possibly turtles (Figure 12.16). The dinosaurs are divided into two main groups (Figure 12.21), the lizard-hipped **saurischians**, which include all the carnivorous theropods and the gigantic sauropods, and the bird-hipped **ornithischians**, which include many of the herbivorous groups including the horned dinosaurs (ceratopsians), the armored dinosaurs (ankylosaurs), and the duck-billed dinosaurs (hadrosaurs, Figure 12.19). Ironically, it is now understood that the lizard-hipped saurischian dinosaurs actually gave rise to the

(a)

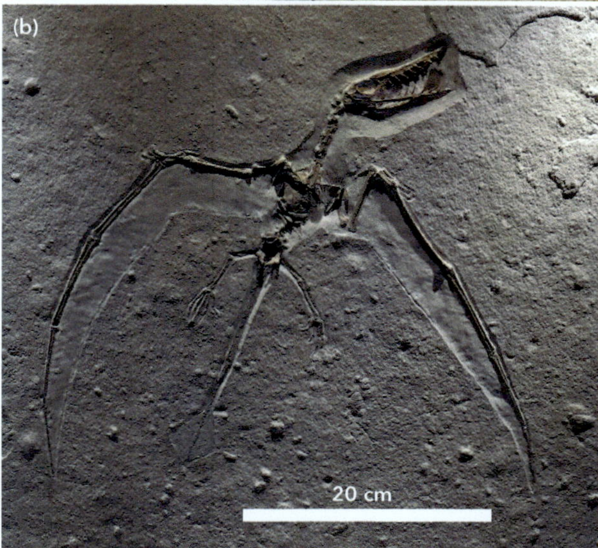

(b)

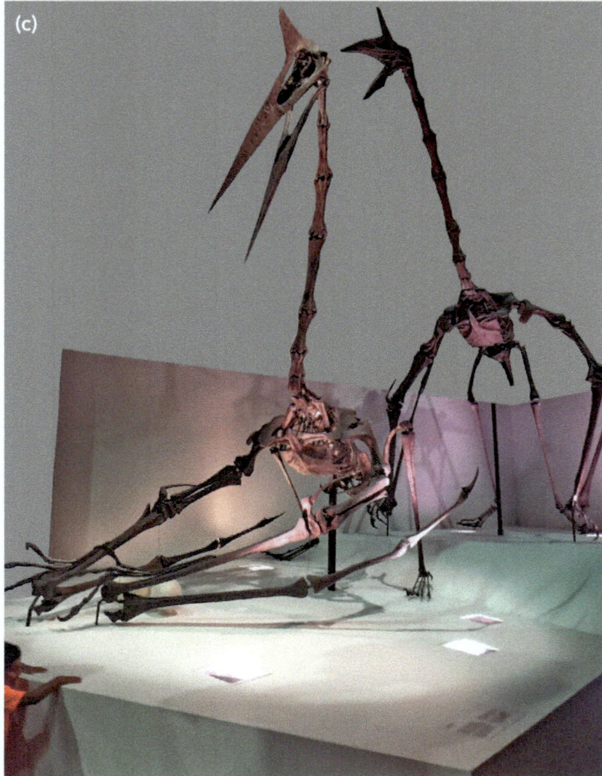

(c)

Figure 12.23 Pterosaurs: (a) *Pterodactyl elegans* and (b) *Rhamphorhynchus*, which were about the size of a seagull, and (c) the

modern birds, and not the bird-hipped ornithischians (Figure 12.21). From the point of view of biological classification, birds are theropods, which are saurischians, which are dinosaurs, which are *Ornithodirans*, which are the variety of archosaurs related to birds. Scientists now refer to birds as "avian dinosaurs" and the traditional dinosaurs as "non-avian dinosaurs." Therefore, it is now argued that the dinosaurs as a clade did not actually become extinct at the end of the Cretaceous even though all non-avian groups within the clade perished (see Chapter 13).

> **KEY POINT**
>
> Dinosaurs and birds are now considered to be part of the same phyletic group (clade).

12.7 Evolution of Feathers

New fossil discoveries of theropod dinosaurs (mostly since about 1996 and mostly in China) show that they evolved feathers long before they developed the ability to fly (Figures 12.21 and 12.22). Feathered dinosaurs include giants like **Yutyrannus huali**, a Cretaceous tyrannosaurid, as well as the smaller **Sinosauropteryx**, which was about the size of a turkey and closely related to the older genus, *Compsognathus* (Figures 12.7 and 12.14). **Archaeopteryx** (which translates as "ancient wing," Figure 12.22(a)) is considered to be one of the true birds, and was found in the 150-million-year-old Jurassic Solnhofen limestone in Germany. There is debate as to how effective a flyer *Archaeopteryx* was, and it is thought that it could only fly in intermittent bursts, like a pheasant or quail, rather than extensive flight over oceans like modern seagulls, eagles, or albatrosses.

Most of the feathered dinosaurs could not actually fly and as a consequence there has been much speculation as to why feathers evolved if not for flight. The most likely theory is that they evolved for thermoregulation, but they may have also played a role in sexual selection, much like is seen in modern peacocks. As an example of this, one of the earliest Cretaceous birds, **Confuciusornis dui** (Figure 12.22(d)), shows evidence of sexual dimorphism, in which males have longer tail feathers, indicating they were likely used to attract females.

It is also clear that wings could not have developed without the earlier evolution of a bipedal gait, which freed the forelimbs from being used in locomotion. In the smaller feathered theropods the forelimbs were adapted into wings that were eventually used for flight.

Figure 12.23 (*cont.*) Late Cretaceous giant *Quetzalcoatlus northropi*. Source: (a) With permission of ROM (Royal Ontario Museum), Toronto, Canada; (b) and (c) With permission from Houston Museum of Natural Science (HMNS).

Fossils of feathers show that they developed initially as simple filaments, and these have been found in both ornithischian and saurischian groups, but in the theropods these developed into full-fledged feathers. The other major group of flying reptiles were the *pterosaurs* (Figure 12.23). Pterosaurs ranged from small animals like Jurassic-age **pterodactyls**, which were about the size of a large seagull, to the Late Cretaceous giant **Quetzalcoatlus**, with a wingspan of about 10 meters. Rare, fossilized skin impressions indicate that pteropods also were covered with filaments, suggesting they were also endothermic, and provide another good example of convergent evolution and diversification between birds and pterosaurs.

KEY POINT

Feathers appeared before they were adapted for flight. They likely were initially used for endothermy and sexual display and only later adapted for flying.

12.8 Gigantism

Not every dinosaur was gigantic, but some approached the size of modern-day jet airplanes (Figure 12.24). From an evolutionary standpoint, one must wonder: what is the advantage of being so large?

12.8.1 Homeothermy, Hollow Bones, and Avian Respiratory Systems

Gigantism may have been an adaptation to homeothermy. Larger animals lose heat much more slowly than small animals because of their lower surface area to volume ratio. However, at some point, if an animal becomes too large, it may be difficult to lose heat, resulting in a risk of overheating and this may put an upper limit on size.

Examination of dinosaur bones, and particularly the giant sauropods, shows that some are hollow, filled with spaces, rather than bone. These spaces are thought to have served as air sacs, balloon-like extensions of the lungs. This accomplished several things: (1) it allows air to interact with a much larger surface areas than lungs alone; (2) it allows oxygen to be extracted during both inhalation and exhalation, such that respiration is far more efficient than in mammals; (3) it enables a higher degree of internal cooling; and (4) the relatively hollow nature of bones lightens the skeleton. All these anatomical features allowed dinosaurs to reach their truly gigantic proportions.

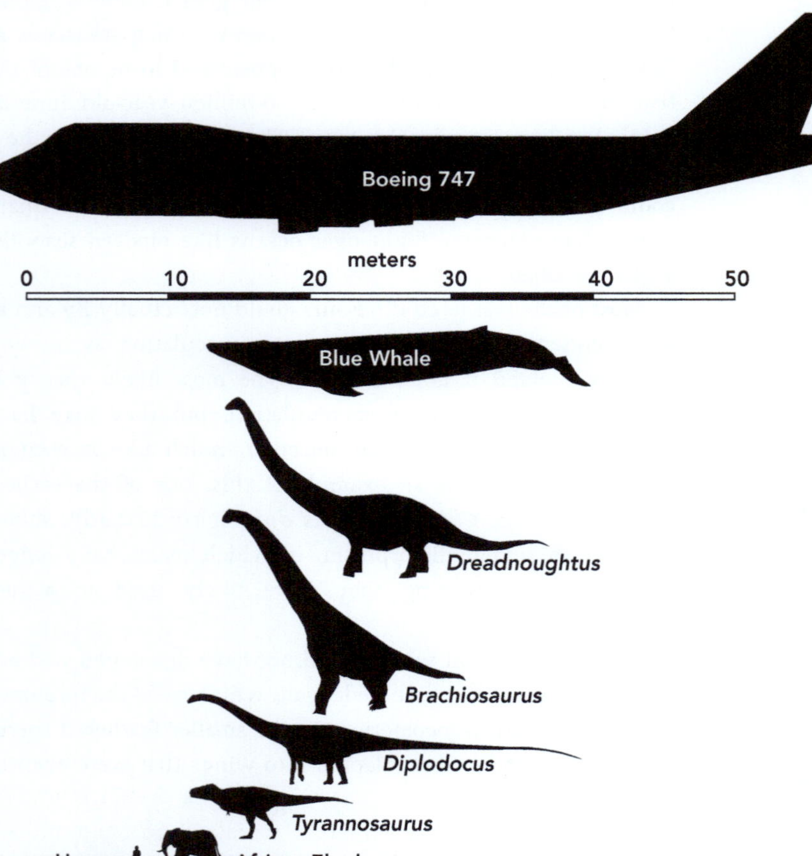

Figure 12.24 Relative size of blue whales, dinosaurs, elephants, humans, and a Boeing 747 Jumbo jet.

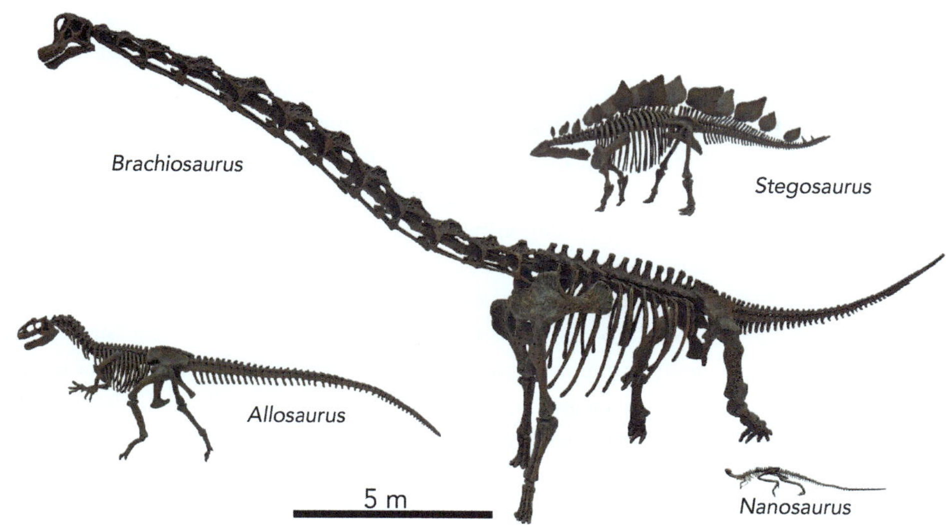

Figure 12.25 Examples of typical Jurassic dinosaurs, *Brachiosaurus*, *Stegosaurus*, *Allosaurus*, and *Nanosaurus*. Source: (*Brachiosaurus*, *Stegosaurus*, *Allosaurus*) Photos by JPB; (*Nanosaurus*) Photo by JPB. With permission from Houston Museum of Natural Science (HMNS).

KEY POINT

Avian respiratory systems helped cool large animals, provided more efficient respiration, and made skeletons lighter, helping drive dinosaurs to gigantism.

12.8.2 Response to Gigantic Predators

The selective pressure that drove herbivores to gigantism was also likely driven by the development of gigantism in the carnivorous theropod dinosaurs. Herbivores would be a fine meal for predators such as *Allosaurus* (a Jurassic predator, Figure 12.25) or **Tyrannosaurus rex** (Figure 12.14), which lived in the Late Cretaceous. The plant-eaters were adapted for eating plants, not fighting with predators, so great size was another adaptation that made sense. Being three to seven times bigger than one's predators makes attacks from the meat-eaters less likely to succeed. Of course, juveniles would be easier prey, so another strategy is to have rapid growth rates such that newborns can achieve large sizes quickly; this also requires a high metabolism. Studies of sauropod bones indicate that they grew relatively fast, certainly significantly faster than lizards, which would support higher metabolic rates.

In response to predation, Jurassic stegosaurs developed long tail spikes (Figure 12.25) and the later ceratopsians evolved large head shields (e.g., Figures 12.11 and 12.19) and impressive horns. These horns may have been used for defense against predators or for fighting for sexual mates. Most sauropods lacked armor, but they were the largest land animals ever seen on Earth and their sheer size would have made them difficult targets for predators (Figure 12.24). In addition, it has been speculated that some groups, and particularly the long-necked diplodocids, may have developed whip tails for protection. The ends of **Diplodocus** tails have very little muscle and are mostly bone and may have been

covered with keratin, the same tissue that covered ceratopsian horns, such as are found in modern rhinoceros. Given their enormous size, when a *Diplodocus* wagged its tail, the tip would have traveled at a speed perhaps exceeding the speed of sound. The supersonic crack may have served as a warning to predators, and there is some speculation that the tail may have been used as a giant whip – woe betide the unwary predator who may have felt its sting. There are also several genera of sauropods that evolved tail clubs and back spines, much like the smaller, and more heavily armored ankylosaurs, which also sported a wide variety of head and body spikes and impressive tail clubs (Figure 12.21). Lastly, herding behavior may also have provided protection especially for juveniles. There is good evidence of herding behavior in most of the larger herbivorous dinosaurs, such as sauropods, hadrosaurs, and ceratopsians, and this may have provided protection for juveniles. There is safety in numbers!

KEY POINT

Gigantism was driven by an increase in the size of predators, triggering a terrestrial evolutionary arms race. In response, herbivores increased in size: sauropods grew too large to be easy prey and may have developed whip tails, ceratopsians developed extensive head armor, and ankylosaurs developed body armor and giant tail spikes, as well as herding behaviors.

12.8.3 Dinosaurs and Plants

The largest dinosaurs, the sauropods, were plant-eaters. It is worth realizing that plants are not actually all that nutritious, so to sustain a homeothermic metabolism for a 50-ton sauropod requires eating a virtual forest for breakfast, lunch, and dinner. One of the requirements that allowed for gigantism must have been the wide availability of vegetation for the sustained herds of large herbivores at the bottom of the food chain.

Figure 12.26 Teeth from (a) *Edmontosaurus*, (b) *Triceratops*, and (c) modern elephant. All are adapted for grinding and chewing in contrast to peg-like teeth of the sauropods (see Figure 12.11) that relied on gastroliths to process low-nutrient vegetation. Source: (*Edmontosaurus*) The Natural History Museum / Alamy Stock Photo; (*Triceratops*) photo by Bradypus, CC BY-SA 3.0 <https://creativecommons.org/licenses/by-sa/3.0>, via Wikimedia Commons; (Elephant) Andrew Sole / Alamy Stock Photo.

A major revolution in plants was the development of fast-growing angiosperms and this may have driven some of the changes in the evolution of herbivores. Before the middle of the Cretaceous Period the dominant terrestrial plants on Earth were plants that reproduced by spores, such as ferns, and the seed-bearing gymnosperms, which grow relatively slowly (see Chapter 11). The Jurassic was dominated by sauropod high browsers that fed on these slow-growing, low-calorie gymnosperms, essentially acting as giant vacuum cleaners. Because of the low nutrition, chewing was largely avoided, and the heavy work of digestion was accomplished using gastroliths and massive stomachs. In the ensuing Cretaceous Period, we see the massive diversification in the fossil record of angiosperms, fast-growing flowering plants (see plant cladogram in Figure 11.1). Angiosperms first appeared in the Cretaceous and are abundant after that, but there is some recent fossil evidence indicating that angiosperms may have appeared in the Jurassic (about 164 Ma). Also, there have been some recent statistical analysis and molecular clock modeling that hypothesize an earlier appearance, perhaps at the end of the Permian Period, 250 million years ago. These newer studies are still controversial, and it seems clear that regardless of a possible early origin, angiosperms did not become dominant until the Cretaceous. In an environment in which dinosaurs are eating plants at a rapid rate, angiosperms have an advantage over gymnosperms in that they grow and reproduce much faster and are more nutritious. The Cretaceous was accompanied by an explosion in the diversity of hadrosaurs, ceratopsians, and ankylosaurs, all grazers that likely took advantage of the expansion of low-to-the-ground, fast-growing angiosperm shrubs versus slow-growing gymnosperm trees. Examination of the teeth of these grazers show arrays of grinding teeth (Figure 12.26), indicating a far greater reliance on chewing compared to the peg-like sauropod teeth (Figure 12.11). Of course, once the angiosperms appeared, they also diversified into larger trees that allowed some of the sauropods to persist, right through to the end of the Cretaceous, but high-browsing sauropods were far less abundant than the herds of grazing dinosaurs.

KEY POINT

The Cretaceous period was marked by the appearance of fast-growing angiosperms, including low shrubs and eventually tall trees. Many of the Cretaceous herbivores, especially hadrosaurs, ankylosaurs, and ceratopsians, were grazers taking advantage of the fast-growing angiosperms.

12.9 Dinosaur Diversity

We now have documented about 2,000 species of dinosaurs, and dinosaurs, as a group were highly diverse, lived across most of the land masses on Earth, and occupied many ecological niches, from jungles to deserts to the high arctic, which, as we will discuss below, was not as cold as today but still would have been dark for half the year. They ranged in size from small animals, not much bigger than a chicken (such as **Compsognathus**, Figure 12.14), to the gigantic sauropods, some of which were heavier than modern jets (Figure 12.24, recall that jets are mostly hollow). In addition to the dinosaurs, the flying pterosaurs (a type of archosaur) and the marine-dwelling ichthyosaurs, **mosasaurs**, and **plesiosaurs** (Figures 12.23 and 12.27) were certainly important members of aerial and marine environments. By almost any measure, the

Figure 12.27 Examples of extinct marine vertebrates: (a) *Tylosaurus*, a type of mosasaur, and *Protostega*, a giant marine turtle; (b) *Ichthyosaurus*; and (c) *Attenborosaurus conybeari*, a type of plesiosaur. Source: (a) Photo by JPB. With permission of Perot Museum, Dallas; (b) and (c) Photos by JPB. By permission of The Trustees of the Natural History Museum, London.

dinosaurs and their avian and marine relatives were hugely successful. They occupied most of the available ecological niches and lived across most of the planet.

KEY POINT

Dinosaurs were enormously successful, occupying a wide diversity of ecological niches across much of planet Earth.

12.10 Mammals in the Shadows

Although the Mesozoic Era is sometimes designated as the "Age of Dinosaurs," mammals and dinosaurs shared the Earth for over 100 million years (Figure 12.28). This section examines what they were doing during all that time.

12.10.1 On Being Small

During this time, mammals took a rather different strategy from dinosaurs. The smallest dinosaurs in the Mesozoic were the **microraptors**, many of them feathered and close relatives of modern birds (e.g., Figures 12.21 and 12.22). The smallest

modern birds are the bee hummingbirds, weighing about 2.6 grams, smaller than your pinky finger, but during the Mesozoic the smallest microraptors were about the size of a pigeon.

Mammals took the low road and evolved to be generally much smaller than dinosaurs. Mesozoic mammals were generally small, the largest about the size of a modern beaver, but most were the size of a rat or shrew. It seems clear that the mammals exploited niches that were closed to the larger dinosaurs. Mammals were largely nocturnal, which explains why most mammals do not have color vision. Color vision does not add much value to a night stalker, and therefore mammals developed keener hearing and smell. Primates are a notable exception, likely representing a later adaptation to life in the trees, where good 3D color vision is essential with a corresponding decrease in our olfactory systems (see Chapter 17). To this day, most mammalian carnivores hunt at dusk rather than midday. Perhaps if you own a dog, you will notice that it sleeps most of the day and is most active in the early evening, and this reflects its Mesozoic nocturnal heritage.

Mammals also developed the ability to hibernate and burrow. These are key advantages to keep out of the way of predators during the day, as well as to survive periods of drought or darkness, like modern bears that hibernate during the winter months. As we will see in Chapter 13, the ability to burrow and hibernate, which was helped by being small, allowed mammals to seek refuge from the calamity that ended the Cretaceous Period.

12.10.2 Mammalian Diversity

Cretaceous mammals were quite diverse and included insectivores, herbivores, omnivores, scavengers, and carnivores, the largest of which have been shown to have preyed on some of the smallest dinosaurs (Figure 12.28). There were even species that glided.

Major anatomical adaptations include simplification of the lower jaw bone, in which many of the lower jaw bones were transformed into the hammer and anvil that make up the ear bones and enhanced hearing (Figure 12.29). The lower jaw in mammals is entirely composed of the dentary jaw bone, which in the synapsid ancestors was the part of the jaw that contained the teeth. Mammals show a much higher level of differentiation in teeth compared to dinosaurs, including front incisors for chopping and cutting, canines for stabbing, and molars for crushing, grinding, and chewing. These enabled more complex chewing and feeding strategies that served increased metabolic rates. Anatomical modifications of the back of the jaw also allows for suckling of milk. Although a few mammals lay eggs, such as the monotremes, which include the duck-billed platypus, most mammals bear their young live.

Mammals were also endothermic and developed fur for thermoregulation. Mammals also, in general, evolved much larger

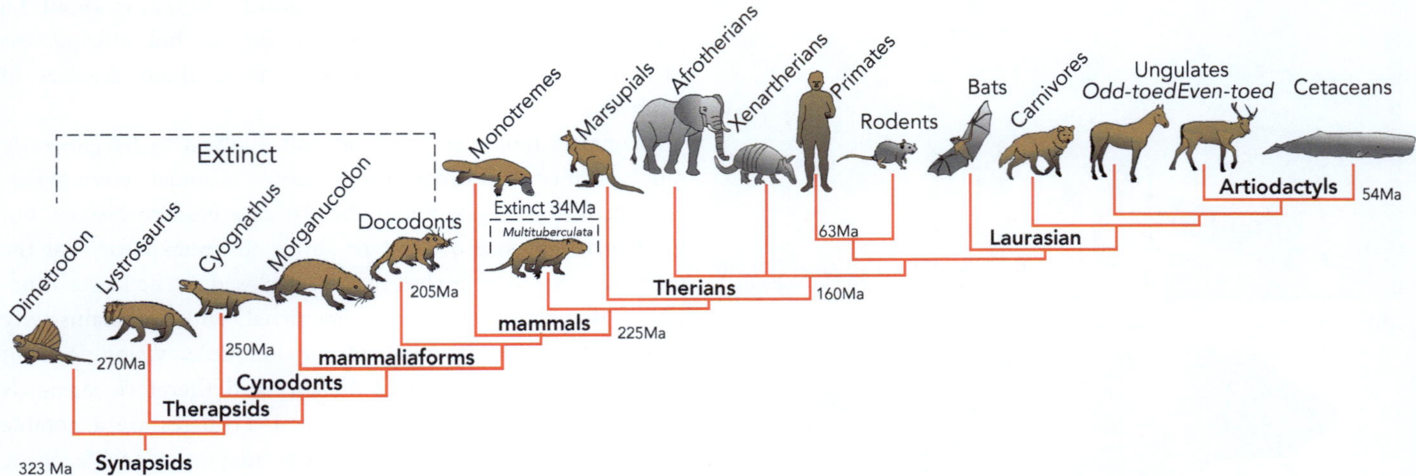

Figure 12.28 Mammal cladogram with approximate dates for appearance and extinctions of key groups. The therians (wild beasts) are the first mammals that bear live young and include the marsupials and placentals. The monotremes lay eggs but are an older sister group of mammals. The most recently evolved mammals are the artiodactyls, which postdate our own group of primates.

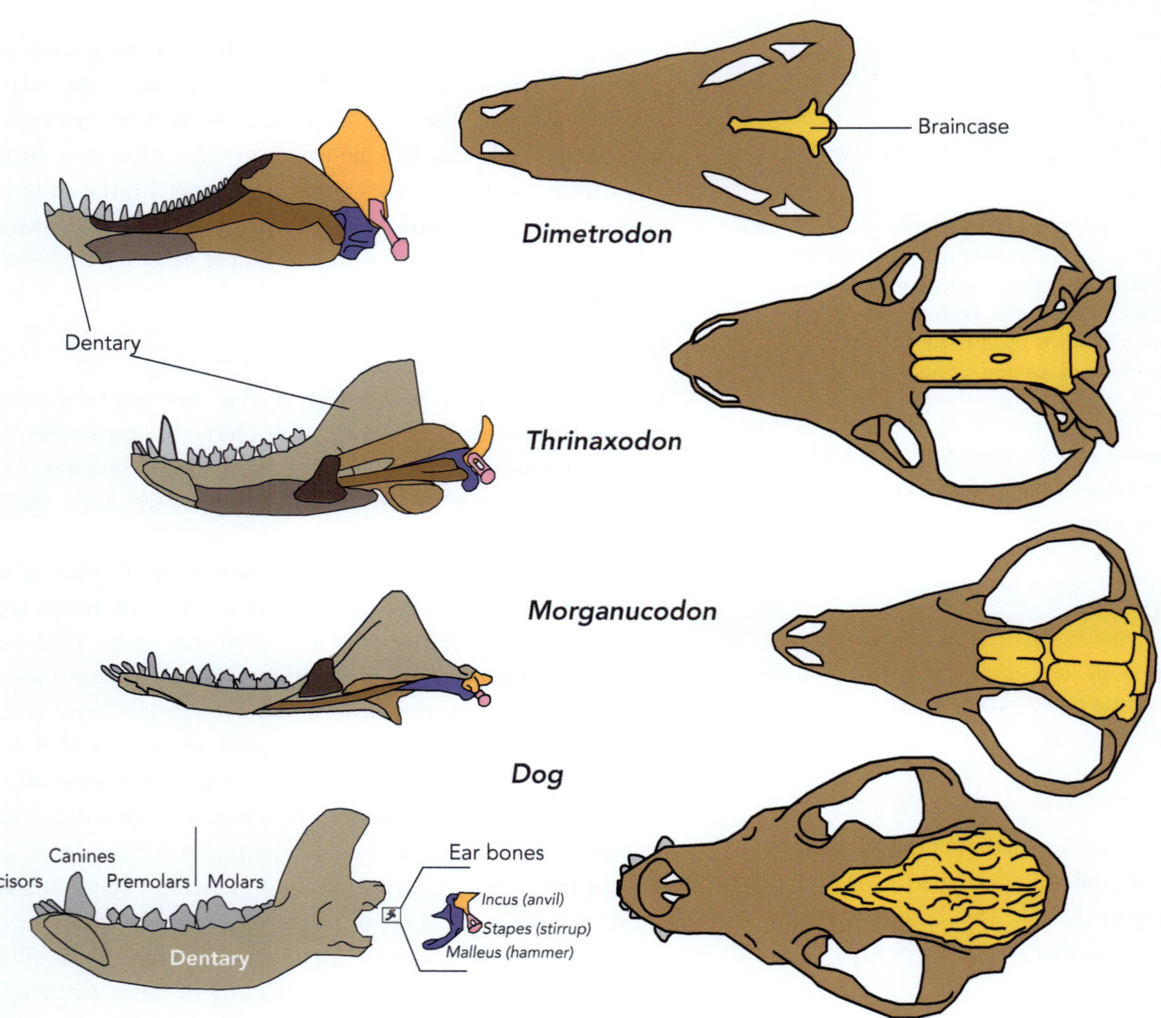

Figure 12.29 Evolution of skulls from early synapsids (e.g., *Dimetrodon*) to modern mammals (e.g., dog). Jaws become simpler, and the rear bones in *Dimetrodon*'s lower skull are eventually adapted to become the inner ear bones in modern mammals. There are massive changes in the relative size of brains versus body size, with much larger ratios in mammals, and an increase in the prefrontal cortex that forms the characteristic bulbous brain tissue. Mammals also show much greater differentiation of teeth that can be used for multi-functions including incisors for cutting, canines for stabbing, and molars for chewing and crushing.

brains for their body sizes, and particularly the prefrontal cortex, and they were almost certainly more intelligent than dinosaurs. However, despite their diversity they were not able to achieve the large body sizes of dinosaurs, and for that a calamity was required.

KEY POINT

Mammals and dinosaurs lived together for over 100 million years. Mammals took a different evolutionary path of small animals that were largely nocturnal, living in the unfilled ecological niches, thereby avoiding the larger areas dominated by dinosaurs.

12.11 End of an Era

There is no evidence of a gradual die-off of dinosaurs at the end of the Mesozoic, despite the fact that the atmosphere was getting a bit cooler. The Mesozoic certainly ended abruptly, and across a single sedimentary surface. Below this, an abundance of fossil dinosaurs can be found but scarcely a single dinosaur bone has been found in the layers above. Worldwide examination of this boundary tells the tale of a global catastrophe that wiped out the non-avian dinosaurs, as well as many other groups, including the ammonites, swimming archosaurs, and flying pterosaurs. Birds, mammals, crocodiles, reptiles, and other species somehow survived, and the causes of this mass extinction is the topic of our next chapter.

12.12 Summary

- The Mesozoic Era was characterized by the transition from arid-hothouse in the Triassic to tropical greenhouse in the Cretaceous Period as Pangea split apart and oceanic sea levels rose.
- The Late Cretaceous was generally humid and characterized by some of the highest sea levels known on Earth. Most of the continents were more or less in their present positions.
- It was initially difficult to reconstruct dinosaurs, as their bones are commonly poorly articulated due to massive cartilage that is not usually fossilized, and the fossil record was very incomplete.
- Triassic trackways showed that the first dinosaurs walked upright, with many bipedal species versus the sprawling gaits of earlier tetrapods. Bipedalism improved dinosaurs' agility and speed.
- Low predator–prey ratios, avian respiratory systems, the appearance of feathers (before they were used for flight), bone structures such as Haversian canals, as well as the recognition of the high agility of theropods suggest that dinosaurs had high metabolic rates and were likely endothermic (i.e., "warm-blooded").
- Feathers evolved in the theropod dinosaurs and some of their sister groups had filaments that could not have been used to fly. The evolution of flight occurred after feathers had already appeared, illustrating the idea that evolution commonly involves adapting older anatomical structures to new functions.
- Birds and dinosaurs are now thought to be closely related. Birds (Aves) evolved from the therapod dinosaurs, which included *T. rex* and *Velociraptor*, the latter of which was certainly feathered, but too large to fly. Smaller winged theropods, including *Archaeopteryx*, illustrate the transition between non-avian dinosaurs and birds.
- Gigantism was partly driven by an evolutionary arms race between giant theropods such as *Allosaurus* and *T. rex* and the giant herbivores such as the sauropods. The increase in temperature and humidity provided the extensive vegetation that the high metabolisms required. Avian respiratory systems, which enable efficient respiration, hollow bones that allowed cooling, as well as high metabolic rates that allowed high growth rates combined to allow the sauropods to become the largest terrestrial animals the Earth has ever seen.
- Dinosaurs were widely distributed and abundant, occupying a wide diversity of lifestyles and behaviors and appear to have been extremely successful. They dominated terrestrial landscapes for over 100 million years.
- Mammals diversified and occupied the unfilled niches by becoming small, and largely nocturnal during the so-called "Age of Dinosaurs."

Key Words

- Zuni sequence
- rudists
- dinosaurs
- *Iguanodon*
- taphonomy
- *Brontosaurus*
- *Camarasaurus*
- sauropod
- ceratopsians
- ankylosaurs
- bipedal
- hadrosaurs
- riparian
- gastroliths
- quadrupedal gait
- *Herrerasaurus*
- theropod
- *Plateosaurus*
- *Marasuchus*
- dinosauromorph
- *Eoraptor*
- ichthyosaurs
- amniotes
- monotremes

- ectotherm
- endotherm
- *Centrosaurus*
- *Triceratops*
- *Edmontosaurus*
- *Deinonychus antirrhopus*
- dromaeosaurids
- velociraptors

- Haversian canals
- archosaur
- saurischians
- ornithischians
- *Yutyrannus huali*
- *Sinosauropteryx*
- *Archaeopteryx*
- *Confuciusornis dui*

- pterodactyls
- *Quetzalcoatlus*
- *Tyrannosaurs rex*
- *Diplodocus*
- *Compsognathus*
- mosasaurs
- plesiosaurs
- microraptors

Further Reading and References

Bakker, R., 1986, *The Dinosaur Heresies: New Theories Unlocking the Mystery of the Dinosaurs and Their Extinction*, Zebra-Kensington.

Brusatte, S., 2018, *The Rise and Fall of the Dinosaurs: A New History of a Lost World*, Mariner Books.

Brusatte, S., 2022, *The Rise and Reign of the Mammals: A New History, from the Shadow of the Dinosaurs to Us*, Harper Collins.

Thulborn, R. A., 1989, The gaits of dinosaurs, in M. J. Benton D. D. Gillette, and M. G. Lockley (eds.), *Dinosaur Tracks and Traces*, Cambridge University Press, p. 39.

Review Questions

1. Describe the general conditions on Earth during the Mesozoic Era, and the transition from arid-hothouse in the Triassic to tropical greenhouse in the Cretaceous Period as Pangea split apart and oceanic sea levels rose.
2. Explain some of the pitfalls in early interpretations of dinosaurs.
3. Describe the first appearance of dinosaurs and their subsequent evolutionary history.
4. Discuss the evidence that dinosaurs were endothermic (i.e., "warm-blooded") including bone structure, predator–prey ratios, and other evidence.
5. Discuss the evolution of feathers and the evidence that they were not initially evolved for flight. In your answer, comment on how *Archaeopteryx* serves as an example of the transition between non-avian dinosaurs and birds.
6. Explain why birds and dinosaurs are considered to be part of the same phyletic group.
7. Explain why so many groups of dinosaurs, and especially the sauropods, evolved to gigantism.
8. Describe some of the adaptations, such as teeth styles, that herbivorous dinosaurs developed to live on plants.
9. Discuss the idea that mammals lived in the "nooks and crannies" of a world dominated by dinosaurs.
10. Why do many mammals lack color vision?
11. What are some of the unique adaptations of mammals?

Tyrannosaurus rex dinosaur escaping the heat and fire of a big meteorite crash. Source: Elena Duvernay / Stocktrek Images / Getty Images.

Chapter 13

The Cretaceous Extinction

The End of the Age of Dinosaurs

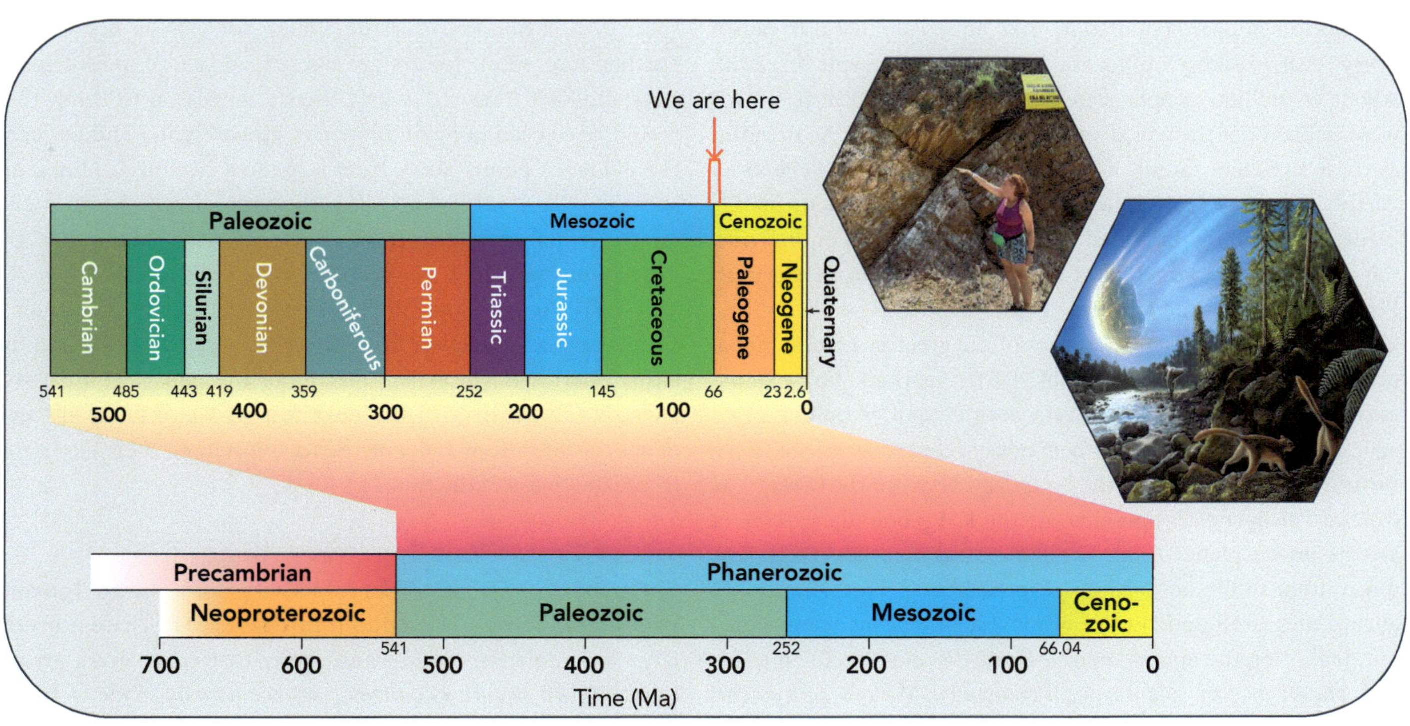

Introduction

This chapter is specifically focused on the events that occurred at the boundary between the Mesozoic and Cenozoic eras, commonly referred to as the K–Pg extinction because it marks the boundary between the Cretaceous (K) and Paleogene (Pg) geologic periods. It is the most recent of the five mass extinctions introduced in Chapter 9 and has been a topic of considerable public interest because it also marks the end of the "Age of Dinosaurs" (see Chapter 12). The two main hypotheses for the cause of the extinction center on a giant asteroid impact and massive volcanic eruptions in India.

13.1 Scales of Natural and Human-Made Disasters versus Mass Extinctions

If you ask Americans what the worst human-made disaster in the past 30 years was, many may mention the events of September 11, 2001, when al-Qaeda operatives hijacked four airliners, three of which were crashed into the World Trade Center in New York City and the Pentagon in Washington, DC. In total about 3,000 people were killed and it caused about $10 billion in property and other economic damage. In contrast, the worst natural disaster in the past 30 years was the Indian Ocean 9.1 magnitude earthquake on December 26, 2004. The resulting series of large waves, or **tsunamis**, it created devastated outlying coastal communities and killed an estimated 230,000 people in 14 countries. The Covid-19 pandemic of 2020–2021 had an even greater death toll, killing over five million people globally. For context, on average about 0.7% of the world's human population dies of mostly natural causes every year, adding up to about 60 million people in 2020. About 60 million people were killed in World War II, which was about 3% of the world's population at the time. As dreadful as these numbers sound, none of them come close to events in Earth's past that have killed up to 96% of all species on the planet, such as we discussed in Chapter 11, which ended with the greatest mass extinction of all time at the end of the Permian Period that ended the Paleozoic Era.

This chapter focuses on another of the greatest natural global catastrophes in the history of Earth marked by a **mass extinction** at the end of the Cretaceous Period, 66 million years ago. This extinction event is commonly referred to as the **K–Pg boundary** event marking the boundary between the Cretaceous (K) and Paleogene (Pg) periods. At the K–Pg boundary, 75% of all species on planet Earth became extinct (see Box 13.1). The devastation to life on Earth appears to have been astoundingly abrupt and profound. Entire groups of organisms were wiped out, including the non-avian dinosaurs, discussed in Chapter 12, and many insects, plants, and mammals. Marine ecosystems were also devastated, with large and diverse groups such as the ammonites and belemnites (Figure 9.29) disappearing completely. These are the extinct relatives of the modern nautilus,

which in turn belong to the phylum Mollusca introduced in Chapter 9. Even groups with eventual survivors, such as the bony fishes and corals, were devastated. Despite the scale of extinctions, a significant number of groups survived, such as omnivorous and scavenging mammals, crocodiles, turtles, fish and amphibians, birds (avian dinosaurs), and many invertebrates (Box 13.1).

The most obvious question about this mass extinction is what caused it? Other questions we might ask are how fast did the extinction occur and how long did it take for ecosystems and life on Earth to recover? Why was the extinction selective, killing off entire groups, yet allowing others to survive? We should also all be interested in knowing the likelihood of such an event occurring again and if there is anything we could do to prevent it. Knowledge of global catastrophes, such as mass extinction events, should provide a perspective for understanding the impact of modern-day global environmental change, especially those that have human-made, or **anthropogenic**, causes (see Chapter 19). We can also compare the K–Pg extinction to the other five mass extinction events, including the end-Permian extinction that was a focus of Chapter 11.

13.2 Cretaceous Extinction Hypotheses

Over the years, many hypotheses have been suggested for the extinction of the dinosaurs. One hypothesis is that environmental changes, such as the gradual cooling of the global climate and falling sea levels in the late Cretaceous, changed habitats, which gradually caused species to go extinct. Another suggests that the flowering plants (angiosperms), which appeared about 100 million years ago, may have poisoned herbivorous dinosaur species to extinction, but the timing for that is not correct. Furthermore, most flowers are not, by and large, poisonous to anything that lives today and there's no reason to think they would have been toxic to dinosaurs. However, by and large, all the evidence points away from a gradual decline in dinosaurs and their ilk leading up to the extinction event because of a more gradual environmental shift, and it appears that the extinction was about as abrupt as it gets: likely less than 100 years. The two most viable hypotheses that can explain such a sudden mass extinction are that it was primarily caused by either enormous volcanism in India or the impact of a massive meteorite in Mexico. In the next section we examine the evidence for volcanism and a meteorite impact and their likely role in causing mass extinction.

13.2.1 Cretaceous Super-Eruptions

Exceptional times of volcanic activity form **large igneous provinces (LIPs)** as introduced in Chapter 11. These are very large accumulations of volcanic rocks that cover areas greater than 100,000 square kilometers and are usually made of basalt, erupted within a geologically brief period, typically a few million years or less. They are thought to be fed by upwelling of the mantle. The Deccan Traps in India form an LIP made up of

BOX 13.1 Who Went Extinct and Who Survived?

Although 75% of all species went extinct at the end of the Cretaceous Period, the extinction was selective. In some cases, entire groups went extinct, whereas in other phyletic groups, variable numbers of species survived. This box summarizes the main groups that went extinct, as well as those that included survivors.

Land Vertebrates

Non-avian dinosaurs: 100% extinction. Note that the dinosaurs lived for millions of years, and many species went extinct long before the K–Pg boundary, which is why Jurassic dinosaurs are different from Cretaceous dinosaurs. However, these previous extinctions were dispersed over time and only affected individual genera and species. They were also largely compensated by changes in the abundance of other species or the appearance of new species so that higher-order clades, particularly the Ornithischian and Saurischian families, persisted throughout the Mesozoic, right up until the K–Pg boundary when all remaining families went extinct (see Chapter 12 for more details).

Turtles: 20% extinction.

Crocodyliforms: about 50% survived.

Lepidosaurians (snakes, lizards, and tuataras): largely survived.

Pterosaurs (non-avian flying reptiles): these were in long decline before the K–Pg event, but the remaining species were all eliminated at the K–Pg boundary.

Mammals: although many species suffered significant extinction, their small size, and burrowing habits, allowed many to survive, particularly those that were omnivorous or scavengers.

Birds (avian dinosaurs): we do not have a detailed fossil record for birds, but post-Cretaceous records indicate that many species went extinct. However, burrowing and aquatic lifestyles allowed enough to survive so that they were able to diversify and become widespread today.

Marine Vertebrates

Mosasaurs and plesiosaurs: 100% extinction (note ichthyosaurs were dominant in the Jurassic Period and went extinct during the mid-Cretaceous, about 25 million years before the K–Pg event). Their extinction may be related to several oceanic anoxic events and associated climatic changes that occurred in the mid-Cretaceous.

Fish: 10% of families of bony fish (teleosts) and sharks went extinct, whereas only about 15% of skates and rays survived.

Marine Invertebrates

Corals: 60–98% extinction.

Ammonites: 100% extinction.

Echinoderms: 35% extinction.

Rudist and inoceramid bivalves: 100% extinction.

Terrestrial Invertebrates

Insects: significant extinction, as determined from the proportion of insect damage observed in fossil leaves. Diversity appears to have remained low for about 1.7 million years after the extinction.

Plants and Fungi

Land plants: about 57% extinction; survival flora included ferns and non-photosynthetic fungi.

---

basaltic lavas (also called **flood basalts**) that presently cover an area of 500,000 km², an area larger than all of the states of California, Nevada, Oregon, and Washington combined or the entire country of Spain. As we mentioned in Chapter 11, the word "trap" is derived from the Swedish word for "stairs" and refers to the step-like terraces made up from individual lava flows that built up the 2,000 m-thick formation. At the time of formation, the Indian plate had not yet collided with the rest of Asia (see Chapter 15) and India was in the southern hemisphere traveling northwest over a mantle plume (Figure 13.1). The basaltic flows may have originally covered an area of 1.5 million km² (Figure 13.1).

The key to our present story is the age of these particular basalts. They began forming about 67.5 million years ago, about one million years *before* the end of the Cretaceous Period. The eruptions involved several pulses of activity with a small pulse at 67.5 Ma, a second large pulse at about 65.8 Ma, and a third and final pulse at about 64.8–64.6 Ma in the ensuing Cenozoic Era. In addition to the lava, these eruptions released significant volumes of volcanic gases, including carbon dioxide (CO_2) and sulfur dioxide (SO_2). SO_2 reacts in the atmosphere to form sun-blocking **aerosols**, suspensions of fine solid particles or liquid droplets that form in the air. These aerosols have the opposite effect of greenhouse gases and likely caused global cooling of about 2 °C. The release of CO_2, as well as acid rain produced when SO_2 dissolves in the atmosphere to form sulfuric acid, would have also caused the oceans to have become more acidic (see Chapter 11) and this would have severely stressed the marine ecosystems of the time. The volume of CO_2 released in the eruptions was relatively low and it is thought that the atmospheric cooling effects of SO_2 outweighed any warming effects from the heat-trapping capacity of the released CO_2 (see

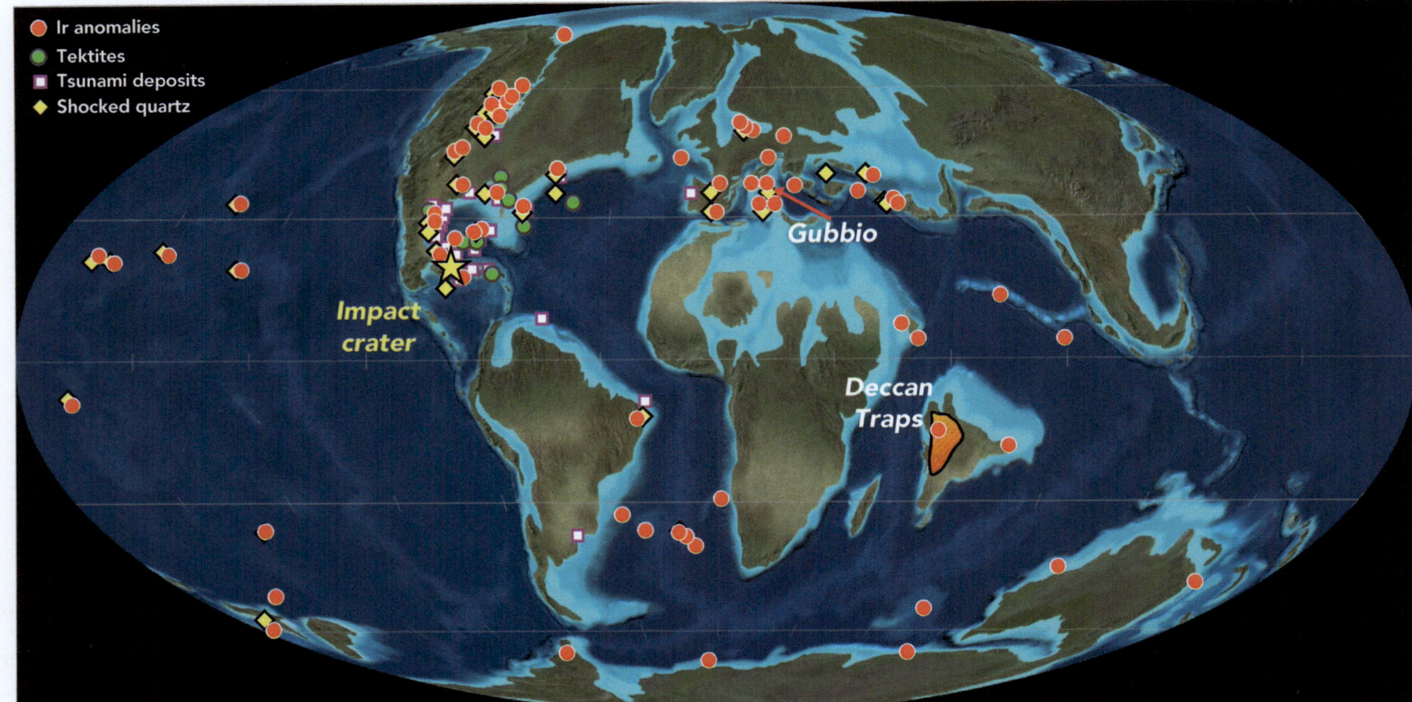

Figure 13.1 Paleogeographic reconstruction of the location of the continents at the end of the Cretaceous Period, at 66 Ma. Also shown are the location of the impact crater in the northern Yucatan (yellow star) and features observed at the stratigraphic boundary between Cretaceous and Paleogene rocks, including an anomalous concentration of iridium (red circles), tektites, or glass spherules produced by the melting of rocks during the impact (green circles), shocked or impact-metamorphosed quartz (yellow diamonds), and sedimentary deposits interpreted to be laid down in giant tsunami waves (white squares). The location of the Deccan Traps, a thick sequence of basalts that erupted in western India at about 66 Ma, is also indicated. Source: © DeepTimeMaps.

Chapter 19). The effects on global ecosystems were certainly significant. Some scientists have proposed that the size and scale of these eruptions may have been sufficient alone to trigger the mass extinction that terminated the Mesozoic Era, in a similar way to the Siberian Traps that triggered the extinction event at the end of the Paleozoic Era (see Chapter 11). There is evidence that the first pulse triggered a global cooling, resulting in diminished sizes of marine plankton, and the mass extinction seems to coincide with the second large pulse. However, as we shall discuss in the next section, volcanoes, lavas, and their gases are only part of the story.

13.2.2 Extraterrestrial Impact

The impact hypothesis suggests that a massive meteorite (asteroid or comet, see Box 13.2) collided with Earth at about 66 Ma and triggered a mass extinction. The evidence for a major impact was discovered by accident using geochemical and geophysical analysis and is the story we tell in this section.

13.2.2.1 Discovering the Iridium Anomaly

In the 1970s, Walter Alvarez, a geology professor at the University of California, Berkeley, was investigating sedimentary rocks at the Mesozoic–Cenozoic boundary. One of his field sites focused on the cream-colored marine limestones, named the Scaglia Formation, exposed in roadcuts just outside the small town of Gubbio, Italy (Figure 13.2). Here the K–Pg boundary is marked by a distinctive, thin clay layer within the Scaglia Formation (Figure 13.2). Alvarez thought that this section of limestone recorded a relatively continuous record of deposition across the boundary without any significant unconformities or changes in the rate of deposition. Consequently, he expected that analysis of the section could help determine the average rate at which the sediments were originally deposited.

One straightforward way to determine the rate of sedimentation is to know the time of deposition of two points in a section of rocks and then divide the thickness between these two rocks (in mm) by the difference in age between the two rocks (in years) to obtain a rate (mm/year). But determining the age of deposition of sedimentary rocks is not often easy, particularity with the precision necessary to make this sort of calculation useful (see Chapter 4).

In the case of the Gubbio section, none of the standard methods for precisely determining the time of deposition (described in Chapter 4) were viable for the sedimentary succession. There were no volcanic layers that could provide absolute ages, the section did not have a good paleomagnetic record, and **biostratigraphy** would not have provided ages that were precise enough to determine sedimentation rates. This required Alvarez and his colleagues to come up with a new method.

BOX 13.2 Asteroids, Comets, Meteors

There is some debate as to whether the K–Pg impactor was an asteroid or a comet.

Asteroids are small, rocky or metallic bodies orbiting the Sun. Most are members of the asteroid belt that lies between the orbits of Mars and Jupiter (see Chapter 3). Comets are small extraterrestrial bodies that contain a significant proportion of ice, but also contain dust and rocky material formed during the early evolution of the solar system. Unlike asteroids, comets have an atmosphere derived from its icy material that also produces the comet's tail. Comets have highly elliptical and more irregular orbits. Some comets originate in the Kuiper Belt beyond Neptune, whereas others originate in the Oort cloud outside the solar system.

When a comet or asteroid, or a fragment of either, enters Earth's atmosphere, the fireball or "shooting star" it produces is called a meteor. In the event that any material makes it to Earth's surface, it is called a meteorite. These bodies can also be distinguished by size: a meteoroid is the broad name given to extraterrestrial objects smaller than 1 meter in diameter that are either ejected from comets or are fragments from an asteroid. The general term impactor is used in astrogeology – the geology of bodies in the solar system – to describe any object that collides with enough force to produce a measurable effect.

Figure 13.2 Photograph of the sedimentary rocks at the K–Pg boundary in Gubbio, Italy.

Recall that they thought they were dealing with a section characterized by a continuous record of sedimentation with a relatively uniform rate. They therefore thought that measuring the concentration of the **iridium** in these rocks could help them work out the rate of sedimentation.

Iridium (Ir) is a member of the platinum group of elements and was present (although in small concentrations) on Earth at the time of its formation. However, because Ir chemically behaves very much like iron, most of the Ir on Earth was drawn into the core, along with iron, during the first few hundred million years of Earth history (see Chapter 3). Ir is therefore *extremely* scarce in terrestrial crustal rocks, with concentrations typically much less than one part per billion (ppb). Conversely, Ir is relatively common in **iron meteorites (chondrites)**, with concentrations of several hundred parts per billion, or up to several thousand times more than in rocks at the Earth's surface (see Box 13.3 for more discussion of the chemical composition of meteorites).

Most meteoroids disintegrate into dust as they pass through our atmosphere (i.e., they do not produce a meteorite; see Box 13.2) and the number of these does not vary much from year to year. Therefore, there is a more-or-less constant rain of extraterrestrial **cosmogenic** dust landing on Earth's surface, which is highly enriched in Ir relative to rocks at the surface.

So, Alvarez and his team reckoned that if there was a constant rain of Ir falling out of the sky and very little Ir coming from anywhere else, the only reason for changes in the concentration of Ir in sedimentary rocks would be the time it took to deposit them. In other words, if it took a long time to deposit a centimeter of rock there would be more Ir in that layer than if the layer was deposited quickly (Figure 13.3). This would not tell the absolute rate (in mm/year) because we are not sure how much Ir enters the atmosphere every year, although we assume that the rate does not change very much. However, this method could pick up relatively subtle variations in sedimentation rate on a fine scale. The Alvarez group was expecting concentrations of Ir on the order of 0.1 parts per billion (ppb) when sedimentation was slow and values at essentially zero when sedimentation was fast.

And so, with these thoughts in mind, the stage was set for the Alvarez team to use this technique to determine the variation of sedimentation rate in the Gubbio section. In the late 1970s it was quite a technical challenge to measure Ir at ppb concentrations. Walter Alvarez sent samples from the Gubbio section to his father, Nobel Prize-winning physicist Luis Alvarez, who had been working on new methods to conduct this type of low-concentration chemical analysis. The results of the analyses are shown in Figure 13.4. Samples from the uppermost Cretaceous rocks at Gubbio have predictably low Ir concentrations (less than 0.1 ppb) but at the K–Pg boundary, which happens to coincide with a layer of clay about 1 cm thick, the Ir concentrations shoot up by a factor of about 300, with concentrations around 9–10 ppb. Above the clay layer, the concentrations of Ir fall back to more normal levels, but not quite as abruptly as the initial increase (Figure 13.4).

Now, following the logic used to set up this series of analyses, the high concentration of Ir in the clay layer would suggest an extreme slow-down in the rate of accumulation of sediment. For example, if the 1 cm of limestone just below the clay layer took 10,000 years to deposit, then the clay layer above would have taken three million years to accumulate. The Alvarez group

BOX 13.3 Siderophile Elements and Meteorites

A key piece of evidence in favor of the giant impact hypothesis for the end K–Pg extinction event is the far higher concentration of iridium (Ir) in meteorites than in the upper crust of Earth. Why is this important? The answer comes from the geochemical characteristics of Ir and the evolution of the meteorites. Ir is a siderophile element, which means it follows iron (Fe) in chemical reactions. Thus, any mineral or rock that has a higher-than-average Fe concentration will also have a higher-than-average Ir concentration. One difference between Fe and Ir is there is much more Fe in the solar system. The initial accretion of interstellar dust that eventually became Earth produced a homogeneous protoplanet. Much was the same for all the other planets, including one that occupied an orbit in between Mars and Jupiter. This planet has been named *Phaeton*, but you may not have heard this name before because Phaeton didn't last long. During the early evolution of the solar system, collisions between planetary bodies of all sizes were common. Today we find between Mars and Jupiter not the large planet Phaeton, but tens of thousands of asteroids and smaller meteoroids (extraterrestrial objects smaller than 1 m in diameter). This is the result of a collision between two bodies that was so energetic that the

pieces were blown so far apart from each other that they were unable to subsequently coalesce into a single planet. This collision took place before Phaeton had evolved enough for the elements to separate out into different layers. These smaller pieces were then too small to evolve in the way Mars, Earth, and Venus did; the asteroids are frozen samples of what planets in our solar system looked like very early in their evolution. Earth also experienced a large collision during this early period. This collision produced our Moon but was not so energetic that it flung material so far away that it could not be brought back together by gravity (see Chapter 3). Thus, after the Moon-producing collision, Earth evolved via igneous differentiation, or the crystallization of different minerals, to have an iron-rich core, a magnesium-rich mantle, and a silicon-rich crust. The formation of the core took several hundred million years after the Moon-forming collision. There wasn't much Ir in the Earth to begin with and now most of it is with most of the Fe in the core. Because Phaeton got broken up before it could undergo this differentiation, the pieces that remain (the meteorites that occasionally come to Earth) have a significantly higher concentration of Ir than crustal rocks on Earth.

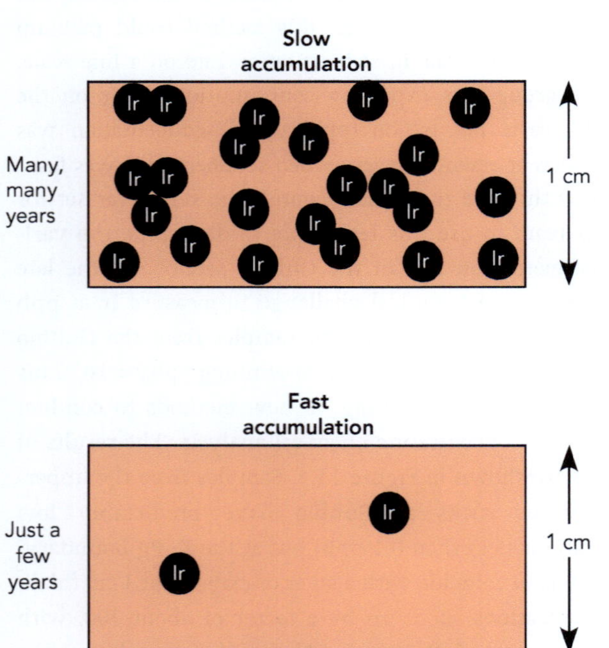

Figure 13.3 Illustration of the relationship between accumulation rate and Ir concentration, assuming a constant flux of Ir. The black circles represent extraterrestrial Ir; the tan background represents a sample of limestone 1 cm thick. If the flux of Ir from meteorites is constant, fast accumulation of the lime mud would leave very little time for Ir to fall from space and therefore, the concentration of Ir in the sample would be low. On the other hand, if the accumulation of the lime mud was slow, more Ir would fall within the time it took to deposit 1 cm of material.

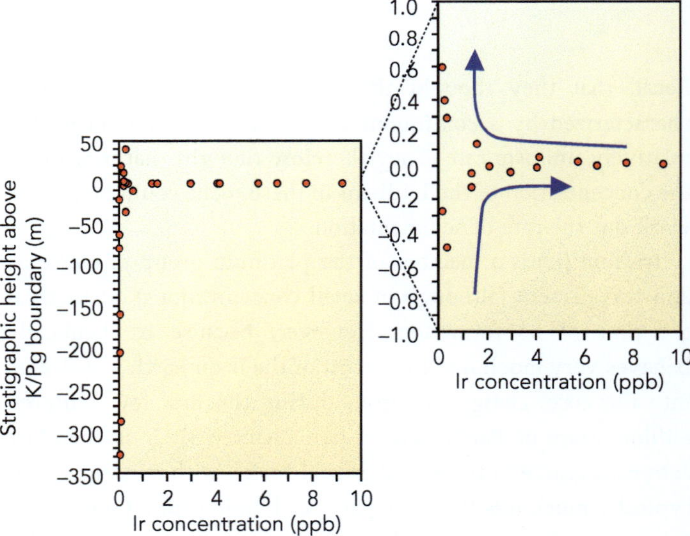

Figure 13.4 The concentration of Ir vs. stratigraphic height at Gubbio, Italy. Source: Plotted from data in Alvarez et al. (1980).

immediately realized that this did not make sense. Based on an understanding of the way sediment accumulates, the accepted magnitude of variation for rates of marine limestone deposition might be as much as a factor of 5 or 10, but not 300. In addition, biostratigraphic analysis showed that the rocks directly above the clay layer weren't millions of years younger, as would be required if sedimentation had fallen to a level slow enough for such a high concentration to occur, assuming a normal background level of Ir accumulation. So, the Alvarez group had to rethink their assumptions. Maybe the reason for the changes in Ir concentration was not because of changes in the rate of accumulation of sediment but because of changes in the rate at which Ir was added to the atmosphere by extraterrestrial material.

They concluded that the easiest way to explain this enrichment of Ir would be from the pulverized remains of an asteroid or comet that produced a large iron meteorite. An extremely compelling aspect of this hypothesis is that it not only explains the extreme Ir enrichment in the clay layer at the K–Pg boundary (and the presence of the clay layer itself), but it also provides a mechanism for the extinction of the dinosaurs and dozens of other life forms at the same time.

A prediction of this **giant-impact hypothesis** is that the Ir enrichment observed at the K–Pg boundary in the Gubbio section in Italy should be found at other locations – an impact that was big enough to kill off a variety of life on all the continents and in all the oceans would not just leave Ir in one place. This led to a worldwide hunt for these Ir anomalies. The search paid off. We now know of dozens of instances of extreme Ir enrichment located precisely at the K–Pg boundary. These sites are found in a wide variety of sedimentary environments from continental to the deep ocean (Figure 13.1); most of these instances of Ir enrichment are associated with a thin layer of clay, as originally observed at Gubbio (Figure 13.2).

Assuming that a 1 cm-thick layer of this Ir-rich clay was deposited everywhere on Earth's surface, and knowing the average Ir concentration in chondritic meteorites, the Alvarez group were able to estimate that the object responsible for the impact and subsequent clay layer must have been about 10 km in diameter.

KEY POINT

The analysis of iridium in sedimentary rocks was pivotal in the development of our understanding of the events at the end of the Cretaceous Period but the initial motivation for this analysis was for another reason. Faced with unexpected data, scientists developed a novel hypothesis, unrelated to the original research.

13.2.2.2 Additional Evidence of an Extraterrestrial Impact

Once an explanation for the Ir anomaly was presented, scientists began looking for other evidence of an extraterrestrial impact at the boundary, and for this they were well rewarded. The K–Pg boundary layer clays also included grains of shocked quartz, microscopic examination of which showed evidence of changes in the crystal structure diagnostic of a sudden and extremely high-pressure but relatively low-temperature impact (Figure 13.5). Quartz crystals do not have natural planes of weakness (cleavage) and most natural quartz grains and crystals show a relatively uniform character under a microscopic thin

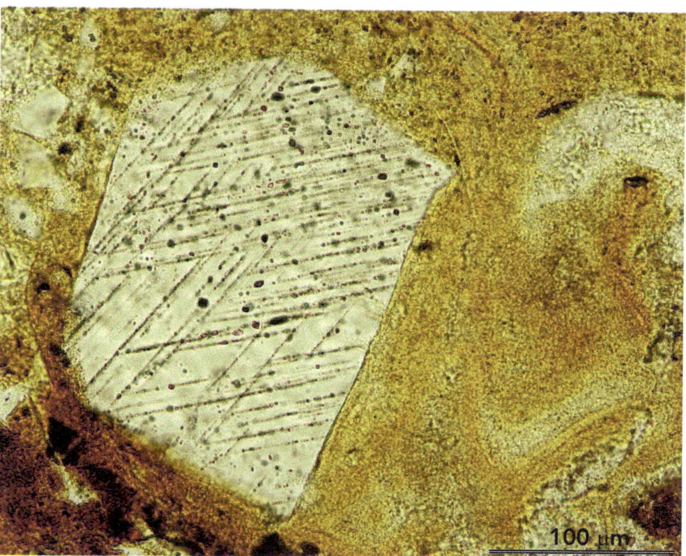

Figure 13.5 Image of shocked quartz in thin section. The grain is approximately 0.5 mm across the lamination seen here are produced when quartz is subjected to impact metamorphism. Source: Martin Schmieder, CC BY 3.0 <https://creativecommons.org/licenses/by/3.0>, via Wikimedia Commons.

section. The shocked quartz crystals contained in the K–Pg boundary clays, however, showed clear evidence of **shock metamorphism**, exhibited as linear streaks in the crystal structure (Figure 13.5). Under most geological processes that metamorphose igneous and sedimentary quartz, the high temperatures result in less **brittle deformation**, and these linear crystal features do not form. Shocked quartz crystals were first discovered associated with the rocks surrounding underground nuclear bomb test sites and later found in the 50,000-year-old Barringer impact crater in northern Arizona. The abundant shocked quartz provided additional evidence of a large impact.

In addition, glass spherules were also found in the clays (Figure 13.6). These spherules were interpreted as **tektites**. Tektites are spherical, elongate or teardrop-shaped glass beads which are formed by falling drops of molten rock that have been melted by an impact. They look similar to molten particles ejected into the air by volcanoes but exhibit a different geochemistry, have a much lower water content, and do not contain micro-crystals, which are common in cooling lava. Whereas Ir anomalies have been observed worldwide, the occurrence of tektites and shocked quartz were found at K–Pg locations in North America and the Caribbean region but were absent in Asia, Africa, and Europe (Figure 13.1).

13.3 The Chicxulub Impact Crater

So, with all that evidence for a massive impact, the next obvious question is, where did it hit? Given that 71% of the Earth is covered by oceans, the likelihood that the impactor that caused the extinction hit a continent, rather than an ocean, is only about 30%. The search for a crater 66 million years old is also hampered by the fact that the surface of Earth is dynamic and undergoes weathering and erosion. Many terrestrial

Figure 13.6 Tektites found from the K–Pg boundary in NW South Dakota. Source: DePalma et al. (2019).

Figure 13.7 (a) and (b) Impact rock. In (c) the green melt phase intrudes gray impact breccia with large white clasts. (d) Shatter cones. (a) and (d) are from the 1.85-billion-year-old Sudbury Crater. Source: (b) and (c) Morgan et al. (2017).

geoscientists working for Pemex, the Mexican national oil company, in their search for buried oil and gas deposits. Pemex drilled an exploration well in the 1950s into the rock in the northern part of the Yucatan Peninsula. They described encountering what they thought at the time was **andesite**, a type of volcanic rock commonly associated with volcanic island arcs along subduction zones (see Chapters 1 and 5). However, it does not make sense to find andesitic volcanic rocks in the tectonically inactive Gulf of Mexico, far from any subduction zone. It wasn't until 1990 that scientists were able to re-examine the samples, and in 2016 a new expedition of the International Ocean Drilling Program (IODP), led by Sean Gulick at the University of Texas and Joanna Morgan at Imperial College London, cored over 1.3 kilometers into the crater. The core samples (Figure 13.7) showed abundant evidence of the impact, including shocked quartz, tektites, a specific rock called **impact breccia** (breccia literally means "broken rock"), melt rock produced by the impact that looks similar to volcanic rocks, as well as shatter cones (Figure 13.7(d)), cone-shaped features uniquely formed by high impacts. Within a decade of hypothesizing that something large had hit the Earth, a candidate for the target location had been identified.

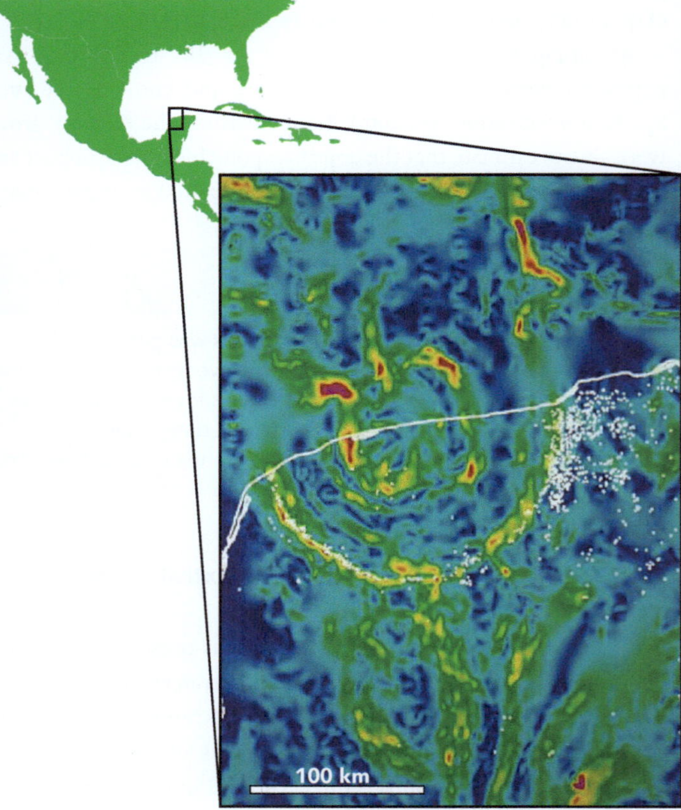

Figure 13.8 Gravity anomaly of the northern Yucatan Peninsula shows a circular feature that is interpreted as the Chicxulub impact crater. White dots are oil wells, drilled by the Pemex oil company. Source: Public domain, via Wikimedia Commons: https://commons.wikimedia .org/wiki/File:Chicxulub-Anomaly-Grav-3.jpg

environments are covered by vegetation and, worse, many surfaces that would have been exposed 66 million years ago are now covered by younger rocks and sediment. And if it hit an ocean, 66 million years has elapsed for the crater to be subducted into the mantle and lost forever.

But, in fact, we think we know where the impact took place. The crater was originally discovered by accident by

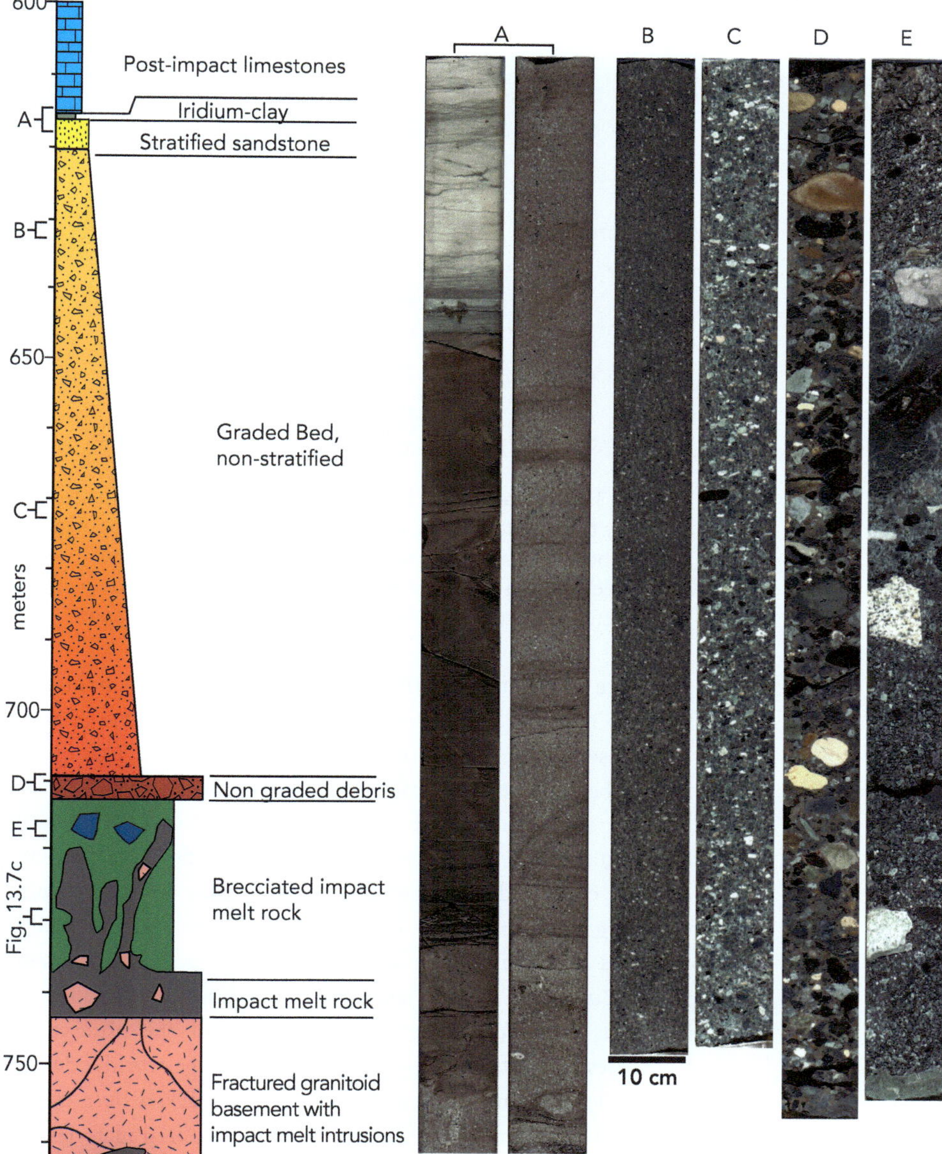

Figure 13.9 IODP core through the Chicxulub peak ring shows fractures and partially melted basement rocks overlain by a 100 m-thick graded bed. The graded beds were deposited as the ocean flowed back into the crater shortly following the impact. Source: Morgan et al. (2017).

Geophysical surveys, including gravity and magnetic data, were soon conducted both onshore and offshore of northern Yucatan. These studies revealed a huge ring-like structure that extended off the shore of the Yucatan Peninsula, near the village of **Chicxulub** (Figure 13.8). The crater is buried by about 1 km of younger Cenozoic rocks. The gravity data showed two rings, the outer one of which is 180 km in diameter. It is now thought that this outer ring marks the location of the inner wall of the crater, which is now estimated to have had a diameter of about 300 km in total. The inner **peak ring** is characteristic of large impacts, and Chicxulub is the only impact crater on Earth with a preserved inner peak ring. The crater is about 20 km deep and indicates that the impact disrupted a significant thickness of the Earth's crust. The 2016 drilling of the peak ring encountered rock that was produced by melting due to the force of the impact (the previously mis-identified andesite), fractured granite, as well as tsunami deposits and other debris associated with the impact (Figure 13.9). Modeling of the impact suggests an event of overwhelming magnitude (Figure 13.10). With a velocity of perhaps 12 km/s the impactor would have caused instantaneous melting of several kilometers of the target rock and vaporization of the impactor within a few seconds. After only one minute, a crater perhaps 80 km in diameter and 20 km deep would have formed but the energy going to produce this monumental feature would rebound and, in a few minutes, the depth of the crater would be substantially smaller.

Chicxulub is one of the third largest impact craters that have been discovered on Earth. The two largest known impact

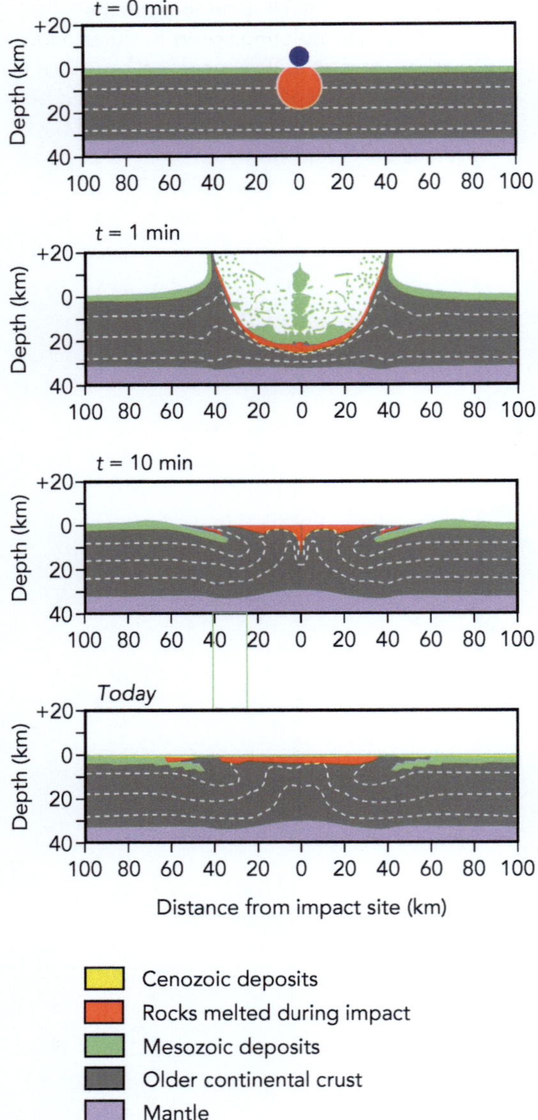

Figure 13.10 Result of numerical modeling of an impactor 10 km hitting continental crust at a velocity of ~12 km/s. Source: Based on Morgan et al. (2016).

Figure 13.11 Aerial view of Barringer Crater, Arizona. This crater was originally thought to be volcanic but investigation of the rocks in and around the crater shows evidence for shock metamorphism associated with the impact of a meteorite. The crater is approximately 1.2 km across. Source: Chris Saulit via Moment / Getty Images

structure and as tektites (Figure 13.6) in other K–Pg boundary sections in the western hemisphere (Figure 13.1). One of the best-preserved sections with these tektites is in Haiti. This material has been dated at 66.04 ± 0.05 Ma. The impact breccia from the Chicxulub crater has been dated at 65.91 ± 0.10 Ma, a result indistinguishable from the tektite value. The synchronicity of these dates and the observation that tektites are only seen at the K–Pg boundary in the Caribbean and North America clearly indicates that the tektites are related to the now-buried crater in the Yucatan. The clay layer lies at the boundary defined by fossils as the end of the Mesozoic Era, and the observation of tektites suggests the clay layer is the same age worldwide. All this evidence points to a single brief event, marked by a many-thousand-fold increase in the flux of extraterrestrial material *and* the extinction of millions of species of organisms, both on land and in the ocean (Box 13.1). The discovery of the impact crater demonstrated that a massive impact occurred and likely played a major role in triggering a mass extinction.

KEY POINT

The initial evidence for the giant-impact hypothesis for the extinction of the dinosaurs came from geochemical and stratigraphic oddities deposited around the globe; the crater produced by the impact was not discovered for many more years because it was buried beneath thousands of meters of rocks deposited over the subsequent 65 million years.

structures were both produced around 2 Ga and include the 1.85 Ga Sudbury Crater in Canada (Figure 13.7). The impacts associated with these craters would have caused devastation even greater than that associated with Chicxulub, but two billion years ago the most complicated life forms on Earth were single celled (see Chapter 7). In contrast, 66 million years ago Earth was rich with multi-celled organisms – trees, plants, and vertebrates (including the dinosaurs). The well-known 40,000-year-old Barringer Crater in Arizona, by comparison, is quite small (about 1.2 km in diameter versus up to 300 km for Chicxulub), lacks a peak ring structure (Figure 13.11), and did not trigger a mass extinction.

A key question in tying the impact that produced the Chicxulub crater to the extinction event is whether they occurred at the same time. The force of the proposed impact would have created new rocks via melting, and these melted and cooled rocks are well represented in the Chicxulub

13.4 Magnitude and Immediate Consequences of the Extraterrestrial Impact

Regardless of whether it was an asteroid or comet, the impact of a 10 km-diameter object is estimated to have been about

10 billion times more powerful than the energy released by the atomic bomb used at Hiroshima at the end of World War II. This impact is estimated to have pulverized, melted, and displaced about 200,000 cubic kilometers of rock, as well as a similar volume of seawater. By comparison, the volume of the impactor itself was only about $500\,km^3$. The superheated ash, dust, and steam cloud injected into the atmosphere would have likely covered the entire Earth for years and possibly decades. In the early days, as this material collapsed back to Earth, larger particles would have been heated to incandescence. This incandescent cloud would have produced an effect quite like a broiler oven and would have literally cooked much of the Earth's surface to several hundred degrees. Much of the material cooled quickly, forming a rain of hot glass now preserved as tektites. The hot ash and glass would have ignited extensive wildfires. Because the impactor hit an area rich in carbonate rocks, these rocks would have been vaporized, causing a corresponding sudden injection of CO_2 into the atmosphere. Analysis of the IODP cores shows a lack of evaporates and carbonates in the impact deposits, even though these are abundant in areas away for the impact area, and this supports the idea that the evaporites and carbonate rocks that were impacted were largely vaporized. Vaporization of the sulfate-rich evaporitic gypsum deposits ($CaSO_4$), which were abundant in the sedimentary layers that were pulverized, also produced large volumes of sulfur dioxide (SO_2), which would have reacted with water in the atmosphere to produce sulfuric acid (H_2SO_4). So, after a few weeks of fires, Earth would have experienced years of cooling, caused by the blocking of sunlight from the ash cloud as well as from acid rain.

The impact also sent massive shock waves through the Earth that would have generated earthquakes and tsunamis, and the actual impact in seawater would have generated a **mega-tsunami**. The mega-tsunami would have caused a huge wave, perhaps 100 times more powerful than that generated by the 2004 Indian Ocean earthquake. Estimates of the magnitude of the wave vary from several tens up to a few hundred meters high. A wave of such magnitude would have traveled hundreds of kilometers inland, inundating land around the Caribbean and beyond. Numerous scientists have since found evidence of distinctive tsunami deposits at the K–Pg boundary, particularly in Argentina, Brazil, the Gulf Coast states of the USA, and other circum-Atlantic locales (Figure 13.1).

The ocean water displaced by the impact soon flowed back into the crater. The IODP core of the peak ring structure shows a 100 m-thick graded bed (Figure 13.9), overlying melted and fractured basement rock, which was formed as the ocean flowed back into the crater and contains clasts and melt rock associated with the impact.

The shock waves of the impact have also been hypothesized to have enhanced the severity of the outpouring of lava from the Deccan Traps in India during the main pulse of volcanic activity. Certainly, the coincidence of increased volcanism and the

impact would have resulted in positive feedback of deleterious effects, such as increased levels of atmospheric SO_2, causing cooling, and increased levels of atmospheric CO_2, causing **oceanic acidification**.

13.4.1 Environmental Collapse

The immediate effects of the impact, including the shock wave, tsunami, acid rain, the hot ash fallout, and subsequent wildfires, would have quickly destroyed vast populations over large areas. Specialized and **endemic species** (i.e., species confined to a very local habitat or area), especially those confined to the areas where there was intense destruction, would have become extinct almost immediately. Acidification of the oceans, and the blocking of sunlight by dust and ash, would have caused severe stress to both marine and terrestrial ecosystems. It is likely that photosynthesis temporarily shut down, probably over a period of several months to a few years, eliminating the base of the food chain. Large herbivorous dinosaurs, including the ceratopsians, ankylosaurs, hadrosaurs, and sauropods (see Chapter 12) could not have survived months, let alone years of starvation (Box 13.1). Starvation would have arrested their metabolic processes and prevented them from reproduction before eventual death, ensuring that there would be no descendants. With the death of the herbivores, carnivorous dinosaurs, including tyrannosaurs and their ilk, would have had a hard time finding a meal and their extinction inevitably followed. The massive size of many of the dinosaurs would also have been a catastrophic liability during such a crisis, not only because of their enormous caloric needs, but also because of their inability to hide or wait out the calamity in a sheltered place.

Cretaceous mammals, in contrast, were small, and lived in the nooks and crannies of a world dominated by dinosaurian giants. In addition to their small size, many mammals were nocturnal (see Chapter 12). Many mammals then, and now, live in burrows that would have provided refuge from the calamity. Dinosaurs were simply too big to dig burrows, but the mammals, living underground, and requiring far less food resources, were better suited to survive the catastrophe. Less specialized **omnivorous**, insectivorous, and scavenging mammals were able to eke out an existence following the extinction event, but none of the purely carnivorous and herbivorous mammals survived.

In the oceans, acidification and loss of sunlight would have severely depleted marine plankton, an important food resource. Ammonites and belemnites appear to have lived in relatively shallow water. Their young would have experienced dissolution of their tiny shells, as well as a massive and abrupt loss of food sources that quickly drove them to extinction. In contrast, some species of octopus, squid, and the shelly nautilus (the closest living relatives to the ammonites) were adapted to deeper water habitats and were less reliant on surface plankton as their food source. These species managed to survive in deeper ocean refuges and are still around today.

The impact also devastated Earth's flora with close to 60% of plant species becoming extinct in North America alone. Birds were also devastated, but enough individuals survived that they were able to recolonize following the catastrophe. In terrestrial environments, freshwater aquatic species, such as turtles, crocodiles, some birds, fish, and amphibians, were also mostly able to survive. It is likely that life in water provided just enough protection for just enough individuals to survive the most devastating effects of the impact. Also, freshwater aquatic species tend to be small and cold-blooded (see Chapter 12) such that their caloric requirements are much lower than land dwellers, allowing them to survive on less food.

KEY POINT

The extinction event at the end of the Cretaceous was remarkably brief, from a geologic perspective. The total disappearance of dinosaurs and many other organisms, on land and in the sea, was completed in perhaps less than 1,000 years.

These biotic changes at the end of the Cretaceous – notably the extinction of the dinosaurs and the rise of mammals – show how natural selection is influenced by environmental changes and how benign traits (such as the ability to burrow and to live underground) can be beneficial, while other traits that were previously beneficial (such as massive size in the dinosaurs) can become a liability when confronted by a catastrophic environmental change.

13.4.2 A Slow Recovery

We can imagine what the world looked like shortly after the impact. There are many fictional accounts of what our world would look like following an all-out nuclear holocaust and, aside from the radioactive fallout, the K–Pg conditions are likely to have been very similar to the "**nuclear winter**" predicted

from such a human-made calamity. Organisms not dependent on photosynthesis, such as fungi, would have quickly re-established themselves and provided a food base for some animals.

A spike in fern pollen has been found in layers just above the K–Pg boundary. Ferns are a common plant that can live in shaded environments and are typically among the first to colonize recently deforested areas. It seems that within a few thousand years of the deforestation caused by the impact-induced wildfires and ensuing "nuclear winter"-like conditions, ferns began the resettlement of the land. These **pioneering communities** have been found all over the world above the K–Pg boundary, including New Zealand, and indicate that the effects of the impact were truly global in extent. The time during which ferns dominated the land lasted at least a thousand years and possibly as long as several tens of thousands of years. Figure 13.12 shows that the Ir spike and fern spike both occur at the K–Pg boundary in the Raton Basin of NE New Mexico, and this dual observation has been repeated in other locations.

One of the hallmarks of mass extinction is that it leaves unoccupied **ecological niches**, or habitats, that can be exploited by survivors. Studies of Paleocene mammal faunas that dominate Earth today showed that they quickly became more widespread and evolved to fill the niches vacated by the extinct dinosaurs. As terrestrial plant communities re-established themselves, with forests replacing ferns, the world once again returned to rather more lush conditions. Bony fishes evolved into the niches vacated by the extinction of Cretaceous aquatic species.

A key question is, how long did it take for life on Earth to recover its former diversity and abundance? Analysis of the age of the oldest fossil-rich Paleocene mammal-bearing sediments indicates that it took at least 185,000 and perhaps as much as 500,000 years for a full recovery to take place. Continued volcanism at the Deccan Traps might have delayed recovery in the

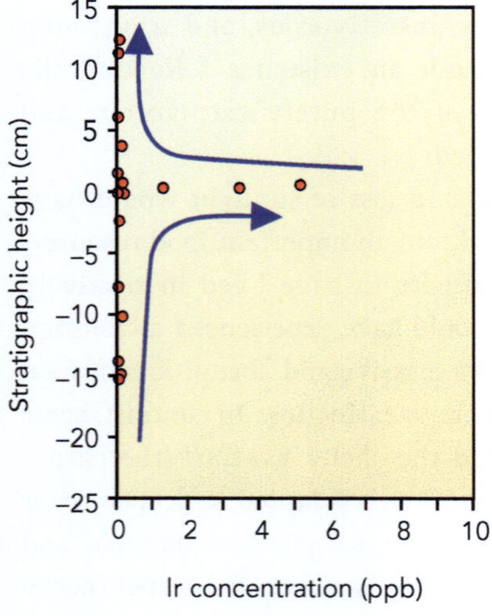
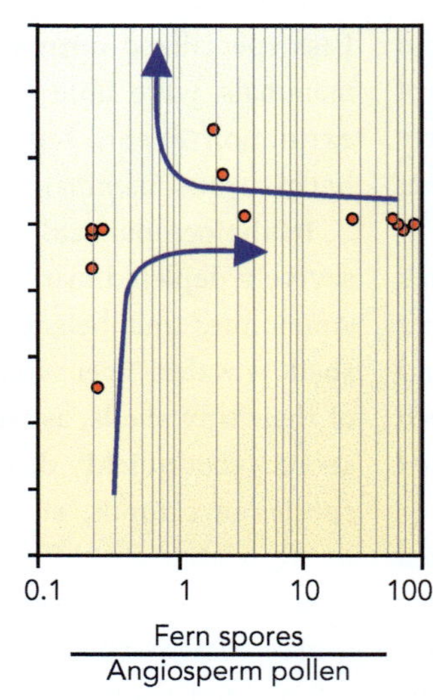

Figure 13.12 Ir concentration and fern spore abundance vs. stratigraphic height at the K–Pg boundary in the Raton Basin, NE New Mexico. These graphs show a clear spike in both Ir and fern spore abundance at the boundary. The Ir concentration then drops off very quickly above the boundary (i.e., after the impact) but the fern spore abundance remains higher after the impact than it was before. Source: Based on Orth et al. (1981).

Early Paleocene Epoch, in similar fashion to continued volcanism of the Siberian Traps during the Early Triassic Period that delayed the recovery following the Permian mass extinction (see Chapter 11)

From the geological perspective, 185,000 years is a relatively short time, given a 4.5-billion-year-old Earth, but, of course, from the human perspective, this is an eternity. For context, few individual humans live longer than 100 years. Human civilizations appear to last a few thousand years, and *Homo sapiens* may have evolved around 250,000 years ago (see Chapter 17). The dinosaurian clade lasted about 160 million years, and in terms of diversity, abundance, and geographical spread, dinosaurs were enormously successful.

13.4.3 Are the Dinosaurs really Extinct?

The final scene of Stephen Spielberg's movie *Jurassic Park* shows a flock of pelicans flying over the Pacific Ocean as the Ingen helicopter escapes Isla Nublar with the exhausted human survivors on board. The scene is quite deliberate. Birds are now widely understood to be the living group most closely related to dinosaurs, to the point that they are commonly referred to as "avian dinosaurs" and are considered part of the **dinosaurian clade** (see Chapter 12).

Cretaceous birds would have been small, with far lower caloric requirements compared to the larger dinosaurs. It is hypothesized that birds had relatively large brains for their body size, and they may have been better able to adapt to the stressful aftermath of the impact. The ability to fly may have provided some advantage, although it did not save the flying pterosaurs (Figure 12.23). Regardless of the myriad of possible hypotheses, birds somehow managed to escape the extinction event that wiped out their fellow non-avian dinosaurs.

13.5 The Origin of the Cretaceous–Paleogene (K–Pg) Extinction

We could now ask, "is the debate about the cause of the K–Pg extinction resolved?" That is, in fact, a very easy question to answer because the answer is always "No," no matter what the topic of scientific inquiry. However, there is consensus among the scientists studying this problem: no scientists dispute the observation of a giant impact crater in the Yucatan Peninsula dated at 66 million years ago, and this almost certainly played a major, if not defining, role in causing the mass extinction at the end of the Cretaceous. But some researchers continue to question aspects of this explanation.

It had been suggested that the timing of the impact was as much as 300,000 years before the extinction of many life forms, which would indicate that the impact could not have been the proximal cause of so much death and has been cited by some to support the importance of Deccan volcanism as a major contributing factor to the extinction instead. However, the uncertainty of most recent dates indicates the impact occurred within 32,000

years of the K–Pg boundary and more recent geochronological analysis of the Deccan Traps showed they were deposited 200,000 years *before* the K–Pg boundary. This highlights the challenge of geochronology. It is not always simple to interpret the dates we obtain. Any alteration or reheating of rocks, however subtle, can alter the results. The extraterrestrial impact event had to be very brief (literally a few minutes or less, although the immediate negative effects likely lasted several years), but the precision of dates for rocks of this age is currently limited to a few tens of thousands of years. The Deccan lavas were likely erupted over periods of hundreds to perhaps tens of thousands of years, causing a more prolonged, but probably lower-level stress on the planet, and dates on these rocks also have uncertainties. As discussed in Chapter 4, dating needs to be precise and, of course, accurate. Assessing the former is easier than the latter. It is likely that new samples and advances in technology will enable scientists to continue to refine these dates in the future.

Some have noted that there is generally a paucity of fossils right at the top of the Cretaceous rock column. If the fires, and the acid rain, and the tsunamis, and the "nuclear winter"-like conditions all attacked the biosphere at once, why isn't there a big pile up of bones at the K–Pg boundary? This may simply reflect the basic problem of paleontology: it's hard to become a fossil. Again, further detailed investigation of more locations will help shed light on this issue. To that end, in 2019 a new fossil discovery was made at the Tanis site, in ancient river-bed deposits of the Hell Creek Formation in Montana. The scientists investigating this site documented a mixture of marine and non-marine fossils, including fossil sturgeons (a type of fish) and paddlefish that contain evidence of glass spherules (i.e., tektites) in their gills. The jumbled nature of the layer in which the fossils are found, and the mixture of marine and non-marine fossils, has been interpreted as a consequence of the tsunami generated by the impact of the asteroid 3,000 kilometers to the south at Chicxulub, bringing seawater far up into the lowland Cretaceous rivers (see Chapter 12).

Other researchers have observed dinosaur fossils in rocks above the K–Pg boundary. At face value, this would indicate that the catastrophe at 66 Ma was not as terrible as described. However, the fossils comprise individual bones, rather than articulated skeletons, and are found with the deposits of ancient river channels. Therefore, these observations of "**Paleogene dinosaurs**" are more likely to be reworked fossils. That is, they are fossils of organisms that did indeed perish at 66 Ma, but their fossilized remains were eroded and redeposited by younger rivers.

So, despite the broad consensus concerning this important and fascinating problem in Earth history, geologists will continue to push on the giant-impact hypothesis to test its strength.

13.6 Conclusion

The story of the K–Pg boundary shows the importance of *serendipity* in science. The measuring of Ir in sedimentary rocks was indeed innovative back in the late seventies, but this innovation was not initially directed towards understanding the extinction of the dinosaurs, nor of finding evidence for giant impacts. This method of using the constant rate of Ir falling to Earth to assess the rate of sedimentation might not have even been considered if other more conventional methods for determining the age of sedimentary rock were more viable at Gubbio. But luck favors the prepared mind. When the data came in, the assumption of a constant Ir flux had to be abandoned and new ideas had to be considered to explain the sudden increase in Ir. The combination of evidence from geochemistry, paleontology, geophysics, and geology has contributed to our current understanding of why so many life forms went extinct about 66 million years ago.

Although Deccan Trap volcanism caused some stress on both the land and oceanic ecosystems, there seems little doubt that an extraterrestrial and extremely rare catastrophe ended their reign. A plot of asteroid size versus the occurrence of impacts shows that impacts of the magnitude of the one that produced the Chicxulub crater are exceedingly rare, occurring only once every 100 million years or so (Figure 13.13). Had the Chicxulub asteroid missed the Earth, the dinosaurs might well still be ruling our biosphere, and it is doubtful that mammals would ever have had their day, and you would not be reading this book.

Should humanity ever experience an event of this magnitude, we have tools that the Cretaceous biota did not. We can detect

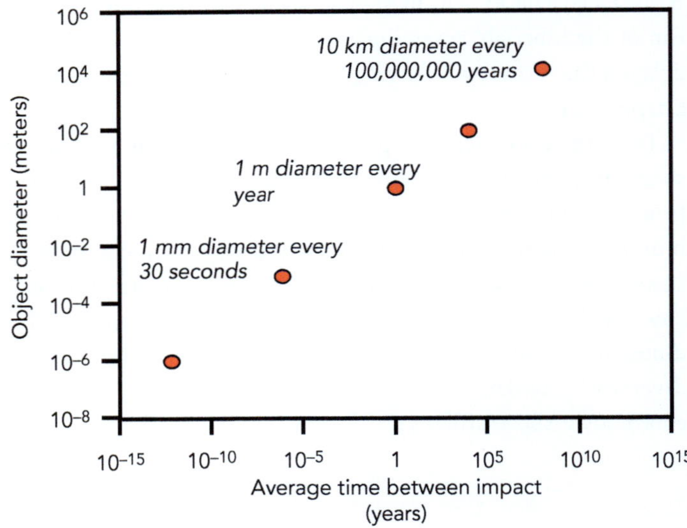

Figure 13.13 The average recurrence time of extraterrestrial impacts of various sizes. Source: Based on De Hon (2023).

asteroids and comets long before they strike our planet. Although we do not have the weaponry to destroy a 10 km asteroid, we know from physics that a small force applied to a body very far away may cause a very slight change in its trajectory, which, after a few million kilometers of continued travel, might be enough to throw an asteroid hurtling towards the Earth off course and stave off a future mass extinction event.

13.7 Summary

- At the K–Pg boundary, 75% of all species became extinct (Box 13.1), including all of the non-avian dinosaurs and many marine species, such as mosasaurs, plesiosaurs, and many invertebrate groups including the ammonites. Aquatic vertebrates, such as amphibians, crocodiles, turtles, and many fish survived, along with burrowing mammals that were also omnivores or scavengers. Avian dinosaurs (i.e., birds) also survived.
- Serendipitous discovery of increased iridium levels at the K–Pg boundary led to the hypothesis that a massive meteorite struck the Earth, triggering the extinction. Eventually, direct evidence of the impact was found in a buried crater near Chicxulub, Mexico.
- The impact had catastrophic environmental impacts, including the release of massive amounts of CO_2 and SO_2, that triggered rapid cooling, as well as dust that blocked sunlight, all of which caused "nuclear winter"-like conditions. The impact also generated massive tsunamis and extensive wildfires, along with the acidification of the oceans. These environmental changes in turn triggered mass extinctions on land and at sea.
- Massive volcanic eruptions from the Deccan Traps in India also contributed large volumes of CO_2, SO_2, and volcanic ash, and enhanced the devastation.

Key Words

- tsunami
- mass extinction
- K–Pg boundary event
- anthropogenic
- large igneous provinces (LIPs)

- flood basalts
- aerosols
- biostratigraphy
- iridium
- iron meteorites (chondrites)

- cosmogenic
- giant-impact hypothesis
- shock metamorphism
- brittle deformation
- tektites

- andesite
- impact breccia
- Chicxulub
- peak ring
- mega-tsunami
- oceanic acidification
- endemic species
- omnivorous
- nuclear winter
- pioneering communities
- ecological niches
- dinosaurian clade
- Paleogene dinosaurs

Further Reading and References

Alvarez, W., 1997, *T. rex and the Crater of Doom*, Princeton University Press.

Alvarez, L. W., Alvarez, W., Asaro, F., and Michel, H. V., 1980, Extraterrestrial cause for the Cretaceous–Tertiary extinction, *Science*, 208(4448), 1095–1108, http://www.jstor.org/stable/1683699.

Courtillot, V., 1999, *Evolutionary Catastrophes: The Science of Mass Extinction*, Cambridge University Press.

David, A., and Fastovsky, D., 2004, Dinosaur extinction, in *The Dinosauria*, 2nd ed., D. B. Weishampel, P. Dodson, and H. Osmólska (eds.), University of California Press, pp. 672–684.

De Hon, R. A., 2023, Meteor impact hazard, in R. Sivanpillai and J. F. Shroder (eds.), *Biological and Environmental Hazards, Risks, and Disasters*, 2nd ed., Elsevier.

DePalma, R. A., Smit, J., Burnham, D. A., et al., 2019, A seismically induced onshore surge deposit at the KPg boundary, North Dakota, *PNAS*, 116(17), 8190–8199, https://doi.org/10.1073/pnas.1817407116.

Hildebrand, A. R., Penfield, G. T. et al., 1991, Chicxulub crater: A possible Cretaceous/Tertiary boundary impact crater on the Yucatán peninsula, Mexico, *Geology*, 19(9), 867–871.

Morgan, J. V., Gulick, S. P. S., Bralower, T., et al., 2016, The formation of peak rings in large impact craters, *Science*, 354, 878–882, https://doi.org/10.1126/science.aah6561.

Morgan, J., Gulick, S., Mellett, C. L., Green, S. L., and the Expedition 364 Scientists, 2017, Chicxulub: Drilling the K-Pg impact crater, *Proceedings of the International Ocean Discovery Program*, 364, https://doi.org/10.14379/iodp.proc.364.2017.

Orth, C. J., Gilmore, J., Knight, J. D., et al., 1981, An iridium abundance anomaly at the palynological Cretaceous-Tertiary boundary in northern New Mexico, *Science*, 214, 1341–1343, https://doi.org/10.1126/science.214.4527.1341.

Rampino, M. R., 2017, *Cataclysms: A New Geology for the Twenty-First Century*, Columbia University Press.

Raup, D., 1986, *The Nemesis Affair*, Norton.

Robertson, D. S., McKenna, M. C., Toon, O.B. et al., 2004, Survival in the first hours of the Cenozoic, *GSA Bulletin*, 116(5–6), 760–768.

Review Questions

1. How does the mass extinction at the end of the Cretaceous compare to recent natural and human-made disasters?
2. What is the evidence for a mass extinction at the end of the Cretaceous?
3. What are the main contending hypotheses proposed to explain the extinction event?
4. What is the geochemical, geological, and geophysical evidence for an asteroid impact at the K–Pg boundary?
5. Why is iridium key in developing the impact hypothesis?
6. Why is there still some controversy around the hypothesis that an asteroid was the main cause of the extinction?
7. What was the role of volcanism in causing the mass extinction?
8. What were the environmental consequences of the impact?
9. What factors determined which organisms survived the extinction?
10. How could dinosaur fossils be found in Paleogene rocks?
11. How long did it take life to recover after the extinction event?
12. What is the frequency of large impacts?

Clearing storm over Flatirons in Boulder, Colorado. Source: beklaus / Getty Images.

Chapter 14

Tectonics of Western North America

A Dynamic Cretaceous and Cenozoic Landscape

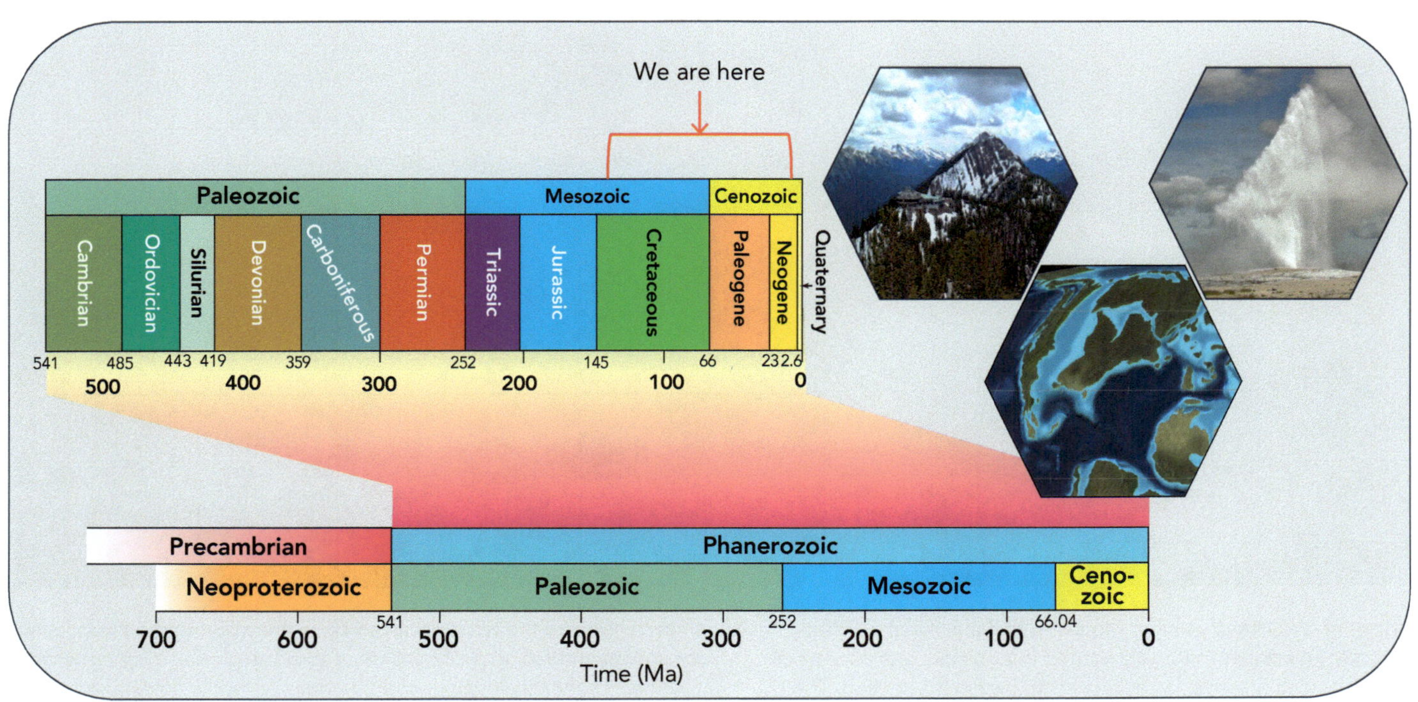

Introduction

The eastern part of North America contains rocks that tell the detailed story of the closing of the Iapetus Ocean during the assembly of the Pangean supercontinent in the Paleozoic that was the subject of Chapter 10. Since about 200 Ma, eastern North America recorded the rifting of Pangea and evolved from a divergent tectonic boundary during the rift phase into its current passive margin at the margin of the modern Atlantic Ocean (see Chapter 5 and Figure 10.2). The western parts of North America contain a geologic record that reflects a similar history but essentially flipped in time. While major orogenic deformation was impacting eastern North America in the Paleozoic, western North America was a passive margin on which a thick sedimentary sequence was deposited on a Proterozoic and older basement. In the Mesozoic, the western margin was transformed into an active margin, producing the **Rocky Mountains**, volcanic and igneous systems including the **Sierra Nevada Batholith**, and the volcanoes of the **Cascade Mountains** associated with tectonic processes that began in the Mesozoic and continue today. The Rocky Mountains stretch from Mexico to Alaska and were formed by a variety of tectonic processes including convergent deformation, extensional deformation, and growth of mountains by intrusion of magma. The region experienced compressional tectonics followed by extensional forces and the result of these varied events can often be seen overprinted in a single area (Figure 14.1). In this chapter we tell the story of the geological processes and events that formed these great physiographic features, focusing on the past 200 million years.

14.1 Suspect Terranes

By about 260 Ma, tectonic movements of the plates had assembled most of the continental land mass into the supercontinent Pangea. With essentially only one continent, most of the rest of Earth was a vast ocean, which geologists refer to as Panthalassa or the paleo-Pacific Ocean because it occupied the same position as the modern Pacific; this ocean covered several tectonic plates, many of which no longer exist on the surface, having been subducted beneath Asia or North America. Examination of the geology of Western North America shows a series of distinctive geological belts that provides the basis for understanding how it formed and its relationship to the adjacent oceanic plates. In the next sections we emphasize how key observations, and particularly the types of rocks, the degree of metamorphism, magmatism, and deformation as well as the use of geochemical and geophysical tools can help understand the history and formative mechanisms of the mountain belts that form western North America.

Although we introduced the idea of **suspect terranes** in discussing the Appalachian orogen in Chapter 10, the concept was first developed by geologists attempting to decipher the complex geology of western North America. Terranes are coherent areas of rocks of regional extent bounded by faults. A given terrane is characterized by a consistent geologic history that is different from the histories of adjacent terranes. One of the hallmarks of suspect terranes is the difficulty of correlating across the faults that bound them. The difficulty of correlating the rocks between varied terranes led to the idea that they did not originate in North America and were therefore suspected (hence the name) of originating somewhere in the Panthalassic Ocean. In western USA, western Canada, and Alaska many terranes have been identified (Figure 14.2). One of the pioneers of this work in western North America was David (Davy) Jones of the United States Geological Survey. Jones followed the lead of earlier scientists like Alfred Wegener and Tuzo Wilson (who worked mostly in eastern North America, see Chapters 5 and 10), and pioneered the use of fossils to decipher the history of

Figure 14.1 Photo of Sulphur Mountain in Banff, Alberta. Tilted Paleozoic limestones, originally formed in a shallow sea, have been uplifted by compressive forces related to Pacific subduction. The U-shaped valley to the left was carved by glaciers during the Pleistocene Ice age that ended about 12,000 years ago. The glacial valley represents erosion of softer shales bounded by ridges of harder limestone.

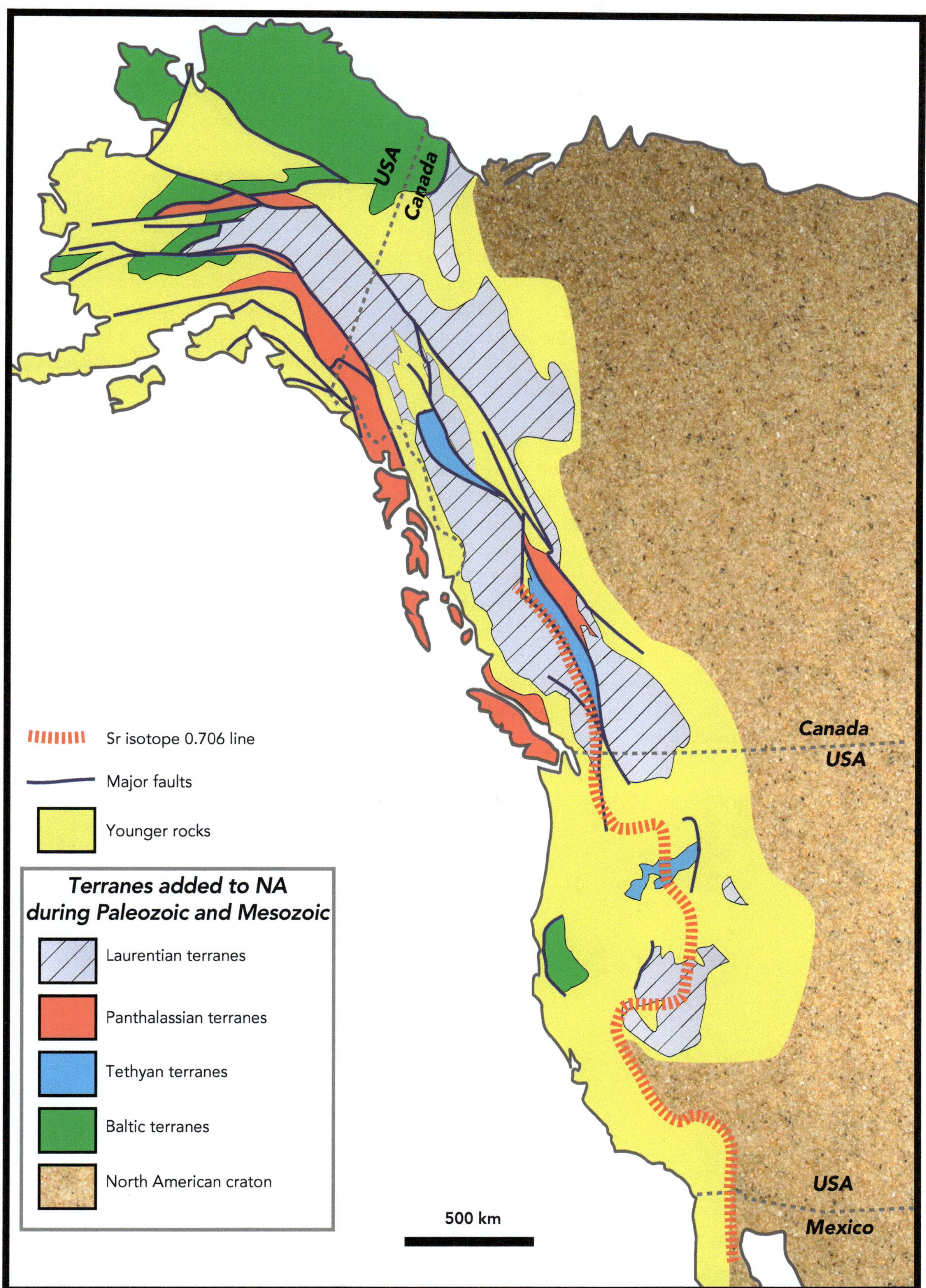

Figure 14.2 Map of western North America showing location of "suspect" terranes, which were formed to the west and south of their present positions and later added to the western margin of North America during the Mesozoic Era. The dotted red line marks the location of igneous rocks with an initial ratio of $^{87}Sr/^{86}Sr$ greater than 0.706 to the east and less than 0.706 to the west. Source: Based on Colpron and Nelson (2009).

plate tectonics and the earlier assembly and breakup of Pangea, working primarily in the 1970s and 1980s. Jones was a paleontologist by training. Many paleontologists are most interested with what fossils can tell us about evolutionary biology (see Chapters 6–9, 11–13), but Jones used his understanding of the significance of fossils, including depositional environment, and paleolatitude, in the service of tectonic analysis to infer major differences that could be linked to significant tectonic dislocation in the history of terranes now juxtaposed in the Rockies.

Some of these blocks show a continental character, suggesting that they represent slivers of continental crust that may have been rifted away from another continent and subsequently carried on an oceanic plate as it drifted. Because many of the continental blocks are thought to have originated by rifting from the periphery of a continent, they are called **pericratonic blocks**. The island of Madagascar, off the coast of east Africa, is a present-day example of a small peri-continental block that was once connected to the African and Indian continents as part of Gondwanaland but rifted away in the Mesozoic. You will recall that some of the blocks in the Appalachian orogen, including Avalonia, were originally attached to Gondwanaland and later rifted away to eventually collide with North America, and these are called the peri-Gondwana terranes.

Other blocks formed in an oceanic setting. These include island arcs and remnants of oceanic crust (i.e., ophiolites introduced in Chapter 5). Recall that much of central Newfoundland consists of island arcs and other rocks that originated in the Iapetus and Rheic oceans. Geologists today think a modern analogue to distribution of the terranes now found in western North America is the region between Australia and Asia (Figure 14.3). This area shows a variety of geologic terranes including island arcs, continental fragments, and oceanic crust. The boundaries between these regions include all possible combinations of convergent, divergent, and strike-slip motion and the change from one type to another often occurs over very short distances. Based on our understanding of the evolution of eastern and western North America, the fate of these many and varied terranes is to ultimately be added to the eastern margin of the Eurasian plate but in different relative positions compared to where they now lie as tectonics accretes these smaller regions on the larger continent.

Based on paleomagnetism, paleontology, geochemistry, and stratigraphy, it is now clear that many terranes now a part of North America were formed in a wide range of geographic origins, including volcanic arcs and sedimentary sequences that formed not far from North America (Laurentian terranes), volcanic and sedimentary sequences that formed far from North America in the Panthalassic Ocean, sequences that formed in the ocean between Asia, Africa, and Antarctica (**Tethyan terranes**, see Chapter 15), and continental fragments from the Baltic (Figure 14.2). These various terranes were carried on the offshore oceanic plates and eventually accreted to Laurentia. Some of these terranes may have accreted to each other as an elongate **ribbon continent** that later accreted to Laurentia. Accretion of these terranes likely did not occur as a simple accordion-like collision. Fossil evidence and paleomagnetic data show that

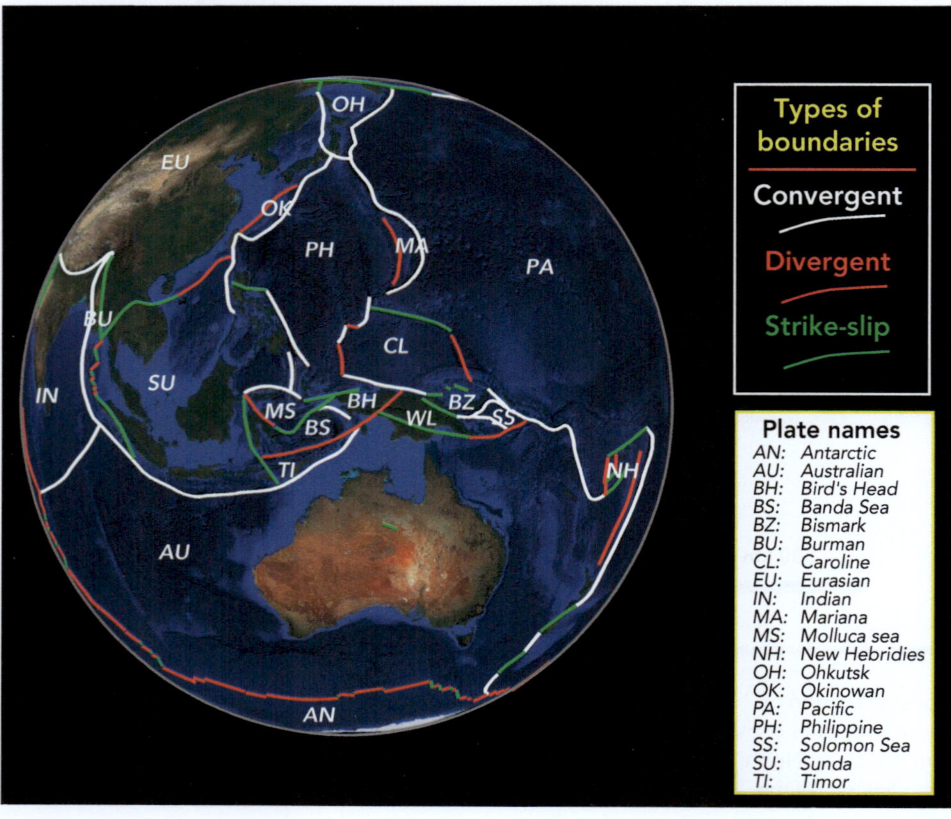

Figure 14.3 The mosaic of plates in the southwest Pacific. As Australia moves north and west, the several smaller plates will be accreted to the continents of Australia and Asia in a fashion seen in the eastern margin of North America in the Paleozoic and the western margin of North America in the Mesozoic. Source: (Globe) imaginima / Getty Images.

many of these terranes experienced up to 4,000 km of northward movement from their latitude of formation, probably along strike-slip faults after accretion to the continent.

KEY POINT

Many terranes of the western part of North America formed far from their present locations and were accreted to the continent in the early Mesozoic.

These various tectonic elements were added to the edge of North America in a complex tectonic amalgamation in the late Paleozoic and early Mesozoic that may have included multiple subduction zones directed both eastward and westward. The timing of accretion of terranes can be understood by the age of igneous intrusions or volcanic rocks that cross-cut and link terranes and dating of contiguous sedimentary strata that overlie adjacent terranes. By around 160 Ma these terranes had mostly coalesced to constitute the upper plate of an east-dipping subduction zone at the western edge of the continent. Subducting beneath North America was an oceanic plate known as the **Farallon plate**. In the middle Jurassic, a continental arc, marked by a chain of volcanoes dotting the southwestern edge of the continent, had developed above the descending plate in what is now California and Baja California (before the development of the Gulf of California). Because the subduction zone at this time dove under the overriding continent at a steep angle, the magmatism and deformation, which thickened the crust, was restricted to an area close to the shoreline. We will see that the location of this volcanic activity would vary greatly over the next 130 million years.

14.1.1 Geochemistry Reveals Crustal Structure

Analysis of the ratio of two of the isotopes of strontium (^{87}Sr and ^{86}Sr) has been used to distinguish igneous rocks that have been derived by partial melting of old continental lithosphere versus direct derivation from younger oceanic lithosphere and mantle. As we discussed in Chapter 4, ^{87}Sr is produced by the radioactive decay of one of the isotopes of rubidium (^{87}Rb). Oceanic lithosphere and the mantle have low concentrations of ^{87}Rb and so, over time, produce small amounts of ^{87}Sr. Thus, the ratio in a magmatic rock of ^{87}Sr (produced by radioactive decay of ^{87}Rb) to ^{86}Sr (a stable, unchanging isotope of Sr) gives us insight into the kind of rocks that were melted to produce the rock in question. In North America, a value of about 0.706 for ^{87}Sr/^{86}Sr has been shown to mark the cut-off between rocks that were derived from old continent versus rocks that were derived from oceanic lithosphere or very young continental crust. Granites with values of ^{87}Sr/^{86}Sr greater than 0.706 were produced by partial melting of old continental lithosphere, which is richer in ^{87}Rb and older than the subducting oceanic plate, so it had more time for ^{87}Sr to accumulate. Regional compilation of the ^{87}Sr/^{86}Sr ratios allow a line to be drawn that marks the western limit of Laurentia, which formed the original North American continental margin in the Paleozoic prior to accretion of the various suspect terranes shown in Figure 14.2. All the igneous rocks to the east of this line have initial ^{87}Sr/^{86}Sr values higher than 0.706, indicating involvement of older Laurentian crust, whereas all the rocks to the west of this line have lower ratios, indicting an origin from partial melting of oceanic lithosphere and mantle associated with convergence at the western margin of the continent.

To the west of the Sr 0.706 line lie the several terranes formed from bits and pieces of the variety of suspect terranes that originated in the Panthalassic Ocean and that were later accreted to North America. Thus, we gain confidence in the exotic formation of these terranes, originally recognized based on structure, paleontology, and paleomagnetism, by using geochemistry to delineate the edge of the main part of North America.

14.2 Foreland Basin

Recall from our discussion of isostasy in Chapter 3 that low-density lithosphere "floats" on a higher-density asthenosphere and when the crust (the upper part of the lithosphere) is thickened by thrusting, the elevation of the surface increases (although not in a 1:1 proportion). However, this is not the only change produced by crustal thickening. The thickening of the crust in the fold and thrust belt will affect adjacent areas not

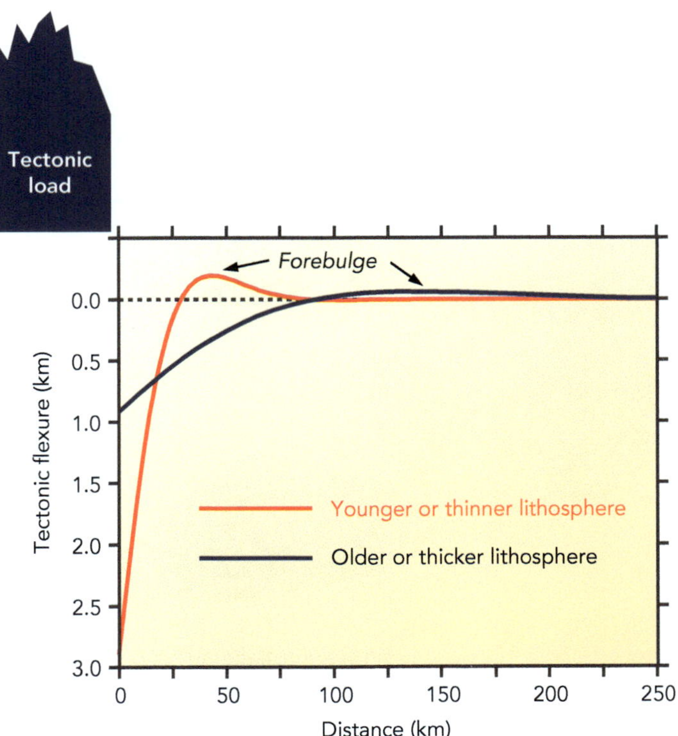

Figure 14.4 Calculation of flexing lithosphere. The flexure illustrated for the two scenarios (young or thin vs. old or thick) assumes the same tectonic load on the edge of the system.

thickened, such that a foreland basin (Figure 5.23) is formed next to the mountain range. Foreland basins are asymmetric depressions that are deepest next to the associated thrust belt. The depth and width of the basin depend on the height and width of the mountain belt, and the flexural rigidity of the lithosphere beneath the foreland basin, which varies with the complexity of its internal structure, its thickness and composition, its age (older lithosphere is usually more rigid), and the heat regime within the basin (Figure 14.4). The mechanics are very similar to a diving board, where the diver standing at the end of the board causes it to flex down. Correspondingly, once the diver leaves the board it bounces back again. Foreland basins can show a similar property due to erosion, although they flex at a much slower rate, typically only a few millimeters per year. Removal of the load by erosion can cause the lithosphere to rebound.

The foreland basin can be divided into sub-basins depending on their location to part of the bending upper plate. Material deposited near the orogenic load are found in the **wedge-top** and **foredeep** basins (Figure 14.5(a)). Farther away from the area of high subsidence, bending of the lithosphere can cause it to bow upwards, forming a high spot, called a **forebulge**, which under some circumstances can be a zone of erosion and not deposition, and a smaller basin away from the load, called a **backbulge** (Figure 14.5(a)). The distance from the area of deepest downward flexure to the crest of the forebulge represents the wavelength of the deformed lithosphere. Weaker (usually younger) lithosphere will produce a foreland basin that is deep and narrow with a pronounced forebulge and backbulge, whereas strong lithosphere will have a wide but shallow foreland basin with a very subdued peripheral bulge. The position of these sub-basins will change with the location or size of the orogenic wedge. Foreland basins can be filled with marine or non-marine deposits, depending on how deep the basin is and whether it is connected to an ocean. The foreland basin in western North America in the Cretaceous was filled with both marine and non-marine strata from the Jurassic to the Paleocene (Figure 14.5(b)). Thickening in the mountain belt can cause additional flexure of the plate, deepening the basin and changing the type of sediments that are deposited. In the case of

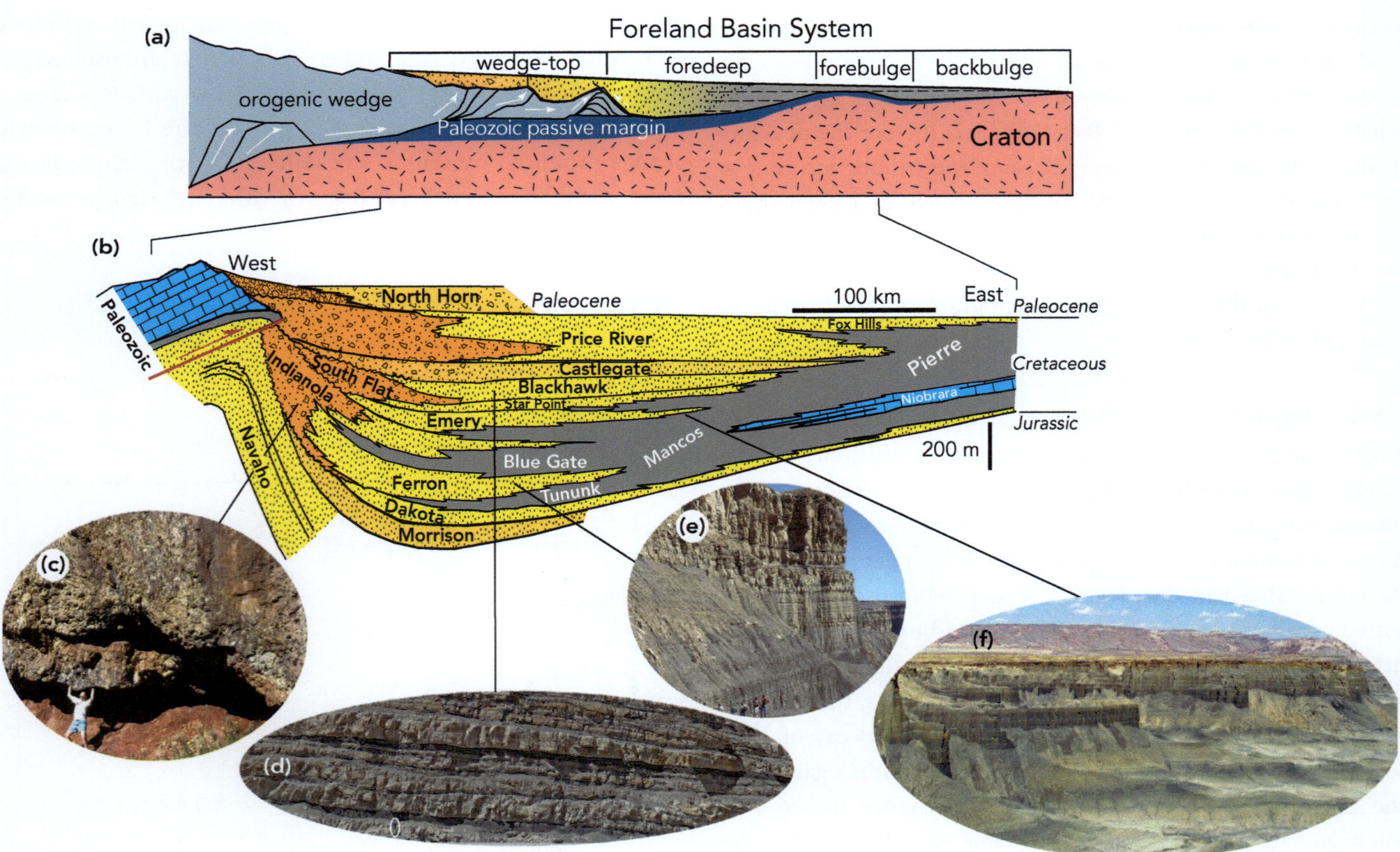

Figure 14.5 The foreland basin system stratigraphy from Jurassic to Paleocene time along a line from eastern Utah in the west to eastern Colorado in the east. (a) Structural cross section; (b) stratigraphic cross section; and (c) Alluvial fan conglomerates of the Indianola mark the most proximal deposits. These types of proximal fans pass into coal and fluvial (river) deposits of the Blackhawk Formation (d). River feed deltas (e) and (f) of the Ferron sandstone and other similar units. Cyclic stacks of sandstones, such as the Ferron, Emery, and Blackhawk, and the intervening Tununk, Mancos Shale, and Bluegate Shales record transgressions and regressions of the Cretaceous Western Interior Seaway. Source: (a) Modified from DeCelles and Giles (1996). With permission from Wiley & Sons; (b) Adapted from original figure by Armstrong (1968). © Geological Society of America. Used with permission.

the **Western Interior Seaway**, we see that in eastern Utah and western Colorado (approximately the center of the basin shown in Figure 14.5(b)) different types of sedimentary rocks are preserved in the strata deposited from 150 to 50 Ma. This is due to changes in the tectonics of the region, producing mountains of different heights (and therefore different tectonic loads, see Figure 14.4) and location.

Part of the subsidence of this basin is also controlled by interactions of the overriding lithosphere and flow of the underlying mantle. If mantle flow results in upwelling this can lessen the subsidence, whereas if downwelling occurs this can increase the subsidence and cause the lithosphere to sink lower into the mantle. These changes in the mantle flow area produced **dynamic topography**, and during the Jurassic to mid-Cretaceous flow was downwards as the mantle was literally sucked down by the subduction of the Farallon plate. This helped maintain the vast seaway that stretched from the Gulf of Mexico to the Arctic, known as the Western Interior Seaway (Figure 14.6).

The raising of sea level that allowed the ocean to make its way across the continent was, as we have seen in rocks of Cambrian and Pennsylvanian age, due to two causes. During most of the Cretaceous Period, both causes were active.

First, during most of the Mesozoic, Earth was experiencing exceptionally warm temperatures, reaching a peak in the mid-Cretaceous that has been referred to as the **Cretaceous Thermal Maximum**. This was a time when global temperatures may have been 5–6 °C warmer than today, leading to the significant

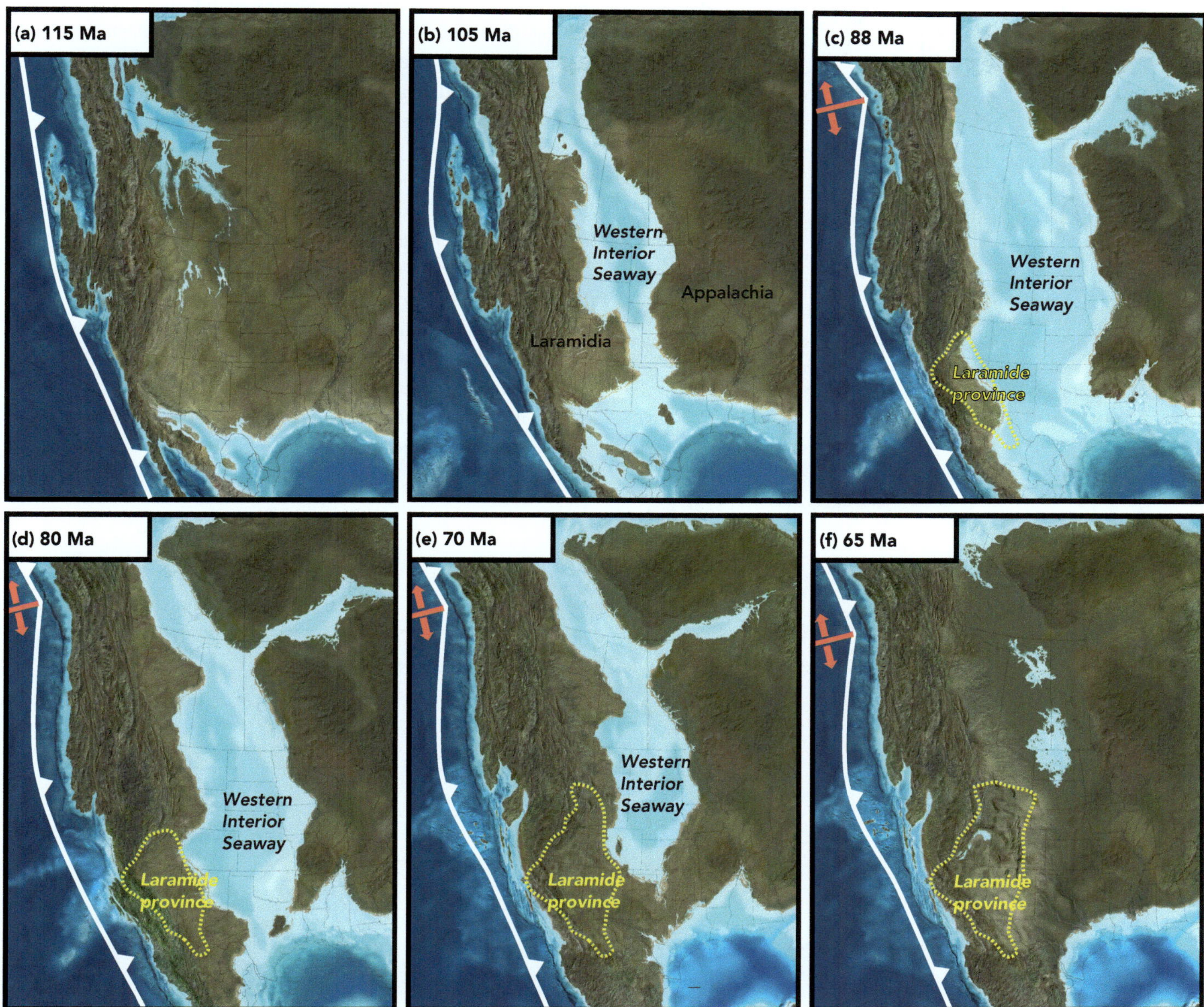

Figure 14.6 The paleogeographic evolution of western North America from 115 to 65 Ma. The white line shows the subduction zone, the red line shows the spreading center, and the yellow dotted line shows the extent of the Laramide deformation province. Source: © DeepTimeMaps.

diminution (if not complete melting) of glaciers worldwide (Figure 14.7).

This had significant effects on the environment and the type of rocks preserved from that time. Because the atmosphere was so warm (average ocean temperatures of over 20 °C or 68°F compared with modern values of around 10 °C or 50 °F, see Figure 14.8), fauna we associate with the tropics today were found in very high latitudes. Today, waters at high latitudes are colder than those in the tropics. In addition, freezing of the ocean surface, such as seen in the Arctic ice sheet, results in the remaining seawater being saltier (because the ice that forms has no salt). The lower temperature and increased salinity results in a greater density and these

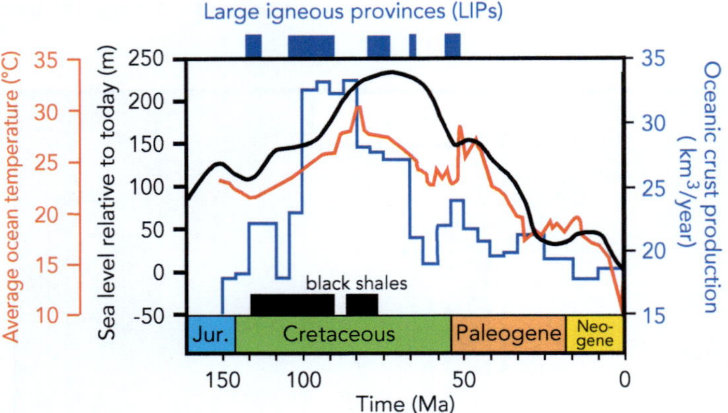

Figure 14.7 The evolution of the average ocean temperature (red), the global rate of production of new oceanic crust (blue), and the level of the sea relative to modern time (black) from 160 Ma to the present. Black and blue bars show times of significant deposition of black shales and eruption of large igneous provinces, respectively. Source: Modified from Marcilly et al. (2022).

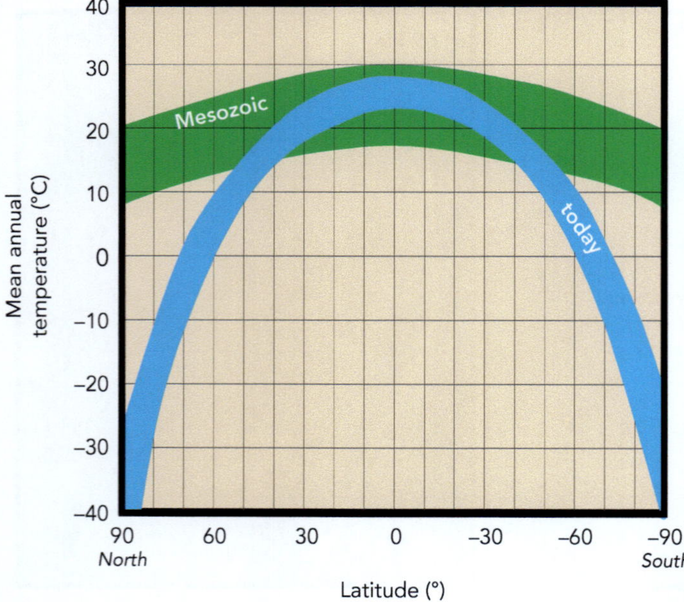

Figure 14.8 The dependence of mean annual surface temperature and latitude for today (blue) and the Mesozoic Era (green). Source: Modified from http://p7griffins-gts.weebly.com/mesozoic-era.html.

dense polar waters tend to sink and flow along the bottom of the ocean, being replaced at the surface by waters flowing from the tropics (Figure 14.9). The high-density water carries with it oxygen from the surface, allowing sediments at the bottom of the ocean to be more easily oxidized. During much of the Mesozoic, and during the Cretaceous Thermal Maximum in particular, the waters near the pole attained temperatures not much colder than in the tropics, preventing freezing (Figure 14.8). This decreased the density difference between polar and tropical waters and therefore the mixing between the top and the bottom of the water column (Figure 14.9). The deep ocean was starved of oxygen and was transformed into an **anerobic** environment in which the sediments remained unoxidized. This produced the organic-rich black shales that are a prominent component of many early to mid-Cretaceous marine sequences (Figure 14.7). These deposits are black because they were not oxidized at the time of deposition, and this allowed for a greater preservation of organic material that would otherwise be destroyed by oxygenation. The organic matter in these Cretaceous black shales was later transformed into oil and gas and are the source rocks of many important oil basins.

The second factor leading to high global sea level occurred because the subduction of oceanic crust of Panthalassa, on the western side of North America, was replaced by new oceanic crust in the Atlantic and modern Pacific Oceans; during the mid-Cretaceous new ocean floor was being produced at a rate about 1.8 times today's rate (Figure 14.7). That young, warm ocean crust was relatively buoyant, riding high on the asthenosphere and this resulted in, on average, a raising of the seafloor and therefore a reduction in the volume of the global ocean basin (Figure 14.10). Not only was the Cretaceous a time of high rates of seafloor spreading, but many large outpourings of basalt (mostly found away from mid-ocean ridges), known as large

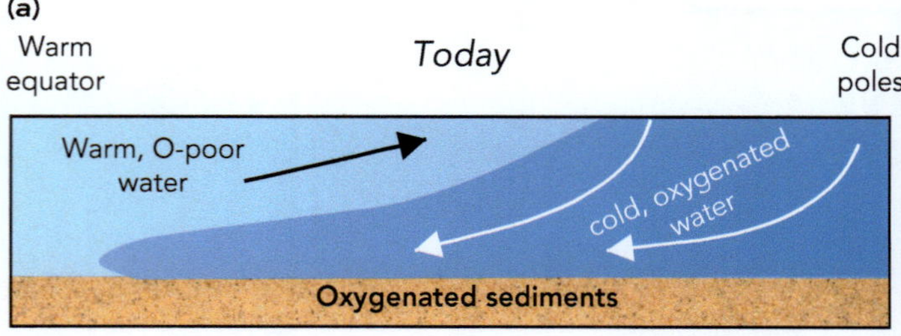

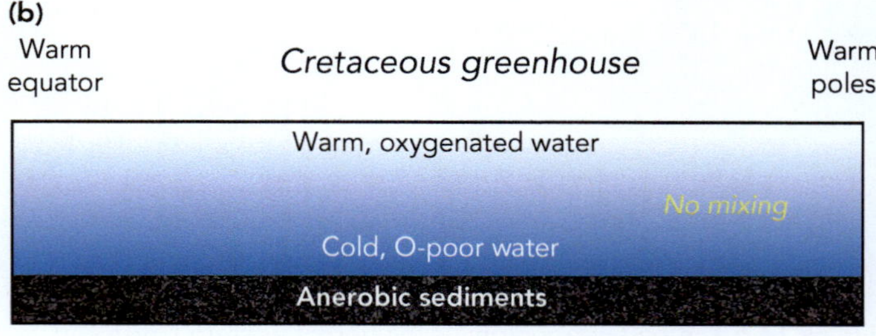

Figure 14.9 Illustration of the global ocean currents prevalent in modern times (a) and the Cretaceous greenhouse time (b).

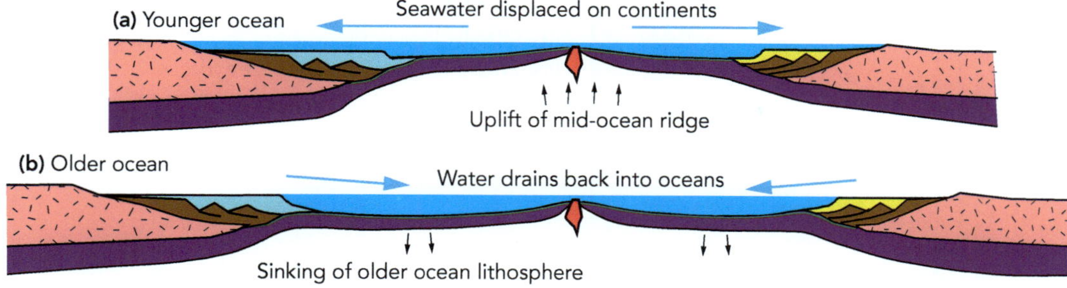

Figure 14.10 Illustration of the effect of new ocean floor on the height of sea level.

igneous provinces (LIPs), also attest to the overall increase of magmatic activity globally during this time (Figure 14.7).

The displaced water from the smaller oceans, combined with substantial melting of continental ice sheets, combined to cause global rise of sea level to more than 200 m above today's level (Figure 14.7), covering many continents, including North America, where the tectonically produced foreland basin was flooded in a configuration referred to as the Cretaceous Western Interior Seaway flanked by **Laramidia** to the west and Appalachia to the east (Figure 14.6). The seaway was flanked by rivers and their shorelines that drained the adjacent Laramidian highlands.

Figure 14.6 shows the geographic evolution of North America from 115 to 65 Ma. We see that the initial encroachment of the sea onto the continent came from both the north and south with the Western Interior Seaway becoming a throughgoing body of water around 105 Ma, which continued for approximately another 30 million years. This evolution of sea level in North America is due to a combination of factors including global climate, warming, increase in the relative rates of seafloor spreading and seafloor subduction, and regional tectonics, which we will discuss in more detail in the next section.

14.3 Subduction Angle Controls Tectonics: The Laramide Orogeny

The subduction of the oceanic Farallon plate began in the Jurassic Period. Like all oceanic plates, the Farallon was produced at a spreading center, in this case, one which was a part of a triple junction far to the west of the western margin of North America; one arm of this triple junction was the spreading center that produced the Pacific plate and the Farallon plate. An example of a modern-day triple junction can be found in the eastern Pacific,

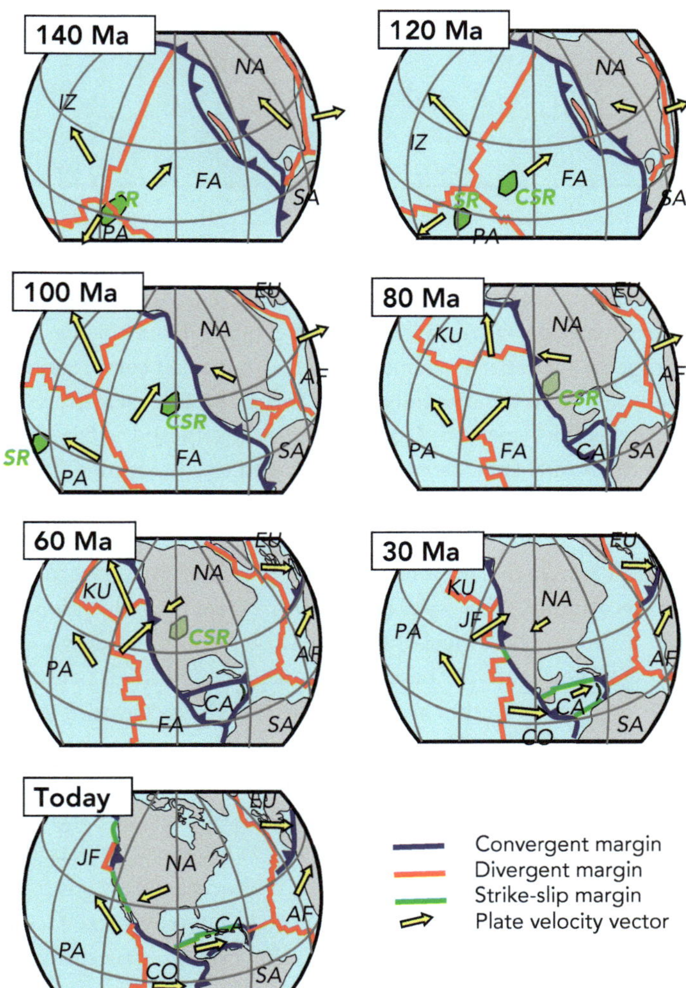

Figure 14.11 Evolution of the tectonic plates surrounding North America from 140 Ma to today. NA – North American plate, IZ – Izanagi plate, FA – Farallon plate, SA – South American plate, PA – Pacific plate, KU – Kula plate, CA – Caribbean plate, CO – Cocos plate, EU – European plate, AF – African plate, SR – Shatsky Rise, CSR – Conjugate Shatsky Rise. Source: Modified from Yonkee and Weil (2015). Copyright (2015). With permission from Elsevier.

where the Pacific, Cocos, and Nazca plates are forming at three spreading centers that come together at a triple point southwest of Central America (Figure 14.11). Geologists refer to deformation across the Rockies related to the subduction of the Farallon plate

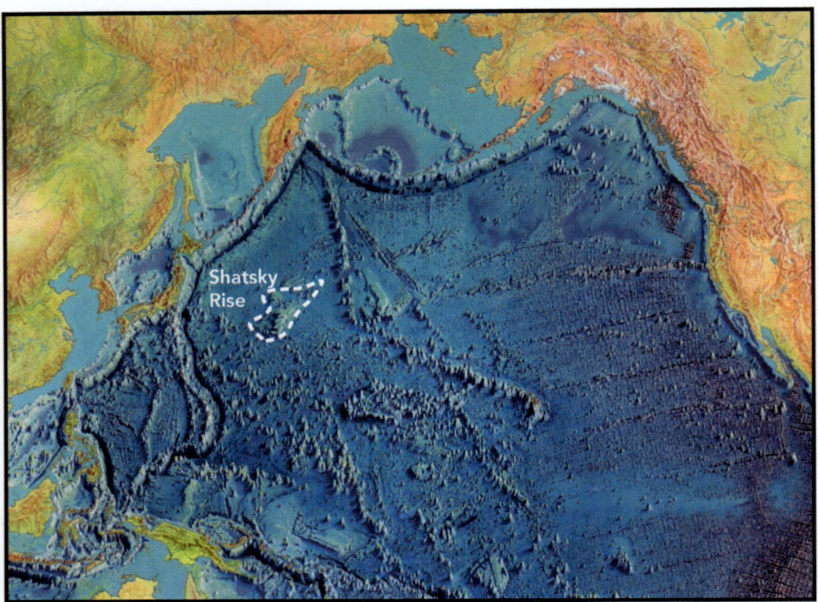

Figure 14.12 Location of the oceanic plateau known as the Shatsky Rise in the western Pacific. Source: Modified from Berann, Heinrich C., Heezen, Bruce C., Tharp, Marie., CC0, via Wikimedia Commons.

during the Cretaceous and early part of the Paleogene Periods as the **Laramide Orogeny**. This name comes from the Laramie Range in southern Wyoming, where large-scale compressional folds and faults that cut deep into Precambrian basement rocks of the North American Craton are clearly exposed. After these compressional structures were described in the Laramie Range, other workers found evidence for similar deformation in other locations in the Rocky Mountains and therefore applied the name Laramide, which means "like Laramie." Afterward any observation of these sorts of structures from many locations in the Rockies, and especially where thrust faults cut down into much older cratonic basement rocks, were called Laramide deformation and the broad region in which we find these structures is called the Laramide province.

An important aspect of the deformation of the North American continent during the Laramide Orogeny started far from the Laurentian continent in the Panthalassic Ocean. The rates of spreading and magmatism at mid-ocean ridges generally do not vary abruptly and the interplay between these factors can have a significant effect on the crustal structure. For example, the mid-Atlantic ridge today has a greater rate of volcanism around Iceland such that the oceanic crust is so thick that it is locally above sea level; this is the result of the intersection of a mid-ocean ridge and a hot spot (see Chapter 5), making this the site of an unusual amount of volcanism and therefore a much thicker than normal oceanic crust. At about 145 Ma, a similar **oceanic plateau** was created at the Pacific–Farallon spreading center (Figure 14.11). The conditions that formed this plateau only existed for a few million years and after magmatism returned to normal this plateau was pulled apart, with half of it moving west on the Pacific plate and the other half moving east on the Farallon plate. The plateau on the Pacific plate can still be seen today on the ocean floor approximately 1,500 km west of

Japan and is known as the **Shatsky Rise** (Figure 14.12) with an area larger than the state of Texas. Because spreading centers are symmetrical, the Shatsky Rise must have had a counterpart or conjugate, on the other side of the spreading center, but this is no longer observable because it has been subducted beneath the North American plate.

KEY POINT

The foreland basin of the Laramide Orogeny in the western United States reflects the interplay between local tectonic activity and global sea-level rise.

An important difference between normal oceanic lithosphere and oceanic lithosphere, which contains a volcanic plateau such as the Shatsky Rise or Iceland, is the average density. The density of the mantle lithosphere is about $3.37\,\text{kg/m}^3$ and the density of oceanic crust is about $2.95\,\text{kg/m}^3$. The thickness of mantle lithosphere is quite uniform at about 90 km, but the thickness of the crust in oceanic lithosphere can vary from as little as 6 km in normal situations to as much as 25 km in the Shatsky Rise in the western Pacific. Large igneous provinces (also known as LIPs) create excess crust, and this can decrease the average density of the oceanic lithosphere by as much as 8% (Figure 5.15) significantly affecting the way an oceanic plate is subducted. In the example in Figure 5.15, the normal lithosphere has a bulk density of $3.23\,\text{kg/m}^3$, whereas the lithosphere with a 24 km-thick oceanic plateau has a bulk density of $3.27\,\text{kg/m}^3$. Such differences can significantly affect the style of subduction of an oceanic plate.

Subduction zones are often the site of volcanism. This occurs not because of the melting of the down-going oceanic plate but

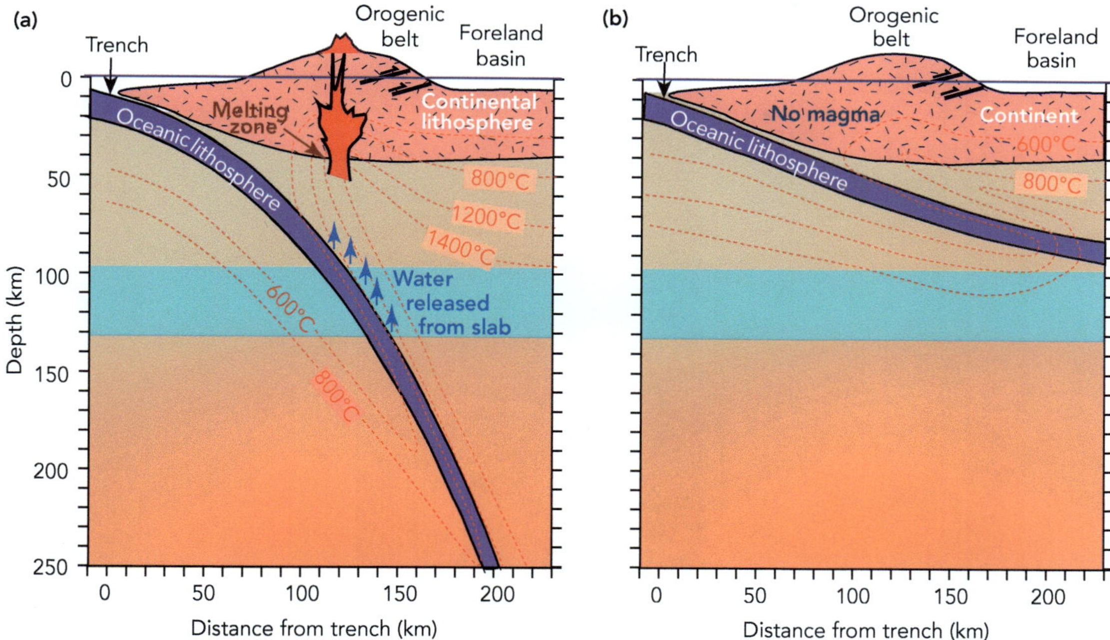

Figure 14.13 Illustration of the process of magma generation in subduction zones. (a) Steep subduction allows the subducting oceanic plate to reach ~110 km only 100–150 km from the trench, causing dehydration. The liberated water from the down-going plate rises into the overlying rocks and causes them to melt because of the change from dry to wet conditions. (b) Shallow subduction moves the locus of dehydration further from the trench and may prevent the oceanic crust reaching depths greater than 120 km, leading to little to no magma generation.

primarily because of dehydration of the down-going plate in which the water released rises into the warm overriding plate, lowering the melting point of the rocks. This dehydration of the subducted slab occurs at a depth of about 110 km (Figure 14.13 (a)). The rising water can decrease the melting temperature of the overlying mantle rocks by as much as 300–400 °C. However, the melting does not occur in the lower part of the overriding plate because the cold down-going plate cools the lower portion of the upper plate. The water must, paradoxically, rise to a point where the temperature is high enough to allow melting. The magmas produced by this melting can rise to become plutonic rocks within the crust or continue all the way to the surface to become volcanoes. Thus, the location of ancient plutonic or volcanic rocks gives us insight as to where the down-going slab passed through a depth of about 110 km. Where this happens relative to the edge of the overriding plate depends on the angle of subduction of the down-going plate (Figure 14.13). If the subduction angle is steep, volcanoes form near the trench, but as the subduction angle decreases, volcanoes are produced further away on the overriding plate. If subduction occurs at a very shallow angle, the oceanic crust may never descend to a depth sufficient to release significant amounts of water, and in those cases magmatism and volcanoes will be absent (Figure 14.13(b)).

Some of the most picturesque examples of subduction-related Plutonism can be found in the Sierra Nevada Range of eastern California. Half Dome in Yosemite National Park is a particularly famous example of granitic rock produced by subduction-related melting (Figure 14.14). These rocks, and others like them in Baja California, were emplaced in the Jurassic and early Cretaceous periods. Much of the fold and thrust belt rocks in western California comprise sediments originally deposited in very deep water that were deposited in the original trench of the subduction zone and which have been subsequently uplifted and plastered along the western edge of the compressed mountain belts. Reconstruction of the paleogeography of this time shows that the Farallon–North America (FA–NA) plate boundary was at the approximate location of the modern shoreline in the Jurassic. Because the plutonic rocks of the Sierra Nevada are only about 200 km from the contemporaneous trench, we can say that the subduction angle of the Farallon plate was around 30° from the horizontal.

Based on plate reconstructions and the current location of the Shatsky Rise, we know that the conjugate to the Shatsky Rise (CSR) on the Farallon plate reached the FA–NA margin about 90 Ma (Figure 14.11). As explained above, this marked a change in the character of the lithosphere being subducted, with the CSR substantially decreasing the density of that part of the Farallon plate entering the trench. This lower density would have caused the angle of subduction to decrease because a lower density lithosphere would be more difficult to subduct as it is more buoyant.

The geology of the upper plate reflects this change, as the location of subduction-related magmatism moved from near the plate margin towards the NE from 90 Ma to about 70 Ma, with

Figure 14.14 Panorama of Yosemite National Park. The famous Half Dome is on the horizon. All rocks in this image are granitic plutons formed due to subduction of the Farallon plate beneath the North American plate at a steep angle. Source: Cavan Images / Getty Images.

magmatism becoming rare and widely dispersed in western North America from around 70 to 50 Ma (Figure 14.15). The paucity of magmatism in the Rockies from around 70 to 50 Ma is likely due to the angle of subduction becoming so shallow that the subducting plate was no longer deep enough to cause dehydration and subsequent melting of the overlying continental lithosphere.

A modern example of this can be seen in the Andes, where different portions of the Nazca plate subduct at different angles beneath South America (Figure 14.16). In some segments, the angle is so shallow that volcanism is absent, whereas in other areas we see many active volcanoes adjacent to the trench. The reason for this segmentation is because the parts of the oceanic Nazca plate contain oceanic plateaus, formed as part of LIPs, that change the average density of the lithosphere and therefore the angle of subduction. For example, the Nazca and Juan Fernandez ridges are regions of anomalously thick oceanic crust, and in the sections of the Andes where these ridges are being subducted there are no active volcanoes (Figure 14.16). In this modern example we have the advantage of active geophysical measurements of the earthquakes that define the distinct Wadati–Benioff zones (Figures 5.13 and 14.16), which is not possible when studying ancient tectonic systems. The confidence we gain from understanding this active system with contemporary geophysics and geology makes it easier to study older tectonic events, such as the Laramide Orogeny.

So, although during the period from about 70 to 50 Ma, volcanism waned in the Laramide province, Figure 14.15(b)

shows that after about 50 Ma, magmatism became more active. This suggests that the angle of subduction once again became steep enough to promote dehydration of the Farallon plate. We will discuss the reasons for this change below.

Igneous activity is very important in understanding the details of an orogenic belt but it's not just the location of magmatic rocks that informs us about this complicated plate interaction of the Laramide Orogeny. As the Farallon plate touched the North American plate, the force of the interaction caused deformation in the overriding plate. This deformation is today manifest in folds and faults (mostly reverse), which produced local uplifts that were eroded to fill nearby basins. As discussed in Chapter 1, analysis of these structures and sedimentary layers can yield information about the timing of initiation of the deformation by taking note of the youngest rocks in an area that are deformed and the oldest that are not. Figure 14.17 shows an example of such deformation from the Persimmon Gap region of Big Bend National Park in southern Texas. Here we see Paleozoic and Cretaceous strata folded and thrust over other Cretaceous rocks. The rocks at the top of the hill (the Glen Rose Formation) are from the Albian stage of the Cretaceous Period (around 110 Ma) and the rocks at the bottom of the hill are from the younger Campanian stage of the Cretaceous (about 73 Ma). We can date the time of initiation of deformation as starting after the formation of the youngest rocks involved. Thus, the rocks at Persimmon Gap were deformed sometime after 73 Ma.

Figure 14.17 shows one of these classic structures in which Paleozoic and Cretaceous rocks are folded and thrust over

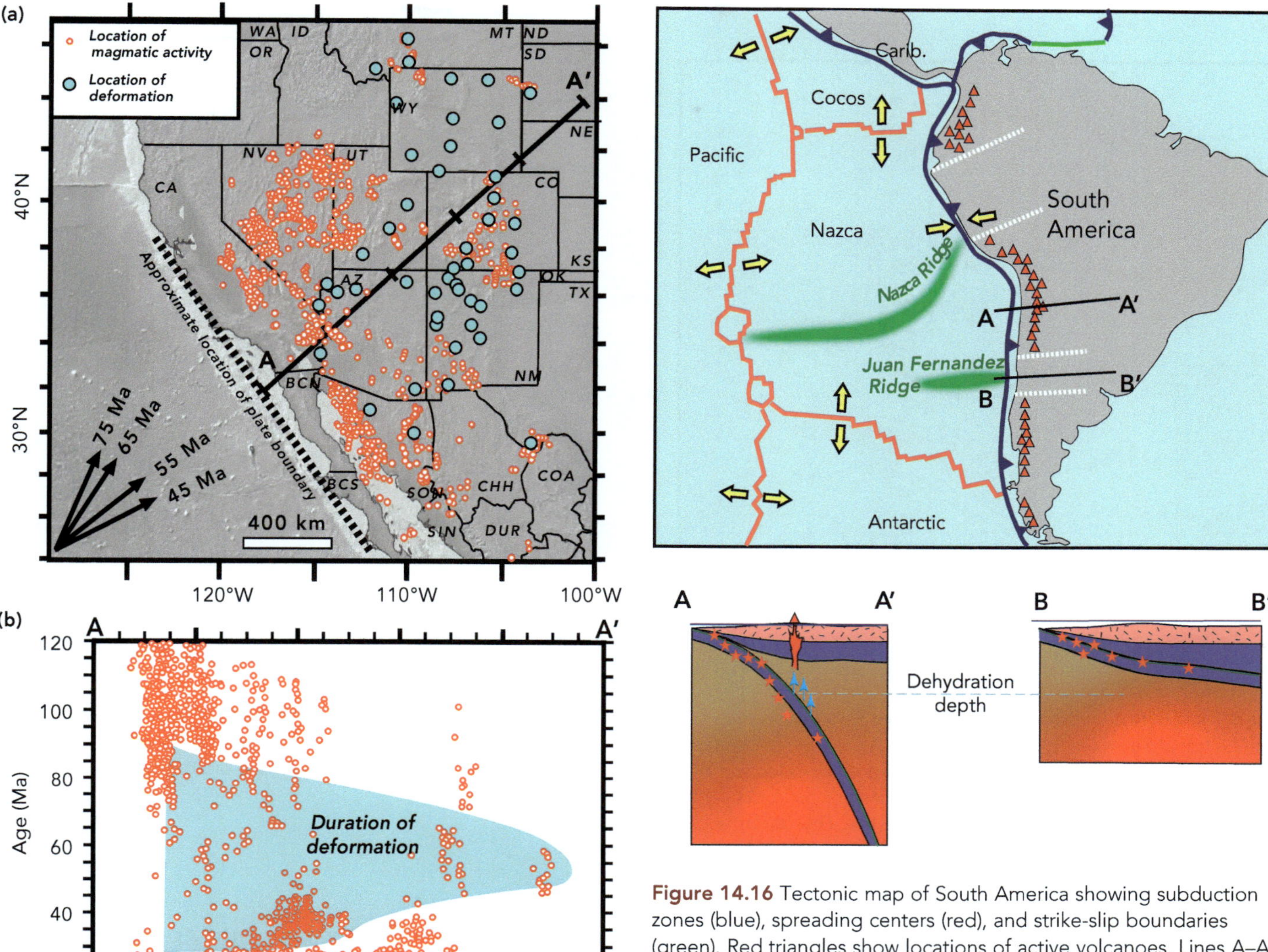

Figure 14.16 Tectonic map of South America showing subduction zones (blue), spreading centers (red), and strike-slip boundaries (green). Red triangles show locations of active volcanoes. Lines A–A' and B–B' show Wadati–Benioff zones for segments of the convergent margin with and without active volcanism, respectively.

Figure 14.15 Geologic consequences of the subduction of the Farallon plate. (a) Map showing locations of igneous rocks (red circles) and layered strata that have deformed by thrusting or folding (blue circles). Black arrows show the direction of convergence between the Farallon and North American plates at the times given. (b) Plot of locations in (a) collapsed onto line A–A' showing the age of individual magmatic rocks and the duration of deformation.

Cretaceous rocks. This example illustrates how it is sometimes difficult to pin down the timing of deformation as all one can say for sure is that the deformation is younger than about 73 Ma, the age of the youngest rocks involved. In other cases, younger, undeformed, rocks are found deposited on top of the Laramide structures, indicating that the age of deformation is older than the age of the youngest rocks. Igneous intrusions provide the same sort of constraint on our estimates. When deformation is not easily bracketed by such cross-cutting relationships, the timing of uplift can be inferred from the age of coarse-grained sedimentary strata deposited in basins adjacent to evolving uplifts.

Putting together this kind of information allows us to unravel the history of the Laramide Orogeny. Figure 14.15 shows locations across the Rockies where estimates can be made regarding the beginning and ending of Laramide deformation. This shows that the duration of the Laramide Orogeny was longer (from about 90 to 30 Ma) in the SW part of the region and much briefer (from about 65 to 55 Ma) in eastern Wyoming and the Black Hills of South Dakota. If we ascribe this sort of deformation to the interaction of the Farallon and North American plates, this pattern also suggests that the angle of subduction of the Farallon plate became shallow enough that deformation (and slight magmatism) can be seen occurring around 55 Ma approximately 2,000 km from the plate boundary in the Black Hills of South Dakota (Figure 14.15). However, the duration of this activity in this northeasternmost part of the Laramide province was quite brief compared to the southwestern part, suggesting that South Dakota was the end of the line for this shallow-angle subduction, and after this time the Farallon plate began to **roll back**, increasing its angle of subduction, and

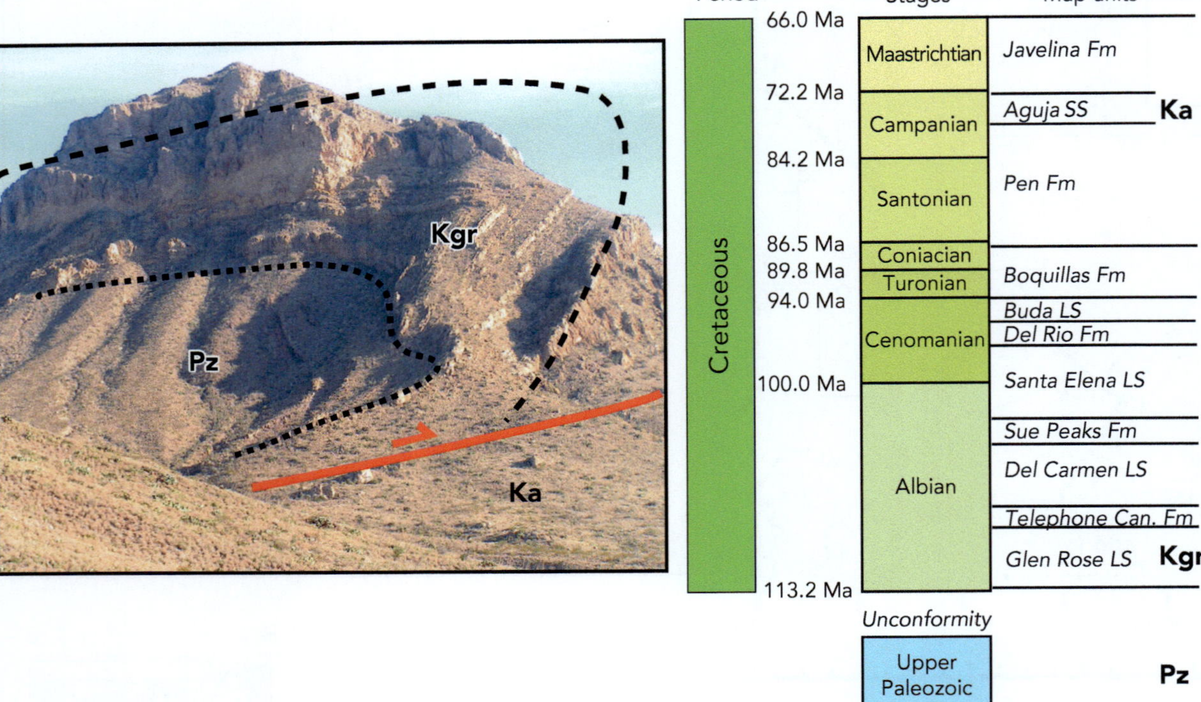

Figure 14.17 Photo of the Laramide fold from Persimmon Gap in Big Bend National Park, Texas. The folded rocks here are the Cretaceous Glen Rose Formation (Kgr), about 110 Ma, and older Paleozoic sedimentary rocks. These units are thrust over younger rocks of the Aguja Sandstone (Ka), deposited at about 73 Ma. This deformation is typical of the Laramide Orogeny in the Rocky Mountains. This allows us to say that at this location the folding and thrusting must have occurred after the deposition of the Aguja, or about 73 Ma. Source: Photo by Mike Murphy.

reducing the space where it and the North American plate were in contact.

This, of course, leads to the question, why would this happen? Figure 14.18 illustrates a model of evolution of the Farallon–North American interaction from 90 to 30 Ma. At 90 Ma, the Farallon slab was subducting at a steep angle, but this changed as the lower density rocks of the CSR were introduced into the subduction zone. The lower bulk density of the thick oceanic plateau caused the Farallon plate to be more difficult to subduct, leading to a shallow angle of subduction. From around 60 to 50 Ma, the Farallon plate attained a status sometimes called **flat-slab subduction** (although "flat" here really applies to the angle of subduction, not the slab itself), allowing interaction between the Farallon plate and the North American plate for up to 2,000 km (Figures 14.15 and 14.18). This explains why Laramide-style deformation is seen from California to South Dakota (Figure 14.15) and why magmatism in the region was subdued in this period compared to the times before and after. However, because these processes were only active for a brief time in the northeastern part of the region and because magmatism becomes more active after 50 Ma, it seems necessary for the angle of subduction to become steep once again. It is thought that although the Farallon lithosphere with the much thicker CSR was initially anomalously buoyant, this condition changed when the basalt (density 2.95 g/cm^3) was metamorphosed by the extreme temperatures and pressures that it was subjected to as it sank into the mantle, transforming it into a denser rock called eclogite (3.45 g/cm^3) (see Chapter 5, and Figures 5.16 and 14.18). This change takes place gradually and it appears that the CSR maintained its lower density for tens of millions of years before it became dense enough to cause the previously more buoyant lithosphere to sink.

Thus, the Laramide Orogeny gives a clear example of how deformation at ocean–continent plate boundaries can be seen over a wide geographic range and that the extent of this range can depend strongly on the angle of subduction, which itself depends on the character of the subducted lithosphere, which can vary depending on the thickness and density of the sinking plate.

14.4 The Impermanence of Plate Margins

Because of the variation of convection in the mantle and the variation inherent in having a sphere made up of several plates moving at different speeds and in different directions, no plate tectonic map of the globe will be static. Many stories in previous and subsequent chapters attest to this. We next consider the change in plate boundary conditions that took place after the Laramide Orogeny, leading to the development of the San Andreas fault system in California and Mexico.

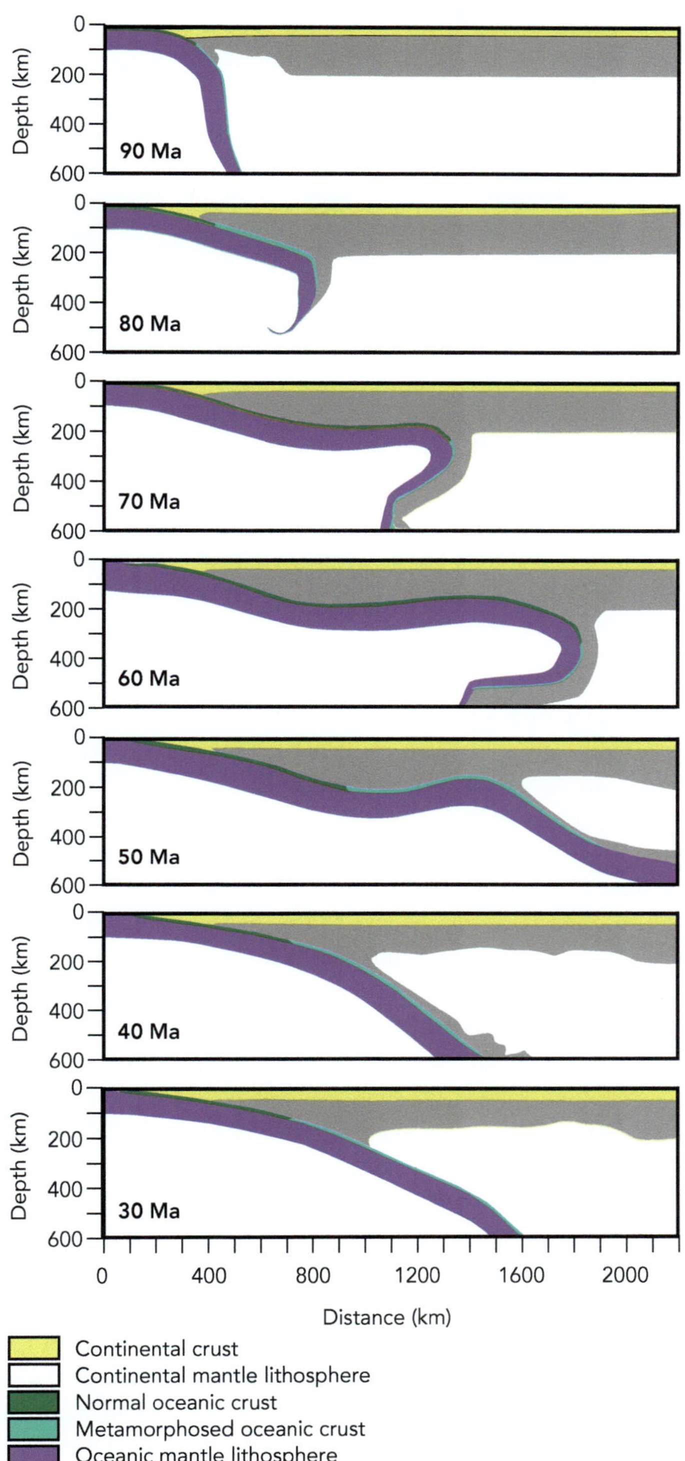

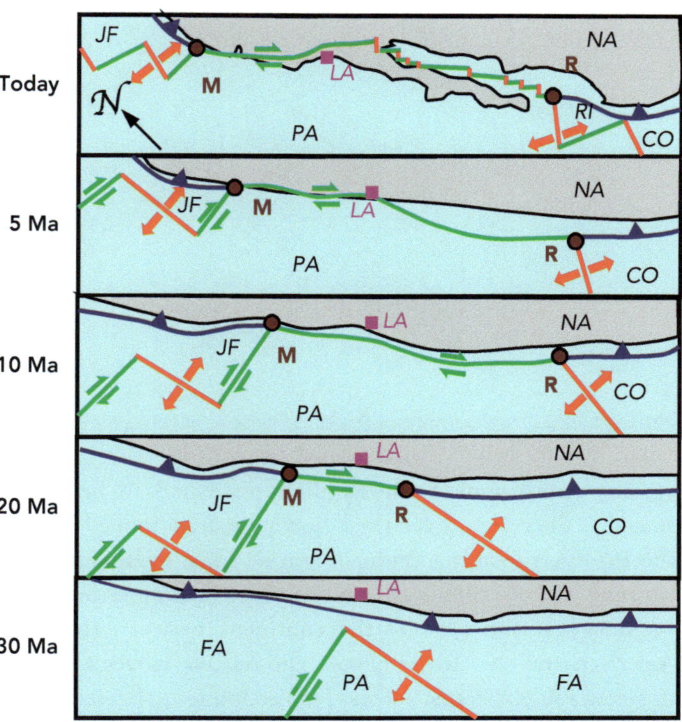

Figure 14.19 Schematic evolution of the plate margin of western North America from 30 Ma to the present. PA – Pacific plate; FA – Farallon plate; NA – North American plate; CO – Cocos plate; JF – Juan de Fuca plate; RI – Rivera plate. Red lines are spreading centers, green lines are strike-slip faults, and blue lines mark subduction zones. Brown dots are triple junctions.

Figure 14.18 Model of subduction of an oceanic plate with a thick oceanic plateau (see Figure 14.13). Source: Modified from Copeland et al. (2017). © Geological Society of America. Used with permission.

14.4.1 Development of the San Andreas Fault System

Although a map of the distribution of tectonic plates (e.g., Figure 5.14) might give one an impression of stability, given the scale of the plates, it must be understood that, because the plates are in motion relative to each other (albeit at a speed of only a few cm per year that makes any movement difficult to appreciate during our lifespans), none of the plate boundaries are permanent in either their location or the type of motion seen across them. This is because gravity-driven convection in the asthenosphere is chaotic and changing (but again, only on long timescales) and because motions of plates on the outside of a sphere are inherently unstable. The western margin of North America in the Cenozoic is an example of this instability.

We've already discussed the concept that the Farallon plate was being subducted beneath North America for most of the Mesozoic and Cenozoic eras and that the relative motion of the western margin of the North American plate and the eastern margin of the Farallon plate were towards each other (Figures 14.11, 14.15, and 14.19). We can think of such motion from the perspective of the site of the formation of the Farallon plate, a mid-ocean spreading center. During parts of the Mesozoic and Cenozoic eras, the Pacific and the Farallon plates were formed with the former moving generally westward and the latter moving generally eastward to be subducted beneath North America. This is a relatively straightforward situation but not one that can be maintained indefinitely, because the positions of all plates that make up the lithosphere are always shifting relative to each other. The key relationship to note in this system is that North America was moving to the west faster than the Pacific plate was moving east (this means the

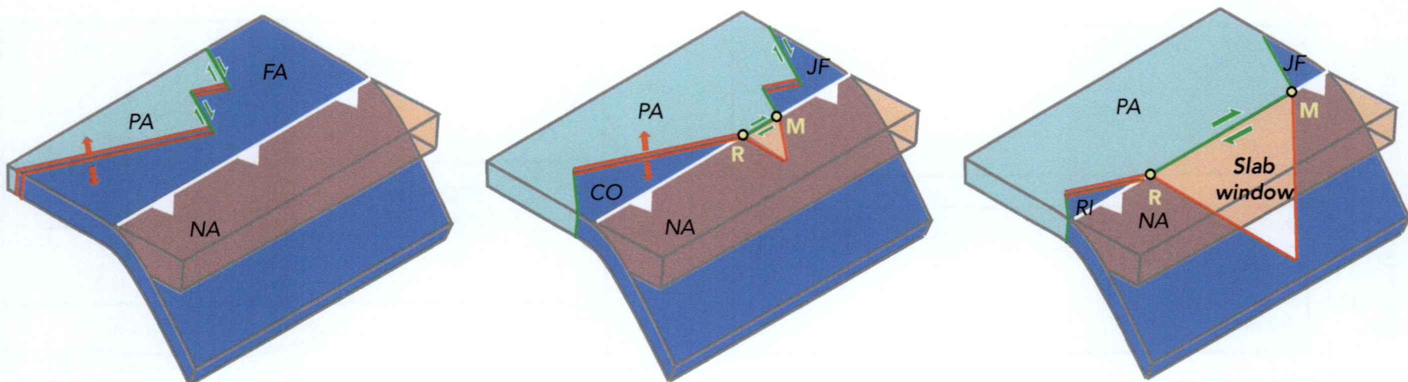

Figure 14.20 Development of the slab window, which developed in conjunction with the evolution of the San Andreas fault system (green line).

surface expression of the Farallon plate had to be getting smaller). Therefore, over time North America moved towards the PA–FA spreading center (Figure 14.19), eventually over-running the Farallon plate entirely, producing a new plate boundary at the edge of the continent between the North American and the Pacific plates. The relative movement across this new PA–NA boundary was (and continues to be) **left-lateral strike-slip** motion; this transform plate boundary evolved from its initiation around 30 Ma into the modern San Andreas fault system (Figure 14.19). The first major consequence of this change was to separate the Farallon plate into two plates north and south of the new strike-slip boundary; today we call these plates the Juan de Fuca plate, named for the Juan de Fuca Strait, of the eastern North Pacific Ocean, between the Olympic Peninsula of Washington, and Vancouver Island, and the **Cocos plate**, named for Cocos Island, approximately 550 km southwest of Costa Rica, in the center of the plate.

The San Andreas fault system developed as a consequence of North America overriding the spreading center between the Pacific and Farallon plates.

At each end of this strike-slip fault is a triple junction. The northern **Mendocino triple junction** is the point where a sub-duction zone (today's Cascadia trench), the San Andreas system, and a strike-slip fault meet and form part of the broader PA–FA divergent system. The southern Rivera triple junction is where the San Andreas system meets the Central American subduction system and a spreading center between the Cocos (a Farallon remnant) and Pacific plates. Because of the geometry of these several plates and their relative motions, the distance between the Mendocino and Rivera triple junctions has increased over the past 30 million years and are now approximately 2,500 km apart (Figure 14.19)

Today, the San Andreas fault system has evolved such that the left-lateral strike-slip plate boundary between the Pacific and North American plates is found on land. The modern-day San Andreas fault hugs the coastline from Cape Mendocino to the San

Francisco Bay area. From there, the fault system moves inland approximately 100 km and parallels the coast until the strike-slip system meets a complex of small spreading centers linked by longer transfer faults in the Gulf of California (Figure 14.20).

14.4.2 The Basin and Range Province

The change from a convergent subduction margin to a left-lateral transform margin in California is not the only major change that took place in western North America as the result of the over-running of the PA–FA spreading center. As the North American plate came in contact with the Pacific plate, the relative motion between NA and the plate to its west changed from convergent to left-lateral strike-slip (Figure 14.20). As mentioned above, this caused the San Andreas fault system to grow over time at the edge of the continent but the consequences of this change in plate geometry were seen beneath the North American plate as well. Prior to 30 Ma, the Farallon plate was being subducted under the western margin of the continent, but as soon as the transform boundary initiated, this subduction stopped. Moreover, as the transform boundary grew, the region over which there was no longer any subduction also grew as the Rivera and Mendocino triple junctions moved in opposite directions (Figures 14.19 and 14.20). However, to the north of the Mendocino triple junction and to the south of the Rivera triple junction, subduction continued. This produced a region between the two subduction zones in which no slab was being placed beneath North America, known as the **slab window** (Figure 14.20) and it has played an important role in the tectonics of western North America since 30 Ma.

As the slab window develops, asthenosphere rises from below to fill the space that would have been occupied by the subducting slab. This material is warmer than the down-going slab on either side and this upwelling eventually warms and weakens the continental crust above. This upwelling does not occur passively but follows the motion of the spreading center, which is now subducted beneath the continent. In other words, although the spreading center is subducted, the spreading doesn't stop. Not immediately, anyway. Thus, the warm and weakened crust beneath the growing slab window experiences extension. The consequence of this is a series of closely spaced normal faults that produce individual mountain ranges in intervening basins throughout essentially all of Nevada and parts of California, Arizona, Utah, New Mexico, and the northern parts of Mexico (Figure 14.21). The extension in some regions exceeds 50 to 100% (Figure 14.22).

The region that is now being pulled apart in extension previously experienced compression during the Laramide Orogeny and before. Therefore, in the **Basin and Range province**, we see the superposition of the effects of an extensional tectonic regime on the products of an earlier tectonic history produced by compressional forces. So, the geologic history of

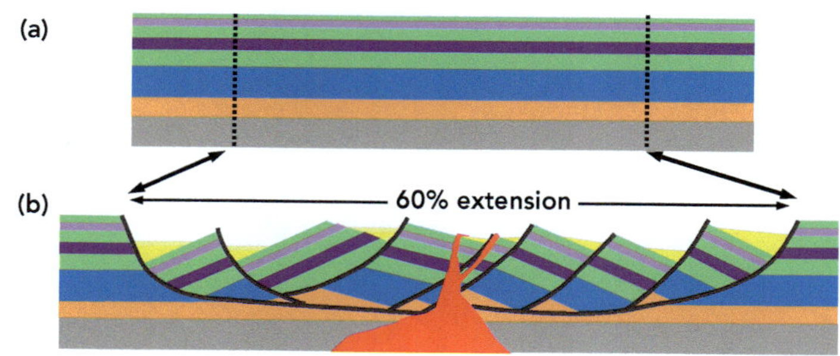

Figure 14.22 Illustration of the formation of Basin and Range topography via extensional faulting. This example starts with simple layer-cake stratigraphy, but most ranges in the Basin and Range physiographic province contain sequences of rocks that were deformed prior to the beginning of extension around 30 Ma, in particular during the Laramide Orogeny (90 to 30 Ma).

ranges in the Basin and Range province is more complicated than shown in Figure 14.23, in which there is assumed to be a simple layer-cake stratigraphy prior to extension.

Over millions of years, this combination of a no-slab window and asthenospheric upwelling has led to the extensive rifting, faulting, and stretching of the crust in western North America, giving rise to the stunning and unique topography that we see today in the Basin and Range province. This is a region in which many geologic stories can be told from studying the rocks in just one place. We used the Grand Canyon as an example of this in Chapter 1, but hundreds of locations in the Basin and Range province contain a Paleozoic sequence of sedimentary rocks deposited in a passive margin after the rifting of the supercontinent Rodinia in the late Proterozoic, and a Mesozoic sequence of sedimentary and volcanic rocks intruded by plutonic rocks. In most places this Paleozoic–Mesozoic package was deformed during compressional tectonics in the Mesozoic and early Cenozoic, and deformed a second time under extension in the Late Cenozoic Era. The basins formed during this younger phase of tectonics are filled with younger sedimentary and volcanic rocks.

14.5 Hot Spots and Plate Motions

In Chapter 5 we introduced the idea of hot spots, or areas where mantle upwelling occurs. One of the best-known active hot spots in the continental USA is the Yellowstone volcanic zone, centered in the northwest corner of the state of Wyoming. Like the Hawaiian–Emperor chain in the Pacific (Figure 5.22), movement of North America over this hot spot has left a trail of volcanic deposits, allowing us to better understand the movement of the overlying plate (Figure 14.23). The Columbia River Basalts form a large igneous province (LIP) that covers over 200,000 square kilometers of northwest USA, including much of Washington, Oregon, Idaho, and Nevada, and they likely extended into western Canada (Figure 14.23). These were the rocks through which the floods starting at Glacial Lake Missoula carved the features studied by J Harlan Bretz discussed in Chapter 2.

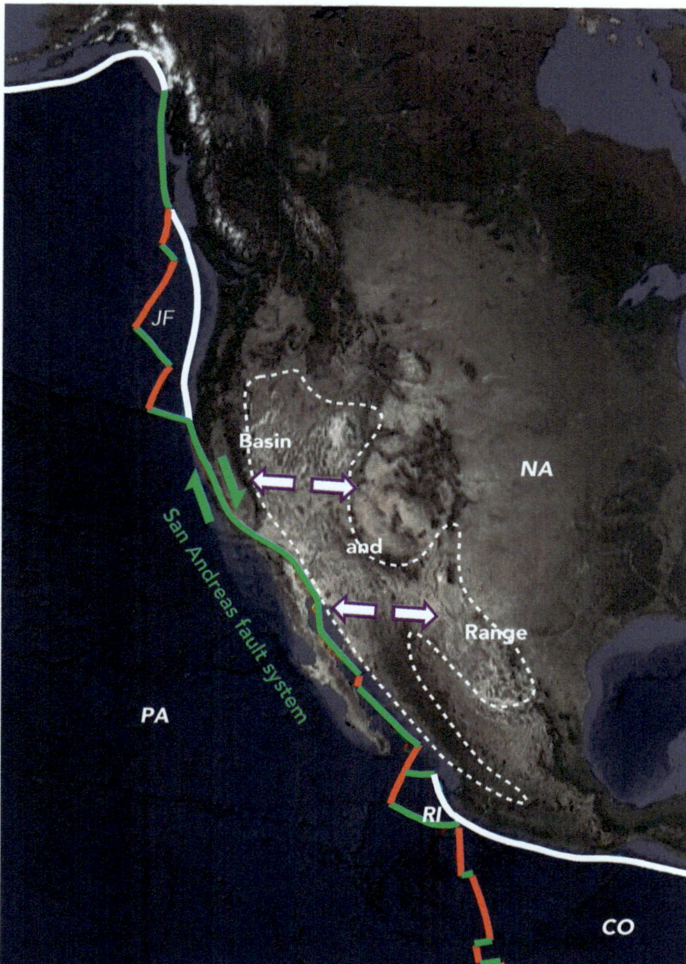

Figure 14.21 Location of the physiographic province known as the Basin and Range. White arrows indicate the regional direction of extension. Lower panel shows the typical landscape of the province.

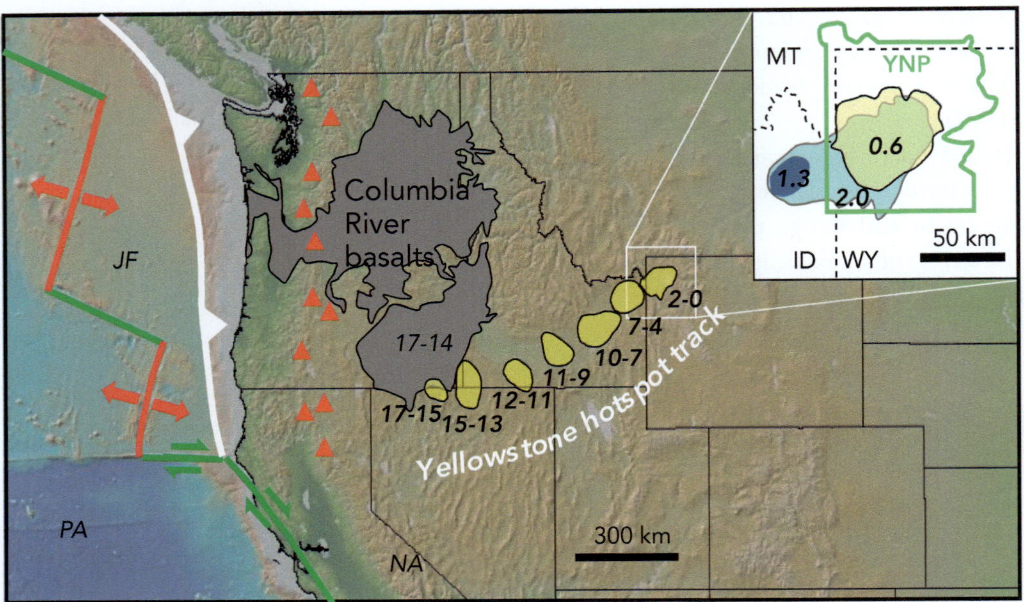

Figure 14.23 Map of volcanic centers associated with the Yellowstone hot spot. Gray region indicates location of lavas of the Columbia River basalts (erupted 17–14 Ma). Yellow ovals show location of rhyolitic volcanic centers with timing of eruption in Ma. White line is the site of the Cascadia subduction zone, green lines are strike-slip plate boundaries, and red triangles show locations of active volcanoes. Inset shows location and ages of eruption (in Ma) of the three youngest calderas of Yellowstone National Park (YNP) region.

(a)

(b)

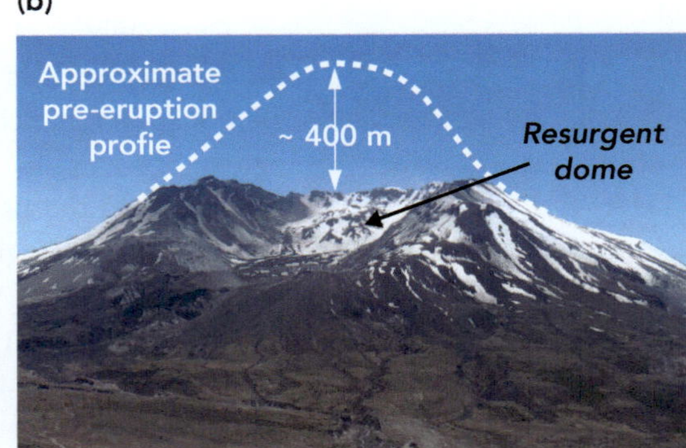

Figure 14.24 Mount St. Helens (a) taken on the morning of May 18, 1980, at the beginning of the main eruption (view from the south). (b) View from the north, taken in June 2019. Source: (a) US Geological Survey; (b) Photo by Peter Copeland.

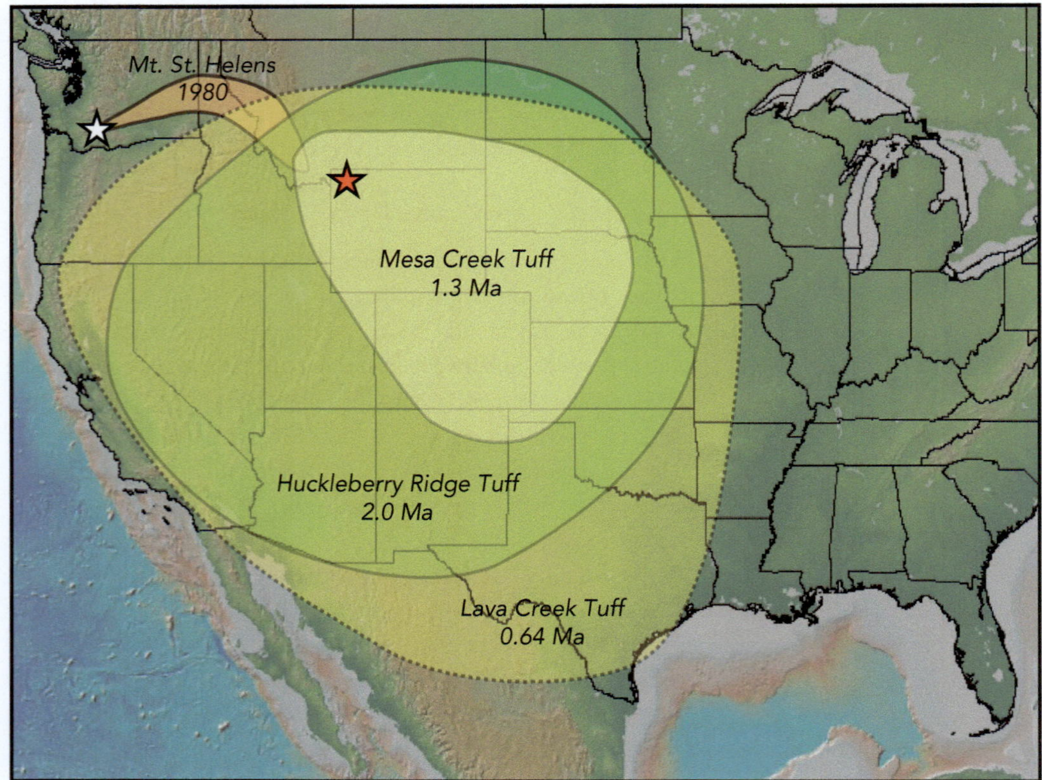

Figure 14.25 Map comparing depositional area for three major eruptions from the Yellowstone volcanic center (red star) to the 1980 eruption of Mount St. Helens (white star).

The Columbia River basalts were erupted mostly as lavas during the interval from 17 to 14 Ma. The chemical composition of these basalts is not consistent with formation in a subduction zone but more like basalts seen at other hot spots such as Hawaii and Iceland. These lavas extend far north of the track defined by younger volcanic rocks aged 17.0 to 0.6 Ma (Figure 14.23), possibly because the magma was deflected to the north by the complex structure of the crust and mantle in this location.

As North America migrated southwestward over the hot spot, there was change in the type of volcanism from more basaltic to more silica-rich **rhyolitic lavas** that have a granitic (i.e., felsic) composition. The change from more basic to more felsic magmatism indicates that the basaltic magma, which was rising to the surface from 17 to 14 Ma to produce extensive basalt flows in Washington and Oregon, began ponding within the continental crust. The residual heat from these mafic magma bodies was enough to melt the surrounding continental rocks, which are mostly made of quartz and feldspar with low melting points. The different composition of the crustal melts produced a different sort of eruptive style. Rather than the relatively gentle eruptions of basaltic lava (as seen in Hawaii), felsic magmas erupt explosively because they are more viscous. The high viscosity means the magma cannot easily flow through small cracks in the rock and eruptions may not occur until the pressure of volatiles (mostly water) accumulating in the magma becomes greater than the strength of the surrounding rocks and they fail catastrophically, producing explosive eruptions. The force of these eruptions often leaves behind a crater known as a **caldera**.

A series of calderas that overlie associated domes, formed by upwelling rhyolitic magmas produced from the heat of the mantle plume, has been mapped and dated showing a progression from activity beginning about 17 Ma in the SW and progressing to the NE, getting younger until the active volcanism in and around Yellowstone National Park (Figure 14.23). The three calderas in this region were formed 2.0, 1.3, and 0.64 million years ago. The eruptions that created these calderas were some of the largest observed in the geologic record. The youngest caldera was formed during an explosive eruption that threw approximately 1,000 cubic kilometers of material into the air. The ash-fall deposits extend from the Pacific coast inland to central USA, including southwestern Minnesota, Iowa, and Missouri, covering an area of about 5 million square kilometers. The caldera covers an area of about 3,000 km^2 (Figure 14.23 inset), which later filled with rhyolitic lava flows that erupted 140 ka and 70 ka.

The pattern in time and space shown by the several volcanic centers shown in Figure 14.23 indicates that North America has been moving southwestward over a mantle plume feeding the Yellowstone region (assuming a fixed-hot-spot model). This is consistent with conclusions about the motion of North America based on magnetic anomalies in the Atlantic and Pacific oceans.

To put into perspective the size of these Yellowstone eruptions (which are so large the system has been given the designation of "supervolcano"), let us compare Yellowstone to the most recent volcanic eruption in the region and the deadliest eruption in the recorded history of North America. On May 18, 1980, at 8:30 a.m., the top of Mount St. Helens in southern Washington was blown off (Figure 14.24), setting off fires, mudslides, and floods, and killing 57 people. The eruption decreased the height of the mountain by over 400 m in just a

Figure 14.26 Volcanic rocks typical of the Yellowstone hot spot. (a) Yellowstone Falls carves into yellow and reddish colored rhyolite layers. (b) Close up of a rhyolitic breccia. (c) Pumice fragments compressed flat from the weight of overlying material.

few minutes. The ash from the eruption covered parts of Washington, Idaho, and Montana to depths of a few centimeters. The total amount of material ejected into the air is estimated to be about $0.5\,km^3$. The eruption of Mount St. Helens illustrates that subduction is still active in the northwest part of the United States, with active volcanoes present from northern California to northern Washington (Figures 14.11, 14.19, and 14.23).

Although the eruption of Mount St. Helens was indeed an impressive event, it was small in comparison with the eruptions known to have been sourced from the Yellowstone region. The deposits from the three most recent eruptions from the Yellowstone volcanic center are named the Huckleberry Ridge, Mesa Falls, and Lava Creek deposits. The last is named for a location with an older lava flow – all of the material from these eruptions was in the form of **pyroclastic** (fire particles) ash ejected in violent eruptions (see Figure 14.25). These eruptions had approximate volumes of 2450, 280, and $1,000\,km^3$, respectively. In other words, these eruptions were between 500 and 5,000 times bigger than the eruption of Mount St. Helens in 1980. Whereas the deposits from Mount St. Helens were just a few centimeters thick at their thickest, deposits of the three Yellowstone eruptions

were in some places as thick as $200\,m$ and the distribution of the ash covered most of the United States west of the Mississippi (Figure 14.25). Outcrops of these rocks are shown in Figure 14.26.

The three most recent major eruptions associated with the Yellowstone hot spot erupted at 2.0, 1.3, and 0.64 million years ago, suggesting a recurrence interval of about 700,000 years. Given that the last eruption was 640,000 years ago, is the region "due" for an eruption sometime soon? Geologically speaking, yes, but this really means some time in the next quarter of a million years, unless the mantle plume feeding this activity changes. Such plumes can persist for many millions of years (e.g., the Hawaiian hot spot seems to have been active for more than 74 million years) but the excessive volcanism that created the Shatsky Rise (Figure 14.12) may have lasted for less time than the Yellowstone system has been active.

KEY POINT

The Yellowstone region, the site of three super-eruptions in the past two million years, is thought to be the result of a mantle plume heating the mantle and crust below.

14.6 Summary

- The Paleozoic passive margin of western North America changed to a series of subduction zones in the Mesozoic, producing magmatism and compressional deformation over an area from California in the west and south across the Rocky Mountains to as far as the Black Hills of South Dakota in the northeast.
- Much of the western margin of North America consists of suspect terranes that were formed in the Panthalassic Ocean as microcontinents or volcanic arc that were later accreted to the continent.
- Igneous rocks with initial $^{87}Sr/^{86}Sr$ ratios greater than ~0.706 are thought to have formed by melting of old continental crust; rocks with values lower than this were derived from the mantle.
- The foreland basin of North America reflects the tectonic load placed on the continental lithosphere by shortening in the Rocky Mountains during the Mesozoic and early Cenozoic.

- Deformation and magmatism in western North America were strongly influenced by the angle of subduction of the Farallon plate from 90 to 30 Ma. The angle started steep, then became shallow, then steep again.
- The sea level during the Cretaceous period was very high due to a combination of greenhouse conditions and rapid seafloor spreading in the Atlantic Ocean.
- The formation of the San Andreas fault is a consequence of North America contacting the Pacific–Farallon spreading center, causing a change in plate geometry, leading to extension in the Basin and Range province.
- As Western North America rode over a hot spot it formed a series of volcanic centers, including the Columbia River Basalts in Oregon and Washington, and culminating in the Yellowstone rhyolites. The change from basaltic to rhylotic lavas reflects partial melting of thicker felsic continental crust in the Yellowstone area, versus direct delivery of mantle plume melts farther west.

Key Words

- Rocky Mountains
- Sierra Nevada Batholith
- Cascade Mountains
- suspect terranes
- pericratonic blocks

- Tethyan terranes
- ribbon continent
- Farallon plate
- wedge-top
- foredeep

- forebulge
- backbulge
- Western Interior Seaway
- dynamic topography
- Cretaceous Thermal Maximum

- anerobic
- Laramide Orogeny
- oceanic plateau
- Shatsky Rise
- roll back
- flat-slab subduction
- left-lateral strike-slip
- Cocos plate
- Mendocino triple junction
- slab window
- Basin and Range province
- rhyolitic lavas
- caldera
- pyroclastic

Further Reading and References

Armstrong, R. L., 1968, Sevier orogenic belt in Nevada and Utah, *GSA Bulletin*, 79(4), 429–458, https://doi.org/10.1130/0016-7606(1968)79[429:SOBINA]2.0.CO;2.

Colpron, M., and Nelson, J. L., 2009, A Palaeozoic Northwest Passage: incursion of Caledonian, Baltican and Siberian terranes into eastern Panthalassa, and the early evolution of the North American Cordillera, *Geological Society, London, Special Publications*, 318, 273–307, https://doi.org/10.1144/SP318.10.

Copeland, P., Currie, C. A, Lawton, T. F., and Murphy, M. A., 2017, Location, location, location: The variable lifespan of the Laramide orogeny, *Geology*, 45(3): 223–226. https://doi:10.1130/G38810.1.

DeCelles, P. G., and Giles, K. A., 1996, Foreland basin systems, *Basin Research*, 8, 105–123, https://doi.org/10.1046/j.1365-2117.1996.01491.x.

Marcilly, C. M., Torsvik, T. H., and Conrad, C. P., 2022, Global Phanerozoic sea levels from paleogeographic flooding maps, *Gondwana Research*, 110, 128–142, https://doi.org/10.1016/j.gr.2022.05.011.

Yonkee, W. A., and Weil, A. B., 2015, Tectonic evolution of the Sevier and Laramide belts within the North American Cordillera orogenic system, *Earth-Science Reviews*, 150, 531–593.

Review Questions

1. What evidence suggests that large parts of the western margin of North America are "suspect terranes" formed far from their current location?
2. How do Sr isotopes tell us something about continental evolution?
3. What factors influence the formation of foreland basins?
4. What conditions led to the widespread deposition of black shales during the Cretaceous Period?
5. Why was the sea level so high in the Cretaceous?
6. How did the Shatsky Rise and its conjugate form? How did this influence the tectonics of North America?
7. Why was magmatism prominent in the Sierra Nevada range in the Jurassic and Cretaceous but shut off from after about 70 Ma?
8. Why was the expression of the Laramide Orogeny in eastern Wyoming and South Dakota so brief, compared with locations to the southwest?
9. How can the modern Andes help us understand the tectonics of the Rocky Mountains from 90 to 30 Ma?
10. How did the San Andreas fault system form?
11. Describe the evolution of the Basin and Range province.
12. What formed the Columbia River Basalts and why are the Yellowstone area volcanic rocks more rhyolitic?

Base Camp Below Makalu Peak.
Source: Charlie Munsey / Getty Images.

Chapter 15

The Indo-Asian Collision

The Formation of the Himalaya

LEARNING OBJECTIVES

- Explain how continent–continent collisions are different from ocean–continent collisions.
- Identify the key evidence that indicates the beginning of the Indo-Asian collision.
- Understand how paleomagnetism played a role in the understanding of the Himalayan orogen.
- Describe the concept of continental escape and how it accommodated continued compression of the Himalaya after the Indian continent had collided with Asia.
- Illustrate the relationship between metamorphism and the major faults of the Himalaya.
- Recall how normal faults develop in a compressional orogeny.
- Explain how we understand the Himalayan mountains by studying the sediments derived from them.

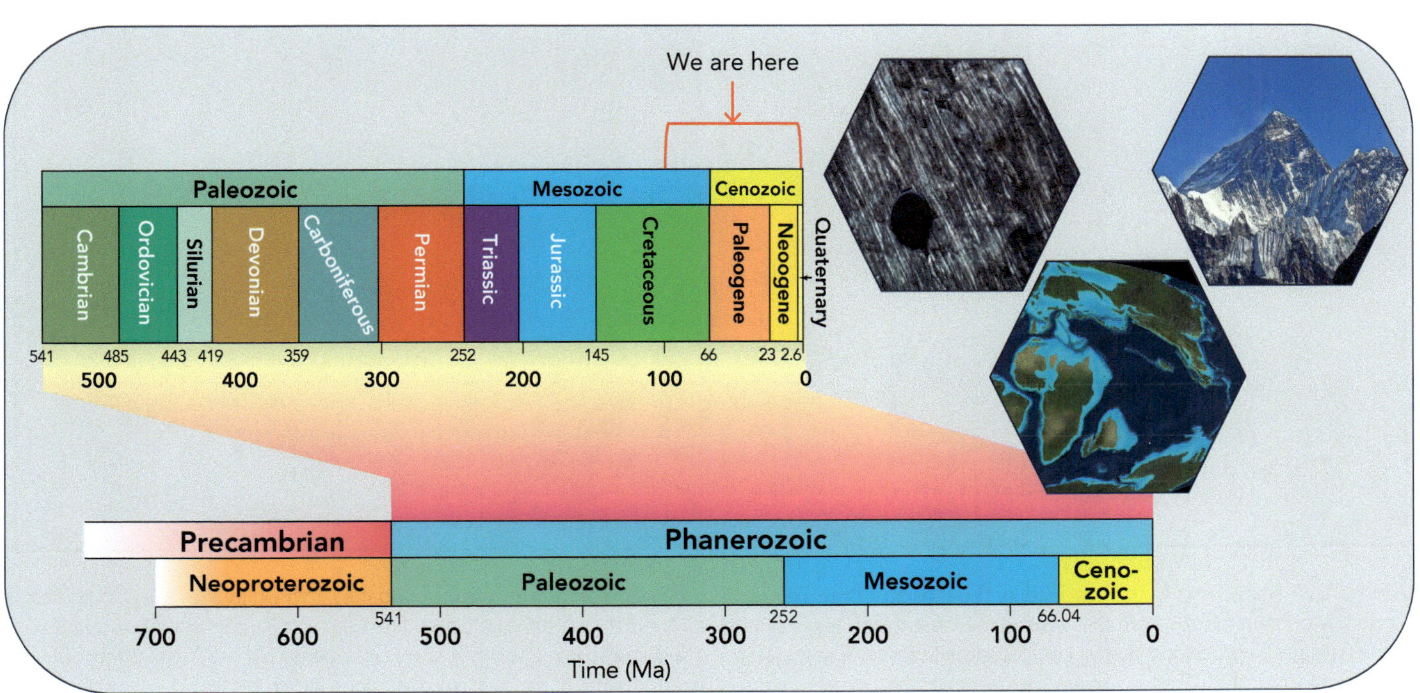

Introduction

The Himalaya is the highest mountain range on Earth, stretching for more than 2,500 km (1,500 miles) with dozens of peaks having elevations over 6,300 m (20,000 feet). At the summit of Mount Everest, the highest point on Earth, at 8850 m (29,029 feet) above sea level, we find limestone, originally deposited below sea level. The great height of the mountains led to our understanding of isostasy via the controversy of topographic compensation debated by Pratt and Airy (see Chapter 3). The first formal geologic study of the region may have been by Swiss geologists Arnold Heim and Augusto Gansser, in 1936, soon after the concept of continental drift was taking hold in Europe (see Chapter 5). Heim and Gansser, in a classic 1939 memoir, identified the major tectonostratigraphic zones and major faults of the Himalaya, which we describe below. Gansser followed this with his seminal work, *Geology of the Himalaya*, in 1964 and continued to publish on the region past his 90th birthday. In the middle part of the twentieth century, much of the mapping of the Himalaya was done by Indian geologists, notably from Kumaun University in Nainital and the Wadiya Institute of Geology in Dehradun, both located in the foothills of the Himalaya.

Despite the general allure of high mountains and the clear significance the Himalaya must have in any recounting of Earth history, the logistics and politics of the area kept the Himalayan Mountains and the adjacent Tibetan plateau relatively unstudied until the 1970s. Since then, as fieldwork became easier, geologists from India, Nepal, Pakistan, and China were joined by collaborators from Europe and North America applying a variety of geologic techniques, new and old, some of which we will discuss in this chapter.

The area of geologic activity includes not just these high mountains but the Tibetan plateau to the north, an area three times the size of Texas with an average elevation of approximately 5,000 m (16,400 ft) above sea level (Figure 15.1). All of this is the consequence of the closing of an ocean, known as **Tethys**, which separated Africa, India, Australia, and other islands to the south from Eurasia to the north. This collision began in the Atlas Mountains of northwest Africa in the Mesozoic and progressed like a closing zipper, eastward through the Pyrenees, Alps, Dinarides, and Carpathian Mountains in Europe, the Caucasus between the Black Sea and the Caspian Sea, the Zagros Mountains in Iran and Iraq, and the Himalayan Mountains stretching from Afghanistan and Pakistan, through India, Nepal, and Bhutan, across southern Tibet and into southwest China (Figure 15.1). In Figure 15.2 we see the movement of the continents from 120 Ma to today, illustrating the closing of the large ocean down to just small fragments of the Tethys Ocean crust, which can be found in parts of the Mediterranean, Black, and Caspian seas today.

The Himalayan segment of this ocean closure is the consequence of a collision between the continents of Asia and India that has been going on for around 50 million years and will be the focus of this chapter.

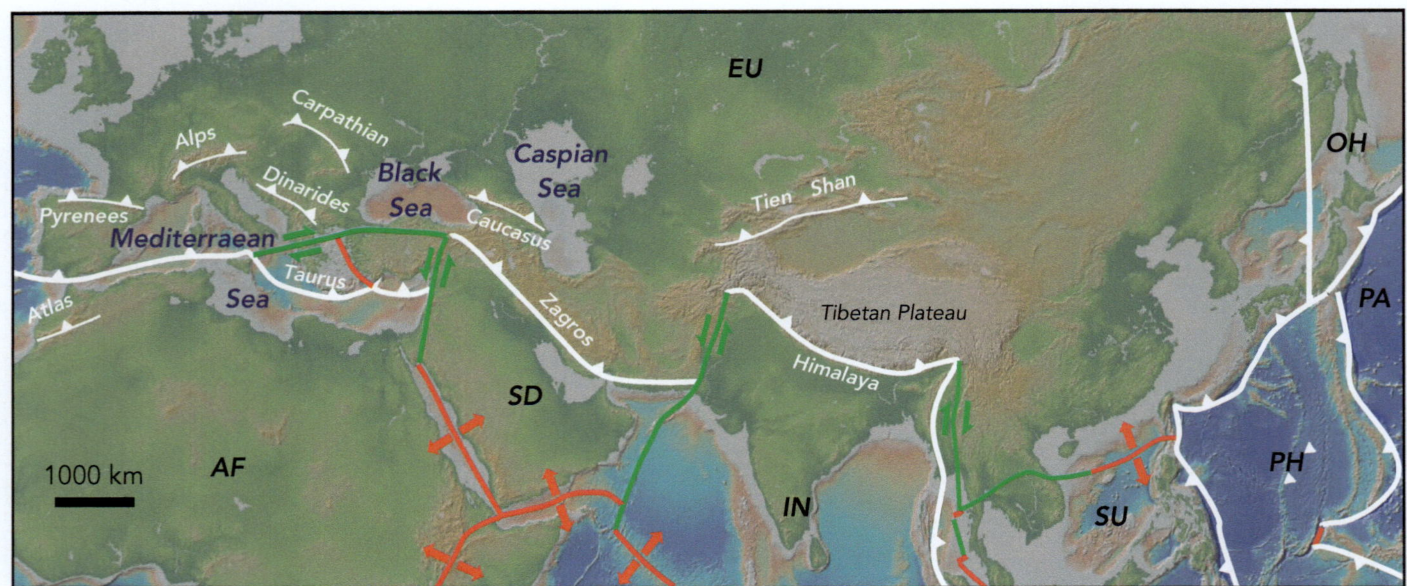

Figure 15.1 Map showing the products of the closing of the Tethys Ocean. Mountain ranges from the Atlas in the west to the Himalaya in the east were all produced during compressional tectonics associated with the closing of the Tethys Ocean. Remnants of this ocean remain in the Mediterranean, Black, and Caspian seas. Bold white lines show modern convergent plate boundaries; thin white lines show locations of intra-plate shortening. Green lines show strike-slip plate boundaries. Red lines show divergent plate boundaries. AF – African plate; SD – Saudi plate. EU – Eurasian plate; IN – Indian plate; SU – Sunda plate; PH – Philippines plate; PA – Pacific plate; OH – Ohkutsk plate.

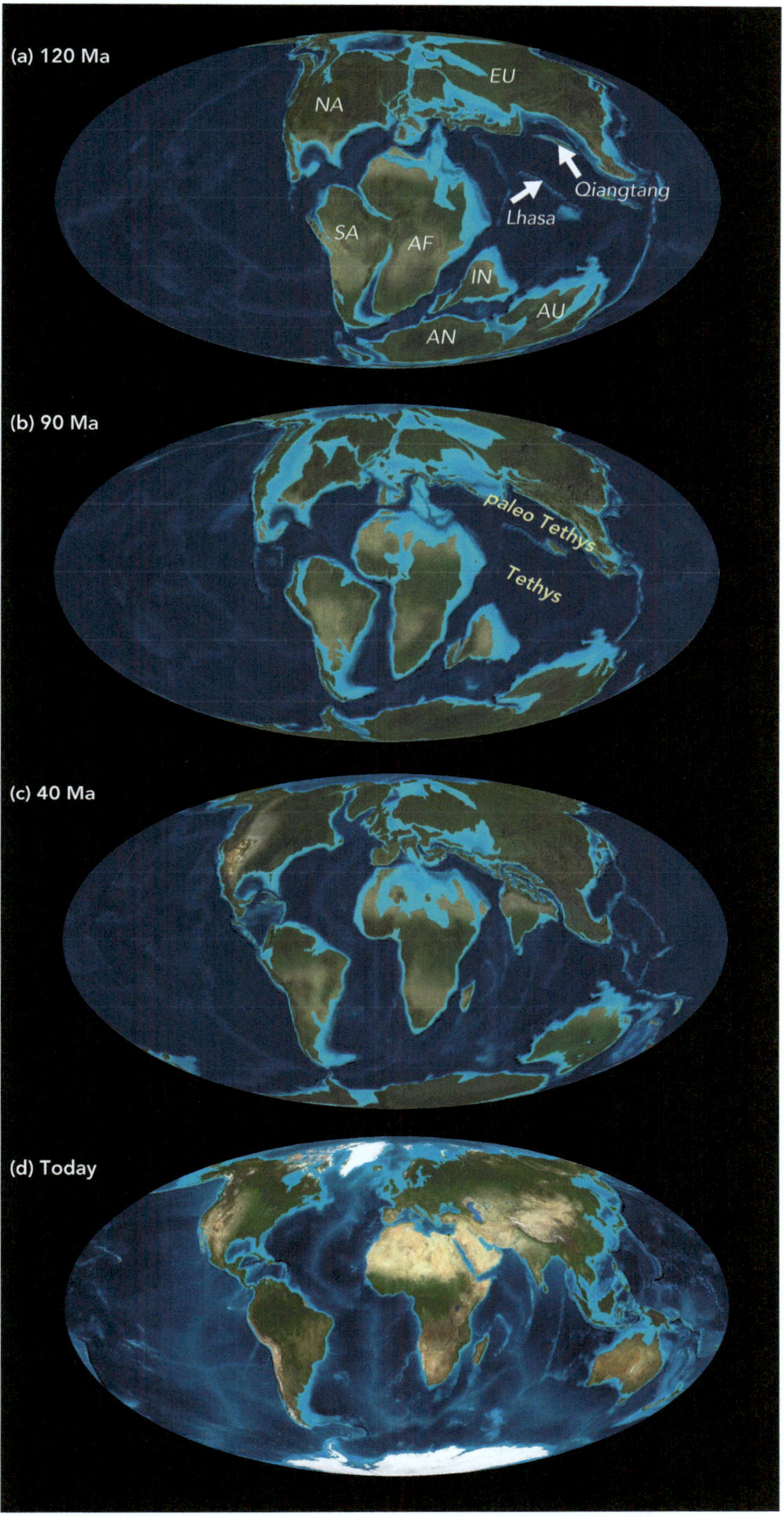

Figure 15.2 Paleogeographic maps at 120, 90, 40, and 0 Ma. Source: © DeepTimeMaps.

15.1 Before the Collision

To best understand this event, it is helpful to first examine the tectonic configuration of the region before collision. The assembly of the Eurasian plate involves more than just the most recent collision of India but rather a sequence of rifting and collision taking place over the Paleozoic and Mesozoic (Figure 15.3). Several continental fragments and island arcs developed during this time and were eventually added to the southern margin of

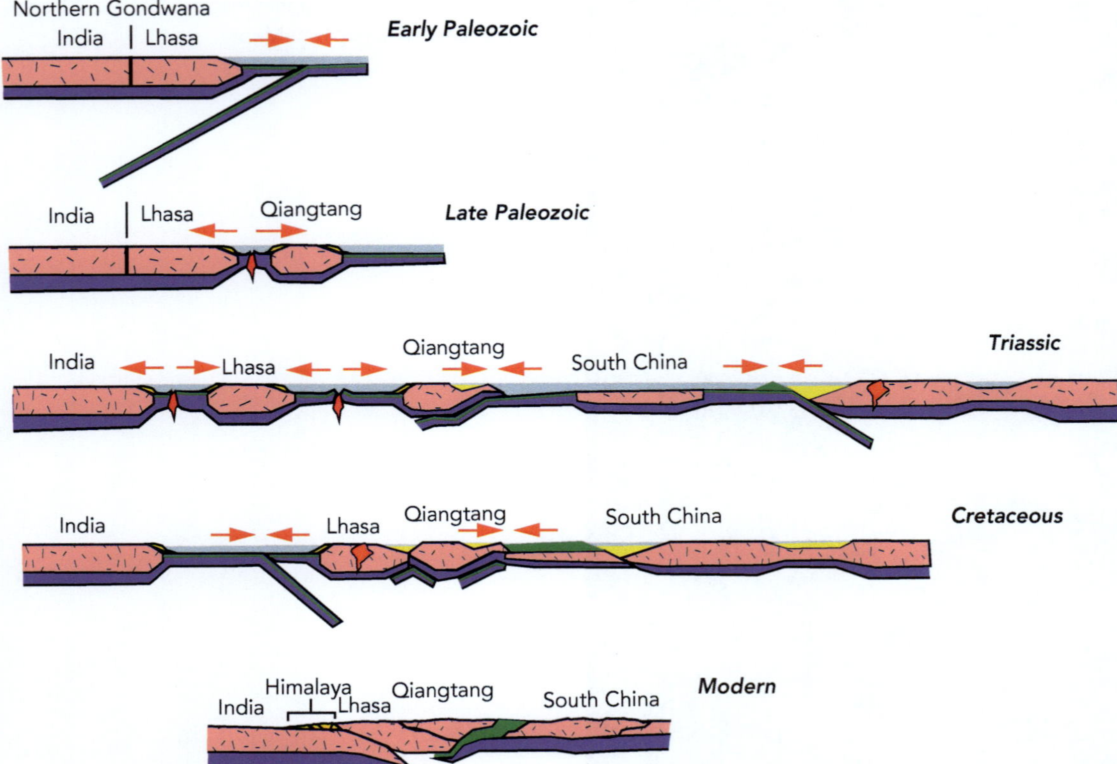

Figure 15.3 Schematic diagram of the assembly of Asia. The assembly of Asia is the product of several continental collisions (and associated destruction of ocean basins) throughout the Phanerozoic.

Asia, producing an Andean-type margin in the Cretaceous in the south (see discussion below). This activity was the prelude to the final closure of the Tethys Ocean with the collision of India and Asia.

15.1.1 Tibet in the Cretaceous

By the end of the Jurassic Period, around 150 Ma, the supercontinent of Pangea had begun to split apart with the opening of the North Atlantic (see Figure 12.2). Around 120 Ma, India, which was originally part of the southern Gondwana supercontinent, broke away from Africa, Antarctica, and Australia via spreading centers on its east and west. This resulted in the birth of the Indian Ocean where the continents separated. To the north lay the ocean geologists call Tethys. This large ocean would close in the same way that the Iapetus Ocean (the precursor to the Atlantic) closed in the Paleozoic, as we discussed in Chapter 10.

The area that would become Tibet was made up of a collection of elongate volcanic island arcs and microcontinents that were converging towards southern Asia, including the Lhasa and Qiangtang terranes shown in Figure 15.2(a). These elongate terranes are very similar to the terranes that converged on western North America at the same time, as discussed in Chapter 14, and are also similar to the older terranes that formed in the Appalachians during the assembly of Pangea, such as Ganderia and Avalonia in the Iapetus and the Rheic oceans, which we reviewed in Chapter 10. In southern Asia, these

several terranes were amalgamated to the southern edge of the continent by the middle of the Cretaceous Period, around 100 Ma.

India separated from Gondwana around 120 Ma and moved northward towards Asia. Significant ocean basins (the paleo-Tethys and Tethys oceans) were consumed by subduction to accommodate the motion of India towards Asia.

15.1.2 The Gangdese Batholith

The southern edge of the Asian continent in the Cretaceous looked a lot like the western edge of South America does today. That is, there was a subduction zone formed where the crust of the Tethys oceanic lithosphere was being subducted beneath the Asian continent. The evidence for this is seen in a variety of geologic observations, including a series of granites and similar rocks in southern Tibet, on the north side of the Himalaya (Figure 15.4(a) and (c)). These rocks are known as the **Gangdese batholith** and were produced in a tectonic setting identical to the Sierra Nevada batholith in California in the Mesozoic (i.e., subduction of relatively old oceanic crust under continental crust with volcanism near the plate margin; see Chapter 14 and Figure 15.5). Today (as in the Sierra Nevada of eastern California) the Gangdese batholith consists almost only

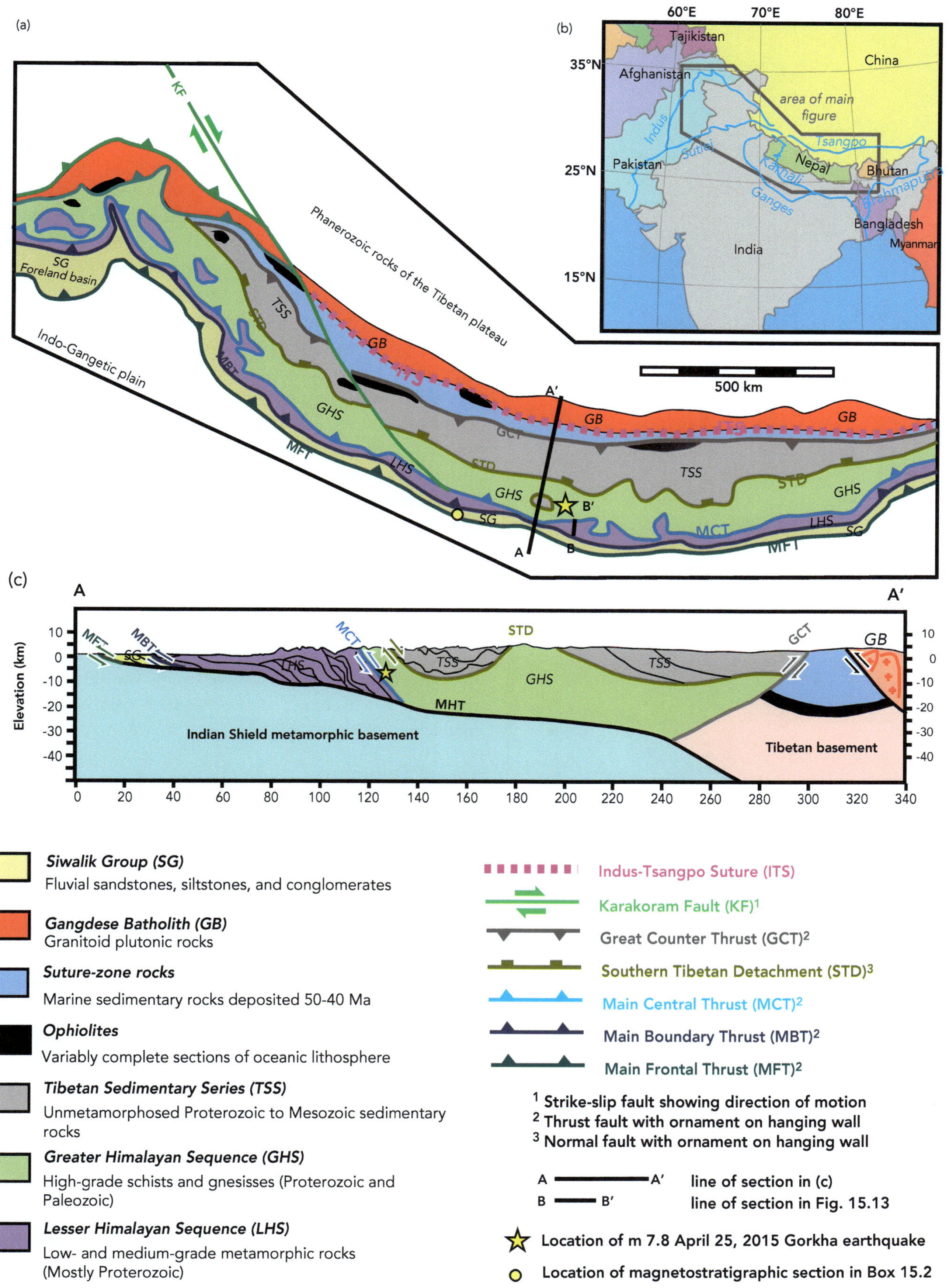

Figure 15.4 Simplified geologic map of the Himalaya. (a) Map showing location of Himalaya in Asia; (b) geologic map of Himalaya and southern Tibet; and (c) geologic cross section along line A–A' in (b).

Figure 15.5 Images of granites from the Gangdese batholith in southern Tibet and the Sierra Nevada batholith in eastern California. (a) Lhasa granite in the background taken in the city of Lhasa, southern Tibet; (b) granites of Yosemite National Park, in the Sierra Nevada mountains of eastern California; and (c) typical mineralogy and texture of rocks from this location. Source: (a) Photo: World Pilgrimage Guide, sacredsites.com; (b) Morten Falch Sortland, Collection: Moment Open, Getty Images; (c) James St. John, CC BY 2.0 <https://creativecommons.org/licenses/by/2.0>, via Wikimedia Commons.

of the plutonic roots of the volcanic arc that once marked the southern edge of Asia.

As the Tethyan plate slid under Asia, it released its fluids and other volatile elements, and these caused partial melting in the overlying Asian plate. The granites are the remnants of a continental volcanic arc that would have looked like modern-day Peru or the Cascades volcanoes in northwest USA. In Tibet and California, almost all the volcanic rocks that were a part of this system have been eroded away and now only the deep plutonic roots of these now-eroded volcanoes are preserved; after subduction ceases, due to a change in tectonic regime, magmatism ceases and erosion takes over. Standing on either batholith today, a geologist can imagine an Andean-type arc that stretched for perhaps 1,500 km. The Gangdese batholith is therefore a testament to the ephemeral nature of any tectonic setting: what was once a continent–ocean collision zone is now the site of an active continent–continent collision. However, as we look along the Tethyan suture we see that the Mediterranean Sea represents the last vestige of this once larger ocean. Along the northern margin, the Mediterranean plate is sliding under parts of Europe, creating the Alps (Figure 15.1). Unlike India, where the ocean has long since vanished via subduction, the Mediterranean oceanic plate is still actively subducting, and devolatilization of the sinking plate results in partial melting that is feeding several active volcanoes, such as Mount Etna, Stromboli, and Vesuvius in Italy, which devastated the Roman towns of Pompeii and Herculaneum in AD 79, and that are still all active today. The absence of active volcanism is a hallmark of the transition from subduction to continental collision.

15.1.3 Ophiolites

As we introduced in Chapter 5 and gave examples of in Chapter 10, ophiolites are remnants of oceanic lithosphere that mark the location of a former ocean. To the north of the high Himalaya a discontinuous group of ophiolites are seen, stretching from southern Tibet to Pakistan (Figure 15.4). Figure 15.6 shows a schematic description of the Xigaze ophiolite, the easternmost ophiolite shown on Figure 15.4.

This and the other examples in Figure 15.4 are all that remain of the several ocean basins shown in Figure 15.2(b). Just as with the ophiolites in the Appalachian Mountains (see Chapter 10), these Tibetan and Himalayan ophiolites are essential in the interpretation of the region and indicate in just a few tens of kilometers the past existence of an ocean once thousands of kilometers wide that marks the line against which the two once separated continents are now juxtaposed.

15.2 The Collision Begins

As India moved northward, the Tethys Ocean became smaller and smaller as oceanic crust was subducted beneath the southern margin of Tibet. At some point, southern Tibet changed

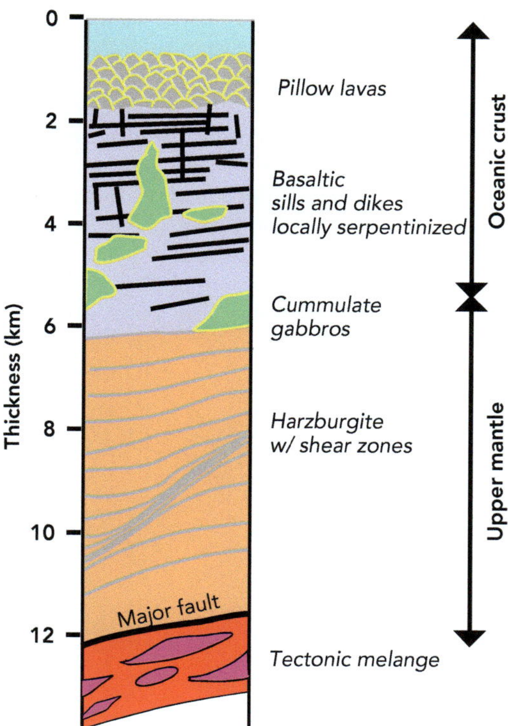

Figure 15.6 Schematic stratigraphy of the Xigaze ophiolite in southern Tibet (the easternmost ophiolites shown on Figure 15.2).

from an ocean–continent convergent margin to a continent–continent convergent margin. Exactly when that occurs rather depends on what definition is used to define the characteristics of the two regimes, but from our perspective today, we must look for changes in the rock record that indicate a change in tectonic setting.

Subduction of the Tethys Ocean brought India and Asia towards each other over thousands of kilometers from the early Cretaceous to the Paleogene. But subduction can only proceed as long as one of the two converging plates can sink into the asthenosphere. This works fine when the down-going plate is oceanic crust because it is denser than the asthenosphere, but as we explained in Chapter 5, continental crust is far too buoyant to subduct. So, when plate tectonic forces move two continents towards each other, the convergence is initially accommodated by subduction of an intervening ocean. When that ocean is consumed, subduction must stop. However, in the case of the collision of India and Asia, convergence did not stop when the subduction did. So, this brings up two important questions: when did subduction stop and how did the collision proceed afterwards?

15.2.1 Change in Velocity of India

As we discussed in Chapter 5, going back at least as far as the work of Alfred Wegener, it has been understood that India was a part of the Gondwanan supercontinent for most of the Paleozoic. Magnetic anomalies in the seafloor indicate that India broke away from Australia, Antarctica, Africa, and Madagascar a bit before 120 Ma (Figure 15.7), beginning its

northward journey towards the eventual collision with southern Eurasia. Because we know the times of magnetic field polarity flips from analysis of magnetic striping patterns (see Chapter 5), we can determine the duration of a particular magnetic epoch (either reversed or normal). The distance between any two boundaries of a magnetic strip on the seafloor is simply read from a map. The rate of spreading for that section of lithosphere is then simply calculated by dividing the distance by the duration. The width and orientation of stripes can change, and this helps decipher the history of plate movement. This, combined with the paleomagnetic pole for rocks of different ages on the continent, allows us to track the position and velocity of India from its separation from Gondwana when it lay in the southern hemisphere, to its initial collision with Asia, to its final position today (Figure 15.7(a)). This shows how the Indian continent was once attached to Antarctica (which was and still is centered on the South Pole) but later passed through the tropics and is now colliding with Asia at 30° N. This trip from near the South Pole to thousands of kilometers north of the equator took more than 100 million years and included a major change in tectonic regime during the journey.

Magnetic anomalies for the Indian Ocean (Figure 15.7(b)) show that the rate of convergence between the Indian continent and Eurasia dropped from about 160 mm/year to around 70 mm/year in the interval from 55 to 40 Ma. At this rate it would take between 70 and 170 years for a chicken to cross a road but from 43 Ma to the present the rate of northward motion of India steadily decreased from 70 to about 40 mm/year (Figure 15.7(c)).

The explanation for the slowing of the motion of India comes from the difficulty in subducting continental lithosphere in a subduction zone as the collision proceeded. With the Tethys oceanic lithosphere fully subducted, the trailing Indian continent that was being tugged along resisted subduction (because of its low density), producing resistance to northward motion. It's easier to subduct oceanic lithosphere under continental lithosphere than it is to jam two continental lithospheres into each other. Thus, the paleomagnetic evidence for a decrease in the northward velocity of India (Figure 15.7(c)) suggests the collision began around 50 Ma and continues to this day. A rate of 40 mm per year may not sound very fast, but the accumulation of motion at this rate over tens of millions of years has produced the Himalaya and the Tibetan plateau, the scale of which are seen nowhere else on Earth today and with few rivals in the past billion years.

15.2.2 Age of Youngest Marine Sedimentary Rocks

There are two key ways in which the sedimentary record has been used to mark the change from subducting Tethys Ocean to continent–continent collision. The first, is the presence of sediments deposited on the Indian plate that were derived from erosion of the Eurasian plate. Deposition routinely takes place where the water meets the shore, but deposition for sand-sized

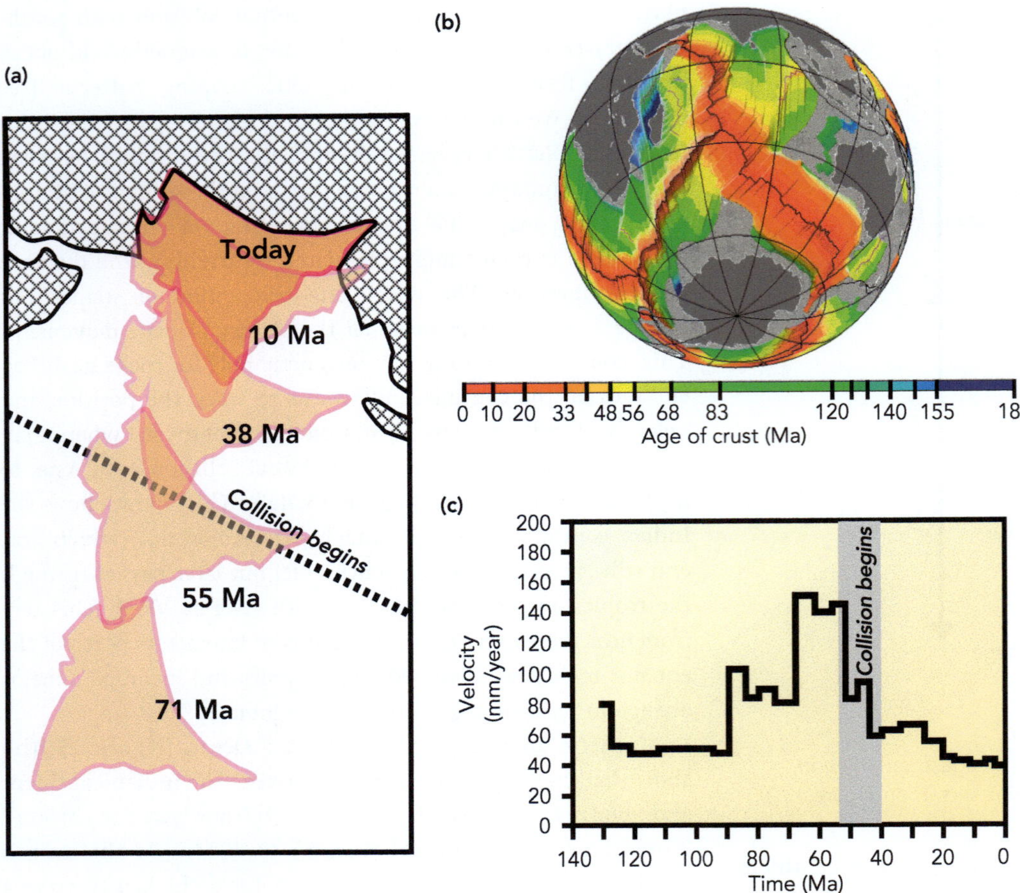

Figure 15.7 The convergence velocity between India and Asia from the beginning of the Cretaceous Period to the present day. (a) Location of India relative to Asia from 71 Ma to present; (b) map of the age of oceanic crust in the Indian Ocean; and (c) velocity of India relative to Asia from 140 Ma to present. Source: (b) Reproduced from Müller et al. (2004).

particles is restricted to a relatively small region relative to the size of most oceans. Sand is big enough to sometimes allow identification of the source, but finer particles, such as clay, are often difficult to trace back to their source.

Around 55 Ma, or at the beginning of the Eocene Epoch of the Paleogene Period, sands with a Eurasian affinity were first deposited on Indian continental crust, indicating the Tethys Ocean must have become so narrow that sediment could be transported across it. We can tell the sediment came from Asia to the north and not India because of the difference in the age of the rocks. Zircons in the sand have been dated (see Chapter 4) and indicate that Asia began to be a significant source for sand deposited on Indian crust around 55 Ma. This indicates the intervening ocean must have been almost gone and very shallow by that time.

Second, the transition from marine sedimentary rocks to either non-deposition or deposition of non-marine rocks would indicate a major tectonic change. Certain rock types, or sequences of types, can be diagnostic of the environment of deposition, and many organisms that eventually become fossils in such rocks either live in salt water or fresh water, but not both. Therefore, it is often a relatively straightforward task to determine if a rock was deposited in the ocean or on land. A survey of sedimentary rocks south of and within the suture zone finds that the oldest non-

marine deposits were formed around 54 Ma and the youngest marine sedimentary rocks were deposited as late as 38 Ma. One reason for such a change could be global sea level dropping (see Chapter 18), but the evidence for the location of sea level at this time from other locations shows that, during the interval from 50 to 40 Ma, there has been very little change since the high stand of the Cretaceous Period (see Chapter 12). However, major drops in the level of the global ocean are observed in rocks deposited at around 31–29 Ma and 15–10 Ma. Therefore, the more likely reason for this change in the environment of deposition of the sands is a local tectonic change caused by the deformation of the Indian plate that lifted the region above global sea level. This tectonic interpretation is strengthened by the observation of angular unconformities in southern Tibet between older marine rocks and younger non-marine rocks.

This tells us that the transition from a time when India was separate from Asia with the intervening Tethys Ocean in between came to an end in the later part of the Eocene Epoch, and no later than around 38 Ma.

15.2.3 Age of Youngest Subduction-Related Magmatism

As discussed above, prior to the beginning of the collision between India and Asia, the convergence between them was accommodated by subduction of oceanic crust. This process leaves a distinct mark in the rock record in the form of igneous rocks produced by subduction. The modern example of an ocean–continent convergence zone is the Andes in western South America, where the oceanic Nazca plate is being subducted beneath South America (see Figures 5.11(b) and 14.16). The most obvious products of this subduction are the volcanoes in Ecuador, Peru, Bolivia, Chile, and Argentina; these volcanoes produce mostly felsic **rhyolites** and andesites (which get their name from the mountains), and such rocks are a signature of subduction zones. The Gangdese batholith (Figure 15.5), described above, is representative of **Andean-type magmatism**. This name comes from the modern example of such ocean–continent tectonics. The Gangdese is a consequence of magmatism associated with the consumption of the Tethys Ocean as India moved toward Asia.

In southern Tibet, the products of subduction form the plutonic rocks of the Gangdese batholith and their few remaining volcanic equivalents. Just as is the case for the Sierra Nevada batholith in eastern California (see Chapter 14), the Gangdese batholith consists of hundreds of individual intrusions, or plutons, and many of these plutons have been isotopically dated, allowing geologists to determine the time and duration of subduction of the Tethys Ocean lithosphere.

Subduction zones are associated with the recycling of oceanic lithosphere into the mantle, and these can introduce specific trace elements to the magmas produced by partial melting of the overlying mantle. Therefore, igneous rocks formed in subduction zones often display higher concentrations of these key elements. Other processes taking place during melting and transport of subduction-related magmas can produce other distinctive chemistries in the plutonic and volcanic rocks formed. Therefore, it is sometimes possible to use variations in major and trace elements in these rocks to determine the tectonic setting in which they formed.

Although magmatism of a wide variety continued in southern Tibet as recently as 17 Ma, a survey of the Gangdese batholith shows that the youngest of the plutons in the region that show a geochemical signature of subduction of Indian oceanic lithosphere are about 40–38 million years old. Because plutonic rocks such as these are indicators of subduction, we can infer that subduction of the Tethys Ocean must have ceased in the later part of the Eocene Epoch, around 40 Ma, and this also coincides with the slowing in convergence rate that we discussed earlier.

A visitor to southern Tibet at 60 Ma would have seen a continental margin similar to the modern day, marked by many active volcanoes, but these would have been shut off by around 38 Ma, after which the volcanoes would have begun eroding away exposing the plutons that represent the cooled deep magma chambers that originally fed the overlying volcanoes.

This points out a key feature of continent–continent collision zones. As we will see, after the oceanic lithosphere that separated the two continents is consumed by subduction, deformation can continue but volcanism cannot.

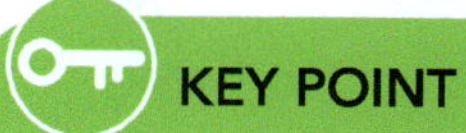

We can date the initiation of the Indo-Asian collision to around 50–40 Ma based on geophysical, geochemical, and sedimentologic data.

15.3 The Collision Continues

So, by several lines of evidence the collision between India and Asia began sometime during the Eocene Epoch. Modern geophysical observations, including the location and magnitude of earthquakes and the tracking of the motion of the Indian continent using GPS measurements, indicate that the collision continues today. Paleomagnetic data from rocks in India shows that the northern edge of the continent was some 2,000 km south of its present location at the time the collision between two continents began. This indicates that although subduction ceased by around 40 Ma, India continued to move north, albeit at a slower pace. A critical question is how can convergence continue if subduction has ceased? Why don't the plates simply lock up? Why doesn't the convergence shift to the boundary between the Indian continent and its now trailing oceanic plate farther south, much like convergence shifted from the Iapetus to the Rheic Ocean during the assembly of Pangea that we discussed in Chapter 10? We will next consider the mechanisms of convergence between India and Eurasia in the absence of subduction. Because much of southern Asia was deformed over this time, it is sometimes best to refer to the convergence between India and Siberia, in the northern part of Asia, because this part of the continent has not been affected by the collision in the south.

The fact that we must travel over 3,000 km away from the suture zone between Indian and Asian lithosphere illustrates how deformation in continent–continent collisions is distributed over a much wider zone than in ocean–continent or ocean–ocean collisions. Figure 15.7(a) shows the position of India at the time of collision, approximately 50 Ma, to be around 2,000 km from the location it occupies today. The convergence of this distance after the cessation of subduction produced this wide zone of deformation.

15.3.1 Continental Escape

When two cars run into each other, one possible result is that each one gets smashed, such that each car is a bit shorter than before the collision. To facilitate this shortening, the cars would have to crumple, making at least some parts now taller than at the beginning. In a plate tectonic collision, the taller parts would be recognized as new mountain belts. However, when two cars collide, deformation of the cars at the point of impact is not the only possible outcome. It's possible for one of the cars to be pushed into a new position as the result of the collision with

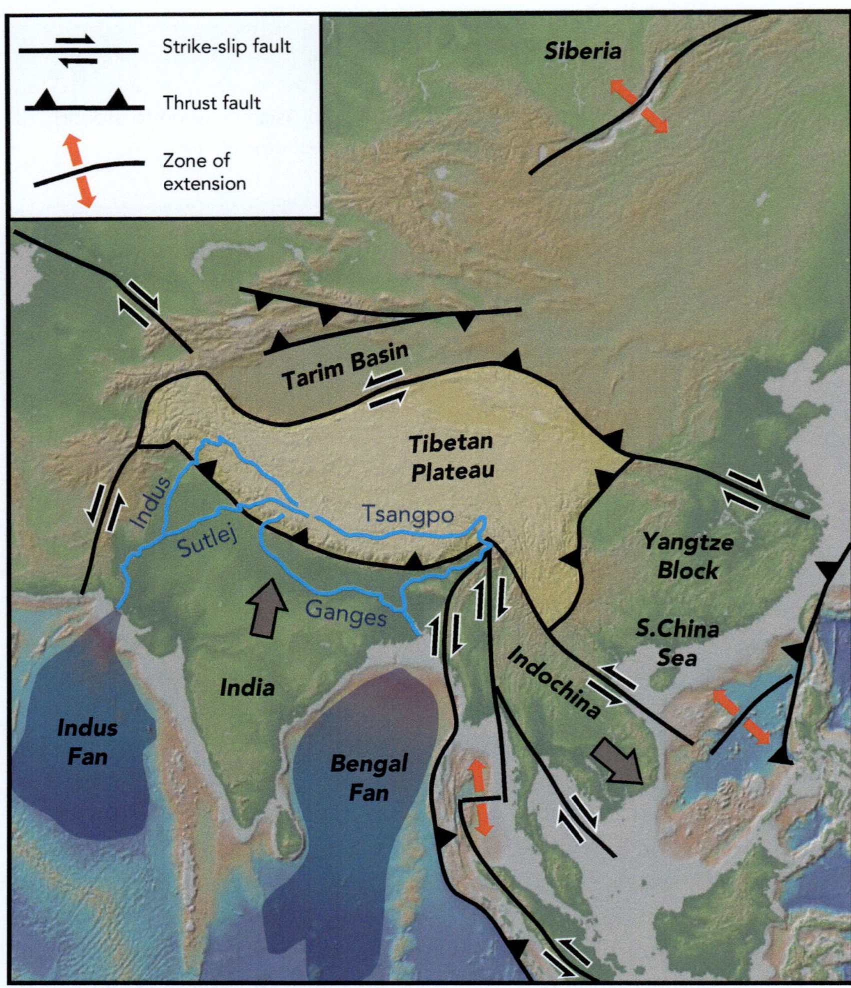

Figure 15.8 Major tectonic elements of the Indo-Asian collision. Gray arrows show direction of motion of tectonic blocks.

only modest deformation of the vehicle, more like striking a billiard ball obliquely. The degree to which the cars might be deformed or displaced might depend on if the road was wet or icy, or the angle of the collision.

In the Indo-Asian collision, the first consequence of the cessation of subduction was not mountain building, but rather a process known as **continental escape**, in which a region moves sideways along large-scale strike-slip faults. In this case the Indochina microcontinent, which includes present-day Vietnam, Laos, Cambodia, Thailand, and Myanmar, moved from its original location in southern Asia southeastward along large-scale strike-slip faults (Figure 15.8).

The sideways translation of Indochina to the east and south caused a reshuffling of parts of southeast Asia. The Yangtze and Indochina blocks (Figure 15.8) moved relative to each other with some rotation, which induced extension or compression between these blocks and adjoining continental and oceanic lithosphere. Extension is best seen in the South China Sea (Figure 15.8), where new ocean crust was formed with a spreading rate around 30 mm/year from 32 to 27 Ma and then more slowly from 27 to 16 Ma (we know the timing from magnetic

anomalies formed in this new oceanic crust during extension). This, and the timing of individual large-scale strike-slip faults (based on cross-cutting relationships) in Indochina and southern China, indicates that continental escape was an important mechanism for the accommodation of the motion between India and Siberia from the early stages of the collision to around 20 Ma. As these blocks slid sideways at rates of 40–50 mm/year, they allowed for continued convergence of mainland India with Asia.

15.3.2 The Collision Continues: The Rise of the Himalaya

Although the sideways translation of modern-day Indochina to the east and south ceased around 25 Ma, the northward motion of India did not. With India and northern Asia continuing to move closer to each other at a rate of about 40 mm/year, another mechanism to accommodate this convergence in southern Asia is required: in this case shortening along major thrust faults. This is recorded in the stratigraphy, structure, and metamorphic record of the Himalaya and is covered in the next section.

15.3.2.1 Stratigraphy and Structure of the Collision Zone

The formation of the highest mountain areas within the Himalaya is a result of compression and uplift of continental blocks, associated with large-scale reverse faults, rather than subduction of oceanic lithosphere. The geology exposed in central Nepal contains four major rock units that show north to south changes in metamorphism, which help us decipher the mechanisms that form high mountains (Figure 15.4).

The northernmost Tethyan sedimentary sequence consists of about 10 kilometers of lightly to unmetamorphosed marine sandstones, shales, and limestones, mostly deposited along the southern passive margin of Asia that originally faced the Tethys Ocean, long before it closed. These sedimentary rocks range in age from Proterozoic to Cretaceous and are similar to age-equivalent rocks found in North America. These rocks, mostly deposited close to or below sea level, form the **Greater Himalayan Sequence**, and are found mostly north of the high peaks of the Himalaya (Figure 15.4).

The Greater Himalayan Sequence consists of about 30 km-thick units of high-grade metamorphic rocks (Figure 15.9) intruded by granites. They form the northernmost and highest part of the Himalaya in central Nepal and show the highest degree of metamorphism, indicating that they were exposed to the highest pressures and temperatures in the central part of the Himalayan orogenic belt (Figure 15.4).

The **Lesser Himalayan Sequence** lies south of the Greater Himalayan Sequence in central Nepal (Figure 15.4). It is composed of less metamorphosed sedimentary and low-grade metamorphic rocks.

The metamorphic rocks of both the Greater and Lesser Himalayan sequences were originally deposited as sediments on the Indian craton when it was still attached to Gondwanaland from the late Proterozoic to early Paleozoic. The difference in their metamorphic assemblages today reflects different levels of burial.

The younger Siwalik Group lies on top of the Indian continent and consists of layers of unmetamorphosed sedimentary rocks, including sandstones, mudstones, and conglomerates (e.g., Figure 15.10), mostly formed in non-marine environments as rivers and alluvial fans (see Figure 1.26 for examples of depositional environments). The sediments were deposited in a **foreland basin**, which is the low area formed by isostatic depression of the Indian lithosphere in front of (i.e., south) of the rising Himalayan Mountains (see Figures 5.23 and 15.4). These sedimentary sequences are a direct consequence of the

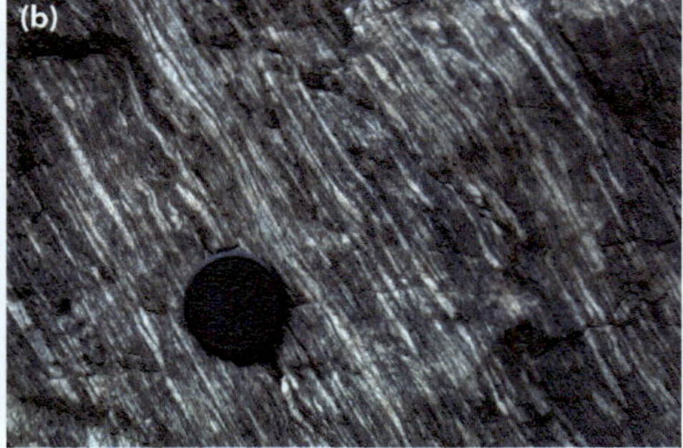

Figure 15.9 Two examples of high-grade metamorphic rocks of the Greater Himalayan Sequence in the field. These rocks were originally sedimentary but were completely transformed during Himalayan metamorphism, creating new minerals and textures. (a) Alternating mafic and felsic gneiss; (b) strongly foliated felsic gneiss. Source: Photos by Alex Robinson.

Figure 15.10 An example of the interbedded sandstones, shales, and conglomerates of the Siwalik basin of Nepal. This material was shed from the Himalaya and deposited in the adjacent foreland basin.

uplift of the Himalaya and, like many of the orogenic episodes that we have discussed in Chapters 10 and 14, represent the sedimentary record of mountain building. Much of the material eroded from the Tethyan sedimentary sequence, and Greater and Lesser Himalayan sequences were dumped into the growing foreland basin as compression, shortening, and crustal thickening progressed.

The structural development of the Himalaya is characterized by a series of large-scale thrust faults and associated folds. As with most fold and thrust belts, the faults tend to develop such that the first major thrust is followed by others that migrate towards the foreland (in the case of the Himalaya, this is to the south), with the locus of convergence moving to the younger faults, forming a series of stacked thrusts. However, older faults can also be reactivated after a newer fault has initiated.

The first thrust to develop in the Himalaya is known as the **Main Central Thrust**. It plays a crucial role in the geology and tectonics of the region (Figure 15.4), and most of the recent earthquakes occur on the Main Central Thrust in the mountains and the Main Frontal Thrust in the foreland (Figure 15.4).

The Main Central Thrust separates the higher-grade metamorphic rocks of the Greater Himalayan Sequence from the lower-grade rocks of the Lesser Himalayan Sequence. The total amount of slip on the thrust system is hundreds of kilometers. Movement on the Main Central Thrust began about 25 Ma. Large earthquakes associated with the Main Central Thrust, such as the 2015 Gorka earthquake (which killed over 9,000 people), have resulted in significant ground ruptures and further displacement along the fault (Box 15.1).

KEY POINT

Since 25 Ma, major thrust faults within the Himalaya have accommodated hundreds of kilometers of motion.

BOX 15.1 The April 25, 2015 Gorkha Earthquake, Central Nepal

Being an active plate margin, the Himalaya are the site of dozens of earthquakes every year. Figure 15.11(a) shows all the earthquakes with magnitude greater than 4.0 in the Indo-Asian collision zone from 1964 to 1995 as well as the location of significant ground ruptures associated with earthquakes since 1505.

The 2015 Gorkha earthquake was a devastating seismic event that occurred on April 25, 2015, in Nepal and the surrounding region. It is the deadliest earthquake to hit Nepal in more than 80 years. The earthquake had a magnitude of 7.8 on the Richter scale. Its epicenter was located in the Gorkha district, approximately 80 kilometers northwest of Nepal's capital, Kathmandu. The initial earthquake was followed by numerous aftershocks (Figure 15.11(a) and (b)), including a major one on May 12, 2015, with a magnitude of 7.3.

The earthquake caused widespread devastation across Nepal and affected neighboring countries, including India, China, and Bangladesh. Thousands of buildings, including historic monuments, temples, and residential structures, were

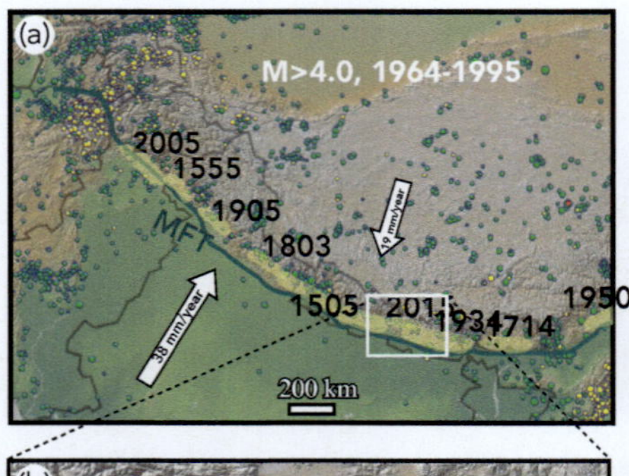

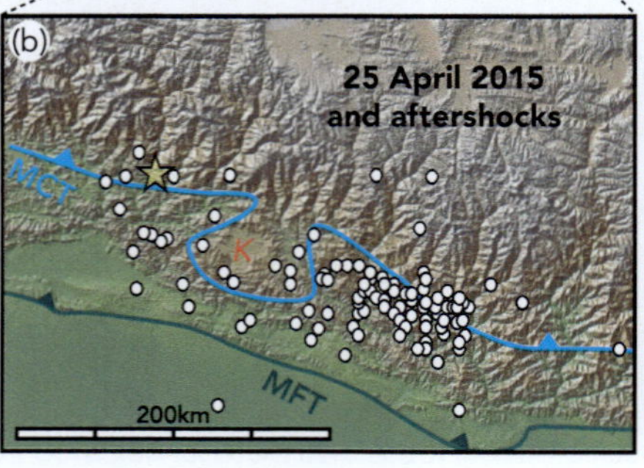

Figure 15.11 (a) Map of earthquakes in the Indo-Asian collision zone of magnitude 4.0 or greater, 1964–1995. Size of symbols is proportional to magnitude; color indicates depth of focus: green = less than 50 km; yellow = 50–250 km; red = greater than 250 km. Also shown in yellow are areas of ground rupture for historical earthquakes. (b) Close up on the area of April 25, 2015, Gorkha earthquake (M = 7.8, yellow star) and subsequent aftershocks greater than M = 4.0. K – Kathmandu. (c) Damage to buildings in Kathmandu. (d) Effect of a landslide in the Langtang valley (approximately 50 km NE of Kathmandu) covering most of a small village. Source: (c) Tom Van Cakenberghe / Getty Images AsiaPac; (d) Punya, CC BY-SA 4.0 <https://creativecommons.org/licenses/by-sa/4.0>, via Wikimedia Commons.

destroyed or severely damaged (Figure 15.11(c)). Entire villages were flattened (Figure 15.11(d)), and infrastructure such as roads and bridges suffered significant destruction. The earthquake resulted in a staggering loss of life. Official figures indicate that over 9,000 people were killed, and more than 22,000 were injured. The actual number of casualties may be higher due to remote and inaccessible areas affected by the earthquake.

The focus of this earthquake (the site of initial rupture) was at a depth of ~8 km near the estimated trace of the Main Central Thrust indicating that although major thrust faults have developed to the south of the Main Central Thrust since its inception around 25 Ma, older faults are still subject to movement in major earthquakes. Figure 15.11(a) shows that almost all earthquakes in this region form at depths less than 50 km; this reflects the lack of significant subduction of continental crust in the collision zone.

The location of the 2015 earthquake was between historic surface-rupturing zones from the 1505 and the 1934 earthquakes. Gaps in the historical record in NW and NE India make these regions more likely for major ruptures in the future.

In Chapter 1, we noted that the high-grade metamorphic rocks in the Grand Canyon were at the bottom, and reflected formation deep in the Earth's crust, in the roots of a long-eroded mountain chain. It is understood that higher-grade metamorphic rocks form at deeper levels in the crust, where the highest temperatures and pressures are realized. Usually, metamorphic grade decreases as one passes structurally and topographically up. One section of the Himalaya in central Nepal, in contrast, shows the opposite. The Greater Himalayan Sequence is composed of a variety of high-grade metamorphic rocks, including gneisses, schists, and marbles. These rocks were originally sedimentary and volcanic rocks that were buried and subjected to high temperatures and pressures during the Himalayan Orogeny. Normally, as rocks are buried deeper and subjected to higher temperatures and pressures, they undergo more intense metamorphism, resulting in the formation of higher-grade rocks. As we introduced in Chapters 1 and 5, in metamorphic terranes the grade of metamorphism can be determined by looking at the minerals that formed at the highest temperature (Box 15.2, Figure 15.12).

In general, metamorphic grade (higher temperature and pressure) increases with depth, so when we see metamorphic rocks at the surface with high-grade mineral assemblages, we understand that such areas must have experienced more erosion than nearby zones with minerals that reflect lower temperatures and pressures, and we expect relatively smooth transitions across metamorphic zones (Box 15.2). One of the most distinctive features of the metamorphic rocks above the Main Central Thrust zone is that the grade of metamorphism *increases* as one moves topographically up. This is referred to as **inverted metamorphism** because usually, the observed sequence of metamorphic rock types is to have the low-grade rocks on top and the high-grade rocks on the bottom of any eroded sequence.

In this case, the apparent inverted metamorphism is the result of the high-grade metamorphic rocks of the Greater Himalayan Sequence having been thrust over lower-grade rocks of the Lesser Himalayan Sequence. Thrusting allowed higher-grade metamorphic rock formed deeper in the crust to be placed over lower-grade rocks formed at an originally shallower level. The assemblages of metamorphic minerals in the Greater Himalayan Sequence indicate they were formed at a depth much greater than the Lesser Himalayan Sequence that are now above. From this we know that there have been many tens of km of movement on the Main Central Thrust that thrust these higher-grade metamorphic rocks on top of the more modestly metamorphosed sequence to the south. This thrusting influenced the uppermost part of the Lesser Himalayan Sequence, which we can see in the texture and mineral assemblage of the Lesser Himalayan Sequence rocks in direct contact with the Greater Himalayan Sequence at the Main Central Thrust.

However, the Main Central Thrust fault (MCT, Figure 15.4) has been active, dormant, and then reactivated several times. Figure 15.13(d) illustrates that the first thrusting of the Greater Himalayan Sequence over the Lesser Himalayan sequence began around 25 Ma but over time, new thrusts developed to the south of the original MCT fault. This action caused rocks that were unmoved below the original MCT fault (i.e., in the original footwall, see Figure 1.35(b)) to now be above the new thrust, becoming part of the hanging wall (Figure 1.35). Figure 15.4 shows the main younger faults that formed south of the MCT (labelled MBT and MFT) with the MBT forming first then the MFT. Since about 15 Ma, these faults, along with faults of lesser displacement, have taken turns accommodating the convergence between India and Asia, which continues at ~4 cm/year (Figure 15.7). Broadly speaking, orogens develop major faults with deformation moving across the region. In the case of the Himalaya, the broad sequence of fault development is from north to south. Reactivation of faults such as the MCT after significant movement has occurred on other faults to the south is called out-of-sequence thrusting.

Although widespread subduction-related magmatism was shut off around 40 Ma, some very localized melting has occurred within the Himalaya during the evolution of the mountains. These small dikes and sills of magma cross cut older units and thus provide a constraint on our understanding of the history of structures such as the Main Central Thrust because these

BOX 15.2 Barrow's Zones and Metamorphic Assemblages

Metamorphic zones are recognized by the mineral assemblages and textures observed in rocks in the field and with the microscope. George Barrow (1854–1932) studied the metamorphic rocks in northeast Scotland and was the first to recognize the order in which minerals appeared in the field (Figure 15.12(a)), which he interpreted to reflect the different metamorphic conditions the rocks experienced. The rocks in NE Scotland were originally shales and he noticed that the lowest grade metamorphic transformation involved clay becoming chlorite. Thus, rocks with chlorite (but not biotite) were designated to be in the *chlorite zone*. As temperature and pressure increased, some of the chlorite is transformed to biotite but by the time garnet is

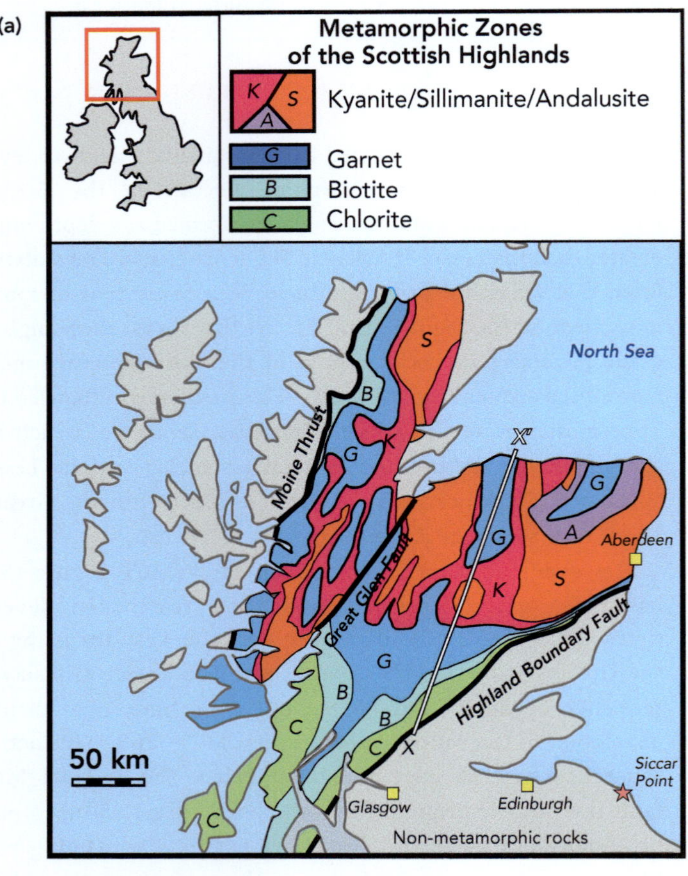

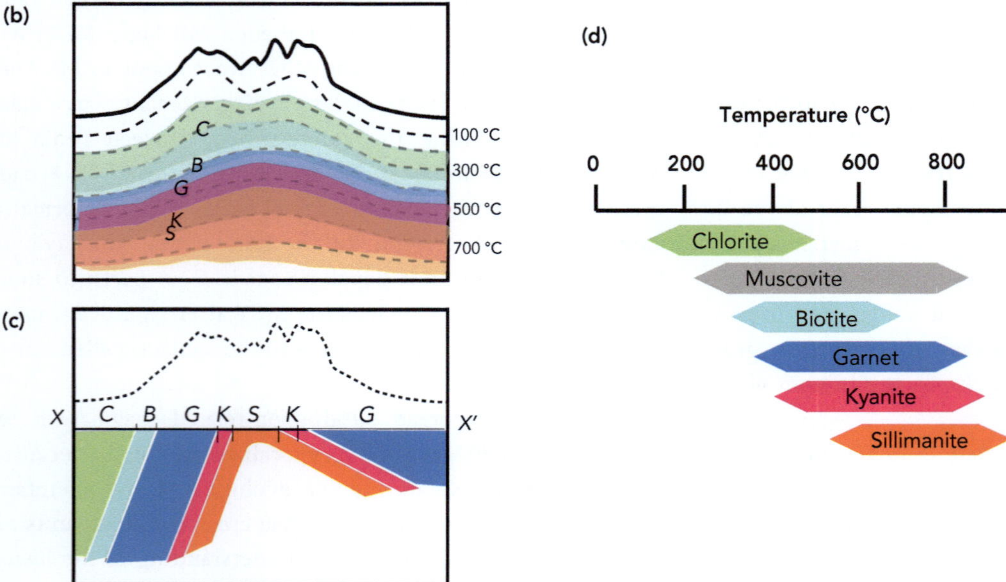

Figure 15.12 (a) Metamorphic zonation map of the Scottish Highlands based on the work of George Barrow in the early twentieth century. (b) Schematic depiction of the pattern of **Barrow's zones** beneath an active mountain range. (c) Cross section X-X' in (a), showing folding of metamorphc zones subsequent to metamorphism. (d) Approximate temperature range for formation of key metamorphic minerals in the metamorphism of shales.

formed all of the chlorite is consumed and as higher metamorphic grades are reached, new minerals are formed at the expense of previously formed minerals (Table 15.1).

The metamorphic zones of the Scottish Highlands mapped by Barrow clearly show the variation in erosion for different areas. The map pattern in Figure 15.12(a) is a result of the original distribution of metamorphic zones as a function of depth (Figure 15.12(b) and (d) and subsequent deformation and erosion (Figure 15.12(c)).

The pattern of metamorphic zonation, originally observed by Barrow in Scotland, can also be seen in other mountain belts around the world including New England. This is, of course, not surprising as Scotland and New England were once part of the same orogenic belt formed when the Iapetus Ocean closed in the Early Paleozoic (see Chapter 10). This pattern of metamorphic assemblages is now referred to as **Barrovian metamorphism**, after Barrow's seminal work in the early twentieth century. By examining the metamorphic assemblage of minerals in a rock, geologists can determine the peak temperature and pressure a metamorphic rock experienced. When placed in a regional context, such information is essential in working out the tectonic history of a region.

Table 15.1 Minerals associated with Barrow's metamorphic zones

Mineral zone	Mineral assemblage
Chlorite	Quartz, plagioclase feldspar, chlorite
Biotite	Quartz, plagioclase feldspar, chlorite, biotite, muscovite
Garnet	Quartz, plagioclase feldspar, biotite, garnet, muscovite
Staurolite	Quartz, plagioclase feldspar, biotite, garnet, staurolite, muscovite
Kyanite	Quartz, plagioclase feldspar, biotite, garnet, kyanite, muscovite
Sillimanite	Quartz, plagioclase feldspar, biotite, garnet, sillimanite, muscovite

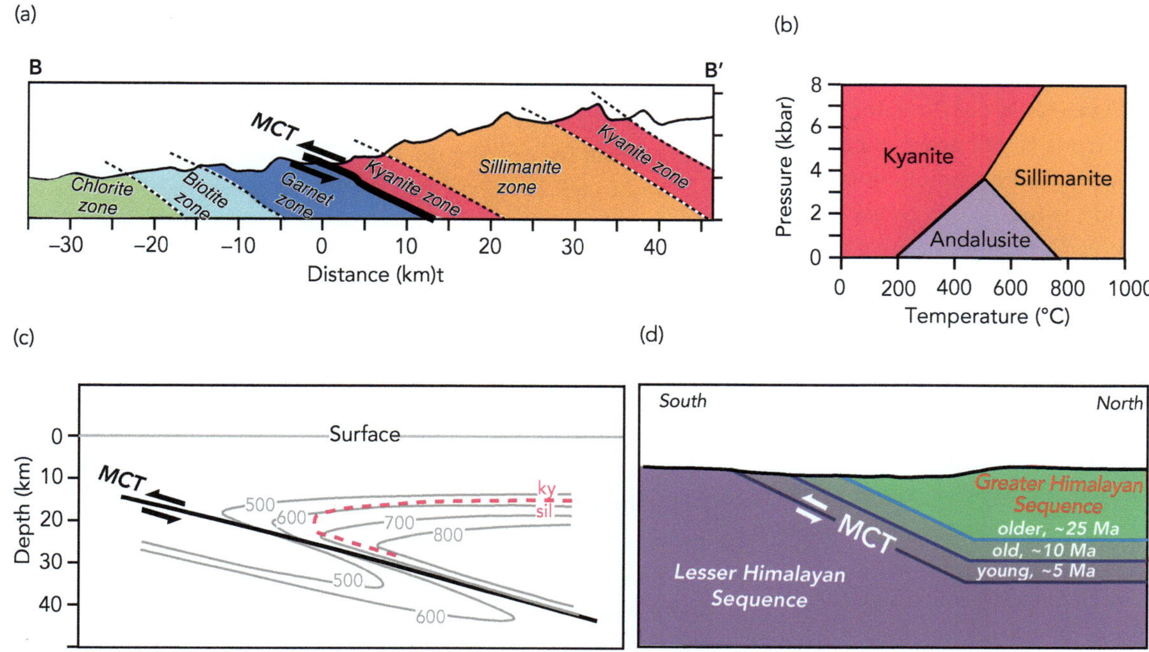

Figure 15.13 (a) Cross section B-B' (see map in Figure 15.4) across the Main Central Thrust (MCT) in central Nepal, showing distribution of metamorphic zones (see Box 15.2). (b) Phase diagram for the polymorphs of Al_2SiO_5, kyanite, sillimanite, and andalusite. (c) Illustration of the folding of isotherms associated with the rapid thrusting along the MCT; red dotted line shows the isograd marking the boundary between the stability of kyanite and sillimanite. (d) Illustration of the development of out-of-sequence thrusting along the MCT, occurring after the development of other major faults to the south (see Figure 15.4). All diagrams with no vertical exaggeration.

igneous dikes can be dated by isotopic methods (see Chapter 4). The oldest dikes that cross cut the Main Central Thrust are around 25 Ma, indicating that the large-scale thrusting in the Himalaya began about that time.

The Main Central Thrust is just one of several major structures that have developed during the growth of the Himalaya. As is common in other convergent tectonic regimes (e.g., western North America in the Cretaceous to Paleogene and western

South America today, see Chapter 14), a fold and thrust belt developed along the collision zone because of the convergence of India and Asia (Figure 15.4). Thrust faults in these environments develop in one location and then later newer faults tend to develop outboard from the mountains, resulting in a shift in the location of the faults that accommodate the compression. In the Himalaya, the Main Central Thrust developed early on, but later thrusts such as the Main Boundary Thrust and Main Frontal Thrust, developed farther south. However, motion on the older faults did not cease altogether when the new faults were formed and the difference in the dip of the faults and the rate of motion on them caused deformation of the intervening rocks in folds and faults, forming features (Figure 15.4(c)).

KEY POINT

Metamorphic assemblages in the rocks of the Himalaya indicate exhumation of most of the mountain range from depths greater than 20 km.

Finally, despite an overall compressional regime, there is one major normal fault seen in the Himalaya (Figure 15.4(c)). Some may wonder, how could a major normal fault be formed in a convergent tectonic regime? The answer comes from understanding that there is a limit to how high thrust sheets can be stacked on top of each other. The mountains initially grow because of lateral compressional forces, but gravity can counteract this as a vertical force. Eventually the thrusting produced a pile of rocks so tall that it tends to fall apart under its own weight. Rocks simply cannot be piled up infinitely to the sky, and it seems from the record in the Indo-Asian collision that the height limit for mountains on Earth is approximately the height of the modern Himalaya. The major normal fault in these mountains, known as the **Southern Tibetan Detachment Fault**, lies to the north of the crest of the range and drops the mostly unmetamorphosed Tethyan sedimentary sequence down on to the high-grade metamorphic rocks of the Greater Himalayan Sequence. This fault, along with other extensional features in southern Tibet, limit the height of the Himalaya and Tibetan plateau. An interplay of thrusting, erosion, and normal faulting have kept it at this height since around 10 Ma.

The structural and metamorphic history of the rocks exposed in the Himalaya indicate this is one of the most significant mountain ranges Earth has ever seen, with likely counterparts in the collisional tectonics operating in eastern North America and western Europe and Africa in the Paleozoic associated with the assembly of Pangea (see Chapter 10).

15.3.2.2 Sandstones Show the Uplift of the Mountains

As with other mountain-building events we have discussed (Chapters 10 and 14), much of what we know about the Himalayan Orogeny comes from study of the sedimentary deposits produced because of the uplift of the mountains. In the Indo-Asian collision, we have the advantage of being able to compare the active mountain belt with the proximal and distal basins filled with Himalayan-derived sediment, but long after the Himalaya is worn down, the record of their presence will be available to read in these sedimentary sequences. We will now discuss two basins associated with the Himalaya, one on land and one in the deep ocean.

As discussed in Chapters 10 and 14, when the lithosphere is loaded by crustal thickening from convergent tectonics, it flexes, producing local topographic highs and lows. The lows adjacent to the mountains are called foreland basins (Figure 5.23). The width and depth of these basins are dependent on the magnitude of the load (i.e., the height of the mountains) and the strength of the lithosphere, which is dependent on the composition and temperature. The Himalayan foreland basin stretches from Pakistan in the west to easternmost India. Today, its width is broadly approximated by the Indo-Gangetic Plain, across which several of the great rivers of the Himalaya flow. The five great rivers of the Himalaya, the Indus, the Sutlej, the Karnali, the Ganges, and the Yarlung-Tsangpo, are all sourced near Mount Kailas in southwestern Tibet (Figure 15.4(b)). The Ganges and the Karnali rivers drain the southern side of the Himalaya, and the Yarlung-Tsangpo, which starts in Tibet on the north side of the Himalaya, turns around the eastern edge of the mountains and becomes the Brahmaputra River, joining with the Ganges in eastern India, eventually flowing into the Bay of Bengal feeding the Bengal Submarine Fan. The Indus and Sutlej rivers drain the western side of the Himalaya (Figures 15.4 and 15.8) and flow in the Arabian sea, feeding the Indus Submarine Fan (Figure 15.8).

We can tell how long the foreland basin has been present by determining the age of the sedimentary strata found within it. In Pakistan, northern India, and Nepal the oldest layers are a part of a sequence of **fluvial** (i.e., deposited by rivers) conglomerates, sandstones, and siltstones called the Dumri Formation. Separated by a fault in some locations, another group of fluvial deposits, known as the Siwalik Group, sits on top of the Dumri Formation. Fluvial deposits are often difficult to date because rivers are high-energy environments, which make preservation of fossils less likely. An even better way of dating sedimentary rocks is to date interbedded volcanic rocks using isotopic geochronology (see Chapter 4), but, unfortunately, this foreland basin is adjacent to a continental collision zone in which volcanism shut off around 40 Ma (as mentioned above, some smallscale plutonic dikes are seen but no related volcanic rocks that would aid in dating these sandstones). In such cases, and in the case of the Himalayan foreland basin, we must find another way to determine the time of deposition of these rocks. The method that has been used to greatest effect in the Himalayan foreland basin has been **magnetostratigraphy**, a technique that utilized the occasional flipping of the polarity of Earth's magnetic field (see Box 15.3). This analysis shows that the basin was first filled around 20 Ma and has experienced relatively continuous

BOX 15.3 Using Magnetostratigraphy to Date Sedimentary Rocks

The minerals in igneous rocks are formed at the same time the rocks they are found in. This means if we can date the formation of the minerals (or at least some of them), we can learn the time of formation of the rock. Sedimentary rocks come in two types. The first forms from the precipitation of minerals from a solution (usually seawater); these include limestones, dolostones, salt, and gypsum. Like with igneous rocks, the newly precipitated minerals are the same age as the rocks they are found in. Unfortunately, these minerals (mostly calcite) usually do not have sufficient concentrations of radioactive parents (most notably potassium and uranium) to make dating them directly with useful precision very easy. The second type of sedimentary rock, detrital rocks (e.g., sandstones, shales) are made of pieces of previously existing rocks, so any dating of minerals in these rocks will only tell us about the age of the rocks that were the source material of the rocks in question, not the age of formation.

So, other methods are commonly needed to determine the age of deposition for sedimentary rocks. Fossils often provide good information in this regard. Most sedimentary rocks are dated using fossils. The absolute age of fossils comes from dating volcanic rocks associated with the fossiliferous deposits. However, in some instances, detrital deposits have neither useful fossils nor interbedded volcanic rocks to help determine the age of deposition. In these cases, another method is needed and one of the alternatives is magnetostratigraphy, which is based on the observations that the Earth's magnetic field occasionally switches polarity from north to south (see discussion in Chapter 5) and that sedimentary deposits (especially fine-grained deposits) record within them the polarity of the magnetic field at the time of deposition.

Because the time the magnetic field stays in one orientation generally lasts several tens of thousands of years (up to many millions of years), analysis of the magnetic information in an individual sample is not enough to determine its age; a sequence of layers deposited over a long time (several million years or more) is generally needed. In the field, geologists will sample layers with a spacing of a few tens of meters, noting the orientation of the sample and its location in the stratigraphic interval in question. The samples are then taken to the lab where the magnetic polarity of the samples can be determined. The results of these analyses are then plotted with respect to stratigraphic thickness measured in the field (Figure 15.14). The thickness of the measured section is then transformed into time by reference to the global **Geomagnetic Time Scale** (Figure 15.14). This timescale has been produced by analyzing thousands of rocks of known age (in many different labs) for their geomagnetic signature. This has identified the times over the past 180 million years or so (data is not so good for older rocks) in which Earth's magnetic field has switched polarity.

Thus, by comparing the pattern of changes in the polarity of our measured section in thickness to the pattern of changes in the global geomagnetic timescale, we can determine the age of deposition of the measured section. This analysis, however, does come with some potential complications. First, we must have some reasonable estimate for the age of the rocks in question. In the example in Figure 15.14, from the Siwaliks of western Nepal, it is clear that the sandstones and mudstones in question were deposited since the growth of the Himalaya and many lines of evidence, stated elsewhere in this chapter, tell us that was around 25 Ma. Therefore, we only must try and match the pattern of polarity changes to a shorter interval of time. Without this, many potential matches might be possible, and the age of the sequence would remain uncertain. Second, if the rate of sediment deposition was highly variable in the measured section, then the pattern in thickness might not match very well the pattern in the global time scale. This problem is exacerbated if there are any (possibly unrecognized) unconformities in the section.

These problems notwithstanding, magnetostratigraphy has proven to be a valuable tool in determining the time of deposition of fossil-poor sequences. This technique has been used extensively in the Siwalik Group because the high-energy fluvial environment is not conducive to fossil preservation; Figure 15.14 shows the results (and Figure 15.4 shows the location) of just one such successful study in the foreland basin of the Himalaya.

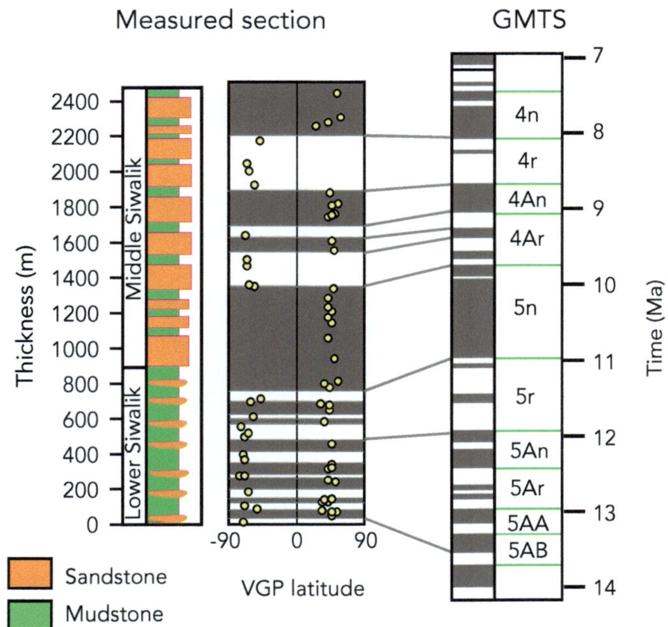

Figure 15.14 Paleomagnetic results from Khutia Khola in western Nepal (Figure 15.4) showing measured section (left), results of paleomagetic Virtual Geomagnetic Poles (center), and comparison to the worldwide Geomagnetic Time Scale (right).

sedimentation since then. This tells us that the Himalayan Mountains must have grown to a thickness sufficient to cause the lithosphere to flex sometime slightly before 20 Ma. This is consistent with the timing of the waning of continental escape. Although the sliding of Indochina to the east and south slowed down around this time, the northward passage of the Indian craton kept moving at much the same pace. This then requires a new mechanism to accommodate the convergence between India and Siberia. One of the important mechanisms that took over after 25–20 Ma was shortening of the Himalaya on a series of large-scale reverse faults described above. These uplifted areas also served as sites of erosion that produced the sediments that filled the adjacent foreland basins. The total thickness of the Dumri plus Siwalik strata reaches up to 7 km. This does not mean there was ever a basin that was 7,000 m deep. The basin evolved with the mountain range and sediments were continuously provided by rivers to keep the basin close to filled as it subsided.

KEY POINT

The foreland basin of the Himalaya has accumulated several kilometers of sediment since around 20 Ma.

However, the sediments derived from erosion of the great mountains of the Himalaya are not just found directly adjacent to the mountains on land. A far greater amount of sediment completely bypassed the foreland basin and was carried to the sea by the Indus, Ganges, and Brahmaputra rivers and deposited on the ocean floor as gigantic submarine fans (Figure 15.8). The **Indus fan** covers an area of 110,000 km² and is over 9 km thick at its thickest. The **Bengal fan** is the largest submarine fan on Earth and is about 3,000 km long, over 1,000 km wide, and about 16.5 km thick. The thickness of the Bengal fan is nearly twice as high as Mount Everest and it covers an area larger than the Indian continent, making it the largest sedimentary feature on Earth today. The volume of the Bengal fan has been estimated between 15 and 20 million cubic kilometers. If we imagine that this material came from the Himalaya, which today has dimensions of about 3,000 km long and about 300 km wide, we can say that the Bengal fan was produced by removing an average of between 16 and 22 km of rock across the Himalaya, and this does not even include the material in the Indus fan or the Siwalik basin. To put it another way, if we took the material in the Bengal fan and sprinkled it over the 48 contiguous United States, it would cover the country to a depth of more than 6,000 m! Removing this much material in this short amount of time requires very fast rates of erosion. Thermochronology of sand grains from strata in the foreland basin and the Bengal fan (Figure 15.15) indicates erosion rates as fast as 5 mm/year in some cases (see Box 15.4).

15.4 The Interplay of Tectonics, Erosion, and Climate

How are high mountains maintained? For high topography to develop, the rate of uplift must exceed the rate of erosion. The Himalaya are almost 9 km above sea level. Significant amounts of sediment derived from the Himalaya have been deposited in the Siwalik Basin, but in addition to this, even more detritus has been transported across the continent to be deposited in near-shore and deep marine environments. Erosion of the Himalayan highlands has produced some of the largest rivers in the world, which have dumped vast piles of sediment at the edges of the Indian continent and onto the adjacent seafloor, forming the world's largest submarine fans (Figure 15.8).

Despite all this erosion, the mountains have continued to grow or maintain their height. This may seem counterintuitive. If mountains are eroded, shouldn't they decrease in height? The answer to this question is, "not necessarily." Recall in Chapter 3 we introduced the concept of isostasy (Figure 3.13). Although we think of continental lithosphere as riding above the mantle, for a typical mountain, its roots (i.e., base of lithosphere) depress and displace the asthenosphere, much like the iceberg in Figure 3.13(d). For a typical mountain the root will be about 5.6 times deeper than the exposed height above sea level. For the Himalaya, the roots extend to at least 50 km below Earth's surface. As erosion occurs at the surface, the load decreases and rebound and uplift of the eroded areas occurs. The surface will not necessarily rebound to the same level as was originally eroded, because mass has been removed, but significant elevation will nevertheless be maintained. In the case of the Himalaya, uplift is also driven by the continuing compression and thickening via thrust faults and folding, and this can cause a complex feedback loop. Compression causes uplift but the uplift is not without limit. As the crust becomes thicker it is hard to lift it higher. Here erosion helps by removing mass, this decreases the weight of the mountains and allows compression and uplift to continue. In many active mountain belts, the rates of uplift may be balanced by the rate of erosion. The Himalaya and the Tibetan plateau are also thick enough to generate internal forces causing the crust to fall apart via normal faults, such as the Southern Tibetan Detachment discussed above. Also, as mountains grow higher, they can become glaciated, and glaciers typically cause far more erosion than is possible with wind and water alone. Many mountainous landscapes, including the Himalaya, reflect a balance of the uplifting tectonic forces versus flattening weathering processes that form part of the rock to sediment cycle. The isostatic rebound that follows erosion also explains how plutonic and highly metamorphosed metamorphic rocks formed within deep mountain roots can eventually be exhumed by erosion and brought to the surface, as we first introduced in Chapter 1 by looking at the Vishnu basement rocks in the floor of the Grand Canyon.

BOX 15.4 Thermochronology of Detrital Minerals: Evidence for Rapid Uplift of the Himalaya

As we discussed in Chapter 4, when a mineral is dated isotopically, the ratio of parent to daughter measured in the lab reflects not the time since the mineral crystallized but rather the time since the isotopic system passed through the closure temperature of the mineral in question. For example, for the decay of ^{40}K to ^{40}Ar, the daughter product (^{40}Ar) is not retained in the minerals muscovite and K-feldspar until the temperature falls to below 400 °C and 200 °C, respectively.

The Himalaya is made up of igneous and metamorphic rocks in which muscovite and K-feldspar are common constituents, and if we assume a geothermal gradient of 30 °C/km, the closure temperatures for these minerals correspond to depths of around 13 and 6 km. So, when we date these minerals using the K–Ar system, we are determining the last time these minerals were at these depths. This is very helpful in tracking the position of rocks relative to the surface over time.

Because muscovite and K-feldspar are common in the mountains, they are also common in the sediments shed off the mountains. Muscovite and K-feldspar have been dated from many sandstones in the Siwalik foreland basin and in the distal parts of the Bengal fan. In these cases, we can date the time of deposition of the sandstone independently (in the Siwaliks by paleomagnetism, see Box 15.3) and in the Bengal fan using marine fossils.

Usually when this work is done, not just one sand grain is dated but dozens from the same sample. Often there is a range

in the measured ages, but the most interesting one in this context is the youngest measured grain. The difference between the youngest observed age for the sand grains in the sandstone and the time of deposition of the material is called the **lag time** (Figure 15.15). This is the time it took for the grain to cool below its closure temperature and later be deposited in the basin. Because there are no volcanoes in the Himalaya, we know that when these minerals were passing through these temperatures, they were deep inside the mountains, not when they were shot out of a volcano in a violent eruption. In almost all the sedimentary basins associated with the Himalaya that have been studied in this way, the lag time for some of the material is very close to zero. That is, we cannot see a difference in the estimates for the time the sand grains passed through the closure temperature (deeper in the mountains) versus the time of deposition of the sand in the basin, given the uncertainty of the two methods.

When geologists study the absolute rates of erosion and deposition, anything above 0.5 mm/year is considered rapid erosion or deposition. In some sedimentary basins the rate of deposition may be as low as 0.01 mm/year. The observations of lag times near zero in many basins associated with the erosion of the Himalaya implies rates of erosion many times more than this. In some parts of the mountain range, estimates (over brief intervals of time) exceed 2 mm/year, and in some instances rates of up to 10 mm/year have been noted.

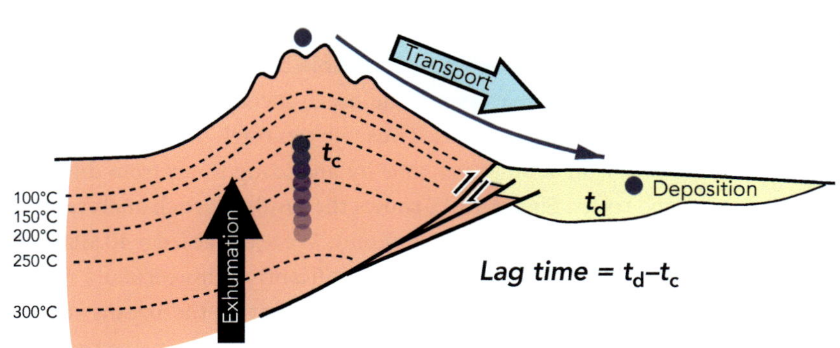

Figure 15.15 Illustration of the use of thermochronology of detrital minerals to assess erosion rates in mountain belts. t_c = time of reaching closure temperature of the system in question (here we illustrate 200 °C), t_d = time of deposition of grain in the foreland basin. Lag time is the difference between t_d and t_c. The smaller the lag time, the more vigorous the rate of exhumation of the mountain belt.

The growth of mountains can also have an important effect on the climate of a region because tall mountains will change the flow of air around the surface, and this will change when and where rain will fall. One effect of mountains on climate seen in many parts of the world is the rain shadow (Figure 15.16). When air currents at the surface encounter a mountain range, they are forced to rise over the mountain and as they do so the air cools. Cooler air has less capacity to hold water so as the air rises, rain, or snow precipitates out of the

clouds. The higher the air rises, the cooler it gets, promoting more precipitation. Thus, as air moves over very high mountains, much of the moisture in the clouds will rain out, leaving the side of the mountains facing the prevailing wind wet and the downwind side dry. The effect can even be exacerbated as the now dry air falls down the lee side of the mountain it heats up, increasing its capacity for holding water, thus we often find desert conditions in the rain shadow of large mountain ranges. As an example, the Atacama Desert, which stretches

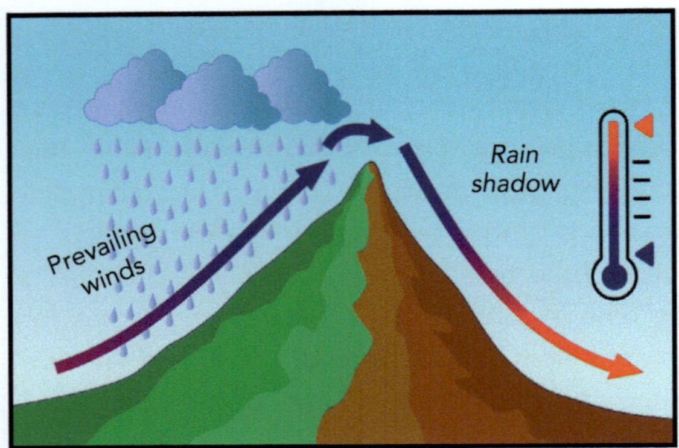

Figure 15.16 Illustration of the rain shadow effect for tall mountains. Source: Tttrung, CC BY-SA 4.0 <https://creativecommons.org/licenses/by-sa/4.0>, via Wikimedia Commons.

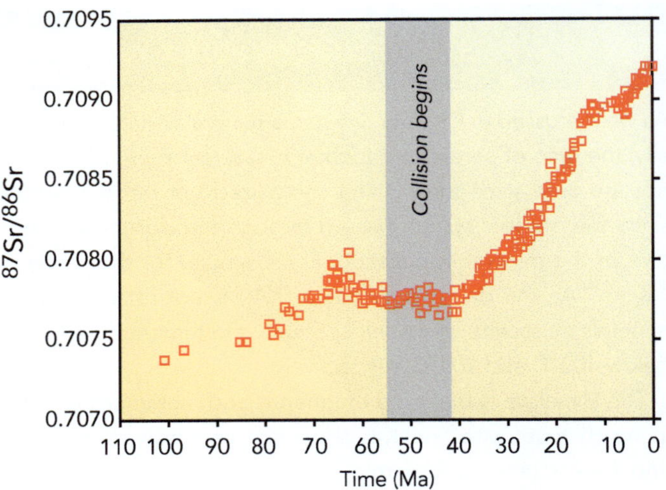

Figure 15.17 Results for ^{87}Sr/^{86}Sr analysis of marine fossils from 100 Ma to the present.

from Bolivia to Peru, is one of the driest areas on Earth and lies on the leeward (dry) side of the Andes.

In the case of the Indo-Asian collision, there is not only a very high mountain range, but the great height of the Tibetan plateau modifies air movement. Because of the high elevation of the plateau (at an average altitude of 5,000 m above sea level) it heats up more rapidly than the surrounding land and ocean during the summer months. This warming air rises, producing a low-pressure system over the plateau. The Indian Ocean does not heat up as much as the land on the plateau, so it experiences relatively higher pressure. This leads to winds moving from the ocean towards the high plateau during the summer. In the winter, the conditions are reversed with cold, dry air flowing away from the plateau. Thus, the Tibetan plateau's high elevation significantly changes the atmospheric circulation patterns of central Asia and the surrounding regions.

The south side of the Himalaya receives over 4,000 mm of rain every year, mostly during the monsoon months of June to September. Such rainfall plays a key role in the erosion of the mountains discussed earlier, but we can see that the rainfall and tectonics are related. The mountains are there because of the tectonics and the rain is focused there because of the topographic gradient created by the structural deformation. The focusing of the main thrust faults, shown in Figure 15.4, and the very high amounts of rain that fall on this region explains why almost all of the Himalaya north of the Main Frontal Thrust (Figure 15.4) is medium- to high-grade metamorphic rocks, whereas the rocks of the Tibetan plateau are mostly unmetamorphosed Phanerozoic sedimentary rocks.

Finally, the effect of the uplift and erosion of the Himalaya and Tibetan plateau can be seen not just locally but globally in the composition of the world's oceans. The river that delivers the most water to the world's oceans is the Amazon in South America, but the river that contributes the most sediment is the Ganges–Brahmaputra draining the central and eastern portions of the Himalaya and southeastern Tibetan plateau. The Indus and Sutlej rivers to the west also add a significant quantity of sediment (Figure 15.8). Sediment delivered to the ocean can dissolve and then change the composition of the water. The sediment

coming from the Himalaya is particularly able to do this because the rocks in the northern part of the Indian plate, deformed and eroded as a result of the collision with Asia, have some particular differences from average crust. The precursors of the rocks in the Himalaya were old and granitic in composition. As discussed in Chapter 14, one of the characteristics of continental crust is a higher concentration of rubidium (Rb) than oceanic crust or the mantle. This is because when rocks melt, Rb is more likely to go into the liquid phase than to persist in the remaining solids. One of the isotopes of Rb, ^{87}Rb, is radioactive and decays to ^{87}Sr, which grows relative to stable, non-radiogenic isotopes of Sr such as ^{86}Sr. Therefore, over time, continental crust will evolve a much higher ^{87}Sr/^{86}Sr ratio than other rocks and detritus from old continental crust will be identifiable by this high ^{87}Sr/^{86}Sr ratio.

Organisms that live in the ocean incorporate oceanic Sr into their shells or skeletons. Thus, we can track the evolution of the ^{87}Sr/^{86}Sr in oceans over time by analyzing shells from different-aged rocks. Figure 15.17 shows the results of such analyses from rocks from 100 Ma to the present. We see that from 40 Ma to the present, this value increased significantly. This certainly reflects an increase in the delivery of sediment to the oceans derived from old continental crust and the timing of the increase corresponds with the initiation of the Indo-Asian collision.

15.5 Conclusion

The Himalaya are the first thing many people think of when they hear the word "mountains" because of their scale. Not only are they the tallest and longest mountain chain today but the Indo-Asian collision zone is one of the largest such features seen on Earth in the past billion years.

We understand the history of this region by modern geophysical observations, study of metamorphic and sedimentary rocks formed during the collision, and by comparing the active collision to similar tectonic regimes of the past including the Appalachians in eastern North America.

15.6 Summary

- India separated from Pangea around 120 Ma and began colliding with Asia around 50 Ma.
- Key features of the change from an ocean–continent tectonic boundary to a continent–continent boundary include the cessation of volcanism and marine sedimentation in the collision zone and a slowing down of the northward velocity of India.
- Before the growth of the Himalaya, continental escape played an important role in accommodating convergence between India and Asia, with parts of Indochina moving to the east and southeast from around 40 to 25 Ma.
- Based on thermochronology of the bedrock, cross-cutting relationships, and sedimentation in the foreland basins, the mountains of the Himalaya did not start to become a major topographic feature until after 25 Ma.
- The Himalaya have not only a local influence of the deformation and sedimentation in Asia but also the chemistry of the oceans and the circulation of the atmosphere.

Key Words

- Tethys Ocean
- Gangdese batholith
- rhyolite
- Andean-type magmatism
- continental escape
- Greater Himalayan Sequence
- Lesser Himalayan Sequence
- foreland basin
- Main Central Thrust
- Barrow's zones
- inverted metamorphism
- isotherms
- Barrovian metamorphism
- Southern Tibetan Detachment Fault
- fluvial
- magnetostratigraphy
- Geomagnetic Time Scale
- Indus fan
- Bengal fan
- lag time

Further Reading and References

Müller, R. D., Roest, W. R., Royer, J.-Y., Gahagan, L. M., and Sclater, J. G., 2004, Digital Isochrons of the World's Ocean Floor, 28 May, https://www.earthbyte.org/Resources/Agegrid/1997/digit_isochrons.html#anchor1131748.

Yin, A., and Harrison, T. M., 2000, Geologic evolution of the Himalayan-Tibetan orogen, *Annual Review of Earth and Planetary Science*, 28, 211–280.

Zhisheng, A., Kutzbach, J. E., Prell, W. L., and Porter, S. C., 2001, Evolution of Asian monsoons and phased uplift of the Himalaya–Tibetan plateau since Late Miocene times, *Nature*, 411, 62–66.

Review Questions

1. Describe the main orogenic events, oceans, and terranes that collided with Asia before the Indian continent collided.
2. What are the several types of data that indicate when the collision between India and Asia began?
3. What is the Tethys Ocean and what role did it play in the evolution of the Himalayan and Tibetan plateau?
4. Why are there many active volcanoes in the Andes Mountains but none in the Himalaya?
5. Did the Himalayan mountains begin to form at the beginning of the collision between India and Asia? What evidence supports this understanding?
6. Explain the distribution of metamorphic rocks in the Himalaya referred to as "inverted metamorphism."
7. Despite being a compressional orogeny, dominated by thrust faults, major normal faults are also found in the Himalaya. How do they form?
8. What is the significance of the Indus and Bengal fans to the history of the Himalaya?

Strait of Gibraltar. Source: Oscar Sánchez Photography / Getty Images.

Chapter 16

The Messinian Crisis

The Great Drying of the Mediterranean Sea

LEARNING OBJECTIVES

- Explain the geological history of the Mediterranean and how it relates to the Tethys Ocean.

- Describe the process of evaporation in the Mediterranean and how it may have occurred due to changes in sea level and climate.

- Understand how diapirs and unconformities can be used to develop hypotheses regarding sea-level changes in the Mediterranean.

- Explain how the Deep Sea Drilling Project provided evidence for the evaporation of the Mediterranean Sea and the formation of evaporites.

- Discuss the stratigraphic evidence for a major sea-level drop associated with the drying of the Mediterranean.

- Discuss the consequences of the Messinian salinity crisis, including faunal exchange and insular dwarfism.

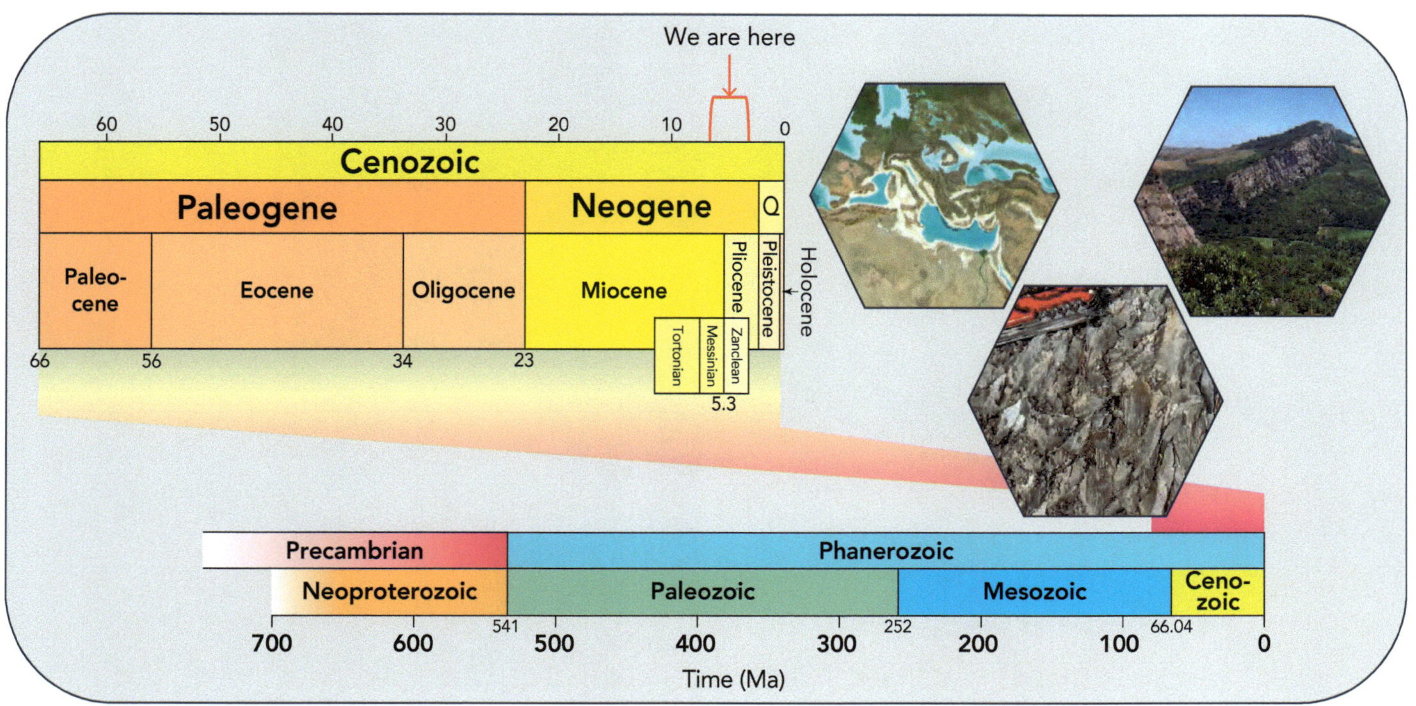

Introduction: The Grandest Canyon

The largest canyon on land today is the Grand Canyon, one of the seven natural wonders of the world, 446 kilometers long, 29 kilometers wide, and 1,857 meters deep, that we reviewed in Chapter 1. Of course, these dimensions pale compared to oceanic basins, but despite their far more extensive dimensions, oceans are always filled with seawater. Or are they? As we have seen in the chapter on plate tectonics, oceans do not last forever. The oldest **lithosphere** at the floors of the modern oceans is Triassic in age (about 180–200 Ma), but there is an exception. The Mediterranean Sea contains a 340-million-year-old remnant of the shrunken Paleozoic Tethys Ocean. Tethys lay on the equator connecting the Atlantic, Indian, and Pacific oceans and originally separated Eurasia from Gondwanaland (Figure 15.2). The Tethys Ocean has been slowly closing, as a consequence of the Alpine to Himalayan Orogeny discussed in Chapter 15, and the Mediterranean Sea, Black Sea, and Caspian Sea (Figure 16.1 and as discussed in Chapter 15) are the last vestiges of this once mighty ocean. The closing of Tethys and joining of continents began changing the climate to more arid conditions, with a corresponding decrease in forests and increase in savannas, which may have driven human evolution (see Chapter 17).

You may notice that the Mediterranean is now almost completely cut off from any other ocean (Figure 16.1). At its western margin it is connected to the Atlantic Ocean through the Strait of Gibraltar and to the northeast it is connected to the Sea of Marmara through the Dardanelles Strait (Figure 16.1). The Sea of Marmara is in turn connected to the Black Sea through the Bosporus Strait, next to which lies the city of Istanbul. The Mediterranean Sea extends for about 3,700 kilometers from the Strait of Gibraltar in the west to Israel in the far east and reaches almost 1,000 kilometers wide from Libya in the south, to Albania in the north (Figure 16.1). The Mediterranean Sea averages about 1,500 m deep and its deepest point is 5,267 m. The Strait of Gibraltar, in contrast, ranges from 300 to 900 m deep and is only 15 km wide. The Bosporus Strait is only 700 m at its narrowest point, and ranges from 11 to 110 m deep.

About 5.9 million years ago, near the end of the **Miocene Epoch**, the connection of the Mediterranean to the Atlantic Ocean closed, likely as a result of tectonic uplift in the area around the Strait of Gibraltar. Despite a number of rivers that drained into the Mediterranean Sea, including the Nile, the Mediterranean is largely bordered by deserts, including the Sahara to the south, and the actual flow of land-derived fresh water into the basin is quite low. Eventually, the rate of evaporation of seawater far exceeded the rate of replenishment of fresh water via rivers and the Mediterranean Sea literally dried up (Figure 16.2). The resulting landscape would have formed one of the largest and deepest terrestrial canyons ever seen on Earth, three times deeper, six times longer and 34 times wider than the Grand Canyon. About 600,000 years after the desiccation, the Atlantic connection was re-established, and the Mediterranean experienced a massive flood that marked the beginning of the **Zanclean Stage**. The great drying out of the Mediterranean is now referred to as the **Messinian salinity crisis** and this chapter reviews the history of the observations that led to the discovery of this great Earth event.

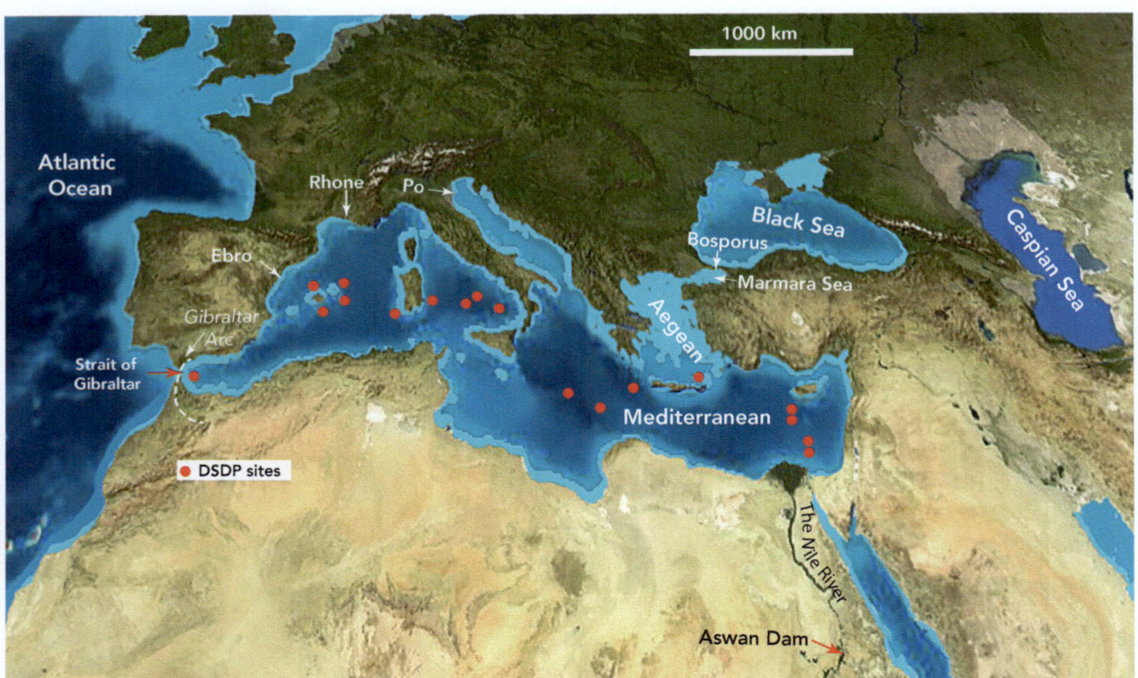

Figure 16.1 Mediterranean area, connections, and key locations of selected major rivers. Red dots indicate location of the Deep Sea Drilling Project (DSDP) holes. Source: © DeepTimeMaps.

Figure 16.2 Paleogeographic reconstruction of the Mediterranean coastlines during the Messinian salinity crisis, about 5.5 Ma BP. Note the closed Gibraltar Strait, the separation of the Black Sea from the Mediterranean Sea, the continental link between Sicily and Africa, and the drying of the Adriatic Sea. Source: © DeepTimeMaps.

KEY POINT

The Mediterranean Sea and linked Caspian and Black seas are elongate remnants of the once larger Tethys Ocean, which has been closing over the past hundred million years or so. Because evaporation exceeds freshwater input, these seas are susceptible to evaporation, but are kept high by input from the Atlantic through narrow straits. Closing of these connections 5.9 million years ago caused the Mediterranean to dry out, initiating the Messinian salinity crisis.

16.1 Evidence of Desiccation from the Deep Sea: Ken Hsu and the *Glomar Challenger*

In 1966, the United States National Science Foundation, along with 22 international partners, instituted the Deep Sea Drilling Project (DSDP), now known as the Integrated Ocean Drilling Program (IODP), to conduct basic research into the overall nature and history of ocean basins. This project was made possible by advances in drilling technology. Since its inception, this project has collected thousands of cores drilled in over 500 locations throughout the world's oceans. In 1970, leg 42 of the DSDP, led by chief scientist Ken Hsu at the Swiss Federal Institute of Technology, collected the first drill core samples from the floor of the Mediterranean Sea, the first of a six-well program. The vessel was the *Glomar Challenger*, a US ship. As we will see, the ultimate interpretation of the data from this cruise came from the collaboration of a geologist, Hsu, geophysicist Bill Ryan, and paleontologist Maria Cita.

The interest in drilling the Mediterranean was based on geophysical surveys, and particularly seismic data (see Box 16.1), that showed several features that were hard to interpret.

Geophysical surveys of the Mediterranean showed a series of strong **seismic reflections** in the sedimentary layers below the surface. The seismic data showed a regional, basin-wide **unconformity** (M and N, Figures 16.5 and 16.6), with a deformed layer in between, but it wasn't clear if the deformed layer was mud or salt. Salt might indicate a major phase of evaporation. Seismic cross sections also showed distinctive "volcano-like" intrusions, called **diapirs** (Figures 16.5 and 16.6). Similar diapirs off the coast of Nigeria and in the Caspian Sea were formed by deformation of muds, but diapirs in the Gulf of Mexico were formed by salt. The drilling was primarily proposed to test whether the Mediterranean sediments contained salt versus mud.

The first core samples penetrated the upper M layers. Despite rather poor core recovery, the scientists found a few bits of **gypsum** mixed with the drilling mud. Gypsum ($CaSO_4 \cdot 2H_2O$) is a mineral uniquely indicative of evaporation of seawater. During further drilling, the scientists were even more surprised to find layers of red-colored gravel and sand. Such a red coloration is typically the result of **hematite** (Fe_2O_3) staining. Hematite (otherwise known as iron oxide) indicates oxidation of iron (i.e., rust) typical of sediments formed on dry land, in rivers and deserts; those formed in the deep sea are typically gray or black.

BOX 16.1 Seismic Surveys

Seismic data involves shooting sound waves into the Earth and observing the reflections that literally bounce of the features below (Figure 16.3). Anytime there is a geologic feature that has a sharp change in density, such as between a layer of hard limestone or softer shale, or an igneous versus sedimentary rock, some of the sound wave energy will bounce back off this surface, much as light is reflected in a mirror producing an image that you can see, and these reflected sound waves can be detected at the surface (Figure 16.4). Although the actual process of collecting and processing the data is quite complex, ultimately an image of the subsurface geology can be produced (Figure 16.5).

In addition to academic purposes, seismic data can be used to detect buried structures that may contain hydrocarbons, and is one of the main tools used in hydrocarbon exploration. One of the challenges is that the vertical scale is the time it takes for a sound wave to leave the surface, hit the reflector, and bounce back (called the **two-way travel time**), but geoscientists want the actual depth, especially when drilling an expensive well (a typical deep water well in the Gulf of Mexico may cost upwards of half a billion dollars). To make this conversion one needs to know the density of the rock, or in marine seismology the density of the water, sediment, and rock, as well as the velocity of the sound wave. If this is integrated with knowledge of the time it takes for the wave to travel, and the velocity of the waves, it is relatively easy to convert the seismic data from time to depth. In Hsu's day (the early 1970s), seismic acquisition was a relatively new technology, data were limited to two-dimensional lines (i.e., vertical cross sections), and data quality was generally not very good and not digital, so there were significant uncertainties regarding the actual nature of geological features seen on the rather poor seismic data. These days, seismic data are collected in three dimensions and processing is far more robust, but

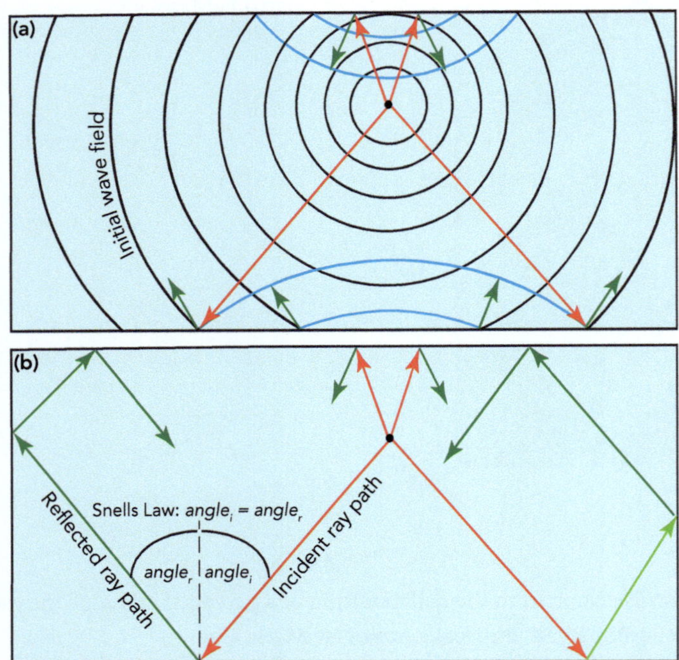

Figure 16.3 Principles of seismic acquisition. (a) If you throw a pebble in a pool, the waves propagate away from the point where the pebble hits the water. The blue lines show reflected waves. (b) Ray paths can be drawn perpendicular to the wave field and if one could detect and map the reflection points, then the geometry of the pool can be reconstructed.

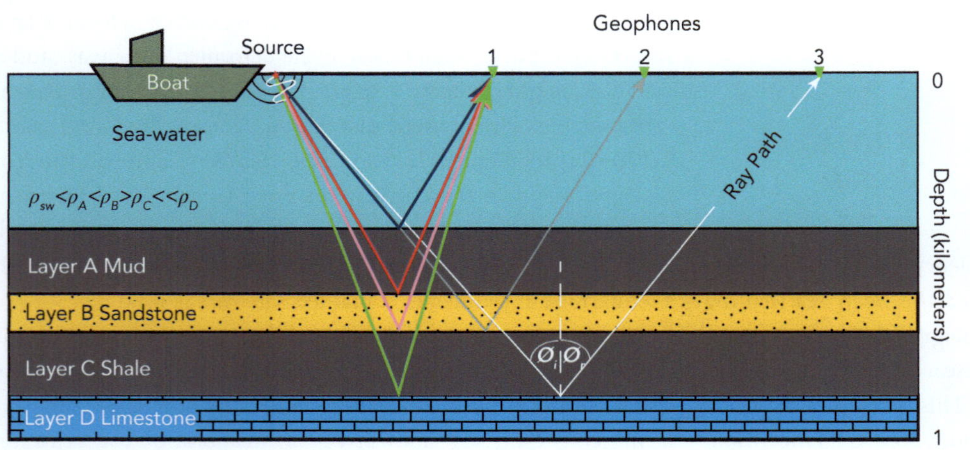

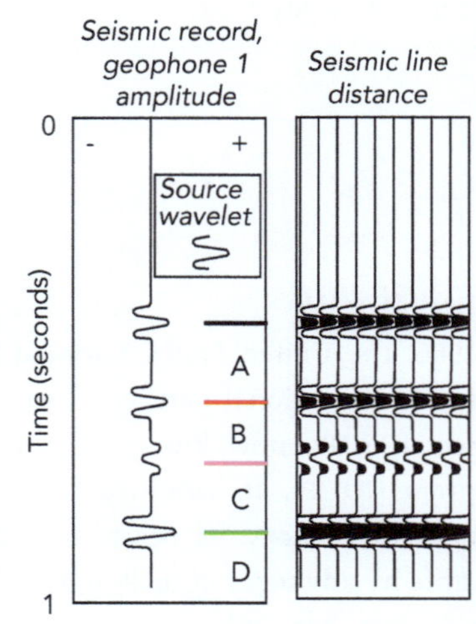

Figure 16.4 In this simple case, the boat produces an energy pulse, which produces sound waves that propagate to the seafloor and penetrate into the buried layers. Some of the energy is reflected back and is "heard" or received by the geophones. Note that it takes more time for the sound waves to travel to a deeper layer and back to any given geophone. The seismic records show the reflected energy (or amplitude of the wave) in so-called "two-way travel time." If one knows the speed at which the sound wave travels through a given layer, which is dependent on the density of the layers (e.g., the velocity of sound in water is about 1,500 m/s), and given the two-way travel time, seismic data can be converted to depth. Individual traces from the different geophones can be gathered and placed side-by-side to yield a typical seismic cross section from which the various geological features may be interpreted.

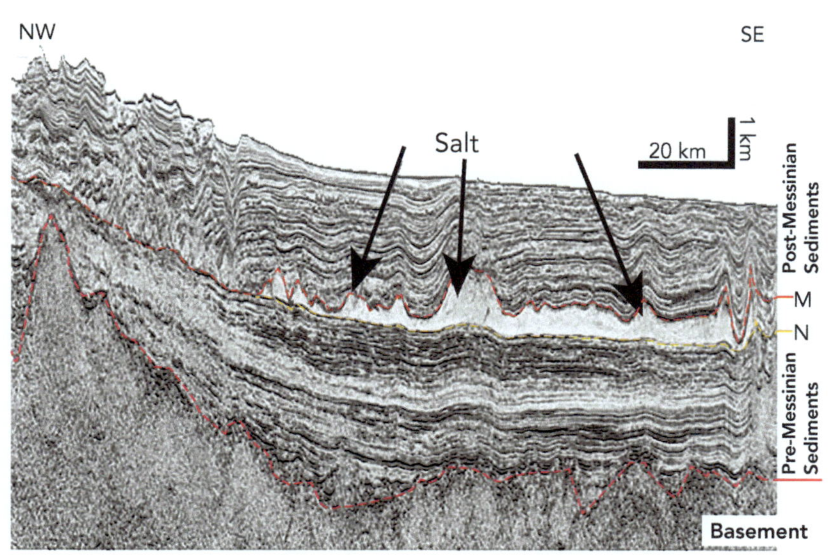

Figure 16.5 Example of a seismic line from the Mediterranean showing salt diapirs and the M and N unconformities associated with the Messinian salt deposits. Source: Adapted from Ryan (2009). Copyright (2009). With permission from John Wiley and Sons.

despite all the improvements, confirmation of what is down there inevitably requires drilling and sampling, with the rocks telling the story.

We already saw in Chapter 13 that the Chicxulub impact crater was imaged with geophysical data by Pemex, the Mexican oil company, in their search for hydrocarbons. They thought this feature was volcanic and might contain oil or gas and drilled into it. It was only later with additional drilling and sampling that the crater hypothesis was successfully tested.

Maria Bianca Cita is an Italian paleontologist who specialized in the study of plankton and was a critical team member aboard the ship. The gravels also contained stunted shallow water mollusks and in these samples, Cita found microscopic, single-celled **benthic foraminifera** including ***Ammonia beccarii*** and ***Elphidium macellum***. These species are typical of organisms that live in coastal brackish waters, rather than deep-sea environments. This invited the question: what were shallow brackish water fossils doing on the floor of the deep Mediterranean Sea, thousands of meters below the present sea level? The combination of shallow fresh-to-brackish-water fossils contained within river gravels and minerals indicative of evaporation suggested a profound change in the environment had occurred and indicated that the Mediterranean Sea was much shallower 5.3 million years ago. This invited more questions: was the Mediterranean shallower prior to 5.3 million years ago with subsequent subsidence and deepening or did the basin go from deep to shallow and then back to deep again?

The layers immediately above the M zone consisted of limy mudstones (termed **marls**) of the Trubi Formation. Here Cita's work again provided critical clues as to the massive changes in sea level that had occurred. She showed that the Trubi included foraminiferal species such as *Oridosalis umbonatus* that are diagnostic of a much deeper-water environment than the foraminifera found in the underlying M layer and are more consistent with the present-day deep water environment of the modern Mediterranean Sea. The geological evidence indicated that the rocks of the M layers were deposited in much shallower brackish to freshwater environments, whereas the overlying Pliocene sediments were deposited in water depths of over 1,000 meters. Moreover, based on the age of the sediments, it looked like these changes in water depth had happened rather quickly, over as little as a few tens of thousands of years. This major change in the environment also coincided with the Miocene–Pliocene boundary.

The term Pliocene was given to this epoch by Charles Lyell, based on the observation that fossil mollusks are very similar to those living today. The word Pliocene literally translates as "more recent." The previous "less recent" Miocene Epoch is characterized by rather different fossils, as is the Oligocene ("few recent") and the Eocene ("dawn of the recent") fossils that characterize those epochs respectively. Clearly something major happened at the end of the Miocene, not only to the Mediterranean Sea but globally.

KEY POINT

The 1970 DSDP mission was designed to test whether diapirs and unconformities in the Mediterranean Sea were the result of salt evaporation. Recovery of evaporitic gypsum and red river gravels supported the hypothesis that the Mediterranean Sea had essentially dried out. Analysis of microfossils also showed alternations from deep marine to brackish conditions, indicative of major changes in sea levels and salinity.

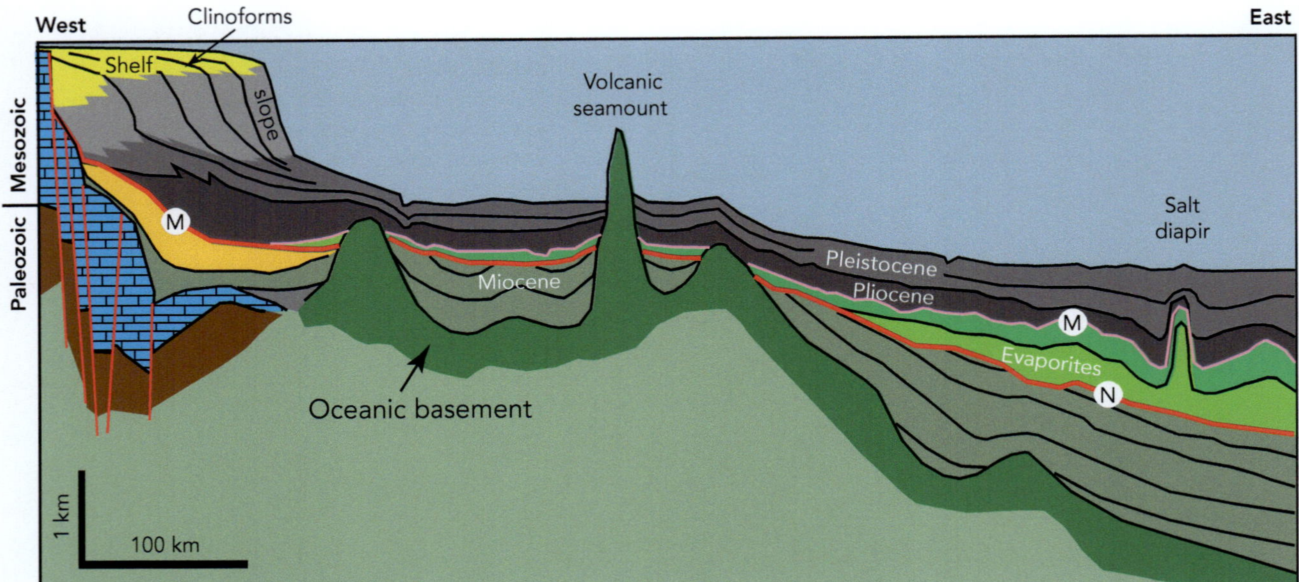

Figure 16.6 Regional west to east cross section from the Ebro delta to the middle of the Balearic Basin. Location of line shown in Figure 16.12. Source: Adapted from Ryan (2009). Copyright (2009). With permission from John Wiley and Sons.

16.2 Formation of Evaporites from Seawater

Before discussing the geological evidence, we need to review the basis for formation of evaporites from seawater. If you have ever tasted seawater, you might think that the only thing dissolved in it is salt (NaCl), whose formal mineral name is **halite**. Perhaps you have noticed a white crust around a shower head or fresh-water tap and have heard about hard versus soft water? This crust represents lime (mostly $CaCO_3$ or $MgCO_3$), which forms "hard" water. Most waters have a host of ions in solution. Seawater contains about 2% chlorine (Cl), 1% sodium (Na), and lesser amounts of magnesium (Mg), sulfur (S), potassium (K), bromine (Br), and carbon (C). Some of these elements occur as ions, such as sulfate (SO_4^{2-}) and carbonate (CO_3^{2-}). Seawater even contains silica (SiO_2). When organisms, such as single-celled *radiolarians* and *diatoms*, and certain species of sponges, build their skeletons, they do so by extracting the dissolved silica to make SiO_2 skeletons. When these organisms die, the skeletons sink to the seafloor, and in some places they concentrate to form siliceous oozes, which can become fossilized to form the rock type chert or flint, which was historically used to make knives, axes, and spearpoints. Other types of organisms, especially corals, clams, mussels, and other bivalves; echinoderms, including starfish and sea urchins; and microscopic single-celled foraminifera and **coccoliths**, make their skeletons of calcium carbonate, including the minerals calcite and aragonite and form the rock called limestone. During the Cretaceous Period, coccoliths were so abundant they formed massive chalk deposits.

As we discussed in Chapter 3, most of the ions and elements in seawater originate from chemical weathering of the land. As a result of significant proportions of these ions being precipitated, either by organisms or by evaporation, the oceans for the most part have a fairly uniform composition, reflecting the balance between new ions delivered by rivers and ions removed by a variety of biogenic and chemical processes.

The proportion of any given element or ion in water is a function of its solubility. If you put a tablespoon of salt, a pearl (made of aragonitic $CaCO_3$), and a quartz crystal in a cup of fresh water, after only a few hours, the salt will have dissolved; a few days later the pearl may have lost a bit of its luster, but the quartz crystal will be unchanged. This reflects the fact that the salt is the most soluble and quartz (SiO_2) the least soluble. If you try the same experiment with common white vinegar, both the salt and pearl will dissolve, which reflects the more potent effects of the acid in dissolving solids; however, the quartz will still remain unchanged.

All seawater experiences some evaporation, but the hydrologic cycle usually ensures that the rate of evaporation is balanced by replenishment from rivers and rain. However, there are places on Earth, especially dry hot areas, where seawater experiences evaporation in excess of replenishment. In these areas, evaporitic precipitates can be deposited forming **sabkhas** (Arabic for "salt flats") such as occur on the shores of the Persian Gulf. There are also a few lakes that are quite salty, one of the most famous being the Great Salt Lake in Utah and the Dead Sea in Israel.

If you take a sample of seawater and evaporate it, the first minerals to precipitate are carbonates (calcite and dolomite), as these are less soluble (Figure 16.7). As evaporation continues, the next major mineral to precipitate is calcium sulfate, forming the mineral gypsum ($CaSO_4\cdot2H_2O$), which has water in its crystal structure. At higher temperatures, **anhydrite** forms ($CaSO_4$), a similar mineral but without the water. Halite precipitates next,

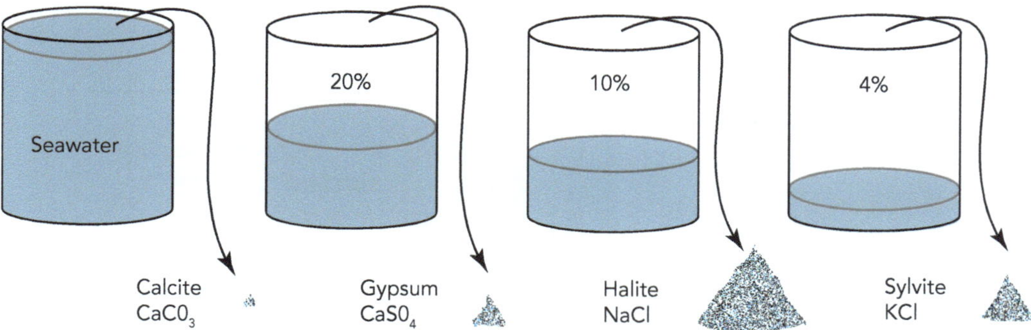

Figure 16.7 The succession of salts precipitated from seawater. Limestone ($CaCO_3$) is the first to precipitate. After 81% of the water has evaporated, gypsum or anhydrite $CaSO_4$ precipitate. Not until 90.5% of the seawater has evaporated does halite precipitate. The bitter magnesium- and potassium-rich salts are the last to precipitate.

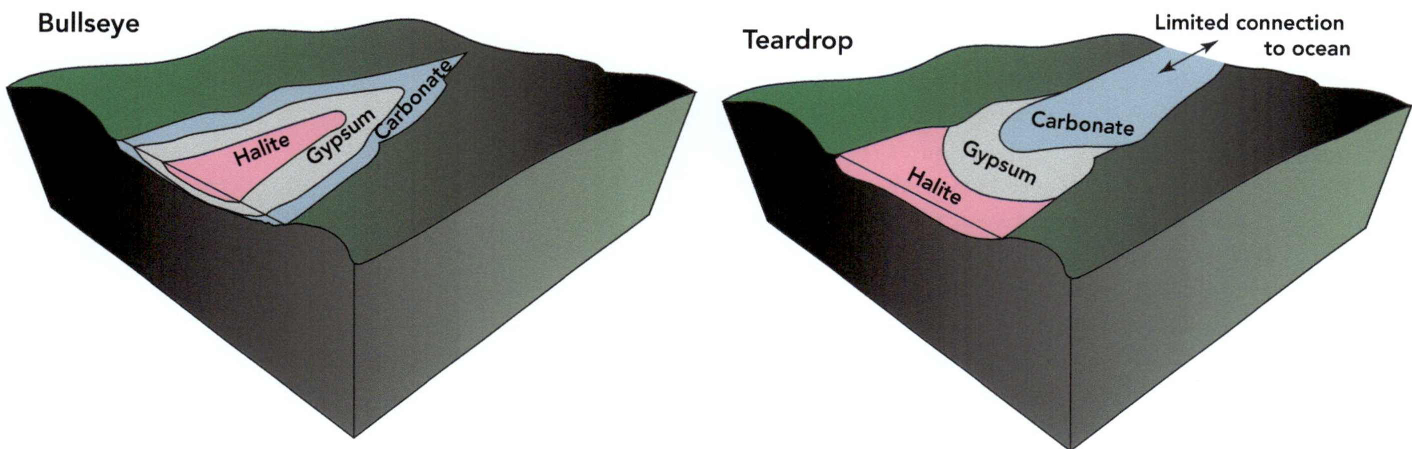

Figure 16.8 Bullseye versus teardrop pattern of evaporite minerals.

showing that the Na and Cl are highly soluble. When almost all the water is gone, potassium salts (KCl), also called bitter salts or sylvite, precipitate. As the seawater evaporites, the remaining water becomes richer in salt, forming a **hypersaline** brine.

In map view, desiccation of a fully enclosed sea results in a bullseye pattern, with the first evaporites found on the margin and later-formed evaporites at the center (Figure 16.8). In contrast, evaporation of a sea that remains partially connected to an ocean will show a teardrop pattern, with earlier evaporite mineral phases deposited close to the area of connection, and later phases at the periphery (Figure 16.8).

KEY POINT

When seawater evaporates, the first minerals to precipitate are the least soluble, and include calcite and dolomite. As the water becomes a brine, gypsum and finally halite precipitate. Depending on the shape of the basin, bullseye or teardrop patterns of evaporite minerals are formed on the surface.

The density of fresh water (i.e., in rivers) is about $1{,}000\,\mathrm{kg/m^3}$ whereas seawater is about $1{,}025\,\mathrm{kg/m^3}$. The density of hypersaline brines can be significantly higher, and the Dead Sea for example has a density of $1{,}240\,\mathrm{kg/m^3}$. The salinity of the Dead Sea is so high that almost nothing can live there, hence the name. Similar increases in salinity can also occur as water freezes, such as in the Arctic Ocean. Freezing of the ocean surface removes fresh water, forming ice, leaving behind seawater that is more salty than normal. These cold salty polar waters are heavier than the underlying ambient seawater and sink to the ocean bottom, forming deep-sea thermohaline currents. During hotter greenhouse periods in Earth history (such as the Cretaceous Period), when polar ice is either absent or very sparse, oceanic circulation can stop completely, and the stagnation can result in global anoxia at the seafloor.

The process of sinking brines led American geologist Robert F. Schmalz to hypothesize in 1969 that evaporites can be deposited in quite deep water, especially if the water becomes "briny" enough that it exceeds the solubility of a given evaporite mineral. This led to a significant controversy as to the relative importance of a deep versus shallow water origin of evaporites in the geological record. Many ancient evaporitic deposits are many hundreds and, in a few places, up to 1,000 m thick. If all evaporites are deposited in shallow water sabkha-type environments it makes it very hard to explain how thick deposits accumulate, as this would require rather exceptional and extremely rapid subsidence of a basin. If someone could find evidence of modern deep water evaporites, this would bolster Schmalz's hypothesis.

KEY POINT

Sinking brines in oversaturated seas can lead to precipitation of evaporites in deep water.

16.3 Stratigraphic Evidence of Mediterranean Evaporation

Evaporites, including salt and gypsum, were well known in outcrops and mines from around the Mediterranean, including the Gessoso-Solfifera Formation in Italy (Figures 16.9 and 16.10). In 1867, Swiss geologist Karl Mayer-Eymar noted late Miocene fossils in the marls that lay in between the Solfifera gypsum layers. He called this the **Messinian Stage**, marking the last stage in the Miocene Epoch (Figure 16.9). Several other European geologists studying these onshore evaporates, which included gypsum and halite, also considered the possibility that the Mediterranean area may have undergone a significant evaporation event. The gypsum deposits show several distinct forms. **Selenitic gypsum** forms as large, radial clusters of crystals (Figure 16.10(b)). The gypsum layers are interbedded with thin marl horizons.

Gypsum ($CaSO_4 \cdot 2H_2O$) that forms in warm shallow sabkhas looks very different and forms a bulbous, nodular texture forming so-called "chicken-wire" gypsum, or as it is buried and loses it water, becomes the mineral anhydrite (Figure 16.10(c)), which represents gypsum, but with the loosely held water driven off during burial.

Drilling and examination of the onshore geology around the Mediterranean suggested that the Messinian event occurred in three main phases (Figure 16.9). The lower phase comprises carbonate-gypsum dominated deposits (known seismically as the N zone, Figures 16.5 and 16.6), the middle phase includes the main salt deposits, and the upper phase consists of carbonate-gypsum-anhydrite deposits of the M zone, also called the Pasquasia Formation in onshore outcrops.

The origin of the selenitic gypsum (Figure 16.10(b)) was long controversial. Modern gypsum evaporites were only found in shallow sabkhas, so the standard interpretation of these older gypsum deposits, using a strictly uniformitarian approach, was that they were shallow water. However, Schmalz suggested that they could be deep water evaporites, which would not require

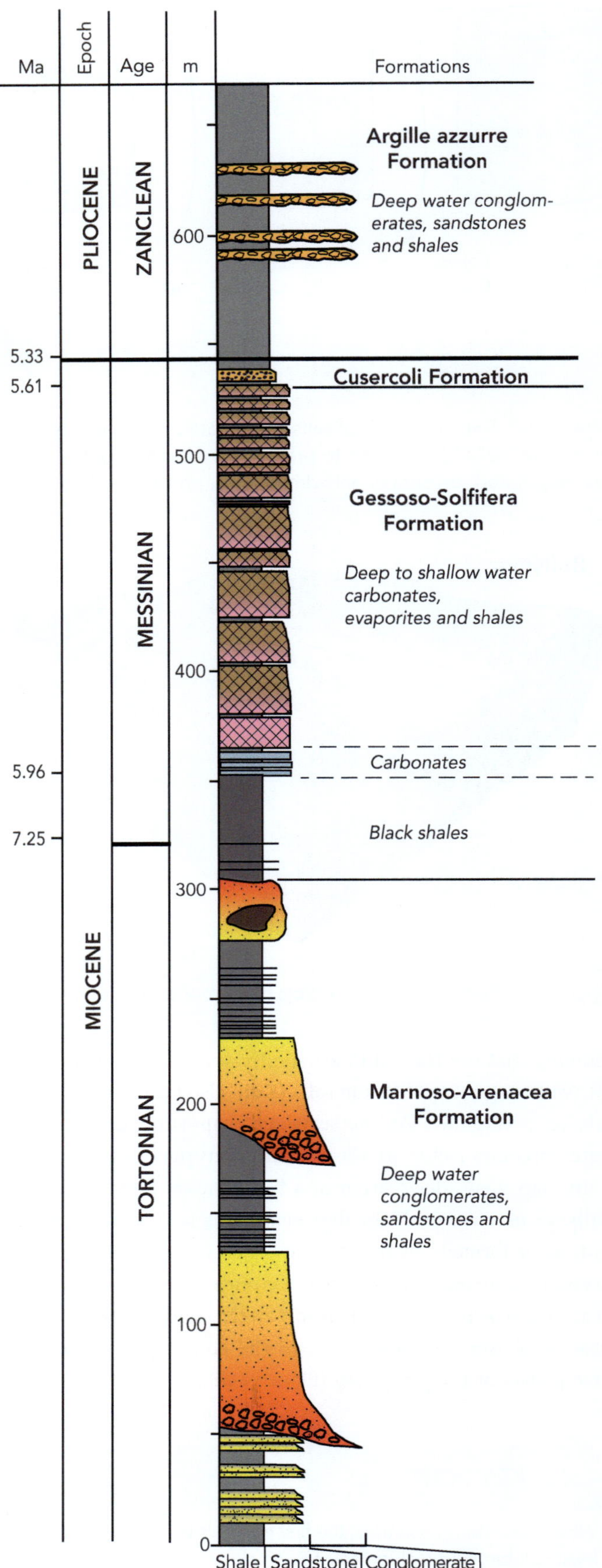

Figure 16.9 Measured section showing the stratigraphy of the Santerno Valley in the Po River plain. Location shown in Figure 16.12. Source: Redrafted from Roveri et al. (2006).

Figure 16.10 (a) Solfifero gypsum overlying deep water sediments, Santerno Valley, Italy. (b) Close up showing bladed selenitic gypsum crystals indicative of evaporation in deep water. (c) Example of chicken-wire anhydrite.

wholesale drawdown of the Mediterranean Sea. More careful examination was required to test the deep versus shallow water evaporite hypothesis. The lower gypsum layers did not show obvious evidence of exposure to the air, such as mudcracks or soil, and the interbeds of marl and their contained fossils indicated deposition underwater. There was also some evidence that the deeper gypsum layers may have been transported into deeper water from a shallower environment, as indicated by deformation structures and brecciation of the rock.

The Eocene foreland basins in Italy and Spain were originally linked to the Mediterranean, but owing to the continued closing are now uplifted and exposed. These foreland basins formed as a consequence of crustal thickening due to compression as Tethys closed (see Figure 15.2). The sedimentary rocks of the Marnoso-Arenacea Formation underlying the Solfifero (Figure 16.9) consist of interbedded conglomerates, sandstones, and shales that show distinctive sedimentary structures such as graded bed and current ripples. These are taken as evidence of deposition into water deeper than 1,000 m by **turbidity currents**, which are sediment-laden flows that commonly form in deep water, supporting the idea that the Mediterranean area had been deep prior to the Messinian. Thin-bedded mudstones (Figure 16.11(a) and (b)) alternate with thicker sandstones (Figures 16.9 and 16.11(c)), which shows evidence of channels interpreted to be due to scour by energetic sand-laden turbidity currents.

Eventually, the salt layer was also sampled by the DSDP drilling, and initially there was also controversy as to the depth at which it formed. The halite crystals contained inclusion of the fluid brines from which they had precipitated. The chemistry and thermodynamics of **fluid inclusions**, literally "bubbles of fluid," trapped in the crystals can be studied to determine the temperature at which the halite formed. The upper halite layers showed large fluctuations in temperature, more characteristic of shallow environments, whereas the lower halites showed evidence of cooler and more consistent temperatures, indicative of formation in a deeper-water setting. The deeper salts also showed a more plate-like crystal structure and are interbedded with **kainite**, a complex sulfate mineral ($KMg(SO_4)Cl·3H_2O$). These halites show no evidence of subaerial exposure and were thus interpreted as being precipitated in relatively deep water. The upper halites, show cracks filled with silt and sand interpreted to be windblown, suggesting that the upper salt was exposed to the air, thus requiring complete desiccation of the sea. The distribution of the evaporites also broadly conformed to the bullseye pattern, where halite pods are surrounded by carbonate and gypsum (Figure 16.12).

At the margins of the Mediterranean, most of the salt is either not preserved or eroded away at the M-unconformity (Figure 16.9) and much of the lower evaporite layer is also eroded. Where erosion is evident, the contact is marked by a clear erosional unconformity (Figures 16.5 and 16.6) and by the

Figure 16.11 (a) Interbedded mudstones and sandstones exposed on the banks of the Santerno Valley lie below the gypsum layers. (b) Close-up view shows deformed ripple-cross laminated sandstone bed, typical of rapid deposition by slowing turbidity currents. These alternate with (c), thickly bedded sandstones. The red dashed line highlights a channel scour, cut by an energetic, sand-laden turbidity current.

river gravels referred to above. It later became clear that there are two major unconformities (M and N), which are superimposed on each other at the basin margin but are separated by the salt layer deeper in the basin (Figures 16.5 and 16.6).

16.4 Evidence of Hypersaline Lakes: The Lago Mare

The upper evaporites comprise mostly anhydrite/gypsum and dolomite. The laminated dolomitic marls alternate with chicken-wire anhydrites and are thought to have been deposited in very shallow water sabkha environments. As we discussed earlier, Dr. Cita showed that they contain fossils characteristic of stagnant, brackish, land-locked briny lakes. The extent of the M layer indicated a very extensive shallow lake, termed the **Lago Mare**, literally "Lake-Sea."

Strontium isotopes of shallower sedimentary basins, on the flanks of the Mediterranean and at higher elevations, such as in Spain and Cyprus, also show a different isotopic composition than the age-equivalent deeper Mediterranean evaporites (Figure 16.13). The initial cycle of evaporites have a $^{87}Sr/^{86}Sr$ ratio of about 0.709, which is close to contemporary seawater, indicating that they formed from evaporation of ocean water that matched global values. The second cycle, however, has a lower ratio of about 0.7086. The areas around the Mediterranean include rocks that have lower $^{87}Sr/^{86}Sr$ ratios, and it is thought that the second cycle of evaporites formed by river waters that eroded these rocks and thus had a correspondingly lower ratio. This is consistent with the Mediterranean being isolated from the Atlantic as it became deeply desiccated, receiving water only from local sources and the global ocean. The Argille Azzurre and Trubi formations show a return to a pre-Messinian, oceanic signature.

> 🔑 **KEY POINT**
>
> As the Mediterranean Sea retreated, it left a series of brackish-water lakes, called the Lago Mare. $^{87}Sr/^{86}Sr$ ratios record a transition from seawater to fresh water delivered by the rivers that drained into these lakes from local sources.

16.5 Plate Tectonics in 1973 and the Depth of the Mediterranean Sea

The theory of plate tectonics, which as we saw in Chapter 5 was only fully accepted in the late 1960s, was still a new idea in 1973. Much was still not known about mechanisms and rates, and especially the rate at which the lithosphere could be uplifted or subsided. The DSDP data was thus a puzzle. In 1973 it was not clear when the Mediterranean Sea formed, and whether it was shallower or deeper in the past. Was it possible that the Mediterranean Sea formed after the deposition of evaporites? This is the shallow–deep hypothesis. Alternatively, could the evaporites have been deposited in deep water, such that the sea never really dried out but somehow became so saturated with solutes that they precipitated in deep water (the deep–deep hypothesis), consistent with the Schmalz model? Or was there a deep Mediterranean Sea that dried out and then flooded again (the deep–shallow–deep hypothesis), as Ken Hsu and his group had suggested?

To answer these questions, it would be critical to sample the sediments *below* the salt layer to see if the older sediments were

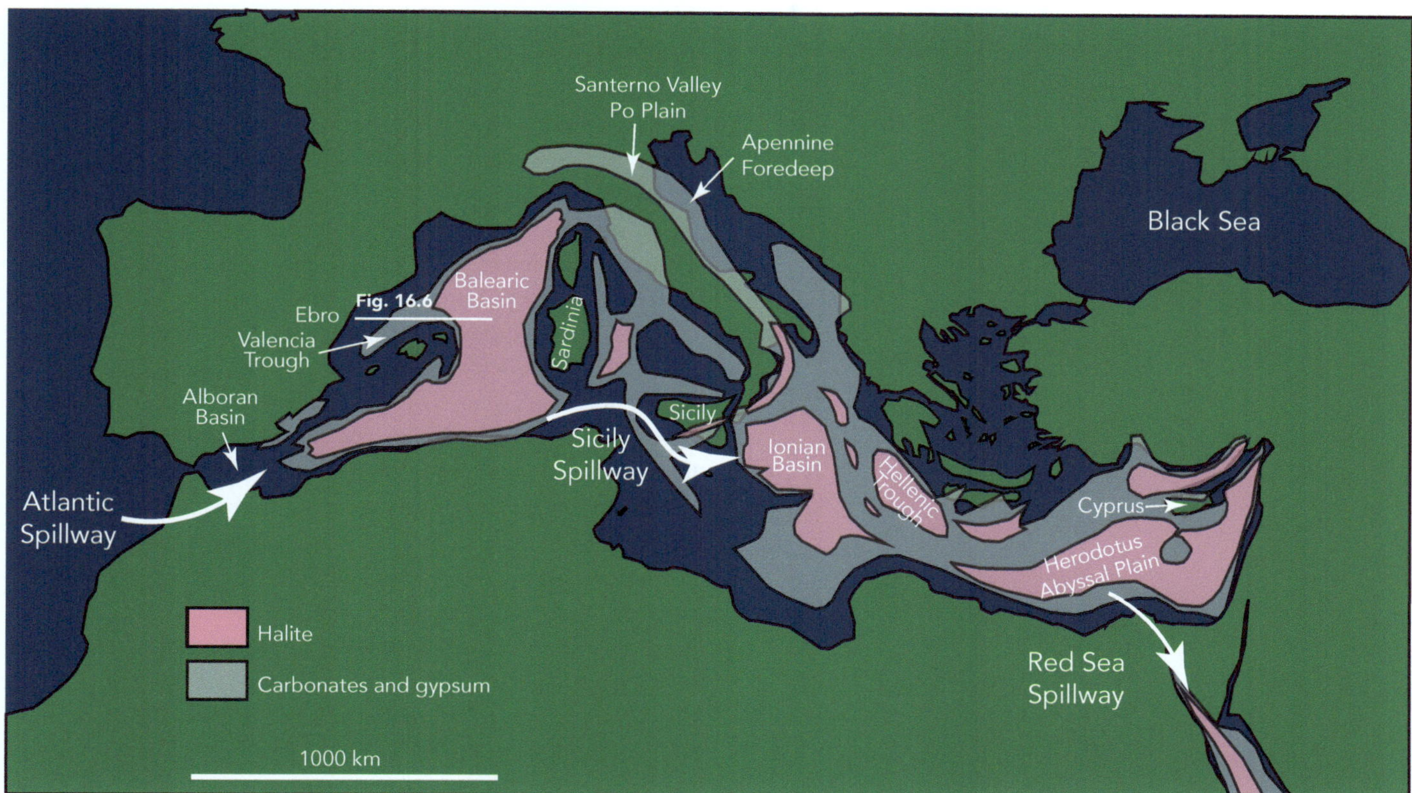

Figure 16.12 Distribution of Mediterranean evaporites showing concentric, bullseye pattern. Connections to the Atlantic Ocean and Red Sea, and spillways between basins within the Mediterranean are highlighted. Map shows location of cross section in Figure 16.6. Source: Adapted from Ryan (2009). Copyright (2009). With permission from John Wiley and Sons.

deposited in deep or shallow water. DSDP results in this regard were very clear. In all areas where the pre-Messinian layers were sampled, there was unequivocal evidence of deep water, and this matched work being done on the outcrops in Italy that showed deep water turbidites below the Solfifero (Figures 16.9 and 16.11).

KEY POINT

Sampling both above and below the Messinian unconformity showed evidence of originally deep water with a shallowing in between. This suggested that tectonics played a relatively minor role in the sea-level fall.

16.6 Unconformities and Incised Valleys: Further Evidence for a Massive Drop in Sea Level

In preparation for constructing the Aswan Dam across the Nile River, in the 1960s, Russian geologists were surprised to drill over 300 meters of loose sediment before hitting bedrock (Figure 16.14(a)). Even stranger, the lower 130 meters consisted of marine Pliocene mudstone, despite the fact that the dam site is almost 1,000 kilometers landward of the present-day shoreline. They had assumed bedrock would be much shallower than this, and never imagined they would encounter any marine sediments. The anomaly required an explanation.

Aswan is about 100 m above sea level, so the deep sedimentary fill indicated that the proto-Nile River cut a canyon 200 m below the modern-day sea level, 1,000 kilometers inland of the present coastline. The lower parts of this canyon were filled with Pliocene marine sedimentary rocks and the upper parts were filled with non-marine deposits (Figure 16.14(a) and (b)). This suggested that the canyon was cut by a river during a time of falling sea level (Figure 16.15) and filled during a later Pliocene event that flooded the valley with seawater, indicating a subsequent rise of sea level. Continued coring and seismic profiling showed that this canyon could be traced for another 1,000 km towards the Mediterranean coast, extending well offshore of the city of Cairo. The mouth of the canyon could be traced to about 3 km depth in the areas offshore of the Egyptian coast (Figures 16.14 and 16.15), where it is buried by younger marine deposits. Subsequent seismic surveys and drilling indicated kilometers-deep incised canyons of a similar age all around the margin of the Mediterranean Sea. Many of the major present-day rivers, including the Ebro in Spain, the Rhone in France, and on the African side the Nile in Egypt, showed deep and wide canyons buried deep below the surface (Figures 16.14 and 16.15). Similar age and scale erosional valleys were also discovered in the Black Sea. Extensive coring showed that many of the canyons were floored by alluvial gravels deposited by

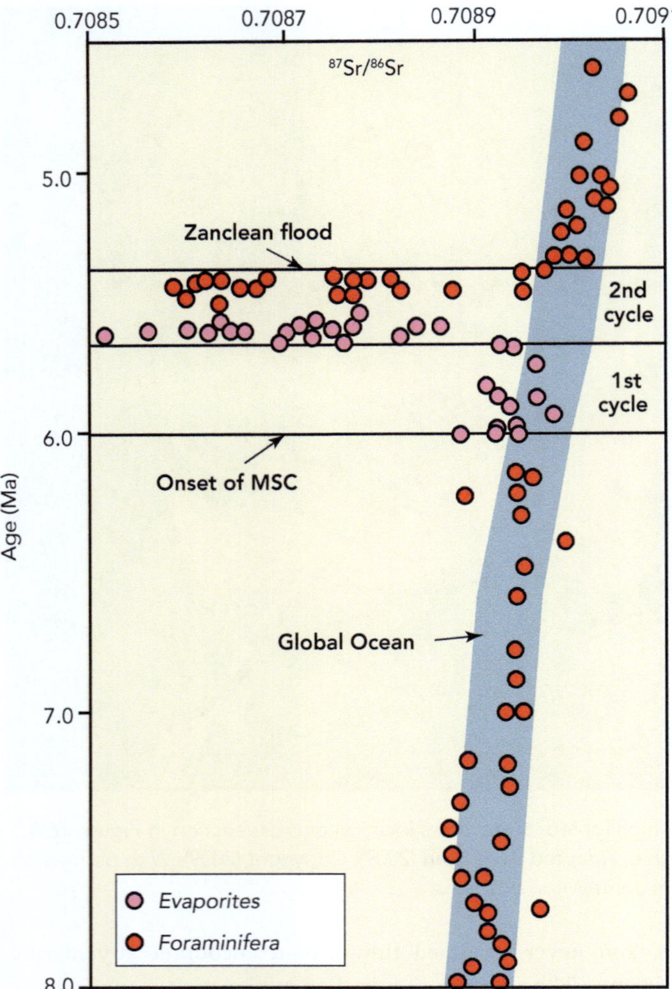

Figure 16.13 Strontium isotopic compositions of Mediterranean carbonates and gypsum as a function of age. The shaded blue belt represents the global ocean trend. At the onset of the Messinian salinity crisis (MSC), the first cycle gypsum deposits lie close to "oceanic" values indicating that the primary supply of water to the Mediterranean brine was from the Atlantic. However, the brines that precipitated the second cycle gypsum are quite different, indicating that the waters were supplied by river input from the surrounding continents. Following the Zanclean flood, $^{87}Sr/^{86}Sr$ ratios return to the normal oceanic average. Source: Adapted from Ryan (2009). Copyright (2009). With permission from John Wiley and Sons.

gravelly rivers that ultimately fed terminal alluvial fans deposited at the mouths of the canyons where they intersected the old seafloor. In many areas these coarse-grained alluvial deposits are overlain by evaporitic salt (Figure 16.14(c)). These observations collectively indicated a massive drop in sea level that exposed the steep margins of the Mediterranean Ocean. These steep margins were eroded into deep valleys and canyons by the rivers (Figure 16.15) that originally drained into the ocean, and are linked to the M unconformity that piqued Ken Hsu's interest and led to the DSDP drilling program to literally look at the rocks associated with the seismic anomalies discussed earlier.

Seismic cross sections (see Box 16.1) also show that the modern-day shelf-to-slope sediments form a distinctive "**clinoform**" geometry characteristic of continental margins that border deep oceans (Figures 16.6 and 16.14(c)). The height of the clinoform can be used to estimate the water depth. The seismic data clearly show these shelf-slope clinoforms built out into a deep-sea basin, but these clinoforms were deposited on top of the Messinian erosional unconformity, which was cut by rivers. This is consistent with the hypothesis that a massive increase in water depth occurred *after* formation of the river valley. This also explains why many of the valley fills are filled with Pliocene marine sediments, recording filling during the transgression associated with the subsequent sea-level rise and flooding of the Mediterranean, following the re-connection with the Atlantic. Unconformities can certainly be caused by tectonic uplift, but this circum-Mediterranean unconformity was just as deeply incised on the Egyptian side, which is a tectonically inactive "passive" margin, as well as the tectonically active northern margins. The regional nature shows that it was caused by a **eustatic** (sea-level) fall rather than a tectonic uplift. Also, the observation of clinoforms in older Miocene strata (Figure 16.14(c)) shows that the Mediterranean was deep both before and after the Messinian event.

🔑 KEY POINT

Observations of kilometer-deep valleys incised around the margins of the Mediterranean Sea provided additional evidence of a kilometers-scale drop of sea level that could not be accounted for by tectonic uplift, further supporting the evaporation hypothesis.

16.7 The Salinity Crisis

Slowly, but surely, the evidence began to build up. The core data showed very shallow to terrestrial, non-marine river and evaporite deposits, recording a kilometers-scale sea-level drop, sandwiched between deep-sea marl deposits, indicating high sea levels. Sea levels associated with the advance and retreat of continental glaciers, such as characterize "icehouse" periods of Earth history, can cause changes on the order of 100 meters, but the Mediterranean looked like it experienced a drop of sea level that was on the order of thousands of meters, an order of magnitude more than could be accounted for by simple climate-driven cycles. This then led to the search for a mechanism that was related to desiccation.

The initial conditions that led to the evaporation included the near isolation of the Mediterranean through the gradual closing of Tethys. Around 7 Ma, a channel existed in present-day Morocco, allowing mixing between the Atlantic and the Mediterranean. Shortening in the western Mediterranean region, including northwest Africa and the southern Iberian Peninsula, uplifted continental fragments sufficient to block

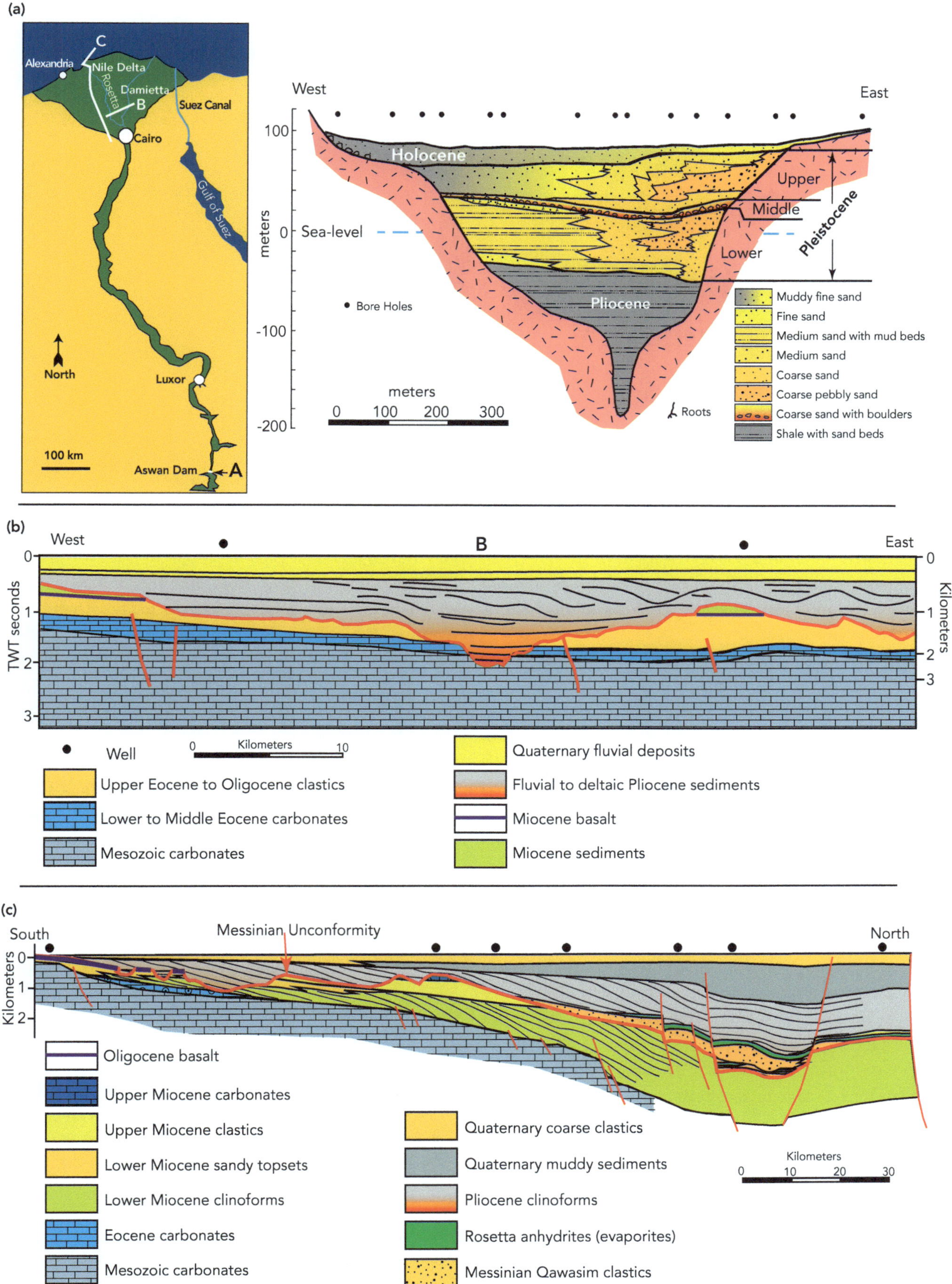

Figure 16.14 (a) Cross section of the Nile Canyon at the Aswan Dam. The canyon is cut to 200 m below sea level during the Messinian drawdown. The lower 120 m is filled with marine Pliocene sediments associated with the Zanclean flood. The upper parts of the fill are

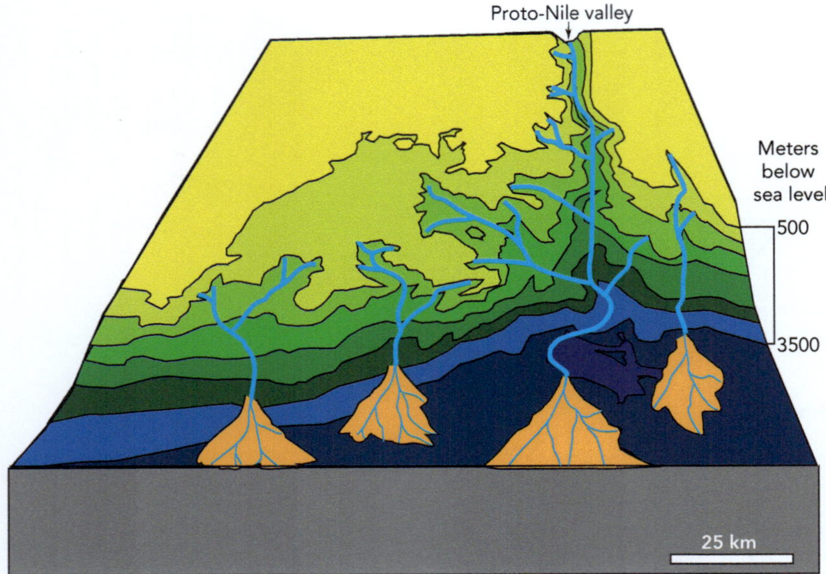

Figure 16.15 Perspective view of Messinian erosional surface in the Nile Delta area, originally mapped by Barber (1981). The proto-Nile River would have cut a canyon over two kilometers deep and deposited coarse alluvial fans and deltas of the Qawasin Formation on the exposed basin floor, as seen in Figure 16.14(c).

communication between the Atlantic Ocean and Mediterranean Sea, producing an isthmus between African and Iberia. This blockage persisted for around 600,000 years until subsidence in the region now occupied by the Strait of Gibraltar allowed Atlantic water to refill the Mediterranean basin in what is known as the Zanclean flood.

The most recent interpretations of the Messinian salinity crisis (Figure 16.16) indicate that the initial phase of evaporation began about 5.96 Ma. At this time, evaporation exceeded the rate of replenishment from the Atlantic as well as freshwater sources, resulting in concentration to the level that carbonate could be deposited. This was accompanied by a moderate drop of sea level and increase in salinity. In nearshore areas, reefs rich in organisms that do not thrive in normal seawater are found. As evaporation continued, the Mediterranean became saltier and selenitic gypsum began to be deposited in deeper water (Figure 16.16(b)). There were indeed deep water evaporites in the initial stages of evaporation, confirming the Schmalz deep water evaporite model. The deepest brines became increasingly enriched in solutes. Orbital changes in climate, which cause glacial cycles, resulted in several gypsum–marl cycles in the lower evaporites, and recent dates shows that these correlate to well-established climate cycles. This allows the timing of the Messinian events to be determined very precisely, down to a few tens of thousands of years. At 5.59 Ma, the connection to the Atlantic was largely closed, and the main drawdown event began (Figure 16.16(c)–(e)). This resulted in desiccation of the Mediterranean Sea, with a drop of water level of over 3,000 meters, formation of the Messinian (M) unconformity (Figure 16.14

(b) and (c)), deposition of river gravels, and shallow water evaporites. As a consequence of the fall in sea level, the rivers draining into the margins of the Mediterranean cut massive canyons, several kilometers deep (Figures 16.14 and 16.15). These side-canyons would have been similar in scale to today's Grand Canyon in Arizona. The main halite deposits are thought to have been deposited in less than 100,000 years in hypersaline seas. The deepest parts of the Mediterranean were left as the Lago Mare. Because they were smaller, the rivers draining into them caused them to become brackish lakes, versus hypersaline seas (Figure 16.16(e)), and some of these were still over 1,000 m deep, but much of the old sea was now a dry desert. The last stages of the crisis, 5.50–5.33 Ma, were marked by cycles of wetting and drying of these lakes, again driven by orbital climate cycles.

At 5.33 Ma, the connection to the Atlantic was re-established through the Strait of Gibraltar and the Mediterranean Sea was re-established. The ensuing Zanclean flood event took place probably over just a few decades, which would have resulted in a sea-level rise on the order of 100 m per year. This is probably one of fastest sea-level rises that has ever occurred in Earth history.

KEY POINT

The salinity crisis occurred in a few stages, with deep water and shallow water evaporite phases. The ensuing rise resulted in the Zanclean flood, probably one of the fastest sea-level rises ever recorded.

Figure 16.14 (*cont.*) non-marine deposits. (b) The cross section across the Nile Delta shows the kilometers-deep incision. (c) The dip cross section shows that the incised canyon erodes into older clinoforms, indicative of deep water. The unconformity can be traced downdip where it feeds coarse sediments of the Qawasim Formation, probably deposited in alluvial fans and overlain by the Rosetta evaporites. Younger Pliocene sediments eventually fill the canyon and build clinoform deposits diagnostic of the return to deep water conditions. Source: (a) Adapted from Chumakov (1973); (b) and (c) Adapted from Barber (1981). Copyright (1981), with permission from Elsevier.

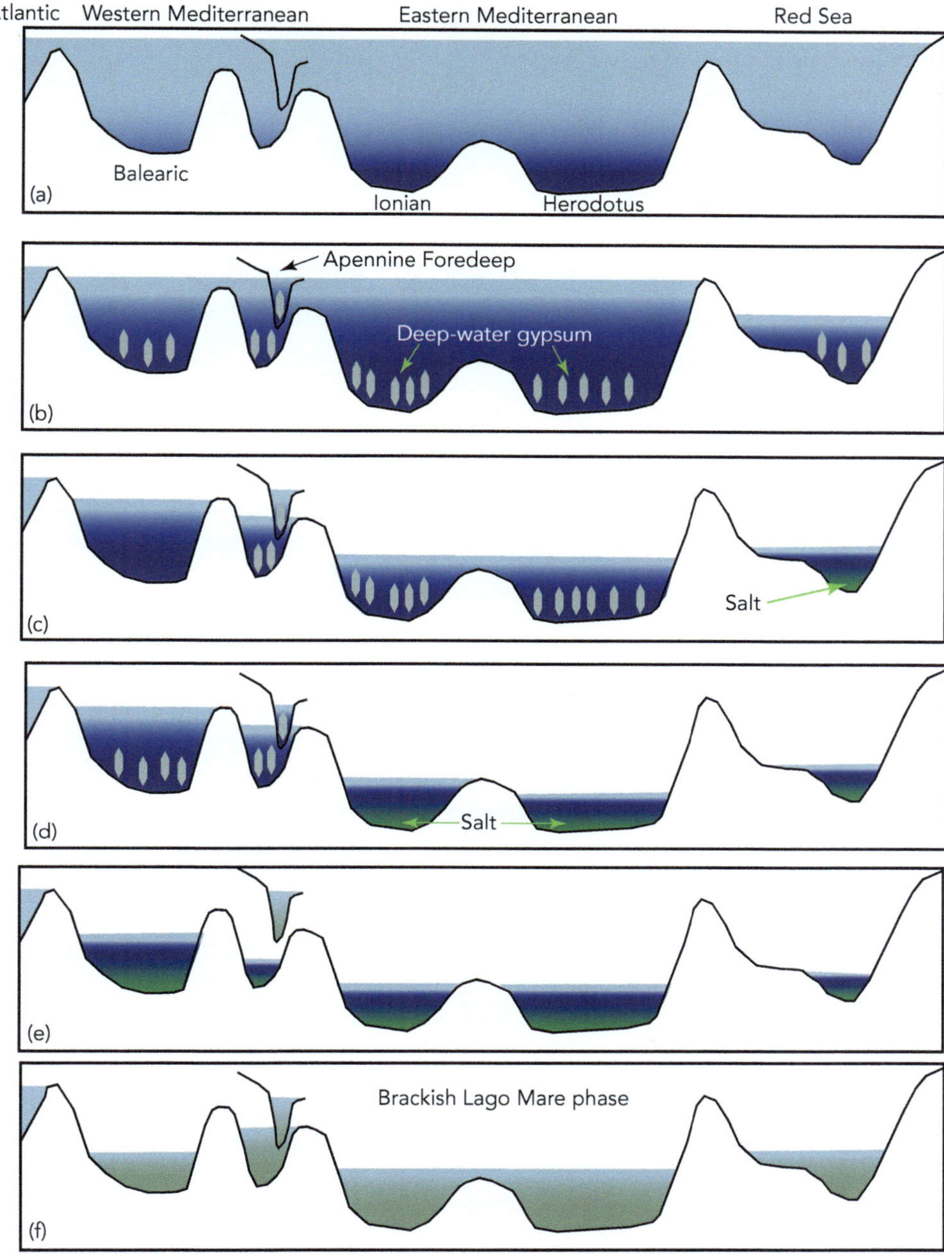

Figure 16.16 Evolution of the Mediterranean during the Messinian salinity crisis. (a) Onset with salinity increasing and deposition of carbonate. (b) Salinity increases further allowing for bottom growth of first cycle selenite deep water gypsum in all circum-Mediterranean satellite basins. The deep basins also accumulate euxinic mud and laminated anhydrite. (c) Evaporation over the whole surface of the Mediterranean and Red seas exceeds the input from the Atlantic combined with rivers plus rain. The Red Sea experiences evaporative drawdown. Evaporative concentration deposits Red Sea salt. The connection to the Apennine foredeep is lost. (d) The Eastern Mediterranean becomes isolated, producing thick salt deposits in the Ionian and Herodotus basins. (e) Atlantic input is now so small that the Western Mediterranean experiences massive halite accumulation in the Balearic Basin. (f) As the Atlantic input stops, each basin becomes occupied by its own Lago Mare lakes whose surfaces rise and fall with short-lived climate cycles. Stages (c)–(d) occur in less than 100 ka. Source: Adapted from Ryan (2009). Copyright (2009). With permission from John Wiley and Sons.

16.8 Tectonic Cause

As we discussed in Chapter 15, the Mediterranean, Black, and Caspian seas are the remnants of the largely closed but once much larger Tethys Ocean. The Strait of Gibraltar region marks the boundary between the African plate, to the south and the Iberian Peninsula to the north, which is part of the European plate. The Gibraltar Arc (Figure 16.1) is linked to an east-dipping oceanic slab (a remnant of Tethys) that is sliding under the Mediterranean. The modern Strait of Gibraltar contains a sill, defined as a high ridge between two basins, in this case the Mediterranean and the Atlantic, that is 280 m deep. As we shall discuss in Chapter 18, climate-driven cycles (i.e., glaciations) can cause up to 110 m of sea-level change, with falls correlating

to glacial periods, but to separate the Atlantic from the Mediterranean would require a drop almost three times this. Therefore, a tectonic mechanism is considered the most likely primary cause.

Compression due to the collision of the African and European plates provides a general cause, but the details have not as yet been fully sorted out. Subduction of the slab associated with the Gibraltar Arc may have caused changes in local uplift, with activation of faults causing local closing of the connection. There are also suggestions that the initial connection, prior to the Messinian salinity crisis, lay south of the present-day Strait of Gibraltar (Figure 16.12), but the Zanclean flood occurred across the modern-day strait.

16.9 Consequences of the Crisis

There were numerous consequences associated with the Messinian salinity crisis. The great drying allowed significant exchange of fauna between Africa and Europe. African animals including hippopotamus, antelopes, and elephants migrated north. When sea level again rose, Mediterranean islands such as Crete, Sicily, Malta, and Cyprus served as refugia. The isolation resulted in **insular dwarfism**, in which dwarf forms of these mammals (Figure 16.17) evolved in response to living in the far more limited environments afforded by the islands. The evolutionary response of the larger animals was to either go extinct or evolve to a smaller size. A similar observation was made in the Cretaceous Period, during which much of central Europe was flooded, creating a series of islands. Dinosaurs found in places such as Germany and Romania, which were then islands, represent much smaller versions of the giants found on continents and indicate that they were forced to evolve to smaller sizes once they were isolated in islands with lower food resources (see Chapter 12).

The water that evaporated from the Mediterranean was likely redistributed in the remaining world's oceans and may have caused a global rise of sea level on the order of 10 m. The combination of uplift in the African continent and global desertification may also have resulted in loss of forests and an increase in the scale of African savannas. As we will see in the next chapter, this roughly coincided with the appearance of the first fully upright hominids, the australopithecines, about five million years ago. The australopithecines evolved into our own genus *Homo*, and developed tools and hunting skills that led to the extinction of many species, including the dwarf mammals that appeared as a consequence of the Messinian events.

Although evaporation and drawdown of the Mediterranean is now one of the best and most recently documented, evaporates are known from many other times in Earth history. One of the requirements is a semi-enclosed, commonly elongate, sea or ocean that can be cut off from the world's oceans. As we saw in Chapter 12, in the Jurassic, the early forming Atlantic looked similar to the modern Red Sea, and there is abundant evidence that it was at times cut off from the worlds' oceans as indicated

Figure 16.17 Dwarf elephants (*Palaeoloxodon falconeri*), found on the islands of Sicily and Malta in the Mediterranean, are smaller than people in the background, reaching about 1 m in height. Source: Photo by Szilas in the Senckenberg Naturmuseum. Own work, Public domain, via Wikimedia Commons.

by Jurassic-age salt deposits on both the North American and African margins. These salts have subsequently floated upwards, forming diapirs that form major traps for hydrocarbon fields in Canada and the Gulf of Mexico.

Paleogeographic maps of the Permian Basin area in the southwestern USA also show that it had a limited connection to global oceans, and there are abundant evaporites in that basin as well. Paleozoic evaporites, such as in the Williston Basin in Montana and Saskatchewan, which produce potash fertilizer from the potassium-rich salts, reflecting extreme evaporation of shallow restricted seas as a consequence of the equatorial position of North America, reviewed in Chapters 10 and 11.

 **KEY POINT**

Drying of the Mediterranean Sea allowed for exchange of fauna between Africa and Eurasia. Following the Zanclean flood, large mammals became isolated on islands such as Crete and Sicily, and experienced insular dwarfism as they evolved to much smaller sizes.

16.10 Summary

- The Mediterranean Sea, which is an elongate remnant of the old Tethys Ocean, has a higher degree of evaporation versus replenishment from freshwater rivers, and maintains its level as a result of input from the Atlantic through the narrow Strait of Gibraltar. About the end of the Miocene Epoch about 5.9 Ma this connection was closed, and the sea experienced a catastrophic drawdown of over 1,000 meters.

- The 1970 Deep Sea Drilling Project was designed to test whether unconformities and diapirs seen in seismic data from the Mediterranean Sea were the result of shale tectonics or salt, the latter of which might be linked to the evaporative drawdown. The deep-sea drilling found deposits of rivers gravels, evaporites, and shallow water fossils on the Mediterranean seafloor, which indicated a major evaporative drawdown event referred to as the Messinian salinity crisis.

- Subsequent drilling far inland around the Aswan Dam in the Nile River valley, as well as circum-Mediterranean seismic lines, discovered a series of smaller canyons around the margin of the Mediterranean basin cut by rivers during a 3,000 m fall of sea level, far too extensive to be the result of better-known glacial cycles.
- The pattern of evaporites shows both teardrop and bullseye patterns, and the early coarse-crystalline selenitic gypsum indicates that the drawdown occurred in cycles, with early precipitation beginning when water was still over 1,000 m deep with later wholesale desiccation.

- At the maximum, the Lago Mare, a series of deep, brackish lakes formed.
- Onshore outcrops in Italy and Spain also show this record of deep–shallow and deep sea-level cycles, but due to continued closing these rocks are now exposed at the surface.
- As a result of the drawdown, large mammals were able to migrate from Africa to Eurasia, but after the Zanclean flood – one of the most rapid sea-level rises in Earth history – large mammals became isolated on islands such as Crete and Sicily and these adapted to small sizes via insular dwarfism.

Key Words

- lithosphere
- Miocene Epoch
- Zanclean Stage
- Messinian salinity crisis
- seismic
- two-way travel time
- seismic reflections
- unconformity
- diapirs
- gypsum

- hematite
- benthic foraminifera
- *Ammonia beccarii*
- *Elphidium macellum*
- marls
- halite
- coccoliths
- sabkhas
- anhydrite
- hypersaline

- Messinian Stage
- selenitic gypsum
- turbidity currents
- fluid inclusions
- kainite
- Lago Mare
- terminal alluvial fans
- clinoform
- eustatic
- insular dwarfism

Further Reading and References

Barber, P. M., 1981, Messinian subaerial erosion of the proto-Nile Delta, *Marine Geology*, 44, 253–272.

Chumakov, I. S., 1973, 44.3. Pliocene and Pleistocene deposits of the Nile Valley in Nubia and Upper Egypt, in *Deep Sea Drilling Project, volume XIII*, https://doi:10.2973/dsdp.proc.13.144-3.1973.

Hsu, K. J., 1982, *The Mediterranean Was a Desert: A Voyage of The Glomar Challenger*, Princeton University Press.

Roveri, M., Manzi, V., Genneri, R., and Lugli, S., 2006, The record of Messinian events in the Northern Apennines for-deep basins, *L'Ateneo parmense. Acta naturalia: organo della Società di medicina e scienze naturali di Parma*, 42(3), 47–125.

Ryan, W. B. F., 2009, Decoding the Mediterranean salinity crisis, *Sedimentology*, 56(1), 95–136.

Review Questions

1. Where is the oldest oceanic lithosphere in the modern oceans?
2. What is the water budget in the Mediterranean?
3. How did the replenishment from the Atlantic Ocean become cut off?
4. What were the observations from deep-sea drilling in the Mediterranean that led to the desiccation hypothesis?
5. What are the types of evidence indicating that evaporites can form either in shallow or deep water?
6. What is the evidence onshore for the Messinian event?
7. What is the significance of finding marine Pliocene shales in the Nile Valley at the Aswan Dam, 1,000 km inland?
8. What is the geophysical evidence, particularly from seismic data, that suggested 3,000 m sea-level fluctuations in the Mediterranean including fall and subsequent rise?
9. How do clinoforms provide evidence that the Mediterranean Sea was deep before the salinity crisis?
10. What is the fossil evidence for these sea-level changes and desiccation?
11. How fast did these sea-level changes occur and what were the main causes?
12. How did the Messinian salinity crisis lead to the evolution of insular dwarf mammals on the Mediterranean islands, such as Sicily, Crete, Malta, and Cyprus?

Australopithecus afarensis,
artwork. Source: Raul Martin / MSF /
Science Photo Library.

Chapter 17

Out of Africa

Human Evolution

LEARNING OBJECTIVES

- Explain why Darwin thought we share a common ancestry with the great apes and why he hypothesized that we first evolved in Africa.
- Explain how cultural and ideological biases led late-twentieth-century scientists to accept the Piltdown forgery in the search for the "missing link."
- Describe the evolution of humans from a common chimp-like ancestor, including some of the major steps in our evolutionary history.
- Explain the evidence for the transition to an upright gait in the early hominids.
- Explain the evolutionary drivers for bipedalism and bigger brains.
- Explain the evidence for the evolution of humans in Africa and their subsequent migrations.
- Explain the concepts of neoteny and paedomorphosis that may have been a mechanism for the evolution of modern humans.

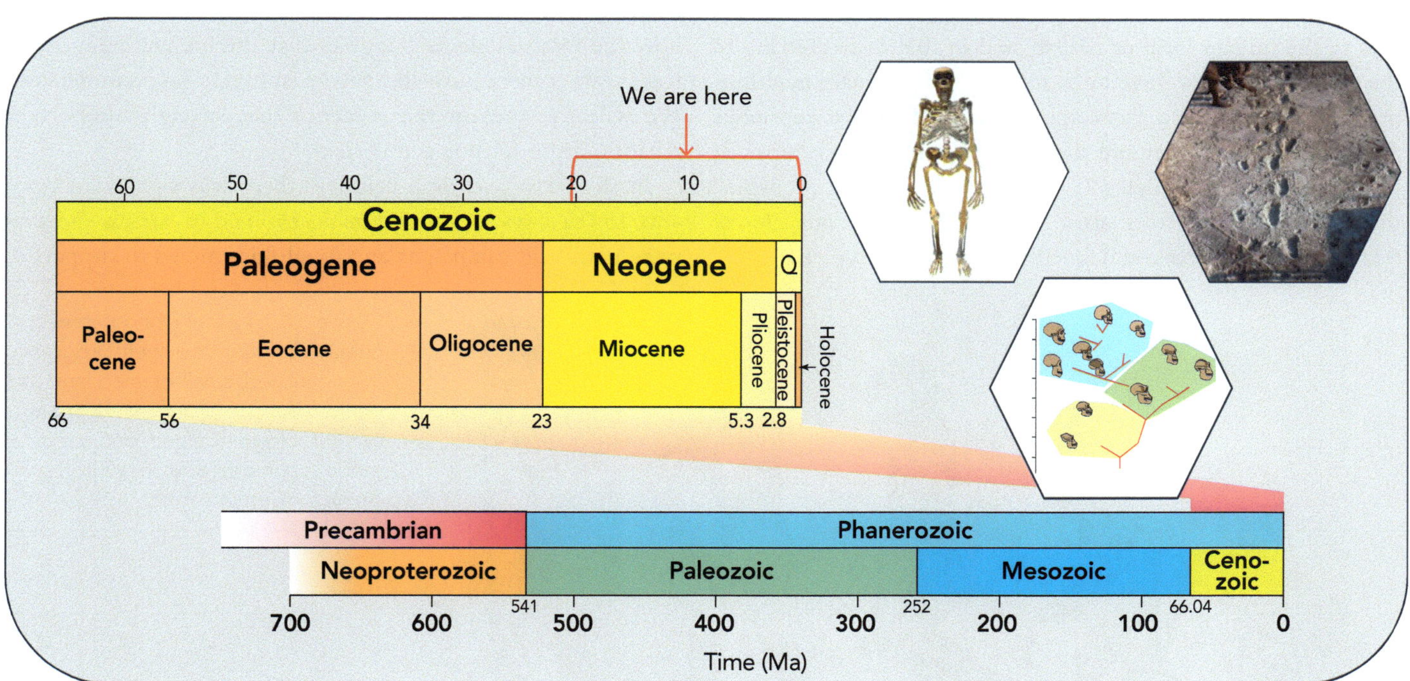

Introduction: Where Did Humans First Evolve?

In Charles Darwin's 1871 book, *The Descent of Man*, a follow-up to his landmark book *On the Origin of Species*, he made it clear that humans were likely an offshoot of "old world **simians**," which include the great apes – chimps, gorillas, and orangutans. The "Old World," refers to Africa, Asia, and Europe, versus the "New World," which refers to the Americas and Oceania. Although Gregor Mendel published his first paper on genetics in 1865, Darwin was not aware of this, so he began his discussion of the evolutionary origin of humans by detailing the physical similarities of humans with apes and mammals. Darwin noted the broad similarities of skeletal organization, embryological similarities, susceptibility to similar diseases, and other anatomical aspects (i.e., homologies discussed in Chapter 6) that indicate their relatedness. He also devoted considerable attention to pointing out the similarities of behavior and mental faculties of humans with the "lower" animals. Darwin was very focused on demonstrating that humans differ from the apes and other animals by degree, rather than by essence. In other words, all the aspects of humans that we might conceive as unique, such as our ability to make tools and our language, can be found, albeit in rudimentary form, in the "lower" animals. Darwin suggested that humans are more similar to the great apes than the great apes are to other mammals such as cats and dogs.

Darwin thus concluded that man must be descended from some "lower" form of primate, but noted that the connecting links (i.e., intermediate species) had not yet been discovered. Darwin made the prediction that some ancient member of the **anthropomorphous** sub-group (i.e., organisms with similarities to the human form or nature, and in this case referring to the great apes) likely gave birth to humans. The main modifications in the transition from ape to human was the enormous expansion of our brains and the development of a fully upright posture. Darwin speculated it was "somewhat more probable" that our early ancestors arose in Africa, where our closest relatives, chimpanzees and gorillas, presently live. He also

pointed out that, in his time, the regions most likely to contain the fossilized remains of our extinct ape-like ancestors had not yet been searched by geologists. Darwin's predictions led to the search for the fossilized remains of our evolutionary ancestors, particularly the "missing link," the hypothetical man-ape that would demonstrate the transition between apes and humans.

KEY POINT

Darwin recognized our close similarities to the great apes, now found only in Africa, and hypothesized that we shared a common ancestry and first evolved in Africa.

17.1 The "Missing Link" and the Ladder of Progress: A Eurocentric View

The concept of a "**missing link**" was commonly used in the eighteenth and nineteenth centuries to refer to a hypothetical creature, intermediate between human and ape, that formed part of an assumed linear evolutionary series of human ancestors. Although the term is no longer used by scientists, the idea that life on Earth forms a **Great Chain of Being**, with humans at the top of an evolutionary ladder, fulfilled eighteenth-century Enlightenment concepts of progress. This in turn built upon Aristotle's theory that life on Earth comprised "higher" and "lower" animals. The idea that evolution is linear and represents a chain or ladder of progress from "lower" **quadruped** (i.e., four-legged), hunched small-brained, and darker-skinned apes to "higher" fully upright, **bipedal**, big-brained, and lighter-skinned humans remains an entrenched icon that is widely used even today (Figure 17.1). In the following sections we will show how the fossil evidence suggests that the human evolutionary tree is rather more bush-like than past investigators understood. We will also present the evidence that strongly supports an African origin for our genus *Homo*.

In the early twentieth century, there was significant resistance to Darwin's idea that humans evolved in Africa, and there were both field campaigns and fossil finds aimed to support the

Figure 17.1 Typical iconography of human evolution as a linear progression from hunched, small-brained, and dark-skinned apes to progressively more upright, bigger-brained, and paler modern humans. Source: Hulton Archive / Getty Images.

idea that humans evolved in Eurasia. Despite a complete lack of any scientific evidence, persistent modern folkloric claims of the existence of ape-men, including the "abominable snowman," or yeti, in the Himalaya and the sasquatch, or bigfoot, in the Pacific Northwest of North America, potentially bolster a Eurasian or even North American origin. The proposal that a "missing link" still lives on Earth today, either in Asia or North America, rather than Africa, is perhaps testimony to the desire to demonstrate that modern humans evolved anywhere but Africa. This may reflect twentieth-century European unease with the idea that black Africans were more likely to have been our progenitors versus lighter-skinned Eurasians. In the early twentieth century, classification of our own species was unclear. Darwin noted that anywhere from 1 to 63 races had been defined by various scientists of the time. After Darwin published *On the Origin of Species* and the *Descent of Man*, although his concept of evolutionary natural selection was accepted, there were significant debates as to whether the various human races reflected descent from one or several common ancestors, and whether white Europeans could trace their ancestry to Africa, as Darwin hypothesized, or Eurasia.

17.2 Piltdown Man: An Example of Twentieth-Century Bias in Established Scientific Views on the Origin of Humans

The strong European-origin bias of the scientific community in the 1920s, was exemplified by one of the most famous hoaxes known as the discovery of "Piltdown man," in an English Quarry in 1912. Amateur archeologist Charles Dawson (Figure 17.2) claimed to have discovered the fossils of the "missing link" in Pleistocene-age gravels near the town of Piltdown, supposedly given to him by a quarry worker. The fossils consisted of a human-like skull and ape-like lower jaw bones.

He presented his fossils at a meeting of the Geological Society of London, and his "missing link" was subsequently named *Eoanthropus dawsoni*, which translates as "Dawson's dawn-man," honoring Dawson as the discoverer. Despite some initial skepticism about the find, Piltdown man was more-or-less accepted as genuine by the English scientific community at the time. In 1938, a memorial stone was placed at the Piltdown discovery site, by Sir Arthur Keith, a well-regarded Scottish anthropologist, a Fellow of the Royal Society, and a key figure in legitimizing Dawson's discovery. Keith was staunchly against the theory that humans evolved in Africa or Asia and was dismissive of the eventual first discoveries of **australopithecines** (unequivocal "missing links") in Africa in 1925, classifying them as mere apes.

In 1931, Keith was a co-presenter with John Walter Gregory (also a member of the Royal Society and a former president of the Geological Society of London) at the annual Conway Hall lecture. Their talk focused on **eugenics,** the idea of controlled selective breeding of humans, such as by sterilization, to improve the population's genetic composition, and promoted the idea that inter-marriage between Caucasian, mongoloid, and negroid races produces inferior progeny. Their arguments supported their perceived need for racial segregation and emphasized their belief in the inferiority of black Africans. These widespread racist views in the scientific community of the time may explain the very negative views on an African origin for humans and the willingness to accept the discovery of Piltdown man as the key English "missing link." This find reinforced the prejudices of the day and attempted to discredit Darwin's out-of-Africa hypothesis.

It wasn't until 1953, two years before Keith's death, that the Piltdown fossils were finally proven to be a hoax. New analysis, including the ability to date the samples, showed that they consisted of a human medieval-age skull, a 500-year-old lower

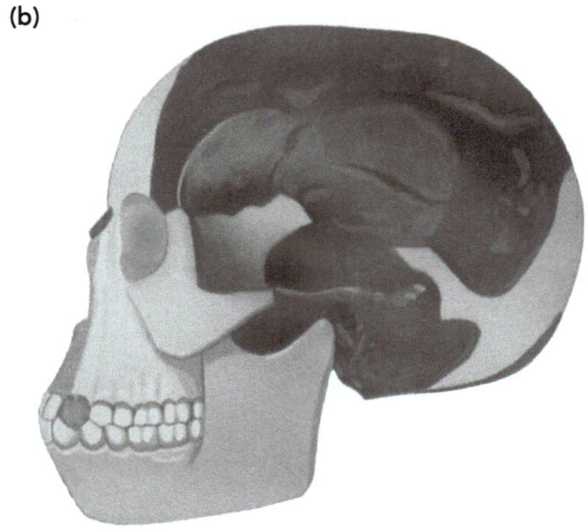

Figure 17.2 (a) 1915 Group portrait of examination of the Piltdown skull. Arthur Keith is seated measuring the skull and wearing a white lab coat, and discoverer Charles Dawson is standing in front of the portrait of Charles Darwin. (b) Reconstruction of the skull. Source: (a) John Cooke, Public domain, via Wikimedia Commons; (b) J. Arthur Thomson, *The Outline of Science*, 1922, Public domain, via Wikimedia Commons.

jaw of an orangutan, and the fossil teeth of a chimpanzee. The bones were stained to make them look like fossils and microscopic examination of the teeth showed that they had been filed to make them look more human-like. It has been suggested that Charles Dawson may have planted the fossils himself, and it has also been speculated that Keith and others may also have been involved in the hoax.

A key point of telling this story is that it shows how a small number of influential Western scientists embraced a hoax that fit their worldview and were able to disseminate their views for several decades. However, it also shows that scientific skeptics eventually won the day, as newer analytical methods were brought to bear to test and debunk a potentially groundbreaking fossil find.

> ### 🔑 KEY POINT
>
> Racist and eugenic ideologies caused resistance to the out-of-Africa hypothesis and may explain why early twentieth-century scientists embraced the Piltdown forgery, which suggested the earliest humans evolved in Britain, rather than Africa.

17.3 The Primate Family Tree

Since the formulation of Darwin's hypothesis, we now have a considerably more robust fossil record that illustrates the origin of primates and humans and their intermediate forms. Humans belong to the order **Primates** (Figure 17.3). The name Primates

reflects the outdated idea that our group belongs to the prime or highest rank of life on Earth representing the culmination of the ladder of progress. The primate clade is divided into two suborders, the "wet-nosed" Strepsirrhini (these include lemurs) and the "dry-nosed" Haplorhini, which includes apes and humans (the **anthropoids**). The anthropoids are in turn divided into the Platyrrhini (literally "flat-nosed") and Catarrhini (literally "narrow-nosed"), which include the "old-world" monkeys (technically the Cercopithecoidea), such as baboons and macaques, and our own clade, **Hominoidea** (Figure 17.3). The Hominoidea are the tailless primates, and include chimpanzees (*Pan*), gorillas, orangutans (*Pongo*), and gibbons (Hylobatidae) (Figure 17.3). As we discussed in Chapter 6, chimps and humans share 99% of their coding DNA, indicating that they recently shared a common ancestor. Humans and gorillas share about 98% of their coding DNA, indicating a common ancestor a little farther back in time.

17.4 The First Primates: Life in the Trees

Fossils with primate-like characteristics have been found in early Cenozoic rocks, dating back to 65 million years ago, and indicate adaptation to an **arboreal** habitat, that is, life in the trees. These adaptations include altered shoulder joints to allow a large range of motion for the arms, dexterous hands and feet useful for grabbing branches, stereoscopic vision, which is required to know if the branch you are reaching for is closer or farther away, and relatively large brains to help navigate the complex environment.

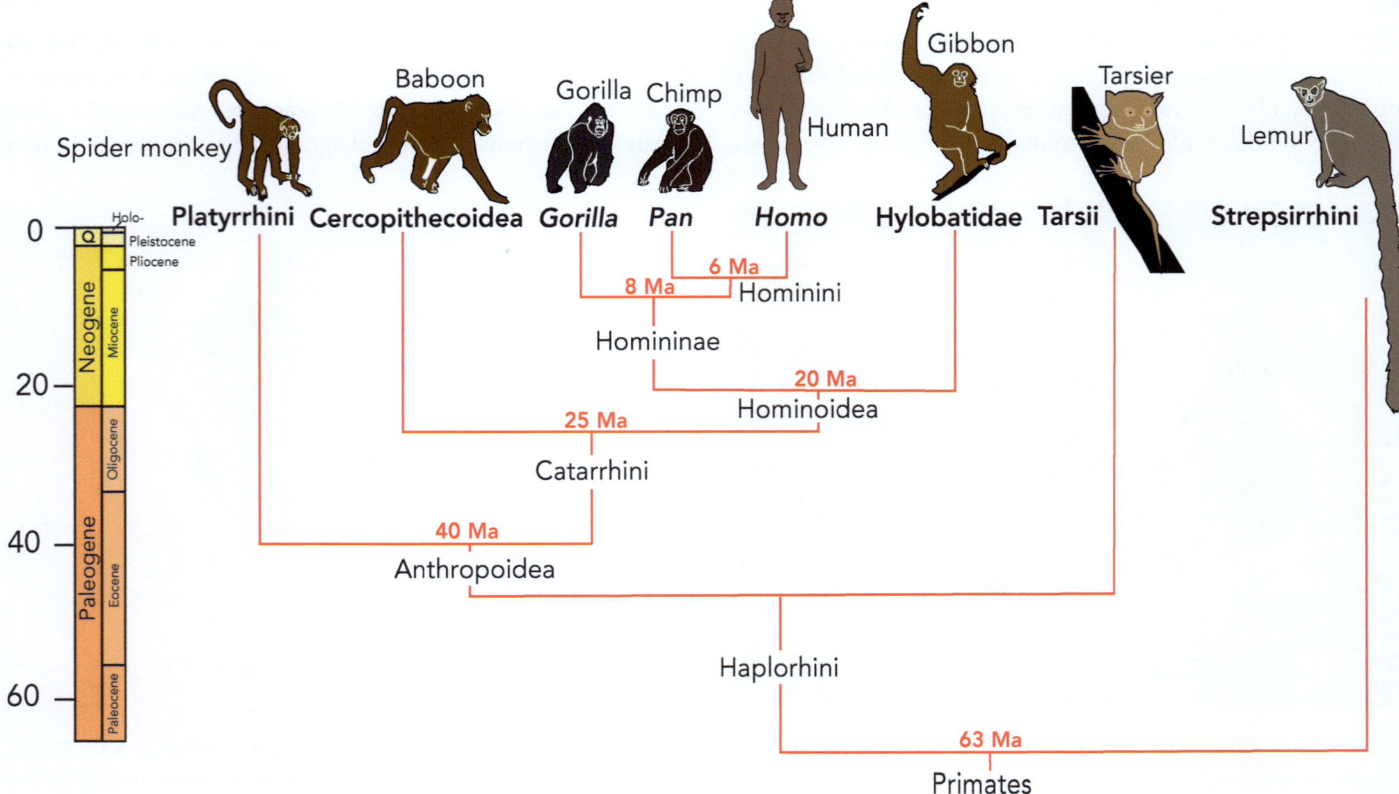

Figure 17.3 Primate family tree with approximate dates of lineage splits. Humans are part of the Catarrhini clade.

Primate fossils are mostly found in rocks that are younger than about 55 Ma (Late Paleocene to Eocene), whereas molecular clocks (discussed in Chapter 6) suggest that they may have evolved in the late Cretaceous, up to 85 million years ago.

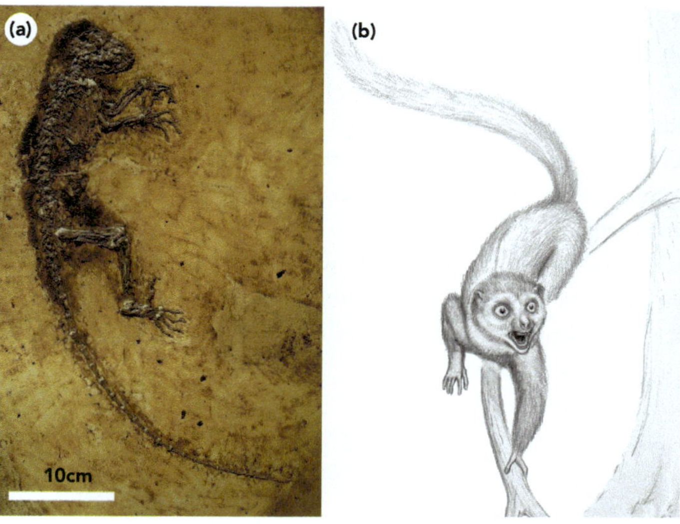

Figure 17.4 *Darwinius masillae*, a Middle Eocene primate. The fossil (a) preserves the outline of the body and fur. Note the opposing large toe and thumb. The reconstruction (b) shows a lemur-like appearance and lacks the flattened face that is characteristic of the anthropoids. Source: (a) Franzen et al. (2009). © 2009; (b) Sketches by Bogdan Bocianowski, https://commons.wikimedia.org/wiki/File:Darwiniusfigs62cropped.jpg, BY CC 2.5.

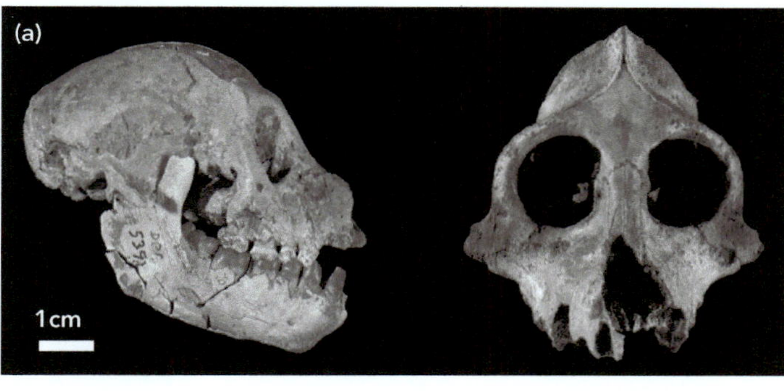

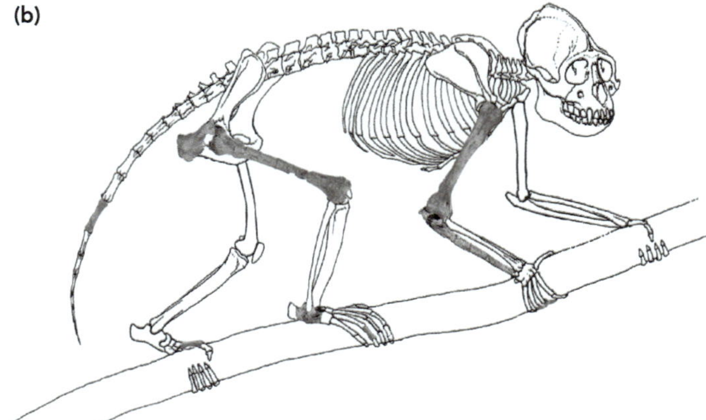

Figure 17.5 (a) Skull fossils and (b) reconstruction of *Aegyptopithecus zeuxis*. One of the oldest anthropoid primates. Source: (b) Used with permission of University of California Press – Books, from Werdelin and Sanders (2010); permission conveyed through Copyright Clearance Center, Inc.

Figure 17.6 Skeletal reconstruction of *Proconsul*, a Miocene ape. Note the lack of a tail but otherwise similarity to the *Aegyptopithecus* in Figure 17.5. Fossil from the Muséum d'Anthropologie, Zurich, Switzerland. Source: Guérin Nicolas (messages), CC BY-SA 3.0 <https://creativecommons.org/licenses/by-sa/3.0>, via Wikimedia Commons.

It has been estimated that perhaps only about 5–7% of all primate species have been preserved in the fossil record, and this likely reflects their preference for a habitat that is not ideal for fossilization. A recent discovery in Messel, Germany provides a rare glimpse of a complete skeleton of a Middle Eocene strepsirrhinid female primate named *Darwinius masillae* (Figure 17.4). She was beautifully preserved within volcanic layers (which allows the fossil to be dated to 47 Ma) deposited in a shallow lake bordered by subtropical rainforests that provided a diet of fruit and leaves. Fossil discoveries in the Fayum depression in Egypt contain good examples of the earliest anthropoid primates, including the 38-million-year-old *Aegyptopithecus zeuxis* (Figure 17.5), which pre-dates the split between hominoid apes and old-world monkeys (cercopithecids, Figure 17.3). Comparison of *Aegyptopithecus* with *Darwinius* (Figures 17.4 and 17.5) shows a dramatic reduction in the length of the nose, and transition to the flatter face that characterizes the hominoids.

The oldest hominoid fossils, discovered in 2013 in Tanzania, include 25-million-year-old teeth and jaw fragments of *Rukwapithecus fleaglei*. Prior to this discovery, the oldest-known ape fossils were about 20 million years old (e.g., *Proconsul*, Figure 17.6). DNA molecular clocks indicate that the first apes likely appeared in the Oligocene Epoch about 25–30 million years ago, and these recent fossil finds are consistent with this. It is generally thought that these earlier hominoids primarily walked on tops of branches, using all four limbs (quadrupedalism), as opposed to swinging underneath primarily using their arms (**brachiation**). Humans are quite capable brachiators, and you can see this if you go to a children's park and see kids swinging from the "monkey bars." The transition from arboreal quadrupedalism to brachiation is a major step in primate evolution and may have led to the loss of a tail. More energy was required in using the arms versus the legs, and as a result, the balancing value of a tail was lost. This transition to brachiation may also have been a prerequisite for later adaptation to a fully bipedal gait in the early hominids.

17.5 Our Greatest Evolutionary Step (Meet Lucy)

Although chimps and gorillas can walk and stand on their feet, they mostly move on all four limbs (Figure 17.7). This is referred to as **facultative bipedalism**, in that they have the facility to walk on two legs but spend most of their time in a quadrupedal stance. Humans and all other hominids are considered to demonstrate **obligate bipedalism**, which is to say we are only able to walk or run on two legs. By 6 to 10 months most babies are crawling, but all parents anticipate the first time their child can walk on two legs and from that point on, most humans never crawl again.

One of the biggest questions in human evolution is, what drove the transition from life in the trees? Holding branches requires both opposable thumbs and toes, but walking upright would be difficult with a splayed-out toe and likely explains why toes migrated to align with the other toes as bipedalism developed. Another question is, what drove the transition to bigger brains? In addition, there is the question of which came first, upright gait or big brains or did both change gradually in tandem? At the start of the twentieth century, only a couple of extinct hominids were known, *Homo neanderthalensis* and *Homo erectus* (Figure 17.8), both of which had relatively large brains.

The common hominid ancestor of our own tribe the **Hominini**, which includes our own genus *Homo*, the australopithecines, and the chimpanzees (genus *Pan*), likely lived around 7 to 8 Ma (Figures 17.3 and 17.8), and the chimp–australopithecine split is thought to have occurred about six million years ago. It is probable that the oldest upright, bipedal hominid was *Sahelanthropus tchadensis* (Figure 17.8). The only fossil remains of this hominid is a skull, but the key observation is that the **foramen magnum**, which is the hole that attaches the skull to the spine, is in a forward position, indicating an upright posture (Figure 17.9). In contrast, note that in obligate quadrupeds, such as a dog, the foramen magnum is at the back of the skull (Figures 17.7 and 17.9).

The first australopithecine (literally "southern ape") fossils were discovered in 1925, in Tuang, South Africa, consistent with Darwin's predictions. Australopithecines were small-brained, and had rather ape-like teeth, albeit with reduced canines (Figure 17.9). A key observation was that the foramen magnum was significantly more central in *Australopithecus*, suggesting that it held its head upright, implying an upright gait (Figure 17.9). The key discovery was a 40% complete 3.6-million-year-old *Australopithecus afarensis* skeleton,

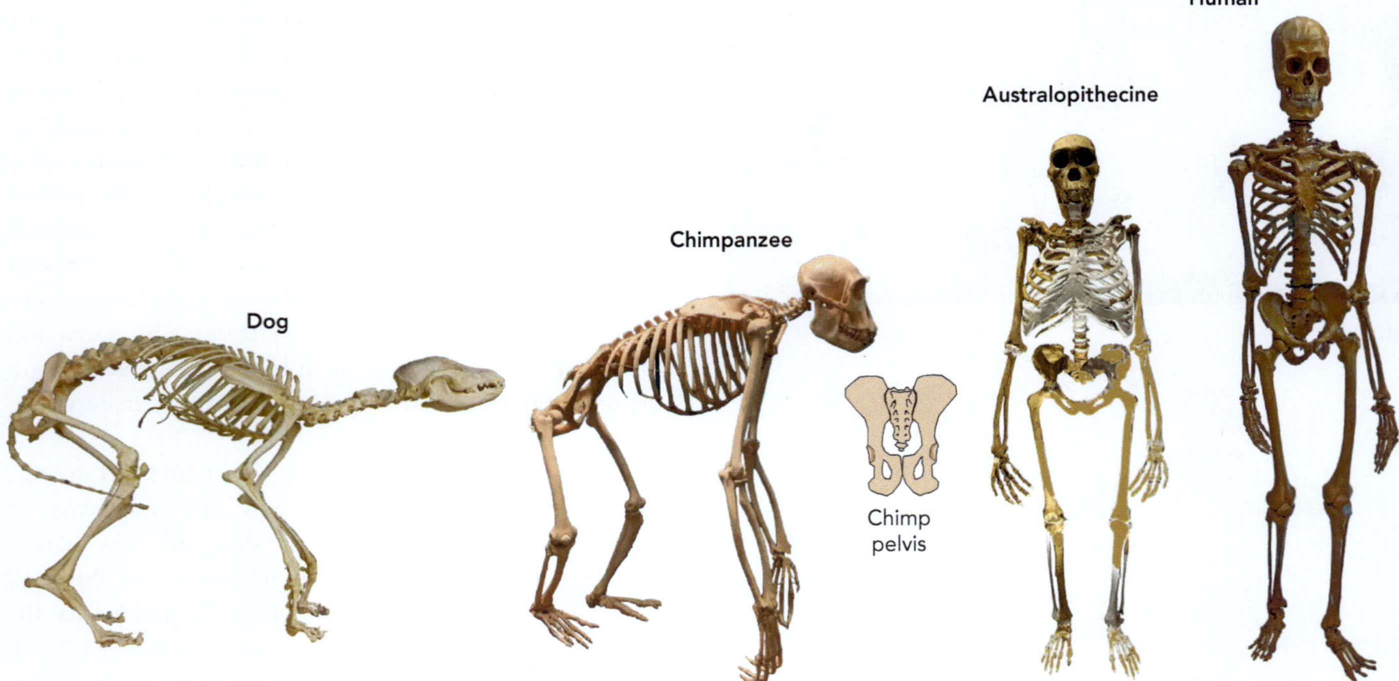

Figure 17.7 Comparison of skeletons of obligate quadrupeds (dogs and chimps) versus obligate bipedalists (australopithecines and humans). In australopithecines and humans, attachment of the spine to the head is underneath the skull versus at the back of the skull in quadrupeds (see Figure 17.9). Also note that the *Australopithecus* pelvis is far more human-like versus the chimp pelvis. Source: (dog) Photo by JPB, McMaster University; (chimpanzee) Photo by JPB. © 2024, Field Museum. All rights reserved; (australopithecine) Photo by JPB. By permission of The Trustees of the Natural History Museum, London; (human) Photo by JPB courtesy of the Indian Museum, Kolkata.

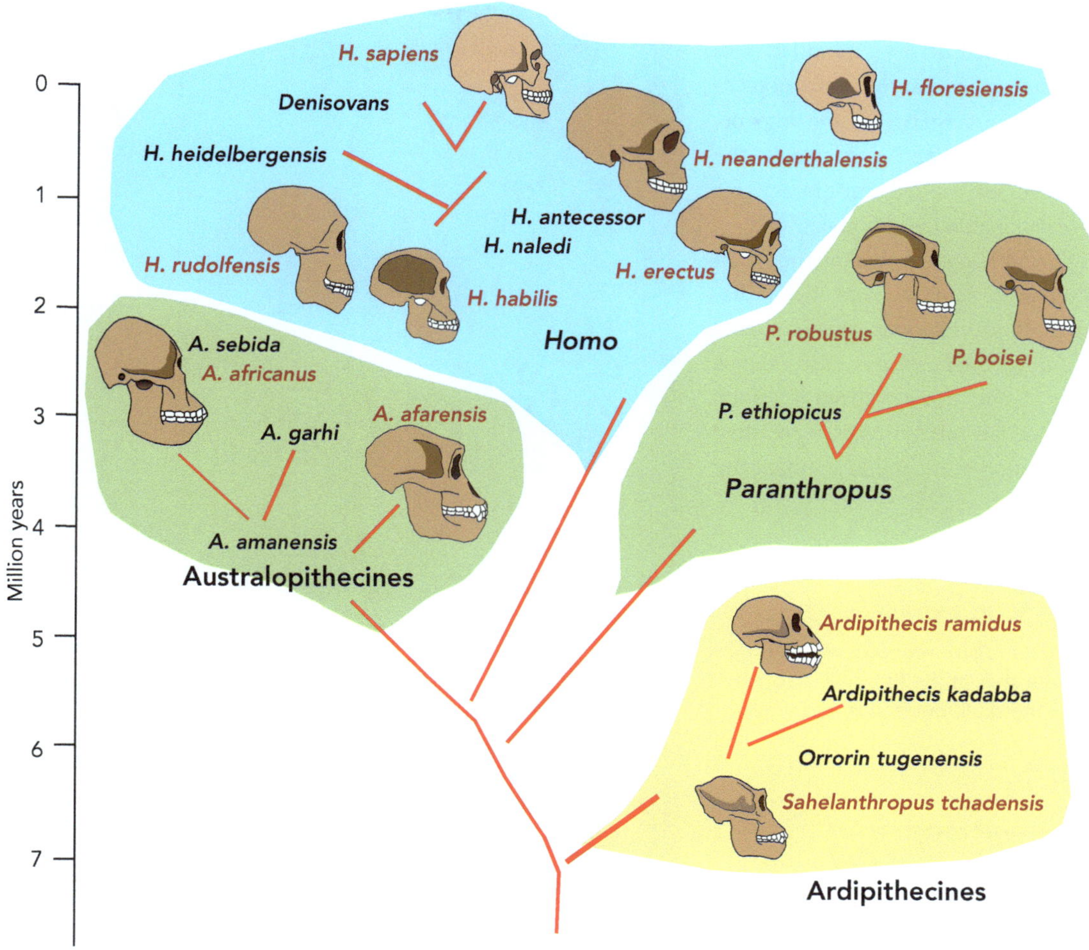

Figure 17.8 Hominin family tree. Red lettered species are adjacent to skull drawings.

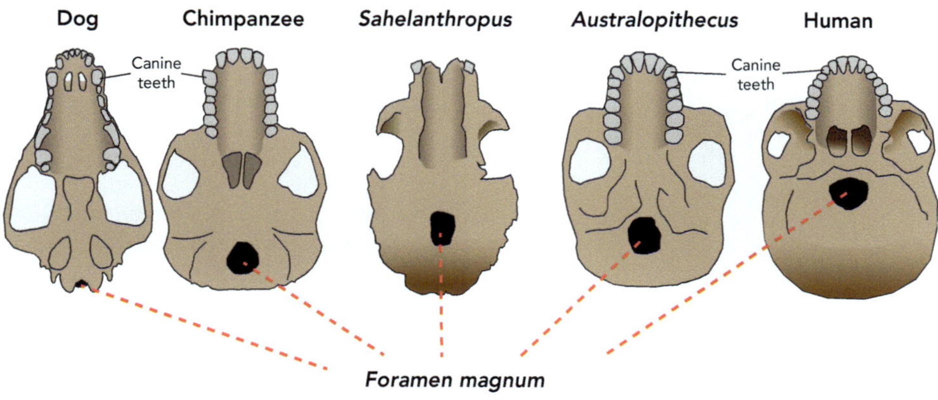

Figure 17.9 The foramen magnum marks the point where the skull attaches to the spine. Farther back, as in dogs and chimpanzees, indicates a four-legged gait, whereas a position in the middle of the skull, such as in the *Sahelanthropus*, *Australopithecus*, and human, indicates upright, bipedal gait. The more central position of the human foramen magnum is largely reflective of a much larger brain case. Note that the *Australopithecus* dentition shows a similar curvature to the chimp, versus the more arcuate teeth pattern in the human skull, but the *Australopithecus* canine teeth are more reduced than the chimp.

found in 1974, nicknamed "Lucy," which included a pelvis and femur (Figure 17.10). Comparison of australopithecine pelvises with humans shows a remarkable similarity (Figure 17.7). Lucy's pelvis is short and bowl-shaped, enabling the gluteal muscles to wrap around the side of the body, providing stability when walking upright. Unlike humans, the big toe is splayed sideways, indicating that Lucy was also good at climbing trees.

In 1976, famed anthropologist Mary Leakey discovered footprints at Laetoli, Tanzania, in the same volcanic layer as the *Australopithecus afarensis* fossils (Figure 17.11). The footprints showed evidence that three individual australopithecines, including two adults and one juvenile, walked across a mudflat. The footprints are consistent with the interpretation that australopithecines walked upright, with modest divergence of the big toe. The anatomical similarities with humans, and particularly the footprints, show that australopithecines walked upright on two legs, although they were still very capable tree climbers.

Since the discovery of Lucy, a 45% complete skeleton of an older bipedal hominin, *Ardipithecus ramidus*, has been discovered (Figure 17.12). At 4.4. million years old, this is almost a million years older than Lucy. *Ardipithecus* also shows evidence of a human-like pelvis, and a centralized foramen magnum, indicating that it was bipedal. Intriguingly, it retained grasping, opposable toes that are much more ape-like indicating it was still well-adapted to climbing trees.

KEY POINT

The first upright hominids showed a massive reorganization of the pelvis, loss of opposing toes, and a shift of the foramen magnum under as opposed to the back of the skull, but brains and teeth remained ape-like.

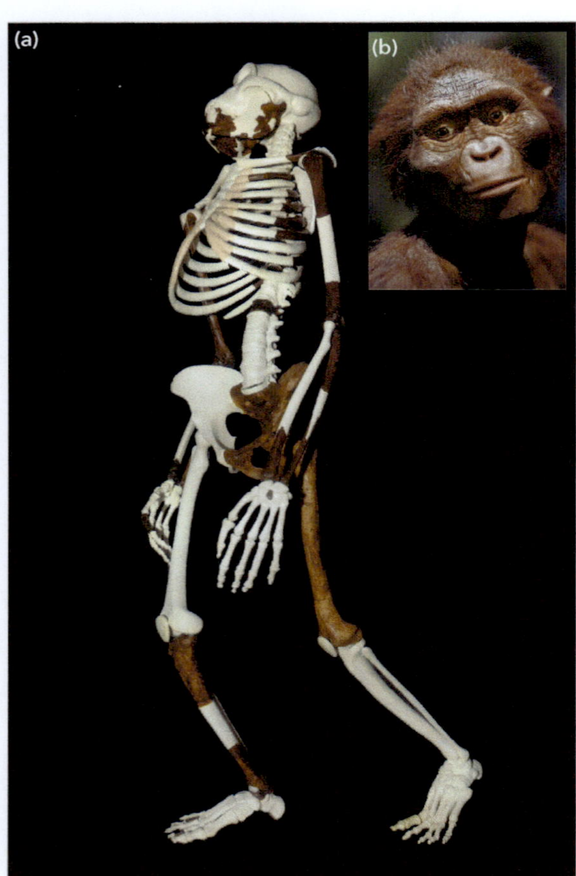

Figure 17.10 *Australopithecine afarensis* (a) reconstruction of skeleton showing upright gait, (b) face that emphasizes ape-like features. Source: (a) Christophe Boisvieux / The Image Bank Unreleased / Getty Images; (b) Bloomberg / Getty Images.

Figure 17.11 3.6 Ma trail of hominid footprints fossilized in volcanic ash, Laetoli, Tanzania in 1978. It shows that hominids had acquired the upright, bipedal gait. The trail probably belongs to *Australopithecus afarensis*. The footprints show a well-developed arch to the foot and no divergence of the big toe. They are of two adults with possibly a third set belonging to a child who walked in the footsteps of one of the adults. The prints to the far right belong to a hipparion, an extinct three-toed horse. Source: John Reader / Science Photo Library.

17.6 Important Environmental Changes

Now that it has been established that upright, bipedal gait evolved before bigger brains, we can return to the question of possible evolutionary drivers for this. The Late Neogene period, during which hominins evolved, was characterized by an overall global cooling, which culminated in the "icehouse" of the Pleistocene ice ages (see Chapter 18). Between 20 and 14 million years ago, during the Miocene Epoch, the planet was about 5 °C hotter than today, but between about 7 Ma to about 3 Ma, global temperatures cooled, settling at about 2 °C above present-day temperatures. During this time, Himalayan uplift was also changing global climate patterns (see Chapter 15). The effect on Africa was an increase in aridity and a change from heavily forested tropical environments to a greater proportion of savannas and grasslands.

It has long been thought that the transition to a greater proportion of grasslands may have been a key driver in the evolution of an upright, bipedal gait. The adaptive advantage of a bipedal gait

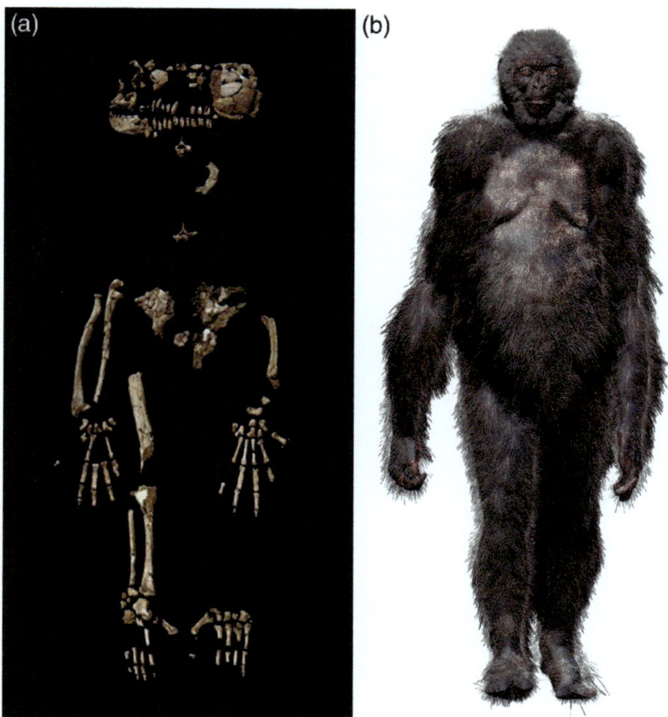

Figure 17.12 *Ardipithecus ramidus*, skeleton and reconstruction. Unlike Lucy, *Ardipithecus* retained opposable toes, indicating adaptation to climbing trees, as well as a bipedal upright gait. Source: (a) Used with permission of American Association for the Advancement of Science, from White et al. (2009); permission conveyed through Copyright Clearance Center, Inc.; (b) John Bavaro Fine Art / Science Photo Library.

may be explained by an improved ability to see over tall savanna grass, to better escape predators, or to walk more efficiently over long distances. Bipedalism would also have freed the hands, improving the ability to make tools for hunting, as well as for gathering food and carrying babies before they can walk on their own. Moreover, an erect posture also exposes less skin to the sun, keeping body temperature cooler in the open terrain of the savanna, versus shaded woods.

Other hypotheses do not require a savanna. For example, one hypothesis is that bipedalism allowed males to provide more food to females and their babies, allowing females to spend less of their time gathering and more of their time nurturing a greater number of offspring. In return for the food provisions gathered and carried by males using their arms and hands that were no longer needed for locomotion, a male would gain greater sexual access to a particular female, ensuring proliferation of his offspring and providing a significant evolutionary advantage. Bipedalism thus may have been a key adaptation in freeing the hands for more food carrying, ensuring more reproductive and consequently evolutionary success.

Many australopithecine fossils have been found in environments that were probably wetter than today's African deserts, and new discoveries of their foot bones show that they were still able to climb trees. *Ardipithecus ramidus,* described above, was discovered in an environment that is thought to be largely

wooded. It clearly retained its adaptations to an arboreal environment, including opposable toes. Given how early this hominid lived, this suggests that bipedalism was not simply driven by loss of forests and increasing grasslands. Other factors, such as improved reproductive success through improved ability to access food resources, may also have been as important.

> **KEY POINT**
>
> Human evolution may have been driven by a loss of forests as the climate cooled, reducing forests and increasing grass and habitats. Bipedalism may have freed hands to do more work, such as finding food and carrying babies, which improved reproductive success.

17.7 How Smart was Lucy? Evolution of the Brain

Clearly, one of the key features of humans is our large brain capacity and a critical question is when this was achieved. Australopithecine brains had a volume of about 400–500 cm^3. *Ardipithecus* (Figure 17.12) also has a small brain, 300–350 cm^3, significantly smaller than *Australopithecus*. Humans are taller than both australopithecines and *Ardipithecus* and have brains that range from 950 to 1,800 cm^3 in volume. *Ardipithecus* brains were only about 20% of the size of a human brain and a bit smaller than a modern chimpanzee. *Australopithecus* brains were about 35% of the size of a human brain and similar in size to modern chimps and gorillas. Therefore, it is likely that *Ardipithecus* and *Australopithecus* were very ape-like in their brain power, supporting the idea that a fully bipedal gait evolved significantly earlier than large brains.

It is key to appreciate that brain size has a strong correlation with body size, such that larger individuals will have larger brains, but this does not mean that they are necessarily any smarter. For example, whales and elephants have brains that are larger than humans', but they are evidently not as intelligent (Figure 17.13). This is so because absolute comparisons of brain size between species are less valuable than comparison of the ratio of brain to body size (Figure 17.13). Dolphins have brains that are about the same size as humans, but some of that large brain services a larger body, rather than problem-solving skills or complex language. Modern humans lie the farthest away from the centerline in Figure 17.13, a consequence of having the largest ratio of brain volume to body mass, whereas the australopithecines were likely closer to chimps in their overall intelligence.

> **KEY POINT**
>
> Rather than absolute brain size it is the ratio of brain size to body size that better determines intelligence. Humans have the largest brain and highest level of intelligence compared to their body size of any other living animal.

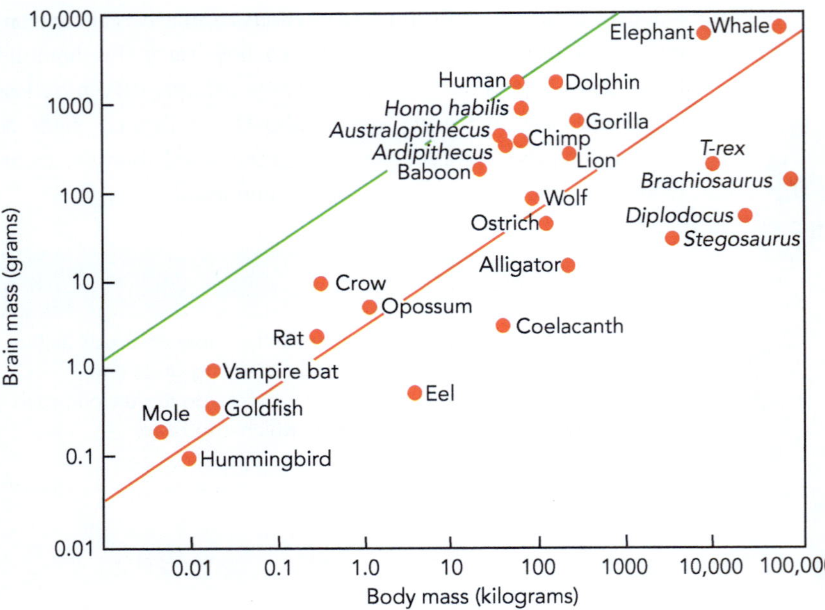

Figure 17.13 Body mass versus brain-size plots. Elephants and whales have brains that are larger than human brains, but the ratio of brain-to-body mass is lower than humans, who lie farthest away from the centerline. Australopithecines and chimps are clearly smarter than wolves, lions, and whales. Dinosaurs (see Chapter 12) lie significantly below the centerline. Source: adapted from Sagan (1977).

17.8 Early Humans and the First Tools

Our genus, *Homo*, is thought to have appeared between about 3.0 to 2.5 million years ago (Figure 17.8), but the fossil record for that time is very sparse. In 2013, well-preserved jaw bones were discovered in Ledi-Geraru, Ethiopia. At 2.8 million years old, these are the oldest fossil record of our own genus. The find also demonstrates that there was significant overlap in time between australopithecines and humans. The earliest species of our genus, including *Homo habilis* (literally "handy man") and *Homo rudolfensis*, lived around 2.1–1.5 Ma (Figure 17.8), although there is significant debate as to whether these are distinct species. *Homo habilis* has a cranial capacity between 550 and 687 cm^3, which is about 50% larger than *Australopithecus*, but half the size of modern human brains. *Homo habilis* is also found with abundant stone tools (Figure 17.14) and was long thought to be the first toolmaker, although examples of simple tools dated to about 3.3 million years old indicate that australopithecines may also have made primitive stone tools. The stone-age (**Paleolithic**) includes the oldest **Oldowan** (also called Mode 1) stone tools, typically comprising cobbles with a few flakes chipped off using another stone (Figure 17.14). This style of tool was replaced by the later, more sophisticated, **Acheulean** technologies (Figure 17.17) showing increasing complexity and thereby suggesting

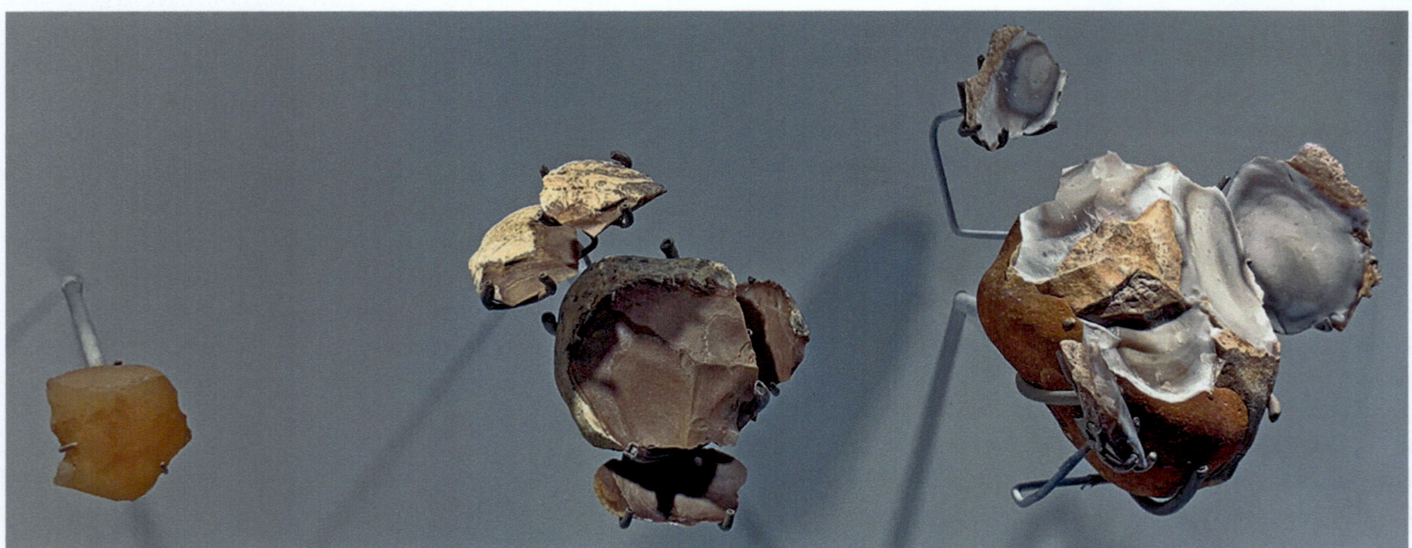

Figure 17.14 Stone tools made by 2.6–1.4-million-year-old humans of the Oldowan style, possibly made by *Homo habilis*. Left, flake; middle, small chopper and flakes; right, large chopper and flakes. Source: Photos by JPB. With permission from Houston Museum of Natural Science (HMNS).

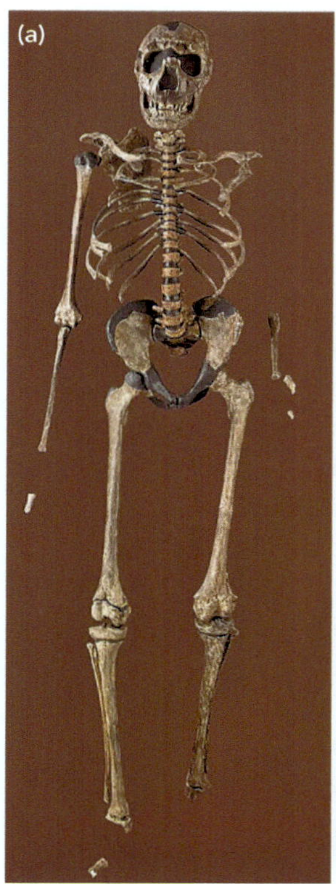

Figure 17.15 Turkana Boy, *Homo erectus* (*ergaster*), (a) skeleton and (b) reconstruction. Skeleton was found on the bank of the Nariokotome River near Lake Turkana, in Kenya. Estimated to be 1.5 to 1.6 million years old. Source: (a) Photo by Chip Clark, Smithsonian Institution, Copyright Smithsonian Institution; (b) © Neanderthal Museum.

increased intelligence associated with larger brains. The Oldowan tools were originally discovered by famed anthropologist Louis Leakey in Olduvai Gorge in Tanzania, where many of the hominid fossils such as *Homo erectus*, *Paranthropus boisei*, and *Homo sapiens* were also discovered.

Although somewhat controversial, about nine species of humans have been proposed (Figure 17.8), some of which have better fossil records than others. One of the earliest species discovered was *Homo erectus* (Figure 17.15), which also included finds in China and Indonesia showing early migration out of Africa nearly two million years ago. Acheulean tools, too, are found in several widely distributed locations, which also supports migration and expansion of *Homo erectus*, although the main concentration of tools discovered is still in Africa and southern Europe. There remains debate as to whether *Homo erectus* evolved in Africa, and later migrated to southeast Asia, or whether an earlier ancestor migrated out of Africa, evolved into *Homo erectus,* and then later migrated back to Africa to give rise to *Homo sapiens*. *Homo ergaster* is a species found exclusively in Africa but is now considered to be the same species as *Homo erectus* and is better thought of as its African variety. *Homo erectus* is thus considered the most likely candidate that gave rise to *Homo sapiens*.

17.9 Neanderthals vs. Sapiens

There are a few more recent "archaic" humans, including **Homo heidelbergensis**, *Homo neanderthalensis* (or "Neanderthals"),

Homo floresiensis, and the **Denisovans** (Figure 17.8). *Homo heidelbergensis* (Figure 17.8) lived from 700,000 to 300,000 years ago and is intermediate between *Homo erectus* and Neanderthals, with a brain about $1,250\,cm^3$, a bit smaller than Neanderthals. The first Neanderthal fossils were found in 1829 but were assumed to be fully human and it wasn't until 1859 that Neanderthals were classified as a separate species. Neanderthals lived from about 400,000 to 40,000 years ago and certainly overlapped with *Homo sapiens*. Neanderthal brains averaged between $1,300–1,600\,cm^3$, which is the same as modern humans. Neanderthals do have some clear anatomical and genetic differences, such as enhanced brow ridges above the eyes, and a stocky body (Figure 17.16). It is not clear how sophisticated the Neanderthals were in terms of intelligence, language, belief systems, or artistic abilities. Fossils of Neanderthals are abundant in Europe and western Asia. They made sophisticated Acheulean stone tools (Figure 17.17), cooked their food, used medicinal plants, crafted simple clothes, and had speech, although it is not known how sophisticated their speech may have been.

Homo floresiensis fossils (Figure 17.8) have been found in a single cave on the island of Flores in Indonesia. Dated at 50,000 years old, they were small (about 1.1 m tall), with brains about $380\,cm^3$, which is chimp-sized, but stone tools indicate they were more intelligent than apes. They have been nicknamed "hobbits," after the characters in the *Lord of the Rings*, due to their diminutive stature. There has been some controversy as to whether this might have been a small group of archaic humans with a genetic or physical disorder that caused dwarfism. The

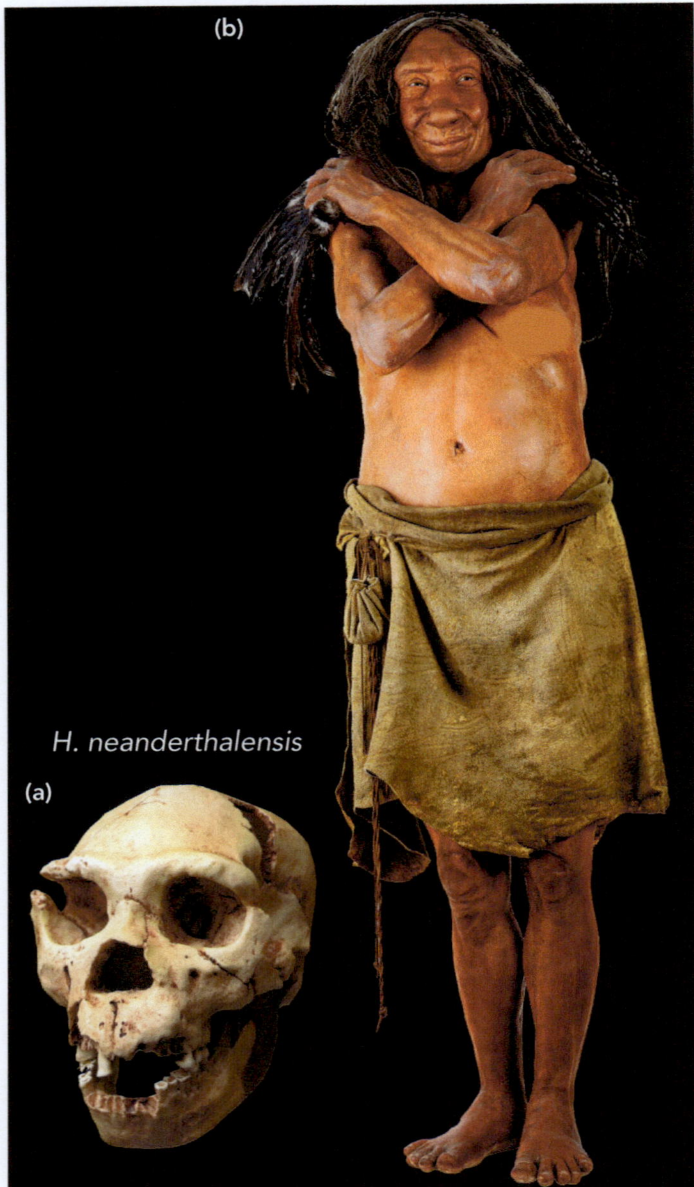

Figure 17.16 *Homo neanderthalensis* (a) skull and (b) reconstruction. Source: (a) ©Neanderthal Museum; (b) Jerónimo Roure Pérez, CC BY-SA 4.0 <https://creativecommons.org/licenses/by-sa/4.0>, via Wikimedia Commons.

Denisovans, discovered in Siberia, with other finds in China, appear to be another late branch of our genus. DNA analysis shows them to be genetically distinct, but the fossil evidence has been insufficient to define an agreed-upon species name. DNA analysis also shows that about 1–4% of non-African *Homo sapiens'* DNA is derived from Neanderthals, indicating that Neanderthals and *Homo sapiens* interbred. Denisovan-derived DNA is found in Melanesians, as well as Australian aborigines.

17.10 *Homo sapiens*

Based on genetic molecular clocks (see Chapter 6), the speciation – the initial split between modern humans and their closest likely ancestors (*Homo heidelbergensis* or *Homo erectus*) – is estimated to have been on the order of a few hundred thousand years ago. Fossil finds of 315,000-year-old archaic *Homo sapiens* have recently been discovered in Morocco (Figure 17.18). These fossils are distinct from Neanderthals but have thicker brow ridges and flatter skulls than modern humans, indicating that significant evolutionary differentiation had yet to occur. The oldest anatomically modern humans have been found in Ethiopia, and date to around 195,000 years. The evidence of complex cultures, as exemplified by cave paintings, musical instruments, and more complex hunting lifestyles, such as fishing, do not really appear until about 50,000 years ago, and there remains a significant gap in our understanding of how modern human behavior and modern human brains appeared.

Anatomically modern *Homo sapiens* fossils older than about 50,000 years are not found anywhere outside of Africa, again supporting the hypothesis that *Homo sapiens* evolved in Africa and migrated from there (Figure 17.19). It thus seems that there were at least two waves of migrations out of Africa: *Homo erectus*, about 1.8 million years ago, and *Homo sapiens*, about 100,000 years ago. These observations support Darwin's original out-of-Africa hypothesis.

KEY POINT

All of the fossil evidence of archaic *Homo sapiens* are found in Africa, supporting the idea that humans evolved in Africa.

17.11 The Naked Ape: Neoteny and Paedomorphosis

The last question we will deal with in this chapter is the evolutionary mechanism that resulted in large brains and loss of fur to produce what author Desmond Morris referred to in his 1967 book **The Naked Ape**. Human brain tissue must be kept within a narrow heat range to avoid cell death, and therefore thermoregulation is critical and comes at a high metabolic cost, as mentioned above. It is assumed that early humans were **diurnal** (i.e., awake during the day), and likely able to hunt during the day, in contrast to most furry predators, such as lions, that would overheat in the midday African sun and were therefore **nocturnal** (i.e., awake during the night). Humans are also able to track and hunt over long periods, but all of this requires the ability to keep cool during the day. That requires sweat, and sweating is enhanced by an increase in sweat glands in furless skin. Humans have many more sweat glands than other primates. The flip side is that loss of fur could be a liability in cooler climates, leading to the hypothesis that the genus *Homo* lived at much lower altitudes than furrier australopithecines, where night temperatures in Africa would have been warm enough for them to survive despite the lack of

Figure 17.17 More modern tools (200–30 ka) likely made by neanderthals and archaic humans. Source: Photo by JPB. With permission from Houston Museum of Natural Science (HMNS).

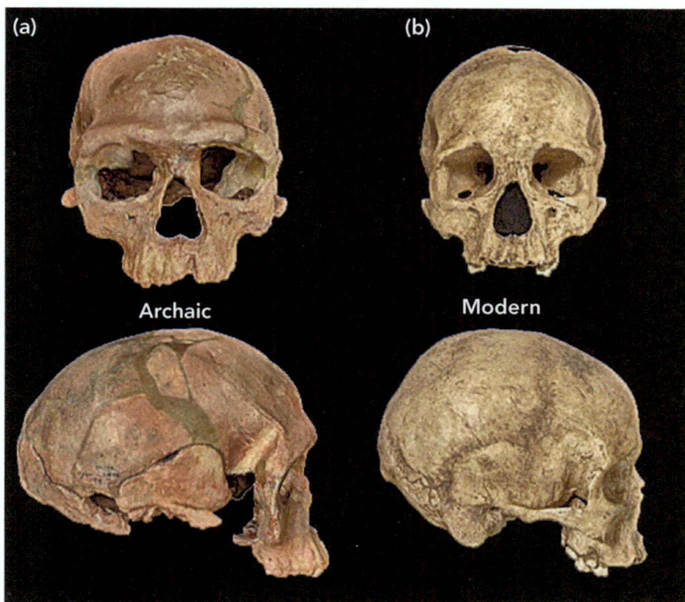

Figure 17.18 Comparison of (a) earliest archaic *Homo sapiens* shows thicker brow ridges and a more flattened skull compared to (b) modern skulls. Source: from Callaway (2017), *Nature*. NHM CC BY.

fur. Ultimately, the ability to control fire and make clothes would eventually solve this problem. It has also been suggested that less fur meant less parasitical lice and may have enhanced sexual attractiveness. Like many aspects of human evolution, there are several theories that have been developed to explain how our various unique traits evolved.

One of the popular theories to explain how our brains and our nakedness evolved is the concept of **neoteny**, that is, the slowing or delaying of the development of an organism that results in **paedomorphosis**, or the retention of juvenile attributes through adulthood. For example, the Mexican axolotl is a salamander, but it retains external gills (Figure 17.20), which most amphibians lose as they pass from larval stage into

adulthood. Many scientists have noticed that juvenile chimps have heads that are much more like humans rather than adult chimps (Figure 17.21). One of the ways to keep a brain big is to delay the development of adult features and corresponding body proportion (e.g., adult chimp in Figure 17.21) by retaining the juvenile proportions into adulthood. This mostly requires changes in the timing of triggering of the genes that code for these changes, either by suppressing or delaying them. Paedomorphic features that humans have, compared to adult apes, include the more ball-shaped skull and consequent large brain (Figures 17.8 and 17.18); thin skull bones with a reduction of the brow ridges that are diagnostic of Neanderthals (Figure 17.16); a flattened and broadened face with larger eyes; and reduced jaws, nose, and teeth. Other juvenile features include a hairless body as well as shorter limbs (Figure 17.16).

Some of these changes have serious consequences. An upright gait required a smaller pelvis than apes (Figure 17.7), but this makes childbearing more difficult, and especially birthing. As an adaptation, human females give birth long before the babies can be independent of the parents. Perhaps you have seen a nature film in which some African savanna animal such as a giraffe gives birth. After a few minutes of wobbling uncertainly on its feet, the baby giraffe is soon walking around, feeding, and playing, exhibiting far more independence than a completely helpless human newborn, unable to walk or feed itself. Typically, children can walk independently and explore their surroundings after two years of age, and of course human children do not become adults (i.e., reaching puberty) until well after their first decade of life. Having such helpless offspring cannot be successful unless parents or guardians are able to protect, nurture and train them, continuously, for at least the first few years of life. There is evidence that humans specifically require the ability to cooperate for complex language to succeed and this may have led to humans being a highly social animal. All these factors were

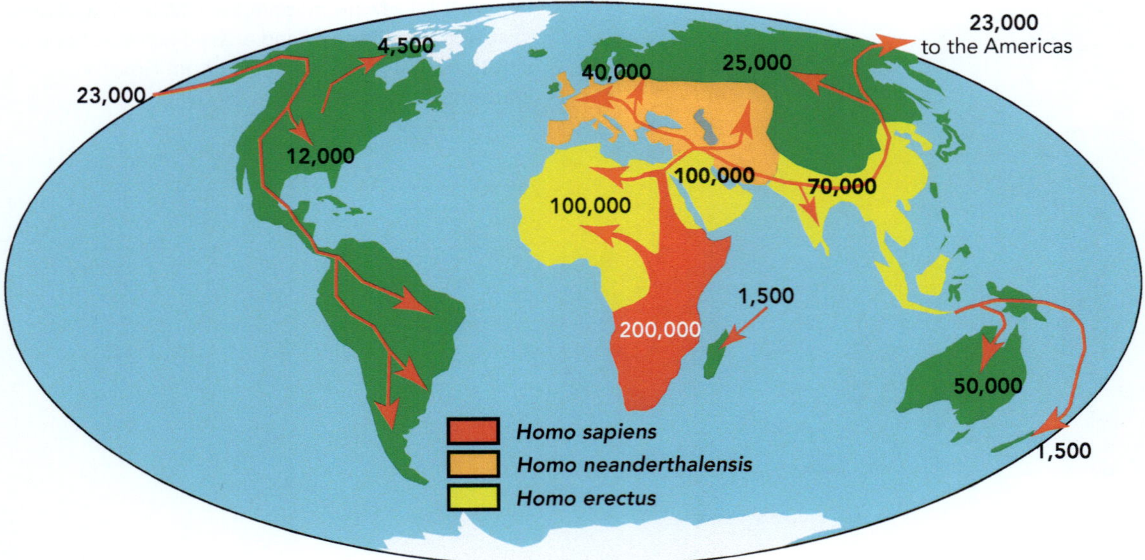

Figure 17.19 Migration of humans. *Homo erectus* migrated throughout Africa and southern Europe and Asia. Neanderthals were confined to Europe and western Asia whereas *Homo sapiens* are distributed worldwide.

Figure 17.20 Adult axolotls retain larval features, including a dorsal fin and external gills. These are usually absent in other species of adult salamanders. Source: M-Production via iStock / Getty Images Plus.

Figure 17.21 Juvenile chimp has a more upright head, large forehead, smaller jaws, larger eyes, and other features compared to the older adult. Juvenile chimp has a more human-like appearance. Source: Reprinted by permission of Springer Nature: The Science of Nature, Über die Urformen der Anthropomorphen und die Stammesgeschichte des Menschenschädels, Adolf Naef, copyright 1926.

probable elements in the evolution of human social units. Because of our large brains, nurturing characteristics, and a language that allows us to impart complex information across generations, we have built complex technological societies, which require enormous investments in time and information to be successful.

KEY POINT

The evolution of big brains and loss of fur may have been a result of neoteny and the slowing of development and paedomorphisis, the consequent retention of juvenile characteristics into adulthood. One consequence is that human babies are essentially helpless, requiring many years of nurturing that would been aided by language and the development of the more complex social behavior that characterizes *Homo sapiens*.

17.12 Conclusion

Although Darwin retained the concept of "higher" and "lower" species, he was quite clear that humans should be placed within a lineage of organic evolution rather than special creation by a divine hand. To that end, in his book *The Descent of Man* he stated,

> . . . we have given to man a pedigree of prodigious length, but not, it may be said, of noble quality. The world, it has often been remarked, appears as if it had long been preparing for the advent of man; and this, in one sense is strictly true, for he owes his birth to a long line of progenitors. If any single

link in this chain had never existed, man would not have been exactly what he now is. Unless we willfully close our eyes, we may, with our present knowledge, approximately recognize our parentage; nor need we feel ashamed of it. The humblest organism is something much higher than the inorganic dust under our feet; and no one with an unbiased mind can study any living creature, however humble, without being struck with enthusiasm at its marvelous structure and properties.

In other words, Darwin recognized that humans have evolved over a long time (*the pedigree of prodigious length*), and that our evolution was by natural selection with many ancestors (*progenitors*) versus a "noble" origin (created by God). Rather than feeling ashamed of having evolved from "humble" biological beginnings, Darwin argued that we should be enthused at the wonder of the diversity of life on Earth. Since Darwin's time, we now have a plethora of fossil evidence that the human clade is richer than we might have expected. Around two million years ago, there was a wider range of hominid species in which *Homo erectus* and robust australopithecines lived at the same time (Figure 17.8), if not in similar places, and as recently as 50,000 years ago, *Homo sapiens*, Neanderthals, Denisovans, and *Homo floresiensis* lived together. Today only one species survives, our own *Homo sapiens*, and despite the abundance and proliferation of our unique species, our hominin clade has not been particularly successful. Only one twig of the numerous branches of the hominin family tree has survived (Figure 17.8). Compare that to many other genera, such as common carpenter ants (genus *Camponatus*) or the common rat (genus *Rattus*), which boast over 40 separate species. It is not known why and how most hominin species became extinct, although competition with *Homo sapiens* might have been an important factor.

17.13 Summary

- Charles Darwin hypothesized that humans likely shared a common ancestor with our closest relatives, chimpanzees and gorillas, and that ancestral humans first evolved in Africa, but paleontological data were sparse in Darwin's time.
- Early twentieth-century resistance to Darwin's out-of-Africa idea based in racist and eugenic ideologies led to the Piltdown man hoax, which supported an English origin for the "missing link" between humans and apes, but this was later debunked.
- The first primates appeared in the Cenozoic and were arboreal. The transition from arboreal quadrupedalism to brachiation led to the loss of a tail, which enabled later adaptation to a fully bipedal gait in the early hominids.
- Fieldwork in Africa showed that the first hominids appeared about six million years ago, and included the australopithecines marked by major anatomical changes in the pelvis, as well as skull attachment to the spine indicating adaptation to obligate bipedalism. Australopithecine brains and teeth, however, were still very ape-like. Bipedalism may have been driven by environmental change, including loss of forests and increase in savannas, although this is controversial.
- The member of our own genus, *Homo*, appeared about 2.5 to 3 million years ago and was marked by larger brains, and tool making, indicating increased intelligence.
- *Homo sapiens* appeared around 250,000 years ago and lived alongside and may have interbred with *Homo neanderthalensis*. Neanderthals had brains as large as *Homo sapiens* and exhibited many human cultural characteristics. No evidence of *Homo sapiens* older than 50,000 years has been found outside Africa, further supporting the out-of-Africa hypothesis.
- The discovery of cave paintings, complex burial sites, and other cultural evidence suggests that the human capacity for language was fully developed by about 50,000 years ago and probably earlier.
- *Homo sapiens*' loss of fur, large brains, and other unique features is interpreted to be the result of neoteny and paedamorphosis, retention of juvenile characteristics through adulthood.

Key Words

- simian
- anthropomorphous
- missing link
- Great Chain of Being
- quadruped
- bipedal
- australopithecine
- eugenics
- primates
- anthropoids
- Hominoidea
- arboreal
- brachiation
- facultative bipedalism
- obligate bipedalism
- *Homo neanderthalensis*
- *Homo erectus*
- Hominini
- foramen magnum
- *Australopithecus afarensis*
- Paleolithic

- Oldowan
- Acheulean
- *Homo heidelbergensis*
- *Homo floresiensis*
- Denisovans
- *Naked Ape*
- diurnal
- nocturnal
- neoteny
- paedomorphosis

Further Reading and References

Callaway, E., 2017, Oldest *Homo sapiens* fossil claim rewrites our species' history, *Nature*, 546, 289–293.

Franzen, J. L., Gingerich, P. D., Habersetzer, J., et al., 2009, Complete primate skeleton from the middle Eocene of Messel in Germany: Morphology and paleobiology, *PLoS One*, 4(5), e5723.

Gould, S. J., 1981, *The Mismeasure of Man*, Norton.

Leakey, R. E., 1981, *The Making of Mankind*, Dutton.

Morris, D., 1967, *The Naked Ape*, Jonathan Cape Publishing.

Naef, A., 1926, Über die Urformen der Anthropomorphen und die Stammesgeschichte des Menschenschädels, *The Science of Nature/Naturwissenschaften*, 14, 472–477, https://doi.org/10.1007/BF01507526.

Sagan, C., 1977, *The Dragons of Eden, Speculations on the Evolution of Human Intelligence*, Random House.

Werdelin, L., and Sanders, W. J., 2010, *Cenozoic Mammals of Africa*, University of California Press.

White, T. D., Asfaw, B., Beyene, Y. et al., 2009, *Ardipithecus ramidus* and the paleobiology of early hominids, *Science*, 326 (5949), 75–86.

Review Questions

1. What are the major steps in the evolution of humans from the first primates? Describe the major anatomical changes and evolutionary drivers.
2. Explain, based on the fossil evidence, how we know which evolved first, big brains or upright posture.
3. What is the evidence that humans and apes share a common ancestor?
4. What is the first evidence for human intelligence?
5. How were australopithecines similar to modern humans and how were they different?
6. What is the evidence that our genus *Homo* first evolved in Africa?
7. Why did Darwin first hypothesize that humans likely originated in Africa?
8. Why has there been resistance to the out-of-Africa hypothesis for the evolution of humans?
9. What were some probable evolutionary advantages of an upright, bipedal gait and what environmental factors may have caused bipedalism?
10. What is the evidence for bipedalism in australopithecines?
11. Explain neoteny and paedomorphism and how they explain how modern humans evolved.
12. Explain how bipedalism, the evolution of language and big brains, and paedomorphism are related.
13. How would you characterize the success of our clade?
14. Does the evidence of the diversity within our clade support the idea that evolution results in a simple linear progression, forming the so-called "Great Chain of Being" or the "Ladder of Progress"?
15. What is the advantage of a big brain? Are there disadvantages?
16. Are big brains alone the best way to estimate the intelligence of a species?

Perito Moreno Glacier, Lago Argentino, Los Glaciares National Park, Patagonia, Argentina. Source: Philipp Rohner / 500px / Getty Images.

Chapter 18

Ice Ages and Sea Level

Quaternary Environmental Change

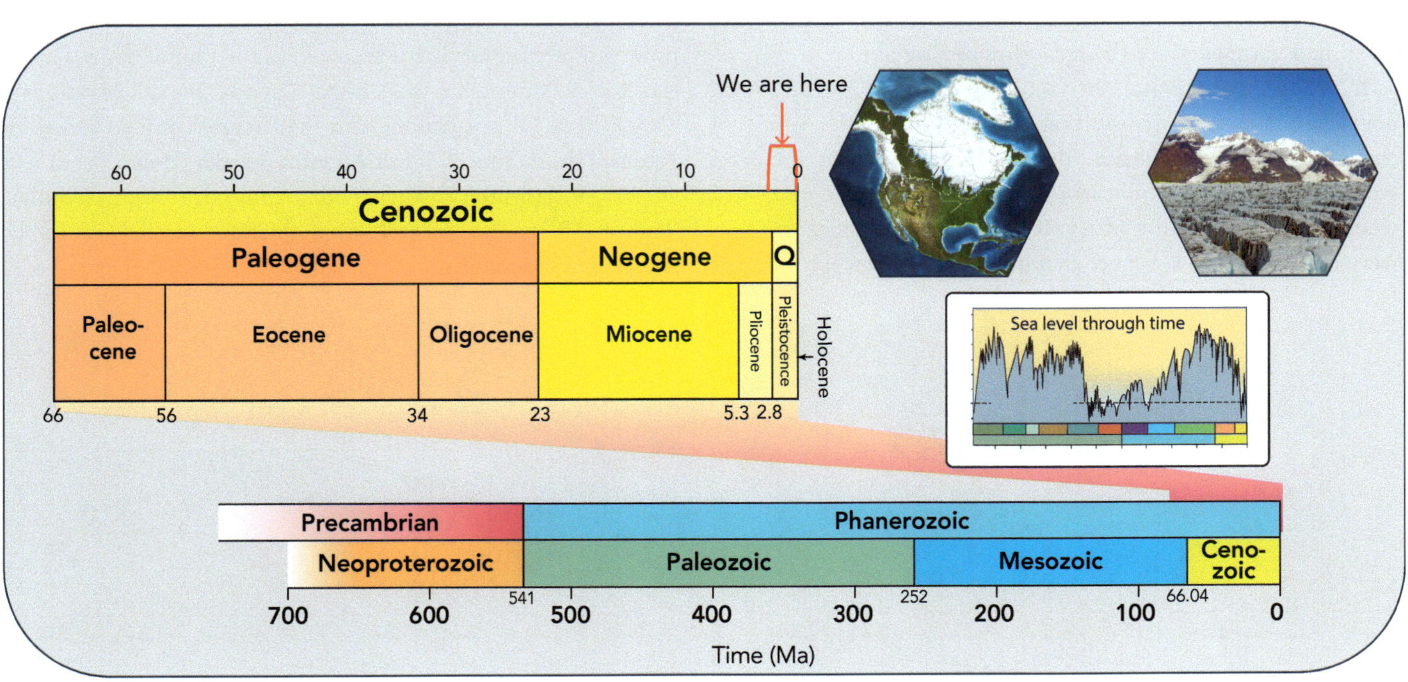

Introduction

Ice ages and sea-level changes have occurred throughout the history of our planet, and both processes have left a clear signature in the rock record. In Chapter 8 we reviewed the history of the idea that Earth was largely frozen, including key evidence of global Neoproterozoic glaciations, referred to as the snowball Earth. In this chapter we begin with a focus on the more recent glaciations that mark the Pleistocene Epoch and the relationship between climate and sea level. We then examine the evidence for sea-level change as recorded in the organization of stratigraphic layers and changes in environments of deposition, emphasizing observations from the rock record, to show that sea levels have also varied, with remarkable periodicity, throughout Earth history and that they record the complex interaction of tectonics and climate.

18.1 From Floods to Ice Ages

Prior to the Enlightenment, Christian scholars sought evidence for the great Noachian deluge, as described in the Book of Genesis, and noted the presence of **diluvium**, coarse sediment typically consisting of boulders that were believed to provide evidence of the great flood. These boulders, now referred to as **erratics**, are pieces of rock that differ in size and type to the rocks native to the area in which they are found (Figure 18.1). In the early nineteenth century, Swiss geologist Ignatz Venetz, a founder of the science of glaciology, explained the dispersal of erratics as being due to glaciers, rather than a great flood. Famed Swiss-American geologist Louis Agassiz further developed the idea that the world had recently experienced ice ages. His ideas were initially met with skepticism, as it was assumed that Earth had experienced steady cooling from its originally molten state, and if they accepted Agassiz's theory, this would require that Earth had warmed since the hypothesized ice age.

In 1864, the self-taught Scottish scientist James Croll developed the first theory that ice ages were the result of changes in Earth's climate, driven by changes in its orbit around the Sun, but his theory was not fully accepted until well into the twentieth century. It also became clear that ice ages were closely tied to sea-level changes, another concept that was kicked around for nearly a century before full acceptance in the 1970s.

Nineteenth-century observations of erratics suggested that the Earth had been glaciated and this led to the hypothesis that orbital changes controlled climate changes and drove glacial cycles.

18.2 Conditions Necessary for Glaciers and Continental Ice Sheets

In this section we review the conditions required for modern glaciers to form, including particulars of location and climate.

18.2.1 Where Glaciers are Found Today

Glaciers can only exist where the summers are cold enough to prevent some snow from melting by the end of the season, allowing it to persist until the following winter. These conditions typically occur at high altitudes and at high (i.e., polar) latitudes. The **snow line** marks the elevation above which temperatures are cold enough to preserve snow and ice through the summer (Figure 18.2). The snow line varies with latitude. For example, the extinct volcano of Mauna Kea, in the tropical island of Hawaii, at 19.8° N latitude, accumulates up to a meter of snow at elevations above 3,700 m in the winter months, but the snow melts in the summer (Figure 18.2(b) and (c)), and thus Mauna Kea does not have any glaciers. Mount Kilimanjaro, at 3° S latitude in Tanzania, in contrast, is permanently frozen above about 5,500 m elevation, and sports a permanent glacier, although it has shrunk from about 11 km^2 to less than 2 km^2 since 1912. More significant glaciers occur at lower elevations at progressively higher latitudes, such as the Columbia Ice Field in the Rocky Mountains of Alberta, Canada (ice at elevations of ~2,000 m at 55° N latitude), and the Patagonian Ice Field in the Andean Mountains of South America (glaciers down to sea level at 49° S latitude). Outside of Antarctica and Greenland (Figure 18.3), the largest accumulation of ice is in the Himalayan mountains.

Figure 18.1 House-sized rocks lie on a glaciated surface, Flat Rock, Newfoundland.

Figure 18.2 Snow lines. (a) Beartooth Mountains, Montana. (b) Mauna Kea, Hawaii. The tree line is marked by the appearance of a reddish-brown color indicating the loss of vegetation. (c) Close-up shows the Mauna Kea astronomical observatory with patches of snow, indicating elevation above the snow line.

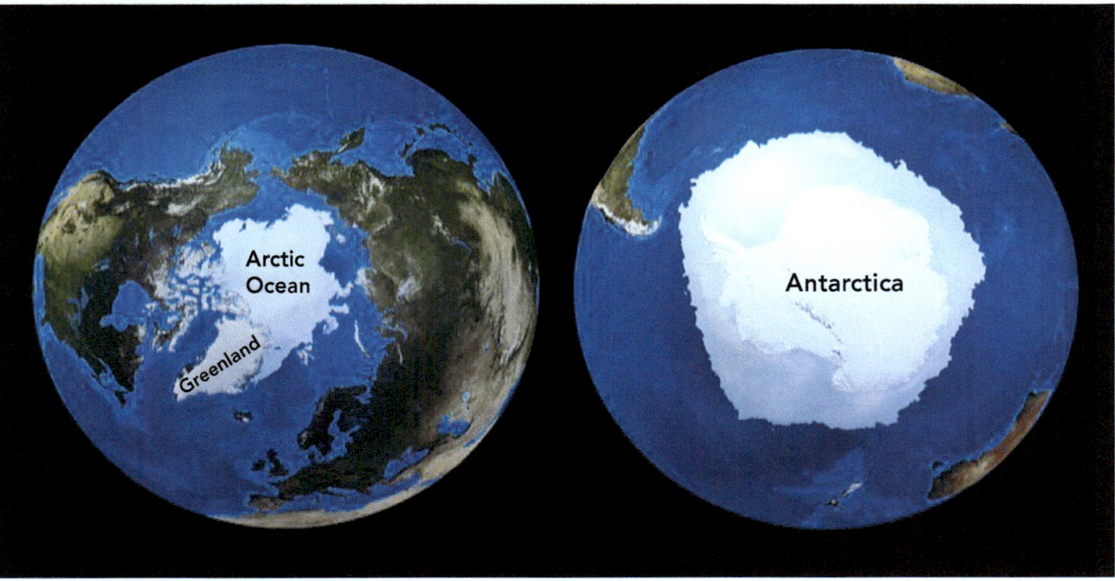

Figure 18.3 View of the poles showing Antarctica (right) and Greenland (left). Note that the Arctic ice cap has shrunk considerably over the last few decades due to global warming (see Chapter 19). Source: NASA Goddard Space Flight Center.

Glaciers are formed at high latitudes and high altitudes where snow and ice accumulation exceeds melting and ablation.

18.2.2 Antarctica and Greenland

The largest ice sheets on land are those that cover most of Greenland and Antarctica (Figure 18.3). The North Pole is indeed covered by ice, but this is ice floating in the Arctic Ocean and it has been estimated that there has been an almost 50% reduction in its late-summer extent since the 1970s. The Arctic ice sheet covers water, not land, whereas 98% of the Antarctic continent is covered by ice, an area of about 14 million square kilometers. The ice here has an average thickness of about 2 km, and the volume of ice is around 26 million cubic kilometers. This is about 60% of the freshwater resources on Earth and if all the Antarctic ice melted, global sea level would rise about 58 m. The Greenland ice sheet is similarly thick and covers an area of almost 2 million square kilometers. If the Greenland ice sheet melted sea level would rise an additional 7 m.

The Antarctic continent has been in a polar position since about 300 Ma and was first glaciated at about that time (see Chapter 11). As discussed in Chapter 10, Antarctica was originally part of the southern Gondwana supercontinent. As Gondwana broke up, Antarctica became increasingly isolated, and by the Neogene the cold **circumpolar current** isolated Antarctica from any warm equatorial waters (note how Antarctica is isolated in Figure 18.3). The modern Antarctic ice sheets began growing in the middle Eocene around 45 Ma and have waxed and waned in size as global climate patterns have shifted. The Greenland ice sheet appears to have been formed around 18 million years ago, based on the observations of ice-rafted sediment seen in cores taken from deep-sea sediment around the continent.

18.3 Continental Glaciers: The Last Icehouse

Eighteen thousand years ago, a significant proportion of the northern hemisphere land masses were covered by kilometers-thick ice sheets (Figure 18.4) at the culmination of the **last glacial maximum** (LGM). The North American ice sheets covered an area of about 17 million km^3, about 20% larger than the Antarctic Ice sheet, and comprised two major sheets, the **Cordilleran Ice Sheet** to the west, and the larger **Laurentide Ice Sheet** to the east (Figure 18.4). The areas that now host every major city in Canada, from St. Johns, Newfoundland to Victoria BC, were buried under several kilometers of ice. The North American ice sheet extended as far south as St. Louis in the USA and covered coastal cities such as Boston and New York (Figure 18.4). The European ice sheet, although smaller

(Figure 18.4(a)), covered most of Northern Europe, including all of Scandinavia, most of the United Kingdom, and much of France, Germany, Poland, and Russia. In addition, glaciers expanded in mountainous regions throughout the world.

During the last glacial maximum much of North America was covered by the kilometers-thick Cordilleran and Laurentide ice sheets.

18.4 Geologic Record of Ice Sheets and How They Move

We have both direct and indirect evidence that the northern-hemisphere ice sheets advanced and retreated many times over the past few million years. There are several factors that allow ice to flow. First, precipitation of snow is uneven, so the thickness of ice sheets can vary with location. If an ice sheet is thicker in one place, gravity may cause it to flow to the areas where it is thinner, especially if the thickness changes result in differential elevation of the top of the ice (see Figure 8.2). More importantly, as ice becomes thicker, the pressure builds up beneath it and if the ice is deposited on a slope, then it will flow down the slope (Figure 8.2). As ice moves, major cracks called **crevasses** may form on the surface (see Figures 8.3 and 18.5). If surface melting occurs during warmer summer periods, water can percolate down these crevasses resulting in a fast-moving ice stream at the base of the glacier, which can cause the glacier to accelerate. During colder periods, the base of glaciers may be dry and show much less movement. Alpine glaciers can move as fast as 30 m/day but typically move at about 25 cm/day (about 1 cm/hour, or 0.00001 km/hour). By comparison, a typical river flows at about 1 m/second (3.6 km/hour).

There are a variety of larger-scale landforms and features that glaciers produce, such as **moraines** and **drumlins** (Figures 18.6 and 18.7), as well as similar-shaped features that can be carved into solid bedrock, such as the elongate Finger Lakes in New York State. Mapping of these landforms, as well as measuring the direction of flow from features such as striations and chatter marks (see Figure 8.5), have been used to determine the direction of flow of the ice sheets that covered North America and Europe. It is worth considering that a 2 km-thick ice sheet exerts a vertical pressure of about 20 million pascals (nearly 3,000 pounds per square inch), or almost 200 times atmospheric pressure. These forces enable glaciers to carve bedrock, erode mountains, and leave a tremendous imprint on the landscapes over which they flow.

Mapping of glacial tills (i.e., direct ice contact deposits, introduced in Chapter 8) and associated landforms also allows us to determine the extent of the ice sheets. Tills have been identified and mapped, and attempts made to correlate them, within and

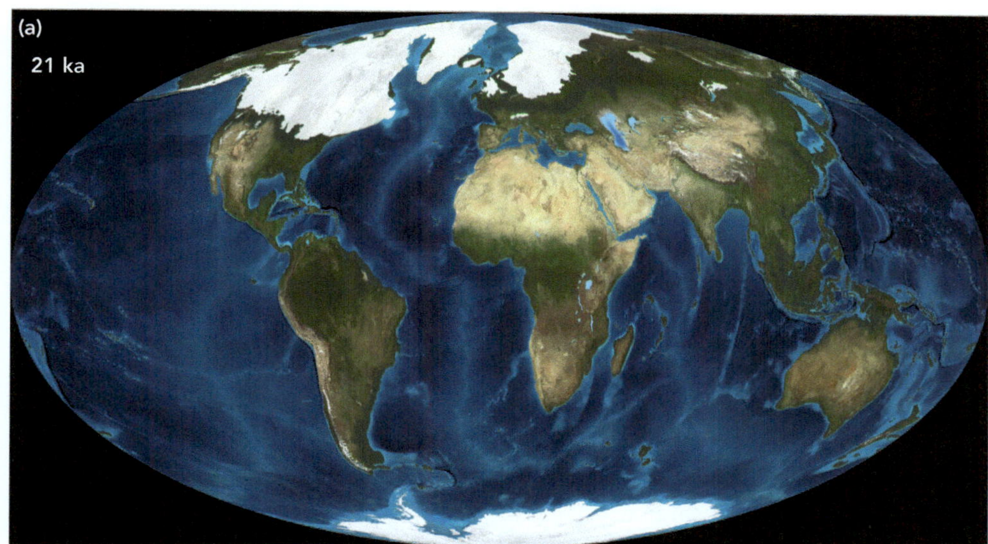

Figure 18.4 (a) Paleogeographic map of the world, at the last glacial maximum (LGM), 21,000 years ago. Both north and south regions were largely covered with extensive continental glaciers. (b) Close-up shows paleogeography at 12 ka at the beginning of glacial retreat with clear subdivision of the Laurentide and Cordilleran ice sheets. Major glacial lakes are also labeled. Source: © DeepTimeMaps.

between continents, to determine the history of glaciations, particularly in the Pleistocene Epoch. Earlier attempts suggested that there were four major phases of glacial growth in North America, and from youngest to oldest were referred to as Wisconsin, Illinoian, Kansan, and Nebraskan, based on the location of the southernmost ice during these intervals. The intervening interglacial periods were also given formal names. Four glacial periods were also recognized in Europe, but given different names (Günz, Mindel, Riss, and Würm), and in the 1970s and 1980s there was a major effort to correlate these glacial cycles using a variety of techniques including chronology by dating intervening volcanic layers, magnetostratigraphy, biostratigraphy, and other techniques. In the end, the notion that there were only four glacial episodes in the Pleistocene broke down, and it was recognized that there were about 11 major glacial episodes. We will return to this theme in Section 18.8.

18.5 Glacial Termini, Great Lakes, and Great Floods

The maximum limit of the advance of the glaciers during the last ice age is marked by the limit of deposits of **terminal moraines** that form at the terminus of a glacier. At the end of these glaciers, giant lakes were formed (Figure 18.4), including Lake Iroquois, which shrank to form Lake Ontario, Lake Agassiz, which shrank to form Lake Winnipeg, Lake Manitoba, and Lakes Winnipegosis and Winnipeg in the Canadian province of Manitoba, and Lake Bonneville, which shrank to the present-day Great Salt Lake in Utah. The Niagara escarpment is a major geographic feature that partly formed at the margin of Lake Iroquois, now Lake Erie and Ontario, separated by Niagara Falls.

Figure 18.5 Crevasses, Alaska. Source: Diego Delso. photo, License CC-BY-SA.

Figure 18.6 Glacial landforms in Manang, Nepal. Lateral and terminal moraine can be seen that formed along the sides and end, respectively of the now retreated glacier. A kettle lake lies behind the terminal moraine that was formed when large blocks of buried ice melted as the glacier retreated, leaving a water-filled depression. Source: Bijaya2043, CC BY-SA 3.0 <https://creativecommons.org/licenses/by-sa/3.0>, via Wikimedia Commons.

Figure 18.7 Oblique air photo of a drumlin field, south of Horicon Marsh, Wisconsin. Note the teardrop-shaped pattern. These formed under the Laurentian Ice Sheet. The elongation indicates the direction that the ice sheets flowed. To give you an idea of scale, the lake in the middle is about 8 km across and in the lower left you can see a road crossing the area. Source: Doc Searls from Santa Barbara, USA, CC BY 2.0 <https://creativecommons.org/licenses/by/2.0>, via Wikimedia Commons.

Eighteen thousand years ago, Glacial Lake Missoula, farther to the west, was dammed at its northwestern margin by an arm of the Cordilleran Ice Sheet, separating it from Glacial Lake Columbia (Figure 18.5 and see also Figure 2.8). This ice dam occasionally failed, releasing massive amounts of water that flowed up to 36 m/second (130 km/hour), carving the Columbia Gorge and forming the Channeled Scablands of eastern Washington. J Harlan Bretz's interpretation of the Scablands as being formed by catastrophic floods was discussed in Chapter 2 as an example of actualism. Bretz's interpretation was initially met by skepticism, as no known floods traveled at that velocity or were known to be of that scale. Similarly, the idea that significant areas of Earth were covered by glaciers seemed to be incredible and did not seem compatible with the purely gradualistic view popular at the time. Eventually scientists converged on the idea that geologic products must always obey the rules of chemistry, physics, and biology. Although the pace of geologic change is mostly slow and steady, which led to the idea of gradualism, Earth sometimes experiences cataclysmic episodes that have an outsized effect on the products relative to their duration, such as the outsize floods of Missoula Lake that carved the Channeled Scablands, or the meteorite impact that triggered the Cretaceous–Paleogene extinction. Given the enormous stretch of geological time, these rare and sometimes catastrophic events have an excessive effect on the history of our planet.

KEY POINT

Mapping of glacial landforms as well as stratigraphic analysis helped understand the extent and timing of most recent glaciations as well as the origin of the modern lakes that are the remnants of much larger glacial lakes.

18.6 Milankovitch Cycles: Orbital Control on Glacial Episodes

Following James Croll and others' theories that the ice ages were controlled by changes in the Earth's orbit, Serbian geophysicist, astronomer, and climatologist Milutin Milankovitch used Newtonian mechanics to calculate the perturbations in Earth's orbit due to the gravitational interaction of the Earth with the Moon, Sun, and the other planets, with Jupiter as the main player, because of its great size. Milankovitch integrated his calculations of orbital perturbations to estimate changes in **insolation** (Figures 18.8 and 18.9), which is a measure of the overall strength of the Sun's incoming radiation. In doing so, Milankovitch developed a unified theory that showed how changes in Earth's orbit cause insolation differences strong enough to drive glacial cycles. The cyclic changes in Earth's orbit, include the degree that its axis of rotation is tilted towards the Sun, termed **obliquity**, the rotation of the axis of rotation, termed **precession**, and the degree to which the elliptical orbit

around the Sun departs from a circle, called the **eccentricity** (Figure 18.8).

Recall that in Chapter 8 we reviewed the hypothesis that in early Earth history, obliquity was much higher (Figure 8.12), and this was briefly considered to explain the equatorial snowball glaciations. Milankovitch showed that the obliquity ranges from 22.1° to 24.5° and changes over a period of 41,000 years. The precessional cycles occur over periods of 19, 22, and 24 thousand years, and the changes in the eccentricity change at 400,000-year (long eccentricity) and 105,000- and 95,000-year (short eccentricity) cycles (Figures 18.8 and 18.9). Changes in the axial tilt and the precession change the insolation (Figures 18.8 and 18.9), and the effect is more pronounced in polar regions such that they become either warmer or colder. The eccentricity changes the distance of the Earth from the Sun (Figure 18.9), which also causes cooling or heating. Colder poles allow build-up of glaciers and warmer climates result in glacial retreat.

However, these cycles interact such that they may be in phase (e.g., cooler, or warmer parts of the different cycles coincide such that Earth is either much colder or much warmer) or out-of-phase, in which case the effects may cancel each other out. This is shown by the solar forcing graph in Figure 18.9, which shows nested cycles of warmer and colder climate. The precessional graph (Figure 18.9) also shows nested cycles (these are technically called "**beats**") in which the three cycles (19, 22, and 24 ka) either reinforce or cancel each other out. The broader "beat" cycle shows a 400,000-year periodicity, similar to the eccentricity in its frequency but different in its cause.

Since Milankovitch completed his calculations in the 1940s, improved estimates of these orbital cycles have been produced, and we can now make predictions of the effect of the **Milankovitch cycles** on climate going back tens of millions of years. Of course, these orbital variations in solar forcing have occurred throughout Earth history, inviting the question of whether there is evidence for Milankovitch cycles further back in the rock record.

KEY POINT

Milutin Milankovitch calculated the cyclic perturbations of Earth's orbit and its control on solar insolation that drives glacial cycles, and these are now referred to as Milankovitch cycles.

18.7 Oxygen Isotope Evidence of Global Climate Cycles and Changes of Sea Level

Geochemical evidence for Milankovitch cycles was found using geochemistry, and particularly oxygen isotopes. Oxygen has two main isotopes, lighter ^{16}O and heavier ^{18}O, so any sample of water (H_2O) will have some combination of the two. When water evaporates, a higher proportion of ^{16}O molecules move

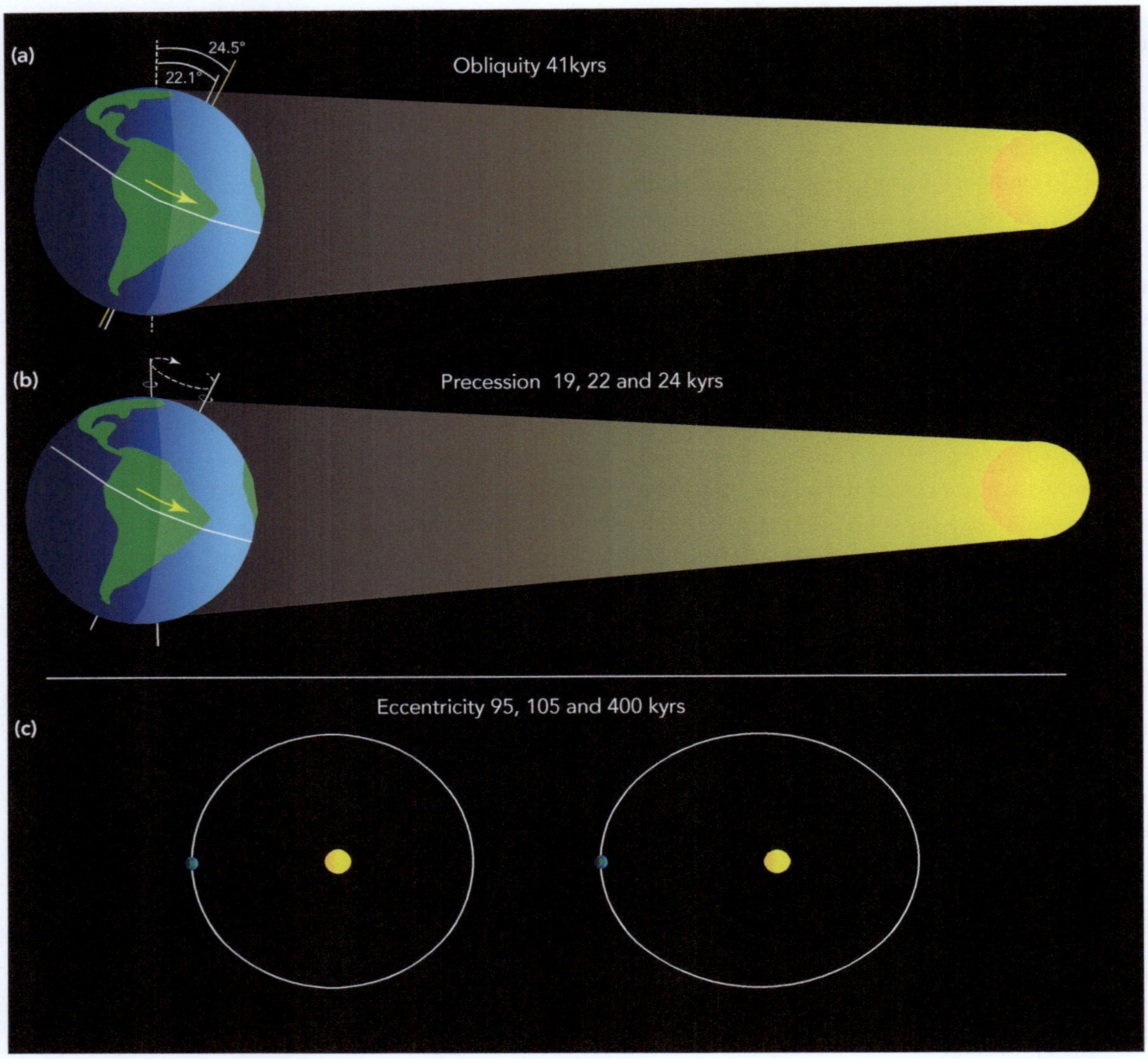

Figure 18.8 Changes in the Earth's orbit include (a) the obliquity, (b) precession, and (c) eccentricity. The more vertical the Earth is, the colder the poles become, causing global cooling. As the Earth's orbit becomes more eccentric it also results in an overall cooler climate as the Sun is farther away.

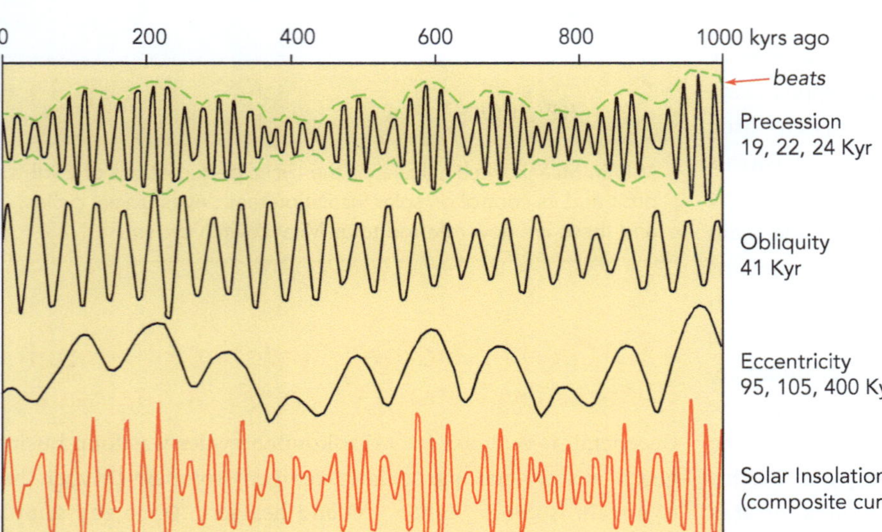

Figure 18.9 The Milankovitch cycles (precession, obliquity, and eccentricity) cause changes in the effect of the Sun's heat. These combine to produce the solar forcing graph, and this generates changes in climate recorded by glaciations. Source: Based on Incredio, CC BY 3.0 <https://creativecommons.org/licenses/by/3.0>, via Wikimedia Commons.

into the vapor phase because it requires less energy to evaporate lighter ^{16}O versus heavier ^{18}O. Consequently, the proportion of heavier ^{18}O will increase in a body of water as more vapor is evaporated from it. In the case of the oceans, $^{18}O/^{16}O$ increases when water is evaporated from the surface of the ocean, but if the water that evaporates to form clouds simply falls back to the surface as rain, and returns to the ocean, then there will be no net change in the isotopic composition of the ocean. However, if the precipitation that forms from these clouds falls as snow and not rain, and that snow forms glaciers that grow from year to year, then the return path of water back to the ocean is interrupted, resulting in two reservoirs of water that are fractionated with respect to oxygen isotopes. Glaciers will have low $^{18}O/^{16}O$ values, as ^{16}O has been preferentially partitioned into the vapor phase as seawater evaporates, and the oceans will have high $^{18}O/^{16}O$ values, and that enrichment will persist until the glaciers melt and return the water with anomalously low $^{18}O/^{16}O$ back to the oceans. Samples with high $^{18}O/^{16}O$ values are said to be isotopically heavy and low values are referred to as isotopically light.

Because the number of glaciers on land is a function of the average temperature of the atmosphere, this gives us a way of measuring the temperature of ancient atmospheres. We can't sample bits of ancient oceans or glaciers, but we can understand the composition of the ocean (and by extension the glaciers) by sampling the remains of organisms that lived in the ocean at various times in the past. In Chapter 19 we will review some of the data derived from actual measurements of gas bubbles trapped in glaciers and retrieved from ice cores.

As we have discussed in earlier chapters, since the Cambrian explosion there have been many marine organisms that build their shells or skeletons out of calcium carbonate ($CaCO_3$). By and large, the shell material will have the same $^{18}O/^{16}O$ value as the ocean that the organism lived in, so by measuring this value in fossil samples we can understand whether or not the organism lived in a time when there were lots of glaciers on land or a time when most of those glaciers had melted and returned that water to the sea. During cooler periods, when there are ice sheets, the ocean has higher ratios

of $^{18}O/^{16}O$, and lower ratios in warmer periods when there is less isotopic fraction in glaciers and the oceans are isotopically lighter. Many of the fossils used for this analysis are also valuable for understanding the age of the rocks in which they are found. Thus, by combining biostratigraphy with geochemistry, we can construct a curve of warming and cooling of the atmosphere, using our proxy of $^{18}O/^{16}O$ in fossils versus time. Going back at least for a few million years, the magnitude of the variations of $^{18}O/^{16}O$ in fossils match exceptionally well to the orbital cycles calculated by Milankovitch, and these have provided an independent test of the idea that orbital cycles control Earth's climate (Figures 18.10 and 18.11). Rather than using the older names of glacial episodes, we now refer to the **Oxygen Isotope Stage** (OIS in Figure 18.10). The recent Wisconsian and Illinoian glaciations are associated with OIS 2–5 and 6 respectively, but the older names correlate to multiple older glaciations and are no longer used. Some of the isotope stages correlate to warmer interglacial periods, such as OIS 1, 5 7, 9, and 11, whereas others are associated with times of glacial maxima, such as OIS 2, 6, and 10 (Figure 18.10).

As water is removed from oceans and stored on land as glaciers this causes falls of sea level. The term **eustasy** refers to global changes of sea level caused either by tectonics, termed **tectono-eustasy**, or by growth and decay of glaciers, termed **glacio-eustasy**. We have already mentioned that melting of the Antarctic and Greenland ice caps could cause 48 m and 7 m of glacio-eustatic sea-level rise, respectively. Conversely, we are interested in knowing how much sea level fell because of the last ice age such that we can better understand the relationship between sea level and climate cycles.

Figure 18.10 illustrates the link to climate changes and sea level. Periods of glacial advance are characterized by falling sea level, and glacial maxima are linked to sea-level lowstands. Periods of melting glaciers cause rising sea level (i.e., transgression) followed by interglacial warm periods marking sea-level highstands. The last glacial maximum (LGM) occurred about 18,000 years ago, at the end of the Pleistocene Epoch (OIS 2 in Figure 18.10). Global sea level was more than 100 meters lower than today, and this period of low sea level, referred to as

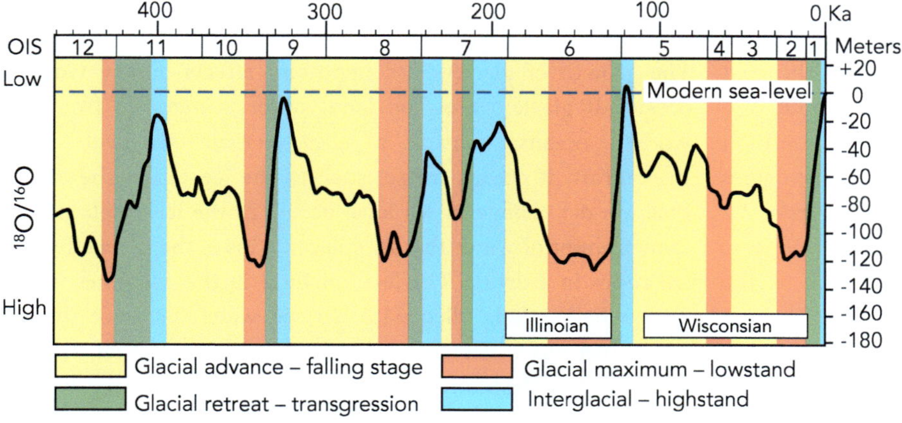

Figure 18.10 Oxygen isotope curve from analysis of deep-sea foraminifera and calculated sea-level curve. OIS numbers 1–12 indicate oxygen isotope stages, where even numbers are lower, odd numbers are higher. Two scales of cyclicity can be seen: a lower-frequency 100 ka cycle, and a lower-amplitude 20 ka cycle. The times of lowest sea level (oxygen isotope stages 10, 8, 6, and 2) are marked by coincidence of the obliquity and short eccentricity cycles. The cycles are also asymmetric, in that the rises of sea level (green transgressions) occur much faster than the falls indicating episodic growth of ice sheets followed by much more rapid melting and sea-level rise.

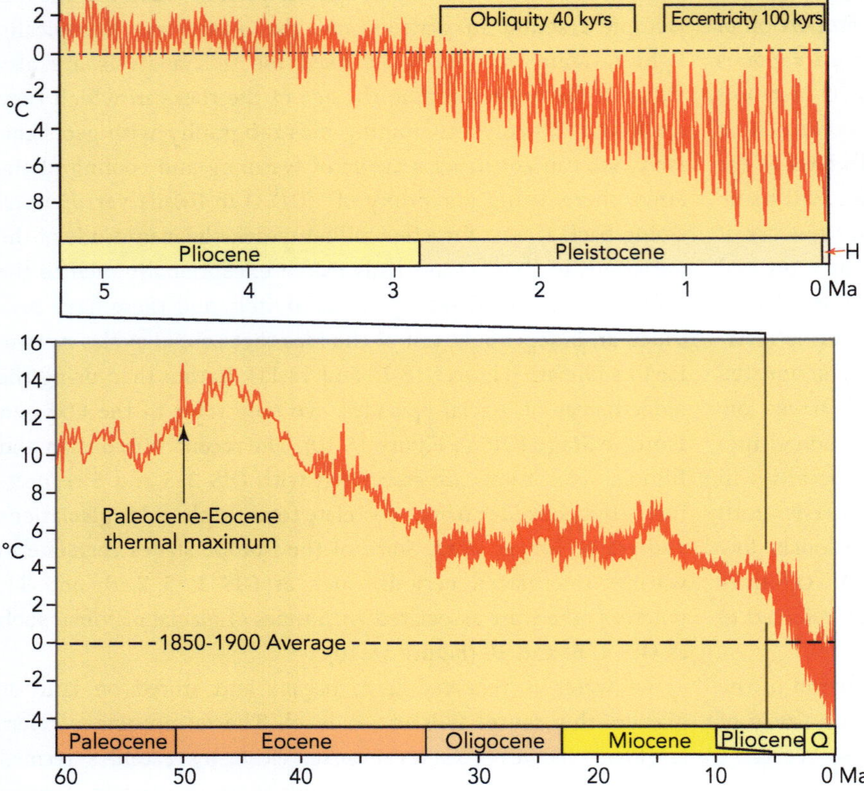

Figure 18.11 Estimated temperature changes over the last 5 and 65 million years. Note that in addition to an overall cooling trend over the last 50 million years, the last million years are dominated by the slower short 100 ka eccentricity cycles, whereas from 2.5 to 1 million years ago the faster 41 ka obliquity cycle was dominant. H – Holocene. Source: (Bottom image) Adapted from Raymo and Lisiecki (2005). Copyright (2005). With permission from John Wiley and Sons.

a **lowstand**, lasted about 10,000 years. In addition to highlighting the lowstands, the isotope data also show that these glacial maxima were preceded by a stepped drop of oxygen isotope values, indicating a step-like build-up of ice sheets and a consequent stepped fall of sea level (Figure 18.10). As the glaciers built up, they experienced shorter intervals of partial glacial retreat that mark the faster obliquity cycles. The glacial maxima, which correlate to the sea-level lowstands, mark the coincidence of the obliquity and short eccentricity cycles. These lowstands were followed by remarkably rapid deglaciations that occurred over time periods of less than 10,000 years.

Going back farther in time (Figure 18.11), there was an exceedingly warm period around 50 million years ago referred to as the **Paleocene/Eocene thermal maximum**, with temperatures 15 °C warmer than today. Over the past 30 million years, global temperatures have been about 5 °C warmer than today. It also appears that the past million years were dominated by the slower short eccentricity cycles with superimposed 20 ka obliquity cycles, whereas from 2.5 to 1 million years ago the faster 40 ka precession cycle was dominant. These changes in the dominance of one of the Milankovitch cycles over another relates back to the way that the Milankovitch cycles interact either constructively or destructively to reinforce or diminish each other at a variety of "beat" timescales. You will also notice

(Figure 18.11) that our current period is actually one of the coldest over the past 65 million years.

KEY POINT

The ratio of $^{18}O/^{16}O$ in the oceans is controlled by the amount of fresh water stored on land as ice and is higher during glacial periods. Measurement of $^{18}O/^{16}O$ in the shells of marine organisms provided an independent test of Milankovitch cycles and led to the development of estimates of global temperature and sea level over geologic time.

18.8 Global Consequences of Continental Glaciation

The growth of glaciers has important effects on the world's oceans. If glaciers grow on land, water is removed from the world's oceans, causing a glacio-eustatic sea-level drop. In contrast, if oceans freeze, such as the ice cap at the North Pole, no net change in sea level occurs as the ice displaces the same amount of water that it replaces. This is the same effect as ice cubes in a drink. The level of fluid in the container is the same whether it is filled with ice and water, or, once the ice melts, just water. However, when ocean water freezes, the ice that forms is fresh and the salt stays in the remaining liquid. The leftover water is both cold and, because of the excess

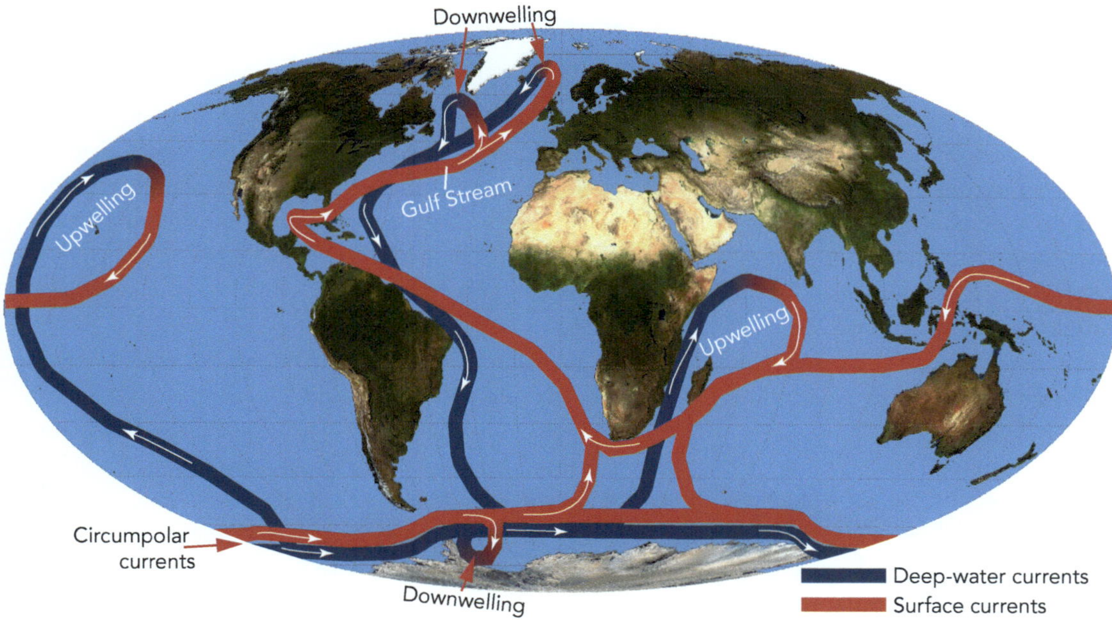

Figure 18.12 Thermohaline oceanic currents. Deep, cold, and salty polar waters (blue lines) descend and flow counterclockwise in the northern oceans and clockwise in the southern oceans. Source: Earth map. Modified from Mdf, Public domain, via Wikimedia Commons.

dissolved salt, denser so it tends to sink. These cold salty descending polar waters flow along the ocean floor and because of the Earth's rotation they flow in a spiral, or *gyre*, resulting in deep **thermohaline** (literally "cold and salty") oceanic currents that also mix the world's oceanic waters and bring oxygenated water into the deep oceans (Figures 14.9 and 18.12). These currents are clockwise in the southern oceans and counterclockwise in the northern oceans.

During glacial periods, most of the world's oceans are well mixed and ocean floors well oxygenated. Upwelling of these currents may occur in equatorial positions such as the Galapagos Islands, bringing nutrient-rich waters that enhance bio-productivity. However, during warmer greenhouse times, such as the Late Cretaceous, a lack of sea ice (and the general elevated temperature of the atmosphere and oceans) can stop these thermohaline currents, resulting in a loss of oxygen to the ocean floors. Several **oceanic anoxic events** have been documented through Earth history and these are commonly marked by the deposition of black-colored shales, rich in organic matter, as the anoxic conditions prevent the growth of oxygen-loving organisms and bacteria that would otherwise feed on this organic material (Figure 14.9). The Burgess Shale, which we reviewed in Chapter 9, is one such example of a richly fossiliferous, organic-rich black shale (Figure 9.15). As these organic-rich sediments are buried, the organic matter heats up and over long periods of time the original proteins, fats, and other organic matter can be broken down into simpler hydrocarbon molecules that form many of the world's largest oil and gas **source rocks** (see Box 18.1). Therefore, petroleum geologists became very interested in the relationship between climate, sea level, and the origin of these source rocks, which we will return to later.

KEY POINT

During cold periods, descending thermohaline ocean currents keep oceans mixed and oxygenated. During periods of extreme warmth, ocean circulation can cease, causing global oceanic anoxia and deposition of organic-rich shales that form hydrocarbon source rocks.

18.9 Sea Level and Sedimentation

Given that Earth has experienced climate change over many timescales, which in turn causes changes in sea level, the next question is what is the direct evidence for this sea-level change, especially if we look at the sedimentary rock record? Above we also mention that at the last glacial maximum, sea level was over 100 m below today's level, and this invites the next question: how do we know this?

In Chapter 1 we reviewed the geology of the Grand Canyon, emphasizing its long geological history as recorded in the rocks exposed along the canyon floor and walls. The various layers record alternations of marine and non-marine conditions punctuated by unconformities that provide clear evidence of transgression, regression with some tectonic episodes of uplift, and erosion. However, in the Grand Canyon, these changes represent tens to hundreds of millions of years, versus the orbital Milankovitch cycles that occur over tens to hundreds of thousands of years and that can cause sea-level changes on the order of tens to a hundred meters.

BOX 18.1 Origin of Oil

Fossil fuels include coal, formed by the compressed remains of swamps and other wetlands, and oil and gas. Oil and gas originate as organic matter, mostly from microscopic plankton and algae, as well as bits and pieces of vegetation that are transported by rivers and deposited into lakes or oceans (Figure 18.13). *Source rocks* refer to mudstones, usually with 1.5% or more of dispersed organic carbon. As organic matter settles on a lake or seafloor, over time it becomes buried within the sediment. Muddy sediments are more likely to eventually

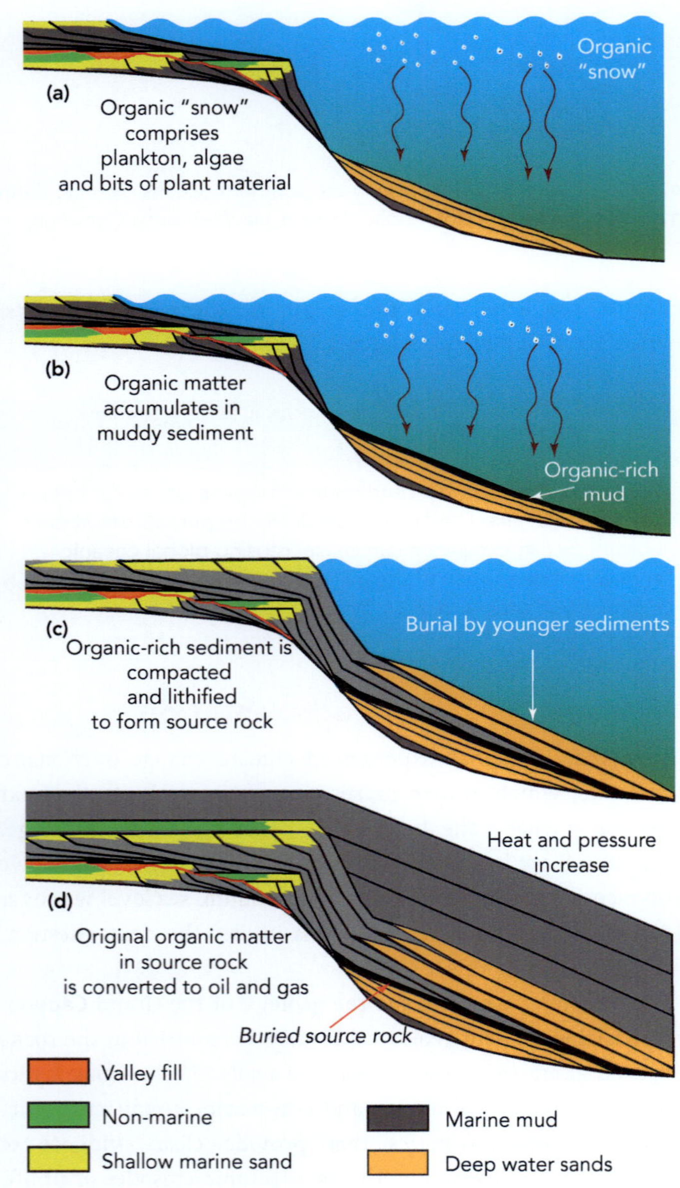

Figure 18.13 Formation of source rocks begins with accumulation of organic matter in sediment. As the organic-rich sediment is buried, the organic matter experiences increasing heat and pressure that gradually transforms it into hydrocarbons (oil and gas).

produce hydrocarbons, as they commonly are less well oxygenated than sandy or gravelly sediment, and the less oxygen the greater the likelihood of preservation of the organic material. In terms of Earth history, predicting or knowing those greenhouse times of excessively high sea levels is of particular interest, as these correlate to the development of worldwide source rocks.

As the source rock is buried, it begins to experience an increase in pressure and temperature (Figure 18.13). At around 60 °C the original organic matter, which can include fats, carbohydrates, proteins, as well as cellulose from plants, begins to break down and petroleum geologists refer to this area in a basin as "the kitchen" (Figure 18.14). The process is similar to cooking, such as when you fry onions and they begin to caramelize, or the darkening of a roast in the oven.

Between about 60 °C and 120 °C the organic matter is converted by **thermogenesis** (or heating) to oil and later gas, and these temperature zones are also referred to as the **oil and gas windows** (Figure 18.14). **Biogenic gas** can also be made by the breakdown of organic matter by anerobic bacteria in the shallow subsurface (Figure 18.14).

Oil and gas are less dense and **immiscible** with water. That is, oil will not dissolve in water, much like an oil and vinegar dressing that will quickly separate if not vigorously mixed. Once petroleum and natural gas molecules begin to form in the source rock, they tend to float upwards because they are less dense than the material that surrounds them (even the water). Of course, all this is happening at the microscopic scale of the tiny pores between the grains or, in the case of clays, flakes of sediment. Regardless of the physical scale, over geologic time, oil and gas does indeed escape from the kitchen. The molecules try to move upwards, but in many cases, they will hit against impermeable layers, or seals, and are forced to move sideways (Figure 18.14). Sometimes they may be able to migrate along faults, fractures, or other zones of increased permeability. Most of the time, the oil and gas molecules will be dispersed, but depending on the structural configuration of the basin, sometimes the hydrocarbons are focused into a *trap*, an area where oil and gas stops migrating and accumulates (Figure 18.14). The most common type of trap is an **anticline**, essentially an upside-down bowl, commonly caused by compression and folding of the rocks, and in the early days of oil and gas exploration, geologists would map anticlines at the surface and drill into the center of these structures. The oil and gas typically seep into and accumulate in **reservoir** rocks, such as sandstones or limestones, that are both porous and permeable (Figure 18.14). If the reservoir layer is overlain by an impermeable seal, then the accumulation can build up.

For most of the history of the oil business, oil has been discovered and produced from conventional high-porosity,

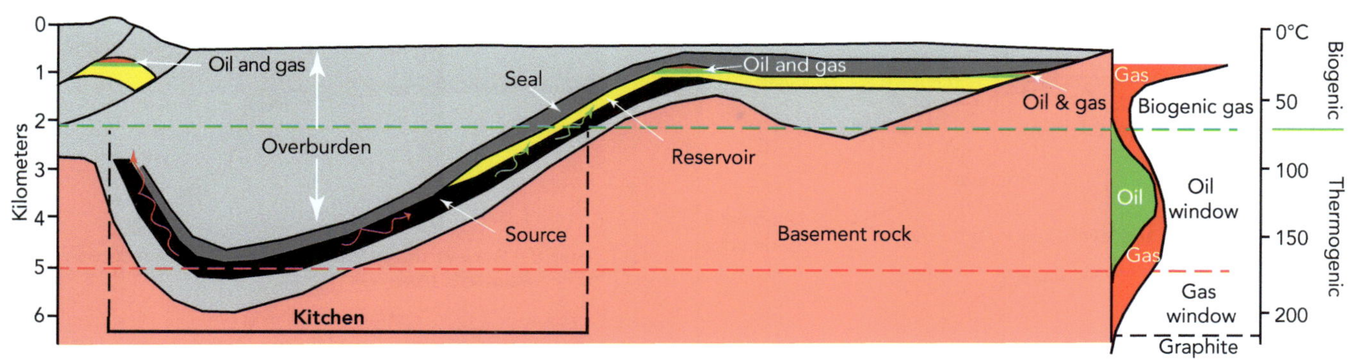

Figure 18.14 A typical petroleum system includes a source rock, which must be buried deep enough in the "kitchen" area of a sedimentary basin for organic matter to be converted to oil and gas. These liquids and gases are unable to mix with water and are less dense so they naturally rise and flow upwards where they may be trapped to form oil or gas accumulations.

high-permeability reservoirs, such as the anticlinal traps described above. However, by the late twentieth century, the USA was running out of conventional domestic reserves and was importing huge amounts of oil from Canada and Mexico. At the beginning of the twenty-first century, US petroleum engineers developed the ability to drill horizontal wells, and with improvements in the ability to fracture rocks, were able to target source rocks directly. Although controversial, this development of **unconventional resources** opened vast new reserves and reduced US dependence on imported hydrocarbons. In the final chapter in this book, we will examine the long-term consequences of hydrocarbon production and use, and in particular its effect on the global carbon cycle and global warming.

In Chapter 11 we reviewed the Late Paleozoic Ice Age and mentioned that Pennsylvanian strata commonly show rapid cyclic alternations of marine and non-marine sedimentary layers termed cyclothems (Figure 11.20). These cyclothems are inferred to be a consequence of glacially driven, sea-level changes associated with the Milankovitch scale orbital cycles and are analogous to the glacial cycles that we have discussed above.

18.9.1 Seismic Stratigraphy and Global Sea-Level Cycles

In the 1970s, scientists at Exxon corporation had collected seismic data from around the world in their search for oil and gas deposits (Box 18.1). In Chapter 16 we showed how seismic lines (see Box 16.1) can image the cross-sectional geometry of a continental margin. In the case of the Mediterranean, two sets of large-scale clinoforms mark the shelf, slope, and deep water environments, indicating that the Mediterranean Sea was quite deep in the Miocene and Pliocene, but as we discussed, these were separated by the regional M unconformity (Figures 16.5 and 16.6), which recorded a massive sea-level drop caused by the drawdown of the Mediterranean Sea during the Messinian salinity crisis. Recall that the clinoforms are formed by the sediments deposited in the shelf, slope, and deep water environments. As sediment is delivered to a continental margin, all else being equal, the sediment will build outwards and the clinoform will correspondingly move seaward (Figure 18.15). If sea-level rises faster than the delivered sediment can continue to build outward, the clinoforms will shift landward, marking a **transgression**. If sea-level falls, as seen in the Mediterranean and shown in Figure 18.15, the clinoforms will shift basinward, marking a **regression**, and rivers may erode across the top of the previous clinoform and can cut an incised valley, or, when the fall is large as it was during the Messinian drawdown, a larger canyon may be cut.

In seismic cross sections, places where reflections are eroded or terminate against each other, termed **lapout**, are used to define unconformity-bounded sequences (Figure 18.16). Once a cross section is interpreted, it can be exploded, with geological time rather than thickness on the vertical axis (Figures 18.16 and 18.17). These **time–stratigraphic diagrams** show how sediments are shifting in time, and particularly if sediments are confined to more distal positions in the basin, marking sea-level lowstands, or if they extend landward, marking sea-level **highstands**. The unconformities are represented by areas that have no rock. Normally, biostratigraphy from wells is used to assign ages to the various stratigraphic sequences. The changes in stratigraphic patterns are controlled by the **subsidence** (i.e., sinking of the land surface either by tectonics or simple sediment compaction) and sea-level changes in the basin. Most passive continental margins experience relatively uniform subsidence, with no uplift, and if the subsidence can be estimated, then any leftover changes can be attributed to sea-level changes.

Using these analytical techniques Exxon scientists, led by geologist Peter Vail and paleontologist Bilal Haq, were able to estimate the age, frequency, and magnitude of sea-level changes in their global seismic data sets (Figure 18.18). When they began

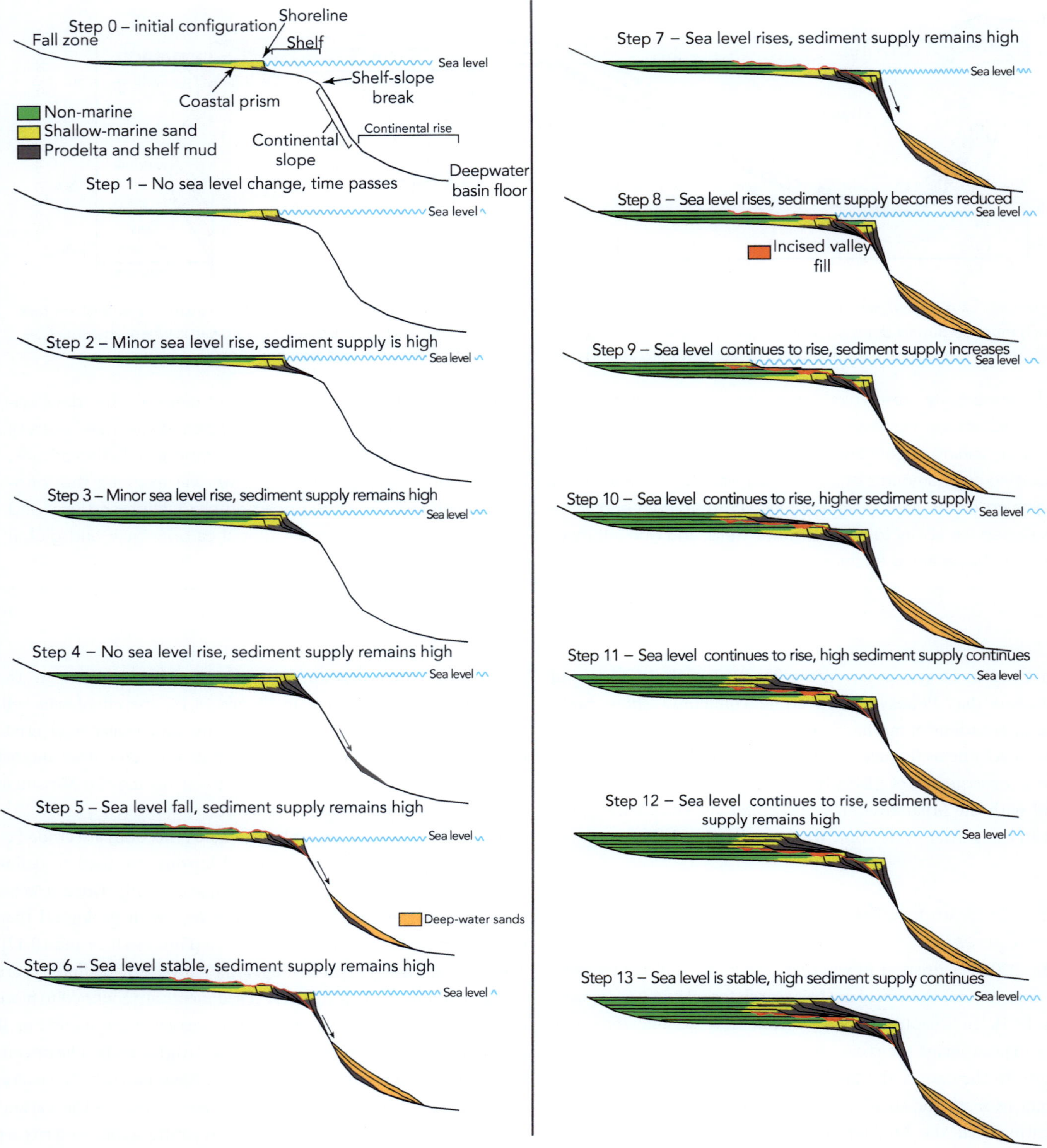

Figure 18.15 Formation of a clinoform during a cycle of high, falling, rising, and high sea level.

comparing the pattern of global cycles in basins from around the world, they noticed remarkable similarity both in the patterns and the age of the cycles and the positions of the unconformities, and realized that these must be the results of global variations in sea level. This led to publication of global changes of sea level throughout the Phanerozoic (Figure 18.18), sometimes referred to as **the Haq curve**.

Of course, Exxon scientists weren't just interested in deciphering the history of global sea level. This was a tool to help them better explore for oil and gas (see Box 18.1). They realized that times of very high sea levels were prone to deposition of source rocks (Figure 18.13), particularly during times of ocean anoxia. Conversely, times when sea level was low were commonly associated with deposition of reservoir rocks, such as

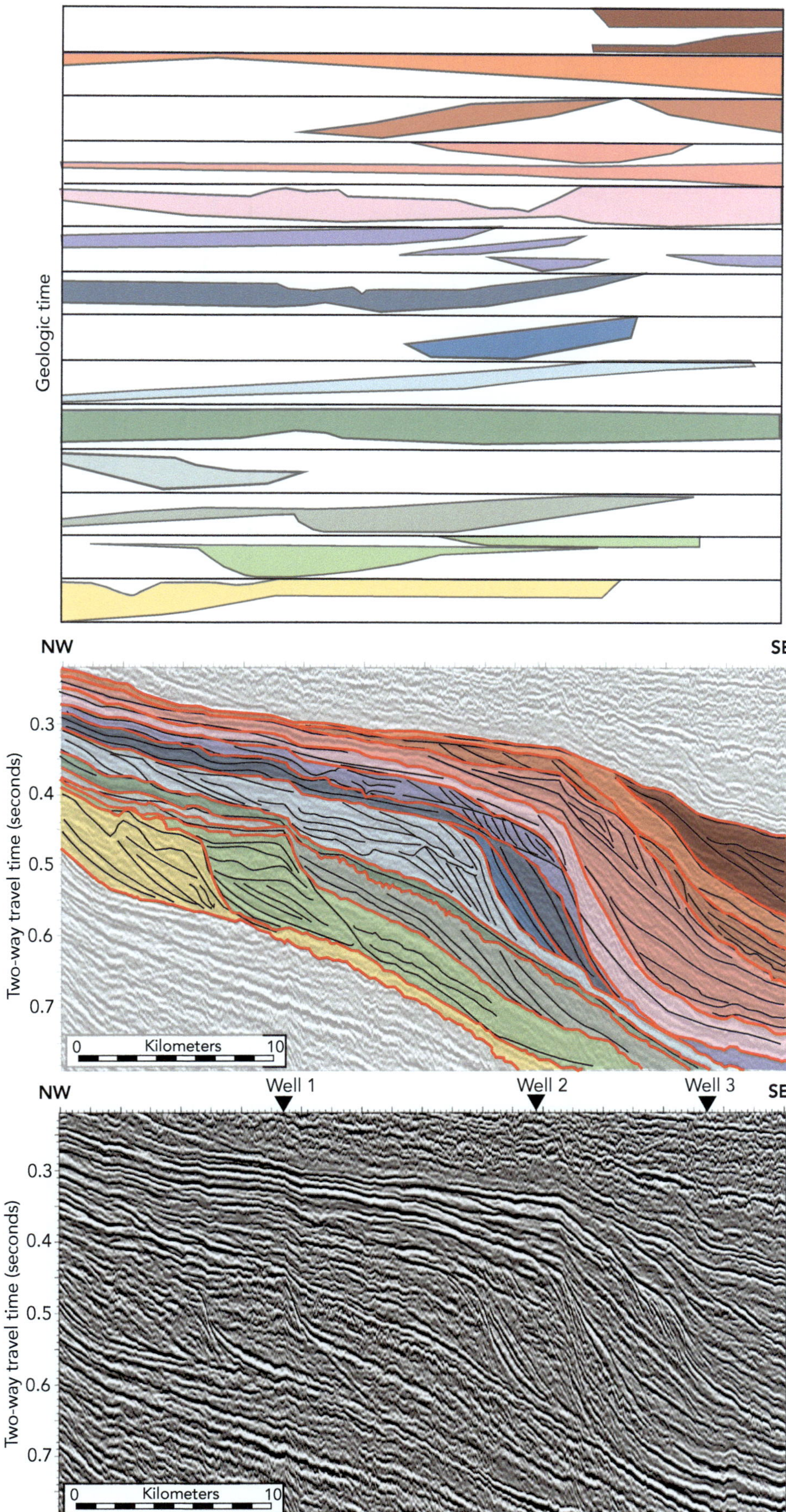

Figure 18.16 Bottom: Seismic cross section, from offshore New Jersey. Middle: Interpretation showing sequences of shelf-slope clinoforms; Top: Time–stratigraphic representation. Source: (middle and bottom) Modified after Miller et al. (2018).

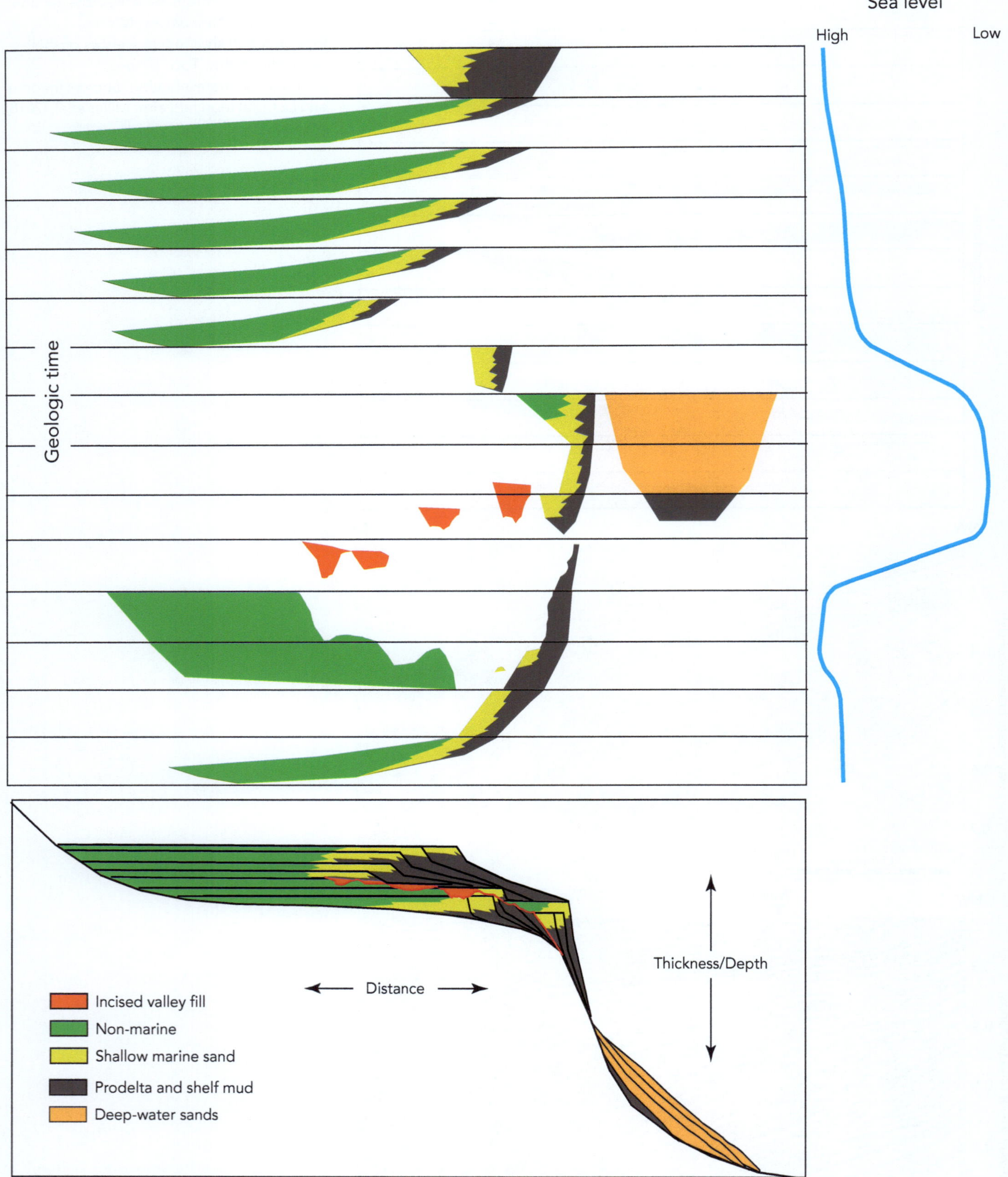

Figure 18.17 Time–stratigraphic analysis of the geologic cross section constructed in Figure 18.14. Such cross sections can be used to infer sea-level change.

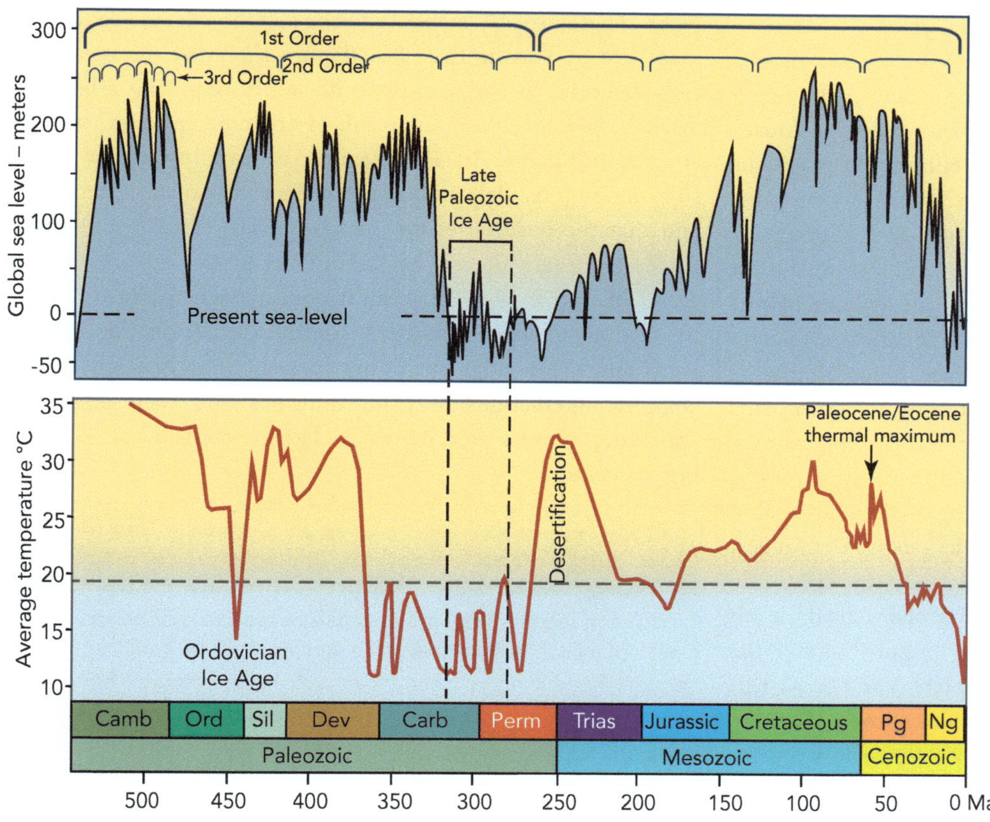

Figure 18.18 Exxon chart of sea level and global temperature showing Phanerozoic global sea-level changes with periods of major glaciations. Broad cycles can be discerned including the longest first-order cycles, related to super continental Wilson cycles, and the tens of millions of years duration second-order cycles. This chart also shows third-order cycles as the shortest, but these consist of shorter-term Milankovitch cycles that are too detailed to show at this scale (see Figures 18.10 and 18.11). Source: Adapted from an image by Robert A. Rohde created from publicly available data and is incorporated into the Global Warming Art project: https://commons .wikimedia.org/wiki/File:Phanerozoic_ Sea_Level.png#file

sands or shallow water limestones (Figures 18.14 and 18.15). Subsequent rises of sea level could place impermeable seals above the reservoirs, forming reservoir–seal pairs.

A limitation of the seismic data is that it does not provide direct information about lithologies, but by integrating layout patterns and inferences of sea level, and identifying major unconformities, Exxon scientists were able to improve the prediction of lithologies encountered while drilling. Given that only about 1 in 10 exploration wells results in a discovery, anything they could do to improve their success rate helped their profitability. Despite their business interest, the Exxon scientists did contribute to our understanding of how sea level has changed over the past half billion years. Eventually both Vail and Haq left Exxon, Vail taking a professorship at Rice University and Haq heading up the Ocean Science research program at the US National Science Foundation, where they continued to make contributions to our understanding of sea level through time. The fundamental contributions of this work was recognized in 2003 when Vail was awarded the Penrose Medal of the Geological Society of America – the same award given to J Harlan Bretz that we discussed in Chapter 2.

KEY POINT

Stratigraphic analysis of seismic lines from around the globe allowed Exxon scientists to estimate changes in global sea level over much of the Phanerozoic.

18.9.2 Sea Level Throughout Earth History

The Haq curve recognized several scales of sea-level cycles over the Phanerozoic (Figure 18.18). Most of these stratigraphic sequences were bounded by disconformities, rather than angular unconformities, suggesting that tectonics was not the most important control. In their landmark 1977 publication (Payton, 1977), they emphasized a glacio-eustatic origin for most of their sea-level cycles. This caused a stir in the academic community, who were concentrating on the relatively new plate tectonic revolution, and there was considerable criticism suggesting that Exxon scientists had largely ignored tectonic controls. Even Pete Vail's supervisor, Larry Sloss, critiqued his own former grad student and his fellow "Neo-neptunists" in a paper entitled a "The tectonic factor in sea level change: A countervailing view," and argued that global cycles were largely tectono-eustatic and that glacio-eustasy played only a secondary role (Sloss, 1991).

18.9.3 Orders of Sea Level and Their Causes

The Haq curve (Figure 18.18) shows several orders of sea-level change from the longest first-order cycles, lasting for 100s of millions of years through second-order cycles that occur over periods of tens of millions of years and third-order cycles of a few million years. Fourth- and fifth-order cycles were not included on the original charts as they were beyond the resolution of seismic data of the time, but they represent the Milankovitch orbital cycles explained above.

18.9.3.1 First-Order Tectono-Eustatic Cycles

There are two first-order cycles marked by peaks in the Paleozoic and Cretaceous through Paleogene. These cycles occur over a period of several hundred million years but what causes these sea-level changes? Could these be climatic in origin or something else? In general, the duration of these two great cycles seem to correlate with the formation and breakup of supercontinents, the Wilson cycles that we introduced in Chapters 5 and 10. In addition to changing the volume of seawater in the oceans by storing it on land in glaciers (glacio-eustasy), tectonics also controls the volume of ocean basins (tectono-eustasy, Figure 14.10). Recall that most modern oceans have very extensive ridges that run along the spreading centers, such as the mid-Atlantic Ridge referred to in Chapter 5 (Figures 5.8 and 5.9). The reason these ridges stand up is that the new oceanic lithosphere is hotter and less dense than older, cooler, and denser lithosphere far away from the spreading centers and this younger less dense lithosphere floats a bit higher on the underlying asthenosphere (Figure 14.10), the isostatic effect that we introduced in Chapters 3 and 5. As this lithosphere cools, it becomes denser and sinks lower, forming the deep abyssal plains that can be seen in the world's oceans. At times in Earth history, seafloor spreading has been more vigorous, resulting in the creation of a greater proportion of young ocean lithosphere. From about 120 to 90 Ma, during the Cretaceous Period, the rate of production of oceanic lithosphere doubled and remained high until about 80 Ma (Figure 14.7). This rapid expansion of the seafloor meant the average temperature of oceanic crust increased, reducing the overall volume of the world's ocean basins. Because the amount of water in the oceans didn't change, the water was displaced onto the continents. Thus, the exceedingly high sea level during the Cretaceous was influenced by tectonics, but this increase in rates of ocean spreading also coincided with a time of generally much higher global temperatures, leading some to claim that during at least parts of the Cretaceous there was no ice on Earth's surface. However, some detailed studies of Cretaceous formations show evidence of Milankovitch-frequency sea-level changes (which imply the existence of glaciers), although the magnitudes are generally tens versus hundreds of meters. It has been suggested that small, ephemeral ice sheets may have existed in Antarctica, primarily because of the high elevation in the interior of the continent, which as we have seen, has lain at the South Pole for most of its history. It has been recently suggested that changes in the proportion of fresh water stored in groundwater aquifers (not in glaciers) may be enough to account for the low-magnitude changes observed, and we now have a new hypothesis of **aquifer-eustasy** to test.

18.9.3.2 Second-Order Tectonic Cycles of Sea Level

The Haq chart (Figure 18.18) also shows a set of second-order cycles that represent durations of several tens of millions of years. These broadly correlate to the tectono-eustatic sequences identified by Sloss in the 1960s (Figure 1.21) that are unequivocally bounded by angular unconformities requiring a tectonic origin. These are likely the result of orogenic episodes, such as covered in Chapter 10. For example, we saw in Chapter 5 that subduction (i.e., slab pull) is the main driver of plate tectonics and that the creation of new oceanic lithosphere at mid-ocean ridges is largely driven by subduction at the other end of a plate. Of course, once an ocean is gone and especially if it ends in continental arc or continent–continent collision, subduction stops and ocean-ridge spreading correspondingly stops. These complex interactions of plate collisions and their effect on spreading centers is thought to largely control the second-order cycles.

KEY POINT

First- and second-order cycles of sea level occur over hundreds to tens of million years respectively and are likely tectonic in origin.

18.9.3.3 Third- and Higher-Order Cycles of Sea Level

Much of the controversy around the original sea-level curves centered on the so-called third-order cycles of a few million years in duration. It is worth recalling that over the past few million years we have an excellent record of glacio-eustatic sea-level changes, and growth and decay of glaciers can cause 100 m of vertical change in sea level over 10,000 years, or about 10 mm/year. Typical rates of tectonic subsidence and uplift are usually much lower, at 1 to 0.1 mm/year. Consequently, we can have some confidence in interpreting high-frequency and high-amplitude fourth- and higher-order cycles as largely glacio-eustatic (or possibly aquifer-eustatic) in origin. The third-order cycles may be primarily an artifact of the low resolution of the seismic data compiled by Exxon in the late 1970s and early 1980s. We introduced the idea that interacting signals of different frequencies can cause beats. This is the essence of the vibrating change in volume that you hear when a pair of guitar strings that are slightly out of tune play the same note simultaneously. The frequency of the vibration is lower than the frequency of the note played. We saw how this can cause either constructive or destructive interference and it may be that the third-order cycles are artifacts of overlapping second-, fourth-, and fifth-order cycles of variable origin.

KEY POINT

There is still uncertainty over the origin of the third-order sea-level cycle, but they likely represent the interaction of tectonic and climatic cycles. The fourth-order and higher cycles are primarily glacio-eustatic in origin.

18.10 Time's Arrow, Time's Cycle

Some events in Earth history are rare, such as the appearance of the first cells or the various mass extinction events, like the meteorite impact 66 million years ago that ended the reign of the dinosaurs, and others recur, such as the Wilson cycles that are associated with the assembly and breakup of supercontinents or the Milankovitch climate cycles that are the focus of this chapter. Reading Earth history yields the idea that some changes, such as evolution, are unidirectional and arrow-like, and of course allow us to order the fossil record by Smith's principle of faunal succession, whereas other processes and events are clearly more cyclic.

In predicting Earth's future, one of the big questions is, what events are cyclic and predictable versus one-offs or extremely rare? Following the discovery of the Chicxulub crater, which confirmed that a meteorite triggered the K–Pg extinction, paleontologists hypothesized that regular disruption of comets in the Oort cloud, which lies outside our solar system, occurs every 26 million years, knocked into the solar system by a renegade Red Dwarf star, called Nemesis. Despite searching for evidence of regular cycles of mass extinction, such as the iridium anomaly associated with the K–Pg event, there has been no luck in demonstrating that any of the other four mass extinctions are associated with meteors, and in retrospect they all now seem to be associated with massive volcanism associated with large igneous provinces. Whether these igneous events are cyclic or random is not known but there is little doubt that there will be catastrophic eruptions in the future – whether these will be in our lifetimes or long after is nearly impossible to predict.

Milankovitch climate cycles are one of the most ideal examples of remarkably predictable cycles. If you examine the oxygen isotope record over the last 450,000 years (Figure 18.10), you will see that the longest lived sea-level highstand, reflecting interglacial warm periods, lasted for a maximum of about 10,000 years. The last glacial retreat and warming began about 12,000 years ago, marking the beginning of the Holocene Period, which is the subject of our final chapter. Right now, we are just about at the interglacial warming peak. It is thus easy to predict, all else being equal, that in about 10,000 years Earth will be well into the next ice age. A big question for humanity is whether anthropogenic warming could counteract or disrupt these natural climate cycles, and we will discuss this in the next and final chapter.

18.11 Summary

- Nineteenth-century observations of erratics led to the idea that Earth had experienced several ice ages, which were hypothesized to be caused by changes in Earth's orbit.
- Currently glaciers are found at high (polar) latitudes and high altitudes, with the largest ice sheets covering Antarctica and Greenland. Melting of these ice sheets could cause 58 and 7 meters of sea-level rise, respectively.
- 18,000 years ago, North America was covered by the kilometers-thick Cordilleran and Laurentide ice sheets. The extent of ice sheets has been determined by mapping and stratigraphic analysis of glacial landforms and deposits and explains the origin of many North American lakes, which are the remnants of the lakes formed at the terminus of the ice sheets as they began to retreat.
- Milutin Milankovitch calculated the changes in solar insolation controlled by the changes in Earth's orbit because of the interaction of the Sun, Moon, and large planets such as Jupiter. These include the 20 ka precession, 40 ka obliquity, and 100 and 400 ka eccentricity cycles that drive changes in Earth's climate and linked glaciations and global sea level.
- Measurements of $^{18}O/^{16}O$ ratios in shells of marine organisms show that high values correspond to ice ages and low sea levels when more of the lighter ^{16}O is tied up in land glaciers, whereas lower values record warmer periods.
- Stratigraphic analysis of seismic sections, particularly from passive margins that do not experience tectonic uplift, allowed Exxon scientists to decipher the global sea level for the Phanerozoic Eon. The slowest 100-million-year duration first-order cycles were associated with the assembly and breakup of supercontinents that mark Wilson Cycles (see Chapter 10). The tens of million-year second-order cycles were interpreted as tectonic in origin and match the Sloss sequences. The third-order cycles may represent the interaction of tectonic and higher-frequency Milankovitch cycles.

Key Words

- diluvium
- erratics
- snow line
- circumpolar current
- last glacial maximum (LGM)
- Cordilleran Ice Sheet
- Laurentide Ice Sheet
- crevasses
- moraines
- drumlins
- terminal moraines
- insolation
- obliquity
- precession
- eccentricity

- beats
- Milankovitch cycles
- Oxygen Isotope Stage (OIS)
- eustasy
- tectono-eustasy
- glacio-eustasy
- lowstand
- Paleocene/Eocene thermal maximum
- thermohaline
- oceanic anoxic events
- source rocks
- thermogenesis
- oil and gas windows
- biogenic gas
- immiscible
- anticline
- reservoir
- unconventional resources
- transgression
- regression
- lapout
- time–stratigraphic diagrams
- highstands
- subsidence
- the Haq curve
- aquifer-eustasy

Further Reading and References

Haq, B. U., Hardenbol, J. A. N., and Vail, P. R., 1987, Chronology of fluctuating sea levels since the Triassic, *Science*, 235(4793), 1156–1167.

Haq, B. U., and Schutter, S. R., 2008, A chronology of Paleozoic sea-level changes, *Science*, 322(5898), 64–68.

Miller, K. G., Lombardi, C. J., Browning, J. V., et al., 2018, Back to basics of sequence stratigraphy: Early Miocene and Mid-cretaceous examples from the New Jersey Paleoshelf, *Journal of Sedimentary Research*, 88(1), 148–176, https://doi.org/10.2110/jsr.2017.73.

Payton, C. E., 1977, *Seismic Stratigraphy—Applications to Hydrocarbon Exploration*, American Association of Petroleum Geologists.

Raymo, M. E., and Lisiecki, L. E., 2005, A Pliocene-Pleistocene stack of 57 globally distributed benthic $\delta^{18}O$ records, *Paleoceanography and Paleoclimatology*, 20(1), https://doi.org/10.1029/2004PA001071.

Sloss, L. L., 1991, The tectonic factor in sea level change: A countervailing view, *Journal of Geophysical Research: Solid Earth*, 96(B4), 6609–6617, https://doi.org/10.1029/90jb00840.

Review Questions

1. What factors are needed for a glaciation to occur?
2. Where do glaciers occur today?
3. What is the snow line?
4. What would be the consequence of melting the Greenland and Antarctic ice sheets?
5. What is some of the key evidence that northern hemisphere continents were covered by ice sheets?
6. How is the extent of glaciations known?
7. How are glaciations and climate cycles related to the Earth's orbit?
8. Explain Milankovitch cycles and their control on solar insolation over time.
9. What is the difference between obliquity, eccentricity, and precession?
10. Describe the relationship between glaciers and sea level.
11. How do oxygen isotopes record changes in climate and sea level?
12. How do climate cycles change ocean circulation and the generation of organic-rich source rocks?
13. Are the relatively cold temperatures of our modern period common in Earth history?
14. What are the origins of first-, second-, and higher-orders of sea-level change?

Wind turbines, West Texas, USA.
Source: mj0007 / Getty Images.

Chapter 19

A Human World

Our Impact in the Holocene

LEARNING OBJECTIVES

- Understand major events in the Holocene, including ice sheet melting, sea-level rise, and climate events.
- Define greenhouse gases and their impact on Earth's surface temperature.
- Explain why the residence time of different greenhouse gases is important in their effects on long-term climate.

- Identify anthropogenic sources of greenhouse gases and their impact on climate and the environment.
- Discuss uncertainties and errors in climate models, and compare anthropogenic effects to events in deep time, including the possibility of a sixth extinction.

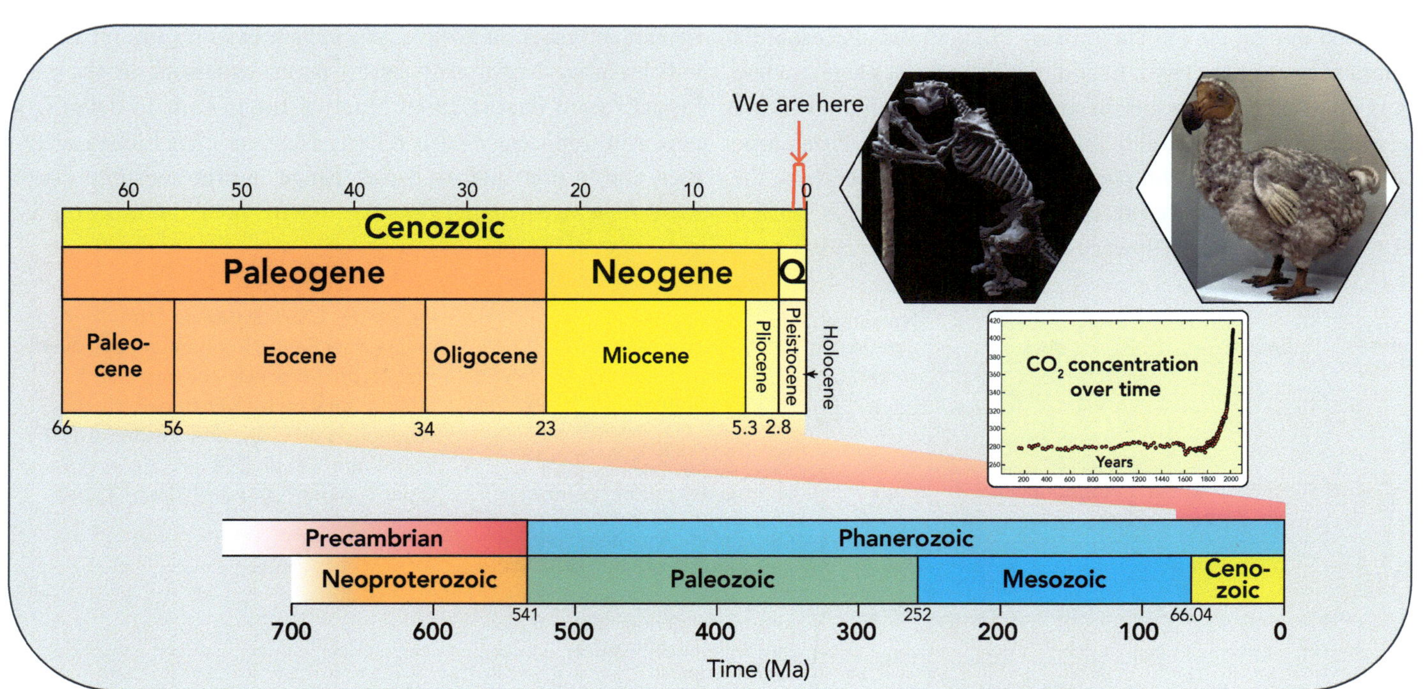

Introduction

So far, we have examined the history of our planet through the lens of a geologist, in which we observe the products captured in the rock record and try to interpret how they originate in the context of the complex interactions of geological processes over long geological timescales. This includes relocation of land masses through plate tectonics over hundreds of million-year Wilson cycles, as discussed in Chapters 5 and 10, to shorter-term sea-level and climate changes that occur over tens of thousands of years associated with orbital cycles reviewed in Chapter 18. We have also reviewed some of the major catastrophes in Earth history, with a focus on the mass extinctions that marked the end of the Paleozoic and Mesozoic eras (Chapters 11 and 13). In this chapter, we consider the idea that human activities over fewer than 300 years are now so profound that they might leave a permanent record in the geology of our planet.

19.1 The Holocene

As we discussed in Chapter 1, geologic time is subdivided into large and small parts (Figures 1.2 and 19.1). Eons are broken into eras, eras are divided into periods, and periods are divided into epochs. Although many of these units were codified in the late nineteenth century, geoscientists have continued to revise the geological timescale, not only with improvements in chronometric techniques, as discussed in Chapter 4, but with new observations and insights. One example is the Cryogenian Period, newly defined in 1990, which we reviewed in Chapter 8.

Currently, we are in the **Holocene Epoch**, which started about 11,650 years BP, marked by the melting and retreat of the Pleistocene ice sheets caused by natural climate warming as discussed in Chapter 18. The melting of ice sheets also caused a significant rise of sea level. In addition to rising sea levels, a few areas of Earth, such as northern Canada and Scandinavia, are experiencing isostatic uplift due to removal of the Pleistocene ice sheets whose weight pushed those areas deeper into the mantle. Now that the ice sheets have melted, isostasy is causing those areas to rebound upwards. In contrast, other areas have been inundated by the Holocene transgression and are sinking below the rising sea.

Although the Holocene has been a relatively warm period between major glaciations, there have been some anomalous temperature excursions of interest including the Medieval Warm Period, which lasted for about 300 years from around 950 to 1250 AD and was followed by a cooler period referred as the **Little Ice Age** (LIA), marked by several cooler pulses between 1300 and 1850 AD. The LIA was not a global event, but there is evidence of a generally cooler climate in the northern hemisphere, where northern latitudes were more severely impacted. The LIA has been attributed to a variety of causes including lower solar radiation, higher volcanic activity, and land-use changes caused by massive loss of human populations because of the Black Plague in Eurasia and the introduction of diseases such as smallpox, cholera, chicken pox, influenza, and many other others into the Americas by Europeans settlers that the people of the "New World" had no immunity against. The idea is that these massive losses of human populations stopped agriculture and allowed forest regrowth, causing CO_2 absorption from the atmosphere and storing carbon into biological pools and promoting cooling of the atmosphere. It has also been suggested that the earlier **Medieval Warm Period** may have been partly caused by anthropogenic deforestation and development of agriculture that promoted large CO_2 release into the atmosphere, causing warming. Thus, there is some evidence that human activity and particularly the development of agriculture may have played a role in changing climate long before the start of the industrial revolution in the 1750s.

In 2008, the stratigraphic commission of the Geological Society of London proposed a new epoch of the Cenozoic Era, the **Anthropocene**, reflecting the idea that the influence of human activities on Earth's atmosphere is so significant that it will leave (or has already left) a permanent mark in the geological record (Figure 19.1). Much of the interest in defining a new Anthropocene Epoch reflects concerns that human activities, and in particular land-use change and the consumption of fossil fuels, is changing the climate with potential catastrophic consequences.

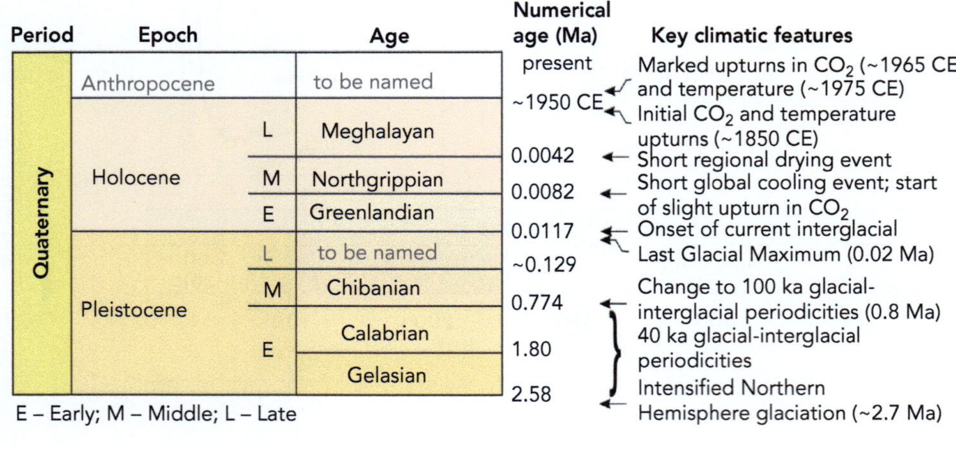

Figure 19.1 Subdivisions of the Quaternary Period and key climatic events. Black type – officially approved and ratified; gray type – yet to be sanctioned officially. Source: Reproduced from Waters et al. (2021). © Colin Waters, Colin Summerhayes and Simon Turner.

Figure 19.2 Atomic bomb test, Bikini Atoll, 1946. Source: Operation_Crossroads_Baker_(wide).jpg, via Wikimedia Commons.

There has been significant debate as to where to draw the line between the proposed Anthropocene and the preceding Holocene Epoch. The appearance of stone tools about three million years ago, discussed in Chapter 17, marks the first geological record of human intelligence and we could define the start of the Anthropocene there, but apart from this important archeological evidence, this advance in human behavior is not obviously associated with any other measurable widespread geologic or biological changes. However, it has been hypothesized that humans may have played a role in the disappearance of woolly mammoths and other megafauna in the Americas at the start of the Holocene. In more recent time, there is abundant documentation of human-caused extinctions within the last few hundred years, such as the disappearance of the dodo and the passenger pigeon. Such changes in the biological record have led some to advocate that this should mark the beginning of the Anthropocene. Others point out the correlation between agriculture and the climate change of the Medieval Warm Period and ensuing Little Ice Age, and suggest this time as the boundary between the Holocene and Anthropocene.

Since the first nuclear tests in 1945 (Figure 19.2), detectable radioactive fallout has been found in sediment layers around the planet, and this has been used to date these young deposits. Several specific markers (e.g., radioactive fallout, plastics, and heavy metals concentrated by mining) may be detectable in sediments, and the appearance of each of these have been separately proposed as a candidate for geological markers that could be used to define the beginning of the Anthropocene. As yet there is no global consensus on the start date of this newly proposed epoch, and it has not yet been formally accepted by the various international committees that approve naming of new time divisions.

KEY POINT

There is some evidence that recent climate events, such as the Medieval Warm Period and the Little Ice Age, may have been partly influenced by humans. This has led to the proposal of a new epoch, the Anthropocene, suggesting that humans may leave a permanent mark in the geological record.

19.2 Earth's Surface Temperature

As we learned in Chapter 18, Earth's temperature varies with its orbital changes, and this causes climate to experience cycles of change over tens of thousands to hundreds of thousands of years. Because of the complex interaction of multiple orbital cycles, at times the Earth is hotter and at other times colder, and these Milankovitch cycles are a major control on recent glacial and **interglacial** periods. The Late Cretaceous and Paleocene–Eocene were both exceptionally hot times in Earth history. Temperatures were likely at least 6 °C warmer than today and may have been up to 14 °C higher. One of the interests in studying these warmer periods is to understand how Earth systems adapt and behave in much warmer conditions, especially given our current concerns about global warming.

Our present Holocene Epoch marks a relatively warm interglacial period, but as we saw in Chapter 18, the pace of Milankovitch cycles virtually guarantees that Earth will experience another ice age in the next 20,000 years or so. Even this time span, although short in geological terms, may feel like an eternity in the context of our own lifetimes. Therefore, most societal concerns with anthropogenic climate change are about what happens in the next 50 to 100 years, versus the inevitable cooling that will likely start to occur in the next 10,000.

As we introduced in Chapter 8, some of the Sun's energy is absorbed by Earth's surface, some is reflected out into space, and some is trapped in the atmosphere. The overall temperature of Earth's surface reflects the balance between incoming solar energy and outgoing heat radiated away from Earth, and the efficiency of the atmosphere in trapping some of the outgoing radiation.

The reflectance, or albedo, of the Earth (see Chapter 8) plays an important role in cooling the Earth, but from the atmospheric perspective, the presence of greenhouse gases is very important because they prevent outgoing radiation of heat to escape and thereby cause the atmosphere to warm up (Figure 19.3). Next, we will further discuss the overall energy budget of Earth, including the incoming solar radiation, which includes short-wavelength light and longer-wavelength heat, and the transfer of heat away from Earth.

19.2.1 Earth's Energy Balance

Because the energy reaching the Earth's surface is balanced by the energy leaving the planet, this helps the Earth to keep a fairly uniform average surface temperature over short periods of time such as days and years (Figure 19.3). As Earth's temperature increases, the outgoing radiation of heat from the Earth also increases. This limits runaway warming and helps maintain a relatively even temperature. However, long-term variations of incoming radiation, as caused by Milankovitch cycles, and changes in the concentration of atmospheric greenhouse gases can cause Earth's climate to shift to a new equilibrium, as observed in the many climate cycles that Earth has experienced over the ages. Given that Earth has experienced significant natural swings in its climate in the past, one could wonder why there is so much concern about climate change today. The concern and apprehension about these changes stems from the fact that the rate of climate change observed in recent decades (Figure 19.4) is much greater than inferred for past warm periods in the geologic record.

KEY POINT

Although Earth's surface temperature is naturally modulated, the average temperature of Earth's atmosphere has increased by 1 °C since the beginning of the industrial revolution. This change is much faster than previous warming events.

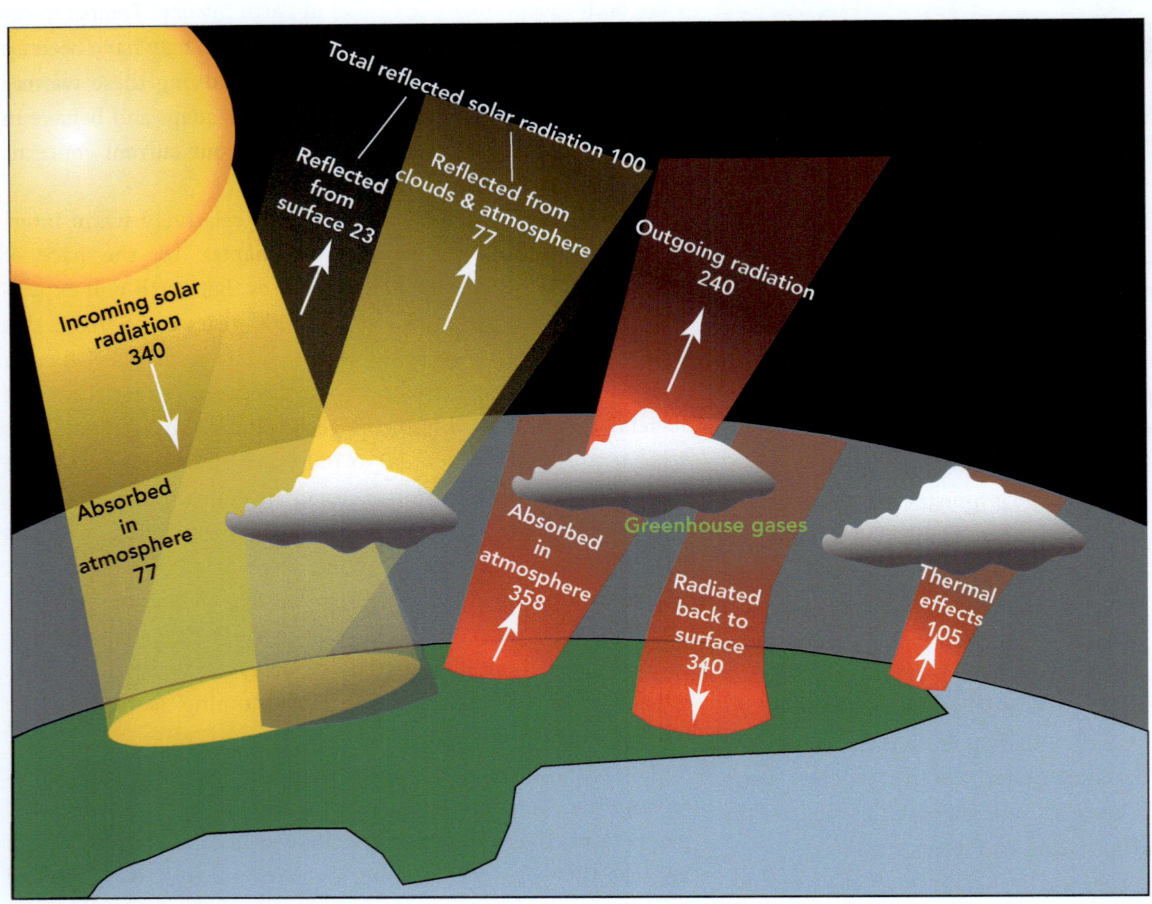

Figure 19.3 Earth's energy budget shows the balance between the incoming radiant energy from the Sun versus the energy that flows from Earth back out to space. The atmosphere absorbs, stores, and re-emits much of the radiation, keeping Earth's temperature in balance. Source: adapted after NASA, Public domain, via Wikimedia Commons.

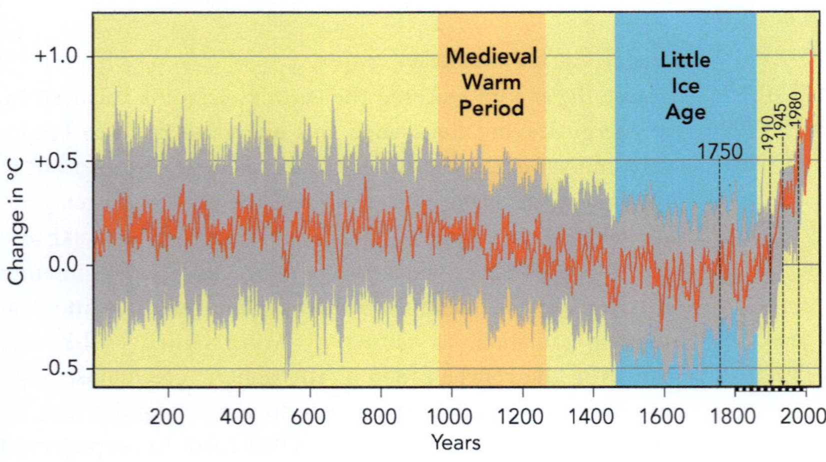

Figure 19.4 Global temperatures over the past 2,000 years. The industrial revolution began in 1750 in the latter part of the Little Ice Age. Source: adapted after NASA, Public domain, via Wikipedia Commons.

19.2.2 The Greenhouse Effect

In 1824 French mathematician and physicist Joseph Fourier first suggested that gases in the atmosphere may trap heat, preventing all the radiant heat energy from leaving the planet. In 1856, amateur American scientist Eunice Newton Foote, and three years later Irish scientist John Tyndall, independently measured the effect of various greenhouse gases and showed that CO_2 was one of the most important. It is doubtful that Tyndall was aware of Foote's contribution, as in her time, women were disallowed from public presentation and her work was presented by Joseph Henry who was the secretary of the Smithsonian Institute. The Swedish chemist Svante Arrhenius (1859–1927) followed up on this research and showed that cutting the concentration of CO_2 then present in the atmosphere by half could trigger another ice age, whereas doubling the concentration of CO_2 could cause warming of 5–6 °C. Arrhenius understood that a cooler atmosphere would hold less water vapor and this water vapor would accumulate as ice and snow at high latitudes and altitudes, increasing Earth's albedo. This would induce positive feedback, enhancing cooling and driving the Earth into another ice age. During Arrhenius' time, industrial production of CO_2 was low compared to today and he was far more concerned about the danger of global cooling than the threat from increasing temperatures. Arrhenius estimated that human-induced warming would likely occur over thousands of years but thought that the effects would be mostly beneficial. The important point here is that climate science and our understanding of the role greenhouse gases play in atmospheric warming is not new and was well established by the nineteenth century. Fourier's hypothesis pre-dates Darwin and represents science that is just as robust as anything else discussed in this book.

KEY POINT

The effect of greenhouse gases, CO_2 in particular, in absorbing thermal energy, was well established in the nineteenth century.

19.2.3 Atmospheric Composition

Our atmosphere is 78.0% nitrogen (N), 21.0% oxygen (O), and 0.9% argon (Ar). The remaining 0.1% is made up of other gases, including carbon dioxide (CO_2), methane (CH_4), water vapor (H_2O), ozone (O_3), nitrous oxides (NO_x), sulfur dioxide (SO_2), chlorofluorocarbons (CFCs), hydrofluorocarbons, and volatile organic compounds (VOCs). Also suspended in the atmosphere are small solid and liquid particles (e.g., fog, dust, and smoke) known as aerosols. The concentration of the major constituents, O, N, and Ar, has little effect on surface temperatures but the presence or absence of the trace components listed above has a disproportionate effect on the temperature of the atmosphere. In other words, although these gases are present in only trace amounts, they are very important to the temperature of – and therefore life on – Earth.

19.2.4 Greenhouse Gases and Their Effects

Greenhouse gases absorb and re-emit energy. Different gases absorb different frequencies of radiation (Figure 19.5) and block some of the outgoing **thermal emission** from the Earth, producing an insulating effect (Figure 19.3). Without this insulating effect, the average temperature of Earth would be much colder, about −15 °C, rather than the present average of +15 °C. The greenhouse gases thus act as an insulating blanket, providing about 30 °C of warming that prevents our world from becoming a permanent icehouse state. For example, the Moon, the same distance from the Sun as Earth but without an atmosphere or ocean, ranges from −183 °C at night to 106 °C in the day and has an average temperature of about −23 °C. Venus, in contrast, has an atmosphere that is over 96% CO_2, and an oven-like surface temperature of over 400 °C! Having an atmosphere, especially one rich in greenhouse gases, creates a highly stabilizing effect on planetary temperatures, but an excess of greenhouse gases, like on Venus, can make it very uncomfortable indeed.

KEY POINT

Greenhouse gases (CO_2, NH_4, H_2O, O_3) provide a thermal blanket around Earth that provides the planet with more than 30 °C warming of the atmosphere.

In addition to the differences in the wavelengths of heat that can be absorbed, the effectiveness and proportion of gases is also variable (Table 19.1). The effect of water vapor, which is also a greenhouse gas, is the strongest, but its distribution and proportion is highly variable, accounting for between 36 and 72% of heat retention in the atmosphere. Water vapor has two effects on climate. Atmospheric H_2O absorbs heat, disallowing the heat to escape to space, which has an insulating effect, but clouds also reflect radiation from the Sun, increasing planetary albedo and thus reducing the heating. Because CO_2 is more efficient in trapping heat than water, it has a significant effect on the temperature of the atmosphere, accounting for between 9 to 26% of heat retention, even though it is present in concentrations of just a few hundred parts per million (ppm). Methane (CH_4) is an even more powerful greenhouse gas but is present today in low enough concentrations such that it accounts for only 4–9% of heat retention. However, because CO_2 and CH_4 as well as other greenhouse gases including nitrous oxides (NOx), O_3, and CFCs are so efficient at trapping heat in the atmosphere, even modest additions to their concentrations can produce significant increases in temperature.

The time a gas molecule remains in the atmosphere before being removed (to rocks, soil, plants, or the ocean) is called the **residence time**. The residence time of greenhouse gases is highly variable and one of the most important factors in controlling the temperature of the atmosphere (Table 19.1). Water vapor has the shortest residence time, averaging about 11 days.

Table 19.1 Greenhouse gas effects

Compounds	Formula	Concentration (ppm)	Residence time (years)	Contribution to heat trapping (%)
Water vapor and clouds	H_2O	10–50,000	0.03	36–72%
Carbon dioxide	CO_2	~410	100	9–26%
Methane	CH_4	~1.8	12	4–9%
Ozone	O_3	2–8	0.3	3–7%
Nitrous oxide	N_2O	0.1	0.02	5–10%

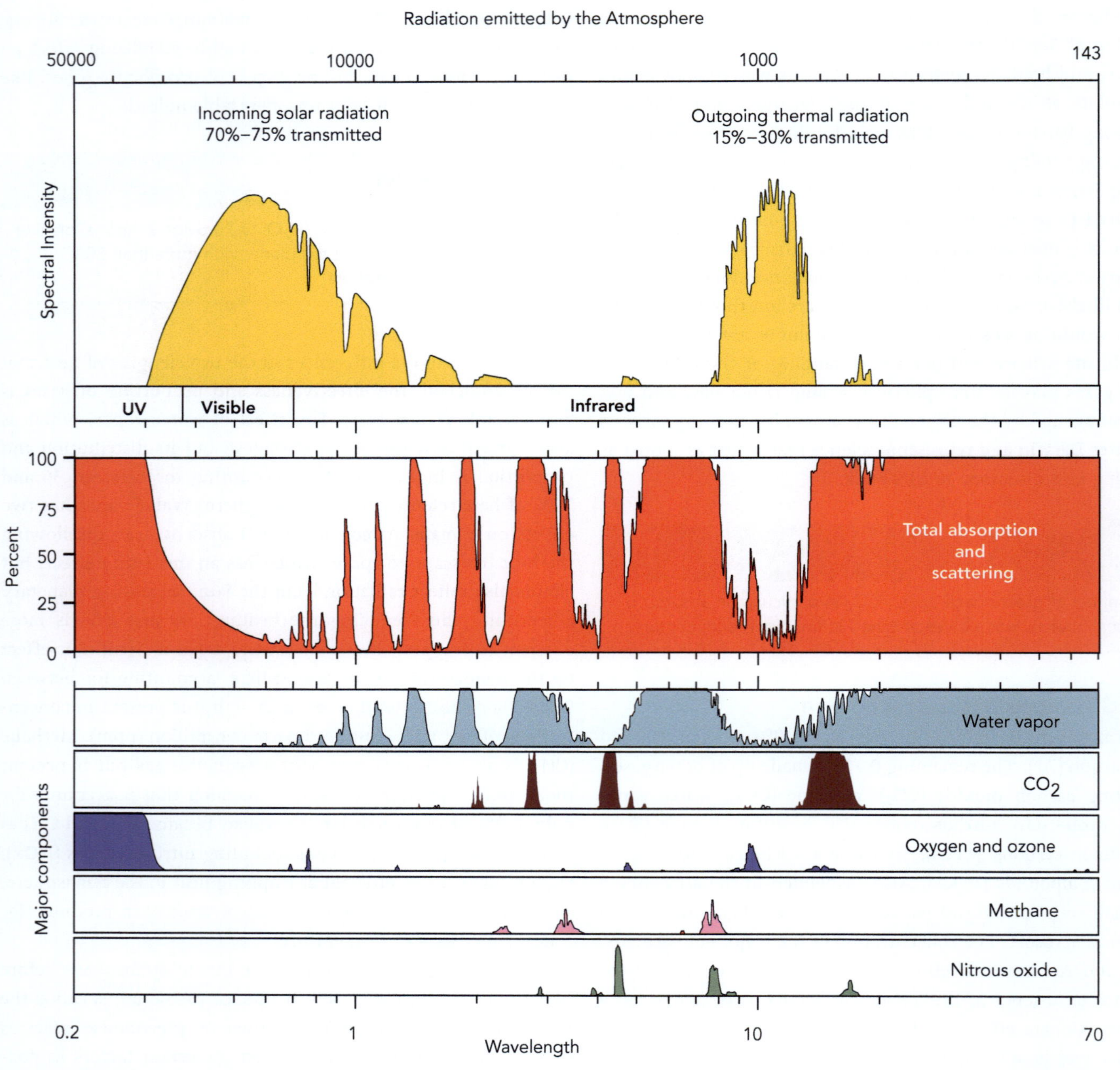

Figure 19.5 Absorption and scattering spectra of the greenhouse gases compared to incoming solar radiation versus outgoing thermal emission from the Earth. Most of the outgoing radiation is within a narrow band of wavelengths not absorbed or scattered by the atmosphere. Source: Figure modified from R. A. Rohde, who created this image for the Global Warming Art project. https://commons.wikimedia.org/wiki/File:Atmospheric_Transmission.png.

You have all seen clouds form overhead, marking a build-up of atmospheric water vapor, which then falls back to Earth's surface as precipitation. This water may then be evaporated back into the atmosphere. However, although water vapor has an amplifying effect on warming, it tends to follow warming trends rather than cause them. As the atmosphere becomes hotter it can hold much more water, and as Earth becomes hotter, evaporation can be more aggressive, putting more water into the atmosphere.

CO_2, in contrast, can remain in the atmosphere from about 300 to 1,000 years. As we discussed in Chapter 8, CO_2 can be removed by weathering of silicate rocks, photosynthesis, and storage in sediment, either as carbonate or organic carbon. Photosynthetic storage is usually good for the lifetime of the plants or algae that store it unless this biological carbon is stored in soil pools or buried and preserved in sediment. Most of the biological carbon is usually quickly released back to the atmosphere by bacterial breakdown of organic matter or through oxidation, such as occurs during forest fires. Only a small fraction of this organic carbon remains in soil for long periods of time. Of course, some trees can live for hundreds of years – similar to the residence time of CO_2 – which is why planting trees to slow deforestation can be important. Organically locked CO_2 can be stored for much longer periods in sediment and sedimentary rocks. Therefore, the rate at which organic CO_2 is buried can make big changes in atmospheric levels. CO_2 can be stored in carbonate rocks for billions of years, but chemical weathering at the surface and metamorphism of carbonates can also release some of this over geologic timescales. The fossil fuels, coal, oil, and gas, contain organic material that has been stored, in many cases, for millions of years or longer. You will recall in Chapter 11 that a massive increase in accumulation of coal deposits in the Carboniferous Period triggered the Late Paleozoic Ice Age, whereas the release of some of the CO_2 stored in that coal by later volcanism in Siberia caused a dramatic warming that triggered the end-Permian extinction event. The industrial revolution was initially fueled by burning coal, later supplemented by oil and natural gas. So, in the past two hundred years, through the burning of fossil fuels, significant amounts of CO_2 have been released into the atmosphere that had been stored in the subsurface for millions of years or more.

KEY POINT

The average time that CO_2 remains in the atmosphere ranges from 300 to 1,000 years; the H_2O atmospheric residence time is only about 11 days. The higher residence time of CO_2 means it has a much-longer-term effect on climate.

Methane has a residence time of about 12 years. Although it is a strong greenhouse gas, its shorter residence time and lower concentration means that it currently plays only a modest role in longer-term climate change. However, as we saw in Chapter 3, early Earth's atmosphere was richer in CO_2 and CH_4, and lacked free oxygen. Higher CO_2 and CH_4 levels were important in preventing Earth's early oceans from freezing, despite a fainter Sun. However, as we discussed in Chapter 7, the rise of photosynthesis drove the Great Oxygenation Event that both oxidized methane and consumed CO_2, reducing their effectiveness in maintaining a warm Earth. This, combined with the dramatic increase in precipitation of carbonate, along with a fainter Sun, helped trigger the Huronian glaciations and the later snowball conditions in the Cryogenian Period, covered in Chapter 8. These examples show that there is strong evidence that changes in the concentration of greenhouse gases over geologic time have had major effects on global climate and have contributed to several of the glaciation and deglaciation stories that we have told so far.

19.2.4.1 Natural Versus Anthropogenic Sources of Greenhouse Gases

In Chapter 11 we noted that volcanoes currently release about 250 million tonnes of CO_2 per year. In contrast, humans emit about 37 billion tonnes of CO_2 per year, and another 13 billion tonnes of CO_2 equivalents in methane and other greenhouse gases. This is 200 times greater than greenhouse gases produced by natural sources. Recall in Chapter 8 that during the snowball Earth periods the typically low rate of volcanic emission meant that it took many millions of years for CO_2 to build up to the levels required for Earth to warm back up again and escape the snowball state. So where do all these anthropogenic greenhouse gases come from?

Methane is a significant byproduct of agriculture, especially from livestock and rice cultivation. Methane is also a common component of hydrocarbon accumulations, mostly trapped deep below the surface of the Earth, although in a few places it naturally escapes in gas seeps. Methane is also found trapped in a frozen form as **gas hydrates** in **permafrost** at high latitudes, such as northern Canada and Siberia, as well as within the sediment deposited on continental shelves of most oceans. Most of this frozen methane originated by bacteria that broke down organic matter hosted in the buried sediment. As the methane rose it froze (recall that temperatures below the surface increase with depth and consequently decrease upwards), at the point where the P–T conditions are ideal the gas combines with water to form a crystalline hydrate. There is evidence that falls in sea level and Arctic warming could cause this methane hydrate to be released as a gas and this may have happened in the past and helped trigger rapid warming events.

CO_2 and methane are also added to the atmosphere because of exploiting fossil fuels. CO_2 is a byproduct of the combustion of fossil fuels and methane is emitted mostly as leakage from gas wells and other sources. A common but now largely banned practice in the hydrocarbon industry was the **flaring** (i.e., burning) of natural gas at the well, because methane is less valuable than oil and sometimes not worth the trouble of storing or taking to market.

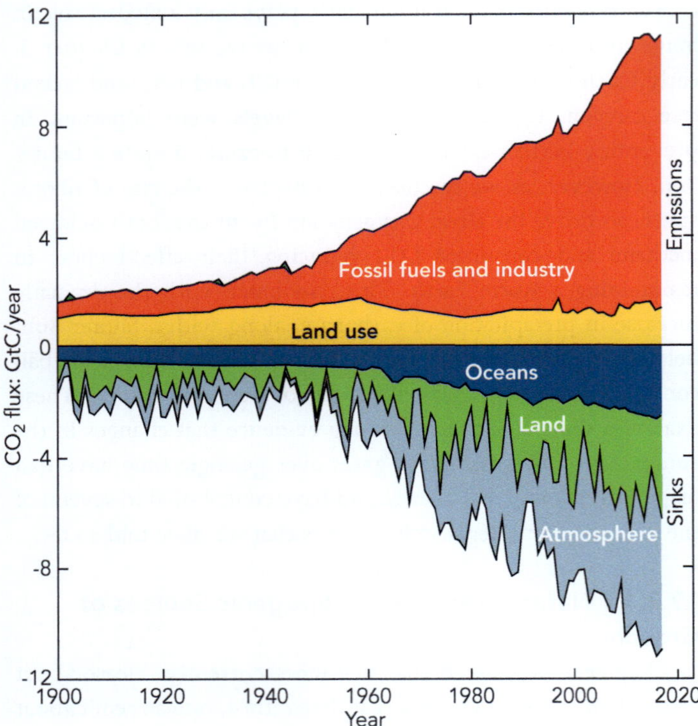

Figure 19.6 Carbon emissions and sinks since 1900. Source: Figure modified from Le Quéré et al. (2018). © Author(s) 2018.

Burning coal produces about twice as much CO_2 as burning the energy equivalent of methane. Of the 37 billion tons of CO_2 emitted annually (Figure 19.6), Asia emits about 53% and Europe and the USA account for about 45%. About 76–87% of all global CO_2 emissions are from fossil fuel use, 9% from agriculture and the remainder from manufacturing such as the cement industry. About 42% of fossil fuels are used for generation of electricity, about 29% in transportation, and the remainder in industrial and residential uses.

19.3 Observations of Change

It is now well documented that concentration of CO_2 in the atmosphere has increased by about 41% since the beginning of the industrial revolution, with the first noticeable increase in about 1750 (Figure 19.7). Evidence for this comes from two sources. The first comes from cores of glacial ice obtained from Antarctica. When snow falls, the snow deposit or layer is full of air. As more snow falls, the ice crystals are compacted but not all the air is expelled. Eventually, these layers of snow and air become glacial ice, and in Antarctica and Greenland there are parts of glaciers that have a record of continuous ice accumulation extending back as far as 800,000 years. These ice deposits can be precisely dated simply by counting the layers, with thicker parts for winter and thin parts for summer. The air trapped in bubbles between the ice crystals is a preserved sample of the atmosphere at the time the snow fell, and these bubbles have been analyzed so that we know the past composition of the atmosphere. Figure 19.7 shows the results for the past 2,000 years. The concentration of CO_2 in these bubbles was stable at about 275 ppm until about 1750. From 1800 to 1950 the concentration increased by about 20% to around 325 ppm. After 1950, our record of CO_2 in the atmosphere is from direct measurement, most notably at the Mauna Loa observatory in Hawaii, chosen for being far away from industrial sources of CO_2. This shows that since 1950, CO_2 has increased by another 25% to about 422 ppm as of 2024. It is increasing by about 2 to 3 ppm per year. This value is higher than at any time in the past two million years. The concentration of

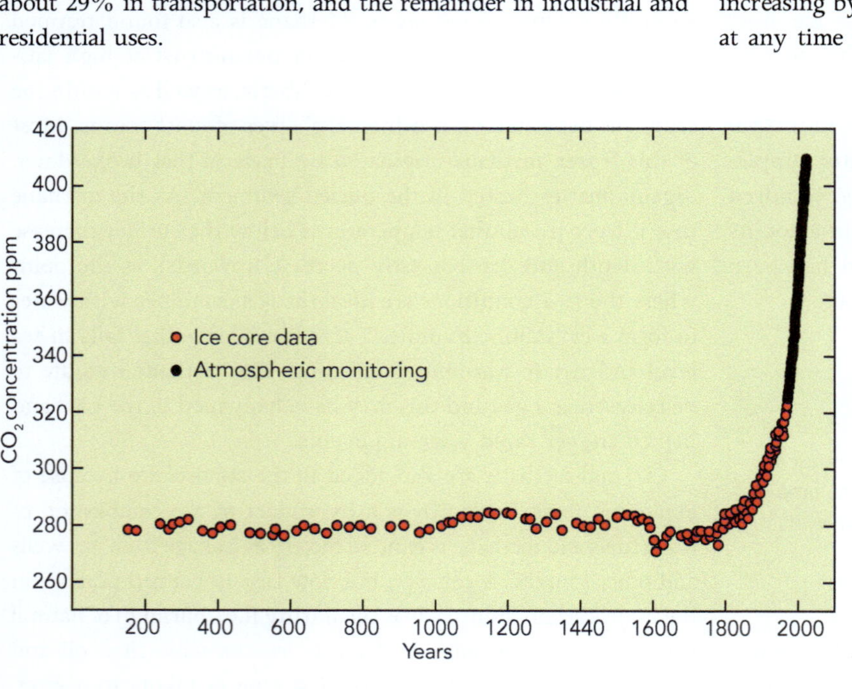

Figure 19.7 CO_2 concentrations for the last 2,000 years. Red dots are data from samples of ice bubbles in ice cores; black dots are data from direct atmospheric monitoring at Mauna Loa in Hawaii. Source: Figure modified from Le Quéré et al. (2018). © Author(s) 2018.

methane has also increased in the past two centuries and this change in CH_4 has an effect that is equivalent to an additional 40 ppm of CO_2.

KEY POINT

We can understand the composition of the atmosphere from many thousands of years ago by sampling air bubbles trapped in glaciers. These show that atmospheric CO_2 has nearly doubled since the beginning of the industrial revolution.

The correlation between CO_2 and average surface temperature is apparent in Figure 19.8, where the correspondence between these two parameters is striking. The temperature record is noisier than the CO_2 changes, but this is expected because surface temperature is dependent on other things. However, the similarity of the two records is consistent with the idea that CO_2 is a strong greenhouse gas. Moreover, the noise is reduced if we consider just the average. The past four decades have had an average surface temperature warmer than any decade since 1859, the year John Tyndall measured the greenhouse gas effects. Since 1880, the average surface temperature has increased more than $1.2\,°C$ (Figures 19.8), with most of that change occurring since 1960.

Ice cores from the Antarctic ice sheets show that CO_2 concentrations correlate to temperature (Figure 19.8). Long-term temperature changes are largely driven by the Milankovitch orbital cycles, as discussed in Chapter 18. In this case, Milankovitch-induced temperature changes of -8 to $0\,°C$ cause relatively small changes in CO_2 of 200–280 ppm with a lag of about 100 years, and this records the interactions in the carbon cycle discussed in Chapter 8.

In addition to the observation that the temperature of Earth's surface has increased in the past century, other changes have been observed that are the expected consequences of rising temperature. In winter, the sea ice in the Arctic Ocean grows and in summer it melts. Satellite observation of the extent of this ice shows that since 1979 the area of ice in March has decreased by about 10% and the area of ice in September has decreased by about 45% (Figures 19.9 and 19.10). The thickness of the ice is also decreasing. If this trend continues, by the mid 2050s there will be no ice in the Arctic Ocean at the end of summer.

The melting of sea ice will not affect global sea level because the ice is already in the ocean. However, the same warming that is melting ice in the ocean is having the same effect on glaciers on land. Since 1920, the area of land covered by snow in the northern hemisphere has decreased by more than 10%. Since 1975, the temperature of the permafrost in Alaska has risen from $1.5\,°C$ below freezing to $0.5\,°C$ *above* freezing. The warming rate is much higher in northern latitudes. Since 1990, the mass of ice in the Antarctic and Greenland ice sheets have decreased by about 7 billion tons. This meltwater will find its way to the sea thus raising the level of water in the ocean. Observations at several locations indicate that average sea level has increased about a quarter of a meter since 1880, with half of that increase occurring since 1975 (Figure 19.9). A significant proportion of this increase in sea level is the result of **steric** effects. As seawater temperature increases, the density of the water decreases and therefore takes up more space, causing expansion of water. During this time there has also been a 400% increase in freshwater storage on land

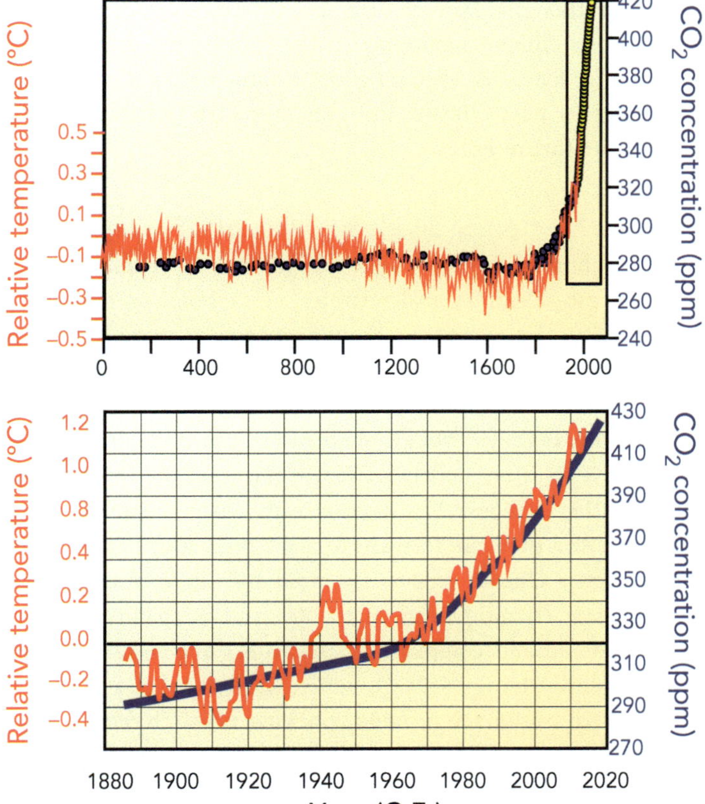

Figure 19.8 The evolution of atmospheric CO_2 concentration and average surface temperature change since 1880. Temperature change is measured relative to the temperature in 1965. Source: Original source from NOAA, redrafted from original by the author.

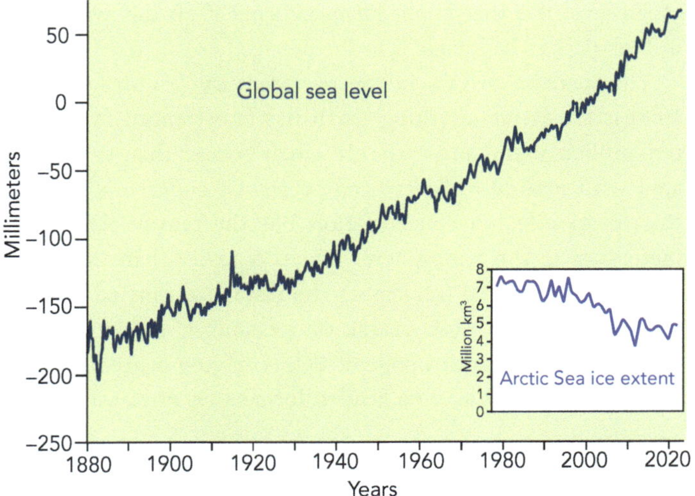

Figure 19.9 Change in global sea level compared to average, 1993–2008. Inset shows decrease in the extent of Arctic Sea ice, 1979–2022. Source: Original source from NOAA, redrafted from original by the author.

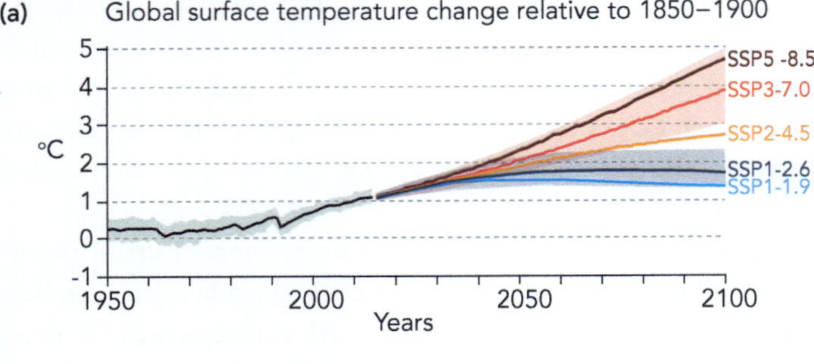

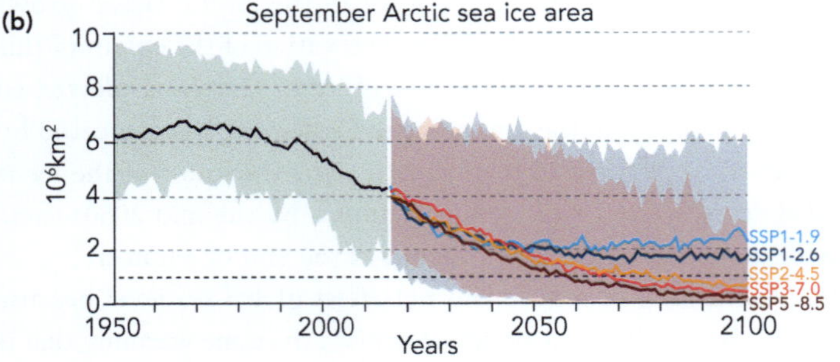

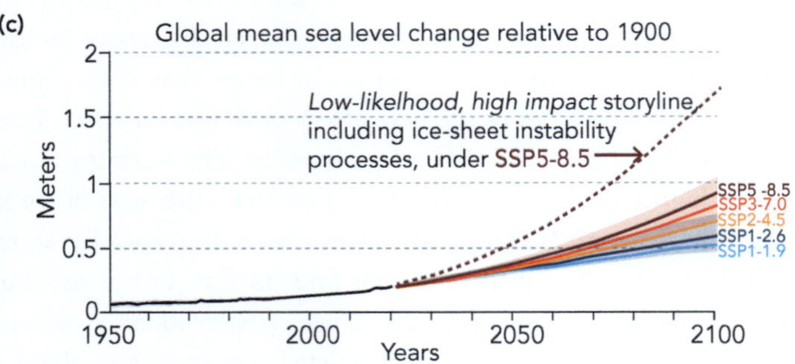

Figure 19.10 Predictions for (a) global average temperature, (b) area of sea ice in the Arctic Ocean in September, and (c) global mean sea level for a series of models based on different input assumptions, called shared socioeconomic pathways, described in Table 19.2. Source: Climate Change 2021: The Physical Science Basis. Working Group I Contribution to the IPCC Sixth Assessment Report.

because of building of dams, and this has mitigated the full effect of sea-level rise that might have occurred if all this water from melted glaciers had made it to the oceans.

The increase in CO_2 in the past century (Figures 19.7 and 19.8) is far above anything Earth has experienced in the past two million years, and certainly much greater than that associated with natural Milankovitch cycles. Of concern is not only this unprecedented concentration but the rate at which it is increasing in the atmosphere, which is faster than the rate at which it is removed (remember the residence time of CO_2 may range from 300 to 1,000 years). Understanding the implications of this increase in anthropogenic CO_2 is driving many to ring the alarm bell that we may be headed for a series of catastrophes.

KEY POINT

Increasing CO_2 levels are causing global warming, sea-level rise, and other climate effects such as flooding and increased intensity of storms.

19.4 Modeling Global Change

Despite our focus in this book on understanding Earth history (i.e., the past), many geoscientists are today involved in prediction of things yet to come, including estimating the risk to the built environment from earthquakes, floods, and fires and the risk to society in general from expected changes in the global climate pattern. Commonly, predictions use some form of model. Modeling the future usually begins by comparing it to the past and the better we understand the past, the greater our confidence that we will be able to anticipate the future.

19.4.1 Climate Versus Weather

Weather refers to meteorological conditions, such as temperature, humidity, precipitation, and wind patterns, of specific locations over timescales from hours to several days, whereas climate refers to averages of these conditions over longer periods of months, years to thousands of years or longer.

Chaotic systems such as weather and climate are inherently difficult to predict, and the further into the future, the more difficult become the predictions. The extreme sensitivity that long-term weather forecasting is subject to is sometimes called the **butterfly effect**, referring to the idea that when a butterfly flaps its wings in Tokyo, the turbulence generated by this small activity may be amplified such that the weather in Denver can be different in two weeks. However, climate modeling, which averages variables over time and space, is becoming better and better at describing changes in past climate and can therefore help us consider what the future holds.

19.4.2 Numerical Models of Chaotic Systems

Although they involve some simplifications, climate models allow researchers to make future predictions or to test the sensitivity of a system to changes in one or more parameters without the noise inherent in natural systems. Climate models, for example, test how changes in CO_2 affect average global temperatures, even though temperature can be highly variable locally. Typically, models are first run to see how well they can mimic the changes that have already occurred over several decades and then run forward to make predictions about future conditions. Climate models are used by universities, government agencies, and private companies, and much of these results are gathered and compiled by the **Intergovernmental Panel on Climate Change (IPCC)**, which was established in 1988 by the United Nations to provide policymakers with regular scientific assessments on the current state of knowledge about climate change.

The IPCC issued its most recent report in 2023 (IPCC, 2023), with over 100 authors from dozens of countries contributing, describing the state of our knowledge of the past and present state of global climate, along with the results of models

Table 19.2 SSPs used in IPCC models for predicting future climate change.

Scenario	Description
SSP1–1.9	*Most optimistic.* This describes a world where global CO_2 emissions are cut to net zero around 2050. Societies switch to more sustainable practices, with focus shifting from economic growth to overall well-being. Investments in education and health go up. Inequality falls.
SSP1–2.6	*Modestly optimistic.* Global CO_2 emissions are cut severely, but not as fast, reaching net-zero after 2050. It imagines the same socioeconomic shifts towards sustainability as SSP1–1.9.
SSP2–4.5	*"Middle of the road."* CO_2 emissions maintain current levels before starting to fall mid-century, but do not reach net-zero by 2100. Socioeconomic factors follow their historic trends. Progress toward sustainability is slow, with development and income growing unevenly.
SSP3–7.0	*Pessimistic.* Emissions and temperatures rise steadily and CO_2 emissions roughly double from current levels by 2100. Countries become more competitive with one another, shifting toward national security, and ensuring their own food supplies.
SSP5–8.5	*Very pessimistic.* Current CO_2 emission levels roughly double by 2050. The global economy grows quickly, but this growth is fueled by exploiting fossil fuels and energy-intensive lifestyles.

Source: Modified from *US News and World Report*, August 9, 2021.

describing various future changes assuming different inputs. The report broadly concludes that:

> Human activities, principally through emissions of greenhouse gases, have unequivocally caused global warming, with global surface temperature reaching 1.1 °C above 1850–1900 in 2011–2020. Global greenhouse gas emissions have continued to increase, with unequal historical and ongoing contributions arising from unsustainable energy use, land use and land-use change, lifestyles and patterns of consumption and production across regions, between and within countries, and among individuals.

The report also describes the results of climate modeling to predict climate changes for the coming century. Inputs to such models are complicated, so the IPCC incorporates a variety of scientific and social factors into individual models that they call **shared socioeconomic pathways** (SSPs). IPCC reports use several different SSPs to model different scenarios for changes in many features of global climate (Table 19.2). These range from the most optimistic, in which CO_2 emissions are cut to net zero by 2050, with societies shifting from placing an emphasis on economic growth towards overall well-being of citizens, to the most pessimistic, where CO_2 output doubles by 2050 and social conditions change very little. Modeling a wide number of different scenarios should allow us to make decisions as to the risks of taking no or little action (pessimistic scenario) versus more aggressive measures at CO_2 mitigation in the more optimistic models.

Some of the outputs of the models based on these different SSP inputs are shown in Figure 19.10. Each model (one for each SSP) has an uncertainty associated with the output, that is, the predicted condition of the future, determined by changing the inputs and observing the sensitivity to the produced outputs. The results in Figure 19.10 shows a most likely central estimate along with a bracket of possible but less likely results for each model.

In these models, a range of global temperature increases of 1.5 to 5.5 °C are predicted, with "middle of the road" models at about 3 °C. The "middle of the road" model predicts practically no sea ice in the Arctic Ocean by 2060 and a global sea-level rise of more than half a meter by 2100 (Figure 19.10). Worst-case models predict a sea-level rise of over a meter.

Tropical cyclones (hurricanes in the Atlantic and typhoons in the Pacific) are predicted to become more intense and hence potentially devastating because their intensity is related to the surface temperature of the ocean. As global temperatures increase so does ocean-water temperature, which promotes stronger cyclones. Although an increase in intensity can be predicted, with varying effects depending on the scenario, it is much harder to predict the magnitude of the increase. Hurricanes range in wind intensity from category 1 to 5 with about a 10% increase between each category. As with sea-level rise, those areas close to sea level, such as the Gulf of Mexico and eastern seaboard of North America, are predicted to experience far more damaging storms with increased global warming.

A rise in the mean level of the oceans of 0.5 m only describes the average water height, not the level at high tide or during storms, which of course could be much higher (depending on the local relief). Therefore, a rise in mean sea level of 0.5 m could mean that many lowland places where hundreds of millions of people now live will become uninhabitable, such as many parts of Florida and many island nations across the world.

Figure 19.11 (a) Healthy Great Barrier Reef, Australia versus (b) bleached Great Barrier Reef, Australia. Source: (a) AndriiSlonchak via iStock / Getty Images Plus; (b) Brett Monroe Garner via Moment / Getty Images.

In addition, increasing temperature increases the humidity in the atmosphere but due to the short residence time of water vapor, virtually guarantees a more vigorous hydrological cycle, with increased and more frequent rainfall in some places. Increased rainfall can also result in a greater likelihood of flooding, as has been seen throughout the world in recent years. This increase in atmospheric water produces *atmospheric rivers*, air masses enriched in water vapor that cause catastrophic rainfall and flooding. However, over much longer geological timescales, increasingly vigorous hydrological cycles can enhance the rate of silicate weathering and consequently CO_2 removal, as was hypothesized to have occurred during the Neoproterozoic, where it was one of the factors that helped drive Earth into a snowball state, as discussed in Chapter 8.

With the global average temperature increasing, the likelihood of prolonged heat waves is also greater. Long spells of temperatures over 100 °F (38 °C) are often accompanied by sparse rainfall, leading to an increased likelihood of forest fires. Bushfires in Australia burned over 180,000 km^2 in 2019–2020, and wildfires in the United States and Canada in 2023 occurred at unprecedented rates and are predicted to become worse in the future.

As CO_2 increases, more is absorbed in the oceans, resulting in acidification. As we saw in Chapter 11, massive increases in CO_2, mostly from the volatilization of coal deposits through volcanism in Siberia, led to catastrophic global warming at the end of the Permian Period. Oceanic acidification caused seawater to become toxic and triggered a mass extinction. Over the past few hundred years the world's oceans have become more acidic. Effects of this acidification include the death of many coral communities (Figure 19.11), and it has been estimated that there has been about 80% loss of coral reefs globally in the past 100 years. Loss of coral obviously affects organisms that depend on reefs, such as reef fish, and their predators, indicating that there is likely to be a collapse of the coral-reef food chain, and this of course could affect marine life more generally. Reefs are particularly sensitive to very small changes in the temperature and chemistry of the oceans and provide a good example that we have likely reached a **tipping point**, where crossing a key threshold causes a quick and catastrophic change.

Polar bears feed mostly on seals that swim in the ocean. They hunt seals by riding on sea ice far from the shore (Figure 19.12)

Figure 19.12 A hungry polar bear hunting for seals on a raft of sea ice. Source: Andreas Weith, File:Endangered_arctic_-_starving_polar_bear.jpg, via Wikimedia Commons.

and must use the ice as rafts because they cannot swim far enough to hunt seals reliably. Polar bears can fast up to 180 days between feeding on seals, but with declining sea ice in the Arctic Ocean (Figure 19.9) hunting seals will become more and more difficult and fasting times will increase for many individuals – to the point of starvation. If these conditions are sustained, this could lead to their extinction. Even the most optimistic models of Arctic Sea ice area over the next few decades (Figure 19.10) foretell a dire future for polar bears.

Depending on the scenario, changes in climate could have a profound effect on agriculture. For example, we are already seeing that harvest times for grapes in Australia and France are now a month earlier than in the early twentieth century, but many other key crops are under pressure. Wheat requires a particular range of temperature to mature to a bountiful crop. The state of Kansas produces as much wheat as almost all the other US states combined. The kind of wheat farmed in Kansas is used to make bread and is called hard red winter wheat because it is planted in November and harvested in the late spring. This crop does best when temperatures are below 75 °F (24 °C), which is typical for western Kansas in the spring. However, climate modeling predicts that the number of days with maximum temperatures above 90 °F (32 °C) rising from about 75 per year in 2000 to 120 (middle of the road) or 150 (pessimistic) per year by the end of century. This would mean much shorter springs in western Kansas, making the winter wheat crop no longer viable because warmer temperatures will result in lower crop yields. Such a change would have significant effects on the economy of Kansas but also on the worldwide food supply. Disruptions in many other aspects of agriculture are likely in a world that warms as predicted by the models in Figure 19.10.

We do not know what the tipping points are for many of the areas of specific concern with respect to climate change, such as the point at which substantial melting of an ice sheet would be triggered, or the point at which we pass from a "background" extinction rate to a mass extinction. The uncertainties are still too high for us to make confident predictions, but we can make evidence-based estimates and, perhaps most importantly, assign probabilities to the risk, following which we can weigh the consequences of inaction versus the cost of mitigation of the risk.

KEY POINT

Climate models are useful for predicting a range of possible changes in climate and other environmental effects that depend on the choices and policies enacted regarding mitigation.

19.5 The Sixth Extinction?

The background extinction rate, also known as the normal extinction rate, refers to the standard rate of extinction in Earth's geological and biological history before humans became a primary contributor to extinctions. In the last few hundred years we have directly caused extinction of many animals, such as the dodo in Mauritius, the great auk on the Funk Islands off the coast of Newfoundland, Canada, and the passenger pigeon (Figure 19.13), which was endemic to the North American continent, and which experienced severe habitat loss due to deforestation and was eventually hunted to death by 1900.

Going back a little further in time, it is now known that humans first entered North America around 23,000 years ago across the land bridge that connected Asia to Alaska, which existed because of lowered sea levels during times of widespread continental glaciation (see Figure 17.19). It is certainly clear from the fossil record that prior to human migration, North America boasted a rich megafauna. This included three species of American elephant, including mastodons and mammoths, large beavers, five species of horses, the North American camel, giant ground sloths, giant armadillos, saber-toothed cats, the dire wolf, short-faced bears, and many others (Figure 19.14). It is thought that hunting by humans may have been a factor in the megafauna extinctions, but there were also some significant Late Pleistocene climate changes, such as the Late Pleistocene **younger Dryas** cooling period, prior to the Holocene warming, that involved significant environmental change.

There is increasing concern among biologists that we may be at the beginning of a sixth, human-induced mass extinction. Recent studies suggest that the number of extinction events for fish, amphibians, reptiles, birds, mammals, and other vertebrates since the beginning of the twentieth century would have taken between 2,000 and 10,000 years at rates of extinction found before the nineteenth century. In addition to outright extinction, many species are seeing massive declines in populations as habitats become more restricted. The big five extinctions were marked by at least 75% extinction of species on Earth. Although none of the recent modeling studies show that we are at that level yet, they do suggest that, unless action is taken, we could reach that level of extinction in as few as 500 years.

Examination of the geological record shows that life on Earth recovered after each of the big five extinctions, indicating that over the long-term life on Earth is indeed resilient. However, the evidence from the rock record also shows that it takes tens of thousands to millions of years for recovery. As with natural climate cycles, these times are an eternity in comparison to our own lifespans, and it is hard to imagine how or if humanity could cope through such a crisis. Although the sixth extinction is not yet a certainty, there is evidence that if the rate of species extinctions and population declines continue at the present rate there is a distinct possibility of catastrophe, even if we can't yet assign the probability.

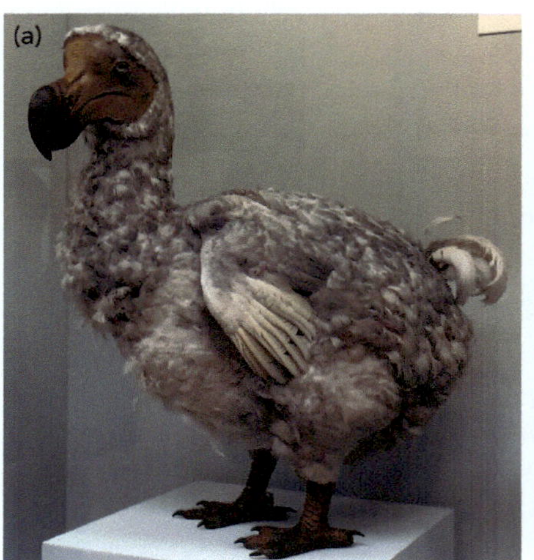

Figure 19.13 Extinct birds: (a) dodo and (b) passenger pigeon. Source: Photos by JPB. By permission of The Trustees of the Natural History Museum, London.

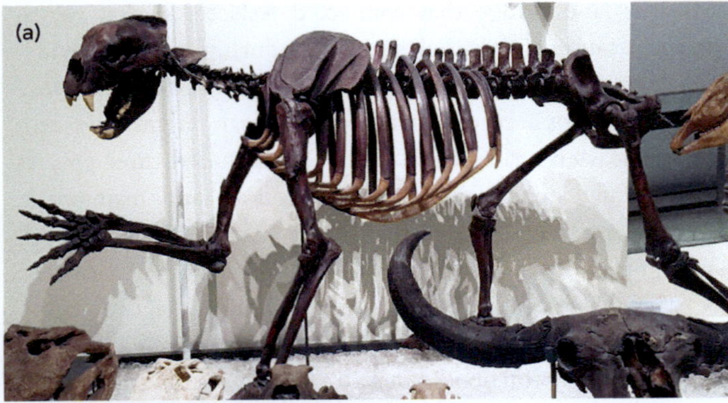

Figure 19.14 Extinct Pleistocene mammals: (a) short-faced bear; (b) giant ground sloth; (c) saber-toothed cat. Source: (a) Photo by JPB. With permission of ROM (Royal Ontario Museum), Toronto, Canada; (b) Photo by JPB. By permission of The Trustees of the Natural History Museum, London; (c) Bone Clones, CC BY-SA 3.0 <https://creativecommons.org/licenses/by-sa/3.0>, via Wikimedia Commons.

KEY POINT

Although the sixth extinction is not yet a certainty, the current stresses on life on Earth suggest a possibility of catastrophe.

19.6 Global Change in a Deep Time Perspective

If there is any rule in Earth history, it is that change is inevitable. Most of the chapters in this book discuss the changes that Earth has experienced. We began this discussion in Chapter 2 with a review of the debates regarding the pace of change. Although the idea that most geological and evolutionary processes are gradual in comparison to human lifespans, we now know that occasionally more catastrophic changes occur that are too fast for species to adapt to, resulting in the mass extinctions that were the focus of Chapters 11 and 13.

In terms of the current concern regarding anthropogenic climate change, climate models provide estimates of the range of probabilities and risks of outcomes that may occur depending on the decisions that we make in the coming decades. We can now begin to compare rates of anthropogenic CO_2 emissions against historical and deep time rates. There is some evidence that current rates may be on the same order as those that caused the greatest mass extinction of all time at the end of the Permian Period, and this should give pause for serious concern.

Since 1750 (i.e., less than 300 years ago), humans have emitted about 1.5 trillion tons of CO_2. This averages about 5.5 billion tons per year, with emissions during the twentieth century averaging four to five times that value. Volcanoes naturally emit about 200 million tons per year, or less than 1% of the current rate of CO_2 emissions by humans. As we saw in Chapter 11, the Siberian trap magmas caused the release of about 5.4 trillion tons of CO_2 over about 30,000 years (not counting all other volcanoes at the time). If the release was at a steady rate, which is highly unlikely, yearly rates would be about 180 million tons of CO_2, not much different than that delivered from volcanoes today. However, it is far more likely that eruptions were pulsed. If we assume that eruptions occurred over a few thousand of those 30,000 years, the Late Permian rates (which seemed to have led to the greatest mass extinction ever seen on Earth) might be comparable to modern anthropogenic rates. This is one of the reasons that there has been so much focus on trying to understand the causes of previous global mass extinctions to help understand the consequences of rapid global change today.

Faced with similar environmental concerns in the recent past, introduction of environmental regulations, such as the 1990 Clean Air Act in the USA, which included cap and trade provisions, has reversed the imminent threat of freshwater acidification from industrial sulfate pollutants. Banning of DDT, a deadly anti-malarial insecticide that was widely used after World War II, reversed the near extinction of bald eagles, brown pelicans, and other birds in the early 1970s. The 1989 ban of ozone-destroying chlorofluorocarbons also avoided a potential atmospheric catastrophe. In 1966, humpback whales numbered around 5,000 and were critically endangered. Banning whale hunting has allowed the humpback whale population to expand to around 80,000 today and they are now designated to be of least concern, a truly remarkable comeback that illustrates the power of effective international cooperation.

KEY POINT

Through scientific analysis and planning, humans have managed to avert some global catastrophes such as the threat of acid rain, destruction of the ozone layer, and extinction of many endangered species.

There seems little doubt in the scientific community that global warming is enhancing the worst effects of events such as floods, storms, wildfires, and sea-level rise. Of course, if we completely shut down all anthropogenic CO_2 emissions, we will not stop hurricanes, floods, or wildfires. Also, even when alternatives are found we will still need vast amounts of energy, agriculture, and technology to sustain what will soon be over eight billion humans that rely on our planet's resources. It is also not just about climate: eight billion humans require land to live on, potable water, and electricity, and these continue to be rare for the nearly one billion humans that live in developing countries on less than $2 a day. Providing sustainable solutions to human energy needs, land and water use, and the continuing goal of lifting the less fortunate out of extreme poverty are challenges that face humanity and must be considered as we balance the environmental versus human costs.

How confident can we be about climate models? In Chapter 2 we emphasized that scientists are in the testing business, not the proving business. In that regard, climate science is no different than all the other sciences. In this chapter we have made the case that climate science has been around since the nineteenth century and is as robust as many of the other branches of science that we largely take for granted. Despite the debates about Newton's gravity versus Einstein's relativity, apples did not suspend themselves in the air awaiting the resolution. The IPCC climate models are very clear about the uncertainties of their prediction and indeed that is why they present a variety of scenarios, each of which has a range of outcomes and assigned probabilities.

Science alone cannot answer the equally important questions regarding actions that we may need to take, but it can tell us that there is measurable and significant risk of anthropogenic-induced climate change, including CO_2 emissions that drive global warming and all the consequences that entails, and that we should certainly consider the cost versus benefits of actions that we may take to mitigate that risk. We also emphasize that the geological record provides a long and reasonably well-understood record of global changes and their consequences that give us some idea of how to evaluate the worst-case scenarios that climate models predict. Perhaps future stratigraphers will agree that the base of the Anthropocene should be coincident with the radioactive fallout from nuclear bombs in the 1940s and 1950s, or perhaps from ash from continental-scale fires of the early to mid twenty-first century, or the last occurrence of coral reefs around 2100. However, perhaps we can still hope that these stratigraphers of the future will find no reason for any boundary at all. The choices we make today will play a major role in whether the rock record tells the story of environmental collapse or not.

KEY POINT

The deep time geological record provides a long-term perspective on global change that could help with the decisions we make now as to how to manage our future.

19.7 Summary

- The Holocene is the warm period that followed the retreat and melting of the last glacial ice sheets, characterized by warmer temperatures and higher sea level. Land masses where glaciers have retreated are experiencing isostatic uplift.
- The newly proposed Anthropocene Epoch recognizes that human activities may leave a permanent mark in the geologic record, including a possible sixth mass extinction. Several markers for the base have been proposed, including the beginning of testing of nuclear weapons, which leaves a distinctive radioactive layer.
- Greenhouse gases, and particularly CO_2, play a major role in mediating Earth's surface temperature, which is controlled by the balance of incoming radiation from the Sun, versus outgoing radiation and temporary trapping in Earth's oceans and buffering by the atmosphere. Current rates of CO_2 increase in the atmosphere may not have been seen on Earth since the end of the Permian Period.
- Since 1750, excessive anthropogenic CO_2 emissions, primarily from fossil fuel consumption, have already caused about 0.8 °C of global warming and 0.3 m of sea-level rise, as well as increased hydrological cycles and stronger storms, and in the absence of mitigation these effects are predicted to increase.
- Current predictions of global warming are based on sophisticated climate models that postulate a variety of possible scenarios (no mitigation of CO_2 versus extensive mitigation) with outcomes that vary in uncertainty.
- Population modeling suggests that the current rate of extinction may be 100 times background levels, suggesting that we are at the beginning of a sixth extinction.

Key Words

- Holocene Epoch
- Little Ice Age
- Medieval Warm Period
- Anthropocene
- interglacial
- thermal emission
- residence time
- gas hydrates
- permafrost
- flaring
- steric
- butterfly effect
- Intergovernmental Panel on Climate Change (IPCC)
- socioeconomic pathways
- tipping point
- younger Dryas

Further Reading and References

IPCC, 2023, *AR6 Synthesis Report: Climate Change 2023*, Intergovernmental Panel on Climate Change, https://www.ipcc.ch/report/sixth-assessment-report-cycle/.

Le Quéré, C., Andrew, R. M., Friedlingstein, P. et al., 2018, Global Carbon Budget 2018, *Earth System Science Data*, 10(4), 2141–2194, https://doi.org/10.5194/essd-10-2141-2018.

Nordhaus, W., 2013, *The Climate Casino, Risk, Uncertainty, and Economics for a Warming World*, Yale University Press.

Wagner, G., and Weitzman., M. L., 2015, *Climate Shock: The Economic Consequences of a Hotter Planet*, Princeton University Press.

Waters, C., Summerhayes, C., and Turner, S., 2021, How humans are influencing climate change and its significance in defining a new geological epoch: The Anthropocene, *Climate of the Future, Climate of the Past, Climate of the Present*, blog, April 26, https://blogs.egu.eu/divisions/cl/2021/04/26/how-humans-are-influencing-climate-change-and-its-significance-in-defining-a-new-geological-epoch-the-anthropocene/.

Review Questions

1. What are some of the major events that define the Holocene Epoch?
2. How has Earth's surface temperature changed over the Holocene and what were the causes of the Medieval Warm Period and the Little Ice Age?
3. Why has an Anthropocene Epoch been proposed and what events are considered candidates for marking the base of this new geologic period?
4. Explain how greenhouse gases and Earth's energy budget control surface temperature.
5. What are the effects of greenhouse gases, and particularly CO_2, on climate, sea level, and hydrological cycles?
6. What are the implications for the future if CO_2 emissions continue at current rates?
7. What are the scenarios used by the IPCC to build their climate models? How likely are the scenarios and the outcomes?
8. What is the evidence for a sixth extinction?
9. How do current rates of CO_2 emission compare with past events in Earth history and particularly the volcanic eruptions that caused the Permian mass extinction?

GLOSSARY

Abiotic: Non-living, formed by a non-biologic process.

Ablation: Loss of ice either by melting or sublimation.

Acadian Orogeny: Mountain building in eastern North America during the Devonian Period.

Acanthostega: One of the amphibious tetrapods that evolved from the lobe-finned fishes.

Acasta Gneiss: The oldest-known rock formation on Earth (~4.1 Ga), named after the Acasta River in the Northwest Territories of Canada.

Accretionary prism: Sedimentary accumulation of detrital material at a trench formed by a subducting oceanic plate.

Acheulean: Paleolithic stone tools, named after the site of St. Acheul on the Somme River in France, which were a revolutionary technological advance over the older Oldowan stone tools.

Acidification: Reduction in the pH of the ocean over an extended period of time, caused primarily by absorption of carbon dioxide from the atmosphere. It was a factor in the mass extinction at the end of the Paleozoic.

Actualism: A branch of uniformitarianism, the view that older geological phenomena can be explained without the constraints of strict gradualism.

Adenosine triphosphate (ATP): Molecule that stores chemical energy produced by mitochondria.

Aerosols: Suspensions of fine solid or liquid particles in the air.

Ages: Subdivisions of periods and epochs in the Geologic Time Scale.

Agglutinated foraminifera: Organisms that survived the Permian extinction because their shells were composed of sediment particles rather than calcite or other secreted substances.

Albedo: The reflectivity of the surface of Earth.

Alleghanian Orogeny: The last major mountain-building event, associated with formation of the Appalachian Mountains in Eastern North America.

Allochthonous: Rocks or sediment formed in one place and later transported to another place, such as by thrusting.

Allopatric speciation: The process where geographic isolation of a population prevents the flow of genes that would normally occur through interbreeding.

Amino acids: The fundamental molecule that serves as the building block for proteins.

Ammonia beccarii: A type of single-celled benthic foraminifera that lives in coastal brackish waters, rather than deep-sea environments.

Ammonites: Extinct group of shelled mollusk related to living octopus and squid.

Amniotes: The group of vertebrates that have developed an amniotic sac, which is required for living and reproducing on dry land.

Amniotic: Semipermeable sac surrounding an embryo that allows mammals and reptiles to reproduce on dry land.

Anapsids: The earliest amphibians and reptiles, which had no holes in their skulls other than eyeholes and nostrils.

Andean-type magmatism: An ocean–continent convergence zone where devolatization of the down-going oceanic plate beneath continental plates causes magmatism and volcanoes.

Andesite: A type of volcanic rock intermediate in composition between basalt and rhyolite that is commonly associated with volcanic island arcs along subduction zones.

Anerobic: Organism that can survive and grow where there is no oxygen.

Angiosperms: Flowering plants, meaning "cased seed," that first appeared in the Cretaceous Period of the Mesozoic Era.

Angular unconformity: A surface between two sequences of layered rocks in which the dip of the units in the two groups of rocks is significantly different.

Anhydrite: One of the major minerals in evaporite deposits ($CaSO_4$), formed when gypsum is buried and loses its water.

Ankylosaurs: Armed dinosaurs belonging to the ornithischian clade.

Annelids: Segmented worms.

Anthozoa: Corals.

Anthropocene: A proposed new epoch of the Cenozoic Era, reflecting the idea that the influence of human activities on Earth's atmosphere is so significant that it will leave (or has already left) a permanent mark in the geological record.

Anthropogenic: Human-made.

Anthropoids: Apes and humans.

Anthropomorphous: Organisms with similarities to the human form or nature.

Anticline: Geological fold in which the oldest rocks are on the inside of the fold.

Appalachia: Ancient land mass comprising Eastern North America separated from the western land mass Laramidia by the Cretaceous Western Interior Seaway.

Appalachian Mountains: Mountain belt that forms the eastern margin of North America, formed through a series of orogenic events associated with closing of Iapetus and its related oceans.

Appalachian Orogeny: The full suite of orogenic events that form the Appalachian Mountains in eastern North America, which include the Taconic, Acadian, Cherokee, and Alleghanian that culminated in the assembly of the Pangean supercontinent.

Appalachians: Mountain range running from the US state of Georgia to the Canadian province of Newfoundland created during the Paleozoic due to collision between North America and Europe, Africa, and smaller continental fragments.

Aquifer-eustasy: Global changes of sea level caused by changes in the proportion of fresh water stored in groundwater aquifers.

Arboreal: Associated with trees.

Archaea, Bacteria, Eukarya: The three major domains into which life is divided.

Archaeopteryx: Feathered dinosaur considered to be one of the oldest true birds.

Archeocyathid: Cambrian reef-building sponges.

Archosaur: Vertebrate group that includes dinosaurs, crocodilians, and possibly turtles, but excludes lizards and snakes.

Aridification: General drying of Earth's climate as rainfall decreases resulting in an increase in deserts (called desertification).

Arthropleura: A 2-meter-long Carboniferous millipede known from the distinctive trackways that it made.

Artificial selection: The selective breeding process, routinely practiced by humans in the development of favored plants and animals. Used by Darwin to explain how evolution occurs via the process of natural selection.

Asteroid belt: A collection of millions of rocky fragments that are the remains of the material that formed the proto-planetary disk that formed our solar system, mostly found between Mars and Jupiter.

Asthenosphere: The mobile, convective layer of the upper mantle that underlies the rigid lithosphere.

Atomic number: The number of protons in the nucleus of an atom.

Australopithecine: Subtribe in the Hominini tribe.

Australopithecus afarensis: One of the longest-lived and best-known early human species.

Autochthonous: Formed in place.

Autotrophic: Mode of feeding in which organisms make their own food using light, water, carbon dioxide, or other chemicals

Avalon zone: See Avalonia.

Avalonia: One of the four major peri-Gondwana terranes in the Appalachian Mountains containing continental crust that originally lay close to the margins of Gondwanaland and later accreted to North America as the Iapetus Ocean closed.

Azoic Eon: Meaning "time without life," assigned to Precambrian rocks before single-celled Precambrian fossils were discovered.

Backbulge: Low area associated with *foreland basin* systems that lies seaward of the *forebulge.*

Barrow's zones: Metamorphic zones that are recognized by the mineral assemblages and textures observed in rocks in the field and with the microscope. Named for British geologist George Barrow (1854–1932), who studied the metamorphic rocks in northeast Scotland and was the first to recognize the order in which minerals appeared in the field, which he interpreted to reflect the different metamorphic conditions the rocks experienced.

Barrovian metamorphism: Pattern of metamorphism named after George Barrow, first recognized in Scotland.

Basalts: Fine-grained silica-poor igneous rocks formed from cooling of mafic lavas.

Basin and Range province: A series of closely spaced normal faults that produced individual mountain ranges in intervening basins, including essentially all of Nevada and parts of California, Arizona, Utah, New Mexico, and the northern parts of Mexico.

Beats: The product of interference of waves of varying frequency. In climate cycles, interactions of 20 ka, obliquity, 41 ka presession, and 100 and 400 ka eccentricity cycles interact to form beats that produce lower frequency climate cycles.

Bengal fan: The largest submarine fan on Earth, produced where the Brahmaputra River joins with the Ganges in eastern India, eventually flowing into the Bay of Bengal. It is about 3,000 km long, over 1,000 km wide, and about 16.5 km thick at its thickest point.

Benthic foraminifera: Single-celled organisms that live on the seafloor rather than free-swimming or floating in the open ocean.

Big Bang: The hypothesis that the presently observed expansion of the universe may be extrapolated backwards to determine when this expansion began, with current data suggesting that the universe formed some 13.77 billion years ago from an exceedingly hot and dense state.

Biogenic gas: Natural gas (mostly methane) produced by bacterial breakdown of organic matter.

Biostratigraphy: Branch of stratigraphy that uses fossils to establish relative ages of sedimentary layers.

Biozones: Periods of time characterized by age-diagnostic fossil species or fossil assemblages.

Bipedal: Two-legged.

Biramous: Limbs that have two branches, such as those common in modern crustaceans.

Blastoids: A type of calcifying stalked echinoderm.

Blastula: The ball of cells that forms in the early stage of cell division in a developing embryo.

Blueschists: A common metamorphic rock associated with sinking oceanic slabs. Named after the common blue metamorphic minerals found within them.

Boulder pavements: Grounding line areas where ice has come in direct contact with the seafloor and later melted leaving behind a layer of boulders.

Brachiation: Locomotion using the arms.

Brachiopods: Phylum of shelled, marine, invertebrate animals that came into existence during the earliest part of the Paleozoic Era.

Brittle deformation: Deformation that occurs when an otherwise elastic material fractures without any apparent sign or little evidence of material deformation prior to failure, resulting in displacement along a fault.

Brontosaurus: Species of sauropod dinosaur.

Bryophtyes: Simple non-vascular plants similar to modern mosses and liverworts.

Bryozoans: Small colonial filter feeding animals that look somewhat like miniature corals.

Butterfly effect: Metaphor for the difficulty of long-term weather forecasting, in which a butterfly flapping its wings at one place on Earth can affect the weather at another place on Earth.

Calcite: The most important mineral of the rock limestone formed of calcium carbonate ($CaCO_3$).

Caldera: The large crater-like depression formed when a volcano erupts and the area above the shallow magma reservoir collapses.

Caledonian: Mountain chain in Scandinavia equivalent in age to the Appalachians in Eastern North America.

Camarasaurus: Species of sauropod dinosaur with a distinctive stubby snout.

Cambrian explosion: Geological time period marked by the appearance of organisms with hard parts.

Cambrian fauna: The oldest of the three major faunal groups of the Phanerozoic Era identified by paleontologist Jack Sepkoski.

Cambroernids: An informal clade of extinct animals marked by coiled bodies and tentacles.

Camera lucida: A traditional device in which the image of a fossil, seen under a microscope, is projected onto a piece of paper allowing the viewer to trace the image.

Canadian Shield: Stable area that forms the interior part of the North American Continent, formed mostly of metamorphic and igneous rocks, also termed a *craton.*

Cap carbonate: Limestone or dolostone commonly found overlaying glacial deposits in the Cryogenian "snowball Earth" Period.

Carapace: A head covering.

Carolina: One of the four major terranes containing continental crust that were accreted to Eastern North America during the formation of the Appalachian Mountains during the Paleozoic Era.

Cascade Mountains: Pacific Northwest mountain range that represents the roots of ancient volcanoes formed by subduction beneath the western margin of North America during the Mesozoic.

Catastrophism: The now defunct idea that major, short-lived events are responsible for many of the major features visible on Earth.

Cenozoic: The youngest Era of geologic time.

Centrosaurus: Species of ceratopsian (horned) dinosaur that formed giant herds.

Cephalopods: A class of invertebrate aquatic animals that includes octopuses, nautiluses, squid, cuttlefish and extinct ammonites and belemnites.

Ceratopsians: Horned dinosaurs, dominant in the Cretaceous Period that include *Triceratops*.

Changsingian Stage: A stage of the Late Permian Period.

Channeled Scablands: A vast area located in Eastern Washington, characterized by unusual surface features, including large, braided channels that cut through basalt bedrock, gravel deposits, giant ripples, cliffs over which waterfalls plunge tens of meters, and large bowl-like depressions formed in the basalt.

Chatter marks: Curved, U-shaped cracks produced by chipping and fracturing of a bedrock surface, as a glacier flows over it.

Chelicerates: A group of arthropods that includes modern organisms such as spiders and scorpions, as well as extinct groups like the giant-clawed eurypterids, which are common in the Silurian Period.

Chemosynthesis: The process of creating food using energy released from chemical reactions.

Chengjiang biota: One of the more recent Konservat lagerstätten discoveries, found in the Maotianshan Shale in China.

Cherokee Orogeny: Mountain-building event caused by accretion of the Carolinian terrane in the Late Ordovician to Early Silurian.

Chicxulub: Site of an impact crater on the Yucatán Peninsula of Mexico, believed to be the location of the giant-impact hypothesis that caused the K–Pg mass extinction.

Chitin: The stiff material arthropods use to make their skeletons, and which also forms the hard beaks on squids and octopuses.

Chlorophyll: The key element in cyanobacterial aerobic photosynthesis.

Choanocytes: Cells that line the inside of sponges that are identical to single-celled choanoflagellates both of which have a collar around the flagellum.

Choanoflagellates: Collared single-celled organisms thought to be the closest ancestor to the metazoan sponges.

Chondrichthyes: The cartilaginous fishes including sharks and rays.

Chromoplasts: Plastid organelles within eukaryotic cells that manufacture and store pigments.

Chromosome: Thread-like structure in the nucleus of animal and plant cells which contains the DNA.

Circumpolar current: An ocean current that isolated Antarctica from any warm equatorial waters by the time of the Neogene Period.

Clade: A statistically defined grouping used to classify organisms that have a shared ancestor, as exhibited by shared anatomical or genetic similarities.

Cladistics: The study of clades.

Cladogram: A graphic representation of how organisms are organized according to their inferred phylogenetic history and the order in which they are related as determined by the addition of newly derived traits.

Clasts: Pieces of broken rock.

Clays: Fine (diameter less than 0.0039 mm), grained muddy sediment formed primarily by clay minerals produced by chemical weathering of crystalline rocks.

Clinoform: Sloping depositional surfaces.

Closure temperature: The temperature at which a radioactive system becomes closed in a mineral (i.e., the temperature below which the daughter products are essentially all retained).

Cnidaria: Sister group to ctenophores, including jellyfish, anemones, and corals. Their fossilized remains have been found in Cambrian rocks, indicating that they appeared at least 540 million years ago and probably earlier.

Coccoliths: Microscopic plates or scales of calcium carbonate that forms chalk, and is formed by disaggregation of coccolithophores, a type of photosynthetic marine plankton.

Cocos plate: One the smaller oceanic plates subducting under North America formed of the originally larger Farallon plate.

Coelacanth: The best-known of the lobe-finned fish. Originally thought to have been extinct since the Cretaceous Period, until its 1938 discovery off the coast of South Africa in the waters of the Indian Ocean.

Coelom: Body cavity.

Compsognathus: Dinosaur species not much bigger than a chicken.

Conformable: Stratigraphic successions in which there is a relatively continuous record of deposition.

Confuciosornis dui: One of the earliest Cretaceous birds.

Conifers: Evergreen trees, which are the main type of gymnosperm alive today.

Conjugation: Transfer of genes directly between cells.

Continental drift: The theory, proposed by German scientist Alfred Wegener (1880–1930), that the location of continents and oceans was not fixed over time.

Continental escape: A tectonic process in which a region moves sideways along large-scale strike-slip faults.

Convection: The mechanism in which mantle material can flow, albeit very slowly, originally proposed by English geologist Arthur Holmes (1890–1965) as the cause of continental drift.

Cooksonia: One the earliest stalked land plants with a spore-producing capsule on top found in rocks of the Silurian Period.

Cordillera: Mountain belt.

Cordilleran Ice Sheet: The smaller of the two major sheets that covered North America at the culmination of the last glacial maximum.

Core: Central part of Earth mostly made of iron and nickel.

Coriolis forcing: The deflection of wind and water flows due to Earth's rotation.

Cosmogenic: That which originates from outside of Earth, such as extraterrestrial, dust, and meteorites.

Cratons: A stable part of the continent, usually composed of high-grade metamorphic and plutonic igneous rocks, commonly Precambrian in age.

Cretaceous Thermal Maximum: Exceptionally warm temperatures that reached a peak in the mid-Cretaceous Period.

Crevasse: Cracks that form in glacier ice when the glacier is put under too much stress for it to deform by flowing.

Crinoids: Group of stalked sessile marine invertebrates, also known as sea lilies, which are related to starfish, sea urchins, and sea cucumbers.

Crown organisms: Ancestral organisms that lie at the top of a clade.

Crown tetrapods: Amphibians, reptiles, birds, and mammals.

Crust: Outermost layer of the Earth that forms part of the rigid lithosphere.

Cryoconite: A dark crust formed by ablation of the ice that would increase the concentration of dust on top of the ice and form local ponds that may have served as refugia for life during the Cryogenian "snowball Earth" Period.

Cryogenian Period: One of the newest identified geological periods (720–635 Ma), when it is thought that ice extended from high latitudes and elevations all the way to sea level in the tropics, covering most of Earth's surface, forming a "snowball Earth."

Ctenophores: Metazoans, also known as "comb jellies," thought to have appeared around the same time as sponges.

Cyanobacteria: Phylum characterized by blue-green colored photosynthetic bacteria.

Cycads: A type of gymnosperm.

Cyclothems: Repeated shallowing-upward stratigraphic cycles of deposition. commonly interpreted to reflect rapid sea-level changes associated with climate cycles.

Cystoids: An extinct calcifying stalked echinoderm, related to the modern *crinoids*.

Cytoskeleton: A network of long fibers that make up the structural framework of a cell.

Daughter element: The element formed when another element undergoes radioactive decay.

Decay constant: The probability of decay of a radioactive element in a given time period.

Decompression melting: Partial melting that is triggered by reduction of the pressure of rocks near their melting point (e.g., when oceanic plates split apart at an oceanic spreading center producing new oceanic crust).

***Deinonychus antirrhopus*:** A small, Early Cretaceous bipedal carnivorous dinosaur with a huge retractable sickle-shaped claw on its foot and stiffening rods along its tail.

Demosponges: One of the marine organisms that dominate Modern fauna.

Denisovans: An apparent late branch of early humans.

Depositional environment: The physical, biological, or geochemical processes by which sediment was transported and deposited, as well as the place of deposition.

Deuterostome: (Literally "second mouth".) Embryos in which the first dimple in embryonic differentiation of the gastrula (hollow ball of cells) becomes the anus and the second dimple the mouth.

Diabase: A fine-grained mafic intrusive igneous rock, and extrusive basalt.

Diamict: (Literally "two mixtures.") Poorly sorted sediment consisting of mud, sand, gravel, and boulders that results when a glacier retreats and especially where it enters the sea.

Diapirs: A dome in which the overlying rocks have been deformed or ruptured by the squeezing out and rising up of underlying mobile material, usually salt or mud.

Diapsids: One of the two main clades of vertebrates characterized by skulls with two additional holes in addition to eyes and nasal passage, the other being the *synapsids*.

***Die eiszeit*:** (Literally "the ice time".) One of the earliest suggestions of a recent global freezing, made by Swiss-American scientist Louis Agassiz, who proposed that Earth had experienced a recent glaciation, initiated by God, that covered the globe.

Diluvium: Coarse sediment, now referred to as alluvial, thought by early Christian scholars to have been the result of Noah's Flood but now ascribed to a glacial origin.

***Dimetrodon*:** A carnivorous, fin-backed synapsid reptile of the Permian Period.

Dinosaurian clade: Approximately 2,000 species of animals that flourished from about 235 million years ago to 66 million years ago, occupying most of the available ecological niches and living across most of the planet.

Dinosauromorph: Archosaur group, closely related and ancestral to the dinosaurs.

Dinosaurs: Group of reptiles of the clade Dinosauria. Translated as "terrible lizard," they were first named by Sir Richard Owen in 1842, based on reconstructions of fossil remains.

***Diplichnites*:** Trace fossils associated with the giant Late Carboniferous centipede *Arthropleura*.

Diploblast: Phyla with two layers of cells.

***Diplodocus*:** Species of sauropod dinosaur.

Disconformities: Major undulating erosional surface resulting from a fall of sea level rather than tectonic uplift and tilting.

Distal: Indicating deposition farther from the sediment source area.

Diurnal: Awake during the day.

Divergent plate boundary: Locations where two plates move in opposite directions away from the boundary. Mature divergent boundaries will evolve into ocean basins.

DNA: Deoxyribonucleic acid, the chromosome molecule that contains the genetic information for an organism.

Domains: The major categories of life, which includes Archaea, Bacteria, and Eukarya.

Dromaeosaurids: Feathered theropod dinosaurs.

Dropstones: A common feature in glaciomarine sediments, which occur where larger clasts, such as pebbles, cobbles, or boulders are rafted out to sea, typically on icebergs or larger ice shelves.

Drumlins: One of a variety of large-scale landforms that glaciers produce.

Ductile: Condition in which rocks can flow without discontinuities while still in the solid state.

***Dunkleostus*:** Devonian fish which grew to over 5 m long.

Dunnage zone: The main deformed remnants of the Iapetus Ocean, and its associated island arcs.

Duplex structures: A structure in which thrust faults merge into flat faults both upward and downward.

Dynamic topography: Changes in the uplift or subsidence of Earth's lithosphere caused by flow in the aesthenospheric mantle.

Eccentricity: The degree to which the Earth's elliptical orbit around the Sun departs from a circle.

Echinodermata: Phylum of marine animals with radial symmetry in their body shapes.

Eclogite: Dense metamorphic rock, rich in garnet, formed by descending ocean slab in the lithosphere.

Ecological niches: A habitat.

Ectotherm: Cold-blooded.

***Edaphosaurus*:** An herbivorous, fin-backed synapsid of the Permian Period.

Ediacaran fauna: Animals that appeared about 25 million years after the Cryogenian Period ended.

Ediacaran Period: Newly named in 2004 after the Ediacara Hills in Australia, the first new geological period (570–543 Ma) declared in 120 years and containing the first unequivocal metazoan fossils.

Edmontosaurus: One of the largest members of the herbivorous hadrosaur family.

Elphidium macellum: A type of single-celled benthic foraminifera that lives in coastal brackish waters, rather than deep-sea environments.

Embryophytes: Land plants.

Endemic species: Species confined to a very local habitat or area.

Endolithic: Within rock.

Endosymbiosis: The process by which one organism lives within the cell of another organism.

Endotherm: Warm-blooded.

Eocrinoids: Primitive echinoderms, common among the Tommotian fauna of the late Ediacaran.

Eolian: Wind-blown desert environment.

Eons: The longest category in the Geologic Time Scale, consisting of the Phanerozoic, Proterozoic, Archean, and Hadean.

Eoraptor: One of the earliest true dinosaurs, relatively small, with a bipedal gait and likely an omnivorous diet.

Epeiric seas: Vast shallow seas that characterized the Cambrian Sauk transgression, which was caused by a global rise in sea level.

Epochs: Subdivisions of Periods in the Geologic Time Scale.

Eras: Subdivisions of Eons in the Geologic Time Scale, including the Paleozoic, Mesozoic, and Cenozoic.

Erratics: Pieces of rock that differ in size and type to the rocks native to the area in which they are found. Formerly referred to as diluvium.

Escape velocity: The velocity needed to overcome the gravitational attraction of a planetary body.

Eugenics: The idea of controlled selective breeding of humans, such as by sterilization, to improve the population's genetic composition.

Eurypterids: A scorpion-like chelicerate predator that appeared during the late Ordovician Period, eventually going extinct at the end of the Permian.

Eustasy: Global changes of sea level.

Eustatic: Sea level.

Eusthenopteron: Lobe-finned fish that evolved into amphibious tetrapods.

Evaporite: Sediments produced by precipitation from seawater that include limestone, dolomite, gypsum, and halite (salt).

Evolution: The observation, based on the rock record, that life on Earth has changed over time, as species become extinct and are often succeeded by new species.

Expanding Earth hypothesis: The proposition that in the Paleozoic Era, the supercontinent Pangea covered most of the Earth's surface, requiring a much smaller Earth that somehow became larger throughout the Mesozoic.

Extremophiles: Organisms that can survive extreme environments.

Extrusive: Igneous rock that was solidified on or above the surface. (Synonymous with volcanic.)

Facultative bipedalism: Having the facility to walk on two legs but spending most of the time in a quadrupedal stance.

Faint young Sun: The Sun during the Neoproterozoic Era, when it produced about 6% less energy than today, resulting in significantly less solar energy reaching Earth.

Farallon plate: An oceanic plate subducting beneath North America during the Cretaceous and early Cenozoic.

Fault: Fracture in rock across which displacement can be observed.

Feldspar: The most widespread silica-rich mineral group, found in most types of rock.

Felsic: Magmas that are rich in feldspar (fel) and silica (si) and form light-colored igneous rocks like rhyolite and granite.

Fetch: Stretch of open water over which the wind blows.

Filamentous bacteria: Bacteria that grow in filamentous colonies.

Fishapods: Transitional forms between fish and tetrapods, reflecting the fact that they enjoyed a largely aquatic lifestyle, but were able to breathe air and crawl out of the water in their search for food and new shallow water habitats.

Flaring: Burning of natural gas.

Flat-slab subduction: Term for the status attained by the Farallon plate from around 60 to 50 Ma, allowing for interaction between it and the North American plate for up to 2,000 km.

Flood basalts: Basaltic lavas.

Fluid inclusions: Bubbles of fluid trapped in crystals.

Fluvial: Pertaining to rivers.

Focus: In seismology, the place inside Earth's crust where an earthquake originates.

Foliation: The laminated structure in metamorphic rocks resulting from segregation of minerals, commonly micas, into layers parallel to the schistosity that is found in slates and schists.

Footwall: The block that lies on the underside of a fault (named because it is where one's feet would be when standing in a tunnel coincident with the fault).

Foramen magnum: The hole that attaches the skull to the spine.

Forebulge: High uplifted area seaward of the foredeep that forms part of a foreland basin system.

Foredeep: Low area adjacent to a fold and thrust belt that forms the proximal part of a foreland basin system.

Foreland basin: Asymmetric depressions that are deepest next to the associated thrust belt.

Formations: The fundamental unit of mapping constituting strata-form bodies of rock having a consistent set of physical characteristics that can be used to distinguish them from adjacent rocks, and which occupy a particular position in a vertical sequence of layers.

Fractional crystallization: A process in which minerals crystallize out of a magma and the resultant crystals and liquid are segregated from each other, producing rocks with new compositions relative to the original.

Frondomorphs: Organisms of the Petalonomae phylum, so named because of their petal- or frond-like forms.

Fusulinids: Giant extinct calcareous single-celled foraminifera common in the Paleozoic Era.

Gabbros: Low silica mafic plutonic rock consisting primarily of calcic plagioclase and clinopyroxene.

Gander zone: Deformed remnants of the Iapetus Ocean that originally lay close to the margins of Gondwanaland.

Ganderia: One of the four major tectonic terranes in the Appalachian Mountains that represent the remnant of the Iapetus Ocean that closed to form Pangea.

Gangdese batholith: A series of granites and similar rocks in southern Tibet, on the north side of the Himalaya.

Gas hydrate: A crystalline solid formed of water and gas, containing large amounts of methane, commonly found in permafrost and in continental ocean margin sediments.

Gastroliths: Stomach stones, used for chewing food that are common in birds and non-avian dinosaurs.

Genomics: The study of all of an organism's genes.

Geochemistry: The study of the chemical aspects of rocks and minerals.

Geomagnetic Time Scale: A chronology based on dated reversals of Earth's magnetic field.

Geosynclinal theory: The theory of continental deformation in which most geologic features were explained by vertical motion. Supplanted by what is now known as plate tectonics in which the most significant motion is horizontal.

Giant-impact hypothesis: The idea that the K–Pg extinction event was caused by a giant extraterrestrial object, either an asteroid or comet, colliding with Earth.

Ginkgos: A type of gymnosperm common in the late Paleozoic Era.

Glacial Lake Missoula: A massive glacial lake that existed at the end of the last ice age, and was located near the site of modern Missoula, Montana.

Glacio-eustasy: Global changes of sea level caused by growth and decay of glaciers and changes in the volume of ocean basins.

Glossopteris: Late Permian ferns that disappeared very soon after the extinction event in the Early Triassic.

Gneisses: High-grade metamorphic rocks.

Gnetophytes: A type of gymnosperm.

Gondwanaland: Supercontinent that existed about 200 million years ago that lay close to the South Pole in which Antarctica was joined to South America, Africa, India, Australia, and the Arabian Peninsula.

Gorgonopsians: Clade of extinct saber-toothed carnivorous therapsid reptiles that were the top predators from the Middle to Upper Permian.

Graben: Low areas, commonly filled with sediments, formed by normal faults commonly associated with extension and rifting.

Granites: Plutonic rock consisting largely of feldspar and quartz.

Great Chain of Being: Scientifically obsolete metaphor for life on Earth as an evolutionary chain or ladder with humans at the top.

Great Ordovician Biodiversification Event: A second radiation and diversification (following the Cambrian explosion), from about 485 to 445 Ma, which saw an increase from a few hundred to around 1500 families in about 50 million years.

Great Oxygenation Event (GOE): An event marking the rise of oxygen in the early Earth's atmosphere, estimated to have begun around 2.5 Ga.

Greater Himalayan Sequence: One of the major tectonic units of the Himalaya consisting of high-grade metamorphic rocks.

Greenhouse: Warmer periods in Earth's history, which alternates with glacial "Icehouse" periods.

Greenhouse gases: Gases that exert a major control on Earth's surface temperature by absorbing heat, consisting of carbon dioxide (CO_2), water vapor (H_2O), and methane (CH_4).

Greenstone belts: Green-colored rock usually composed of metamorphosed basalts and minor sedimentary rocks, and common in the Precambrian.

Grenville Orogeny: A major Mesoproterozoic mountain-building event along the eastern margin of Laurentia associated with the assembly of the supercontinent Rodinia.

Grounding line: Marks the area where submarine tills can be deposited as glacial ice sheets come into contact with the seafloor.

Groups: Stratigraphic unit consisting of several Formations grouped into a large-scale unit for the purpose of naming.

Grypania: Curved, tubular fossils regarded by some to be the oldest eukaryotes.

Gymnolaemata: A class of bryozoans, marine invertebrate organisms that dominate the Modern fauna.

Gymnosperms: The first seed-bearing plants, meaning "naked seed," referring to the fact that their seeds do not have a covering.

Gypsum: A calcium sulfate mineral uniquely indicative of evaporation of seawater ($CaSO_4 \bullet H_2O$).

Hadrosaurs: Duck-billed sauropod with extensive head ornaments originally hypothesized to have been used as some sort of aqualung, but now thought to have probably served as amplifiers for generating low loud sounds.

Half-life: The time it takes for a given amount of radioactive isotope to decrease by half.

Halite: Formal mineral name of salt (NaCl).

Halophiles: One of the first modern archaea, which are able to live in areas of high salt-content.

Hanging wall: The overhanging wall of a fault (named because that is where one would hang a lantern).

Harzburgite: A mafic igneous rock made up of olivine and pyroxene.

Haversian canals: Microscopic tubes in bones that host blood and nerve vessels. Increased densities of Haversian canals may be correlated to "warm-bloodedness."

Heat-producing elements: Radioactive elements, such as uranium, that decay releasing heat in the process.

Hematite: The principal ore of iron, the oxidation of which causes the red coloration typical of sediments formed on dry land, rivers, deserts as well as rust.

Hemes: The components of molecules that include iron and form part of the hemoglobin that gives blood its red color.

Hemoglobin: A red-colored protein used to transport oxygen in blood.

Herrerasaurus: One of the earliest Triassic theropod dinosaurs.

Heterotrophic origin of life theory: The idea that life evolved in the "primordial soup" of a shallow freshwater pond.

Hexactinellids: Glass sponges.

Hexapoda: Six-legged insects that were among the first land-dwelling arthropods.

Highstands: Times when sea levels are at their highest.

Holocene Epoch: The current epoch, beginning about 11,650 years BP, marked by the melting and retreat of the Pleistocene ice sheets caused by natural climate warming.

Holoptychius: A type of lobe-finned fish.

Hominini: Taxonomic tribe that includes *Homo*, the australopithecines, and the chimpanzees.

Hominoidea: Tailless primate superfamily that includes gibbons, chimps, orangutans, gorillas, and humans.

Homo erectus: One of the earliest human species discovered.

Homo floresiensis: Extinct species of "archaic" humans that inhabited the island of Flores, Indonesia.

Homo heidelbergensis: Extinct species of "archaic" humans that lived from 700,000 to 300,000 years ago and is intermediate between *Homo erectus* and Neanderthals.

Homo neanderthalensis: Extinct species of "archaic" humans.

Homology: Similar anatomical or genetic structures attributable to a common evolutionary origin.

Humber Zone: Westernmost zone that forms the marks the boundary of the Appalachian Mountains and which represents the ancient eastern margin of Laurentia.

Huronian Glaciation: Period characterized by glacial deposits lasting from 2.4–2.1 Ga.

Hydrological cycles: The continuous circulation of water in the Earth–Atmosphere system, including evaporation, transpiration, condensation, precipitation, and runoff.

Hydrothermal vents: Openings in the deep seafloor where hot gases and fluids are vented and around which life can proliferate.

Hylonomus lyelli: One of the oldest fossil reptiles, discovered by Canadian geologist John William Dawson, along the sea cliffs in northwest Nova Scotia in a hollowed-out stem of a fossil *Lycopsid* tree found in the Carboniferous Joggins Formation.

Hypercapnia: Extinction mechanism caused by high CO_2 levels in seawater that inhibits metabolic processes and suffocates marine organisms.

Hypersaline: Water with a very high salinity level.

Hypocenter: The location below Earth's surface where an earthquake is initiated.

Hypoxia: Loss of oxygen.

Iapetus: Name given to now-vanished Paleozoic ocean that was destroyed during the assembly of the supercontinent Pangea, and remnants of which are found in the Appalachian Mountains. Named after the father of Atlas, the namesake of the Atlantic Ocean.

Icehouse: Glacial periods in Earth's history, which alternate with warmer "greenhouse" periods.

Ichthyosaurs: Marine-dwelling reptiles, which lived at the same time as dinosaurs and evolved from land-dwelling reptiles.

Ichthyostega: One of the Devonian amphibious tetrapods that evolved from the lobe-finned fishes.

Igneous: Related to rocks formed from magma.

Igneous dikes: A crack in rock that is filled with magma.

Iguanodon: Late Jurassic to Early Cretaceous herbivorous dinosaur, a stem ancestor related to the hadrosaurs (duck-billed dinosaurs).

Immiscible: Liquids not forming a homogeneous mass when mixed.

Impact breccia: Broken rock formed by meteorite impacts.

Impact hypothesis: The tilts of Earth and Uranus were caused by major impacts that tilted the planets as well as caused the formation of their moons.

Imperfect adaptations: Less well-adapted traits in some organisms that can be traced back to ancestral organisms in which the traits played a different role.

Indus fan: The dominant sedimentary feature of the Arabian Sea, located between the Arabian, Eurasian, and Indian plates. It covers an area of $110,000 \, km^2$ and reaches over 9 km thick.

Infaunal: Marine animals that live in sediment.

Inflationary epoch: The time, lasting from 10^{-36} to 10^{-32} seconds after the Big Bang, when the universe expanded in scale by a factor of 10^{+78}.

Inostrancevia: An extinct saber-toothed gorgonopsian, common in the Permian Period.

Insolation: A measure of the overall strength of the Sun's incoming radiation.

Insular dwarfism: Evolutionary trend in which larger mainland animals become smaller over time as a result of isolation in a small area, such as an island.

Interglacial: A period of milder climate between two glacial periods.

Intergovernmental Panel on Climate Change (IPCC): An organization established in 1988 by the United Nations to provide policymakers with regular scientific assessments on the current state of knowledge about climate change.

Intrusive: Igneous rocks formed from magmas that cool and solidify beneath the surface. (Synonymous with *plutonic*.)

Inverted metamorphism: When the grade of metamorphism increases as one moves topographically upward.

Iridium: An element that is relatively common in meteorites, but extremely scarce in terrestrial crustal rocks, the high concentrations of which provide evidence for the end-Cretaceous meteorite hypothesis.

Iron meteorites (chondrites): Meteorites rich in iron and nickel and thought to resemble Earth's core in composition.

Island arcs: A curved chain of volcanic islands typically formed above a subduction zone.

Isochron diagram: Graph used to determine the age of cogenetic minerals in which there exists significant amounts of daughter product in the minerals at the time of crystallization.

Isostasy: Concept that explains the heights of mountains and the depths of oceans primarily as a function of differences in density and buoyancy between the rigid lithosphere and mobile asthenosphere.

Isotherms: Lines of equal temperature.

Isotopes: Different types of atoms of a given element, referred to by the total number of protons, which is always the same for a given element, plus the number of neutrons, which may vary. Some, but not all, isotopes are also radioactive.

Kainite: A complex sulfate mineral.

Konservat lagerstätten: A sedimentary rock that preserves soft parts of animals.

K–Pg boundary event: Mass extinction that occurred at the boundary between the Mesozoic and Cenozoic eras.

Lag time: The difference between the youngest observed age for sand grains in a sand or sandstone and the time of deposition of the material.

Lago Mare (lake sea): Large brackish to hypersaline lakes that remained after evaporation of the Mediterranean Ocean during the Messinian salinity crisis, 5.3 million years ago.

Lapout: Places in seismic cross sections where reflections are eroded or terminate against each other.

Laramide Orogeny: Deformation across the Rocky Mountains related to the flat-slab subduction of the Farallon plate during the Cretaceous and early part of the Paleogene periods characterized by deep-seated basement faulting.

Laramidia: Landmass that lay on the western margin of the Cretaceous Western Interior Seaway.

Large igneous provinces (LIPs): Very large accumulations of volcanic rocks that cover areas greater than 100,000 square kilometers and are usually made of basalt, erupted within a geologically brief period, typically a few million years or less.

Last glacial maximum (LGM): A period of glacial advance, ending about 18,000 years ago, when a significant proportion of the northern hemisphere land masses were covered by kilometers-thick ice sheets.

Late Heavy Bombardment (LHB): Period from 4.1 to 3.9 billion years ago when planetary bodies in the solar system were characterized by extensive bombardment by asteroids.

Latent plastic virtue: According to pre-scientific theory, a force inherent in Nature that causes a transformation of rocks into forms that resemble living things – what we now call fossils – but that these forms were not actually related to ancient life.

Laurentia: Term for the ancient continent of North America formed prior to the

Appalachian Orogeny that now forms the Canadian Shield.

Laurentide Ice Sheet: The larger of the two major sheets that covered North America at the culmination of the *last glacial maximum*.

Laurussia: Landmass formed about 400 Ma, when Laurentia, Baltica, and Siberia joined together.

Lava: Liquid magma that flows at the surface.

Left-lateral strike-slip: Strike-slip fault in which when standing looking towards the fault, the movement of the opposing block is to the left.

Lesser Himalayan Sequence: The less metamorphosed sedimentary and low-grade metamorphic rocks that lie south of the Greater Himalayan Sequence in central Nepal.

Lherzolites: A mafic igneous rock made up of olivine and pyroxene.

Lignin: A rigid organic molecule organized into pipe-like structures that are used to transport water and minerals throughout a plant.

Limiting factors: A range of factors such as temperature, oxygen levels, food resources, etc., to which any given species on Earth is typically adapted.

Lineation: A preferred orientation of elongate minerals seen in metamorphic rocks such as gneisses.

Lithosphere: The rigid outer layer of Earth comprising the crust and upper (lithospheric) mantle that form the world's plates.

Little Ice Age: A relatively cooler period of the Holocene, between 1300 and 1850 AD, impacting northern latitudes more severely. Attributed to a variety of causes including lower solar radiation, higher volcanic activity, and land-use changes caused by massive loss of human populations due to disease.

Lobe-finned fish: Fish that evolved into the first amphibious land vertebrates during the Devonian Period.

Lobopodia: Extinct phylum of lobe-footed worms.

Lowstand: Period of low sea level.

Lycopsid: Vascular plants that dominated the Carboniferous Period and became extinct as the Late Paleozoic climate became more erratic.

Lystrosaurus: The most common of the Late Permian herbivorous synapsids and one of the few terrestrial vertebrates that survived the end-Permian mass extinction.

Mafic: Magmas that are rich in magnesium (Ma) and iron (Fe) and low in silica that form dark-colored igneous rocks like basalts and gabbros.

Magma: Naturally occurring silicate liquid from which igneous rocks are derived.

Magnetic anomalies: Observation of rocks with a uniform paleomagnetic signature that pass into areas with a different signature (i.e., reversed polarity), then switch back. Magnetic striping in ocean crust showed that they formed as Earth's magnetic field switched from normal to reversed and back again many times.

Magnetic surveys: Surveys that measure magnetic signature of rocks.

Magnetostratigraphy: A method of determining the age of deposition, which is based on the observations that the Earth's magnetic field occasionally switches polarity from north to south, and that some rocks record within them the polarity of the magnetic field at the time of formation.

Main Central Thrust: A major crustal-scale south-vergent thrust fault along the Himalaya.

Malacostracan: Crustaceans, such as lobsters, shrimps, and crabs.

Mantle: Silicate-rich layer of Earth between the crust and core.

Marasuchus: A Late Triassic fully bipedal animal, classified as a dinosauromorph archosaur, closely affiliated to the dinosaurs.

Marinoan Epoch: Glacial epoch that lasted for about 15 million years (650–635 Ma), following the Sturtian Epoch, which lasted for about 59 million years (718–659 Ma) associated with the snowball Earth.

Marls: Limey mudstones.

Mass extinction: A natural disaster that causes massive loss of species on a global scale.

Medieval Warm Period: A period of the Holocene, which may have been partly caused by anthropogenic deforestation and development of agriculture, that promoted large CO_2 release into the atmosphere, causing warming.

Megacheira: An extinct class of early arthropod found in the Burgess Shale called the Megacheira, meaning "great hands."

Meganeura: An extinct giant dragonfly that had a wingspan of about 68 cm, a size that reflected the high oxygen levels in the late Paleozoic.

Mega-tsunami: A very large wave created by a large displacement of material into a body of water.

Meguma: One of the four major Paleozoic terranes forming the Appalachian orogen that accreted during the assembly of Pangea.

Meltwater lid: Relatively fresh water over 500 m in thickness caused by deglaciation triggered by dramatic warming of Earth at the end of the Cryogenian (snowball Earth) Period.

Members: Smaller-scale subdivisions of Formations.

Mendocino triple junction: The point where a subduction zone (today's Cascadia trench), the San Andreas fault system, and a strike-slip fault between the Juan de Fuca and Pacific plates meet.

Mesozoic: Era of geologic time that includes the Triassic, Jurassic, and Cretaceous periods.

Messinian salinity crisis: The great drying out of the Mediterranean Sea that occurred near the end of the Messinian Stage, about 5.9 Ma.

Messinian Stage: The last stage in the Miocene Epoch that lasted from 7.24 to 5.33 Ma.

Metamorphic: Related to rocks transformed by pressure and temperature.

Metazoan: Multi-celled animals.

Methanogenesis: The reduction of carbon dioxide commonly associated with anerobic bacteria that produces energy and methane as a byproduct.

Microraptors: The smallest dinosaurs in the Mesozoic, many of them feathered and close relatives of modern birds.

Mid-ocean ridges: Underwater mountain ranges with a deep canyon in the middle of the otherwise high-standing feature that mark areas where seafloor spreading occurs.

Milankovitch cycles: High-frequency changes in climate that occur at frequencies of 20,000; 40,000; 100,000 and 400,000 years caused by periodic changes in Earth's orbit, largely related to extraterrestrial forces of the Sun, Moon and Jupiter, which are the main drivers for major glacial periods. Named for the Serbian geophysicist, astronomer, and climatologist Milutin Milankovitch.

Milky Way: The spiral galaxy that includes our solar system.

Miocene Epoch: The fourth of the five epochs which constitute the Paleogene Period.

Missing link: Hypothetical intermediate organisms that demonstrate the transition between species such as the "ape-man" that links apes and humans. Outdated concept assumes that evolution of life follows a linear *Great Chain of Being* versus newer concepts that the history of life is more bush- or tree-like.

Mitochondria: Organelle found in eukaryotic cells that control respiration and energy production.

Modern fauna: One of the three major groups of Phanerozoic Era fauna identified by paleontologist Jack Sepkoski.

Molecular clocks: A technique which estimates the time at which two groups last shared a common ancestor by dividing their genetic distance by the general rate of genetic change.

Mollusks: Along with trilobites, one of the dominant bilateral animal groups of the Paleozoic Era, featuring an external shell that contains a soft-bodied invertebrate.

Monomers: Simple molecules consisting of a single part.

Monoplacophorans: Among the rarest members of the phylum Mollusca, recognized among the Tommotian fauna of the late Ediacaran.

Monotremes: An order of egg-laying mammals that include the duck-billed platypus.

Moraines: One of a variety of large-scale landforms that form by deposition of sediment at the margins of glaciers.

Mosasaurs: Large, marine-dwelling reptiles common in seas of the Late Cretaceous.

Mount Narryer Quartzite: Rocks located in the Jack Hills of Western Australia, containing detrital zircons indicating that Earth is older than 4.4 Ga.

Mutations: Anatomical changes caused by changes in an organism's genes.

Myriapoda: Many-legged arthropods, including centipedes and millipedes, that were among the earliest land-dwelling invertebrates.

***Naked Ape*:** Title of the 1967 book by zoologist Desmond Morris, referring to the evolutionary loss of fur, and increase in brain size that ultimately resulted in humans.

Nama Group: One of the three major Ediacaran faunal assemblages, found in the desert outcrops of Namibia.

Natural disaster: A highly harmful impact on life that is not human-made.

Natural selection: Process of evolution in which traits in an organism that increase the likelihood of healthy offspring are favored.

Nebular hypothesis: The widely accepted idea that the planets grow around eddies in a flat proto-planetary disk that rotated about the evolving star at the center and are initially formed from solar nebulae that form after explosion of giant stars.

Negative feedback loop: Interactions of a natural processes that inhibit one-way change. For example, the weathering and storage of CO_2 in sediment disallows catastrophic build-up that might otherwise cause runaway warming.

Nektonic: Free-swimming fauna.

Neoteny: The slowing or delaying of the development of an organism that results in the retention of juvenile attributes through adulthood.

Neptunism: The theory, advanced by Abraham Gottlob Werner (1749–1817), that almost all the rocks and minerals of Earth's crust were precipitated from retreat and evaporation of a universal ocean.

Nocturnal: Animals whose lifestyle is based on being awake during the night.

Non-conformity: Surface where sedimentary rocks overlie igneous or metamorphic rocks.

Non-vascular plants: Plants without a vascular system that lack xylem and phloem.

Normal faults: Fault in which the *hanging wall* moves down relative to the *footwall*.

Nuclear winter: The environmental devastation predicted to follow a worldwide nuclear disaster.

Obduct: Thrusting of rocks on top of other rocks, antonym of *subduct*.

Obligate bipedalism: Only able to walk or run on two legs.

Obliquity: The degree that Earth's axis of rotation is tilted towards the Sun.

Oceanic acidification: Acidification of the ocean by addition of CO_2 or other acidic compounds thought to be a driver of many marine extinctions.

Oceanic anoxic events: Periods through Earth history which are rich in organic matter, but in which lack of aquatic oxygen prevents the growth of oxygen-loving organisms and bacteria that would otherwise feed on the organic matter.

Oceanic plateau: Area of thicker ocean crust commonly formed by large igneous provinces associated with hot spots.

Oil and gas windows: Areas in sedimentary basins where heat and pressure convert organic material into oil and gas.

Oldowan: Paleolithic stone tools, typically comprising cobbles with a few flakes chipped off using another stone.

Olivine: An olive-colored mafic mineral that is the primary component of the upper mantle.

Omnivorous: Eating food of both plant and animal origin.

Onychophora: Also known by the common name Velvet Worms, a modern phylum discovered in the Burgess Shale.

Open system: System in which matter and/or energy can be exchanged with the external environment, in contrast to a closed system, which exchanges neither matter nor energy. Earth is open as it receives energy from the Sun. In isotope geochemistry a system is open when the daughter products of radioactive decay can easily leave the mineral (usually because of elevated temperature)

Organelles: Subcellular structures, such as mitochondria or chloroplasts in Eukaryotes that have one or more specific jobs to perform in the cell.

Ornithischians: A group of so-called "Bird Hipped" dinosaurs that actually did not gave rise to the birds.

Orogenesis: The process of mountain building by tectonic forces.

Orogeny: Mountain-building event usually involving compressional tectonics.

Osmosis: The passive transport of nutrients by diffusion across a membrane.

Osmotrophic: Exchanging food and oxygen with the environment through osmosis.

Osteichthyes: Bony fishes.

Ostracodes: Microscopic crustaceans that live in a hinged, clam-like shell.

Oxygen Isotope Stage (OIS): Glacial or interglacial stages revealed by oxygen-isotope analysis of sediment and fossils.

Paedomorphosis: Evolutionary process in which juvenile attributes are retained through adulthood.

Paleocene/Eocene thermal maximum: An exceedingly warm period around 50 million years ago, with temperatures 15 °C warmer than today.

Paleocurrents: Sedimentological features contained in sedimentary deposits that enable the direction of movement of the sediment and associated fluid or air at the time of deposition to be determined.

Paleogene dinosaurs: Hypothetical idea that a few dinosaurs survived the K–Pg extinction. All are likely fossils reworked and redeposited in younger sediment.

Paleolithic: Stone-age.

Paleomagnetic field: Earth's magnetic field in the past.

Paleomagnetism: The study of ancient magnetic fields locked into rocks.

Paleontology: The study of fossils.

Paleopole: The position of a magnetic pole of Earth at a particular geological time.

Paleosols: Ancient soil deposits.

Paleozoic: First era of the Phanerozoic Eon.

Paleozoic fauna: The middle of the three major faunal groups of the Phanerozoic Era identified by paleontologist Jack Sepkoski.

Panarthropods: A superclade including Arthropoda, Tardigrada, and Onychophora as well as several extinct groups found in the Cambrian Burgess Shale.

Pangea: Supercontinent (translated as "whole mother-Earth land") that existed at the end of the Paleozoic Era, in which most of Earth's continents were joined together.

Panthalassa: A paleo-Pacific Ocean that existed during the Cambrian Period, and which was connected to the carbonate rocks on the western margin of North America.

Paraconformity: A type of unconformity where the layers above and below a contact appear to have no break, but in which examinations of fossils indicate a gap in time.

Parent element: The radioactive element from which a daughter element is produced by radioactive decay.

Partial melting: The dominant magma-producing process produced by partial versus complete melting of solid rocks.

Passive continental margin: Edge of a continent formed after continents rift apart; site of accumulation of thick sedimentary sequences.

Peak ring: A complex type of impact crater with multiple rings indicative of relatively large impacts.

Pederpes: An early Carboniferous tetrapod that lived between 348 and 347 Ma during the Late Mississippian, making it one of the earliest known animals capable of a partly terrestrial lifestyle.

Pericratonic blocks: Blocks that originated by rifting from the periphery of a continent.

Peridot: Igneous rocks made up primarily of olivine.

Periglacial: Peripheral to a glacier.

Peri-Laurentian arcs: The more common name for the island arcs formed on the periphery of Laurentia in the Iapetus Ocean.

Periods: Subdivisions of eras in the Geologic Time Scale.

Permafrost: Ground that remains completely frozen for at least two consecutive years.

Petrified: Organic material that has been turned into rock over time commonly forming fossils.

Photoautotrophic: Organisms that use sunlight to manufacture food.

Photoheterotrophic: Organisms that use sunlight as well as other nutrients as food intake.

Photosynthesis: Process that uses sunlight, water, and CO_2 to make food producing O_2 as a byproduct.

Phyletic gradualism: Gradual evolutionary change of a species over time.

Phylogenetic: Study of the evolutionary history of an organism based on morphologies, molecular information, and fossil records.

Pillow lavas: Basalts erupted under water forming semi-spherical shapes resembling pillows.

Pioneering communities: The first species to flourish following an extinction event.

Placozoa (Flat animals): The simplest animals but with a poor fossil record.

Planetesimals: "Tiny planets," several kilometers across, that were formed when small bodies, a few meters across, began colliding and coalescing with each other during the early history of the solar system.

Planktonic: Free-floating fauna.

Plastids: Type of membrane-bound organelle found in all plants that commonly contain chlorophyll.

Plate tectonics: The concept that Earth's lithosphere is divided into a series of rigid plates that move relative to each other, producing all major physiographic features of Earth.

Plateosaurus: Late Triassic species of bipedal herbivorous dinosaur closely related to the quadrupedal sauropods.

Plesiosaurs: Large, long-necked, marine-dwelling reptiles that arose in the Triassic Period.

Plutonic: Igneous rocks formed by magmas that cool and solidify beneath the surface. (Synonymous with *intrusive*.)

Plutonism: The theory that the formation of volcanic rocks must be related to heat in Earth's interior.

Polychaete: A type of segmented marine worm.

Polydactyl: Animals with extra fingers or toes. Early tetrapods had more than five digits and these have been largely reduced to the five that characterize most modern terrestrial vertebrates.

Polymers: Molecules that form an elongate chain.

Porifera: Sponges. The earliest and simplest metazoans with a well-preserved fossil record.

Positive feedback loop: Processes in which interactions cause net one-way change that can lead to "runaway" conditions.

Pre-biotic: Before the advent of life on Earth.

Precession: The cyclic change in the direction of Earth's rotational axis, typically at a 41,000-year cycle.

Primates: Order of mammals that includes humans. The name reflects the outdated idea that this group belongs to the prime or highest rank of life on Earth representing the culmination of the ladder of progress and the *Great Chain of Being*.

Primordial soup: Related to the idea that life evolved in a shallow freshwater pond rich in pre-biotic chemicals such as ammonia, carbon dioxide, and phosphate.

Principle of faunal succession: The observation, advanced by William Smith (1769–1839), that fossils succeed each other vertically in a specific and reliable order and can furthermore be identified over wide horizontal distances allowing layers to be correlated and regionally mapped.

Principle of lateral continuity: The third of Steno's stratigraphic laws, stating that previously continuous strata are now discontinuous because of subsequent erosion.

Principle of original horizontality: The first of Steno's stratigraphic laws, stating that sedimentary layers are broadly laid parallel to the surface, because sedimentary particles are deposited under the influence of gravity.

Principle of superposition: The second of Steno's laws, stating that in an undisturbed sequence of layered sediments or rocks, the oldest layers will be at the bottom and the youngest layers will be at the top.

Proboscis: Elongated appendage that extends from the head of an animal.

Proteinoids: Protein-like molecules formed abiotically from amino acids which may have been precursors to the first living cells.

Protolith: Original rock.

Protoplanets: Early accumulation of rock and dust during the coalescing of the solar system; many protoplanets came together to form the planets we observe today.

Protostome: Embryos in which the first dimple in embryonic differentiation of the gastrula (hollow ball of cells) becomes the mouth.

Proximal: Indicating deposition closer to the sediment source area.

Pterodactyls: A winged reptile or pterosaur, about the size of a large seagull.

Punctuated equilibrium: The idea that the evolution of organisms occurs in brief spurts, or punctuations, when stresses in the environment reach a tipping point beyond which an organism cannot adapt and survive.

Pyroclastic: Redeposited particles produced by fire, such as are common in explosive volcanoes.

Pyroxene: A mafic mineral group common in igneous rock.

Quadruped: Four-legged animal.

Quadrupedal gait: Walking using all four limbs.

Quetzalcoatlus: A giant extinct pterosaur (winged reptile) that lived at the end of the Cretaceous Period with a wingspan of about ten meters.

Radioactive decay: The process, discovered by Henri Becquerel in 1895, by which an element can spontaneously transform into a different element.

Radiodontia: Meaning "radial teeth," the name of the group that includes the circular-mouthed anomalocarids.

Reaction centers: The molecular structures that use sunlight in metabolic processes.

Refugia: Environments where survivors are able to take refuge.

Regression: The lowering of sea level.

Reguibat Promontory: The first point of the Gondwana continent to collide with Laurussia, an event associated with the closing of the Rheic Ocean from about 340 to about 280 Ma.

Releasing bend: Bend associated with a strike-slip fault that creates extension.

Reservoir: An area where porous and permeable rocks accumulate oil and gas from anticlines that have been opened by drilling.

Residence time: The time a chemical species remains in the atmosphere or oceans before being removed to another reservoir (to rocks, soil, plants, or the ocean).

Restraining bend: Bend associated with a strike-slip fault that creates compression.

Reverse faults: Fault in which the hanging wall moves up relative to the *footwall* (see also *thrust faults*).

Rheic Ocean: A new ocean that formed at about the time that Iapetus was closing, caused when the Carolinia, Avalonia, and Meguma terranes rifted away from the western margin of Gondwanaland at various times during the Cambrian to Ordovician periods.

Rheology: The way material deforms under varying temperature and pressure.

Rhyniognatha hirsti: One of the oldest land-dwelling arthropods, originally thought to be the first winged insect, but may have been a *myriapod.*

Rhyolite: The extrusive equivalent of granite.

Rhyolitic lavas: Lavas that have a granitic (i.e., felsic) composition.

Ribbon continent: An elongate land mass containing continental lithosphere.

Ridge push: Tectonic force that moves oceanic plates by magma addition at mid-ocean ridges.

Rifting: Area where extension occurs forming normal faults and a series of high areas (horsts) and low areas (grabens). Common at extensional plate boundaries.

Riparian: Pertaining to the banks of a body of water.

Rock cycle: The transformation of one kind of rocks (igneous, metamorphic, sedimentary) into another via erosion, burial, melting, and solid-state phase changes.

Rocky Mountains: Mountain range stretching from Mexico to Alaska, which was formed by a variety of tectonic processes including convergent deformation, extensional deformation, and growth of mountains by intrusion of magma.

Rodinia: A supercontinent that existed during the Proterozoic Eon.

Roll back: Process in which steeply dipping subducting oceanic lithosphere moves away from the overlying plate as it sinks into the asthenosphere. It can cause extension in the overlying oceanic plate.

Romer's Gap: The lack of vertebrate fossils during the 15-million-year period that recorded the transition from fish to land vertebrates. Named for the noted vertebrate paleontologist Alfred Sherwood Romer.

Rudists: A group of bivalve mollusks which were the major reef builders in the Cretaceous Period.

Runaway albedo: An event in which colder air produced more snow, which produced colder air leading to more ice in the oceans, leading to colder oceans and so on. An example of a *positive feedback loop.*

Sabkhas: Arabic for salty tidal flats.

Sandstones: Sedimentary rocks made up primarily of grains between 0.016 and 2 mm in diameter.

Saurischians: Dinosaurs characterized by so-called "lizard-hips," but which gave rise to the birds.

Sauropod: Long-necked herbivorous dinosaurs which were the largest land animals to ever inhabit the Earth.

Schists: A medium-grade metamorphic rock that has flaky mica crystals.

Sclerites: Scale-like coverings.

Sedimentary: Related to rocks from sediment.

Sedimentary basins: Low areas on the surface of the Earth, such as a lake or ocean, where sinking (or subsidence) of the surface occurs that can be filled with sediment.

Seismic: Caused by or related to an earthquake.

Seismic reflection: Sound waves bouncing back after being shot into the Earth.

Seismometers: The instruments used in seismology, the branch of geophysics that studies the way energy from earthquakes passes through the various layers of Earth.

Selenitic gypsum: Large, radial clusters of gypsum crystals indicative of evaporation in relatively deep water.

Serpentine: Fibrous mineral that makes up the hydrated ultramafic igneous rock serpentinite.

Sessile: Organisms that lived attached to or partly buried in the sediment.

Shales: Clay-rich sedimentary rocks that are fissile.

Shatsky Rise: Plateau on the Pacific plate, seen today on the ocean floor approximately 1,500 km west of Japan.

Shock metamorphism: Metamorphism characteristic of shocks such as produced during impacts of meteorites.

Siberian Traps: A massive volcanic deposit, formed of basaltic and rhyolitic lava associated with a larger igneous province that covered much of Siberia and likely triggered the end-Permian mass extinction.

Sierra Nevada Batholith: California mountain range comprising plutonic rocks that represents the roots of ancient volcanoes formed by subduction beneath the western margin of North America during the Mesozoic.

Simian: The great apes, including chimps, gorillas, and orangutans.

Sinosauropteryx: Feathered dinosaurs, about the size of a turkey and closely related to the older genus, *Compsognathus*.

Slab pull: The force produced by the densification and metamorphism of subducting oceanic lithosphere that drives *plate tectonics*.

Slab window: A region between two subduction zones in which no slab is being subducted.

Slate: A metamorphosed shale.

Slushball: The idea that during the Cryogenian Period (snowball Earth), Earth consisted of low-latitude glaciations but with extended periods of open ocean that would have served as refugia for existing life on Earth.

Snow line: The elevation above which temperatures are cold enough to preserve snow and ice through the summer.

Snowball Earth: The idea that, during the Cryogenian Period, ice extended from high latitudes and elevations all the way to sea level in the tropics, covering most of Earth's surface.

Socioeconomic pathways: Climate change scenarios of predicted socioeconomic global changes that would affect future greenhouse gas emissions.

Sorting: The range of different clast sizes found in sediment or sedimentary rock.

Source rocks: Organic-rich sediments that when buried can release and serve as the source of hydrocarbons that may flow away from the source rock and be trapped in oil or gas fields.

Southern Tibetan Detachment Fault: The major normal fault in the Himalaya.

Spicules: Protective needles found on many sponges.

Spiracles: Openings along the sides of insects, which connect to a system of tubes that convey oxygen through their bodies, allowing them to "breathe" without lungs.

Stasis: A period of inactivity or lack of change.

Stegocephalia: Stem-tetrapods that represent transitional forms between fish and tetrapods, reflecting the fact that they enjoyed a largely aquatic lifestyle, but were able to breathe air and crawl out of the water in their search for food and new shallow water habitats.

Stem organisms: Ancestral organisms that lie at the base of a clade.

Stem-tetrapods: The early four-limbed vertebrates that were the ancestors of the four-legged crown tetrapods that we know today as the amphibians, reptiles, birds, and mammals.

Steno's laws: The basic principles for understanding the order of sedimentary layering, introduced in 1669 by the Danish scientist Nicolas Steno.

Sterane: Hydrocarbon compounds diagnostic of the breakdown of cholesterol, found in animals but not fungi or plants.

Steric: Sea-level changes cause by heating or cooling of seawater that in turn causes volume changes.

Strata: Layers of rock.

Stratigraphy: The study of layering of sedimentary rocks, usually at a regional scale.

Stratovolcanoes: Cone-shaped andesitic volcanoes that are common on the upper plates of subduction zones.

Striations: Scratch marks made by pieces of rock stuck at the bottom of a glacier as they are dragged over a rocky surface.

Strike-slip: The fault that marks the boundary between two plates that slide past one another.

Stromatolites: Thinly laminated, bulbous mounds, interpreted to be formed by colonies of sticky bacterial mats that covered the ancient seafloor and are the most common type of fossils in the Precambrian.

Stromatoporoid: Calcareous sponges that were important reef builders in much of the Paleozoic.

Structural geology: The study of deformation of rocks.

Sturtian Epoch: Glacial epoch that lasted for about 59 million years (718–659 Ma), followed by an interglacial period of about 9 million years and the younger and shorter-lived Marinoan Epoch lasted for about 15 million years (650–635 Ma), associated with snowball Earth.

Subduction zones: Areas where oceanic lithosphere is lost by sinking into the aesthenospheric mantle.

Subsidence: Sinking of the ground because of underground material movement.

Supernova explosions: Astronomical explosions in which giant stars (between 5 and 250 times the mass of the Sun) die. These can form solar nebulae that can in turn form new solar systems (see *nebular hypothesis*).

Survival of the fittest: The observation that competition for food and living space results in the elimination of those least well-adapted to their environment.

Suspect terranes: Relatively homogeneous geological provinces transported on adjacent plates and accreted onto a usually larger continent.

Suwanee terrane: A fifth terrane associated with the Appalachian orogen that was inferred and later identified based on subsurface investigations of the geology underlying the younger coastal plain sedimentary rocks that covered it.

Sympatric speciation: Speciation that occurs within large interbreeding populations that may be driven by external forces (for antonym see *allopatric speciation*).

Synapsids: One of the two main clades of vertebrates with a single hole in the skull (excluding the eye sockets and nasal opening), the other being the *diapsids*, which have two extra holes.

Taconic arcs: A series of volcanic island arcs, relatively close to Laurentia, that formed above several subduction zones during the closing of Iapetus.

Taconic Orogeny: An Early Paleozoic mountain-building event named after the Taconic Mountains in New York State.

Taphonomy: The study of the process of fossilization.

Tardigrada: Modern phylum of eight-legged segmented micro-animals, also known by the common name "water bears."

Tectono-eustasy: Global changes of sea level caused by tectonics.

Tektites: Spherical, elongate, or teardrop-shaped glass beads which are formed by falling drops of molten rock that have been melted by an impact.

Telomere: Distinctive sequence of DNA molecules at the ends of chromosomes.

Terminal moraines: Deposits that form at the terminus of a glacier.

Tethyan terranes: Tectonic terranes formed in the Tethys and related oceans that were later accreted to Asia prior to the collision with India.

Tethys Ocean: The ocean that originally separated Africa, India, Australia, and other islands to the south from Eurasia to the north.

The Age of Fishes: Term used to describe the Devonian Period in which fishes flourished.

The Haq curve: Graph of global changes of sea level throughout the Phanerozoic, named after paleontologist Bilal Haq.

Thermal emission: Heat emitted from Earth.

Thermogenesis: Origin by heating.

Thermohaline: Cold and salty, typically used in reference to the process that drives oceanic currents.

Thermophiles: One of the first modern Archaea, which thrive in areas of high heat.

Theropod: The classic bipedal carnivorous dinosaur.

Thin section: A fragment of rock or mineral ground to paper thinness for viewing under a microscope.

Thrust faults: A surface across which layers are moved upwards and on top of each other (see *reverse faults*).

Tiktaalik: One of the earliest lobe-finned fishes that has some capability of land-dwelling and which evolved into amphibious tetrapods.

Till: Usually poorly sorted sediments deposited directly by glaciers.

Time–stratigraphic diagrams: Diagrams that depict rock sections in time and space.

Tipping point: The point at which crossing a key threshold causes a quick and catastrophic change.

Tommotian fauna: "Small shelly" fauna that characterizes the Early Cambrian.

Trace fossils: Petrified impressions, such as footprints or crawling marks, left by organisms as they move, rest, infiltrate the sediment or otherwise behave, though no parts of the organism remain.

Tracheophytes: Fossilized land plants with full-fledged vascular systems.

Transduction: The direct transfer of genetic material from one microorganism to another.

Transform boundary: Strike-slip faults developed perpendicular to mid-ocean ridges that accommodate extension.

Transformation: When genetic material is directly absorbed by a cell and incorporated into its own DNA.

Transgression: Migration of the shoreline landward usually related to a rise of sea level.

Triceratops: Last surviving species and largest of the horned ceratopsian dinosaurs.

Trilobites: A class of arthropods that lived in the Paleozoic Era.

Triple junction: A point where three plate boundaries are connected.

Triploblasts: Phyla in which the embryo develops three cell layers.

Tsunami: Large waves commonly formed by marine earthquakes or large extraterrestrial impacts.

Turbidity currents: Rapid, downhill flow of water and sediment slurries caused by increased density due to high amounts of sediment.

Two-way travel time: In seismology, the time it takes for a sound wave to leave the surface, hit the reflector, and bounce back.

Tyrannosaurus rex: Bipedal carnivorous dinosaur that lived at the very end of the dinosaurs' reign in the Late Cretaceous.

Ultramafic: Igneous rocks very rich in iron and magnesium.

Unconformity: A surface between rock layers across which there is evidence of a significant break in the record of deposition, commonly marked by some erosion.

Unconventional resources: The use of controversial new techniques by United States petroleum engineers to target source rocks directly.

Uniformitarianism: The idea that analysis and understanding of modern processes is critical to understanding Earth's past. Commonly expressed by the term "The present is the key to the past."

Urkontinent: The name given, by German scientist Alfred Wegener (1880–1930), to a proposed supercontinent that contained several of the modern continents.

Vascular systems: Pipe-like structures used to transport water and minerals throughout a plant, which enables them to live farther away from wetland environments, and allowed extensive expansion across the world's land areas.

Vascularization: Tubes for blood vessels.

Velociraptors: Feathered theropod dinosaur.

Vermiform: Wormlike.

Vestigial anatomical structures: Parts of an organism's anatomy which were greatly reduced during its evolution.

Volcanic: Igneous rock that was solidified on or above the surface. (Synonymous with *extrusive*.)

Wedge-top: Wedge-shaped basin commonly formed above a fold and thrust belt and part of a foreland basin system.

Western Interior Seaway: Low area that was flooded due to down-warping of the foredeep during the Cretaceous Period and extended from Alaska to the Gulf of Mexico, separating North America into two land masses, *Appalachia* to the east and *Laramidia* to the west.

White Sea Group: One of the three major Ediacaran faunal assemblages, found in northeast Russia.

Wilson cycles: Named after the Canadian geophysicist J. Tuzo Wilson, the hypothesis that the Earth experiences cycles of continent assembly and breakup that occur over a few hundred million years.

Yavapai Orogeny: Period of mountain building during the Proterozoic Eon associated with formation of the now-eroded Vishnu Mountains that underlie the Grand Canyon.

Younger Dryas: A cooling period, prior to Holocene warming, that involved significant environmental change.

Yutyrannus huali: Giant, feathered dinosaur.

Zanclean Stage: The earliest age of the Pliocene, which began with the massive flooding of the Mediterranean, 600,000 years after it dried up.

Zircon: ($ZrSiO_4$) A favored mineral for determining the formation age of igneous rocks, because it has one of the highest closure temperatures and is highly resistant to subsequent weathering.

Zuni sequence: Term coined in 1963 by Northwestern University geology professor Lawrence Sloss, referring to the sedimentary sequences associated with high sea levels that developed during the Cretaceous Period.